STUDENT'S SOLUTIONS MANUAL

CHARLES ODION
Houston Community Colle

JAMES J. BALL
Indiana State University

MATHEMATICS WITH APPLICATIONS AND FINITE MATHEMATICS WITH APPLICATIONS

NINTH EDITION

Margaret L. Lial
American River College

Thomas W. Hungerford
Saint Louis University

John P. Holcomb, Jr.
Cleveland State University

PEARSON

Addison
Wesley

Boston San Francisco New York
London Toronto Sydney Tokyo Singapore Madrid
Mexico City Munich Paris Cape Town Hong Kong Montreal

Reproduced by Pearson Addison-Wesley from electronic files supplied by the authors.

Copyright © 2007 Pearson Education, Inc.
Publishing as Pearson Addison-Wesley, 75 Arlington Street, Boston, MA 02116.

ISBN   0-321-33595-3

10 11 12 BRR 12 11 10 09

TABLE OF CONTENTS

Chapter 1: Algebra and Equations

Section 1.1 The Real Numbers

1. True. This statement is true, since every integer can be written as the ratio of the integer and 1.

   For example, $5 = \dfrac{5}{1}$.

3. Answers vary with the calculator, but $\dfrac{2,508,429,787}{798,458,000}$ is the best.

5. $6(t + 4) = 6t + 6 \cdot 4$

   This illustrates the distributive property.

7. $0 + (-7) = -7 + 0$

   This illustrates the commutative property of addition.

9. Answer varies.

11. $\begin{aligned} -3(p + 5q) &= -3[-2 + 5(4)] \\ &= -3(-2 + 20) \\ &= -3(18) \\ &= -54 \end{aligned}$

13. $\begin{aligned} \dfrac{q + r}{q + p} &= \dfrac{4 + (-5)}{4 + (-2)} \\ &= \dfrac{-1}{2} \\ &= -\dfrac{1}{2} \end{aligned}$

15. Let $r = 1.5$.
    $\begin{aligned} \text{APR} &= 12r \\ &= 12(1.5) \\ &= 18\% \end{aligned}$

17. Let $\text{APR} = 9$.
    $\begin{aligned} \text{APR} &= 12r \\ 9 &= 12r \\ \dfrac{9}{12} &= r \\ \dfrac{3}{4} &= r \\ r &= .75\% \end{aligned}$

19. $\begin{aligned} 3 - 4 \cdot 5 + 5 &= 3 - 20 + 5 \\ &= -17 + 5 \\ &= -12 \end{aligned}$

21. $8 - 4^2 - (-12)$
    Take powers first.
    $= 8 - 16 - (-12)$
    Then add and subtract in order from left to right.
    $= 8 - 16 + 12$
    $= -8 + 12$
    $= 4$

23. $-(3 - 5) - \left[2 - \left(3^2 - 13\right) \right]$
    Take powers first.
    $= -(3 - 5) - [2 - (9 - 13)]$
    Work inside brackets and parentheses.
    $= -(-2) - [2 - (-4)]$
    $= 2 - [2 + 4]$
    $= 2 - 6$
    $= -4$

25. $\dfrac{2(-3) + \frac{3}{(-2)} - \frac{2}{(-\sqrt{16})}}{\sqrt{64} - 1}$

    Work above and below fraction bar. Take roots.
    $= \dfrac{2(-3) + \frac{3}{(-2)} - \frac{2}{(-4)}}{8 - 1}$
    Do multiplications and divisions.
    $= \dfrac{-6 - \frac{3}{2} + \frac{1}{2}}{8 - 1}$
    Add and subtract.
    $= \dfrac{-\frac{12}{2} - \frac{3}{2} + \frac{1}{2}}{7}$
    $= \dfrac{-\frac{14}{2}}{7}$
    $= \dfrac{-7}{7}$
    $= -1$

27. $\dfrac{2040}{523}, \dfrac{189}{37}, \sqrt{27}, \dfrac{4587}{691}, 6.735, \sqrt{47}$

29. 12 is less than 18.5.
    $12 < 18.5$

31. x is greater than or equal to 5.7.
    $x \geq 5.7$

33. z is at most 7.5.
    $z \leq 7.5$

35. $(-8, -1)$

This represents all real numbers between -8 and -1, not including -8 and -1. Draw parentheses at -8 and -1 and a heavy line between them. The parentheses at -8 and -1 show that neither of these points belongs to the graph.

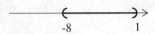

37. $[-2, 2)$

This represents all real numbers between -2 and 2, including -2, but not including 2.
Draw a bracket at -2, a parenthesis at 2, and a heavy line between them.

39. $(-2, 3]$

All real numbers x such that $-2 < x \le 3$
Draw a heavy line from -2 to 3. Use a parenthesis at -2 since it is not part of the graph. Use a bracket at 3 since it is part of the graph.

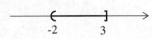

41. $(3, \infty)$

This represents all real numbers greater than 3. Draw a parenthesis at 3 and a heavy line to the right.

43. $(-\infty, -2]$

This represents all real numbers less than or equal to -2. Draw a bracket at -2 and a heavy line to the left.

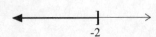

45. $r > 2.7$

The percentage increase in the CPI was greater than 2.7 in 1995, 1996, 2000, and 2001. Therefore, there are 4 years that satisfy this inequality.

47. $r \le 2.7$

The percentage increase in the CPI was less than or equal to 2.7 in 1997, 1998, 1999, 2002, 2003, and 2004. Therefore, there are 6 years that satisfy this inequality.

49. $r > 4$

There is no year in which the percentage increase in the CPI was greater than 4. Therefore, there are 0 years that satisfy this inequality.

51. a. Steffi Graf's height in inches is
$5(12) + 9 = 69$ in.

$$B = \frac{.455W}{(.0254H)^2}$$
$$= \frac{.455(119)}{(.0254(69))^2}$$
$$= \frac{54.145}{(1.7526)^2}$$
$$\approx 17.6$$

b. No, Steffi Graf's body mass index falls below the desirable range.

53. a. Tiger Wood's height in inches is
$6(12) + 2 = 74$ in.

$$B = \frac{.455W}{(.0254H)^2}$$
$$= \frac{.455(180)}{(.0254(74))^2}$$
$$= \frac{81.9}{(1.8796)^2}$$
$$\approx 23.2$$

b. Yes, Tiger Wood's body mass index falls within the desirable range.

55. A wind at 20 miles per hour with a $30°$ temperature has a wind-chill factor of $17°$.
A wind at 10 miles per hour with a $-10°$ temperature has a wind-chill factor of $-28°$.
$|17 - (-28)| = |17 + 28| = 45°$

57. A wind at 35 miles per hour with a $-30°$ temperature has a wind-chill factor of $-69°$.
A wind at 15 miles per hour with a $-20°$ temperature has a wind-chill factor of $-45°$.
$|-69 - (-45)| = |-69 + 45| = |-24| = 24°$

59. $|8| - |-4| = 8 - (4) = 4$

61. $-|-4|-|-1-14| = -(4)-|-15|$
$$= -(4)-(15)$$
$$= -19$$

63. $|5| \underline{\quad} |-5|$
$5 \underline{\quad} 5$
$5 = 5$

65. $|10-3| \underline{\quad} |3-10|$
$|7| \underline{\quad} |-7|$
$7 \underline{\quad} 7$
$7 = 7$

67. $|-2+8| \underline{\quad} |2-8|$
$|6| \underline{\quad} |-6|$
$6 \underline{\quad} 6$
$6 = 6$

69. $|3-5| \underline{\quad} |3|-|5|$
$|-2| \underline{\quad} 3-5$
$2 \underline{\quad} -2$
$2 > -2$

71. When $a < 7$, $a - 7$ is negative.
So $|a-7| = -(a-7) = 7-a$.

73. No, it is not always true that $|a+b| = |a|+|b|$. For example, let $a = 1$ and $b = -1$. Then,
$|a+b| = |1+(-1)| = |0| = 0$, but
$|a|+|b| = |1|+|(-1)| = 1+1 = 2$.

75. $|2-b| = |2+b|$ only when $b = 0$. Then each side of the equation is equal to 2.

77. The statement is true for 2000, 2003, and 2004. For example, for the year 2000,
$$|1,250,000,000 - 2,000,000,000| = |-750,000,000|$$
$$= 750,000,000$$

79. Let x be the percentage of international students from India. Then $x > 13\%$.

81. Let x be the rank of UNC. Then $x > 25$.

Section 1.2 Polynomials

1. $11.2^6 \approx 1,973,822.685$

3. $\left(-\dfrac{18}{7}\right)^6 \approx 289.0991339$

5. -3^2 is negative, whereas $(-3)^2$ is positive. Both -3^3 and $(-3)^3$ are negative.

7. $4^2 \cdot 4^3 = 4^{2+3} = 4^5$

9. $(-6)^2 \cdot (-6)^5 = (-6)^{2+5} = (-6)^7$

11. $\left[(5u)^4\right]^7 = (5u)^{4 \cdot 7} = (5u)^{28}$

13. degree 4; coefficients: 6.2, –5, 4, –3, 3.7; constant term 3.7.

15. Since the highest power of x is 3, the degree is 3.

17. $\left(3x^3 + 2x^2 - 5x\right) + \left(-4x^3 - x^2 - 8x\right)$
$$= \left(3x^3 - 4x^3\right) + \left(2x^2 - x^2\right) + (-5x - 8x)$$
$$= -x^3 + x^2 - 13x$$

19. $\left(-4y^2 - 3y + 8\right) - \left(2y^2 - 6y + 2\right)$
$$= \left(-4y^2 - 3y + 8\right) + \left(-2y^2 + 6y - 2\right)$$
$$= -4y^2 - 3y + 8 - 2y^2 + 6y - 2$$
$$= \left(-4y^2 - 2y^2\right) + (-3y + 6y) + (8 - 2)$$
$$= -6y^2 + 3y + 6$$

21. $\left(2x^3 + 2x^2 + 4x - 3\right) - \left(2x^3 + 8x^2 + 1\right)$
$$= \left(2x^3 + 2x^2 + 4x - 3\right) + \left(-2x^3 - 8x^2 - 1\right)$$
$$= 2x^3 + 2x^2 + 4x - 3 - 2x^3 - 8x^2 - 1$$
$$= \left(2x^3 - 2x^3\right) + \left(2x^2 - 8x^2\right) + (4x) + (-3 - 1)$$
$$= -6x^2 + 4x - 4$$

23. $-9m\left(2m^2 + 6m - 1\right)$
$$= (-9m)\left(2m^2\right) + (-9m)(6m) + (-9m)(-1)$$
$$= -18m^3 - 54m^2 + 9m$$

25. $(3z+5)\left(4z^2 - 2z + 1\right)$
$$= (3z+5)\left(4z^2\right) + (3z+5)(-2z) + (3z+5)(1)$$
$$= 12z^3 + 20z^2 - 6z^2 - 10z + 3z + 5$$
$$= 12z^3 + 14z^2 - 7z + 5$$

27. $(6k-1)(2k+3)$
$= (6k)(2k-3) + (-1)(2k+3)$
$= 12k^2 - 18k - 2k - 3$
$= 12k^2 - 20k - 3$

29. $(3y+5)(2y+1)$
Use FOIL.
$= 6y^2 + 3y + 10y + 5$
$= 6y^2 + 13y + 5$

31. $(9k+q)(2k-q)$
$= 18k^2 - 9kq + 2kq - q^2$
$= 18k^2 - 7kq - q^2$

33. $(6.2m - 3.4)(.7m + 1.3)$
$= 4.34m^2 + 8.06m - 2.38m - 4.42$
$= 4.34m^2 + 5.68m - 4.42$

35. $5k - [k + (-3 + 5k)]$
$= 5k - [6k - 3]$
$= 5k - 6k + 3$
$= -k + 3$

37. $R = 5(1000x) = 5000x$
$C = 150,000 + 2250x$
$P = (5000x) - (150,000 + 2250x)$
$\quad = 2750x - 150,000$

39. $R = 7.50(1000x) = 7500x$
$C = 250,000 + (-3x^2 + 3480x - 325)$
$\quad = -3x^2 + 3480x + 249,675$
$P = (7500x) - (-3x^2 + 3480x + 249,675)$
$\quad = 3x^2 + 4020x - 249,675$

41. a. According to the bar graph, the number of on-line households in 1999 was about 102,000,000.

b. Let $x = 0$.
$-.47x^2 + 19.12x + 101.55$
$= -.47(0)^2 + 19.12(0) + 101.55$
$= 101.55$
The number of on-line households in 1999 was 101,550,000.

43. a. According to the bar graph, the number of on-line households in 2002 was about 157,600,000.

b. Let $x = 3$.
$-.47x^2 + 19.12x + 101.55$
$= -.47(3)^2 + 19.12(3) + 101.55$
$= 154.68$
The number of on-line households in 2002 was 154,680,000.

45. Let $x = 6$.
$-.47x^2 + 19.12x + 101.55$
$= -.47(6)^2 + 19.12(6) + 101.55$
$= 199.35$
The number of on-line households in 2005 will be about 199,350,000.

47. Let $x = 9$.
$-.47x^2 + 19.12x + 101.55$
$= -.47(9)^2 + 19.12(9) + 101.55$
$= 235.56$
The number of on-line households in 2008 will be about 235,560,000.

49. Let $x = 10$.

$1 - .0058x - .00076x^2$
$= 1 - .0058(10) - .00076(10)^2$
$= .866$

51. Let $x = 22$.

$1 - .0058x - .00076x^2$
$= 1 - .0058(22) - .00076(22)^2$
$= .505$

53. a. Calculate the volume of the Great Pyramid when $h = 200$ feet, $b = 756$ feet and $a = 314$ feet.
$$V = \frac{1}{3}(200)\left(314^2 + (314)(756) + 756^2\right)$$
$\approx 60,501,067$ cubic feet

b. When $a = b$, the shape becomes a rectangular box with a square base, with volume $b^2 h$.

55. a Some or all of the terms may drop out of the sum, so the degree of the sum could be 0, 1, 2, or 3 or no degree (if one polynomial is the negative of the other).

b Some or all of the terms may drop out of the difference, so the degree of the difference could be 0, 1, 2, or 3 or no degree (if they are equal).

c Multiplying a degree 3 polynomial by a degree 3 polynomial results in a degree 6 polynomial.

Section 1.3 Factoring

1. $12x^2 - 24x = 12x \cdot x - 12x \cdot 2$
$$= 12x(x-2)$$

3. $r^3 - 5r^2 + r = r(r^2) - r(5r) + r(1)$
$$= r(r^2 - 5r + 1)$$

5. $6z^3 - 12z^2 + 18z$
$$= 6z(z^2) - 6z(2z) + 6z(3)$$
$$= 6z(z^2 - 2z + 3)$$

7. $3(2y-1)^2 + 7(2y-1)^3$
$$= (2y-1)^2(3) + (2y-1)^2 \cdot 7(2y-1)$$
$$= (2y-1)^2[3 + 7(2y-1)]$$
$$= (2y-1)^2(3 + 14y - 7)$$
$$= (2y-1)^2(14y - 4)$$
$$= 2(2y-1)^2(7y-2)$$

9. $3(x+5)^4 + (x+5)^6$
$$= (x+5)^4 \cdot 3 + (x+5)^4(x+5)^2$$
$$= (x+5)^4\left[3 + (x+5)^2\right]$$
$$= (x+5)^4(3 + x^2 + 10x + 25)$$
$$= (x+5)^4(x^2 + 10x + 28)$$

11. $x^2 + 5x + 4$
4 and 1 are factors of 4 that add to 5.
$$x^2 + 5x + 4 = (x+1)(x+4)$$

13. $x^2 + 7x + 12$
3 and 4 are factors of 12 that add to 7.
$$x^2 + 7x + 12 = (x+3)(x+4)$$

15. $z^2 + 10z + 24$
6 and 4 are factors of 24 that add to 10.
$$z^2 + 10z + 24 = (z+4)(z+6)$$

17. $2x^2 - 9x + 4 = (2x-1)(x-4)$

19. $15p^2 - 23p + 4 = (3p-4)(5p-1)$

21. $4z^2 - 16z + 15 = (2z-5)(2z-3)$

23. $6x^2 - 5x - 4 = (2x+1)(3x-4)$

25. $10y^2 + 21y - 10 = (5y-2)(2y+5)$

27. $6x^2 + 5x - 4 = (2x-1)(3x+4)$

29. $3a^2 + 2a - 5 = (3a+5)(a-1)$

31. $x^2 - 81 = x^2 - (9)^2 = (x+9)(x-9)$

33. $9p^2 - 12p + 4$
$$= (3p)^2 - 2(3p)(2) + 2^2$$
$$= (3p-2)^2$$

35 $r^2 + 3rt - 10t^2 = (r-2t)(r+5t)$.

37. $m^2 - 8mn + 16n^2$
$$= (m)^2 - 2(m)(4n) + (4n)^2$$
$$= (m-4n)^2$$

39. $4p^2 - 9 = (2p)^2 - 3^2$
$$= (2p+3)(2p-3)$$

41. $3x^2 - 24xz + 48z^2$
$$= 3(x^2 - 8xz + 16z^2)$$
$$= 3\left[x^2 - 2x(4z) + (4z)^2\right]$$
$$= 3(x-4z)^2$$

43. $a^2 + 4ab + 5b^2$
We are looking for two positive numbers that are factors of 5 and have the sum of 4. Since there are none, the polynomial cannot be factored.

45. $-x^2 + 7x - 12 = (-x+4)(x-3)$
or $(x-4)(-x+3)$

47. $3a^2 - 13a - 30$
Trying various possibilities, we see that
$3a^2 - 13a - 30 = (3a+5)(a-6)$.

49. $21m^2 + 13mn + 2n^2 = (7m + 2n)(3m + n)$

51. $20y^2 + 39yx - 11x^2$
$= (4y - x)(5y + 11x)$

53. $y^2 - 4yz - 21z^2 = (y - 7z)(y + 3z)$

55. $121x^2 - 64 = (11x)^2 - 8^2$
$= (11x + 8)(11x - 8)$

57. $5m^3(m^3 - 1)^2 - 3m^5(m^3 - 1)^3$

Factor out $m^3(m^3 - 1)^2$.

$= m^3(m^3 - 1)^2 \left[5 - 3m^2(m^3 - 1) \right]$

Factor difference of two cubes.

$= m^3 \left[(m - 1)(m^2 + m + 1) \right]^2 \left[5 - 3m^5 + 3m^2 \right]$

$= m^3(m - 1)^2(m^2 + m + 1)^2 \left(5 + 3m^2 - 3m^5 \right)$

59. $a^3 - 216 = a^3 - (6)^3$
$= (a - 6)\left(a^2 + 6a + 36 \right)$

61. $8r^3 - 27s^3 = (2r)^3 - (3s)^3$
$= (2r - 3s) \cdot \left[(2r)^2 + (2r)(3s) + (3s)^2 \right]$
$= (2r - 3s)\left(4r^2 + 6rs + 9s^2 \right)$

63. $64m^3 + 125$
$= (4m)^3 + (5)^3$
$= (4m + 5)\left[(4m)^2 - (4m)(5) + (5)^2 \right]$
$= (4m + 5)\left(16m^2 - 20m + 25 \right)$

65. $1000y^3 - z^3$
$= (10y)^3 - (z)^3$
$= (10y - z)\left[(10y)^2 + (10y)(z) + (z)^2 \right]$
$= (10y - z)\left(100y^2 + 10yz + z^2 \right)$

67. $x^4 + 5x^2 + 6 = \left(x^2 + 2 \right)\left(x^2 + 3 \right)$

69. $b^4 - b^2 = b^2(b + 1)(b - 1)$

71. $x^4 - x^2 - 12 = \left(x^2 + 2 \right)\left(x^2 - 2 \right)\left(x^2 + 3 \right)$

73. $16a^4 - 81b^4 = (2a + 3b)(2a - 3b)\left(4a^2 + 9b^2 \right)$

75. $x^8 + 8x^2 = x^2\left(x^6 + 8 \right) = x^2 \left(\left(x^2 \right)^3 + 2^3 \right)$
$= x^2\left(x^2 + 2 \right)\left(x^4 - 2x^2 + 4 \right)$

77. $6x^4 - 3x^2 - 3 = \left(2x^2 + 1 \right)\left(3x^2 - 3 \right)$ is not the

correct complete factorization because $3x^2 - 3$ contains a common factor of 3. This common factor should be factored out as the first step. This will reveal a difference of two squares, which requires further factorization. The correct factorization is

$6x^4 - 3x^2 - 3 = 3\left(2x^4 - x^2 - 1 \right)$
$= 3\left(2x^2 + 1 \right)\left(x^2 - 1 \right)$
$= 3\left(2x^2 + 1 \right)(x + 1)(x - 1)$

79. $(x + 2)^3 = (x + 2)(x + 2)^2$
$= (x + 2)\left(x^2 + 4x + 4 \right)$
$= x^3 + 4x^2 + 2x^2 + 8x + 4x + 8$
$= x^3 + 6x^2 + 12x + 8,$

which is not equal to $x^3 + 8$. The correct factorization is $x^3 + 8 = (x + 2)(x^2 - 2x + 4)$.

Section 1.4 Rational Expressions

1. $\dfrac{8x^2}{56x} = \dfrac{x \cdot 8x}{7 \cdot 8x} = \dfrac{x}{7}$

3. $\dfrac{25p^2}{35p^3} = \dfrac{5 \cdot 5p^2}{7p \cdot 5p^2} = \dfrac{5}{7p}$

5. $\dfrac{5m + 15}{4m + 12} = \dfrac{5(m + 3)}{4(m + 3)} = \dfrac{5}{4}$

7. $\dfrac{4(w - 3)}{(w - 3)(w + 6)} = \dfrac{4}{w + 6}$

9. $\dfrac{3y^2 - 12y}{9y^3} = \dfrac{3y(y - 4)}{3y\left(3y^2 \right)}$
$= \dfrac{y - 4}{3y^2}$

11. $\dfrac{m^2 - 4m + 4}{m^2 + m - 6} = \dfrac{(m - 2)(m - 2)}{(m + 3)(m - 2)} = \dfrac{m - 2}{m + 3}$

13. $\dfrac{x^2 + 4x - 5}{x^2 - 1} = \dfrac{(x + 5)(x - 1)}{(x + 1)(x - 1)} = \dfrac{x + 5}{x + 1}$

15. $\dfrac{4p^3}{49}\cdot\dfrac{7}{2p^2}=\dfrac{2p\cdot2p^2\cdot7}{7\cdot7\cdot2p^2}$

$=\dfrac{7\cdot2p^2\cdot2p}{7\cdot2p^2\cdot7}$

$=\dfrac{2p}{7}$

17. $\dfrac{21a^5}{14a^3}\div\dfrac{8a}{12a^2}=\dfrac{21a^5}{14a^3}\cdot\dfrac{12a^2}{8a}$ Invert second fraction.

$=\dfrac{3\cdot7\cdot4\cdot3\cdot a^4\cdot a^3}{2\cdot7\cdot4\cdot2\cdot a^4}$ Multiply.

$=\dfrac{7\cdot4\cdot a^4\cdot3\cdot3\cdot a^3}{7\cdot4\cdot a^4\cdot2\cdot2}$

$=\dfrac{9a^3}{4}$ Lowest terms

19. $\dfrac{2a+b}{3c}\cdot\dfrac{15}{4(2a+b)}=\dfrac{(2a+b)\cdot15}{(2a+b)\cdot12c}$

$=\dfrac{15}{12c}$

$=\dfrac{5}{4c}$

21. $\dfrac{15p-3}{6}\div\dfrac{10p-2}{3}$

$=\dfrac{15p-3}{6}\cdot\dfrac{3}{10p-2}$

$=\dfrac{3(5p-1)\cdot3}{3\cdot2\cdot2\cdot(5p-1)}$

$=\dfrac{3(5p-1)\cdot3}{3(5p-1)\cdot2\cdot2}=\dfrac{3}{4}$

23. $\dfrac{9y-18}{6y+12}\cdot\dfrac{3y+6}{15y-30}$

$=\dfrac{9(y-2)}{6(y+2)}\cdot\dfrac{3(y+2)}{15(y-2)}$

$=\dfrac{27(y-2)(y+2)}{90(y+2)(y-2)}=\dfrac{27}{90}=\dfrac{3}{10}$

25. $\dfrac{4a+12}{2a-10}\div\dfrac{a^2-9}{a^2-a-20}$

$=\dfrac{4a+12}{2a-10}\cdot\dfrac{a^2-a-20}{a^2-9}$

$=\dfrac{4(a+3)}{2(a-5)}\cdot\dfrac{(a-5)(a+4)}{(a+3)(a-3)}$

$=\dfrac{4(a+3)(a-5)(a+4)}{2(a-5)(a+3)(a-3)}$

$=\dfrac{2(a+4)}{a-3}$

27. $\dfrac{k^2-k-6}{k^2+k-12}\cdot\dfrac{k^2+3k-4}{k^2+2k-3}$

$=\dfrac{(k-3)(k+2)}{(k+4)(k-3)}\cdot\dfrac{(k+4)(k-1)}{(k+3)(k-1)}$

$=\dfrac{(k-3)(k+2)(k+4)(k-1)}{(k+4)(k-3)(k+3)(k-1)}$

$=\dfrac{k+2}{k+3}$

29. To find the least common denominator for two fractions, factor each denominator into prime factors, multiply all unique prime factors raising each factor to the highest frequency it occurred.

31. $\dfrac{3}{5z}-\dfrac{1}{3z}$

The common denominator is $15z$.

$=\dfrac{3\cdot3}{5z\cdot3}-\dfrac{1\cdot5}{3z\cdot5}$

$=\dfrac{9}{15z}-\dfrac{5}{15z}$

$=\dfrac{4}{15z}$

33. $\dfrac{r+2}{3}-\dfrac{r-2}{3}$

$=\dfrac{(r+2)-(r-2)}{3}$

$=\dfrac{r+2-r+2}{3}$

$=\dfrac{4}{3}$

35. $\dfrac{4}{x}+\dfrac{1}{5}$

The common denominator is $5x$.

$= \dfrac{4\cdot 5}{x\cdot 5}+\dfrac{1\cdot x}{5\cdot x}$

$= \dfrac{20}{5x}+\dfrac{x}{5x}$

$= \dfrac{20+x}{5x}$

37. $\dfrac{1}{m-1}+\dfrac{2}{m}$

The common denominator is $m(m-1)$.

$= \dfrac{m\cdot 1}{m\cdot(m-1)}+\dfrac{(m-1)\cdot 2}{(m-1)\cdot m}$

$= \dfrac{m}{m(m-1)}+\dfrac{2(m-1)}{m(m-1)}$

$= \dfrac{m+2(m-1)}{m(m-1)}$

$= \dfrac{m+2m-2}{m(m-1)}=\dfrac{3m-2}{m(m-1)}$

39. $\dfrac{8}{3(a-1)}+\dfrac{3}{a-1}$

The common denominator is $3(a-1)$.

$= \dfrac{8}{3(a-1)}+\dfrac{3\cdot 3}{3(a-1)}$

$= \dfrac{8}{3(a-1)}+\dfrac{9}{3(a-1)}$

$= \dfrac{8+9}{3(a-1)}=\dfrac{17}{3(a-1)}$

41. $\dfrac{2}{5(k-2)}+\dfrac{5}{4(k-2)}$

The common denominator is $20(k-2)$.

$= \dfrac{8}{20(k-2)}+\dfrac{25}{20(k-2)}$

$= \dfrac{8+25}{20(k-2)}=\dfrac{33}{20(k-2)}$

43. $\dfrac{2}{x^2-4x+3}+\dfrac{5}{x^2-x-6}$

$= \dfrac{2}{(x-3)(x-1)}+\dfrac{5}{(x-3)(x+2)}$

The common denominator is
$(x-3)(x-1)(x+2)$

$= \dfrac{2(x+2)}{(x-3)(x-1)(x+2)}+\dfrac{5(x-1)}{(x-3)(x+2)(x-1)}$

$= \dfrac{2(x+2)+5(x-1)}{(x-3)(x+2)(x-1)}$

$= \dfrac{2x+4+5x-5}{(x-3)(x-1)(x+2)}$

$= \dfrac{7x-1}{(x-3)(x-1)(x+2)}$

45. $\dfrac{2y}{y^2+7y+12}-\dfrac{y}{y^2+5y+6}$

$= \dfrac{2y}{(y+4)(y+3)}-\dfrac{y}{(y+3)(y+2)}$

The common denominator is
$(y+4)(y+3)(y+2)$.

$= \dfrac{2y(y+2)}{(y+4)(y+3)(y+2)}-\dfrac{y(y+4)}{(y+4)(y+3)(y+2)}$

$= \dfrac{2y(y+2)-y(y+4)}{(y+4)(y+3)(y+2)}$

$= \dfrac{2y^2+4y-y^2-4y}{(y+4)(y+3)(y+2)}$

$= \dfrac{y^2}{(y+4)(y+3)(y+2)}$

47. $\dfrac{3k}{2k^2+3k-2}-\dfrac{2k}{2k^2-7k+3}$

$= \dfrac{3k}{(2k-1)(k+2)}-\dfrac{2k}{(2k-1)(k-3)}$

The common denominator is
$(2k-1)(k+2)(k-3)$.

$= \dfrac{3k(k-3)}{(2k-1)(k+2)(k-3)}-\dfrac{2k(k+2)}{(2k-1)(k+2)(k-3)}$

$= \dfrac{3k(k-3)-2k(k+2)}{(2k-1)(k+2)(k-3)}$

$= \dfrac{3k^2-9k-2k^2-4k}{(2k-1)(k+2)(k-3)}$

$= \dfrac{k^2-13k}{(2k-1)(k+2)(k-3)}$

49. $\dfrac{1+\frac{1}{x}}{1-\frac{1}{x}}$

Multiply both numerator and denominator of this complex fraction by the common denominator, x.

$$\frac{x\left(1+\frac{1}{x}\right)}{x\left(1-\frac{1}{x}\right)} = \frac{x \cdot 1 + x\left(\frac{1}{x}\right)}{x \cdot 1 - x\left(\frac{1}{x}\right)}$$

$$= \frac{x+1}{x-1}$$

51. $\dfrac{\frac{1}{x+h}-\frac{1}{x}}{h} = \dfrac{\frac{x-(x+h)}{x(x+h)}}{h}$

The common numerator is $x(x+h)$

$$= \frac{\frac{x-x-h}{x(x+h)}}{h} = \frac{\frac{-h}{x(x+h)}}{h}$$

$$= \frac{-h}{x(x+h)} \div h = \frac{-h}{x(x+h)} \cdot \frac{1}{h}$$

$$= \frac{-1}{x(x+h)} \quad \text{or} \quad -\frac{1}{x(x+h)}$$

53. a. $\dfrac{x^2-10x+25}{x^3-15x^2+50x} = \dfrac{(x-5)(x-5)}{x(x^2-15x+50)}$

$$= \frac{(x-5)(x-5)}{x(x-10)(x-5)}$$

$$= \frac{x-5}{x(x-10)}$$

$$= \frac{x-5}{x^2-10x}$$

b. Let $x = 11$.

$$\frac{11-5}{(11)^2-10(11)} = \frac{6}{121-110} = \frac{6}{11}$$

$$\frac{6}{11}(60) \approx 32.7 \text{ sec}$$

Let $x = 15$.

$$\frac{15-5}{(15)^2-10(15)} = \frac{10}{225-150} = \frac{10}{75} = \frac{2}{15}$$

$$\frac{2}{15}(60) = 8 \text{ sec}$$

Let $x = 20$.

$$\frac{20-5}{(20)^2-10(20)} = \frac{15}{400-200} = \frac{15}{200} = \frac{3}{40}$$

$$\frac{3}{40}(60) = 4.5 \text{ sec}$$

55. a. Let $x = 25$. Then

$$\frac{.072(25)^2+.744(25)+1.2}{25+2} = 2.4$$

The ad cost $2.4 million in 2005

b. Let $x = 45$. Then

$$\frac{.072(45)^2+.744(45)+1.2}{45+2} = 3.84$$

Let $x = 40$. Then

$$\frac{.072(40)^2+.744(40)+1.2}{40+2} = 3.48$$

The cost of an ad will not reach $4 million in 2020.

Section 1.5 Exponents and Radicals

1. $\dfrac{7^5}{7^3} = 7^{5-3} = 7^2 = 49$

3. $(4c)^2 = 4^2 c^2 = 16c^2$

5. $\left(\dfrac{2}{x}\right)^5 = \dfrac{2^5}{x^5} = \dfrac{32}{x^5}$

7. $(3u^2)^3(2u^3)^2 = (27u^6)(4u^6) = 108u^{12}$

9. $6^{-1} = \dfrac{1}{6^1} = \dfrac{1}{6}$

11. $2^{-5} = \dfrac{1}{2^5} = \dfrac{1}{32}$

13. $-7^{-4} = -\dfrac{1}{7^4} = -\dfrac{1}{2401}$

15. $(-y)^{-3} = \dfrac{1}{(-y)^3} = -\dfrac{1}{y^3}$

17. $\left(\dfrac{1}{7}\right)^{-3} = \left(\dfrac{7}{1}\right)^3 = 7^3 = 343$

19. $\left(\dfrac{4}{3}\right)^{-2} = \left(\dfrac{3}{4}\right)^2 = \dfrac{9}{16}$

21. $\left(\dfrac{a}{b^3}\right)^{-1} = \dfrac{b^3}{a}^1 = \dfrac{b^3}{a}$

23. $49^{\frac{1}{2}} = 7$ because $7^2 = 49$.

25. $(5.71)^{\frac{1}{4}} = (7.51)^{.25} \approx 1.55$ Use a calculator.

27. $27^{\frac{2}{3}} = \left(27^{\frac{1}{3}}\right)^2 = 3^2 = 9$

29. $-64^{\frac{2}{3}} = -\left(64^{\frac{1}{3}}\right)^2 = -(4)^2 = -16$

31. $\left(\dfrac{8}{27}\right)^{-\frac{4}{3}} = \left(\dfrac{27^{\frac{1}{3}}}{8^{\frac{1}{3}}}\right)^4 = \left(\dfrac{3}{2}\right)^4 = \dfrac{3^4}{2^4} = \dfrac{81}{16}$

33. $\dfrac{4^{-2}}{4^3} = 4^{-2} \cdot 4^{-3} = 4^{-5} = \dfrac{1}{4^5}$

35. $4^{-3} \cdot 4^6 = 4^3 = 64$

37. $8^{\frac{2}{3}} \cdot 8^{-\frac{1}{3}} = 8^{\frac{1}{3}} = 2$

39. $\dfrac{8^9 \cdot 8^{-7}}{8^{-3}} = 8^9 \cdot 8^{-7} \cdot 8^3 = 8^5$

41. $\dfrac{9^{-\frac{5}{3}}}{9^{\frac{2}{3}} \cdot 9^{-\frac{1}{5}}} = 9^{-\frac{5}{3}} \cdot 9^{-\frac{2}{3}} \cdot 9^{\frac{1}{5}} = 9^{-\frac{7}{3}} \cdot 9^{\frac{1}{5}}$

$= 9^{-\frac{32}{15}}$

$= \dfrac{1}{9^{\frac{32}{15}}}$

43. $\dfrac{z^6 \cdot z^2}{z^5} = \dfrac{z^8}{z^5} = z^{8-5} = z^3$

45. $\dfrac{2^{-1}\left(p^{-1}\right)^3}{2p^{-4}} = \dfrac{2^{-1}p^{-3}}{2^1 p^{-4}} = 2^{-1-1}p^{-3-(-4)}$

$= 2^{-2}p^1 = \dfrac{1}{2^2} \cdot p = \dfrac{p}{4}$

47. $\left(q^{-5}r^3\right)^{-1} = q^5 r^{-3}$

$= q^5 \cdot \dfrac{1}{r^3}$

$= \dfrac{q^5}{r^3}$

49. $\left(2p^{-1}\right)^3 \cdot \left(5p^2\right)^{-2} = 2^3\left(p^{-1}\right)^3(5)^{-2}\left(p^2\right)^{-2}$

$= 2^3\left(p^{-3}\right)\dfrac{1}{5^2}\left(p^{-4}\right)$

$= 2^3 \dfrac{1}{p^3} \dfrac{1}{5^2} \dfrac{1}{p^4}$

$= \dfrac{8}{25p^7}$

51. $(2p)^{\frac{1}{2}} \cdot \left(2p^3\right)^{\frac{1}{3}}$

$= 2^{\frac{1}{2}}p^{\frac{1}{2}} \cdot 2^{\frac{1}{3}} \cdot \left(p^3\right)^{\frac{1}{3}}$

$= 2^{\frac{1}{2}}2^{\frac{1}{3}}p^{\frac{1}{2}}p^1$

$= 2^{\frac{3}{6}}2^{\frac{2}{6}}p^{\frac{1}{2}}p^{\frac{2}{2}}$

$= 2^{\frac{5}{6}}p^{\frac{3}{2}}$

53. $p^{\frac{2}{3}}\left(2p^{\frac{1}{3}} + 5p\right)$

$= p^{\frac{2}{3}}\left(2p^{\frac{1}{3}}\right) + p^{\frac{2}{3}}(5p)$

$= 2p + 5p^{\frac{5}{3}}$

55. $\dfrac{\left(x^2\right)^{1/3}\left(y^2\right)^{2/3}}{3x^{2/3}y^2} = \dfrac{(x)^{2/3}(y)^{4/3}}{3x^{2/3}y^2}$

$= \dfrac{1}{3y^{2-4/3}} = \dfrac{1}{3y^{2/3}}$

57. $\dfrac{(7a)^2(5b)^{3/2}}{(5a)^{3/2}(7b)^4} = \dfrac{7^2a^2 5^{3/2}b^{3/2}}{5^{3/2}a^{3/2}7^4 b^4}$

$= \dfrac{a^{2-\frac{3}{2}}}{7^2 b^{4-\frac{3}{2}}}$

$= \dfrac{a^{1/2}}{49b^{5/2}}$

59. $x^{1/2}\left(x^{2/3} - x^{4/3}\right) = x^{1/2}x^{2/3} - x^{1/2}x^{4/3}$

$= x^{7/6} - x^{11/6}$

61. This is a difference of two squares.

$\left(x^{1/2} + y^{1/2}\right)\left(x^{1/2} - y^{1/2}\right) = \left(x^{1/2}\right)^2 - \left(y^{1/2}\right)^2$

$= x - y$

63. $(-3x)^{\frac{1}{3}} = \sqrt[3]{-3x}$, (f)

65. $(-3x)^{-\frac{1}{3}} = \dfrac{1}{(-3x)^{\frac{1}{3}}} = \dfrac{1}{\sqrt[3]{-3x}}$, (h)

67. $(3x)^{\frac{1}{3}} = \sqrt[3]{3x}$, (g)

69. $(3x)^{-\frac{1}{3}} = \dfrac{1}{(3x)^{\frac{1}{3}}} = \dfrac{1}{\sqrt[3]{3x}}$, (c)

71. $\sqrt[3]{125} = 125^{\frac{1}{3}} = 5$

73. $\sqrt[4]{625} = 625^{\frac{1}{4}} = 5$

75. $\sqrt[7]{-128} = (-128)^{\frac{1}{7}} = -2$

77. $\sqrt[3]{81} \cdot \sqrt[3]{9} = \sqrt[3]{729} = 9$

79. $\sqrt{81-4} = \sqrt{77}$

81. $\sqrt{81} - \sqrt{4} = 9 - 2 = 7$

83. $\sqrt{8}\sqrt{96} = \sqrt{8}\sqrt{8 \cdot 12} = \sqrt{8}\sqrt{8}\sqrt{12} = 16\sqrt{3}$

85 $\sqrt{75} + \sqrt{192} = 5\sqrt{3} + 8\sqrt{3} = 13\sqrt{3}$

87. $\left(\sqrt{2}+3\right)\left(\sqrt{2}-3\right)$
Use FOIL.
$= \sqrt{2}\left(\sqrt{2}\right) - 3\sqrt{2} + 3\sqrt{2} - 3(3)$
$= 2 - 9$
$= -7$

89. Use FOIL.

$\left(\sqrt{3}+4\right)\left(\sqrt{5}-4\right) = \sqrt{15} - 4\sqrt{3} + 4\sqrt{5} - 16$

91. $\dfrac{3}{1-\sqrt{2}} = \dfrac{3}{1-\sqrt{2}} \cdot \dfrac{1+\sqrt{2}}{1+\sqrt{2}}$
$= \dfrac{3\left(1+\sqrt{2}\right)}{(1)^2 - \left(\sqrt{2}\right)^2}$
$= \dfrac{3\left(1+\sqrt{2}\right)}{1-2} = \dfrac{3\left(1+\sqrt{2}\right)}{-1}$
$= -3\left(1+\sqrt{2}\right) = -3 - 3\sqrt{2}$

93. $\dfrac{4-\sqrt{2}}{2-\sqrt{2}} = \dfrac{4-\sqrt{2}}{2-\sqrt{2}} \cdot \dfrac{2+\sqrt{2}}{2+\sqrt{2}}$
$= \dfrac{8+4\sqrt{2}-2\sqrt{2}-\sqrt{2}\left(\sqrt{2}\right)}{(2)^2 - \left(\sqrt{2}\right)^2}$
$= \dfrac{8+4\sqrt{2}-2\sqrt{2}-2}{4-2}$
$= \dfrac{6+2\sqrt{2}}{2} = \dfrac{2\left(3+\sqrt{2}\right)}{2}$
$= 3 + \sqrt{2}$

95. $\dfrac{2-\sqrt{3}}{2+\sqrt{3}} = \dfrac{2-\sqrt{3}}{2+\sqrt{3}} \cdot \dfrac{2+\sqrt{3}}{2+\sqrt{3}}$
$= \dfrac{4-3}{4+2\sqrt{3}+2\sqrt{3}+\sqrt{3}\sqrt{3}}$
$= \dfrac{1}{7+4\sqrt{3}}$

97. $x = \sqrt{\dfrac{kM}{f}}$

Note that because x represents the number of units to order, the value of x should be rounded to the nearest integer.

a. $k = \$1, f = \$500, M = 100,000$
$x = \sqrt{\dfrac{1 \cdot 100,000}{500}} = \sqrt{200} \approx 14.1$
The number of units to order is 14.

b. $k = \$3, f = \$7, M = 16,700$
$x = \sqrt{\dfrac{3 \cdot 16,700}{7}} \approx 84.6$
The number of units to order is 85.

c. $k = \$1, f = \$5, M = 16,800$
$x = \sqrt{\dfrac{1 \cdot 16,800}{5}} = \sqrt{3360} \approx 58.0$
The number of units to order is 58.

99. Let $x = 12$. Then $111.371(12)^{.19234} \approx 179.61$
The CPI for the first half of 2002 was approximately 179.61

101. Let $x = 15.5$. Then $111.371(15.5)^{.19234} \approx 188.68$
The CPI for the last half of 2005 was approximately 188.68

103. Let $x = 15$. Then $2260.323(15)^{.59743} \approx 11,397$
The number of kidney transplants in 1995 was approximately 11,397.

105. Let $x = 22$. Then $2260.323(22)^{.59743} \approx 14,328$
The number of kidney transplants in 2002 was approximately 14,328.

107. Let $x = 24$. Then $2260.323(24)^{.59743} \approx 15,092$
The number of kidney transplants in 2004 was approximately 15,092.

109. a. $\dfrac{12.8}{1+15.3817(x^{-.844})}$

In 2000, $\dfrac{12.8}{1+15.3817(10^{-.844})} \approx 3.996$.

About 3.996 million SUV's were sold in 2000.

In 2003, $\dfrac{12.8}{1+15.3817(13^{-.844})} \approx 4.629$.

About 4.629 million SUV's were sold in 2003.

In 2005, $\dfrac{12.8}{1+15.3817(15^{-.844})} \approx 4.991$.

About 4.991 million SUV's were sold in 2005.

b. The following table shows that sales would reach 6,000,000 in 22 years. This corresponds to the year 2012.

X	Y1
16.000	5.1580
17.000	5.3163
18.000	5.4668
19.000	5.6102
20.000	5.7470
21.000	5.8776
22.000	**6.0026**

X=22

According to this approximation, sales will not reach 9,000,000 in the next 40 years as shown in the following table.

X	Y1
67.000	8.8742
68.000	8.9082
69.000	8.9415
70.000	8.9741
71.000	9.0061
72.000	9.0376
73.000	9.0684

Y1=9.0061419492

109. Continued

The next 40 years corresponds to the year 2030 and from the table below, when $x = 40$, the sales are about 7.60 million.

X	Y1
39.000	7.5362
40.000	**7.6023**
41.000	7.6665
42.000	7.7289
43.000	7.7896
44.000	7.8486
45.000	7.9061

X=40

Section 1.6 First-degree Equations

1.
$$3x + 8 = 20$$
$$3x + 8 - 8 = 20 - 8$$
$$3x = 12$$
$$\frac{1}{3}(3x) = \frac{1}{3}(12)$$
$$x = 4$$

3.
$$.6k - .3 = .5k + .4$$
$$.6k - .5k - .3 = .5k - .5k + .4$$
$$.1k - .3 = .4$$
$$.1k - .3 + .3 = .4 + .3$$
$$.1k = .7$$
$$\frac{.1k}{.1} = \frac{.7}{.1}$$
$$k = 7$$

5.
$$2a - 1 = 4(a + 1) + 7a + 5$$
$$2a - 1 = 4a + 4 + 7a + 5$$
$$2a - 1 = 11a + 9$$
$$2a - 2a - 1 = 11a - 2a + 9$$
$$-1 = 9a + 9$$
$$-1 - 9 = 9a + 9 - 9$$
$$-10 = 9a$$
$$\frac{-10}{9} = \frac{9a}{9}$$
$$-\frac{10}{9} = a$$

7.
$$2[x-(3+2x)+9]=3x-8$$
$$2(x-3-2x+9)=3x-8$$
$$2(-x+6)=3x-8$$
$$-2x+12=3x-8$$
$$12=5x-8$$
$$20=5x$$
$$4=x$$

9. $\dfrac{3x}{5}-\dfrac{4}{5}(x+1)=2-\dfrac{3}{10}(3x-4)$

Multiply both sides by the common denominator, 10.

$$10\left(\dfrac{3x}{5}\right)-10\left(\dfrac{4}{5}\right)(x+1)$$
$$=(10)(2)-(10)\left(\dfrac{3}{10}\right)(3x-4)$$
$$2(3x)-8(x+1)=20-3(3x-4)$$
$$6x-8x-8=20-9x+12$$
$$-2x-8=32-9x$$
$$-2x+9x=32+8$$
$$7x=40$$
$$\dfrac{1}{7}(7x)=\dfrac{1}{7}(40)$$
$$x=\dfrac{40}{7}$$

11.
$$\dfrac{5y}{6}-8=5-\dfrac{2y}{3}$$
$$6\left(\dfrac{5y}{6}-8\right)=6\left(5-\dfrac{2y}{3}\right)$$
$$6\left(\dfrac{5y}{6}\right)-6(8)=6(5)-6\left(\dfrac{2y}{3}\right)$$
$$5y-48=30-4y$$
$$9y-48=30$$
$$9y=78$$
$$y=\dfrac{78}{9}$$
$$y=\dfrac{26}{3}\quad\text{Lowest terms}$$

13.
$$\dfrac{m}{2}-\dfrac{1}{m}=\dfrac{6m+5}{12}$$
$$12m\left(\dfrac{m}{2}-\dfrac{1}{m}\right)=12m\left(\dfrac{6m+5}{12}\right)$$
$$(12m)\dfrac{m}{2}-(12m)\dfrac{1}{m}=m(6m)+m(5)$$
$$6m^2-12=6m^2+5m$$
$$-12=5m$$
$$\dfrac{1}{5}(-12)=\dfrac{1}{5}(5m)$$
$$-\dfrac{12}{5}=m$$

15. $\dfrac{4}{x-3}-\dfrac{8}{2x+5}+\dfrac{3}{x-3}=0$

Multiply each side by the common denominator, $(x-3)(2x+5)$.

$$(x-3)(2x+5)\dfrac{4}{x-3}$$
$$-(x-3)(2x+5)\dfrac{8}{2x+5}$$
$$+(x-3)(2x+5)\dfrac{3}{x-3}$$
$$=(x-3)(2x+5)0$$
$$4(2x+5)-8(x-3)+3(2x+5)=0$$
$$8x+20-8x+24+6x+15=0$$
$$6x+59=0$$
$$6x=-59$$
$$x=-\dfrac{59}{6}$$

17. $\dfrac{3}{2m+4} = \dfrac{1}{m+2} - 2$

$\dfrac{3}{2(m+2)} = \dfrac{1}{m+2} - 2$

$2(m+2)\ \dfrac{3}{2(m+2)}$

$= 2(m+2)\ \dfrac{1}{m+2} - 2(m+2)(2)$

$3 = 2 - 4(m+2)$

$3 = 2 - 4m - 8$

$3 = -6 - 4m$

$3 + 6 = -4m$

$9 = -4m$

$m = -\dfrac{9}{4}$

19. $9.06x + 3.59(8x - 5) = 12.07x + .5612$

$9.06x + 28.72x - 17.95 = 12.07x + .5612$

$9.06x + 28.72x - 12.07x = 17.95 + .5612$

$25.71x = 18.5112$

$x = \dfrac{18.5112}{25.71}$

$x = .72$

21. $\dfrac{2.63r - 8.99}{1.25} - \dfrac{3.90r - 1.77}{2.45} = r$

Multiply by $(1.25)(2.45)$.

$(2.45)(2.63r - 8.99) - (1.25)(3.90r - 1.77)$

$= (1.25)(2.45)r$

$6.4435r - 22.0255 - 4.875r + 2.2125$

$= 3.0625r$

$6.4435r - 4.875r - 3.0625r$

$\qquad = 22.0255 - 2.2125$

$-1.494r = 19.813$

$r \approx -13.26$

23. $4(a + x) = b - a + 2x$

$4a + 4x = b - a + 2x$

$4a = b - a - 2x$

$5a - b = -2x$

$\dfrac{5a - b}{-2} = \dfrac{-2x}{-2}$

$-\dfrac{5a - b}{2} = x$ or $x = \dfrac{b - 5a}{2}$

25. $5(b - x) = 2b + ax$

First, use the distributive property.

$5b - 5x = 2b + ax$

$5b = 2b + ax + 5x$

$3b = ax + 5x$

Now use the distributive property on the right.

$3b = (a + 5)x$

$\dfrac{3b}{a + 5} = \dfrac{(a + 5)x}{a + 5}$

$\dfrac{3b}{a + 5} = x$

27. $PV = k$ for V

$\dfrac{1}{P}(PV) = \dfrac{1}{P}(k)$

$V = \dfrac{k}{P}$

29. $V = V_0 + gt$ for g

$V - V_0 = gt$

$\dfrac{V - V_0}{t} = \dfrac{gt}{t}$

$\dfrac{V - V_0}{t} = g$

31. $A = \dfrac{1}{2}(B + b)h$ for B

$A = \dfrac{1}{2}Bh + \dfrac{1}{2}bh$

$2A = Bh + bh$ Multiply by 2.

$2A - bh = Bh$

$\dfrac{2A - bh}{h} = \dfrac{Bh}{h}$ Multiply by $\dfrac{1}{h}$.

$\dfrac{2A - bh}{h} = B$

33. $|2h - 1| = 5$

$2h - 1 = 5$ or $2h - 1 = -5$

$2h = 6$ or $2h = -4$

$h = 3$ or $h = -2$

35. $|6 + 2p| = 10$

$6 + 2p = 10 \quad \text{or} \quad 6 + 2p = -10$

$2p = 4 \quad \text{or} \quad 2p = -16$

$p = 2 \quad \text{or} \quad p = -8$

37. $\left|\dfrac{5}{r - 3}\right| = 10$

$\dfrac{5}{r - 3} = 10$

$5 = 10(r - 3)$

$5 = 10r - 30$

$5 + 30 = 10r$

$35 = 10r$

$\dfrac{35}{10} = r$

$\dfrac{7}{2} = r$

or

$\dfrac{5}{r - 3} = -10$

$5 = -10(r - 3)$

$5 = -10r + 30$

$5 - 30 = -10r$

$-25 = -10r$

$\dfrac{-25}{-10} = r$

$\dfrac{5}{2} = r$

39. $|3y - 2| = |4y + 5|$

$3y - 2 = 4y + 5 \quad \text{or} \quad 3y - 2 = -(4y + 5)$

$-2 = y + 5 \qquad\qquad 3y - 2 = -4y - 5$

$-7 = y \qquad\qquad\qquad 7y - 2 = -5$

$\qquad\qquad\qquad\qquad\qquad 7y = -3$

$y = -7 \quad \text{or} \quad y = -\dfrac{3}{7}$

41. When $C = -10$,

$F = \dfrac{9(-10) + 160}{5}$

$= 14$

43. When $C = 18$,

$F = \dfrac{9(18) + 160}{5}$

$= 64.4$

45. $y = .79x + 3.93$

Substitute 8.67 for y.

$8.67 = .79x + 3.93$

$4.74 = .79x$

$6 = x$

Therefore, the federal deficit will be $8.67 billion in 2006.

47. $y = .79x + 3.93$

Substitute 12.62 for y.

$12.62 = .79x + 3.93$

$8.69 = .79x$

$11 = x$

Therefore, the federal deficit will be $12.62 billion in 2011.

49. $E = 73.04x + 625.6$

Substitute $991 in for E.

$991 = 73.04x + 625.6$

$365.4 = 73.04x$

$5.00 = x$

The health care expenditures were at $991 billion in 1995.

51. $E = 73.04x + 625.6$

Substitute $1794.25 in for E.

$1794.25 = 73.04x + 625.6$

$1168 = 73.04x$

$16.00 = x$

The health care expenditures were at $1794.25 billion in 2006.

53. $-2.1977(x - 2001) = y - 12.94$

Substitute 50 for y and solve for x.

$-2.1977(x - 2001) = 50 - 12.94$

$-2.1977x + 4397.598 = 50 - 12.94$

$-2.1977x = 37.06$

$-2.1977x = -4360.538$

$x = \dfrac{-4360.538}{-2.1977}$

$= 1984.1371$

50% of workers were covered in 1984.

55. $-2.1977(x - 2001) = y - 12.94$

Substitute 15 for y and solve for x.

$-2.1977(x - 2001) = 15 - 12.94$
$-2.1977x + 4397.598 = 15 - 12.94$
$-2.1977x = 2.06$
$-2.1977x = -4395.538$
$$x = \frac{-4395.538}{-2.1977}$$
$$= 2000.0628$$

15% of workers were covered in 2000.

57. $P = 2.831x + 224.361$

Substitute 280.980 for P and solve for x.
$280.980 = 2.831x + 224.361$
$56.619 = 2.831x$
$$\frac{56.619}{2.831} = x$$
$$20 \approx x$$

The United States had a population of 280,980,000 in 2000.

59. $P = 2.831x + 224.361$

Substitute 300.798 for P and solve for x.
$300.798 = 2.831x + 224.361$
$76.437 = 2.831x$
$$\frac{76.437}{2.831} = x$$
$$27 = x$$

The United States will have a population of 300,798,000 in 2007.

61. $f = 800, n = 18, q = 36$

$$u = f \cdot \frac{n(n+1)}{q(q+1)}$$

$$= 800 \cdot \frac{18(19)}{36(37)}$$

$$= 800 \cdot \frac{342}{1332}$$

$$\approx 205.41$$

The amount of unearned interest is $205.41.

63. a. $k = .132 \dfrac{B}{W}$

$k = .132 \dfrac{20}{75} = .0352$

b. $R = kd$
$R = .0352(.42)$
$\approx .015$ or 1.5%

63. Continued

c. Number of cases of cancer

$$= \frac{R \cdot 5000}{72}$$

$$= \frac{.015 \cdot 5000}{72}$$

$$\approx 1 \text{ case}$$

65. Let x represent the amount invested at 4%. Then $20,000 - x$ is the amount invested at 6%. Since the total interest is $1040,
$.04x + .06(20,000 - x) = 1040$
$.04x + 1200 - .06x = 1040$
$-.02x + 1200 = 1040$
$-.02x = -160$
$x = 8000$

She invested $8000 at 4%.

67. Let x represent the amount invested at 4%. $20,000 invested at 5% (or .05) plus x dollars invested at 4% (or .04) must equal 4.8% (or .048) of the total investment (or $20,000 + x$). Solve this equation.
$.05(20,000) + .04x = .048(20,000 + x)$
$1000 + .04x = 960 + .048x$
$1000 = 960 + .008x$
$40 = .008x$
$5000 = x$

$5000 should be invested at 4%.

69. Let x represent the distance between the two cities. Since $d = rt$, $t = \dfrac{d}{r}$. Make a table.

	Distance	Rate	Time
Going	x	50	$\dfrac{x}{50}$
Returning	x	55	$\dfrac{x}{55}$

Since the total traveling time was 32 hr,

$$\frac{x}{50} + \frac{x}{55} = 32$$

$$550\left(\frac{x}{50} + \frac{x}{55}\right) = 550(32)$$

$$11x + 10x = 17,600$$

$$21x = 17,600$$

$$x \approx 838.10$$

The distance between the two cities is approximately 840 mi.

71. Let x = the number of liters of 94 octane gas;
200 = the number of liters of 99 octane gas;
200 + x = the number of liters of 97 octane gas.
$$94x + 99(200) = 97(200 + x)$$
$$94x + 19,800 = 19,400 + 97x$$
$$19,800 - 19,400 = 97x - 94x$$
$$400 = 3x$$
$$\frac{400}{3} = x$$

Thus, $\dfrac{400}{3}$ liters of 94 octane gas are needed.

73. Let x = number of miles driven
$$20 + .18x = 45.56$$
$$.18x = 25.56$$
$$x = 142$$
In order for the car rental firms to break even, you must drive 142 miles in a day.

75. $y = 10(x - 65) + 50$
Substitute 100 in for y.
$$100 = 10(x - 65) + 50$$
$$50 = 10(x - 65)$$
$$5 = x - 65$$
$$70 = x$$
Paul was driving 70 mph.

77. Let x represent the width. Since the

length is 3 cm less than twice the width, the length is $2x - 3$.
The perimeter of a rectangle is twice the length plus twice the width. The perimeter is 54 cm, so the equation is $2x + 2(2x - 3) = 54$.
   Solve the equation.
$$2x + 2(2x - 3) = 54$$
$$2x + 4x - 6 = 54$$
$$6x - 6 = 54$$
$$6x = 60$$
$$x = 10$$
The width is 10 cm. The length is
$2(10) - 3 = 20 - 3 = 17$ cm.
Since $2(10) + 2(17) = 20 + 34 = 54$, as stated in the original problem, the solution checks.

79. Let x = the length of the shortest side,
$2x$ = the length of the second side, and
$x + 7$ = the length of the third side.
The perimeter, 27, is the sum of the lengths of the three sides, so
$$x + 2x + x + 7 = 27$$
$$4x + 7 = 27$$
$$4x = 20$$
$$x = 5$$
The length of the shortest side is 5 cm.

Section 1.7 Quadratic Equations

1. $(x + 4)(x - 14) = 0$
$$x + 4 = 0 \quad \text{or} \quad x - 14 = 0$$
$$x = -4 \quad \text{or} \quad x = 14$$
The solutions are –3 and 12.

3. $x(x + 6) = 0$
$$x = 0 \quad \text{or} \quad x + 6 = 0$$
$$x = -6$$

5. $3z^2 = 9z$
$$3z^2 - 9z = 0$$
$$3z(z - 3) = 0$$
$$z = 0 \quad \text{or} \quad z - 3 = 0$$
$$z = 3$$

7. $y^2 + 15y + 56 = 0$
$(y + 7)(y + 8) = 0$ Factor
$y + 7 = 0$ or $y + 8 = 0$ Zero-factor property
$y = -7$ or $y = -8$ Solve
The solutions are –7 and –8.

9. $$2x^2 = 7x - 3$$
$$2x^2 - 7x + 3 = 0$$
$$(2x - 1)(x - 3) = 0$$
$$2x - 1 = 0 \quad \text{or} \quad x - 3 = 0$$
$$x = \frac{1}{2} \quad \text{or} \quad x = 3$$

The solutions are $\dfrac{1}{2}$ and 3.

11.
$$6r^2 + r = 1$$
$$6r^2 + r - 1 = 0$$
$$(3r - 1)(2r + 1) = 0$$
$$3r - 1 = 0 \quad \text{or} \quad 2r + 1 = 0$$
$$r = \frac{1}{3} \quad \text{or} \quad r = -\frac{1}{2}$$

The solutions are $\frac{1}{3}$ and $-\frac{1}{2}$.

13.
$$2m^2 + 20 = 13m$$
$$2m^2 - 13m + 20 = 0$$
$$(2m - 5)(m - 4) = 0$$
$$2m - 5 = 0 \quad \text{or} \quad m - 4 = 0$$
$$m = \frac{5}{2} \quad \text{or} \quad m = 4$$

The solutions are $\frac{5}{2}$ and 4.

15.
$$m(m + 7) = -10$$
$$m^2 + 7m + 10 = 0$$
$$(m + 5)(m + 2) = 0$$
$$m + 5 = 0 \quad \text{or} \quad m + 2 = 0$$
$$m = -5 \quad \text{or} \quad m = -2$$
The solutions are -5 and -2.

17.
$$9x^2 - 16 = 0$$
$$(3x + 4)(3x - 4) = 0$$
$$3x + 4 = 0 \quad \text{or} \quad 3x - 4 = 0$$
$$3x = -4 \qquad\qquad 3x = 4$$
$$x - \frac{4}{3} \quad \text{or} \quad x = \frac{4}{3}$$
The solutions are $-\frac{4}{3}$ and $\frac{4}{3}$.

19. $16x^2 - 16x = 0$
$$16x(x - 1) = 0$$
$$16x = 0 \quad \text{or} \quad x - 1 = 0$$
$$x = 0 \quad \text{or} \quad x = 1$$
The solutions are 0 and 1.

21. $(r - 2)^2 = 7$
Use the square root property to solve this equation.
$$r - 2 = \sqrt{7} \quad \text{or} \quad r - 2 = -\sqrt{7}$$
$$r = 2 + \sqrt{7} \quad \text{or} \quad r = 2 - \sqrt{7}$$

We abbreviate the solutions as $2 \pm \sqrt{7}$.

23. $(4x - 1)^2 = 20$
Use the square root property.
$$4x - 1 = \sqrt{20} \quad \text{or} \quad 4x - 1 = -\sqrt{20}$$
$$4x - 1 = 2\sqrt{5} \quad \text{or} \quad 4x - 1 = -2\sqrt{5}$$
$$4x = 1 + 2\sqrt{5} \quad \text{or} \quad 4x = 1 - 2\sqrt{5}$$

The solutions are abbreviated as $\frac{1 \pm 2\sqrt{5}}{4}$.

25. $2x^2 + 7x + 1 = 0$
Use the quadratic formula with $a = 2$, $b = 7$, and $c = 1$.
$$x = \frac{-b \pm \sqrt{b^2 - 4ac}}{2a}$$
$$= \frac{-7 \pm \sqrt{7^2 - 4(2)(1)}}{2(2)}$$
$$= \frac{-7 \pm \sqrt{49 - 8}}{4}$$
$$x = \frac{-7 \pm \sqrt{41}}{4}$$

The solutions are $\frac{-5 + \sqrt{41}}{4}$ and $\frac{-5 - \sqrt{41}}{4}$, which are approximately $-.1492$ and -3.3508.

27. $4k^2 + 2k = 1$
Rewrite the equation in standard form.
$$4k^2 + 2k - 1 = 0$$
Use the quadratic formula with $a = 4$, $b = 2$, and $c = -1$.
$$k = \frac{-2 \pm \sqrt{2^2 - 4(4)(-1)}}{2(4)}$$
$$= \frac{-2 \pm \sqrt{4 + 16}}{8}$$
$$= \frac{-2 \pm \sqrt{20}}{8}$$
$$= \frac{-2 \pm 2\sqrt{5}}{8}$$
$$= \frac{2\left(-1 \pm \sqrt{5}\right)}{2 \cdot 4}$$
$$k = \frac{-1 \pm \sqrt{5}}{4}$$

The solutions are $\frac{-1 + \sqrt{5}}{4}$ and $\frac{-1 - \sqrt{5}}{4}$, which are approximately $.309$ and $-.809$.

29. $5y^2 + 5y = 2$

$5y^2 + 5y - 2 = 0$

$a = 5, b = 5, c = -2$

$y = \dfrac{-5 \pm \sqrt{5^2 - 4(5)(-2)}}{2(5)}$

$= \dfrac{-5 \pm \sqrt{25 + 40}}{10}$

$= \dfrac{-5 \pm \sqrt{65}}{10}$

$= \dfrac{-5 \pm \sqrt{65}}{10}$

The solutions are $\dfrac{-5 + \sqrt{65}}{10}$ and $\dfrac{-5 - \sqrt{65}}{10}$, which are approximately .3062 and −1.3062.

31. $6x^2 + 6x + 4 = 0$

$a = 6, b = 6, c = 4$

$x = \dfrac{-6 \pm \sqrt{6^2 - 4(6)(4)}}{2(6)}$

$= \dfrac{-6 \pm \sqrt{36 - 96}}{12}$

$x = \dfrac{-6 \pm \sqrt{-60}}{12}$

Because $\sqrt{-60}$ is not a real number, the given equation has no real number solutions.

33. $2r^2 - 3r - 5 = 0$

$r = \dfrac{-(3) \pm \sqrt{9 - 4(2)(-5)}}{2(2)} = \dfrac{-3 \pm \sqrt{49}}{4}$

$= \dfrac{-3 \pm 7}{4}$

$r = \dfrac{-3 + 7}{4} = \dfrac{4}{4} = 1$ or

$r = \dfrac{-3 - 7}{4} = \dfrac{-10}{4} = \dfrac{-5}{2}$

The solutions are $-\dfrac{5}{2}$ and 1.

35. $2x^2 - 7x + 30 = 0$

$x = \dfrac{-(-7) \pm \sqrt{49 - 4(2)(30)}}{2(2)}$

$= \dfrac{7 \pm \sqrt{-191}}{4}$

Since $\sqrt{-191}$ is not a real number, there are no real solutions.

37. $1 + \dfrac{7}{2a} = \dfrac{15}{2a^2}$

To eliminate fractions, multiply both sides by the common denominator, $2a^2$.

$2a^2 + 7a = 15$

$2a^2 + 7a - 15 = 0$

$(2a - 3)(a + 5) = 0$

$2a - 3 = 0$ or $a + 5 = 0$

$2a = 3$

$a = \dfrac{3}{2}$ or $a = -5$

The solutions are $\dfrac{3}{2}$ and −5.

39. $25t^2 + 49 = 70t$

$25t^2 - 70t + 49 = 0$

$b^2 - 4ac = (-70)^2 - 4(25)(49)$

$= 4900 - 4900$

$= 0$

The discriminant is 0.

There is one real solution to the equation.

41. $13x^2 + 24x - 5 = 0$

$b^2 - 4ac = (24)^2 - 4(13)(-5)$

$= 836$

The discriminant is positive.

There are two real solutions to the equation.

43. $4.42x^2 - 10.14x + 3.79 = 0$

$x = \dfrac{-(-10.14) \pm \sqrt{(-10.14)^2 - 4(4.42)(3.79)}}{2(4.42)}$

$\approx \dfrac{10.14 \pm 5.9843}{8.84}$

$\approx .4701$ or 1.8240

45. $7.63x^2 + 2.79x = 5.32$

$7.63x^2 + 2.79x - 5.32 = 0$

$x = \dfrac{-2.79 \pm \sqrt{(2.79)^2 - 4(7.63)(-5.32)}}{2(7.63)}$

$\approx \dfrac{-2.79 \pm 13.0442}{15.26}$

≈ -1.0376 or $.6720$

47. a. Let $d = 36$.

It takes 1.5 seconds for the object to fall 36 feet.

b. Let $d = 36 \cdot 5 = 180$.
$$180 = 16t^2$$
$$\frac{180}{16} = t^2$$
$$\sqrt{\frac{180}{16}} = t$$
$$3.35 \approx t$$
It takes about 3.35 seconds for the object to fall 180 feet.

49. a. Let $D = 1$.
$$1 = .0031x^2 - .291x + 7.1$$
$$0 = .0031x^2 - .291x + 6.1$$
Store $\sqrt{b^2 - 4ac} = \sqrt{(-.291)^2 - 4(.0031)(6.1)}$
$\approx .0951$ in your calculator.
By the quadratic formula, $x \approx 62$ or $x \approx 32$. People of about 32 or 62 years of age have a driver fatality rate of about 1 death per 1000.

b. Let $D = 3$.
$$3 = .0031x^2 - .291x + 7.1$$
$$0 = .0031x^2 - .291x + 4.1$$
Store $\sqrt{b^2 - 4ac} = \sqrt{(-.291)^2 - 4(.0031)(4.1)}$
$\approx .1840$ in your calculator.

By the quadratic formula, $x \approx 77$ or $x \approx 17$. People of about 77 or 17 years of age have a driver fatality rate of about 3 deaths per 1000.

51. a. Let $N = 1000$ (in millions).

$$1000 = .68x^2 + 3.8x + 24$$
$$0 = .68x^2 + 3.8x - 976$$

Store $\sqrt{b^2 - 4ac} = \sqrt{3.8^2 - 4(.68)(-976)}$
≈ 51.6639 in your calculator.

By the quadratic formula, $x \approx 35$.
There were approximately one billion airline passengers in 1985.

51. Continued

b. Let $N = 2000$ (in millions).
$$2000 = .68x^2 + 3.8x + 24$$
$$0 = .68x^2 + 3.8x - 1976$$
Store
$$\sqrt{b^2 - 4ac} = \sqrt{3.8^2 - 4(.68)(-1976)}$$
≈ 73.4109 in your calculator.
By the quadratic formula, $x \approx 51$.
There were approximately 2 billion airline passengers in 2001.

53. a. The eastbound train travels at a speed of $x + 20$.

b. The northbound train travels a distance of $5x$ in 5 hours.
The eastbound train travels a distance of $5(x + 20)$ in 5 hours.

c.

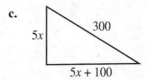

By the Pythagorean theorem,
$$(5x)^2 + (5x + 100)^2 = 300^2$$

d. Expand and combine like terms.
$$25x^2 + 25x^2 + 1000x + 10,000 = 90,000$$
$$50x^2 + 1000x - 80,000 = 0$$
Factor out the common factor, 50, and divide both sides by 50.
$$50(x^2 + 20x - 1600) = 0$$
$$x^2 + 20x - 1600 = 0$$
Store
$$\sqrt{b^2 - 4ac} = \sqrt{20^2 - 4(1)(-1600)} \approx 82.4621$$
in your calculator. By the quadratic formula, $x \approx 31.23$ or $x \approx -51.23$. Since x cannot be negative, then the speed of the northbound train is $x \approx 31.23$ mph, and the speed of the eastbound train is $x + 20 \approx 51.23$ mph.

55. a. Let x represent the length. Then, $\dfrac{300-2x}{2}$ or $150 - x$ represents the width.

b. Use the formula for the area of a rectangle.
$$LW = A$$
$$x(150 - x) = 5000$$

c. $150x - x^2 = 5000$
Write this quadratic equation in standard form and solve by factoring.
$$0 = x^2 - 150x + 5000$$
$$x^2 - 150x + 5000 = 0$$
$$(x - 50)(x - 100) = 0$$
$$x - 50 = 0 \quad \text{or} \quad x - 100 = 0$$
$$x = 50 \quad \text{or} \qquad x = 100$$
Choose $x = 100$ because the length is the larger dimension. The length is 100 m and the width is
$$150 - 100 = 50 \text{ m}.$$

57. Let x = the width of the uniform strip around the rug $(0 < x < 6)$.

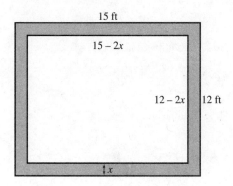

The dimensions of the rug are $15 - 2x$ and $12 - 2x$. The area, 108, is the length times the width. Solve the equation.

$$(15 - 2x)(12 - 2x) = 108$$
$$180 - 54x + 4x^2 = 108$$
$$4x^2 - 54x + 72 = 0$$
$$2x^2 - 27x + 36 = 0$$
$$(x - 12)(2x - 3) = 0$$

$$x - 12 = 0 \quad \text{or} \quad 2x - 3 = 0$$
$$x = 12 \quad \text{or} \qquad x = \frac{3}{2}$$

Discard $x = 12$ since $0 < x < 6$.

If $x = \dfrac{3}{2}$, then

57. Continued

$$15 - 2x = 15 - 2\left(\frac{3}{2}\right) = 12$$

and $12 - 2x = 12 - 2\left(\dfrac{3}{2}\right) = 9$.

The dimensions of the rug should be 9 ft by 12 ft.

59. a.
$$h = 64t - 16t^2$$
$$64 = 64t - 16t^2 \quad \text{Let } h = 64$$
$$16t^2 - 64t + 64 = 0$$
$$t^2 - 4t + 4 = 0$$
$$(t - 2)^2 = 0$$
$$t - 2 = 0$$
$$t = 2$$
The ball will reach 64 ft in 2 sec.

b.
$$h = 64t - 16t^2$$
$$28 = 64t - 16t^2 \quad \text{Let } h = 28.$$

$$16t^2 - 64t + 28 = 0$$
$$4t^2 - 16t + 7 = 0$$
$$(2t - 7)(2t - 1) = 0$$
$$2t - 7 = 0 \quad \text{or} \quad 2t - 1 = 0$$
$$t = \frac{7}{2} \quad \text{or} \qquad t = \frac{1}{2}$$
The ball will reach 28 ft at both $\dfrac{1}{2}$ sec and $\dfrac{7}{2}$ sec.

c. It reaches the given height twice, once on the way up and once on the way down. (Note that it reaches 64 feet only once because that is the maximum height.)

61.
$$S = \frac{1}{2}gt^2 \text{ for } t$$
$$2S = gt^2$$
$$\frac{2S}{g} = t^2$$
$$\pm\sqrt{\frac{2S}{g}} = t$$
$$\pm\frac{\sqrt{2Sg}}{g} = t$$

63.

$$L = \frac{d^4 k}{h^2} \text{ for } h$$

$$Lh^2 = d^4 k$$

$$h^2 = \frac{d^4 k}{L}$$

$$h = \pm\sqrt{\frac{d^4 k}{L}} \cdot \frac{\sqrt{L}}{\sqrt{L}} = \pm\frac{\sqrt{d^4 kL}}{L}$$

$$h = \pm\frac{d^2\sqrt{kL}}{L}$$

65.

$$P = \frac{E^2 R}{(r+R)^2} \text{ for } R$$

$$P(r+R)^2 = E^2 R$$

$$P(r^2 + 2rR + R^2) = E^2 R$$

$$Pr^2 + 2PrR + PR^2 = E^2 R$$

$$PR^2 + (2Pr - E^2)R + Pr^2 = 0$$

Solve for R by using the quadratic formula with $a = P$, $b = 2Pr - E^2$, and $c = Pr^2$.

$$R = \frac{-(2Pr - E^2) \pm \sqrt{(2Pr - E^2)^2 - 4P \cdot Pr^2}}{2P}$$

$$= \frac{-2Pr + E^2 \pm \sqrt{4P^2 r^2 - 4PrE^2 + E^4 - 4P^2 r^2}}{2P}$$

$$= \frac{-2Pr + E^2 \pm \sqrt{E^4 - 4PrE^2}}{2P}$$

$$= \frac{-2Pr + E^2 \pm \sqrt{E^2(E^2 - 4Pr)}}{2P}$$

$$R = \frac{-2Pr + E^2 \pm E\sqrt{E^2 - 4Pr}}{2P}$$

67. a. Let $x = z^2$.

$$x^2 - 2x = 15$$

b.
$$x^2 - 2x = 15$$
$$x^2 - 2x - 15 = 0$$
$$(x-5)(x+3) = 0$$
$$x = 5 \text{ or } x = -3$$

c. Let $z^2 = 5$.

$z = \pm\sqrt{5}$, that is, $z = \sqrt{5}$ or $z = -\sqrt{5}$.

69. $2q^4 + 3q^2 - 9 = 0$

Let $u = q^2$; then $u^2 = q^4$.

$$2u^2 + 3u - 9 = 0$$
$$(2u-3)(u+3) = 0$$
$$2u - 3 = 0 \quad \text{or} \quad u + 3 = 0$$
$$u = \frac{3}{2} \quad \text{or} \quad u = -3$$
$$q^2 = \frac{3}{2} \quad \text{or} \quad q^2 = -3$$
$$q = \pm\sqrt{\frac{3}{2}} \quad \text{or} \quad q = \pm\sqrt{-3}$$

$\pm\sqrt{-3}$ are not real numbers.

$$q = \pm\frac{\sqrt{3}}{\sqrt{2}} \cdot \frac{\sqrt{2}}{\sqrt{2}} = \pm\frac{\sqrt{6}}{2}$$

The solutions are $\pm\frac{\sqrt{6}}{2}$.

71. $z^4 - 3z^2 - 1 = 0$

Let $x = z^2$; then $x^2 = z^4$.

$$x^2 - 3x - 1 = 0$$

By the quadratic formula, $x = \frac{3 \pm \sqrt{13}}{2}$.

$$\frac{3 + \sqrt{13}}{2} = z^2$$

$$\pm\sqrt{\frac{3 + \sqrt{13}}{2}} = z$$

Chapter 1 Review Exercises

1. 0 and 6 are whole numbers.

2. $-12, -6, -\sqrt{4}(=-2), 0$, and 6 are integers.

3. $-12, -6, -\dfrac{9}{10}, -\sqrt{4}(=-2), 0, \dfrac{1}{8}$, and 6 are rational numbers.

4. $-\sqrt{7}, \dfrac{\pi}{4}, \sqrt{11}$ are irrational numbers.

5. $9[(-3)4] = 9[4(-3)]$
   Commutative property of multiplication

6. $7(4 + 5) = (4 + 5)7$
   Commutative property of multiplication

7. $6(x + y - 3) = 6x + 6y + 6(-3)$
   Distributive property

8. $11 + (5 + 3) = (11 + 5) + 3$
   Associative property of addition

9. x is at least 9.
   $x \geq 9$

10. x is negative.
    $x < 0$

11. $-7, -3, -2, 0, \pi, 8$
    (π is about 3.14.)

12. $-\dfrac{5}{4}, -\dfrac{2}{3}, -\dfrac{3}{8}, \dfrac{1}{2}, \dfrac{5}{6}$

13. $|6 - 4| = 2, -|-2| = -2, |8 + 1| = |9| = 9$,
    $-|3 - (-2)| = -|3 + 2| = -|5| = -5$
    Since $-5, -2, 2, 9$ are in order, then
    $-|3 - (-2)|, -|-2|, |6 - 4|, |8 + 1|$ are in order.

14. $-\left|\sqrt{16}\right| = -4, -\sqrt{8}, \sqrt{7}, \left|-\sqrt{12}\right| = \sqrt{12}$

15. $-|-5| + |3| = -5 + 3 = -2$

16. $|-6| + |-9| = 6 + 9 = 15$

17. $7 - |-8| = 7 - 8 = -1$

18. $|-2| - |-7 + 3| = 2 - |-4|$
    $= 2 - 4$
    $= -2$

19. $x \geq -3$
    Start at -3 and draw a line to the right. Use a bracket at -3 to show that -3 is a part of the graph.

20. $-4 < x \leq 6$
    Put a parenthesis at -4 and a bracket at 6. Draw a line segment between these two endpoints.
    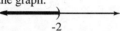

21. $x < -2$
    Start at -2 and draw a line to the left. Use a parenthesis at -2 to show that -2 is not a part of the graph.

22. $x \leq 1$
    Start at 1 and draw a line to the left. Use a bracket to show that 1 is a part of the graph.

23. $(-6 + 3 \cdot 5)(-2) = (-6 + 15)(-2)$
    $= 9(-2) = -18$

24. $-4(-8 - 9 \div 3) = -4(-8 - 3)$
    $= -4(-11)$
    $= 44$

25. $\dfrac{-8 + (-6)(-3) \div 9}{6 - (-2)} = \dfrac{-8 + 18 \div 9}{6 + 2}$
    $= \dfrac{-8 + 2}{8} = \dfrac{-6}{8} = -\dfrac{3}{4}$

26. $\dfrac{20 \div 4 \cdot 2 \div 5 - 1}{-9 - (-3) - 12 \div 3} = \dfrac{5 \cdot 2 \div 5 - 1}{-9 - (-3) - 4}$
    $= \dfrac{10 \div 5 - 1}{-9 + 3 - 4}$
    $= \dfrac{2 - 1}{-6 - 4}$
    $= \dfrac{1}{-10} = -\dfrac{1}{10}$

27. $\left(3x^4 - x^2 + 5x\right) - \left(-x^4 + 3x^2 - 6x\right)$
    $= \left(3x^4 - x^2 + 5x\right) + \left(x^4 - 3x^2 + 6x\right)$
    $= 3x^4 - x^2 + 5x + x^4 - 3x^2 + 6x$
    $= \left(3x^4 + x^4\right) + \left(-x^2 - 3x^2\right) + (5x + 6x)$
    $= 4x^4 - 4x^2 + 11x$

28. $\left(-8y^3 + 8y^2 - 3y\right) - \left(2y^3 - 4y^2 - 12\right)$

$= \left(-8y^3 + 8y^2 - 3y\right) + \left(-2y^3 + 4y^2 + 12\right)$

$= -8y^3 + 8y^2 - 3y + -2y^3 + 4y^2 + 12$

$= \left(-8y^3 - 2y^3\right) + \left(8y^2 + 4y^2\right) - 3y + 12$

$= -10y^3 + 12y^2 - 3y + 12$

29. $-2\left(q^4 - 3q^3 + 4q^2\right) + 4\left(q^4 + 2q^3 + q^2\right)$

$= -2q^4 + 6q^3 - 8q^2 + 4q^4 + 8q^3 + 4q^2$

$= \left(-2q^4 + 4q^4\right) + \left(6q^3 + 8q^3\right) + \left(-8q^2 + 4q^2\right)$

$= 2q^4 + 14q^3 + \left(-4q^2\right)$

$= 2q^4 + 14q^3 - 4q^2$

30. $5\left(3y^4 - 4y^5 + y^6\right) - 3\left(2y^4 + y^5 - 3y^6\right)$

$= 15y^4 - 20y^5 + 5y^6 - 6y^4 - 3y^5 + 9y^6$

$= 14y^6 - 23y^5 + 9y^4$

31. $(4z + 2)(3z - 2)$

$= 12z^2 - 8z + 6z - 4$

$= 12z^2 - 2z - 4$

32. $(8p - 4)(5p + 2)$

$= 40p^2 + 16p - 20p - 8$

$= 40p^2 - 4p - 8$

33. $(4k - 3h)(4k + 3h) = (4k)^2 - (3h)^2$

$= 16k^2 - 9h^2$

34. $(2r - 5y)(2r + 5y) = (2r)^2 - (5y)^2$

$= 4r^2 - 25y^2$

35. $(6x + 3y)^2$

$= (6x)^2 + 2(6x)(3y) + (3y)^2$

$= 36x^2 + 36xy + 9y^2$

36. $(2a - 5b)^2$

$= (2a)^2 - 2(2a)(5b) + (5b)^2$

$= 4a^2 - 20ab + 25b^2$

37. $2kh^2 - 4kh + 5k = k\left(2h^2 - 4h + 5\right)$

38. $2m^2n^2 + 6mn^2 + 16n^2 = 2n^2\left(m^2 + 3m + 8\right)$

39. $3a^4 + 13a^3 + 4a^2 = a^2\left(3a^2 + 13a + 4\right)$

$= a^2(3a + 1)(a + 4)$

40. $24x^3 + 4x^2 - 4x = 4x\left(6x^2 + x - 1\right)$

$= 4x(3x - 1)(2x + 1)$

41. $10y^2 - 11y + 3 = (2y - 1)(5y - 3)$

42. $8q^2 + 3m + 4qm + 6q$

$= \left(8q^2 + 4qm\right) + (3m + 6q)$ Group the terms.

$= 4q(2q + m) + 3(m + 2q)$ Factor each group.

$= 4q(2q + m) + 3(2q + m)$ Factor out $2q + m$.

$= (2q + m)(4q + 3)$

43. $4a^2 - 20a + 25$

$= (2a)^2 - 2(2a)(5) + 5^2$

$= (2a - 5)^2$

44. $36p^2 + 12p + 1$

$= (6p)^2 + 2(6p)(1) + 1^2$

$= (6p + 1)^2$

45. $144p^2 - 169q^2$

$= (12p)^2 - (13q)^2$

$= (12p + 13q)(12p - 13q)$

46. $81z^2 - 25x^2 = (9z)^2 - (5x)^2$

$= (9z + 5x)(9z - 5x)$

47. $8y^3 - 1$

$= (2y)^3 - 1^3$

$= (2y - 1)\ \left[(2y)^2 + 2y(1) + 1^2\right]$

$= (2y - 1)\left(4y^2 + 2y + 1\right)$

48. $125a^3 + 216$

$= (5a)^3 + (6)^3$

$= (5a + 6)\ \left[(5a)^2 - 5a(6) + 6^2\right]$

$= (5a + 6)\left(25a^2 - 30a + 36\right)$

49. $\dfrac{3x}{5} \cdot \dfrac{35x}{12} = \dfrac{3x \cdot 35x}{5 \cdot 12}$

$= \dfrac{3 \cdot 5 \cdot 7x^2}{4 \cdot 5 \cdot 3}$

$= \dfrac{7x^2}{4}$

50. $\dfrac{5k^2}{24} - \dfrac{70k}{36} = \dfrac{5k^2 \cdot 3}{24 \cdot 3} - \dfrac{70k \cdot 2}{36 \cdot 2}$

$\qquad = \dfrac{15k^2}{72} - \dfrac{140k}{72}$

$\qquad = \dfrac{15k^2 - 140k}{72}$

$\qquad = \dfrac{5k(3k - 28)}{72}$

51. $\dfrac{c^2 - 3c + 2}{2c(c-1)} \div \dfrac{c-2}{8c}$

$\qquad = \dfrac{(c-1)(c-2)}{2c(c-1)} \cdot \dfrac{8c}{(c-2)}$

$\qquad = \dfrac{8c(c-1)(c-2)}{2c(c-1)(c-2)}$

$\qquad = \dfrac{8}{2} = 4$

52. $\dfrac{p^3 - 2p^2 - 8p}{3p\left(p^2 - 16\right)} \div \dfrac{p^2 + 4p + 4}{9p^2}$

$\qquad = \dfrac{p\left(p^2 - 2p - 8\right)}{3p(p+4)(p-4)} \cdot \dfrac{9p^2}{(p+2)(p+2)}$

$\qquad = \dfrac{p(p-4)(p+2) \cdot 9p^2}{3p(p+4)(p-4)(p+2)(p+2)}$

$\qquad = \dfrac{3p(p-4)(p+2) \cdot 3p^2}{3p(p-4)(p+2) \cdot (p+4)(p+2)}$

$\qquad = \dfrac{3p^2}{(p+4)(p+2)}$

53. $\dfrac{2y - 10}{5y} \cdot \dfrac{20y - 25}{14}$

$\qquad = \dfrac{2(y-5)}{5y} \cdot \dfrac{5(4y-5)}{14}$

$\qquad = \dfrac{2 \cdot 5 \cdot (y-5)(4y-5)}{2 \cdot 5 \cdot 7y}$

$\qquad = \dfrac{(y-5)(4y-5)}{7y}$

54. $\dfrac{m^2 - 2m}{15m^3} \cdot \dfrac{5}{m^2 - 4}$

$\qquad = \dfrac{m(m-2)}{5 \cdot 3m^3} \cdot \dfrac{5}{(m+2)(m-2)}$

$\qquad = \dfrac{5m(m-2) \cdot 1}{5m(m-2) \cdot 3m^2(m+2)}$

$\qquad = \dfrac{1}{3m^2(m+2)}$

55. $\dfrac{2m^2 - 4m + 2}{m^2 - 1} \div \dfrac{6m + 18}{m^2 + 2m - 3}$

$\qquad = \dfrac{2\left(m^2 - 2m + 1\right)}{(m+1)(m-1)} \cdot \dfrac{m^2 + 2m - 3}{6m + 18}$

$\qquad = \dfrac{2(m-1)^2}{(m+1)(m-1)} \cdot \dfrac{(m+3)(m-1)}{6(m+3)}$

$\qquad = \dfrac{2(m-1)(m+3) \cdot (m-1)^2}{2(m-1)(m+3) \cdot 3(m+1)}$

$\qquad = \dfrac{(m-1)^2}{3(m+1)}$

56. $\dfrac{x^2 + 6x + 5}{4\left(x^2 + 1\right)} \cdot \dfrac{2x(x+1)}{x^2 - 25}$

$\qquad = \dfrac{(x+5)(x+1) \cdot 2x(x+1)}{4\left(x^2 + 1\right) \cdot (x+5)(x-5)}$

$\qquad = \dfrac{2(x+5) \cdot x(x+1)^2}{2(x+5) \cdot 2\left(x^2 + 1\right)(x-5)}$

$\qquad = \dfrac{x(x+1)^2}{2\left(x^2 + 1\right)(x-5)}$

57. $\dfrac{6}{15z} + \dfrac{2}{3z} - \dfrac{9}{10z}$

$\qquad = \dfrac{6 \cdot 2}{15z \cdot 2} + \dfrac{2 \cdot 10}{3z \cdot 10} - \dfrac{9 \cdot 3}{10z \cdot 3}$

$\qquad = \dfrac{12}{30z} + \dfrac{20}{30z} - \dfrac{27}{30z}$

$\qquad = \dfrac{5}{30z} = \dfrac{1}{6z}$

58. $\dfrac{5}{y-3} - \dfrac{4}{y} = \dfrac{5(y)}{(y-3)(y)} - \dfrac{4(y-3)}{y(y-3)}$

$\qquad = \dfrac{5y - 4(y-3)}{y(y-3)}$

$\qquad = \dfrac{5y - 4y + 12}{y(y-3)}$

$\qquad = \dfrac{y + 12}{y(y-3)}$

59. $\dfrac{2}{5q}+\dfrac{10}{7q}=\dfrac{2\cdot 7}{5q\cdot 7}+\dfrac{10\cdot 5}{7q\cdot 5}$

$\qquad\qquad =\dfrac{14}{35q}+\dfrac{50}{35q}$

$\qquad\qquad =\dfrac{64}{35q}$

60. $125^{\frac{2}{3}}=(125^2)^{\frac{1}{3}}=(15,625)^{\frac{1}{3}}=25$

$\quad$ or $125^{\frac{2}{3}}=(125^{\frac{1}{3}})^2=(5)^2=25$

61. $6^{-3}=\dfrac{1}{6^3}$ or $\dfrac{1}{216}$

62. $10^{-2}=\dfrac{1}{10^2}$ or $\dfrac{1}{100}$

63. $-8^0=-\left(8^0\right)=-1$

64. $-3^{-1}=-\left(3^{-1}\right)=-\dfrac{1}{3^1}=-\dfrac{1}{3}$

65. $-\dfrac{6}{5}^{-2}=-\dfrac{5}{6}^2$

$\qquad\qquad =\dfrac{(-5)^2}{(-6)^2}$

$\qquad\qquad =\dfrac{5^2}{6^2}$ or $\dfrac{25}{36}$

66. $\dfrac{3}{2}^{-3}=\dfrac{2}{3}^3$

$\qquad\qquad =\dfrac{(2)^3}{3^3}$

$\qquad\qquad =\dfrac{2^3}{3^3}$ or $\dfrac{8}{27}$

67. $4^6\cdot 4^{-3}=4^{6+(-3)}=4^3$

68. $7^{-5}\cdot 7^{-2}=7^{-5+(-2)}=7^{-7}=\dfrac{1}{7^7}$

69. $\dfrac{9^{-4}}{9^{-3}}=9^{-4-(-3)}=9^{-4+3}=9^{-1}=\dfrac{1}{9}$

70. $\dfrac{5^{-2}}{5^3}=5^{-2-3}=5^{-5}=\dfrac{1}{5^5}$

71. $\dfrac{9^4\cdot 9^{-5}}{\left(9^{-2}\right)^2}=\dfrac{9^4\cdot 9^{-5}}{9^{-4}}$

$\qquad\qquad =9^{4-5-(-4)}=9^3$

72. $\dfrac{k^4\cdot k^{-3}}{\left(k^{-2}\right)^{-3}}=\dfrac{k}{k^6}=\dfrac{1}{k^5}$

73. $4^{-1}+2^{-1}=\dfrac{1}{4}+\dfrac{1}{2}$

$\qquad\qquad =\dfrac{1}{4}+\dfrac{2}{4}=\dfrac{3}{4}$

74. $5^{-2}+5^{-1}=\dfrac{1}{5^2}+\dfrac{1}{5}$

$\qquad\qquad =\dfrac{1}{25}+\dfrac{1}{5}$

$\qquad\qquad =\dfrac{1}{25}+\dfrac{5}{25}$

$\qquad\qquad =\dfrac{6}{25}$

75. $125^{\frac{2}{3}}=\left(125^{\frac{1}{3}}\right)^2=5^2=25$

76. $128^{\frac{3}{7}}=\left(128^{\frac{1}{7}}\right)^3=2^3=8$

77. $9^{-\frac{5}{2}}=\dfrac{1}{9^{\frac{5}{2}}}=\dfrac{1}{\left(9^{\frac{1}{2}}\right)^5}=\dfrac{1}{3^5}$

78. $\dfrac{144}{49}^{-\frac{1}{2}}=\dfrac{49}{144}^{\frac{1}{2}}=\dfrac{7}{12}$

79. $\dfrac{5^{\frac{1}{3}}5^{\frac{1}{2}}}{5^{\frac{3}{2}}}=5^{\frac{1}{3}+\frac{1}{2}-\frac{3}{2}}$

$\qquad\qquad =5^{\frac{2}{6}+\frac{3}{6}-\frac{9}{6}}$

$\qquad\qquad =5^{-\frac{4}{6}}=5^{-\frac{2}{3}}$

$\qquad\qquad =\dfrac{1}{5^{\frac{2}{3}}}$

80. $\dfrac{2^{\frac{3}{4}}\cdot 2^{-\frac{1}{2}}}{2^{\frac{1}{4}}}=\dfrac{2^{\frac{1}{4}}}{2^{\frac{1}{4}}}=1$

81. $\left(3a^2\right)^{\frac{1}{2}}\cdot\left(3^2 a\right)^{\frac{3}{2}}=3^{\frac{1}{2}}a\cdot 3^3 a^{\frac{3}{2}}$

$\qquad\qquad =3^{\frac{1}{2}+3}a^{1+\frac{3}{2}}$

$\qquad\qquad =3^{\frac{7}{2}}a^{\frac{5}{2}}$

82. $(4p)^{\frac{2}{3}} \cdot \left(2p^3\right)^{\frac{3}{2}}$

$= 4^{\frac{2}{3}} p^{\frac{2}{3}} \cdot 2^{\frac{3}{2}} \cdot p^{\frac{9}{2}}$

$= \left(2^2\right)^{\frac{2}{3}} p^{\frac{2}{3}} \cdot 2^{\frac{3}{2}} p^{\frac{9}{2}}$

$= 2^{\frac{4}{3}} \cdot 2^{\frac{3}{2}} p^{\frac{2}{3}} p^{\frac{9}{2}}$

$= 2^{\frac{17}{6}} p^{\frac{31}{6}}$

83. $\sqrt[3]{27} = 3$

84. $\sqrt[6]{-64}$ is not a real number.

85. $\sqrt{54} = \sqrt{9 \cdot 6} = \sqrt{9} \cdot \sqrt{6} = 3\sqrt{6}$

86. $\sqrt{63} = \sqrt{9} \cdot \sqrt{7} = 3\sqrt{7}$

87. $\sqrt[3]{54 p^3 q^5} = \sqrt[3]{27 \cdot 2 p^3 q^3 q^2}$

$= \sqrt[3]{27 p^3 q^3} \cdot \sqrt[3]{2q^2}$

$= 3pq\sqrt[3]{2q^2}$

88. $\sqrt[4]{64 a^5 b^3} = \sqrt[4]{16 a^4} \cdot \sqrt[4]{4ab^3}$

$= 2a\sqrt[4]{4ab^3}$

89. $\sqrt{\dfrac{5n^2}{6m}} = \dfrac{n\sqrt{5}}{\sqrt{6m}} \cdot \dfrac{\sqrt{6m}}{\sqrt{6m}}$

$= \dfrac{n\sqrt{30m}}{6m}$

90. $\sqrt{\dfrac{3x^3}{2z}} = \sqrt{\dfrac{3xx^2}{2z}} \cdot \dfrac{\sqrt{2z}}{\sqrt{2z}}$

$= \dfrac{x\sqrt{3x}\sqrt{2z}}{2z} = \dfrac{x\sqrt{6xz}}{2z}$

91. $2\sqrt{3} - 5\sqrt{12} = 2\sqrt{3} - 5\sqrt{4 \cdot 3}$

$= 2\sqrt{3} - 5 \cdot 2\sqrt{3}$

$= 2\sqrt{3} - 10\sqrt{3}$

$= -8\sqrt{3}$

92. $8\sqrt{7} + 2\sqrt{63} = 8\sqrt{7} + 2\sqrt{9 \cdot 7}$

$= 8\sqrt{7} + 6\sqrt{7}$

$= 14\sqrt{7}$

93. $\left(\sqrt{6} - 1\right)\left(\sqrt{6} + 1\right) = \left(\sqrt{6}\right)^2 - 1^2$

$= 6 - 1$

$= 5$

94. $\left(\sqrt{5} - \sqrt{3}\right)\left(\sqrt{5} + \sqrt{3}\right) = \left(\sqrt{5}\right)^2 - \left(\sqrt{3}\right)^2$

$= 5 - 3$

$= 2$

95. $\dfrac{\sqrt{2}}{1 + \sqrt{3}} = \dfrac{\sqrt{2}\left(1 - \sqrt{3}\right)}{\left(1 + \sqrt{3}\right)\left(1 - \sqrt{3}\right)}$

$= \dfrac{\sqrt{2} - \sqrt{6}}{1 - 3} = \dfrac{-\sqrt{2} + \sqrt{6}}{2}$

96. $\dfrac{4 + \sqrt{2}}{4 - \sqrt{5}} = \dfrac{\left(4 + \sqrt{2}\right)}{\left(4 - \sqrt{5}\right)} \cdot \dfrac{\left(4 + \sqrt{5}\right)}{\left(4 + \sqrt{5}\right)}$

$= \dfrac{16 + 4\sqrt{2} + 4\sqrt{5} + \sqrt{10}}{16 - \left(\sqrt{5}\right)^2}$

$= \dfrac{16 + 4\sqrt{2} + 4\sqrt{5} + \sqrt{10}}{11}$

For Exercises 97–100, use the formula

$$\begin{array}{c}\text{Amount for} \\ \text{large state}\end{array} = \dfrac{E_{\text{large}}}{E_{\text{small}}}^{\frac{3}{2}} \times \begin{array}{c}\text{Amount for} \\ \text{small state}\end{array}$$

and the amount for a small state is $1,000,000.

97. $E_{\text{large}} = 27$, $E_{\text{small}} = 3$

$\begin{array}{c}\text{Amount for} \\ \text{large state}\end{array} = \dfrac{27}{3}^{\frac{3}{2}} \times 1,000,000$

$= 9^{\frac{3}{2}} \times 1,000,000$

$= 27 \times 1,000,000$

$= \$27,000,000$

98. $E_{\text{large}} = 31$, $E_{\text{small}} = 6$

$\begin{array}{c}\text{Amount for} \\ \text{large state}\end{array} = \dfrac{31}{6}^{\frac{3}{2}} \times 1,000,000$

$= 5.16667^{\frac{3}{2}} \times 1,000,000$

$= 11.74399 \times 1,000,000$

$= \$11,743,989.79$

99. $E_{\text{large}} = 55$, $E_{\text{small}} = 5$

$\begin{array}{c}\text{Amount for} \\ \text{large state}\end{array} = \dfrac{55}{5}^{\frac{3}{2}} \times 1,000,000$

$= 36.48287269 \times 1,000,000$

$= \$36,482,872.69$

100. $E_{\text{large}} = 34$, $E_{\text{small}} = 10$

$$\text{Amount for large state} = \frac{34}{10}^{\frac{3}{2}} \times 1,000,000$$

$$= 6.26929 \times 1,000,000$$

$$= \$6,269,290.23$$

101. $2x - 5(x-3) = 3x + 9$

$2x - 5x + 15 = 3x + 9$

$-3x + 15 = 3x + 9$

$6 = 6x$

$1 = x$

102. $4y + 9 = -3(1 - 2y) + 5$

$4y + 9 = -3 + 6y + 5$

$4y + 9 = 2 + 6y$

$-2y + 9 = 2$

$-2y = -7$

$y = \frac{7}{2}$

103. $\dfrac{2z}{5} - \dfrac{4z-3}{10} = \dfrac{-z+1}{10}$

$4z - (4z - 3) = -z + 1$ Multiply by 10.

$3 = -z + 1$

$z = -2$

104. $\dfrac{p}{p+2} - \dfrac{3}{4} = \dfrac{2}{p+2}$

$4(p+2)\,\dfrac{p}{p\,2} - 4(p+2)\,\dfrac{3}{4}$

$= 4(p+2)\,\dfrac{2}{p\,2}$

$4p - 3p - 6 = 8$

$p = 14$

105. $\dfrac{2m}{m-3} = \dfrac{6}{m-3} + 4$

$2m = 6 + 4(m-3)$

$2m = 6 + 4m - 12$

$2m = 4m - 6$

$6 = 2m$

$3 = m$

Because $m = 3$ would make the denominators of the fractions equal to 0, making the fractions undefined, the given equation has no solution.

106. $\dfrac{15}{k+5} = 4 - \dfrac{3k}{k+5}$

$(k+5)\,\dfrac{15}{k+5} = (k+5)4 - (k+5)\,\dfrac{3k}{k+5}$

$15 = 4k + 20 - 3k$

$-5 = k$

If $k = -5$, the fractions would be undefined, so the given equation has no solution.

107. $6ax - 1 = x$

$6ax - x = 1$

$(6a-1)x = 1$

$\dfrac{(6a-1)x}{6a-1} = \dfrac{1}{6a-1}$

$x = \dfrac{1}{6a-1}$

108. $6x - 5y = 4bx$

$6x - 4bx = 5y$

$(6 - 4b)x = 5y$

$\dfrac{(6-4b)x}{6-4b} = \dfrac{5y}{6-4b}$

$x = \dfrac{5y}{6-4b}$

109. $\dfrac{2x}{3-c} = ax + 1$

$(3-x)\,\dfrac{2x}{3-c} = (3-c)(ax+1)$

$2x = 3ax + 3 - acx - c$

$c - 3 = 3ax - acx - 2x$

$c - 3 = x(3a - ac - 2)$

$x = \dfrac{c-3}{3a - ac - 2}$

110. $b^2 x - 2x = 4b^2$

$\left(b^2 - 2\right)x = 4b^2$

$\dfrac{\left(b^2 - 2\right)x}{b^2 - 2} = \dfrac{4b^2}{b^2 - 2}$

$x = \dfrac{4b^2}{b^2 - 2}$

111. $|m - 5| = 9$

$m - 5 = 9$ or $m - 5 = -9$

$m = 14$ or $m = -4$

The solutions are 14 and −4.

112. $|4-x|=12$

$4-x=12$ or $4-x=-12$

$-x=8$ or $-x=-16$

$x=-8$ or $x=16$

The solutions are -8 and 16.

113. $\left|\dfrac{2-y}{5}\right|=8$

$\dfrac{2-y}{5}=8$ or $\dfrac{2-y}{5}=-8$

$5\,\dfrac{2-y}{5}=5(8)$ or $5\,\dfrac{2-y}{5}=-5(-8)$

$2-y=40$ or $2-y=-40$

$-y=38$ or $-y=-42$

$y=-38$ or $y=42$

The solutions are -38 and 42.

114. $|4k+1|=|6k-3|$

$4k+1=6k-3$ or $4k+1=-(6k-3)$

$4k+1=6k-3$ or $4k+1=-6k+3$

$-2k+1=-3$ or $10k+1=3$

$-2k=-4$ or $10k=2$

$k=2$ or $k=\dfrac{1}{5}$

The solutions are 2 and $\dfrac{1}{5}$.

115. a. Let $D=259$ (in millions).

$259=3.7x+111$

$148=3.7x$

$40=x$

Demand will reach 259 million barrels per day in 2010.

b. Let $D=300$ (in millions).

$300=3.7x+111$

$189=3.7x$

$51.08=x$

Demand will reach 300 million barrels per day in 2021.

116. Let $x=$ the original price.

$x-.15x=306$

$.85x=306$

$x=360$

The original price of the laser printer was $360.

117. a. Let $C=10.124$ (in thousands).

$10.124=1.1953x+5.343$

$4.781=1.1953x$

$4=x$

Costs reached $10,124$ in 2004.

b. Let $C=13.710$ (in thousands).

$13.710=1.1953x+5.343$

$8.367=1.1953x$

$7=x$

Costs reached $13,710$ in 2007.

118. Let $x=$ the rate of interest on $750.

Using the interest formula $A=P(1+rt)$,

$500(1+.12t)+250(1+.18t)=750(1+xt)$

$500+60t+250+45t=750+750xt$

$750+105t=750+750xt$

$105=750x$

$.14=x$

Borrowing $750 at a 14% annual interest results in the same total amount of annual interest.

119. Let $x=$ the amount invested at 8%.

Then $100,000-x=$ the amount invested at 5%;

$.08x=$ the interest from 8% investment;

$.05(100,000-x)=$ the interest from 5% investment.

$.08x+.05(100,000-x)=6800$

$.08x+5000-.05x=6800$

$.03x=1800$

$x=60,000$

$100,000-x=40,000$

The firm should invest $60,000 at 8% and $40,000 at 5%.

120. Let $x=$ the number of pounds of chocolate hearts;

$30-x=$ the number of pounds of candy kisses.

$5x+3.50(30-x)=4.50(30)$ Total cost

$5x+105-3.5x=135$

$1.5x=30$

$x=20$

Use 20 lb of hearts and 10 lb of kisses for the mix.

121. $x^2-6x=4$

$x^2-6x-4=0$

$b^2-4ac=(-6)^2-4(1)(-4)=36+16=52$

The discriminant is positive.

There are two real solutions to the equation.

122. $-3x^2+5x+2=0$

$b^2-4ac=(5)^2-4(-3)(2)=25+24=49$

The discriminant is positive.

There are two real solutions to the equation.

123. $4x^2 - 12x + 9 = 0$

$b^2 - 4ac = (-12)^2 - 4(4)(9) = 144 - 144 = 0$

The discriminant is 0.

There is one real solution to the equation.

124. $5x^2 + 2x + 1 = 0$

$b^2 - 4ac = 2^2 - 4(5)(1) = 4 - 20 = -16$

The discriminant is negative.

There are no real solutions to the equation.

125. $x^2 + 3x + 5 = 0$

$b^2 - 4ac = 3^2 - 4(1)(5) = 9 - 20 = -11$

The discriminant is negative.

There are no real solutions to the equation.

126. $(b+7)^2 = 5$

Use the square root property to solve this quadratic equation.

$b + 7 = \sqrt{5}$ or $b + 7 = -\sqrt{5}$

$b = -7 + \sqrt{5}$ or $b = -7 - \sqrt{5}$

The solutions are $-7 + \sqrt{5}$ and $-7 - \sqrt{5}$, which we abbreviate as $-7 \pm \sqrt{5}$.

127. $(2p+1)^2 = 7$

Solve by the square root property.

$2p + 1 = \sqrt{7}$ or $2p + 1 = -\sqrt{7}$

$2p = -1 + \sqrt{7}$ or $2p = -1 - \sqrt{7}$

$p = \dfrac{-1 + \sqrt{7}}{2}$ or $p = \dfrac{-1 - \sqrt{7}}{2}$

The solutions are $\dfrac{-1 \pm \sqrt{7}}{2}$.

128. $2p^2 + 3p = 2$

Write the equation in standard form and solve by factoring.

$2p^2 + 3p - 2 = 0$

$(2p - 1)(p + 2) = 0$

$2p - 1 = 0$ or $p + 2 = 0$

$p = \dfrac{1}{2}$ or $p = -2$

The solutions are $\dfrac{1}{2}$ and -2.

129. $2y^2 = 15 + y$

Write the equation in standard form and solve by factoring.

$2y^2 - y - 15 = 0$

$(y - 3)(2y + 5) = 0$

$y = 3$ or $y = -\dfrac{5}{2}$

The solutions are 3 and $-\dfrac{5}{2}$.

130. $x^2 - 2x = 2$

Write the equation in standard form.

$x^2 - 2x - 2 = 0$

Use the quadratic formula, with $a = 1$, $b = -2$, and $c = -2$.

$x = \dfrac{2 \pm \sqrt{(-2)^2 - 4(-2)}}{2}$

$= \dfrac{2 \pm \sqrt{12}}{2} = \dfrac{2 \pm 2\sqrt{3}}{2}$

$x = 1 \pm \sqrt{3}$

The solutions are $1 \pm \sqrt{3}$.

131. $\qquad r^2 + 4r = 1$

$r^2 + 4r - 1 = 0$

$r = \dfrac{-4 \pm \sqrt{16 + 4}}{2}$

$= \dfrac{-4 + \sqrt{20}}{2}$

$= \dfrac{-4 \pm 2\sqrt{5}}{2}$

$r = -2 \pm \sqrt{5}$

The solutions are $-2 \pm \sqrt{5}$.

132. $\qquad 2m^2 - 12m = 11$

$2m^2 - 12m - 11 = 0$

$a = 2, b = -12, c = -11$

$m = \dfrac{12 \pm \sqrt{144 - 4(2)(-11)}}{2(2)}$

$= \dfrac{12 \pm \sqrt{232}}{4}$

$= \dfrac{12 \pm 2\sqrt{58}}{4}$

$m = \dfrac{6 \pm \sqrt{58}}{2}$

The solutions are $\dfrac{6 \pm \sqrt{58}}{2}$.

133.
$$9k^2 + 6k^2 = 2$$
$$9k^2 + 6k - 2 = 0$$
$$a = 9, \ b = 6, \ c = -2$$
$$k = \frac{-6 \pm \sqrt{36 + 72}}{18}$$
$$= \frac{-6 \pm \sqrt{108}}{18}$$
$$= \frac{-6 \pm \sqrt{36 \cdot 3}}{18}$$
$$= \frac{-6 \pm 6\sqrt{3}}{18} = \frac{6\left(-1 \pm \sqrt{3}\right)}{6 \cdot 3}$$
$$k = \frac{-1 \pm \sqrt{3}}{3}$$

The solutions are $\dfrac{-1 \pm \sqrt{3}}{3}$.

134.
$$2a^2 + a - 15 = 0$$
$$(2a - 5)(a + 3) = 0$$
$$2a - 5 = 0 \quad \text{or} \quad a + 3 = 0$$
$$a = \frac{5}{2} \quad \text{or} \quad a = -3$$

The solutions are $\dfrac{5}{2}$ and -3.

135.
$$12x^2 = 8x - 1$$
$$12x^2 - 8x + 1 = 0$$
$$(2x - 1)(6x - 1) = 0$$
$$2x - 1 = 0 \quad \text{or} \quad 6x - 1 = 0$$
$$x = \frac{1}{2} \quad \text{or} \quad x = \frac{1}{6}$$

The solutions are $\dfrac{1}{2}$ and $\dfrac{1}{6}$.

136.
$$2q^2 - 11q = 21$$
$$2q^2 - 11q - 21 = 0$$
$$(2q + 3)(q - 7) = 0$$
$$2q + 3 = 0 \quad \text{or} \quad q - 7 = 0$$
$$q = -\frac{3}{2} \quad \text{or} \quad q = 7$$

The solutions are $-\dfrac{3}{2}$ and 7.

137.
$$3x^2 + 2x = 16$$
$$3x^2 + 2x - 16 = 0$$
$$(3x + 8)(x - 2) = 0$$
$$3x + 8 = 0 \quad \text{or} \quad x - 2 = 0$$
$$x = -\frac{8}{3} \quad \text{or} \quad x = 2$$

The solutions are $-\dfrac{8}{3}$ and 2.

138.
$$6k^4 + k^2 = 1$$
$$6k^4 + k^2 - 1 = 0$$

Let $p = k^2$, so $p^2 = k^4$.
$$6p^2 + p - 1 = 0$$
$$(3p - 1)(2p + 1) = 0$$
$$3p - 1 = 0 \quad \text{or} \quad 2p + 1 = 0$$
$$p = \frac{1}{3} \quad \text{or} \quad p = -\frac{1}{2}$$

If $p = \dfrac{1}{3}$, $\quad k^2 = \dfrac{1}{3}$
$$k = \pm\sqrt{\frac{1}{3}}$$
$$= \pm\frac{\sqrt{3}}{3}$$

If $p = -\dfrac{1}{2}$, $k^2 = -\dfrac{1}{2}$ has no real number solution.

The solutions are $\pm\dfrac{\sqrt{3}}{3}$.

139.
$$21p^4 = 2 + p^2$$
$$21p^4 - p^2 - 2 = 0$$

Let $u = p^2$; then $u^2 = p^4$.
$$21u^2 - u - 2 = 0$$
$$(3u - 1)(7u + 2) = 0$$
$$3u - 1 = 0 \quad \text{or} \quad 7u + 2 = 0$$
$$x = \frac{1}{3} \quad \text{or} \quad x = -\frac{2}{7}$$
$$p^2 = \frac{1}{3} \quad \text{or} \quad p^2 = -\frac{2}{7}$$

If $x = -\dfrac{2}{7}$, $p^2 = -\dfrac{2}{7}$ has no real number solution.
$$p = \pm\frac{1}{\sqrt{3}} = \pm\frac{\sqrt{3}}{3}$$

The solutions are $\pm\dfrac{\sqrt{3}}{3}$.

140.
$$2x^4 = 7x^2 + 15$$
$$2x^4 - 7x^2 - 15 = 0$$

Let $p = x^2$, so $p^2 = x^4$.

$$2p^2 - 7p - 15 = 0$$
$$(2p + 3)(p - 5) = 0$$
$$2p + 3 = 0 \quad \text{or} \quad p - 5 = 0$$
$$p = -\frac{3}{2} \quad \text{or} \quad p = 5$$

If $p = -\frac{3}{2}$, then $x^2 = -\frac{3}{2}$ has no real number solution.

If $p = 5$, then
$$x^2 = 5$$
$$x = \pm\sqrt{5}$$

The solutions are $\pm\sqrt{5}$.

141.
$$3m^4 + 20m^2 = 7$$
$$3m^4 + 20m^2 - 7 = 0$$

Let $u = m^2$.

$$3u^2 + 20u - 7 = 0$$
$$(3u - 1)(u + 7) = 0$$
$$3u - 1 = 0 \quad \text{or} \quad u + 7 = 0$$
$$u = \frac{1}{3} \quad \text{or} \quad u = -7$$
$$m^2 = \frac{1}{3} \quad \text{or} \quad m^2 = -7$$

For $m^2 = -7$, there is no real number solution.

$$m = \pm\frac{1}{\sqrt{3}} = \pm\frac{\sqrt{3}}{3}$$

The solutions are $\pm\dfrac{\sqrt{3}}{3}$.

142.
$$3 = \frac{13}{z} + \frac{10}{z^2}$$
$$3z^2 = 13z + 10 \quad \text{Multiply by } z^2.$$
$$3z^2 - 13z - 10 = 0$$
$$(3z + 2)(z - 5) = 0$$
$$3z + 2 = 0 \quad \text{or} \quad z - 5 = 0$$
$$z = -\frac{2}{3} \quad \text{or} \quad z = 5$$

The solutions are $-\dfrac{2}{3}$ and 5.

143. $p = \dfrac{E^2 R}{(r + R)^2}$ for r.

$$p(r + R)^2 = E^2 R$$
$$p\left(r^2 + 2rR + R^2\right) = E^2 R$$
$$pr^2 + 2rpR + R^2 p = E^2 R$$
$$pr^2 + 2rpR + R^2 p - E^2 R = 0$$

Use the quadratic formula to solve for r.

$$r = \frac{-2pR \pm \sqrt{4p^2 R^2 - 4p\left(R^2 p - E^2 R\right)}}{2p}$$
$$r = \frac{-2pR \pm \sqrt{4pE^2 R}}{2p}$$
$$= \frac{-pR \pm E\sqrt{pR}}{p}$$

144. $p = \dfrac{E^2 R}{(r + R)^2}$ for E.

$$p(r + R)^2 = E^2 R$$
$$E^2 = \frac{p(r + R)^2}{R}$$
$$E = \pm\sqrt{\frac{p(r + R)^2}{R}}$$
$$E = \frac{\pm(r + R)\sqrt{pR}}{R}$$

145. $K = s(s - a)$ for s.
$$K = s^2 - as$$
$$s^2 - as - K = 0$$

Use the quadratic formula.

$$s = \frac{a \pm \sqrt{a^2 - 4(-K)}}{2}$$
$$= \frac{a \pm \sqrt{a^2 + 4K}}{2}$$

146. $kz^2 - hz - t = 0$ for z.

Use the quadratic formula with $a = k$, $b = -h$, and $c = -t$.

$$z = \frac{-(-h) \pm \sqrt{(-h)^2 - 4(k)(-t)}}{2k}$$
$$= \frac{h \pm \sqrt{h^2 + 4kt}}{2k}$$

147. a. Let $x = 3$ (in millions).

$$N = .364(3)^2 - 2.296(3) + 3.783$$
$$= .171 \times 1,000,000$$
$$= 171,000$$

In 2003, there were 171,000 subscribers.

b. Let $N = 9$ (in millions).

$$9 = .364x^2 - 2.296x + 3.783$$
$$0 = .364x^2 - 2.296x - 5.217$$
$$x = 8.08$$

Subscribers will reach 9 million in 2008.

148. Let $x =$ the width of the walk.

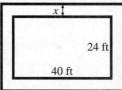

The area $= (24 + 2x)(40 + 2x) - 24(40)$
$$= 960 + 48x + 80x + 4x^2 - 960$$
$$= 4x^2 + 128x$$

To use all of the cement,
$$740 = 4x^2 + 128x$$
$$4x^2 + 128x - 740 = 0$$
$$x^2 + 32 - 185 = 0$$
$$(x + 37)(x - 5) = 0$$
$$x = -37 \text{ or } x = 5$$

The width cannot be negative, so the solution is 5 feet.

149. Let $x =$ the dimension of the playground that does not lie along the side of the building.

The area is 11,250 square meters so
$$x(325 - 2x) = 11,250$$
$$325x - 2x^2 = 11,250$$
$$2x^2 - 325x + 11,250 = 0$$
$$x = \frac{325 \pm \sqrt{325^2 - 4 \cdot 2 \cdot 11,250}}{2 \cdot 2}$$
$$x = \frac{325 \pm \sqrt{15,625}}{4}$$
$$x = \frac{325 \pm 125}{4}$$
$$x = 112.5 \text{ or } x = 50.$$

If $x = 112.5$, then $325 - 2x = 100$.
If $x = 50$, then $325 - 2x = 225$.
The width is 100 m and the length is 112.5 m or the width is 50 m and the length is 225 m.

150.

Let s be the speed of the car going north. Then the speed of the car going west is $s + 10$. Use $d = rt$ with $t = 1$ to label the distances shown above.
By the Pythagorean theorem,
$$s^2 + (s + 10)^2 = 50^2$$
$$s^2 + s^2 + 20s + 100 = 2500$$
$$2s^2 + 20s - 2400 = 0$$
$$s^2 + 10s - 1200 = 0$$
$$(s - 30)(s + 40) = 0$$
$$s = 30 \text{ or } s = -40.$$

Reject –40 as a possible solution because speed cannot be negative. The speed of the car headed north is 30 mph and the speed of the car headed west is 40 mph.

Case 1 Consumers Often Defy Common Sense

1. The total cost to buy and run this refrigerator for x years is $700 + 85x$.

2. The total cost to buy and run this refrigerator for x years is $1000 + 25x$.

3. Over 10 years, the $700 refrigerator costs
   $700 + 85(10) = 1550$ or $1550,
   and the $1000 refrigerator costs
   $1000 + 25(10) = 1250$ or $1250.
   The $700 refrigerator costs $300 more over
   10 years.

4. The total costs for the two refrigerators will be
   equal when $700 + 85x = 1000 + 25x$

   $$60x = 300$$

   $$x = 5$$

   The costs will be equal in 5 years.

Chapter 2: Graphs, Lines, and Inequalities

Section 2.1 Graphs, Lines, and Inequalities

1. $(1,-2)$ lies in quadrant IV

   $(-2,1)$ lies in quadrant II

   $(3,4)$ lies in quadrant I

   $(-5,-6)$ lies in quadrant III

3. $(1,-3)$ is a solution to $3x - y - 6 = 0$ because $3(1) - (-3) - 6 = 0$ is a true statement.

5. $(3, 4)$ is not a solution to $\begin{aligned} 2(0) + x &= 4 \\ x &= 4 \end{aligned}$ because

   $(3-2)^2 + (4+2)^2 = 37$, not 6.

7. $4y + 3x = 12$
   Find the y-intercept. If $x = 0$,
   $$4y = -3(0) + 12$$
   $$4y = 12$$
   $$y = 3$$
   The y-intercept is 3.
   Next find the x-intercept. If $y = 0$,
   $$4(0) + 3x = 12$$
   $$3x = 12$$
   $$x = 4$$
   The x-intercept is 4.
   Using these intercepts, graph the line.

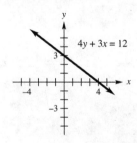

9. $8x + 3y = 12$
   Find the y-intercept. If $x = 0$,
   $$3y = 12$$
   $$y = 4$$
   The y-intercept is 4.
   Next, find the x-intercept. If $y = 0$,
   $$8x = 12$$
   $$x = \frac{12}{8} = \frac{3}{2}$$
   The x-intercept is $\frac{3}{2}$.
   Using these intercepts, graph the line.

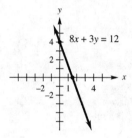

11. $x = 2y + 3$
    Find the y-intercept. If $x = 0$,
    $$0 = 2y + 3$$
    $$2y = -3$$
    $$y = -\frac{3}{2}$$
    The y-intercept is $-\frac{3}{2}$.
    Next, find the x-intercept. If $y = 0$,
    $$x = 2(0) + 3$$
    $$x = 3$$
    The x-intercept is 3.
    Using these intercepts, graph the line.

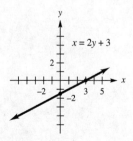

13. The x-intercepts are where the lines cross the x-axis, at -1 and 3.5. The y-intercepts are where the lines cross the y-axis, at 1.

15. The x-intercepts are at -2, 0, and 2. The y-intercept is at 0.

17. $3x + 4y = 12$
To find the x-intercept, let $y = 0$:
$3x + 4(0) = 12$
$$3x = 12$$
$$x = 4$$
The x-intercept is 4.
To find the y-intercept, let $x = 0$:
$-3 + (-6) < m + 6 + (-6) < 2 + (-6)$
$$-9 < m < -4$$
The y-intercept is 3.

19. $2x - 3y = 24$
To find the x-intercept, let $y = 0$:
$2x - 3(0) = 24$
$$2x = 24$$
$$x = 12$$
The x-intercept is 12.
To find the y-intercept, let $x = 0$:
$2(0) - 3y = 24$
$$-3y = 24$$
$$y = -8$$
The y-intercept is -8.

21. $y = x^2 - 9$
To find the x-intercepts, let $y = 0$:
$0 = x^2 - 9$
$$x^2 = 9$$
$$x = \pm\sqrt{9} = \pm 3$$
The x-intercepts are 3 and -3.
To find the y-intercept, let $x = 0$:
$y = 0 - 9$
$$y = -9$$
The y-intercept is -9.

23. $y = x^2 + x - 20$
To find the x-intercepts, let $y = 0$:
$0 = x^2 + x - 20$
$0 = (x + 5)(x - 4)$
$x = -5, \text{or } 4$
The x-intercepts are -5 and 4.
To find the y-intercept, let $x = 0$:
$y = 0^2 + 0 - 20$
$y = -20$
The y-intercept is -20.

25. $y = 2x^2 - 5x + 7$
To find the x-intercepts, let $y = 0$:
$0 = 2x^2 - 5x + 7$

This equation does not have real solutions and so there are no x-intercepts.
To find the y-intercept, let $x = 0$:
$y = 2(0)^2 - 5(0) + 7$
$y = 7$
The y-intercept is 7.

27. $y = x^2$
x-intercept: $0 = x^2$
$\qquad\qquad\quad x = 0$
y-intercept: $y = 0$

x	y
-2	4
-1	1
0	0
1	1
2	4

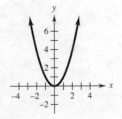

29. $y = x^2 - 3$

x-intercepts: $0 = x^2 - 3$

$x^2 = 3$

$x = \pm\sqrt{3}$

y-intercepts: $y = 0^2 - 3$

$y = -3$

x	y
-3	6
-1	-2
0	-3
1	-2
3	6

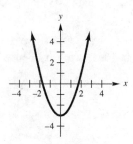

31. $y = x^2 - 6x + 5$

x-intercept:
$0 = x^2 - 6x + 5$
$0 = (x-1)(x-5)$

$x = 1, 5$

y-intercept: $y = (0)^2 - 6(0) + 5$

$y = 5$

x	y
-2	8
-1	2
0	5
1	2
-2	8

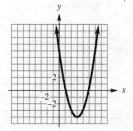

33. $y = x^3$

x-intercept: $0 = x^3$

$x = 0$

y-intercept: $y = 0^3$

$y = 0$

x	y
-2	-8
-1	-1
0	0
1	1
2	8

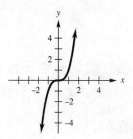

35. $y = x^3 + 1$

x-intercept: $0 = x^3 + 1$

$x^3 = -1$

$x = \sqrt[3]{-1} = -1$

y-intercept: $y = 0^3 + 1$

$y = 1$

x	y
-2	-7
-1	0
0	1
1	2
2	9

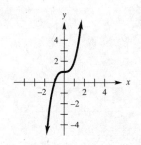

37. $y = \sqrt{x+4}$

    x-intercept: $0 = \sqrt{x+2}$

                $0 = x+2$

                $x = -2$

    y-intercept: $y = \sqrt{0+2} = \sqrt{2}$

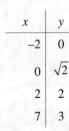

x	y
-2	0
0	$\sqrt{2}$
2	2
7	3

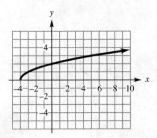

39. $y = \sqrt{4 - x^2}$

    x-intercept: $0 = \sqrt{4 - x^2}$

                    $0 = 4 - x^2$

                    $x^2 = 4$

                    $x = \pm\sqrt{4} = \pm 2$

    y-intercept: $y = \sqrt{4} = 2$

x	y
-2	0
-1	$\sqrt{3}$
0	2
1	$\sqrt{3}$
2	0

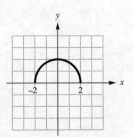

41. Fargo, about 2:00PM; Seattle about 5:00PM.

43. From 11:00AM to 6:00PM.

45. (a) About \$1,250,000   (b) \$1,750,000

    (c) About \$4,250,000

47. (a) About \$500,000 (b) about \$1,000,000.

    (c) About \$1,500,000.

49. About 12 seconds.

51. (a) In year 5, about \$750; year 15, about \$600;Year 25, about \$300

    (b) During the 22nd year.

53. From about the mid-2003 through 2008.

55. About 14% in mid-2004

57. Men, about \$695; women, about \$520

59. About \$635 in 1994

61. Men, about \$10; women, about \$140

63. $y = x^2 + x + 1$

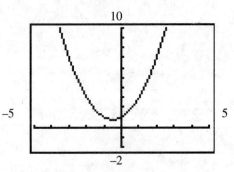

65. $y = (x - 3)^3$

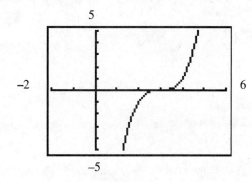

67. $y = x^3 - 3x^2 + x - 1$

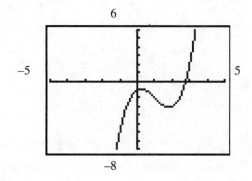

69. $y = x^4 - 2x^3 + 2x$

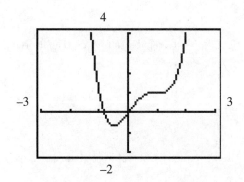

The "flat" part of the graph near $x = 1$ looks like a horizontal line segment, but it is not. The y values change as you trace along the segment.

71. $x \approx -1.1038$

73. $x \approx 2.1017$

75. $x \approx -1.7521$.

77. $r \approx 4.6580$ in.

79. 2004

Section 2.2 Equations of Lines

1. Through (2, 5) and (0, 8)

$$\text{slope} = \frac{\Delta y}{\Delta x}$$
$$= \frac{8 - 5}{0 - 2}$$
$$= \frac{3}{-2} = -\frac{3}{2}$$

3. Through (−4, 14) and (3, 0)

$$\text{slope} = \frac{14 - 0}{-4 - 3} = \frac{14}{-7} = -2$$

5. Through the origin and (−4, 10), the origin has coordinate (0, 0).

$$\text{slope} = \frac{10 - 0}{-4 - 0}$$
$$= \frac{10}{-4} = -\frac{5}{2}$$

7. Through (−1, 4) and (−1, 6)

$$\text{slope} = \frac{6 - 4}{-1 - (-1)}$$
$$= \frac{2}{0}$$

not defined

The slope is undefined.

9. $b = 5, m = 4$
$y = mx + b$
$y = 4x + 5$

11. $b = 1.5, m = -2.3$
$y = mx + b$
$y = -2.3x + 1.5$

13. $b = 4, \ m = -\dfrac{3}{4}$
$y = mx + b$
$y = -\dfrac{3}{4}x + 4$

15. $2x - y = 9$

Rewrite in slope-intercept form.

$-y = -2x + 9$

$y = 2x - 9$

$m = 2.$

$b = -9.$

17. $6x = 2y + 4$

Rewrite in slope-intercept form.

$2y = 6x - 4$

$y = 3x - 2$

$m = 3.$

$b = -2.$

19. $6x - 9y = 16$

Write in slope-intercept form.

$-9y = -6x + 16$

$9y = 6x - 16$

$y = \dfrac{2}{3}x - \dfrac{16}{9}$

$m = \dfrac{2}{3}.$

$b = -\dfrac{16}{9}.$

21. $2x - 3y = 0$

Rewrite in slope-intercept form.

$3y = 2x$

$y = \dfrac{2}{3}x$

$m = \dfrac{2}{3}.$

$b = 0.$

23. $x = y - 5$

Rewrite in slope-intercept form.

$y = x + 5$

$m = 1, b = 5$

25. (a) Largest value of slope is at C.

(b) Smallest value of slope is at B.

(c) Largest absolute value is at B

(d) Closest to 0 is at D

27. $2x - y = -2$

Find the x-intercept by setting $y = 0$ and solving for x:

$2x - 0 = -2$

$2x = -2$

$x = -1$

Find the y-intercept by setting $x = 0$ and solving for y:

$2(0) - y = -2$

$-y = -2$

$y = 2$

Use the points $(-1, 0)$ and $(0, 2)$ to sketch the graph:

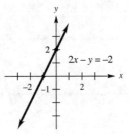

29. $2x + 3y = 4$

Find the x-intercept by setting $y = 0$ and solving for x:

$2x + 3(0) = 4$

$2x = 4$

$x = 2$

Find the y-intercept by setting $x = 0$ and solving for y:

$2(0) + 3y = 4$

$3y = 4$

$y = \dfrac{4}{3}$

Use the points $(2, 0)$ and $\left(0, \dfrac{4}{3}\right)$ to sketch the graph:

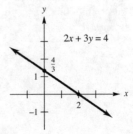

31. $4x - 5y = 2$

Find the x-intercept, by setting $y = 0$ and solving for x:

$$4x - 5(0) = 2$$
$$4x = 2$$
$$x = \frac{1}{2}$$

Find the y-intercept by setting $x = 0$ and solving for y:

$$4(0) - 5y = 2$$
$$-5y = 2$$
$$y = -\frac{2}{5}$$

Use the points $\left(\frac{1}{2}, 0\right)$ and $\left(0, -\frac{2}{5}\right)$ to sketch the graph:

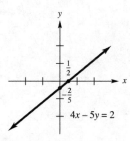

33. For $4x - 3y = 6$, solve for y.

$$y = \frac{4}{3}x - 2$$

For $3x + 4y = 8$, solve for y.

$$y = -\frac{3}{4}x + 2$$

The two slopes are $\frac{4}{3}$ and $-\frac{3}{4}$. Since

$$\left(\frac{4}{3}\right)\left(-\frac{3}{4}\right) = -1,$$

the lines are perpendicular.

35. For $3x + 2y = 8$, solve for y.

$$y = -\frac{3}{2}x + 4$$

For $6y = 5 - 9x$, solve for y.

$$y = -\frac{3}{2}x + \frac{5}{6}$$

Since the slopes are both $-\frac{3}{2}$, the lines are parallel.

37. For $4x = 2y + 3$, solve for y.

$$y = 2x - \frac{3}{2}$$

For $2y = 2x + 3$, solve for y.

$$y = x + \frac{3}{2}$$

Since the two slopes are 2 and 1, the lines are neither parallel nor perpendicular.

39. Triangle with vertices $(9, 6)$, $(-1, 2)$ and $(1, -3)$:

a. Slope of side between vertices $(9, 6)$ and $(-1, 2)$:

$$m = \frac{6 - 2}{9 - (-1)} = \frac{4}{10} = \frac{2}{5}$$

Slope of side between vertices $(-1, 2)$ and $(1, -3)$:

$$m = \frac{2 - (-3)}{-1 - 1} = \frac{5}{-2} = -\frac{5}{2}$$

Slope of side between vertices $(1, -3)$ and $(9, 6)$:

$$m = \frac{-3 - 6}{1 - 9} = \frac{-9}{-8}$$

Yes, the triangle is a right triangle.

b. The sides with slopes $\frac{2}{5}$ and $-\frac{5}{2}$ are

perpendicular, because $\frac{2}{5}\left(-\frac{5}{2}\right) = -1$.

41. Use point-slope form with

$$(x_1, y_1) = (-1, 2), \ m = -\frac{2}{3}$$
$$y - y_1 = m(x - x_1)$$
$$y - 2 = -\frac{2}{3}(x - (-1))$$
$$y - 2 = -\frac{2}{3}(x + 1)$$
$$y - 2 = -\frac{2}{3}x - \frac{2}{3}$$
$$y = -\frac{2}{3}x + \frac{4}{3} \text{ or } 3y = -2x + 4$$

43. $(x_1, y_1) = (-2, -2), \ m = 4$

$$y - y_1 = m(x - x_1)$$
$$y - (-2) = 4(x - (-2))$$
$$y + 2 = 4(x + 2)$$
$$y = 4x + 6$$

45. $(x_1, y_1) = (8, 2), m = 0$

$$y - y_1 = m(x - x_1)$$
$$y - 2 = 0(x - 8)$$
$$y = 2$$

47. Since the slope is undefined, the equation is of a vertical line through $(6, -5)$.

$$x = 6$$

49. Through $(-1, 1)$ and $(2, 7)$
Find the slope.

$$m = \frac{7 - 1}{2 - (-1)} = \frac{6}{3} = 2$$

Use the point-slope form with $(2, 7) = (x_1, y_1)$.

$$y - y_1 = m(x - x_1)$$
$$y - 7 = 2(x - 2)$$
$$y - 7 = 2x - 4$$
$$y = 2x + 3$$

51. Through $(1, 2)$ and $(3, 9)$
Find the slope.

$$m = \frac{9 - 2}{3 - 1} = \frac{7}{2}$$

Use the point-slope form with $(1, 2) = (x_1, y_1)$.

$$y - y_1 = m(x - x_1)$$
$$y - 2 = \frac{7}{2}(x - 1)$$
$$y - 2 = \frac{7}{2}x - \frac{7}{2}$$
$$y = \frac{7}{2}x - \frac{3}{2}$$

53. Through the origin with slope 5.

$$(x_1, y_1) = (0, 0) ; m = 5$$
$$y - y_1 = m(x - x_1)$$
$$y - 0 = 5(x - 0)$$
$$y = 5x$$

55. Through $(6, 8)$ and vertical.

$$(x_1, y_1) = (6, 8) ; m = \text{undefined}$$
$$x = 6$$

57. Through $(3, 4)$ and parallel to $4x - 2y = 5$.

$$(x_1, y_1) = (3, 4)$$
$$4x = 2y + 5$$
$$2y = 4x - 5$$
$$y = 2x - \frac{5}{2} \quad m = 2$$
$$y - y_1 = m(x - x_1)$$
$$y - 4 = 2(x - 3)$$
$$y - 4 = 2x - 6$$
$$y = 2x - 2$$

59. x-intercept 6; y-intercept -6
Through the points $(6, 0)$ and $(0, -6)$.

$$m = \frac{0 - (-6)}{6 - 0} = \frac{6}{6} = 1$$
$$(x_1, y_1) = (6, 0)$$
$$y - y_1 = m(x - x_1)$$
$$y - 0 = 1(x - 6)$$
$$y = x - 6$$

61. Through $(-1, 3)$ and perpendicular to the line through $(0, 1)$ and $(2, 3)$.

$$m_1 = \frac{1 - 3}{0 - 2} = \frac{-2}{-2} = 1$$
$$m_2 = \frac{-1}{1} = -1$$
$$(x_1, y_1) = (-1, 3)$$
$$y - y_1 = m(x - x_1)$$
$$y - 3 = -1(x - (-1))$$
$$y - 3 = -x - 1$$
$$y = -x + 2$$

63. Let cost $x = 12{,}482$ and n years $= 10$. Find D.

$$D = \left(\frac{1}{n}\right)x$$
$$D = \left(\frac{1}{10}\right)(12{,}482)$$
$$D = 1248.2$$

The depreciation is \$1248.20.

65. Let cost $x = 145{,}000$ and n years $= 28$. Find D.

$$D = \left(\frac{1}{n}\right)x$$
$$D = \left(\frac{1}{28}\right)(145{,}000)$$
$$D = 5178.57 \text{ (to the nearest cent)}$$

The depreciation is \$5178.57.

67. a. Let $x = 13$.
$$y = 136.25(13) + 1443.75$$
$$= 3215(1,000,000)$$
$$3,215,000,000$$

b. Let $x = 15$.
$$y = 136.25(15) + 1443.75$$
$$= 3487.5(1,000,000)$$
$$= 3,487,500,000$$

c. 2015

69. a. Let $x = 9$.
$$y = 535.43(9) + 5362.21$$
$$y = 10181.08$$
$$y = 10181.08(1,000,000)$$
$$y = \$10,181,000,000$$

There were approximately $10,181,000,000 in box-office receipts in 2004.

b. Let $y = 14,000$.
$$14000 = 535.43x + 5362.21$$
$$8637.79 = 535.43x$$
$$16 \approx x$$
$$1995 + 16 = 2011$$
Box-office receipts will reach $14,000,000,000 in 2011.

71. a. The given data is represented by the points $(0, 60)$ and $(7, 65)$.

b. Find the slope.
$$m = \frac{65 - 60}{7 - 0} = \frac{5}{7}$$
Use the point-slope form with $(0, 60) = (x_1, y_1)$.
$$y - y_1 = m(x - x_1)$$
$$y - 60 = \frac{5}{7}(x - 0)$$
$$y - 60 = \frac{5}{7}x$$
$$y = \frac{5}{7}x + 60$$

71. Continued

c. Let $x = 3$.
$$y = \frac{5}{7}(3) + 60$$
$$= 62.142857$$

The number of vehicles produced in 2003 was approximately 62,142,857.

d. Let $y = 68$.
$$68 = \frac{5}{7}x + 60$$
$$476 = 5x + 420$$
$$x \approx 11$$
In 2011, vehicle production will reach 68 million.

73. a. Let (x_1, y_1) be $(0, 1.1)$ and (x_2, y_2) be $(23, 5.5)$.
Find the slope.
$$m = \frac{5.5 - 1.1}{23 - 0} = \frac{4.4}{23} = .191$$
Use the point-slope form with $(0, 1.1)$.
$$y - 1.1 = .191(x - 0)$$
$$y - 1.1 = .191x$$
$$y = .191x + 1.1$$

b. Let $x = 33$.
$$y = .191(33) + 1.1$$
$$= 6.303 + 1.1$$
$$= 7.403$$
The number of cohabitating adults in 2010 will be about 7.4 million.

c. In 2003

75. a. The slope in degrees C per year is $\frac{.3}{10} = .03$.
Use the point-slope form with $(0, 15)$.
$$y - 15 = .03(t - 0)$$
$$y - 15 = .03t$$
$$y = .03t + 15$$

b. Let $y = 19$.
$$19 = .03t + 15$$
$$4 = .03t$$
$$133 \approx t$$
In 2103, the average global temperature will rise to 19°C.

77. a. The slope of $-.01786$ indicates that on average, the 5000-meter run is being run $.01786$ seconds faster every year. It is negative because the times are generally decreasing as time progresses.

b. 12.99 minutes

c. World War II occurred during the years 1940 and 1944, and no Olympic Games were held.

Section 2.3 Linear Models

1. a. Let (x_1, y_1) be $(32, 0)$ and (x_2, y_2) be $(68, 20)$.
Find the slope.
$$m = \frac{20 - 0}{68 - 32} = \frac{20}{36} = \frac{5}{9}$$
Use the point-slope form with $(32, 0)$.
$$y - 0 = \frac{5}{9}(x - 32)$$
$$y = \frac{5}{9}(x - 32)$$

b. Let $x = 50$.
$$y = \frac{5}{9}(50 - 32)$$
$$= \frac{5}{9}(18)$$
$$= 10°C$$
Let $x = 75$.
$$y = \frac{5}{9}(75 - 32)$$
$$= \frac{5}{9}(43)$$
$$\approx 23.89°C$$

3. $F = 867°$
$$C = \frac{5}{9}(867 - 32)$$
$$C = \frac{5}{9}(835)$$
$$C \approx 463.89°C$$

5. Let (x_1, y_1) be $(0, 130.7)$ and (x_2, y_2) be $(15, 190.7)$.
Find the slope.
$$m = \frac{190.7 - 130.7}{15 - 0} = \frac{60}{15} = 4$$
Use the point-slope form with $(0, 130.7)$.
$$y - 130.7 = 4(x - 0)$$
$$y - 130.7 = 4x$$
$$y = 4x + 130.7$$
To estimate the CPI in 1995, let $x = 14$.
$$y = 4(14) + 130.7$$
$$= 186.7$$
To estimate the CPI in 2003, let $x = 18$.
$$y = 4(18) + 130.7$$
$$= 202.7$$

7. Let (x_1, y_1) be $(9, 82)$ and (x_2, y_2) be $(13, 152)$.
Find the slope.
$$m = \frac{152 - 82}{13 - 9} = \frac{70}{4} = 17.5$$
Use the point-slope form with $(9, 82)$.
$$y - 82 = 17.5(x - 9)$$
$$y - 82 = 17.5x - 157.5$$
$$y = 17.5x - 75.5$$
To estimate the amount of imports in 2006, let $x = 16$.
$$y = 17.5(16) - 75.5$$
$$= \$204.5 \text{ billion}$$

9. Find the slope of the line.
$$(x_1, y_1) = (0, 244)$$
$$(x_2, y_2) = (3, 245)$$
$$m = \frac{y_2 - y_1}{x_2 - x_1} = \frac{245 - 244}{3 - 0} = \frac{1}{3}$$
Use the slope and one point to find the equation of the line.
$$y - y_1 = m(x - x_1)$$
$$y - 244 = \frac{1}{3}(x - 0)$$
$$y - 244 = \frac{1}{3}x$$
$$y = \frac{1}{3}x + 244$$
To estimate the expenditures in 2007,
Let $x = 7$.
$$y = \frac{1}{3}(7) + 244$$
$$y \approx \$246.33 \text{ billions}$$

11. a. $y = .5x + 1.5$

Data Point (x, y)	Model Point (x, y)	Residual $p - y$	Squared Residual $(p - y)^2$
(1, 2)	(1, 2)	0	0
(2, 2)	(2, 2.5)	–.5	.25
(3, 3)	(3, 3)	0	0
(4, 3)	(4, 3.5)	–.5	.25
(5, 5)	(5, 4)	1	1

$y = x$

Data Point (x, y)	Model Point (x, y)	Residual $p - y$	Squared Residual $(p - y)^2$
(1, 2)	(1, 1)	1	1
(2, 2)	(2, 2)	0	0
(3, 3)	(3, 3)	0	0
(4, 3)	(4, 4)	–1	1
(5, 5)	(5, 5)	0	0

sum = 0 – .5 + 0 – .5 + 1 = 0
sum = 1 + 0 + 0 + (–1) + 0 = 0

b. For $y = .5x + 1.5$,
sum = 0 + .25 + 0 + .25 + 1 = 1.5.
For $y = x$,
sum = 1 + 0 + 0 + 1 + 0 = 2.

c. $y = .5x + 1.5$ is a better fit for the data because the sum of the squares of its residuals is smaller than the sum for $y = x$.

13. Plot the points.

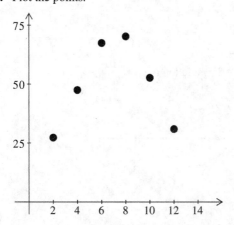

13. Continued

Visually, a straight line looks to be a poor model for the data. Also, the coefficient of correlation is $r \approx .1$, which indicates that the regression line is a poor fit for the data.

15. a. Let (x_1, y_1) be (0, 559) and (x_2, y_2) be (42, 240).
Find the slope.
$$m = \frac{240 - 559}{42 - 0} \approx -7.5952$$
Use the point-slope form with (0, 559).
$$y - 559 = -7.5952(x - 0)$$
$$y = -7.5952x + 559$$
Using a graphing calculator, the regression-line model is
$y = -7.7135x + 563.22$

b. Let $x = 48$.
$y = -7.5952(48) + 559 \approx 194.43$
   (Two-point model)
$y = -7.7135(48) + 563.22 = 192.97$
   (regression-line model)

17. a. Let (x_1, y_1) be (150, 5000) and (x_2, y_2) be (450, 9500).
Find the slope.
$$m = \frac{9500 - 5000}{450 - 150} = 15$$
Use the point-slope form with
(150, 5000).
$$y - 5000 = 15(x - 150)$$
$$y = 15x + 2750$$
Using a graphing calculator, the regression-line model is
$y = 14.9x + 2820$.

b. Using the Two-point model:
Let $x = 150$ ft^2.
$y = 15(150) + 2750 = 5000$
Let $x = 280$ ft^2.
$y = 15(280) + 2750 = 6950$
Let $x = 420$ ft^2.
$y = 15(420) + 2750 = 9050$
Using the regression-line model:
Let $x = 150$ ft^2.
$y = 14.9(150) + 2820 = 5055$
Let $x = 280$ ft^2.
$y = 14.9(280) + 2820 = 6992$
Let $x = 420$ ft^2.
$y = 14.9(420) + 2820 = 9078$

17. Continued

    c. Using the two-point model:

    Let $x = 235$ ft^2.

    $y = 15(235) + 2750 = 6275$

    Using the regression-line model:

    Let $x = 235$ ft^2.

    $y = 14.9(235) + 2820 = 6321.5$

    Adam should choose an air conditioner with a BTU of 6500 if air conditioners are available only with the BTU choices from the table.

19. a. The data points corresponding to the graph are: (4, 1.1), (5, 1.3), (6, 1.8), (7, 2.3), (8, 2.5), (9, 3.1), (10, 3.9), (11, 3.8), (12, 4), (13, 4.4).

    b. Using a graphing calculator, the regression-line model is $y = .3877x - .477$

    c. Let $x = 18$. Then

    $y = .3877(18) - .477$

      $= 6.502$

    Intel will spend about \$6.5 billion.

21. a. Using a graphing calculator, the regression-line model is $y = .03119x + .5635.$

    b. Let $x = 12$, then $y \approx .938$ which is about .938 billion. Now le $x = 16$, then $y \approx 1.06$ which is about 1.06 billion.

23 a. Using a graphing calculator, the regression-line model is $y = 11.1143x + 340.81$; 396,000; 508,000; 619,000

    b. 2037; 2060

25. a. Using a graphing calculator, the regression-line model is $y = 34.9125x + 277.75.$

    b. Yes, this model fits the data fairly well because the coefficient of correlation $r \approx .986,$ which is a value close to 1.

    c. About \$941.09 billions

    d. In 2016

Section 2.4 Linear Inequalities

1. Use brackets if you want to include the endpoint, and parentheses.

3. $-8x \leq 32$

Multiply both sides of the inequality by $-\frac{1}{8}$.

Since this is a negative number, change the direction of the inequality symbol.

$-\frac{1}{8}(-8k) \geq -\frac{1}{8}(32)$

The solution is $[-4, \infty)$.

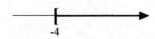
-4

5. $\qquad -2b > 0$

$-\frac{1}{2}(-2b) < -\frac{1}{2}(0)$

$\qquad b < 0$

The solution is $(-\infty, 0)$. To graph this solution, put an open circle at 0 and draw an arrow extending to the left.

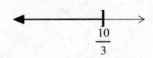
0

7. $\qquad 3x + 4 \leq 14$

$\qquad 3x + 4 - 4 \leq 14 - 4$

$\qquad\qquad 3x \leq 10$

$\qquad \frac{1}{3}(3x) \leq \frac{1}{3}(10)$

$\qquad\qquad x \leq \frac{10}{3}$

The solution is $\left(-\infty, \frac{10}{3}\right]$.

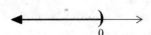
$\frac{10}{3}$

9. $-5 - p \geq 3$

$-5 + 5 - p \geq 3 + 5$

$-p \geq 8$

$(-1)(-p) \leq (-1)(8)$

$p \leq -8$

The solution is $(-\infty, -8]$.

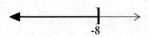

11. $7m - 5 < 2m + 10$

$5m - 5 < 10$

$5m < 15$

$\dfrac{1}{5}(5m) < \dfrac{1}{5}(15)$

$m < 3$

The solution is $(-\infty, 3)$.

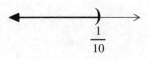

13. $m - (4 + 2m) + 3 < 2m + 2$

$m - 4 - 2m + 3 < 2m + 2$

$-1 - m < 2m + 2$

$-m - 2m < 2 + 1$

$-3m < 3$

$-\dfrac{1}{3}(-3m) > -\dfrac{1}{3}(3)$

$m > -1$

$2p - (3 - p) \leq -7p - 2$

$2p - 3 + p \leq -7p - 2$

$3p - 3 \leq -7p - 2$

$10p - 3 \leq -2$

$10p \leq 1$

$p \leq \dfrac{1}{10}$

The solution is $\left(-\infty, \dfrac{1}{10}\right]$.

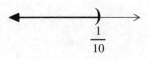

15. $-2(3y - 8) \geq 5(4y - 2)$

$-6y + 16 \geq 20y - 10$

$16 + 10 \geq 20y + 6y$

$26 \geq 26y$

$1 \geq y$

or $y \leq 1$

The solution is $(-\infty, 1]$.

17. $3p - 1 < 6p + 2(p - 1)$

$3p - 1 < 6p + 2p - 2$

$-1 + 2 < 6p + 2p - 3p$

$1 < 5p$

$\dfrac{1}{5} < p$

or $p > \dfrac{1}{5}$

The solution is $\left(\dfrac{1}{5}, \infty\right)$.

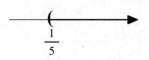

19. $-7 < y - 2 < 5$

$-7 + 2 < y < 5 + 2$

$-5 < y < 7$

The solution is $(-5, 7)$.

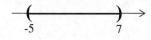

21. $8 \leq 3r + 1 \leq 16$

$8 - 1 \leq 3r \leq 16 - 1$

$7 \leq 3r \leq 15$

$\dfrac{7}{3} \leq r \leq 5$

21. Continued

The solution is $\left[\dfrac{7}{3}, 5\right]$.

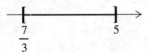

23.

$$-4 \le \dfrac{2k-1}{3} \le 2$$

$$-4(3) \le 3\left(\dfrac{2k-1}{3}\right) \le 2(3)$$

$$-12 \le 2k-1 \le 6$$

$$-12+1 \le 2k \le 6+1$$

$$-11 \le 2k \le 7$$

$$-\dfrac{11}{2} \le k \le \dfrac{7}{2}$$

The solution is $\left[-\dfrac{11}{2}, \dfrac{7}{2}\right]$.

25.

$$\dfrac{3}{5}(2p+3) \ge \dfrac{1}{10}(5p+1)$$

$$10 \cdot \dfrac{3}{5}(2p+3) \ge 10 \cdot \dfrac{1}{10}(5p+1)$$

$$6(2p+3) \ge 5p+1$$

$$12p+18 \ge 5p+1$$

$$12p-5p \ge -18+1$$

$$7p \ge -17$$

$$p \ge -\dfrac{17}{7}$$

The solution is $\left[-\dfrac{17}{7}, \infty\right)$.

27. $x \ge 2$

29. $-3 < x \le 5$

31. a. Let x represent the number of milligrams per liter of lead in the water.

b. 5% of .040 is .002.
$$.040 - .002 \le x \le .040 + .002$$
$$.038 \le x \le .042$$

c. Since all the samples had a lead content less than or equal to .042 mg per liter, all were less than .050 mg per liter and did meet the federal requirement.

33. a. The six income ranges are:
$$0 < x \le 7300$$
$$7300 < x \le 29{,}700$$
$$29{,}700 < x \le 71{,}950$$
$$71{,}950 < x \le 150{,}150$$
$$150{,}150 < x \le 326{,}450$$
$$x > 326{,}450$$

b. The inequalities that give the tax range in dollars for each of the six income ranges are:
$$0 < T \le 730$$
$$730 < T \le 4090$$
$$4090 < T \le 14{,}652.50$$
$$14{,}652.50 < T \le 36{,}548.50$$
$$36{,}548.50 < T \le 94{,}727.50$$
$$T > 94{,}727.50$$

35. $|m| < 2$

$-2 < m < 2$

In interval notation, this solution is written $(-2, 2)$.

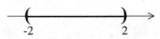

37. $|a| < -2$

Since the absolute value of a number is never negative, the inequality has no solution.

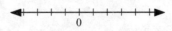

39. $|2x+5| < 1$

$$-1 < 2x+5 < 1$$
$$-1-5 < 2x < 1-5$$
$$-6 < 2x < -4$$
$$-3 < x < -2$$

The solution is $(-3, -2)$.

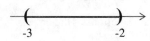

41. $|3z+1| \geq 4$

$3z+1 \geq 4$	or	$3z+1 \leq -4$
$3z \geq 4-1$		$3z \leq -4-1$
$3z \geq 3$		$3z \leq -5$
$z \geq 1$		$z \leq -\dfrac{5}{3}$

The solution is $\left(-\infty, -\dfrac{5}{3}\right]$ or $[1, \infty)$.

43. $\left|5x+\dfrac{1}{2}\right| - 2 < 5$

$$\left|5x+\dfrac{1}{2}\right| < 7$$
$$-7 < 5x+\dfrac{1}{2} < 7$$
$$-7-\dfrac{1}{2} < 5x < 7-\dfrac{1}{2}$$
$$-\dfrac{15}{2} < 5x < \dfrac{13}{2}$$
$$-\dfrac{15}{2}\cdot\dfrac{1}{5} < x < \dfrac{13}{2}\cdot\dfrac{1}{5}$$
$$-\dfrac{3}{2} < x < \dfrac{13}{10}$$

The solution is $\left(-\dfrac{3}{2}, \dfrac{13}{10}\right)$.

45. $|T-49| \leq 20$

$$-20 \leq T-49 \leq 20$$
$$29 \leq T \leq 69$$

47. $|T-62| \leq 19$

$$-19 \leq T-62 \leq 19$$
$$43 \leq T \leq 81$$

49. $|R_L - 26.75| \leq 1.42$

$|R_E - 38.75| \leq 2.17$

a. $R_L - 26.75 \leq \pm 1.42$
$$25.33 \leq R_L \leq 28.17$$
$$R_E - 38.75 \leq \pm 2.17$$
$$36.58 \leq R_E \leq 40.92$$

b. $225(25.33) \leq T_L \leq 225(28.17)$
$$5699.25 \leq T_L \leq 6338.25$$
$$225(36.58) \leq T_E \leq 225(40.92)$$
$$8238.5 \leq T_E \leq 9207$$

51. $16.4x + 121 = 203$
$$16.4x = 82$$
$$x = 5$$
$$16.4x + 121 = 285$$
$$16.4x = 164$$
$$x = 10$$

Between the years of 2005 and 2010, the number of internet users was between 203 and 285 million.

53. $-.8x + 70.5 = 54.5$
$$-.8x = -16$$
$$x = 20$$
$$-.8x + 70.5 = 42.5$$
$$-.8x = -28$$
$$x = 35$$

For ages 20 through 35, the average remaining life expectancy for a male is between 54.5 and 42.5 years.

55. If 1/5 of a mile is 30 cents, then a mile is $1.50. Let x represent the number of miles traveled.
$$5 \leq 2 + 1.50x \leq 11$$
$$3 \leq 1.50x \leq 9$$
$$2 \leq x \leq 6$$
You can travel 2 to 6 miles in a taxi in New York City.

57. $C = 50x + 6000$; $R = 65x$

To at least break even, $R \geq C$.

$$65x \geq 50x + 6000$$
$$65x - 50x \geq 6000$$
$$15x \geq 6000$$
$$x \geq 400$$

The number of units of wire must be in the interval $[400, \infty)$.

59. $C = 85x + 1000$; $R = 105x$

$$R \geq C$$
$$105x \geq 85x + 1000$$
$$105x - 85x \geq 1000$$
$$20x \geq 1000$$
$$x \geq \frac{1000}{20}$$
$$x \geq 50$$

x must be in the interval $[50, \infty)$.

61. $C = 1000x + 5000$; $R = 900x$

$$R \geq C$$
$$900x \geq 1000x + 5000$$
$$900x - 1000x \geq 5000$$
$$-100x \geq 5000$$
$$x \leq \frac{5000}{-100}$$
$$x \leq -50$$

It is impossible to break even.

63. If x is within 4 units of 2, then the distance from x to 2 is less than or equal to 4.

$$|x - 2| \leq 4$$

65. If z is no less than 2 units from 12, then the distance from z to 12 is greater than or equal to 2.

$$|z - 12| \geq 2$$

Section 2.5 Polynomial and Rational Inequalities

1. $(x + 4)(2x - 3) \leq 0$

Solve the corresponding equation.

$$(x + 4)(2x - 3) = 0$$
$$x + 4 = 0 \quad \text{or} \quad 2x - 3 = 0$$
$$x = -4 \qquad x = \frac{3}{2}$$

Note that because the inequality symbol is "$\leq$," -4 and $\frac{3}{2}$ are solutions of the original inequality. These numbers separate the number line into three regions.

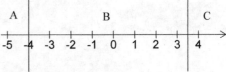

In region A, let $x = -6$:

$(-6 + 5)[2(-6)-3] = 15 > 0$.

In region B, let $x = 0$:

$(0 + 5)[2(0) - 3] = -15 < 0$.

In region C, let $x = 2$:

$(2 + 5)[2(2) - 3] = 7 > 0$.

The only region where $(x + 5) \cdot (2x - 3)$ is negative is region B, so the solution is $\left[-4, \frac{3}{2}\right]$.

To graph this solution, put brackets at -4 and $\frac{3}{2}$ and draw a line segment between these two endpoints.

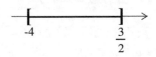

3. $r^2 + 4r > -3$

Solve the corresponding equation.

$$r^2 + 4r = -3$$
$$r^2 + 4r + 3 = 0$$
$$(r+1)(r+3) = 0$$
$$r+1 = 0 \quad \text{or} \quad r+3 = 0$$
$$r = -1 \quad \text{or} \quad r = -3$$

Note that because the inequality symbol is ">," −1 and −3 are not solutions of the original inequality.

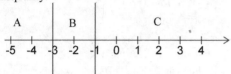

In region A, let $r = -4$:

$(-4)^2 + 4(-4) = 0 > -3$.

In region B, let $r = -2$:

$(-2)^2 + 4(-2) = -4 < -3$.

In region C, let $r = 0$:

$0^2 + 4(0) = 0 > -3$.

The solution is $(-\infty, -3)$ or $(-1, \infty)$.

To graph the solution, put a parenthesis at −3 and draw an arrow extending to the left, and put a parenthesis at −1 and draw an arrow extending to the right.

5. $4m^2 + 7m - 2 \leq 0$

Solve the corresponding equation.

$$4m^2 + 7m - 2 = 0$$
$$(4m-1)(m+2) = 0$$
$$4m-1 = 0 \quad \text{or} \quad m+2 = 0$$
$$m = \frac{1}{4} \quad \text{or} \quad m = -2$$

Note that $\frac{1}{4}$ and −2 are solutions of the original inequality.

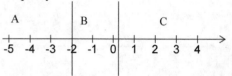

In region A, let $m = -3$:

$4(-3)^2 + 7(-3) - 2 = 13 > 0$.

In region B, let $m = 0$:

$4(0)^2 + 7(0) - 2 = -2 < 0$.

In region C, let $m = 1$:

$4(1)^2 + 7(1) - 2 = 9 > 0$.

The solution is $\left[-2, \frac{1}{4} \right]$.

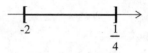

7. $4x^2 + 3x - 1 > 0$

Solve the corresponding equation.

$$4x^2 + 3x - 1 = 0$$
$$(4x-1)(x+1) = 0$$
$$4x-1 = 0 \quad \text{or} \quad x+1 = 0$$
$$x = \frac{1}{4} \quad \text{or} \quad x = -1$$

Note that $\frac{1}{4}$ and −1 are not solutions of the original inequality.

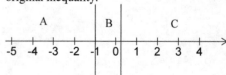

7. Continued

In region A, let $x = -2$:
$4(-2)^2 + 3(-2) - 1 = 9 > 0$.
In region B, let $x = 0$:
$4(0)^2 + 3(0) - 1 = -1 < 0$.
In region C, let $x = 1$:
$4(1)^2 + 3(1) - 1 = 6 > 0$.

The solution is $(-\infty, -1)$ or $\left(\dfrac{1}{4}, \infty\right)$.

9. $x^2 \leq 36$

Solve $x^2 = 36$

$\qquad x = \pm 6$

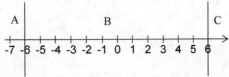

For region A, let $x = -7$:
$(-7)^2 = 49 > 36$.
For region B, let $x = 0$:
$0^2 = 0 < 36$.
For region C, let $x = 7$:
$7^2 = 49 > 36$.
Both endpoints are included. The solution is
$[-6, 6]$.

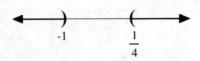

11. $p^2 - 16p > 0$

Solve $p^2 - 16p = 0$

$\qquad p(p - 16) = 0$

$p = 0$ or $p = 16$

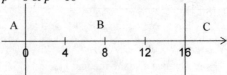

For region A, let $p = -1$:
$(-1)^2 - 16(-1) = 17 > 0$.
For region B, let $p = 1$:
$1^2 - 16(1) = -15 < 0$.
For region C, let $p = 17$:
$17^2 - 16(17) = 17 > 0$.
The solution is $(-\infty, 0)$ or $(16, \infty)$.

13. $x^3 - 9x \geq 0$

Solve the corresponding equation.

$\qquad x^3 - 9x = 0$

$\qquad x(x^2 - 9) = 0$

$x(x + 3)(x - 3) = 0$

$x = 0$ or $x = -3$ or $x = 3$

Note that 0, –3, and 3 are all solutions of the original inequality.

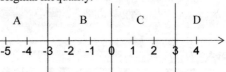

In region A, let $x = -4$:
$(-4)^3 - 9(-4) = -28 \leq 0$.
In region B, let $x = -1$:
$(-1)^3 - 9(-1) = 8 \geq 0$.
In region C, let $x = 1$:
$(1)^3 - 9(1) = -8 \leq 0$
In region D, let $x = 4$:
$4^3 - 9(4) = 28 \geq 0$.
The solution is $[-3, 0]$ or $[3, \infty)$.

15. $(x+6)(x+1)(x-4) \geq 0$

Solve the corresponding equation.

$(x+6)(x+1)(x-4) = 0$

$x+6 = 0$ or $x+1 = 0$ or $x-4 = 0$

$x = -6$ or $x = -1$ or $x = 4$

Note that -6, -1 and 4 are all solutions of the original inequality.

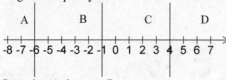

In region A, let $x = -7$:

$(-7+6)(-7+1)(-7-4) = -1(-6)(-11)$

$= -66 < 0$

In region B, let $x = -2$:

$(-2+6)(-2+1)(-2-4) = 4(-1)(-6)$

$= 24 > 0$

In region C, let $x = 0$:

$(0+6)(0+1)(0-4) = 6(1)(-4) = -24 < 0$

In region D, let $x = 5$:

$(5+6)(5+1)(5-4) = 11(6)(1) = 66 > 0$

The solution is $[-6, -1]$ or $[4, \infty)$.

17. $(x+3)\left(x^2 - 2x - 3\right) < 0$

Solve the corresponding equation.

$(x+3)\left(x^2 - 2x - 3\right) = 0$

$(x+3)(x+1)(x-3) = 0$

$x+3 = 0$ or $x+1 = 0$ or $x-3 = 0$

$x = -3$ or $x = -1$ or $x = 3$

Note that -3, -1 and 3 are not solutions of the original inequality.

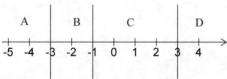

In region A, let $x = -5$:

$(-5+3)\left[(-5)^2 - 2(-5) - 3\right]$

$= -2(25+10-3) = -64 < 0$

In region B, let $x = -2$:

$(-2+3)\left[(-2)^2 - 2(-2) - 3\right]$

$= 1(4+4-3) = 1(5) = 5 > 0$

In region C, let $x = 0$:

$(0+3)\left[0^2 - 2(0) - 3\right] = 3(-3) = -9 < 0$

In region D, let $x = 4$:

$(4+3)\left[(4)^2 - 2(4) - 3\right]$

$= 7(16-8-3) = 7(5) = 35 > 0$

The solution is $(\infty, -3)$ or $(-1, 3)$.

19. $6k^3 - 5k^2 < 4k$

Solve the corresponding equation.

$6k^3 - 5k^2 = 4k$

$6k^3 - 5k^2 - 4k = 0$

$k\left(6k^2 - 5k - 4\right) = 0$

$k(3k-4)(2k+1) = 0$

$k = 0$ or $k = \dfrac{4}{3}$ or $k = -\dfrac{1}{2}$

Note that 0, $\dfrac{4}{3}$, and $-\dfrac{1}{2}$ are not solutions of the original inequality.

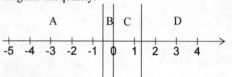

In region A, let $k = -1$:

$6(-1)^3 - 5(-1)^2 = -11$

$4(-1) = -4$

$-11 < -4$

In region B, let $k = -\dfrac{1}{4}$:

$6\left(-\dfrac{1}{4}\right)^3 - 5\left(-\dfrac{1}{4}\right)^2 = -\dfrac{13}{32}$

$4\left(-\dfrac{1}{4}\right) = -1$

$-\dfrac{13}{32} > -1$

In region C, let $k = 1$:

$6(1)^3 - 5(1)^2 = 1$

$4(1) = 4$

$1 < 4$

In region D, let $k = 10$:

$6(10)^3 - 5(10)^2 = 5500$

$4(10) = 40$

$5500 > 40$

The given inequality is true in regions A and C.

The solution is $\left(-\infty, -\dfrac{1}{2}\right)$ or $\left(0, \dfrac{4}{3}\right)$.

21. The inequality $p^2 < 16$ should be rewritten as $p^2 - 16 < 0$ and solved by the method shown in this section for solving quadratic inequalities. This method will lead to the correct solution $(-4, 4)$. The student's method and solution are incorrect.

23. To solve $.5x^2 - 1.2x < .2$,
write the inequality as $.5x^2 - 1.2x - .2 < 0$.
Graph the equation $y = .5x^2 - 1.2x - .2$.

Enter this equation as y_1 and use $-3 \le x \le 3$ and $-3 \le y \le 3$. On the CALC menu, use "zero" to find the x-values where the graph crosses the x-axis. These values are $x = -.1565$ and $x = 2.5565$. The graph is below the x-axis between these two values. The solution of the inequality is $(-.1565, 2.5565)$.

25. To solve $x^3 - 2x^2 - 5x + 7 \ge 2x + 1$,
write the inequality as $x^3 - 2x^2 - 7x + 6 \ge 0$.
Graph the equation $y = x^3 - 2x^2 - 7x + 6$.

Enter this equation as y_1 and use $-4 < x < 4$ and $-10 < y < 10$. On the CALC menu, use "zero" to find the x-values where the graph crosses the x-axis. These values are $x = -2.2635$, $x = .7556$ and $x = 3.5079$. The graph is above the x-axis to the right of -2.2635 and to the left of $.7556$, and to the right of 3.5079. The solution of the inequality is $[-2.2635, .7556]$ or $[3.5079, \infty)$.

27. To solve $2x^4 + 3x^3 < 2x^2 + 4x - 2$,
write the inequality as
$2x^4 + 3x^3 - 2x^2 - 4x + 2 < 0$.
Graph the equation $y = 2x^4 + 3x^3 - 2x^2 - 4x + 2$.

Enter this equation as y_1 and use $-2 < x < 2$ and $-2 < y < 8$. On the CALC menu, use "zero" to find the x-values where the graph crosses the x-axis. These values are $x = .5$ and $x = .8393$. The graph is below the x-axis to the right of $.5$ and to the left of $.8393$. The solution of the inequality is $(.5, .8393)$.

29. $\dfrac{r-4}{r-1} \ge 0$

Write the corresponding equation.

$\dfrac{r-4}{r-1} = 0$

The quotient can change sign only when the numerator is 0 or the denominator is 0. The numerator is 0 when $r = 4$. The denominator is 0 when $r = 1$. Note that 4 is a solution of the original inequality, but 1 is not.

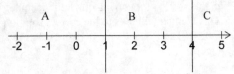

29. Continued

In region A, let $r = 0$:
$\dfrac{0-4}{0-1} = 4 \ge 0$.

In region B, let $r = 2$:
$\dfrac{2-4}{2-1} = -2 \le 0$.

In region C, let $r = 5$:
$\dfrac{5-3}{5-1} = \dfrac{1}{2} \ge 0$.

The given inequality is true in regions A and C, so the solution is $(-\infty, 1)$ or $[4, \infty)$.

31. $\dfrac{a-2}{a-5} < -1$

Write the corresponding equation and get one side equal to 0.

$$\frac{a-2}{a-5} = -1$$

$$\frac{a-2}{a-5} + 1 = 0$$

$$\frac{a-2}{a-5} + \frac{a-5}{a-5} = 0$$

$$\frac{2a-7}{a-5} = 0$$

The numerator is 0 when $a = \dfrac{7}{2}$. The denominator is 0 when $a = 5$. Note that $\dfrac{7}{2}$ and 5 are not solutions of the original inequality.

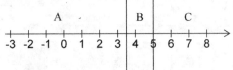

In region A, let $a = 0$:
$\dfrac{0-2}{0-5} = \dfrac{2}{5} > -1$.

In region B, let $a = 4$:
$\dfrac{4-2}{4-5} = \dfrac{2}{-1} < -1$.

In region C, let $a = 10$:
$\dfrac{10-2}{10-5} = \dfrac{8}{5} > -1$.

The solution is $\left(\dfrac{7}{2}, 5\right)$.

33. $\dfrac{1}{p-2} < \dfrac{1}{3}$

Write the corresponding equation and get one side equal to 0.

$$\frac{1}{p-2} = \frac{1}{3}$$

$$\frac{1}{p-2} - \frac{1}{3} = 0$$

$$\frac{3-(p-2)}{3(p-2)} = 0$$

$$\frac{3-p+2}{3(p-2)} = 0$$

$$\frac{5-p}{3(p-2)} = 0$$

The numerator is 0 when $p = 5$. The denominator is 0 when $p = 2$. Note that 2 and 5 are not solutions of the original inequality.

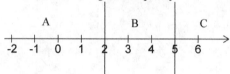

In region A, let $p = 0$:
$$\frac{1}{0-2} = -\frac{1}{2} < \frac{1}{3}.$$
In region B, let $p = 3$:
$$\frac{1}{3-2} = 1 > \frac{1}{3}.$$
In region C, let $p = 6$:
$$\frac{1}{6-2} = \frac{1}{4} < \frac{1}{3}.$$
The solution is $(-\infty, 2)$ or $(5, \infty)$.

35. $\dfrac{5}{p+1} > \dfrac{12}{p+1}$

Write the corresponding equation and get one side equal to 0.

$$\frac{5}{p+1} = \frac{12}{p+1}$$

$$\frac{5}{p+1} - \frac{12}{p+1} = 0$$

$$\frac{-7}{p+1} = 0$$

The numerator is never 0. The denominator is 0 when $p = -1$. Therefore, in this case, we separate the number line into only two regions.

35. **Continued**

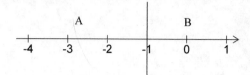

In region A, let $p = -2$:
$$\frac{5}{-2+1} = -5$$
$$\frac{12}{-2+1} = -12$$
$$-5 > -12$$
In region B, let $p = 0$:
$$\frac{5}{0+1} = 5$$
$$\frac{12}{0+1} = 12$$
$$12 > 5$$
Therefore, the given inequality is true in region A. The only endpoint, -1, is not included because the symbol is ">." Therefore, the solution is $(-\infty, -1)$.

37. $\dfrac{x^2 - x - 6}{x} < 0$

Solve the corresponding equation.

$$\frac{x^2 - x - 6}{x} = 0$$

$$x^2 - x - 6 = 0 \quad \text{or} \quad x = 0$$

$$(x-3)(x+2) = 0 \quad \text{or} \quad x = 0$$

$$x - 3 = 0 \quad \text{or} \quad x + 2 = 0 \quad \text{or} \quad x = 0$$

$$x = 3 \quad \text{or} \quad x = -2 \quad \text{or} \quad x = 0$$

Note that -2, 0 and 3 are not solutions of the original inequality.

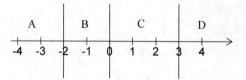

In region A, let $x = -3$:
$$\frac{(-3)^2 - (-3) - 6}{-3} = \frac{9+3-6}{-3} = \frac{6}{-3} = -2 < 0.$$
In region B, let $x = -1$:
$$\frac{(-1)^2 - (-1) - 6}{-1} = \frac{1+1-6}{-1} = \frac{-4}{-1} = 4 > 0$$

37. Continued

In region C, let $x = 1$:

$$\frac{1^2 - 1 - 6}{1} = -6 < 0.$$

In region D, let $x = 4$:

$$\frac{4^2 - 4 - 6}{4} = \frac{16 - 10}{4} = \frac{6}{4} = \frac{3}{2} > 0.$$

The solution is $(-\infty, -2)$ or $(0, 3)$.

39. To solve

$$\frac{2x^2 + x - 1}{x^2 - 4x + 4} \leq 0,$$

break the inequality into two inequalities

$$2x^2 + x - 1 \leq 0 \text{ and } x^2 - 4x + 4 \leq 0.$$

Graph the equations

$$y = 2x^2 + x - 1 \text{ and}$$

$$y = x^2 - 4x + 4.$$

Enter these equations y_1 and y_2, and use $-3 < x < 3$ and $-2 < y < 2$. On the CALC menu, use "zero" to find the x-values where the graphs cross the x-axis. These values for y_1 are $x = -1$ and $x = .5$.

The graph of y_1 is below the x-axis to the right of -1 and to the left of .5. The graph of y_2 is never below the x-axis. The solution of the inequality is $[-1, .5]$.

41. $p = 3x^2 - 35x + 50$

The company makes a profit when

$$3x^2 - 35x + 50 > 0.$$

Solve the corresponding equation.

$$3x^2 - 35x + 50 = 0$$

$$(3x - 5)(x - 10) = 0$$

$$x = \frac{5}{3} \text{ or } x = 10$$

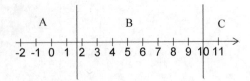

Test a number from each region in the original inequality.

For region A, let $x = 0$.

$$3(0)^2 - 35(0) + 50 = 50 > 0$$

For region B, let $x = 2$.

$$3(2)^2 - 35(2) + 50 = -8 < 0$$

For region C, let $x = 11$.

$$3(11)^2 - 35(11) + 50 = 28 > 0$$

41. Continued

The numbers in regions A and C satisfy the inequality. The company makes a profit when the amount spent on advertising in hundreds of thousands of dollars is in the interval $\left(0, \frac{5}{3}\right)$ or $(10, \infty)$, that is, $0 < x < \frac{5}{3}$ and $x > 10$.

43. $P = -x^2 + 250x - 15,000$

The complex makes a profit when

$$-x^2 + 250x - 15,000 > 0.$$

Solve the corresponding equation.

$$-x^2 + 250x - 15,000 = 0$$

$$-\left[x^2 - 250x + 15,000\right] = 0$$

$$-(x - 100)(x - 150) = 0$$

$$x = 100 \text{ or } x = 150$$

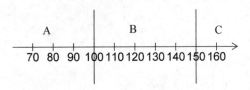

In region A, let $x = 99$:

$$-(99)^2 + 250(99) - 15,000 = -51 < 0.$$

In region B, let $x = 101$:

$$-(101)^2 + 250(101) - 15,000 = 49 > 0.$$

In region C, let $x = 151$:

$$-(151)^2 + 250(151) - 15,000 = -51 < 0.$$

The complex makes a profit when the number of units rented is between 100 and 150, exclusive, or when x is in the interval $(100, 150)$.

45. $26x^2 - 62x + 60 \geq 400\,(x \geq 1)$ Solve the corresponding equation.

$$26x^2 - 62x + 60 - 400 = 0$$
$$26x^2 - 62x - 340 = 0$$
$$(2x - 10)(13x + 34) = 0$$

$$x = 5, \text{ or } x = -\frac{34}{13}.$$

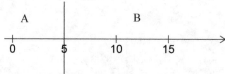

In region A, let $x = 0$:

$26(0)^2 - 62(0) + 60 = 60 \leq 400$

In region B, let $x = 10$:

$26(10)^2 - 62(10) + 60 = 2040 \geq 400$

The given inequality is true in region B and so revenue will exceed $400,000,000 from 2005 to present.

47. $.005x^2 - .052x + 4.7 > 5$

Use a graphing calculator to solve the corresponding equation. $.005x^2 - .052x + 4.7 = 5$
$x = -4.130$ and 14.530

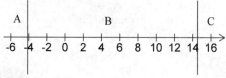

In region A, let $x = -5$,

$.005(-5)^2 - .052(-5) + 4.7 > 5$

$5.085 > 5$

In region B , let $x = 0$.

$.005(0)^2 - .052(0) + 4.7 > 5$

$4.7 > 5$

In region C, let $x = 15$.

$.005(15)^2 - .052(15) + 4.7 > 5$

$5.045 > 5$

Since $x = 0$ corresponds to the year 2000, there will be more than 5 million patients from mid 2014 ($x = 14.530$) onward.

Chapter 2 Review Exercises

1. $y = x^2 - 2x - 5$

$(-2, 3): \; 3 \overset{?}{=} (-2)^2 - 2(-2) - 5$

$\qquad 30 \; 4 + 4 - 5$

$\qquad 3 = 3$

$(0, -5): \; -5 \overset{?}{=} 0^2 - 2(0) - 5$

$\qquad -5 \overset{?}{=} -5$

$(2, -3): \; -3 \overset{?}{=} 2^2 - 2(2) - 5$

$\qquad -3 \overset{?}{=} 4 - 4 - 5$

$\qquad -3 \neq -5$

$(3, -2): \; -2 \overset{?}{=} 3^2 - 2(3) - 5$

$\qquad -2 \overset{?}{=} 9 - 6 - 5$

$\qquad -2 = -2$

$(4, 3): \; 3 \overset{?}{=} 4^2 - 2(4) - 5$

$\qquad 3 \overset{?}{=} 16 - 8 - 5$

$\qquad 3 = 3$

$(7, 2): \; 2 \overset{?}{=} 7^2 - 2(7) - 5$

$\qquad 2 \overset{?}{=} 49 - 14 - 5$

$\qquad 2 \neq 30$

Solutions are $(-2, 3)$, $(0, -5)$, $(3, -2)$, $(4, 3)$.

2. $x - y = 5$

$(-2, 3): \; -2 - 3 \overset{?}{=} 5$

$\qquad -5 \neq 5$

$(0, -5): \; 0 - (-5) \overset{?}{=} 5$

$\qquad 5 = 5$

$(2, -3): \; 2 - (-3) \overset{?}{=} 5$

$\qquad 5 = 5$

$(3, -2): \; 3 - (-2) \overset{?}{=} 5$

$\qquad 5 = 5$

$(4, 3): \; 4 - 3 \overset{?}{=} 5$

$\qquad 1 \neq 5$

$(7, 2): \; 7 - 2 \overset{?}{=} 5$

$\qquad 5 = 5$

Solutions are $(0, -5)$, $(2, -3)$, $(3, -2)$, $(7, 2)$.

3. $5x - 3y = 15$

First, we find the y-intercept. If $x = 0$,
$y = -5$, so the y-intercept is -5. Next we find the
x-intercept. If $y = 0$, $x = 3$, so the x-intercept is 3.
Using these intercepts, we graph the line.

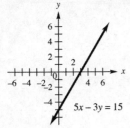

4. $2x + 7y - 21 = 0$

First we find the y-intercept. If $x = 0$,
$y = 3$, so the y-intercept is 3. Next we find the

x-intercept. If $y = 0$, $x = \dfrac{21}{2}$, so the

x-intercept is $\dfrac{21}{2}$. Using these intercepts, we

graph the line.

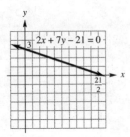

5. $y + 3 = 0$

The equation may be rewritten as $y = -3$. The
graph of $y = -3$ is a horizontal line with y-
intercept of -3.

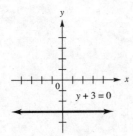

6. $y - 2x = 0$

First, we find the y-intercept. If $x = 0$, $y = 0$, so
the y-intercept is 0. Since the line passes through
the origin, the x-intercept is also 0. We find
another point on the line by arbitrarily choosing a
value for x. Let $x = 2$. Then
$y - 2(2) = 0$, or $y = 4$. The point with coordinates
$(2, 4)$ is on the line. Using this point and the
origin, we graph the line.

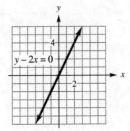

7. $y = .25x^2 + 1$

First we find the y-intercept. If $x = 0$,
$y = .25(0)^2 + 1 = 1$, so the y-intercept is 1. Next
we find the x-intercepts. If $y = 0$,
$0 = x^2 + 1$
$x^2 = -1$
$x = \sqrt{-1}$, not a real number.
There are no x-intercepts.
Make a table of points and plot them.

x	$.25x^2 + 1$
-4	5
-2	2
0	1
2	2
4	5

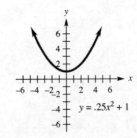

8. $y = \sqrt{x+4}$

Make a table of points and plot them.

x	$\sqrt{x+4}$
−4	0
−3	1
0	2
5	3

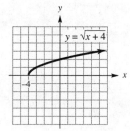

9. a. When was the temperature over 55°?
About 11:30 A.M. to about 7:30 P.M.

b. When was the temperature below 40°?
From midnight until about 5 A.M., and after about 10:30 P.M.

10. At noon in Bratenahl the temperature was about 57°. When is the temperature in Greenville 57°? When the temperature in Bratenahl is 50°, or at about 10:30 A.M. and 8:30 P.M.

11. Answers vary. A possible answer is "rise over run".

12. Through (−1, 4) and (2, 3)

$$\text{slope} = \frac{\Delta y}{\Delta x} = \frac{3-4}{2-(-1)}$$

$$= \frac{-1}{3} = -\frac{1}{3}$$

13. Through (5, −3) and (−1, 2)

$$\text{slope} = \frac{2-(-3)}{-1-5}$$

$$= \frac{5}{-6} = -\frac{5}{6}$$

14. Through (7, −2) and the origin
The coordinates of the origin are (0, 0).

$$\text{slope} = \frac{-2-0}{7-0} = -\frac{2}{7}$$

15. Through (8, 5) and (0, 3)

$$\text{slope} = \frac{3-5}{0-8}$$

$$= \frac{-2}{-8} = \frac{1}{4}$$

16. $2x + 3y = 30$
First we solve for y.

$$3y = -2x + 30$$

$$y = -\frac{2}{3}x + 10$$

When the equation is written in slope-intercept form, the coefficient of x gives the slope. The slope is $-\frac{2}{3}$.

17. $4x - y = 7$
First we solve for y.

$$-y = -4x + 7$$

$$y = 4x - 7$$

The coefficient of x gives the slope, so the slope is 4.

18. $x + 5 = 0$
The graph of $x = -5$ is a vertical line. Therefore, the slope is undefined.

19. $y = 3$
The graph of $y = 3$ is a horizontal line. Therefore, the slope is 0.

20. Parallel to $3x + 8y = 0$
To find the slope of the given line, we solve for y.

$$8y = -3x$$

$$y = -\frac{3}{8}x$$

The slope is the coefficient of x, $-\frac{3}{8}$. A line parallel to this line has the same slope, so the slope of the parallel line is also $-\frac{3}{8}$.

21. Perpendicular to $x = 3y$
To find the slope of the given line, we solve for y.

$$y = \frac{1}{3}x$$

The slope of this line is the coefficient of x, $\frac{1}{3}$.

The slope of a perpendicular line is the negative reciprocal of this slope, so the slope of the perpendicular line is −3.

22. Through $(0, 5)$ with $m = -\dfrac{2}{3}$

Since $m = -\dfrac{2}{3} = \dfrac{-2}{3}$, we start at the point with

coordinates $(0, 5)$ and move 2 units down and 3
units to the right to obtain a second point on the
line. Using these two points, we graph the line.

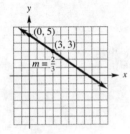

23. Through $(-4, 1)$ with $m = 3$

Since $m = 3 = \dfrac{3}{1}$, we start at the point with

coordinates $(-4, 1)$ and move 3 units up and 1 unit
to the right to obtain a second point on the line.
Using these two points, we graph the line.

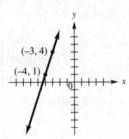

24. Answers vary. One example is:
You need two points; one point and the slope; the
y-intercept and the slope.

25. Through $(5, -1)$, slope $\dfrac{2}{3}$

Use the point slope form with $x_1 = 5$, $y_1 = -1$,

and $m = \dfrac{2}{3}$.

$$y - y_1 = m(x - x_1)$$
$$y - (-1) = \dfrac{2}{3}(x - 5)$$
$$y + 1 = \dfrac{2}{3}x - \dfrac{10}{3}$$

Multiplying by 3 gives
$$3y + 3 = 2x - 10$$
$$3y = 2x - 13$$

26. $(8, 0)$, $m = -\dfrac{1}{4}$

$$y - 0 = -\dfrac{1}{4}(x - 8)$$
$$4y = -1(x - 8)$$
$$4y = -x + 8$$

27. Through $(5, -2)$ and $(1, 3)$
$$m = \dfrac{3 - (-2)}{1 - 5} = \dfrac{5}{-4} = -\dfrac{5}{4}$$
$$y - 3 = -\dfrac{5}{4}(x - 1)$$
$$4(y - 3) = -5(x - 1)$$
$$4y - 12 = -5x + 5$$
$$4y = -5x + 17$$

28. $(2, -3)$ and $(-3, 4)$
$$m = \dfrac{-3 - 4}{2 - (-3)} = -\dfrac{7}{5}$$
$$y - (-3) = -\dfrac{7}{5}(x - 2)$$
$$5(y + 3) = -7(x - 2)$$
$$5y + 15 = -7x + 14$$
$$5y = -7x - 1$$

29. Undefined slope, through $(-1, 4)$
This is a vertical line. Its equation is
$x = -1$.

30. Slope 0, $(-2, 5)$
This is a horizontal line. Its equation is
$y = 5$.

31. x-intercept -3, y-intercept 5
Use the points $(-3, 0)$ and $(0, 5)$.
$$m = \frac{5-0}{0-(-3)} = \frac{5}{3}$$
$$y = \frac{5}{3}x + 5$$
$$3y = 5(x+3)$$
$$3y = 5x + 15$$

32. x-intercept 3, y-intercept 2.
Use the points $(3, 0)$ and $(0, 2)$.
$$m = \frac{2-0}{0-3} = -\frac{2}{3}$$
$$y = -\frac{2}{3}x + 2$$
$$3y = 3\left(-\frac{2}{3}x + 2\right)$$
$$3y = -2x + 6$$
$$2x + 3y = 6$$

Answer option (d).

33. a. Let (x_1, y_1) be $(0, 25.3)$ and (x_2, y_2) be $(12, 22.5)$.
Find the slope.
$$m = \frac{22.5 - 25.3}{12 - 0} = \frac{-2.8}{12}$$
$$= -\frac{28}{120}$$
$$= -\frac{7}{30}$$
Use the point-slope form with $(0, 25.3)$.
$$y - 25.3 = -\frac{7}{30}(x - 0)$$
$$y - 25.3 = -\frac{7}{30}x$$
$$y = -\frac{7}{30}x + 25.3$$

b. The slope is negative because the percentage of smokers is decreasing.

c. Let $x = 18$.
$$y = -\frac{7}{30}(18) + 25.3$$
$$= 21.1\%$$

34. a. Let (x_1, y_1) be $(0, 72.5)$ and (x_2, y_2) be $(12, 68.7)$.
Find the slope.
$$m = \frac{68.7 - 72.5}{12 - 0} = \frac{-3.8}{12}$$
$$= -\frac{38}{120}$$
$$= -\frac{19}{60}$$
Use the point-slope form with $(0, 72.5)$.
$$y - 72.5 = -\frac{19}{60}(x - 0)$$
$$y = -\frac{19}{60}x + 72.5$$

b.

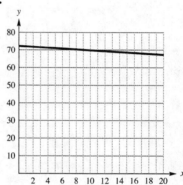

c. Let $x = 20$.
$$y = -\frac{19}{60}(20) + 72.5$$
$$= 66.2\%$$

35. a. Let (x_1, y_1) be $(8, 13.00)$ and (x_2, y_2) be $(13, 15.35)$.
Find the slope.
$$m = \frac{15.35 - 13.00}{13 - 8} = \frac{2.35}{5} = .47$$
Use the point-slope form with $(8, 13.00)$.
$$y - 13.00 = .47(x - 8)$$
$$y = .47x + 9.24$$

b. Using a graphing calculator, the least squares regression line is $y = .4785x + 9.188$.

c. Let $x = 10$. For the two-point model,
$$y = .47(10) + 9.24$$
$$\approx \$13.94$$

$$\$14.00 - \$13.94 = \$.06$$

35. Continued

For the regression model,
$y = .4785(10) + 9.188$

$\approx \$13.97$

$\$14.00 - \$13.97 = \$.03$

The two-point model is off by $.06 and the regression model is off by $.03.

d. Let $x = 18$. Two-point model:
$y = .47(18) + 9.24$

$\approx \$17.70$

Regression model:
$y = .4785(18) + 9.188$

$\approx \$17.80$

36. Solve the inequality:
$6.82 < .14125x + 2.3 < 10.21$

$6.82 < .14125x + 2.3 < 10.21$

$4.52 < .14125x < 7.91$

$32 < x < 56$

Medicare costs will be between 6.82% and 10.21% in 2032 – 2056.

37. a. Using a graphing calculator, the least squares regression line is
$y = 606.5386x + 14,863.20$.

b. The coefficient of correlation is $r \approx .98$. Since this value is so close to 1, the regression line is a good fit for the data.

c. Let $x = 18$ (for 2008).
$y = 606.5386(18) + 14,863.20$

$\approx 25,781$

38. a. Using a graphing calculator, the least squares regression line is
$y = 7.8475x + 11.895$

b. The coefficient of correlation is $r \approx .99$. Since this value is so close to 1, the regression line is a good fit for the data.

c. Let $x = 11$ (for 2001).
$y = 7.8475(11) + 11.895$

$\approx \$98,217,500,000$

39.

$-6x + 3 < 2x$

$-6x + 6x + 3 < 2x + 6x$

$3 < 8x$

$\dfrac{3}{8} < \dfrac{8x}{8}$

$\dfrac{3}{8} < x$

or $\quad x > \dfrac{3}{8}$

The solution is $\left(\dfrac{3}{8}, \infty\right)$.

40.

$12z \geq 5z - 7$

$12z - 5z \geq 5z - 5z - 7$

$7z \geq -7$

$\dfrac{7z}{7} \geq \dfrac{-7}{7}$

$z \geq -1$

The solution is $[-1, \infty)$.

41.

$2(3 - 2m) \geq 8m + 3$

$6 - 4m \geq 8m + 3$

$6 - 4m - 8m \geq 8m - 8m + 3$

$6 - 12m \geq 3$

$6 - 6 - 12m \geq 3 - 6$

$12m \geq -3$

$\dfrac{-12m}{-12} \leq \dfrac{-3}{-12}$

$m \leq \dfrac{1}{4}$

The solution is $\left(-\infty, \dfrac{1}{4}\right]$.

42. $6p - 5 > -(2p + 3)$

$6p - 5 > -2p - 3$

$8p - 5 > -3$

$8p > 2$

$\dfrac{8p}{8} > \dfrac{2}{8}$

$p > \dfrac{1}{4}$

The solution is $\left(\dfrac{1}{4}, \infty\right)$.

43. $-3 \leq 4x - 1 \leq 7$

$\quad -2 \leq 4x \leq 8$

$\quad -\dfrac{1}{2} \leq x \leq 2$

The solution is $\left[-\dfrac{1}{2}, 2\right]$.

44. $\quad 0 \leq 3 - 2a \leq 15$

$\quad 0 - 3 \leq 3 - 3 - 2a \leq 15 - 3$

$\quad\quad -3 \leq -2a \leq 12$

$\quad\quad \dfrac{-3}{-2} \geq \dfrac{-2a}{-2} \geq \dfrac{12}{-2}$

$\quad\quad\quad \dfrac{3}{2} \geq a \geq -6$

The solution is $\left[-6, \dfrac{3}{2}\right]$.

45. $|b| \leq 8$

$\quad -8 \leq b \leq 8$

The solution is $[-8, 8]$.

46. $|a| > 7$

$\quad a < -7 \text{ or } a > 7$

The solution is $(-\infty, -7)$ or $(7, \infty)$.

47. $|2x - 7| \geq 3$

$\quad 2x - 7 \leq -3 \quad \text{ or } \quad 2x - 7 \geq 3$

$\quad\quad 3x \leq 4 \quad\quad \text{ or } \quad\quad 2x \geq 10$

$\quad\quad\quad x \leq 2 \quad\quad \text{ or } \quad\quad\quad x \geq 5$

The solution is $(-\infty, 2]$ or $[5, \infty)$.

48. $|4m + 9| \leq 16$

$\quad -16 \leq 4m + 9 \leq 16$

$\quad -25 \leq 4m \leq 7$

$\quad -\dfrac{25}{4} \leq m \leq \dfrac{7}{4}$

The solution is $\left[-\dfrac{25}{4}, \dfrac{7}{4}\right]$.

49. $|5k + 2| - 3 \leq 4$

$\quad\quad |5k + 2| \leq 7$

$\quad\quad -7 \leq 5k + 2 \leq 7$

$\quad\quad -9 \leq 5k \leq 5$

$\quad\quad -\dfrac{9}{5} \leq k \leq 1$

49. Continued

The solution is $\left[-\dfrac{9}{5}, 1\right]$.

50. $|3z - 5| + 2 \geq 10$

$\quad |3x - 5| \geq 8$

$\quad 3z - 5 \leq -8 \quad \text{ or } \quad 3z - 5 \geq 8$

$\quad\quad 3z \leq -3 \quad \text{ or } \quad\quad 3z \geq 13$

$\quad\quad\quad z \leq -1 \quad \text{ or } \quad\quad\quad z \geq \dfrac{13}{3}$

The solution is $(-\infty, -1]$ or $\left[\dfrac{13}{3}, \infty\right)$.

51. The inequalities that represent the weight of pumpkin that he will not use are $x < 2$ or $x > 10$. This is equivalent to the following inequalities:

$x - 6 < 2 - 6 \text{ or } x - 6 > 10 - 6$

$x - 6 < -4 \text{ or } x - 6 > 4$

$|x - 6| > 4$

Choose answer option (d).

52. Let n = number of milligrams per liter.

$|n - 40| \leq .05$

53. a. Let (x_1, y_1) be $(0, 415)$ and (x_2, y_2) be $(10, 645)$.

Find the slope

$m = \dfrac{645 - 415}{10 - 0} = \dfrac{230}{10} = 23$

Use the point-slope form with $(0, 415)$.

$y - 415 = 23(x - 0)$

$\quad\quad y = 23x + 415$

b. Let $y = 599$.

$599 = 23x + 415$

$184 = 23x$

$\quad x = 8$

Until 1998, the energy production will be less than 599 Mtoe.

c. Let $y = 714$.

$714 = 23x + 415$

$299 = 23x$

$\quad x = 13$

After 2003, the energy production will be more than 714 Mtoe.

54. Let m = number of miles driven.

$$75 > 50 + .05m$$

$$25 > .05m$$

$$m < 500$$

Since m must be a positive number,
$0 \le m < 500$.

55. $r^2 + r - 6 < 0$

Write the corresponding equation and solve to determine values of r where the polynomial changes sign.

$$r^2 + r - 6 = 0$$

$$(r + 3)(r - 2) = 0$$

$$r = -3 \text{ or } r = 2$$

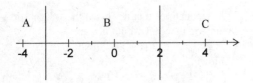

For region A, test -4:
$(-4)^2 + (-4) - 6 = 6 > 0$.
For region B, test 0:
$0^2 + 0 - 6 = -6 < 0$.
For region C, test 3:
$3^2 + 3 - 6 = 6 > 0$.
The solution is $(-3, 2)$.

56. $y^2 + 4y - 5 \ge 0$

Solve the corresponding equation.

$$y^2 + 4y - 5 = 0$$

$$(y + 5)(y - 1) = 0$$

$$y = -5 \text{ or } y = 1$$

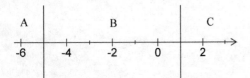

For region A, test -6:
$(-6)^2 + 4(-6) - 5 = 7 > 0$.
For region B, test 0:
$0^2 + 4(0) - 5 = -5 < 0$.
For region C, test 2:
$2^2 + 4(2) - 5 = 7 > 0$.
Both endpoints are included because the inequality symbol is "$\ge$." The solution is $(-\infty, -5]$ or $[1, \infty)$.

57. $2z^2 + 7z \ge 15$

Write the corresponding equation and solve it.

$$2z^2 + 7z = 15$$

$$2z^2 + 7z - 15 = 0$$

$$(2z - 3)(z + 5) = 0$$

$$z = \frac{3}{2} \text{ or } z = -5$$

These numbers are solutions of the inequality because the inequality symbol is "$\ge$."

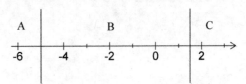

For region A, test -6:
$2(-6)^2 + 7(-6) = 30 > 15$.
For region B, test 0:
$2 \cdot 0^2 + 7 \cdot 0 = 0 < 15$.
For region C, test 2:
$2 \cdot 2^2 + 7 \cdot 2 = 22 > 15$.

The solution is $(-\infty, -5]$ or $\left[\dfrac{3}{2}, \infty\right)$.

58. $3k^2 \le k+14$

Solve the corresponding equation.

$$3k^2 = k+14$$

$$3k^2 - k - 14 = 0$$

$$(3k-7)(k+2) = 0$$

$$k = \frac{7}{3} \text{ or } k = -2$$

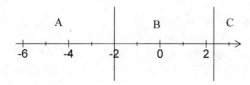

For region A, test −3:

$$3(-3)^2 = 27$$

$$-3+14 = 11$$

$$27 > 11$$

For region B, test 0:

$$3(0)^2 = 0$$

$$0+14 = 14$$

$$0 < 14$$

For region C, test 3:

$$3(3)^2 = 27$$

$$3+14 = 17$$

$$27 > 17$$

The given inequality is true in region B and at

both endpoints, so the solution is $\left[-2, \dfrac{7}{3}\right]$.

59. $(x-3)\left(x^2 + 7x + 10\right) \le 0$

Write the corresponding equation and solve it.

$$(x-3)\left(x^2 + 7x + 10\right) = 0$$

$$(x-3)(x+2)(x+5) = 0$$

$$x - 3 = 0 \quad \text{or} \quad x+2 = 0 \quad \text{or} \quad x+5 = 0$$

$$x = 3 \quad \text{or} \quad x = -2 \quad \text{or} \quad x = -5$$

Note that −5, −2, and 3 are solutions of the original inequality.

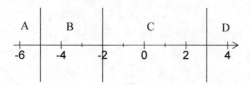

59. Continued

In region A, let $x = -6$:

$$(-6-3)\left((-6)^2 + 7(-6) + 10\right)$$

$$= -9(36 - 42 + 10) = -9(4) = -36 < 0$$

In region B, let $x = -3$:

$$(-3-3)\left((-3)^2 + 7(-3) + 10\right)$$

$$= -6(9 - 21 + 10) = -6(-2) = 12 > 0.$$

In region C, let $x = 0$:

$$(0-3)\left(0^2 + 7(0) + 10\right) = -3(10) = -30 < 0.$$

In region D, let $x = 4$:

$$(4-3)\left(4^2 + 7(4) + 10\right)$$

$$= 1(16 + 28 + 10) = 54 > 0.$$

The solution is $(-\infty, -5]$ or $[-2, 3]$.

60. $(x+4)\left(x^2 - 1\right) \ge 0$

Write the corresponding equation and solve it.

$$(x+4)\left(x^2 - 1\right) = 0$$

$$(x+4)(x+1)(x-1) = 0$$

$$x + 4 = 0 \quad \text{or} \quad x+1 = 0 \quad \text{or} \quad x-1 = 0$$

$$x = -4 \quad \text{or} \quad x = -1 \quad \text{or} \quad x = 1$$

Note that −4, −1, and 1 are solutions of the original inequality.

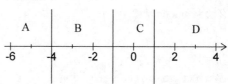

In region A, let $x = -5$:

$$(-5+4)\left((-5)^2 - 1\right) = -1(24) = -24 \le 0.$$

In region B, let $x = -2$:

$$(-2+4)\left((-2)^2 - 1\right) = 2(3) = 6 > 0.$$

In region C, let $x = 0$:

$$(0+4)\left(0^2 - 1\right) = 4(-1) = -4 < 0.$$

In region D, let $x = 2$:

$$(2+4)\left(2^2 - 1\right) = 6(3) = 18 > 0.$$

The solution is $[-4, -1]$ or $[1, \infty)$.

61. $\dfrac{m+2}{m} \le 0$

Write $\dfrac{m+2}{m} = 0$.

The quotient changes sign when

$m+2 = 0$ or $m = 0$

$m = -2$ or $m = 0$

–2 is a solution of the inequality, but the inequality is undefined when $m = 0$, so the endpoint 0 must be excluded.

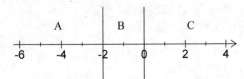

For region A, test –3:

$\dfrac{-3+2}{-3} = \dfrac{1}{3} > 0$.

For region B, test –1:

$\dfrac{-1+2}{-1} = -1 < 0$.

For region C, test 1:

$\dfrac{1+2}{1} = 3 > 0$.

The solution is $[-2, 0)$.

62. $\dfrac{q-4}{q+3} > 0$

Write the corresponding equation.

$\dfrac{q-4}{q+3} = 0$

The numerator is 0 when $q = 4$. The denominator is 0 when $q = -3$.

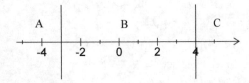

62. Continued

For region A, test –4:

$\dfrac{-4-4}{-4+3} = \dfrac{-8}{-1} = 8 > 0$.

For region B, test 0:

$\dfrac{0-4}{0+3} = -\dfrac{4}{3} < 0$.

For region C, test 5:

$\dfrac{5-4}{5+3} = \dfrac{1}{8} > 0$.

The inequality is true in regions A and C, and both endpoints are excluded. Therefore, the solution is $(-\infty, -3)$ or $(4, \infty)$.

63. $\dfrac{5}{p+1} > 2$

Write the corresponding equation and get one side equal to 0.

$\dfrac{5}{p+1} = 2$

$\dfrac{5}{p+1} - 2 = 0$

$\dfrac{5 - 2(p+1)}{p+1} = 0$

$\dfrac{3 - 2p}{p+1} = 0$

This quotient changes sign when

$3 - 2p = 0$ or $p + 1 = 0$

$p = \dfrac{3}{2}$ or $p = -1$

Neither of these numbers is a solution of the inequality.

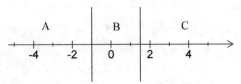

In region A, test –2:

$\dfrac{5}{-2+1} = -5 < 2$.

For region B, test 0:

$\dfrac{5}{0+1} = 5 > 2$.

For region C, test 2:

$\dfrac{5}{2+1} = \dfrac{5}{3} < 2$.

The solution is $\left(-1, \dfrac{3}{2}\right)$.

64. $\dfrac{6}{a-2} \le -3$

Write the corresponding equation and get one side equal to 0.

$$\frac{6}{a-2} = -3$$

$$\frac{6}{a-2} + 3 = 0$$

$$\frac{6+3(a-2)}{a-2} = 0$$

$$\frac{3a}{a-2} = 0$$

The numerator is 0 when $a = 0$. The denominator is 0 when $a = 2$.

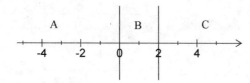

For region A, test -1:

$\dfrac{6}{-1-2} = -2 \ge -3$.

For region B, test 1:

$\dfrac{6}{1-2} = -6 \le -3$.

For region C, test 3:

$\dfrac{6}{3-2} = 6 \ge -3$.

The given inequality is true in region B. The endpoint 0 is included because the inequality symbol is "$\le$." However, the endpoint 2 must be excluded because it makes the denominator 0. The solution is $[0, 2)$.

65. $\dfrac{2}{r+5} \le \dfrac{3}{r-2}$

Write the corresponding equation and get one side equal to 0.

$$\frac{2}{r+5} = \frac{3}{r-2}$$

$$\frac{2}{r+5} - \frac{3}{r-2} = 0$$

$$\frac{2(r-2)-3(r+5)}{(r+5)(r-2)} = 0$$

$$\frac{2r-4-3r-15}{(r+5)(r-2)} = 0$$

$$\frac{-r-19}{(r+5)(r-2)} = 0$$

The quotient changes sign where

$-r-19 = 0$ or $r+5 = 0$ or $r-2 = 0$

$r = -19$ or $r = -5$ or $r = 2$

-19 is a solution of the inequality, but the inequality is undefined when $r = -5$ or $r = 2$.

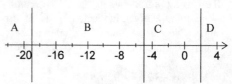

For region A, test -20:

$\dfrac{2}{-20+5} = -\dfrac{2}{15} \approx -.13$ and

$\dfrac{3}{-20-2} = -\dfrac{3}{22} \approx -.14$

Since $-.13 > -.14$, -20 is not a solution of the inequality.

For region B, test -6:

$\dfrac{2}{-6+5} = -2$ and $\dfrac{3}{-6-2} = -\dfrac{3}{8}$.

Since $-2 < -\dfrac{3}{8}$, -6 is a solution.

For region C, test 0:

$\dfrac{2}{0+5} = \dfrac{2}{5}$ and $\dfrac{3}{0-2} = -\dfrac{3}{2}$.

Since $\dfrac{2}{5} > -\dfrac{3}{2}$, 0 is not a solution.

For region D, test 3:

$\dfrac{2}{3+5} = \dfrac{1}{4}$ and $\dfrac{3}{3-2} = 3$.

Since $\dfrac{1}{4} < 3$, 3 is a solution. The solution is $[-19, -5)$ or $(2, \infty)$.

66. $\dfrac{1}{z-1} > \dfrac{2}{z+1}$

Write the corresponding equation and get one side equal to 0.

$$\frac{1}{z-1} = \frac{2}{z+1}$$

$$\frac{1}{z-1} - \frac{2}{z+1} = 0$$

$$\frac{(z+1) - 2(z-1)}{(z-1)(z+1)} = 0$$

$$\frac{3-z}{(z-1)(z+1)} = 0$$

The numerator is 0 when $z = 3$. The denominator is 0 when $z = 1$ and when $z = -1$. These three numbers, -1, 1, and 3, separate the number line into four regions.

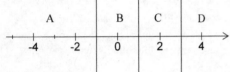

Testing a point in each of these four regions will show that the given inequality is true in regions A and C. Neither endpoint is included in the solution. Thus, the solution is $(-\infty, -1)$ or $(1, 3)$.

67. $p = .06x^2 - 12.5x + 926$

Find x when $p > 500$.
Solve the corresponding equation.

$$500 = .06x^2 - 12.5x + 926$$

$$0 = .06x^2 - 12.5x + 426$$

Using the quadratic formula with $a = .06$, $b = -12.5$, and $c = 426$, $x \approx 42.9$ or $x \approx 165.4$, which is in the future (later than 2000), and so is extraneous.

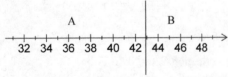

Test a number from each region in the original inequality.
For region A, let $x = 0$.
$.06(0)^2 - 12.5(0) + 926 = 926 > 500$
For region B, let $x = 50$.
$.06(50)^2 - 12.5(50) + 926 \approx 451 < 500$
The population of Cleveland was above 500,000 from 1950 to late 1992.

Case 2 Using Extrapolation to Predict Life Expectancy

1. Since $x = 0$ corresponds to the year 1900, enter the following data into a computing device.

x	y
50	71.3
60	73.1
70	74.7
80	77.4
85	78.2
90	78.8
9.5	78.9

Here is a screen capture.

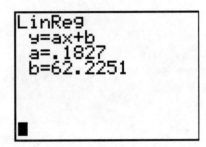

The model is verified.

2. Let $x = 1900$.
$y = .1827(0) + 62.2251 \approx 62.2$. This is about 62.2 years.

3. 62.2 years is significantly higher than 48.3 years. This poor prediction isn't surprising, since we were extrapolating far beyond the range of the original data.

4.

Year of birth (x = 0 is 1900)	Table Value	Predicted Value	Residual
50	71.3	71.4	–.1
60	73.1	73.2	–.1
70	74.7	75.0	–.3
80	77.4	76.8	.6
85	78.2	77.8	.4
90	78.8	78.7	.1
95	78.9	79.6	–.7

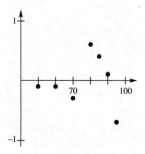

5. It's not clear that any simple smooth function will fit this data. It would be extremely difficult, if not impossible, to predict the life expectancy for females born in 2010.

6. You'll get 0 slope and 0 intercept, because you've already subtracted out the linear component of the data.

7. They used a regression equation of some kind to predict this value!

Chapter 3: Functions and Graphs

Section 3.1 Functions

1.
$$\begin{array}{c|ccccccc} x & 3 & 2 & 1 & 0 & -1 & -2 & -3 \\ \hline y & 9 & 4 & 1 & 0 & 1 & 4 & 9 \end{array}$$

   This rule defines y as a function of x because each value of x determines one and only one value of y.

3. The rule $y = x^3$ defines y as a function of x because each value of x determines one and only one value of y.

5. The rule $x = |y + 2|$ does not define y as a function of x because some values of x determine two values for y. For example, if $x = 4$,

   $4 = |y + 2|$

   $y + 2 = 4$ or $y + 2 = -4$

   $\quad y = 2$ or $\quad\quad y = -6$

7. The rule $y = \dfrac{-1}{x - 1}$ defines y as a function of x because each value of x determines one and only one value of y.

9. $f(x) = 4x - 1$

   The domain of f is all real numbers since x may take on any real-number value. Therefore, the domain is $(-\infty, \infty)$.

11. $f(x) = x^4 - 1$

    The domain of f is all real numbers since x may take on any real-number value. Therefore, the domain is $(-\infty, \infty)$.

13. $f(x) = \sqrt{-x} + 3$

    In order to have $\sqrt{-x}$ be a real number, we must have $-x \geq 0$ or $x \leq 0$. Thus, the domain is all nonpositive real numbers, or $(-\infty, 0]$.

15. $f(x) = \dfrac{1}{x - 2}$

    Since the denominator cannot be zero, $x \neq 2$. Thus, the domain is all real numbers except 2, which in interval notation is written $(-\infty, 2)$ or $(2, \infty)$.

17. $g(x) = \dfrac{x^2 + 4}{x^2 - 4}$

    Solve $x^2 - 4 = 0$ and exclude the solutions from the domain because the solutions make the denominator equal to 0.

    $x^2 - 4 = 0$

    $(x - 2)(x + 2) = 0$

    $x - 2 = 0$ or $x + 2 = 0$

    $x = 2$ or $x = -2$

    The domain of f is all real numbers except $x = \pm 2$ since x cannot take on $x = \pm 2$. Therefore, the domain is $(-\infty, -2)$ or $(-2, 2)$ or $(2, \infty)$.

19. $h(x) = \dfrac{\sqrt{x + 4}}{x^2 + x - 12}$

    Solve $x^2 + x - 12 = 0$ and exclude the solutions from the domain since x cannot take on these numbers.

    $x^2 + x - 12 = 0$

    $(x - 3)(x + 4) = 0$

    $x - 3 = 0$ or $x + 4 = 0$

    $x = 3$ or $x = -4$

    For $\sqrt{x + 4}$,

    $x + 4$ must be positive or 0. That is $x + 4 > 0$ or $x > -4$.

    Therefore, the domain is x such that $x \neq 3$ and $x > -4$.

21. $g(x) = \begin{cases} \dfrac{1}{x} & \text{if } x < 0 \\ \sqrt{x^2 + 1} & \text{if } x \geq 0 \end{cases}$

    For $\dfrac{1}{x}$ if $x < 0$, x can take on any real number.

    For $\sqrt{x^2 + 1}$, x can take on any real number.

    The domain is all real numbers or $(-\infty, \infty)$.

23. $f(x) = 8$

For any value of x, the value of $f(x)$ will always be 6. (This is a constant function).

a. $f(4) = 8$

b. $f(-3) = 8$

c. $f(2.7) = 8$

d. $f(-4.9) = 8$

25. $f(x) = 2x^2 + 4x$

a. $f(4) = 2(4^2) + 4(4)$
$= 2(16) + 16$
$= 32 + 16$
$= 48$

b. $f(-3) = 2(-3)^2 + 4(-3)$
$= 2(9) + (-12)$
$= 18 - 12$
$= 6$

c. $f(2.7) = 2(2.7)^2 + 4(2.7)$
$= 2(7.29) + 10.8$
$= 14.58 + 10.8$
$= 25.38$

d. $f(-4.9) = 2(-4.9)^2 + 4(-4.9)$
$= 2(24.01) + (-19.6)$
$= 48.02 - 19.6$
$= 28.42$

27. $f(x) = \sqrt{x+3}$

a. $f(4) = \sqrt{4+3} = \sqrt{7} \approx 2.6458$

b. $f(-3) = \sqrt{-3+3} = \sqrt{0} = 0$

c. $f(2.7) = \sqrt{2.7+3} = \sqrt{5.7} \approx 2.3875$

d. $f(-4.9) = \sqrt{-4.9+3} = \sqrt{-1.9}$ not defined

29. $f(x) = \left| x^2 - 6x - 4 \right|$

a. $f(4) = \left| 4^2 - 6(4) - 4 \right|$
$f(4) = \left| 16 - 24 - 4 \right|$
$f(4) = \left| -12 \right| = 12$

b. $f(-3) = \left| (-3)^2 - 6(-3) - 4 \right|$
$= \left| 9 + 18 - 4 \right|$
$= 23$

c. $f(2.7) = \left| (2.7)^2 - 6(2.7) - 4 \right|$
$= \left| 7.29 - 16.2 - 4 \right|$
$= 12.91$

d. $f(-4.9) = \left| (-4.9)^2 - 6(-4.9) - 4 \right|$
$= \left| 24.01 + 29.4 - 4 \right|$
$= \left| 49.41 \right|$
$= 49.41$

31. $f(x) = \dfrac{\sqrt{x-1}}{x^2 - 1}$

a. $f(4) = \dfrac{\sqrt{4-1}}{4^2 - 1} = \dfrac{\sqrt{3}}{15} \approx .1155$

b. $f(-3) = \dfrac{\sqrt{-3-1}}{(-3)^2 - 1} = \dfrac{\sqrt{-4}}{8}$ not defined

c. $f(2.7) = \dfrac{\sqrt{2.7-1}}{2.7^2 - 1} = \dfrac{\sqrt{1.7}}{6.29} \approx .2073$

d. $f(-4.9) = \dfrac{\sqrt{-4.9-1}}{(-4.9)^2 - 1}$
$= \dfrac{\sqrt{-5.9}}{23.01}$ not defined

33. $g(x) = \begin{cases} x^2 & \text{if } x < 2 \\ 5x - 7 & \text{if } x \geq 2 \end{cases}$

(a) $g(4) = 5(4) - 7 = 13$

(b) $g(-3) = (-3)^2 = 9$

(c) $g(2.7) = 5(2.7) - 7 = 6.5$

(d) $g(-4.9) = (-4.9)^2 = 24.01$

35. $f(x) = 6 - x$

    a. $f(p) = 6 - p$

    b. $f(-r) = 6 - (-r)$
            $= 6 + r$

    c. $f(m + 3) = 6 - (m + 3)$
                 $= 6 - m - 3$
                 $= 3 - m$

37. $f(x) = \sqrt{4 - x}$

    a. $f(p) = \sqrt{4 - p} \;\; (p \le 4)$

    b. $f(-r) = \sqrt{4 - (-r)}$
             $= \sqrt{4 + r} \;\; (r \ge -4)$

    c. $f(m + 3) = \sqrt{4 - (m + 3)}$
                $= \sqrt{4 - m - 3}$
                $= \sqrt{1 - m} \;\; (m \le 1)$

39. $f(x) = x^3 + 1$

    a. $f(p) = p^3 + 1$

    b. $f(-r) = (-r)^3 + 1 = -r^3 + 1$

    c. $f(m + 3) = (m + 3)^3 + 1$
                 $= m^3 + 9m^2 + 27m + 27 + 1$
                 $= m^3 + 9m^2 + 27m + 28$

41. $f(x) = \dfrac{3}{x - 1}$

    a. $f(p) = \dfrac{3}{p - 1} \;\; (p \ne 1)$

    b. $f(-r) = \dfrac{3}{-r - 1}$ or $-\dfrac{3}{r + 1} \;\; (r \ne -1)$

    c. $f(m + 3) = \dfrac{3}{(m + 3) - 1}$
                 $= \dfrac{3}{m + 2} \;\; (m \ne -2)$

43. $f(x) = 2x - 4$

$$\frac{f(x + h) - f(x)}{h}$$
$$= \frac{[2(x + h) - 4] - (2x - 4)}{h}$$
$$= \frac{2x + 2h - 4 - 2x + 4}{h}$$
$$= \frac{2h}{h} = 2$$

45. $f(x) = x^2 + 1$

$$\frac{f(x + h) - f(x)}{h}$$
$$= \frac{(x + h)^2 + 1 - \left(x^2 + 1\right)}{h}$$
$$= \frac{x^2 + 2hx + h^2 + 1 - x^2 - 1}{h}$$
$$= \frac{2hx + h^2}{h}$$
$$= 2x + h$$

47. $g(x) = 3x^4 - x^3 + 2x$

X	Y1
3.5000	414.31
3.9000	642.51
4.3000	954.73
4.7000	1369.5
5.1000	1907.1
5.5000	2589.8

Y1=414.3125

49.

$$T(x) = \begin{cases} .03x & \text{if } 0 \le x \le 5000 \\ 150 + .04(x - 5000) & \text{if } 5000 < x \le 10,000 \\ 350 + .05(x - 10,000) & \text{if } x > 10,000 \end{cases}$$

a. $.03(4750) = \$142.50$

b. $350 + .05(27,950 - 10,000) = \$1247.50$

c. $150 + .04(9320 - 5000) = \$322.80$

51. $p(x) = \dfrac{x-1}{x}$

a. $p(9) = \dfrac{9-1}{9} = .8888 \text{ or } 89\%$

b. $p(1.9) = \dfrac{1.9-1}{1.9} = .4736 \text{ or } 47\%$

53. $h(x) = .4018x^2 + 2.039x + 50$

a. Let $x = 6$ for 2000.

$h(6) = .4018(6)^2 + 2.039(6) + 50$

$\qquad = 76.699$

$\qquad = 76.699(1000)$

$\qquad = 76,699$

Let $x = 10$ for 2004.

$h(10) = .4018(10)^2 + 2.039(10) + 50$

$\qquad = 110.57$

$\qquad = 110.57(1000)$

$\qquad = 110,570$

b. Let $x = 14$ for 2008

$h(14) = .4018(14)^2 + 2.039(14) + 50$

$\qquad = 157.2988$

$\qquad = 157.2988(1000)$

$\qquad = 157,300$

55. a. $g(x) = -.583x^2 + 38.786x^2 + 12.44x + 363.471$

X	Y1
0.00	363.47
1.00	414.11
2.00	538.83
3.00	734.12
4.00	996.50
5.00	1322.4
6.00	1708.5

Y1◻-.583X^3+38....

b.

X	Y1
.50	379.31
1.50	467.43
2.50	627.87
3.50	857.14
4.50	1151.7
5.50	1508.2
6.50	1922.9

Y1◻-.583X^3+38....

57. At noon, $t = 0$.
Distance traveled $= 475 \cdot t$
Distance from Seattle:
$f(t) = 2000 - 475t$

59. a. $C(x) = 125x + 36,000$

b. Let $x = 600$.
$C(600) = 125(600) + 36,000$
$C(600) = \$111,000$
To find the cost per bicycle, divide \$111,000
by the number of bicycles. The cost per bike
is $\dfrac{111,000}{600} = \$185$

Section 3.2 Graphs of Functions

1. $f(x) = -.5x + 2$
   The graph is a straight line with slope $-.5$ and y-intercept 2.

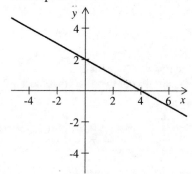

$$f(x) = -.5x + 2$$

3. $f(x) = \begin{cases} x+3 & \text{if } x \le 1 \\ 4 & \text{if } x > 1 \end{cases}$

   Graph the line $y = x + 3$ for $x \le 1$.
   Graph the horizontal line $y = 4$ for $x > 1$.

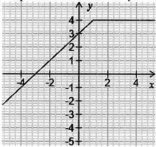

$$f(x) = \begin{cases} x+3 & \text{if } x \le 1 \\ 4 & \text{if } x > 1 \end{cases}$$

5. $f(x) = \begin{cases} 4-x & \text{if } x \le 0 \\ 3x+4 & \text{if } x > 0 \end{cases}$

   Graph the line $y = 4 - x$ for $x \le 0$.
   Graph the line $y = 3x + 3$ for $x > 0$.

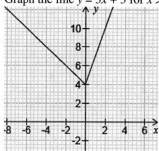

$$f(x) = \begin{cases} 4-x & \text{if } x \le 0 \\ 3x+4 & \text{if } x > 0 \end{cases}$$

7. $f(x) = \begin{cases} |x| & \text{if } x < 2 \\ -2x & \text{if } x \ge 2 \end{cases}$

   Rewrite the function as

   $$f(x) = \begin{cases} -x & \text{if } x \le 0 \\ x & \text{if } 0 < x \le 2 \\ -x & \text{if } x \ge 2 \end{cases}$$

   Graph the line $y = -x$ for $x \le 0$.
   Graph the line $y = x$ for $0 < x \le 2$.
   Graph the line $y = -x$ for $x \ge 2$.

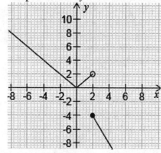

$$g(x) = \begin{cases} x+5 & \text{if } x \le 1 \\ 2-3x & \text{if } x > 1 \end{cases}$$

9. $f(x) = |x - 4|$

Using the definition of absolute values gives

$$f(x) = \begin{cases} x-4 & \text{if } x-4 \geq 0 \\ -(x-4) & \text{if } x-4 < 0 \end{cases}$$

or

$$f(x) = \begin{cases} x-4 & \text{if } x \geq 4 \\ -x+4 & \text{if } x < 4 \end{cases}$$

We graph the line $y = x - 4$ with slope 1 and y-intercept −4 for $x \geq 4$. We graph the line $y = -x + 4$ with slope −1 and y-intercept 4 for $x < 4$. Note that these partial lines meet at the point (4, 0).

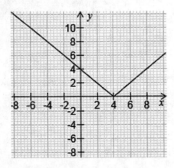

$$f(x) = |x - 4|$$

11. $f(x) = |3 - 3x|$

$$f(x) = \begin{cases} 3-3x & \text{if } 3-3x \geq 0 \\ -(3-3x) & \text{if } 3-3x < 0 \end{cases}$$

or

$$f(x) = \begin{cases} 3-3x & \text{if } x \leq 1 \\ -3+3x & \text{if } x > 1 \end{cases}$$

Graph the line $y = 3 - 3x$ for $x \leq 1$.
Graph the line $y = 3x - 3$ for $x > 1$.
The graph consists of two lines that meet at $(1, 0)$.

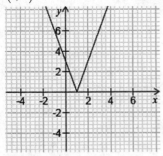

$$f(x) = |3 - 3x|$$

13. $f(x) = -|x - 1|$

$$f(x) = \begin{cases} -(x-1) & \text{if } x-1 \geq 0 \\ -[-(x-1)] & \text{if } x-1 < 0 \end{cases}$$

or

$$f(x) = \begin{cases} -x+1 & \text{if } x \geq 1 \\ x-1 & \text{if } x < 1 \end{cases}$$

Graph the line $y = -x + 1$ for $x \geq 1$.
Graph the line $y = x - 1$ for $x < 1$.
The graph consists of two lines that meet at $(1, 0)$.

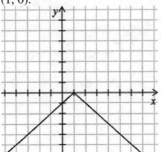

$$f(x) = -|x - 1|$$

15. $g(x) = |x - 2| + 3$

$$g(x) = \begin{cases} x+1 & \text{if } x \geq 2 \\ -x+5 & \text{if } x < 2 \end{cases}$$

Graph the line $y = x + 1$ for $x \geq 2$.
Graph the line $y = -x + 5$ for $x < 2$.
The graph consists of two lines that meet at (2, 3).

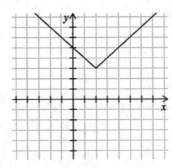

$$g(x) = |x - 2| + 3$$

17. $f(x) = [x - 3]$

For x in the interval $[0, 1)$, the value of $[x - 3] = -3$. For x in the interval $[1, 2)$, the value of $[x - 3] = -2$. For x in the interval $[2, 3)$, the value of $[x - 3] = -1$, and so on. The graph consists of a series of line segments. In each case, the left endpoint is included, and the right endpoint is excluded.

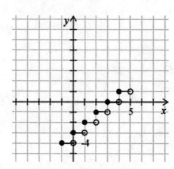

$f(x) = [x - 3]$

19. $g(x) = [-x]$

For x in the interval $(0, 1]$, the value of $[-x] = -1$. For x in the interval $(1, 2]$, the value of $[-x] = -2$. For x in the interval $(2, 3]$, the value of $[-x] = -3$, and so on. The graph consists of a series of line segments. In each case, the left endpoint is excluded, and the right endpoint is included.

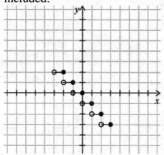

$g(x) = [-x]$

21. $f(x) = [x] + [-x]$

Make a table of intervals and their endpoints:

x	$[x]$	$[-x]$	$f(x)$
-3	-3	3	0
$(-3, -2)$	-3	2	-1
-2	-2	2	0
$(-2, -1)$	-2	1	-1
-1	-1	1	0
$(-1, 0)$	-1	0	-1
0	0	0	0
$(0, 1)$	0	-1	-1
1	1	-1	0
$(1, 2)$	1	-2	-1
2	2	-2	0
$(2, 3)$	2	-3	-1
3	3	-3	0
$(3, 4)$	3	-4	-1

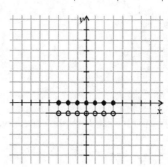

$f(x) = [x] + [-x]$

23. $f(x) = 3 - 2x^2$

Make a table of values and plot the corresponding points.

x	$f(x) = 3 - 2x^2$
−3	−15
−2	−5
−1	1
0	3
1	1
2	−5
3	−15

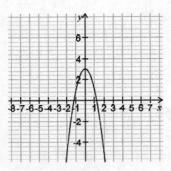

$$f(x) = 3 - 2x^2$$

25. $h(x) = \dfrac{x^3}{10} + 2$

Make a table of values and plot the corresponding points.

x	$h(x) = \frac{x^3}{10} + 2$
−4	−4.4
−3	−.7
−2	1.2
−1	1.9
0	2
1	2.1
2	2.8
3	4.7

25. Continued

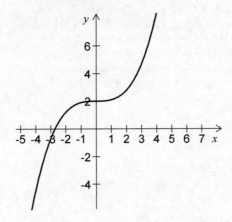

$$h(x) = \frac{x^3}{10} + 2$$

27. $g(x) = \sqrt{-x}$

Make a table of values and plot the corresponding points. The function is not defined for $x > 0$.

x	$g(x) = \sqrt{-x}$
−9	3
−4	2
−1	1
0	0

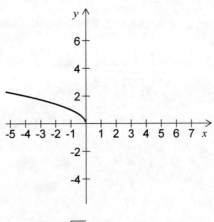

$$g(x) = \sqrt{-x}$$

29. $f(x) = \sqrt[3]{x}$

Make a table of values and plot the corresponding points.

x	$f(x) = \sqrt[3]{x}$
-8	-2
-1	1
0	0
1	1
8	2

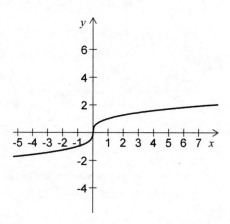

$$f(x) = \sqrt[3]{x}$$

31. $f(x) = \begin{cases} x^2 & \text{if } x < 2 \\ -2x + 2 & x \geq 2 \end{cases}$

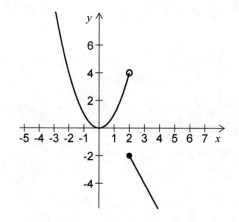

33. Every vertical line intersects this graph in at most one point, so this is the graph of a function.

35. A vertical line intersects the graph in more than one point, so this is not the graph of a function.

37. Every vertical line intersects this graph in at most one point, so this is the graph of a function.

39.

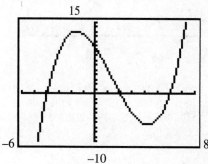

41.

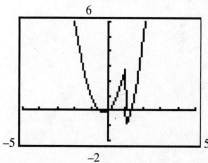

To avoid having the vertical line at $x = 1$, put the calculator in dot mode to obtain the following.

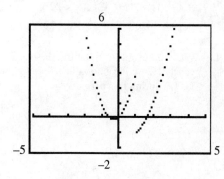

43. Draw the graph and locate the x– intercepts.

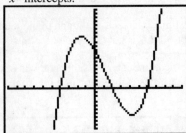

The x – intercepts are –4, 2, 6.

45. Using the maximum/minimum finder, there is a peak (.5078, .3938); and valleys at (– 1.9826, –4.2009) and (3.7248, –8.7035).

47. a.

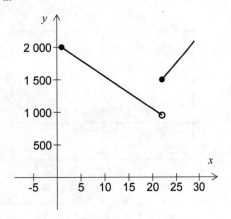

$$f(x) = \begin{cases} -50x + 2050 & \text{if } 1 \le x < 22 \\ 84x - 348 & \text{if } 22 \le x < 29 \end{cases}$$

b. When adjusted for inflation, the maximum yearly IRA contribution fell from $2000 to $1000 during this period.

49. $f(x) = \begin{cases} 6.5x & \text{if } 0 \le x \le 4 \\ -5.5x + 48 & \text{if } 4 < x \le 6 \\ -30x + 195 & \text{if } 6 < x \le 6.5 \end{cases}$

a. We graph the line $y = 6.5x$ for x in [0, 4]. We graph the line $y = -5.5x + 48$ for x in (4, 6]. We graph the line $y = -30x + 195$ for x in (6, 6.5].

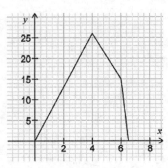

b. The highest point on the graph occurs when x = 4.
$f(4) = 6.5(4) = 26$
The value x = 4 corresponds to the beginning of February when the depth of the snow is 26 inches.

c. The value x = 0 corresponds to the beginning of October. The value x = 6.5 corresponds to mid-April. The snow begins in October and ends in mid-April.

51. a. Find the equations of lines for points
(0, 15.1) and (25, 47.5) and points (25, 47.5)
and (53, 297.1).
Find the slope for the line between (0, 15.1)
and (25, 47.5).

$$m = \frac{47.5 - 15.1}{25 - 0} = 1.296$$
$$(y - 15.1) = 1.296(x - 0)$$
$$y = 1.296x + 15.1$$

Rounding the coefficients to one decimal
place gives us $y = 1.3x + 15.1$.
Find the slope for the line between (25, 47.5)
and (53, 297.1).

$$m = \frac{297.1 - 47.5}{53 - 25} = 8.91$$
$$(y - 47.5) = 8.91(x - 25)$$
$$y = 8.91x - 175$$

Rounding the coefficients to one decimal
place gives us $y = 8.9x - 175$.
Therefore, the piecewise function that
models this data is:

$$f(x) = \begin{cases} 1.3x + 15.1 \text{ if } 0 \le x \le 25 \\ 8.9x - 175.3 \text{ if } x > 25 \end{cases}$$

b.

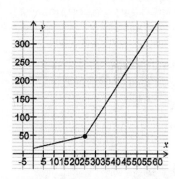

c. Let $x = 2000 - 1950 = 50$
$8.9(50) - 175 = 270$

d. Let $x = 2009 - 1950 = 59$
$8.9(59) - 175 = 350.1$

53. a. No. The graph of the CPI for all items is
always above the x-axis, so the percent of
change is always positive; this means that the
CPI is always increasing.

b. From early 1997 to mid 1998, mid 2001 to
mid 2002.

c. From 1990 to early 1997; from mid 1998 to
mid 2001; mid 2002 to 2003. The percentage
change was positive during these periods,
which means that the CPI was increasing.

55. a.

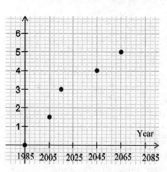

b. Given (2005, 1.5) and (2085, 5). Find the
slope.

$$m = \frac{5 - 1.5}{2085 - 2005} = \frac{3.5}{80} = .044$$

(to the nearest thousandth)
Write an equation for the line.
Let $(2005, 1.5) = (x_1, y_1)$

$$y - y_1 = m(x - x_1)$$
$$y - 1.5 = .044(x - 2005)$$
$$y - 1.5 = .044x - 88.22$$
$$y = .044x - 86.72$$
or $\quad y = .044x - 86.7$

c. Replace y with $f(x)$.
$f(x) = .044x - 86.7$

d. $f(2065) = .044(2065) - 86.7$
$= 4.16$

The table value is 4. This agrees fairly
closely. Therefore, the function
$f(x) = .044x - 86.7$
adequately describes this function.

57. a. According to the graph, $f(2000) = 33$ and
$f(2004) = 37$.

b. The figure has vertical line segments, which can't be part of the graph of a function. To make the figure into the graph of *f*, delete the vertical line segments, then for each horizontal segment of the graph, put a closed dot on the left end and an open-circle dot on the right end.

59. The charges are based on the time rounded up to the next whole hour.

a. 2 hours
$25 + 2(2) = 29$,
so the charge is $29.

b. 1.5 hours
The renter will be charged for 2 hours, so again the charge is $29.

c. 4 hours
$25 + 4(2) = 33$,
so the charge is $33.

d. 3.7 hours
The renter will be charged for 4 hours, so again the charge is $33.

e.

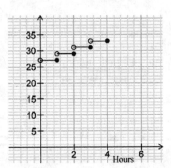

Note that for each step, the left endpoint is not included and the right point is included.

61. Make table of values:

Time	t	$g(t)$	slope
midnight	0	1,000,000	increasing
noon	1	?	decreasing
4:00 P.M.	1	?	increasing
9:00 P.M.	2	?	vertical

There are many correct answers, including:

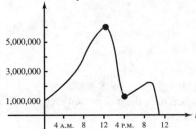

Section 3.3 Applications of Linear Functions

1. The marginal cost is $1, while the fixed cost is $12. Let $C(x)$ be the cost of renting a saw for x hours. Then $C(x) = 1x + 12$ or $C(x) = x + 12$, where x is the number of hours.

3. Let x = the number of hours and $C(x)$ = the total charge for x hours. Then $C(x) = .50x + 2$

5. Fixed cost, $200, 50 items cost $2000 to produce. Since the fixed cost is $200,
$$C(x) = mx + 200$$
$$C(50) = 2000.$$
Therefore,
$$2000 = m(50) + 200$$
$$50m = 1800$$
$$m = 36.$$
$$C(x) = 36x + 200$$

7. Marginal cost, $120; 100 items cost $15,800 to produce.
$$C(x) = mx + b, \ m = 120$$
$$C(100) = 15,800 = 120(100) + b$$
$$b = 15,800 - 12,000 = 3800$$
$$C(x) = 120x + 3800$$

9.

$$\overline{C}(50) = \frac{C(50)}{50}$$

$$= \frac{12(50) + 1800}{50}$$

$$= \frac{2400}{50}$$

$$= \$48$$

$$\overline{C}(500) = \frac{C(500)}{500}$$

$$= \frac{12(500) + 1800}{500}$$

$$= \frac{7800}{500}$$

$$= \$15.60$$

$$\overline{C}(1000) = \frac{C(1000)}{1000}$$

$$= \frac{12(1000) + 1800}{1000}$$

$$= \frac{13,800}{1000}$$

$$= \$13.80$$

11.

$$\overline{C}(200) = \frac{C(200)}{200}$$

$$= \frac{6.5(200) + 9800}{200}$$

$$= \frac{11,100}{200}$$

$$= \$55.50$$

$$\overline{C}(2000) = \frac{C(2000)}{2000}$$

$$= \frac{6.5(2000) + 9800}{200}$$

$$= \frac{22,800}{2000}$$

$$= \$11.40$$

$$\overline{C}(5000) = \frac{C(5000)}{5000}$$

$$= \frac{6.5(5000) + 9800}{5000}$$

$$= \frac{42,300}{5000}$$

$$= \$8.46$$

13. a. Let $(x_1, y_1) = (0, 15350)$ and let $(x_2, y_2) = (4, 9910)$

$$m = \frac{9910 - 15350}{4 - 0} = -1360$$

$$y - 15350 = -1360(x - 0)$$

$$f(x) = -1360x + 15350$$

b. $f(x) = -1360(4) + 15,350 = \$7190$

c. Because the slope is -1360, the car is depreciating at a rate of $1360 per year.

15. a. Let $(x_1, y_1) = (0, \ 120,000)$ and $(x_2, y_2) = (8, \ 25,000)$.

$$m = \frac{25000 - 120,000}{8 - 0} = -11,875$$

$$y - y_1 = m(x - x_1)$$

$$y - 120,000 = -11,875(x - 0)$$

$$y - 120,000 = -11,875x$$

$$y = -11,875x + 120,000$$

$$f(x) = -11,875x + 120,000$$

b. The domain ranges from 0 to 8 years, that is [0, 8].

c. Let $x = 6$. Then

$$f(6) = -11,875(6) + 120,000$$

$$f(6) = \$48,750$$

The machine will be worth $48,750 in 6 years.

17. a. The fixed costs are
$C(0) = \$42.5(0) + 80,000 = \$80,000$

b. The slope of $C(x) = 42.5x + 80,000$ is 42.5, so the marginal cost is $42.50.

c. The cost of producing 1000 books is
$C(1000) = 42.5(1000) + 80,000 = \$122,500$.
The cost of producing 32,000 books is
$C(32,000) = 42.5(32,000) + 80,000$

$$= \$1,440,000.$$

17. Continued

d. The average cost per book when 1000 are produced is:

$$\overline{C}(1000) = \frac{C(1000)}{1000}$$

$$= \frac{42.5(1000) + 80,000}{1000}$$

$$= \$122.50$$

The average cost per book when 32,000 are produced is,

$$\overline{C}(32,000) = \frac{C(32,000)}{32,000}$$

$$= \frac{42.5(32,000) + 80,000}{32,000}$$

$$= \$45$$

19. a. Let (x_1, y_1) be $(100, 11.02)$ and (x_2, y_2) be $(400, 40.12)$.
Find the slope.

$$m = \frac{40.12 - 11.02}{400 - 100} = \frac{29.1}{300} = .097$$

Use the point-slope form with $(100, 11.02)$.

$$y - 11.02 = .097(x - 100)$$

$$y = .097 + 1.32$$

$$C(x) = .097 + 1.32$$

b. The total cost of producing 1000 cups is
$C(1000) = .097(1000) + 1.32 = \$98.32$.

c. The total cost of producing 1001 cups is
$C(1001) = .097(1001) + 1.32$
$= \$98.42$.

d. The marginal cost of the 1001st cup is
$C(1001) - C(1000) = 98.417 - 98.32 = \$.097$, which is about 9.7¢.

e. The slope of $C(x) = .097x + 1.32$ is .097, so the marginal cost is \$.097 or 9.7 cents.

21. The total fixed cost for the company per year is given by
6.15(300,000) = \$1,84,5000, and the revenue per year is given by 2.0337(300,000) = \$610,110.
The total monthly revenue is given by the function $R(x) = 2.0337x + 1,845,000$.

23. a. $C(x) = 10x + 500$

b. $R(x) = 35x$

c. $P(x) = R(x) - C(x)$
$= 35x - (10x + 500)$
$= 25x - 500$

d. The profit on 100 items is,
$P(100) = 25(100) - 500 = \$2000$

25. a. $C(x) = 18x + 250$

b. $R(x) = 28x$

c. $P(x) = R(x) - C(x)$
$= 28x - (18x + 250)$
$= 10x - 250$

d. The profit on 100 items is,
$P(100) = 10(100) - 250 = \$750$

27. a. $C(x) = 12.5x + 18,000$

b. $R(x) = 25x$

c. $P(x) = R(x) - C(x)$
$= 25x - (12.5x + 18,000)$
$= 12.5x - 18,000$

d. The profit on 100 items is,
$P(100) = 12.5(100) - 18,000$
$= -\$16,750$ (loss)

29. $2x - y = 7$ and $y = 8 - 3x$
Solve for y.
$2x - y = 7$
$y = 2x - 7$
Set the two equations equal and solve for x.
$2x - 7 = 8 - 3x$
$5x = 15$
$x = 3$
Substitute x into one equation to find y.
$y = 8 - 3(3) = -1$
The lines intersect at $(3, -1)$.

31. $y = 3x - 7$ and $y = 7x + 4$
Set the two equations equal and solve for x.
$$3x - 7 = 7x + 4$$
$$-4x = 11$$
$$x = -\frac{11}{4}$$
Substitute x into one equation to find y.
$$y = 3\left(-\frac{11}{4}\right) - 7 = -\frac{33}{4} - \frac{28}{4} = -\frac{61}{4}$$
The lines intersect at $\left(-\frac{11}{4}, -\frac{61}{4}\right)$.

33. a.
$$R = 125x \text{ and}$$
$$C = 100x + 5000$$
$$R = C$$
$$125x = 100x + 5000$$
$$25x = 5000$$
$$x = 200$$

The break-even point is $x = 200$ or 200,000 policies.

b. To graph the revenue function, graph $y = 125x$. If $x = 0$, $y = 0$. If $x = 20$, $y = 2500$. Use the point $(0, 0)$ and $(20, 2500)$ to graph the line. To graph the cost function, graph $y = 100x + 5000$. If $x = 0$, $y = 5000$. If $x = 30$, $y = 8000$. Use the points $(0, 5000)$ and $(30, 8000)$ to graph the line.

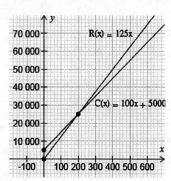

c. From the graph, when $x = 100$, cost is $15,000 and revenue is $12,500.

35. Given $r = .21x$ and $p = .084x - 1.5$

a. Cost equals revenue – profit, so
$$c(x) = r(x) - p(x)$$
$$c(x) = .21x - (.084x - 1.5)$$
$$c(x) = .21x - .084x + 1.5$$
$$c(x) = .126x + 1.5$$

b. $c(7) = .126(7) + 1.5 = 2.382$
The cost of producing 7 units is $2.382 million.

c. At the break-even point, profit $p(x) = 0$, so
$$c(x) = r(x)$$
$$.126x + 1.5 = .21x$$
$$1.5 = .084x$$
$$x \approx 17.857$$
The break-even point occurs at about 17.857 units.

37. $c(x) = 80x + 7000$; $r(x) = 95x$
$$95x = 80x + 7000$$
$$15x = 7000$$
$$x \approx 467$$

The break-even point is about 467 units. Do not produce the item since $467 > 400$.

39. $c(x) = 125x + 42,000$; $r(x) = 165.5x$
$$125x + 42,000 = 165.5x$$
$$42,000 = 40.5x$$
$$x \approx 1037$$
The break-even point is about 1037 units. Produce the item since $1037 < 2000$.

41. The break-even point is about $(2000, \$20/\text{hr.})$

43. a. Net import: Let (x_1, y_1) be $(0, 7.2)$ and

(x_2, y_2) be $(13, 11.2)$.

Find the slope.

$$m = \frac{11.2 - 7.2}{13 - 0} \approx .3077$$

Use the point-slope form with $(0, 7.2)$.

$(y - 7.2) = .3077(x - 0)$

$y = .3077x + 7.2$

$f(x) = .3077x + 7.2$

Domestic: Let (x_1, y_1) be $(0, 9.7)$ and

(x_2, y_2) be $(13, 8.4)$.

Find the slope.

$$m = \frac{8.4 - 9.7}{13 - 0} \approx -.1$$

Use the point-slope form with $(0, 9.7)$.

$(y - 9.7) = -.1(x - 0)$

$y = -.1x + 9.7$

$g(x) = -.1x + 9.7$

b.

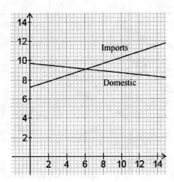

c. The intersection point is about (6.1320, 9.0868). This means the investment in both countries was about $9,086,800 billion in late 1996.

45. On the supply curve, when $q = 20$, $p = 140$. The point (20, 140) is on the graph. When 20 items are supplied, the price is $140.

47. The two curves intersect at the point (10, 120). The equilibrium supply and equilibrium demand are both $q = 10$ or 10 items.

49. $p = 16 - \dfrac{5}{4}q$

a. If $q = 0$, $p = 16$, so for a demand of 0 units, the price is $16.

b. If $q = 4$,

$$p = 16 - \frac{5}{4}(4) = 16 - 5 = 11,$$

so for a demand of 4 units, the price is $11.

c. If $q = 8$,

$$p = 16 - \frac{5}{4}(8) = 16 - 10 = 6,$$

so for a demand of 8 units, the price is $6.

d. From (c), if $p = 6$, $q = 8$, so at a price of $6, the demand is 8 units.

e. From (b), if $p = 11$, $q = 4$, so at a price of $11, the demand is 4 units.

f. From (a), if $p = 16$, $q = 0$, so at a price of $16, the demand is 0 units.

g.

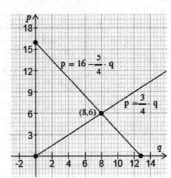

h. $p = \dfrac{3}{4}q$

If $p = 0$, $q = 0$, so at a price of $0, the supply is 0 units.

49. Continued

i. If $p = 10$,

$$10 = \frac{3}{4}q$$

$$\frac{4}{3}(10) = \frac{4}{3}\left(\frac{3}{4}\right)q$$

$$\frac{40}{3} = q$$

When the price is \$10, the supply is $\frac{40}{3}$ units.

j. If $p = 20$,

$$20 = \frac{3}{4}q$$

$$\frac{4}{3}(20) = \frac{4}{3}\left(\frac{3}{4}q\right)$$

$$\frac{80}{3} = q$$

When the price is \$20, the supply is $\frac{80}{3}$ units.

k. See (g).

l. The two graphs intersect at the point (8, 6). The equilibrium supply is 8 units.

m. The equilibrium price is \$6.

51. a. To graph $p = \left(\frac{2}{5}\right)q$: When $q = 0$, $p = 0$, and when $q = 5$, $p = 2$. Use the points (0, 0) and (5, 2). To graph $p = 100 - \left(\frac{2}{5}\right)q$: When $q = 0$, $p = 100$, and when $q = 20$, $p = 92$. Use the points (0, 100) and (20, 92).

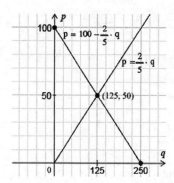

b. The two graphs intersect at the point (125, 50). The equilibrium demand is 125 units.

c. The equilibrium price is 50 cents.

d.

$$100 - \frac{2}{5}q > \frac{2}{5}q$$

$$100 > \frac{2}{5}q + \frac{2}{5}q$$

$$100 > \frac{4}{5}q$$

$$\frac{5}{4}(100) > \frac{5}{4}\left(\frac{4}{5}q\right)$$

$$125 > q$$

$$q < 125$$

Demand exceeds supply when q is in the interval [0, 125).

53. Total cost increases when more items are made (because it includes the cost of all previously made items), so the graph cannot move downward. No; the average cost can decrease as more items are made, so its graph may move downward.

Section 3.4 Quadratic Functions

1. $f(x) = x^2 - 3x - 12$

   This function is written in the form
   $y = ax^2 + bx + c$ with $a = 1$, $b = -3$ and the
   parabola opens upward since
   $a = 1 > 0$.

3. $h(x) = -3x^2 + 14x + 1$

   The parabola opens downward since
   $a = -3 < 0$.

5. $g(x) = 2.9x^2 - 12x - 5$

   The parabola opens upward since
   $a = 2.9 > 0$

7. $f(x) = -3(x - 4)^2 + 5$

   This function is written in the form
   $y = a(x - h)^2 + k$ with $a = -3$,
   $h = 4$, and $k = 5$. The vertex of the parabola, (h, k),
   is (4, 5). The parabola opens downward since
   $a = -3 < 0$.

9. $h(x) = 5(x + 6)^2 - 7$

   This function is written in the form
   $y = a(x - h)^2 + k$ with $a = 5$,
   $h = -6$, and $k = -7$. The vertex of the parabola,
   (h, k), is $(-6, -7)$. The parabola opens upward
   since
   $a = 5 > 0$.

11. $g(x) = 2.1(x - 3.5)^2 - 9$

    This function is written in the form
    $y = a(x - h)^2 + k$ with $a = 2.1$,
    $h = 3.5$, and $k = -9$. The vertex of the parabola, $(h,$
    $k)$, is $(3.5, -9)$. The parabola opens upward since
    $a = 2.1 > 0$.

13. I

15. K

17. J

19. F

21. vertex (1, 2); point (5, 6)

    $f(x) = a(x - h)^2 + k$
    $f(x) = a(x - 1)^2 + 2$
    Use (5, 6) to find a.
    $6 = a(5 - 1)^2 + 2$
    $4 = 16a$
    $a = \dfrac{1}{4}$
    $f(x) = \dfrac{1}{4}(x - 1)^2 + 2$

23. vertex (-1, -2); point (1, 2)

    $f(x) = a(x - h)^2 + k$
    $f(x) = a(x + 1)^2 - 2$
    Use (1, 2) to find a.
    $2 = a(1 + 1)^2 - 2$
    $4 = 4a$
    $a = 1$
    $f(x) = (x + 1)^2 - 2$

25. vertex (0, 0); point (2, 12)

    $f(x) = a(x - h)^2 + k$
    $f(x) = a(x - 2)^2 - 4$

    Use (2, 12) to find a.

    $12 = a(2 - 0)^2 + 0$
    $12 = 4a$
    $a = 3$
    $f(x) = 3(x - 0)^2$
    $f(x) = 3x^2$

27. $f(x) = x^2 - 8x + 4$

    $a = 1$, $b = -8$

    To find the vertex of the function, use the vertex
    formula

    $x = -\dfrac{b}{2a}$ and $y = f\left(-\dfrac{b}{2a}\right)$.

    $x = -\dfrac{(-8)}{2(1)} = 4$

    $y = f(4)$
    $f(4) = 4^2 - 8(4) + 4 = -12$

    The vertex is (4, -12).

29. $h(x) = 2x^2 + 12x - 3$

$a = 2, \ b = 12$

To find the vertex of the function, use the vertex formula

$$x = -\frac{b}{2a} \text{ and } y = f\left(-\frac{b}{2a}\right).$$

$$x = -\frac{(12)}{2(2)} = -3$$

$$y = f(-3)$$

$$f(4) = 2(-3)^2 + 12(-3) - 3 = -21$$

The vertex at $(-3, -21)$.

31. $g(x) = -x^2 + x$

$a = -1, \ b = 1$

To find the vertex of the function, use the vertex formula

$$x = -\frac{b}{2a} \text{ and } y = f\left(-\frac{b}{2a}\right).$$

$$x = -\frac{(1)}{2(-1)} = \frac{1}{2}$$

$$y = f\left(\frac{1}{2}\right)$$

$$f\left(\frac{1}{2}\right) = -\left(\frac{1}{2}\right)^2 + \frac{1}{2} = \frac{1}{4}$$

The vertex at $\left(\frac{1}{2}, \frac{1}{4}\right)$.

33. $f(x) = -3x^2 + 4x + 5$

$a = -3, \ b = 4$

To find the vertex of the function, use the vertex formula

$$x = -\frac{b}{2a} \text{ and } y = f\left(-\frac{b}{2a}\right).$$

$$x = -\frac{(4)}{2(-3)} = \frac{2}{3}$$

$$y = f\left(\frac{2}{3}\right)$$

$$f\left(\frac{2}{3}\right) = -3\left(\frac{2}{3}\right)^2 + 4\left(\frac{2}{3}\right) + 5 = \frac{19}{3}$$

The vertex is at $\left(\frac{2}{3}, \frac{19}{3}\right)$.

35. $f(x) = 3(x - 2)^2 - 3$

To find the x-intercepts, set $f(x) = 0$ and then solve for x.

$$0 = 3(x - 2)^2 - 3$$

$$3(x - 2)^2 = 3$$

$$(x - 2)^2 = 1$$

$$x - 2 = \pm 1$$

$$x = 1 \ \text{ or } \ x = 3.$$

The x-intercepts are $(1, 0)$ and $(3, 0)$.
Let $x = 0$ to find the y-intercept.

$$f(x) = 3(0 - 2)^2 - 3 = 9$$

The y-intercept at $(0, 9)$.

37. $f(x) = (x+2)^2$

$\qquad y = (x+2)^2$

$\qquad y = 1(x+2)^2 + 0$

The vertex is at the point with coordinates $(-2, 0)$. The axis is the line with equation $x = -2$.

x	-1	0	1
y	1	4	9

Plot these points and use the axis of symmetry to find corresponding points on the other side of the axis. Connect the points with a smooth curve

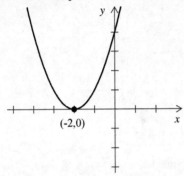

$(-2,0)$

$f(x) = (x+2)^2$

39. $f(x) = (x+2)^2$

$\qquad y = (x+2)^2$

$\qquad y = 1(x+2)^2 + 0$

The vertex is at the point with coordinates $(-2, 0)$. The axis is the line with equation $x = -2$.

x	-1	0	1
y	1	4	9

Plot these points and use the axis of symmetry to find corresponding points on the other side of the axis. Connect the points with a smooth curve.

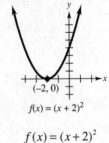

$(-2, 0)$

$f(x) = (x + 2)^2$

$f(x) = (x+2)^2$

41. $f(x) = (x-1)^2 - 3$

$\qquad y = (x-1)^2 - 3$

$\qquad y = 1(x-1)^2 - 3$

The vertex is at the point with coordinates $(1, -3)$. The axis is the line with equation $x = 1$.

x	2	3	4
y	-2	1	6

Use the axis of symmetry to find corresponding points on the other side of the axis.

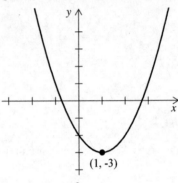

$(1, -3)$

$f(x) = (x-1)^2 - 3$

43. $f(x) = x^2 - 4x + 6$

$\qquad y = x^2 - 4x + 6$

$\qquad = (x^2 - 4x) + 6$

$\qquad = (x^2 - 4x + 4 - 4) + 6$

$\qquad = (x^2 - 4x + 4) + (-4 + 6)$

$\qquad = (x-2)^2 + 2$

The vertex is at the point with coordinates $(2, 2)$. The axis is the line with equation $x = 2$.

x	3	4	5
y	3	6	11

Use the axis of symmetry to find corresponding points on the other side of the axis.

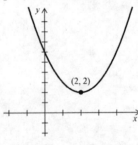

$(2, 2)$

$f(x) = x^2 - 4x + 6$

$f(x) = (x-2)^2 + 2$

45. $f(x) = 2x^2 - 4x + 5$

$$y = 2x^2 - 4x + 5$$
$$= 2(x^2 - 2x) + 5$$
$$= 2(x^2 - 2x + 1 - 1) + 5$$
$$= 2(x^2 - 2x + 1) + 2(-1) + 5$$
$$= 2(x^2 - 2x + 1) + (-2 + 5)$$
$$= 2(x - 1)^2 + 3$$

The vertex is at the point with coordinates (1, 3). The axis is the line with equation $x = 1$.

x	2	3	4
y	5	11	21

Use the axis of symmetry to find corresponding points on the other side of the axis.

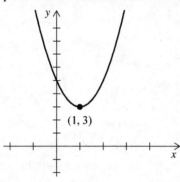

$$f(x) = 2x^2 - 4x + 5$$
$$f(x) = 2(x - 1)^2 + 3$$

47. $f(x) = .0031x^2 - .291x + 7.1$ Let x be the age of the driver. Find the x-value of the vertex. Because the graph is a parabola which opens upward, the rate is lowest at the vertex. For the quadratic function,
$$f(x) = .0031x^2 - .291x + 7.1,$$
$$a = .0031$$
$$b = -.291$$

$$x = -\frac{(-.291)}{2(.0031)} = 46.94$$

The rate is lowest at about age 47.

49. $h(t) = 80t - 16t^2$

a. The maximum height is the y-value of the quadratic function

$h(t) = 80t - 16t^2$. We know that $\begin{matrix} a = -16 \\ b = 80 \end{matrix}$

Use the vertex formula

$$x = -\frac{b}{2a}, \quad y = f\left(-\frac{b}{2a}\right)$$

$$t = -\frac{80}{2(-16)} = 2.5$$

$$y = h(2.5)$$

$$y = 80(2.5) - 16(2.5)^2 = 100$$

The maximum height attained is 100 feet.

b. The object hits the ground when the height is zero. Set $h(t) = 0$ and solve for t.

$$h(t) = 80t - 16t^2 = 0$$
$$80t - 16t^2 = 0$$
$$16t(5 - t) = 0$$
$$16t = 0 \quad \text{or} \quad 5 - t = 0$$
$$t = 0 \quad \text{or} \quad t = 5$$

It takes 5 seconds for the object to hit the ground.

51. (a)-(d)

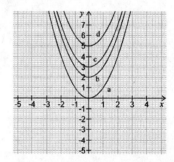

(e) The graph of $f(x) = x^2 + c$ is the graph of $k(x) = x^2$ shifted c units upward.

53. (a)-(d)

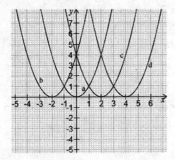

(e) The graph of $f(x) = (x+c)^2$ is the graph of $k(x) = x^2$ shifted c units to the left. The graph of $f(x) = (x-c)^2$ is the graph of $g(x) = x^2$ shifted c units to the right.

Section 3.5 Applications of Quadratic Functions

1. $C(x) = x^2 - 40x + 405$

a. $C(15) = (15)^2 - 40(15) + 405 = 30$
It costs $30 per box to make 15 boxes per day.
$C(18) = (18)^2 - 40(18) + 405 = 9$
It costs $9 per box to make 18 boxes per day.
$C(30) = (30)^2 - 40(30) + 405 = 105$
It cost $105 per box to make 30 boxes per day.

b.

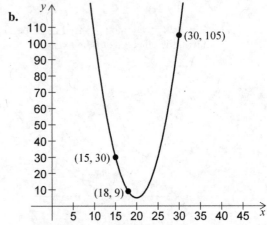

c. The minimum point on the graph corresponds to the number of boxes that will make the cost per box as small as possible. Complete the square to find the vertex.

$C(x) = x^2 - 40x + 405$
$= (x^2 - 40x + 400 - 400) + 405$
$= (x^2 - 40x + 400) + (-400 + 405)$
$= (x - 20)^2 + 5$

Vertex: (20, 5)

d. She should make 20 boxes per day at a cost of $5 per box.

3. $y = -x^2 + 20x - 60$

a. $y = -\left(x^2 - 20x\right) - 60$

$\quad = -\left(x^2 - 20x + 100 - 100\right) - 60$

$\quad = -\left(x^2 - 20x + 100\right) + 100 - 60$

$\quad y = -(x - 10)^2 + 40$

The graph of this parabola opens downward, so the maximum occurs at the vertex, (10, 40). The maximum firing rate will be reached in 10 milliseconds.

b. When $x = 10$, $y = 40$, so the maximum firing rate is 40 responses/millisec.

5. $P(x) = -2x^2 + 60x - 120$

a. If $P(x) = 0$, then

$0 = -2x^2 + 60x - 120$

$0 = x^2 - 30x + 60$

$x = \dfrac{-(-30) \pm \sqrt{(-30)^2 - 4(1)(60)}}{2(1)}$

$\quad = \dfrac{30 \pm \sqrt{900 - 240}}{2}$

$\quad = \dfrac{30 \pm \sqrt{660}}{2}$

$x = \dfrac{30 + \sqrt{660}}{2} \approx 27.8$

or $x = \dfrac{30 - \sqrt{660}}{2} \approx 2.2$.

Therefore, $P(x) > 0$ if $2.2 < x < 27.8$.
The largest number of cases she can sell and still make a profit is 27.

b. There are many possibilities. For example, she might have to buy additional machinery or pay high overtime wages in order to increase product, thus increasing costs and decreasing profits.

c. $P(x) = -2x^2 + 60x - 120$

$\quad = -2\left(x^2 - 30\right) - 120$

$\quad = -2\left(x^2 - 30x + 225 - 225\right) - 120$

$\quad = -2\left(x^2 - 30x + 225\right) + 450 - 120$

$P(x) = -2(x - 15)^2 + 330$

Since the vertex is at (15, 330), she should sell 15 cases to maximize her profit.

7. Given

supply: $p = \dfrac{1}{5}q^2$;

demand: $p = -\dfrac{1}{5}q^2 + 40$.

For demand, make the following table by first choosing values for p and then computing q in the equation.

$p = -\dfrac{1}{5}q^2 + 40$.

p	10	20	30	40
q	12.2	10	7.1	0

For supply, make the following table by first choosing values for p and then computing q in the equation

$p = \dfrac{1}{5}q^2$.

p	5	10	20	30
q	5	7.1	10	12.2

Answer parts (a)-(h) by referring to the appropriate table above.

a. About 12 books are demanded.

b. 10 books are demanded.

c. About 7 books are demanded.

d. No books are demanded.

e. 5 books are supplied.

f. About 7 books are supplied.

g. 10 books are supplied.

h. About 12 books are supplied.

i. Using the values from the tables, plot points and graph the functions.

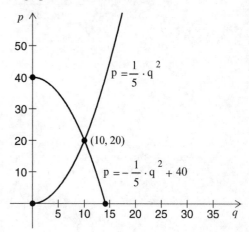

9. $p = 640 - 5q^2$

a. If $q = 0$

$$p = 640 - 5(0)^2 = 640$$

so the price is $640.

b. If $q = 5$

$$p = 640 - 5(5)^2 = 515$$

so the price is $515.

c. If $q = 10$

$$p = 640 - 5(10)^2 = 140$$

so the price is $140.

d. $p = 640 - 5q^2$

q	0	4	8	10
p	640	560	320	140

$p = 5q^2$

q	0	4	8	10
p	0	80	320	500

Graph these two parabolas on the same axes.

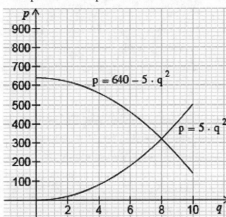

e. demand = supply

$$640 - 5q^2 = 5q^2$$
$$640 = 10q^2$$
$$q^2 = 64$$
$$q = \pm 8$$

A negative value of q is not meaningful. The equilibrium supply is 8 hundreds or 800 units.

f. If $q = 8$

$$p = 5(p)^2 = 320$$

The equilibrium cost is $320.

11. Set $p = 45q$ and $p = -q^2 + 10,000$ equal.

$$45q = -q^2 + 10,000$$

Write this quadratic equation in standard form and solve using the quadratic formula.

$$q^2 + 45q - 10,000 = 0$$

$$q = \frac{-45 \pm \sqrt{45^2 - 4(1)(-10,000)}}{2(1)}$$

$$= \frac{-45 \pm \sqrt{42025}}{2}$$

$$= \frac{-45 \pm 205}{2}$$

$$q = \frac{-45 + 205}{2} = 80 \text{ or } q = \frac{-45 - 205}{2} = -125$$

Since q cannot be negative, $q = 80$.
Since $p = 45(80) = 3600$ and
$p = -(80)^2 + 10,000 = 3600$, the equilibrium quantity is $q = 80$ units and the equilibrium price is $p = \$3600$.

13. Set $p = q^2 + 20q$ and $p = -2q^2 + 10q + 3000$ equal.

$$q^2 + 20q = -2q^2 + 10q + 3000$$

Write this quadratic equation in standard form and solve using the quadratic formula.

$$3q^2 + 10q - 3000 = 0$$

$$q = \frac{-10 \pm \sqrt{10^2 - 4(3)(-3000)}}{2(3)}$$

$$= \frac{-10 \pm \sqrt{36,100}}{6}$$

$$= \frac{-10 \pm 190}{6}$$

$$q = \frac{-10 - 190}{6} = \frac{-200}{6} = -\frac{100}{3} \text{ or}$$

$$q = \frac{-10 + 190}{6} = \frac{180}{6} = 30$$

Since q cannot be negative, $q = 30$.
Since $p = 30^2 + 20(30) = 1500$ and
$p = -2(30)^2 + 10(30) + 3000 = 1500$, the equilibrium quantity is $q = 30$ units and the equilibrium price is $p = \$1500$.

15. Set the revenue function equal to the cost function and solve for x.

$$200x - x^2 = 70x + 2200$$

$$0 = x^2 - 130x + 2200$$

By the quadratic formula,

$$x = \frac{-(-130) \pm \sqrt{(-130)^2 - 4(1)(2200)}}{2(1)}$$

$$= \frac{130 \pm \sqrt{8100}}{2}$$

$x = 20$ or $x = 110$

Since these equations are only valid for $0 \le x \le 100$, 20 is the number of units needed to break even.

17. Set the revenue function equal to the cost function and solve for x.

$$400x - 2x^2 = -x^2 + 200x + 1900$$

$$-x^2 + 200x - 1900 = 0$$

By the quadratic formula,

$$x = \frac{-200 \pm \sqrt{(200)^2 - 4(-1)(-1900)}}{2(-1)}$$

$$= \frac{-200 \pm \sqrt{32400}}{-2}$$

$x = 10$ or $x = 190$

Since these equations are only valid for $0 \le x \le 100$, 10 is the number of units needed to break even.

19. a. Let x be the number of unsold seats. Then the number of people flying is $100 - x$, and the price per ticket is $200 + 4x$. The total revenue is

$$R(x) = (200 + 4x)(100 - x)$$

$$= 20,000 - 200x + 400x - 4x^2$$

$$R(x) = 20,000 + 200x - 4x^2$$

b. $$R(x) = -4(x^2 - 50x) + 20,000$$

$$= -4(x^2 - 50x + 625 - 625) + 20,000$$

$$= -4(x - 25)^2 + 2500 + 20,000$$

$$R(x) = -4(x - 25)^2 + 22,500$$

The vertex is (25, 22,500).

X	5	15	25	35	45
$R(x)$	20,900	22,100	22,500	22,100	20,900

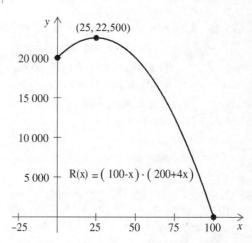

c. The maximum revenue occurs at the vertex, when 25 seats are unsold.

d. $R(25) = 22,500$, so the maximum revenue is $22,500.

21. Let x represent the number of weeks she should wait. Let R represent her revenue in dollars per hog.

$$R(x) = (90 + 5x)(.88 - .02x)$$

$$= 79.2 - 1.8x + 4.4x - .1x^2$$

$$= (-1.x^2 + 2.6x) + 79.2$$

$$= -1.(x^2 - 26x) + 79.2$$

$$= -.1(x^2 - 26x + 169 - 169) + 79.2$$

$$= -.1(x^2 - 26x + 169) + 16.9 + 79.2$$

$$R(x) = -1.(x - 13)^2 + 96.1$$

The vertex is (13, 96.1). She should wait 13 weeks and will receive $96.10/hog.

23.   a. $g(x) = a(x-h)^2 + k$

$g(x) = a(x-0)^2 + 2.1$

$3.1 = a(4-0)^2 + 2.1$

$3.1 = 16a + 2.1$

Solve for a.

$3.1 = 16a + 2.1$

$16a = 1$

$a = \dfrac{1}{16}$

$a = .0625$

$g(x) = .0625x^2 + 2.1$

b. $g(6) = .0625(6)^2 + 2.1 = \$4.35$ billion

$g(8) = .0625(8)^2 + 2.1 = \$6.1$ billion.

25.   a. $f(x) = a(x-h)^2 + k$

$f(x) = a(x-7)^2 + 3820$

Plug in the data point (12, 6538).

$6538 = a(13-7)^2 + 3820$

$6538 = 36a + 3820$

$36a = 2718$

$a = 75.5$

Therefore, $f(x) = 75.5(x-7)^2 + 3820$

b. $x = 15$ corresponds to 2005.

$f(x) = 75.5(15-7)^2 + 3820 = \$8652$

$x = 19$ corresponds to 2009.

$f(x) = 75.5(19-7)^2 + 3820 = \$14,692$

27.
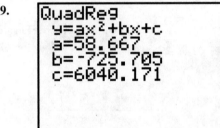

$g(x) = .02857x^2 + .1457x + 2.0771$

$g(6) = .02857(6)^2 + .1457(6) + 2.0771$

$g(6) = \$3.98$ billion in 2006.

$g(8) = .02857(8)^2 + .1457(8) + 2.0771$

$g(8) = \$5.07$ billion in 2008,

27.   Continued
Somewhat lower than those in Exercise 23.

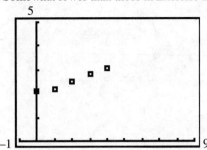

29.
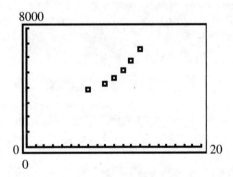

a. The model is

$f(x) = 58.667x^2 - 725.7x + 6040.17$.

Since r^2 is very close to 1, the model is a good one.

b. \$8355;   \$13,431, lower than those in Exercise 25.

31. Let x = each of equal sides of the rectangle and
y = the side opposite the river.
The perimeter of the rectangle is
$2x + y = 320$. Solve for y and get
$y = 320 - 2x$.

The area of the rectangle is the length multiplied by the width.

$A = xy$

$A = x(320 - 2x)$

$A = 320x - 2x^2$

$A(x) = -2x^2 + 320x$

$a = -2$

$b = 320$

The x value of the vertex leads to the maximum value of the area because the graph opens downward.

$x = -\dfrac{b}{2a} = -\dfrac{320}{2(-2)}$

$x = 80$

$y = 320 - 2(80)$

$y = 160$

The dimensions of the rectangle are 80 feet by 160 feet.

33. $R(x) = 400x - 2x^2$
$C(x) = 200x + 2000$

a. $400x - 2x^2 = 200x + 2000$

$0 = 2x^2 - 200x + 2000$

$0 = x^2 - 100x + 1000$

$x = \dfrac{100 \pm \sqrt{(-100)^2 - 4(1)(1000)}}{2(1)}$

$= \dfrac{100 \pm \sqrt{6000}}{2}$

$= \dfrac{100 \pm 20\sqrt{15}}{2}$

$= 50 \pm 10\sqrt{15}$

$x \approx 88.7$ and 11.3

b. $P(x) = R(x) - C(x)$

$= (400x - 2x^2) - (200x + 2000)$

$= -2x^2 + 200x - 2000$

$= -2(x^2 - 100x + 2500) - 2000 + 5000$

$= -2(x - 50)^2 + 3000$

The vertex is at (50, 3000). Profit is maximum when $x = 50$.

33. Continued

c. From the vertex, the maximum profit is $3000.

d. Since the break-even points are 88.7 and 11.3, a loss will occur if $x < 11.3$ or $x > 88.7$.

e. A profit will occur for $11.3 < x < 88.7$.

Section 3.6 Polynomial Functions

1. $f(x) = x^4$

First we find several ordered pairs.

x	−3	−2	−1	0	1	2	3
y	81	16	1	0	1	16	81

Plot these ordered pairs and draw a smooth curve through them.

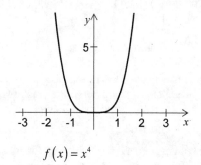

$$f(x) = x^4$$

3. $h(x) = -.2x^5$

First find several ordered pairs

x	−3	−2	−1	0	1	2	3
y	48.6	6.4	.2	0	−.2	−6.4	−48.6

To graph the function, plot these ordered pairs, and draw a smooth curve through them.

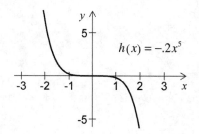

5. a. Yes. The graph could possibly be the graph of some polynomial function because it moves sharply away from the x-axis at the far left and far right.

b. No. The total number of peaks and valleys on the graph of a polynomial function of degree n is at most $n - 1$. Since the graph has 4 peaks and valleys, this could not be the graph of a polynomial function of degree 3.

c. No. Since the graph has 4 peaks and valleys, it could not be the graph of a polynomial function of degree 4.

d. Yes. Since the graph has 4 peaks and valleys and the opposite ends of the graph move sharply away from the x-axis in the opposite direction, it could be the graph of a polynomial function of degree 5.

7. a. Yes. The graph could possibly be the graph of some polynomial function because it moves sharply away from the x-axis as $|x|$ gets large.

b. No. Since it has three peaks and valleys, it could not be the graph of a polynomial function of degree 3 (since $n - 1 = 3 - 1 = 2$).

c. Yes. Since it has three peaks and valleys, it could be the graph of a polynomial function of degree 4 (since $n - 1 = 4 - 1 = 3$).

d. No. The graph could not be the graph of a polynomial function of degree 5 because the opposite ends of the graph move sharply away from the x-axis in the same direction, indicating an even degree.

9. $f(x) = x^3 - 7x - 9$

Since $f(x)$ is degree 3, it has at most 2 peaks and valleys. When $|x|$ is large, the graph resembles the graph of x^3. Therefore, since the y-intercept is -9, D is the graph of the function.

11. $f(x) = x^4 - 5x^2 + 7$.

Since $f(x)$ is degree 4, it has at most 3 peaks and valleys. When $|x|$ is large, the graph resembles the graph of x^4. Therefore, since the y-intercept is 7, B is the graph of the function.

13. $f(x) = .7x^5 - 2.5x^4 - x^3 + 8x^2 + x + 2$

Since $f(x)$ is degree 5, it has at most 4 peaks and valleys. When $|x|$ is large, the graph resembles the graph of $.7x^5$. Therefore, since the y-intercept is 2, E is the graph of the function.

15. $f(x) = (x + 2)(x - 3)(x + 4)$

First, find x-intercepts by setting $f(x) = 0$ and solving for x.

$(x + 2)(x - 3)(x + 4) = 0$

$x = -2$ or $x = 3$ or $x = -4$

These three numbers divide the x-axis into four regions.

Choose an x-value in each region as a test number and compute the function value.

Region	Test No.	Value of $f(x)$	Graph
$x < -4$	-5	-24	below x-axis
$-4 < x < -2$	-3	6	above x-axis
$-2 < x < 3$	0	-24	below x-axis
$3 < x$	4	48	above x-axis

Use this information, the x-intercept, and the fact that the graph can have a total of at most 2 peaks and valleys to sketch the graph.

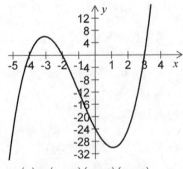

$$f(x) = (x + 2)(x - 3)(x + 4)$$

17. $f(x) = x^2(x-2)(x+3)$

First, find x-intercepts by setting $f(x)$ and solving for x.

$x^2(x-2)(x+3) = 0$

$x = 0$ or $x = 2$ or $x = -3$

These three numbers divide the x-axis into four regions.

Choose an x-value in each region as a test number and compute the function value.

Region	Test No.	Value of $f(x)$	Graph
$x < -3$	-4	96	above x-axis
$-3 < x < 0$	-2	-16	below x-axis
$0 < x < 2$	1	-4	below x-axis
$2 < x$	3	54	above x-axis

Use this information, the x-intercepts and the fact that the graph can have a total of at most 3 peaks and valleys to sketch the graph.

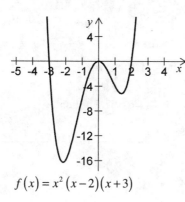

$f(x) = x^2(x-2)(x+3)$

19. $f(x) = x^3 + x^2 - 6x$

$\quad = x(x^2 + x - 6)$

$\quad = x(x+3)(x-2)$

x-intercepts: $x = 0, -3, 2$

Region	Test No.	Value of $f(x)$	Sign
$x < -3$	-4	-24	$-$
$-3 < x < 0$	-1	6	$+$
$0 < x < 2$	1	-4	$-$
$2 < x$	3	18	$+$

19. Continued

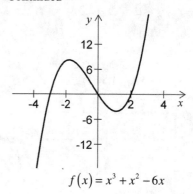

$f(x) = x^3 + x^2 - 6x$

21. $f(x) = x^3 + 3x^2 - 4x$

$\quad = x(x^2 + 3x - 4)$

$\quad = x(x+4)(x-1)$

x-intercepts: $x = 0, -4, 1$

Region	Test No.	Value of $f(x)$	Sign
$x < -4$	-5	-30	$-$
$-4 < x < 0$	-1	6	$+$
$0 < x < 1$	$\frac{1}{2}$	$-\frac{9}{8}$	$-$
$1 < x$	2	12	$+$

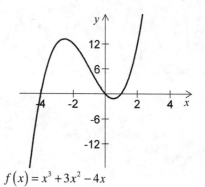

$f(x) = x^3 + 3x^2 - 4x$

23. To graph
$$g(x) = x^3 - 3x^2 - 4x - 5,$$
enter the function as Y_1. There are many possible viewing windows that will show the complete graph. One such window is $-3 \le x \le 5$ and $-20 \le y \le 5$.

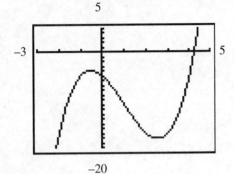

25. To graph
$$f(x) = 2x^5 - 3.5x^4 - 10x^3 + 5x^2 + 12x + 6 \text{ enter}$$
the function as Y_1. There are many possible viewing windows that will show the complete graph. One such window is $-3 \le x \le 4$ and $-35 \le y \le 20$.

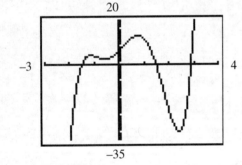

27. a. For a 20% tax rate the revenue is,
$$f(20) = \frac{20(20-100)(20-160)}{240}$$
$$= \frac{224,000}{240}$$
$$\approx \$933.33 \text{ billion}$$

b. For a 40% tax rate the revenue is,
$$f(40) = \frac{40(40-100)(40-160)}{240}$$
$$= \frac{288,000}{240}$$
$$= \$1200 \text{ billion}$$

c. For a 50% tax rate the revenue is,
$$f(50) = \frac{50(50-100)(50-160)}{240}$$
$$= \frac{275,000}{240}$$
$$\approx \$1145.8 \text{ billion}$$

d. For a 70% tax rate the revenue is,
$$f(70) = \frac{70(70-100)(70-160)}{240}$$
$$= \frac{189,000}{240}$$
$$= \$787.5 \text{ billion}$$

e.

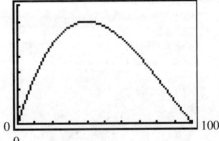

29. $P(t) = t^3 - 18t^2 + 81t$

$P(t) = t\left(t^2 - 18t + 81\right)$

$P(t) = t(t-9)^2$

Because t is time in years from the date of the first reading, $t \geq 0$.

a. $P(0) = 0(0-9)^2 = 0$

$P(3) = 3(3-9)^2 = 108$

$P(7) = 7(7-9)^2 = 28$

$P(10) = 10(10-9)^2 = 10$

b. Notice that $P(9) = 0$. Using this value and the values from (a), plot points and graph the function for $t \geq 0$.

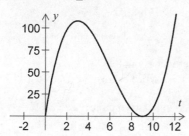

$P(t) = t^3 - 18t^2 + 81t$

c. From the graph, the pressure is increasing for years 0 to 3 and from the ninth year on, and decreasing for years 3 to 9.

31. a. 2008 corresponds to $x = 8$ and 2020 corresponds to $x = 20$.

$g(x) = -.00096x^3 - .1x^2 + 11.3x + 1274$

$g(8) = -.00096(8)^3 - .1(8)^2 + 11.3(8) + 1274$

$g(8) \approx 1357.500$

$g(20) = -.00096(20)^3 - .1(20)^2 + 11.3(20) + 1274$

$g(20) \approx 1452.300$

In 2008, the population will be 1,357,500,000 and 1,452,300,000 in 2020.

b. When x is large, the graph must resemble the graph of $y = -.00096x^3$, which drops down forever at the far right. This would mean that China's population would become 0 at some point.

33. a. 20

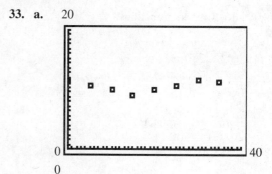

b.

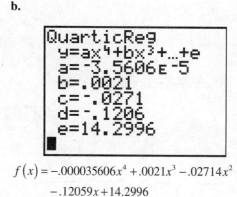

$f(x) = -.000035606x^4 + .0021x^3 - .02714x^2$

$- .12059x + 14.2996$

c. 20

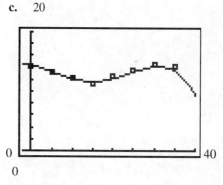

The graph fits reasonably well.

d. Substitute $x = 27$ into $f(x)$ to obtain

$f(27) \approx 13.6704$. According to the model, the enrollment was about 13.7 million in 2002. Next, substitute $x = 31$ into $f(x)$ to obtain

$f(31) \approx 14.1579$. The enrollment is about 14.2 million.

e. Using the minimum finder, enrollment was lowest in late 1989.

Section 3.7 Rational Functions

1. $f(x) = \dfrac{1}{x+5}$

Vertical asymptote:

If $x + 5 = 0$, $x = -5$, so the line $x = -5$ is a vertical asymptote.

Horizontal asymptote:

$$y = \frac{1}{x+5} = \frac{0x+1}{1x+5}$$

The line $y = \dfrac{a}{c} = \dfrac{0}{1}$ or $y = 0$ is a horizontal asymptote.

x	-8	-7	-6	-5.5	-4.5	-4	-3	-2
y	$-\frac{1}{3}$	$-\frac{1}{2}$	-1	-2	2	1	$\frac{1}{2}$	$\frac{1}{3}$

Using these points and the asymptotes, we graph the function.

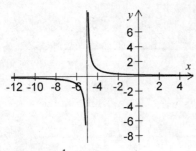

$$f(x) = \frac{1}{x+5}$$

3. $f(x) = \dfrac{-3}{2x+5}$

Vertical asymptote:

If $2x + 5 = 0$, $x = -\dfrac{5}{2}$, so the line $x = -\dfrac{5}{2}$ is a vertical asymptote.

Horizontal asymptote:

$$y = -\frac{3}{2x+5} = \frac{0x-3}{2x+5}$$

The line $y = \dfrac{0}{2}$ or $y = 0$ is a horizontal asymptote.

x	-5	-4	-3	-2	-1	0
y	$\frac{3}{5}$	1	3	-3	-1	$-\frac{3}{5}$

Using these points and the information above, we graph the function.

3. Continued

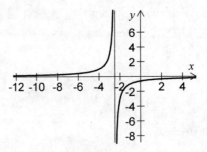

$$f(x) = \frac{-3}{2x+5}$$

5. $f(x) = \dfrac{3x}{x-1}$

Vertical asymptote:

If $x - 1 = 0$, $x = 1$, so the line $x = 1$ is a vertical asymptote.

Horizontal asymptote:

$$y = \frac{3x+0}{1x-1}$$

The line $y = \dfrac{3}{1}$ or $y = 3$ is a horizontal asymptote.

x	-3	-2	-1	0	2	3	4	5
y	$\frac{9}{4}$	2	$\frac{3}{2}$	0	6	$\frac{9}{2}$	4	$\frac{15}{4}$

Using these points and the information above, we graph the function.

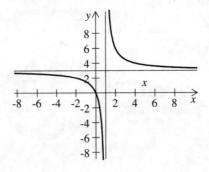

$$f(x) = \frac{3x}{x-1}$$

7. $f(x) = \dfrac{x+1}{x-4}$

Vertical asymptote:

If $x - 4 = 0$, $x = 4$, so the line $x = 4$ is a vertical asymptote.

Horizontal asymptote:

$$y = \dfrac{1x+1}{1x-4}$$

The line $y = \dfrac{1}{1}$ or $y = 1$ is a horizontal asymptote.

x	0	1	2	3	5	6	7	8
y	$-\frac{1}{4}$	$-\frac{2}{3}$	$-\frac{3}{2}$	-4	6	$\frac{7}{2}$	$\frac{8}{3}$	$\frac{9}{4}$

Using these points and the information above, we graph the function.

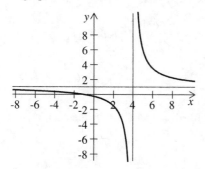

$$f(x) = \dfrac{x+1}{x-4}$$

9. $f(x) = \dfrac{2-x}{x-3}$

Vertical asymptote:

If $x - 3 = 0$, $x = 3$, so the line $x = 3$ is a vertical asymptote. Horizontal asymptote:

$$y = \dfrac{-1x+2}{1x-3}$$

The line $y = -\dfrac{1}{1}$ or $y = -1$ is a horizontal asymptote.

x	-1	0	1	2	4	5	6	7
y	$-\frac{3}{4}$	$-\frac{2}{3}$	$-\frac{1}{2}$	0	-2	$-\frac{3}{2}$	$-\frac{4}{3}$	$-\frac{5}{4}$

Using these points and the information above, we graph the function.

9. **Continued**

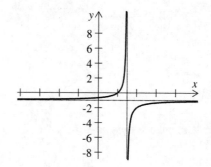

$$f(x) = \dfrac{2-x}{x-3}$$

11. $f(x) = \dfrac{2x-1}{4x+2}$

Vertical asymptote:

If $4x + 2 = 0$, $x = -\dfrac{1}{2}$, so the line $x = -\dfrac{1}{2}$ is a vertical asymptote. Horizontal asymptote:

$$y = \dfrac{2x-1}{4x+2}$$

The line $y = \dfrac{2}{4}$ or $y = \dfrac{1}{2}$ is a horizontal asymptote.

x	-4	-3	-2	-1	0	1	2	3
y	$\frac{9}{14}$	$\frac{7}{10}$	$\frac{5}{6}$	$\frac{3}{2}$	$-\frac{1}{2}$	$\frac{1}{6}$	$\frac{3}{10}$	$\frac{5}{14}$

Using these points and the information above, we graph the function.

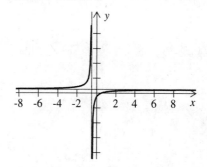

$$f(x) = \dfrac{2x-1}{4x+2}$$

13. $h(x) = \dfrac{x+1}{x^2+3x-4}$

To find the vertical asymptotes, set the denominator
equal to 0 and solve.

$$x^2 + 3x - 4 = 0$$
$$(x+4)(x-1) = 0$$
$$x = -4 \quad x = 1$$

To find horizontal asymptotes, divide the
numerator and the
denominator by x^2.

$$\frac{\frac{x}{x^2}+\frac{1}{x^2}}{\frac{x^2}{x^2}+\frac{3x}{x^2}-\frac{4}{x^2}} = \frac{\frac{1}{x}+\frac{1}{x^2}}{1+\frac{3}{x}-\frac{4}{x^2}}$$

As $|x|$ gets very large, the numerator gets close to
0 and the
denominator gets close to 1, so the function has a
horizontal asymptote at $y = 0$. Find the intercepts.

If $x = 0$, $y = -\dfrac{1}{4}$.

If $y = 0$,

$$0 = \frac{x+1}{x^2+3x-4}$$
$$0 = x+1$$
$$x = -1$$

Make a table of values.

x	-7	-5	-4.5	-3.9	-3	-2
y	$-\frac{1}{4}$	$-\frac{2}{3}$	-1.27	5.92	$.5$	$\frac{1}{6}$

x	-1	0	$.5$	1.5	2	5
y	0	$-\frac{1}{4}$	$-\frac{2}{3}$	$\frac{10}{11}$	$.5$	$\frac{1}{6}$

Using these points and the information above,
graph the function.

$$h(x) = \frac{x+1}{x^2+3x-4}$$

15. $f(x) = \dfrac{x^2+1}{x^2-1}$

To find the vertical asymptotes, set the
denominator equal to 0 and solve.

$$x^2 - 1 = 0$$
$$(x+1)(x-1) = 0$$
$$x = -1 \quad \text{or} \quad x = 1$$

To find horizontal asymptotes, divide the
numerator and the denominator by x^2.

$$\frac{\frac{x^2}{x^2}+\frac{1}{x^2}}{\frac{x^2}{x^2}-\frac{1}{x^2}} = \frac{1+\frac{1}{x^2}}{1-\frac{1}{x^2}}$$

As $|x|$ gets very large, the function approaches 1.
So the function has a horizontal asymptote at
$y = 1$.
Find the intercepts.

If $x = 0$, $\dfrac{0+1}{0-1} = -1$, the y-intercept is -1.

If $y = 0$

$$0 = \frac{x^2+1}{x^2-1}$$
$$x^2 = -1$$

is undefined, so there is no x-intercept.
Make a table of values.

x	-3	-2	-1.5	$-.5$	0	$.5$	1.5	2	3
y	$1\frac{1}{4}$	$1\frac{2}{3}$	2.6	$-1\frac{2}{3}$	-1	$-1\frac{2}{3}$	2.6	$1\frac{2}{3}$	$1\frac{1}{4}$

Use these points and the information above, graph
the function.

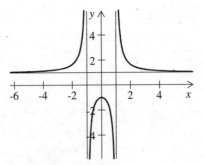

$$f(x) = \frac{x^2+1}{x^2-1}$$

17. $f(x) = \dfrac{x-3}{x^2+x-2}$

$\qquad = \dfrac{(x-3)}{(x+2)(x-1)}$

If $(x+2)(x-1) = 0$, then $x = -2$ or $x = 1$. The lines $x = -2$ and $x = 1$ are vertical asymptotes.

19. $g(x) = \dfrac{x^2+2x}{x^2-4x-5}$

$\qquad = \dfrac{x(x+2)}{(x-5)(x+1)}$

If $(x-5)(x+1) = 0$, then $x = 5$ or $x = -1$. The lines $x = 5$ and $x = -1$ are vertical asymptotes.

21. $f(x) = \dfrac{4.3x}{100-x}$

a. $f(50) = \dfrac{4.3(50)}{100-50}$

$\qquad = 4.3$ or $\$4300$

b. $f(70) = \dfrac{4.3(70)}{100-70}$

$\qquad = 10.03333$ or $\$10,033.33$

c. $f(80) = \dfrac{4.3(80)}{100-80}$

$\qquad = 17.2$ or $\$17,200$

d. $f(90) = \dfrac{4.3(90)}{100-90}$

$\qquad = 38.7$ or $\$38,700$

e. $f(95) = \dfrac{4.3(95)}{100-95}$

$\qquad = 81.7$ or $\$81,700$

f. $f(98) = \dfrac{4.3(98)}{100-98}$

$\qquad = 210.7$ or $\$210,700$

g. $f(99) = \dfrac{4.3(99)}{100-99}$

$\qquad = 425.7$ or $\$425,700$

h. Since $f(100)$ is undefined, all the pollutant cannot be removed according to his model.

21. Continued

i.

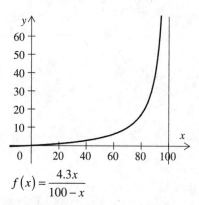

$f(x) = \dfrac{4.3x}{100-x}$

23. a. Since x represents the size of a generation, x cannot be a negative number. The domain of x is $[0, \infty)$.

b. Let $\lambda = a = b = 1$ and $x \geq 0$.

$f(x) = \dfrac{x}{1+x}$

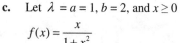

c. Let $\lambda = a = 1$, $b = 2$, and $x \geq 0$

$f(x) = \dfrac{x}{1+x^2}$

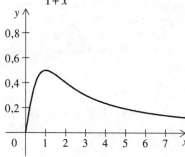

d. Increasing b makes the next generation smaller when this generation is larger.

25. $W = \dfrac{S(S-A)}{A} = \dfrac{3(3-A)}{A}$

a. If $A = 1$, $W = \dfrac{3(3-1)}{1} = 6$

The waiting time is 6 min.

b. If $A = 2$, $W = \dfrac{3(3-2)}{2} = 1.5$

The waiting time is 1.5 min.

c. If $A = 2.5$, $W = \dfrac{3(3-2.5)}{2.5} = .6$

The waiting time is .6 min.

d. The vertical asymptote occurs when the denominator is zero or $A = 0$.

e. Use the vertical asymptote and the values found in (a), (b), and (c) to graph the function for $0 < A \leq 3$.

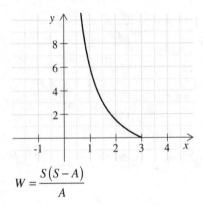

$W = \dfrac{S(S-A)}{A}$

f. When $A < 3$, W is decreasing. The waiting time approaches 0 as A approaches 3. The formula does not apply for $A > 3$ because there will be no waiting if people arrive more than 3 minutes apart

27. $y = \dfrac{900,000,000 - 30,000x}{x + 90,000}$

x	0	10,000	20,000	30,000
y	10,000	6000	2727	0

The maximum number of red tranquilizers is 30,000. The maximum number of blue ones is 10,000.

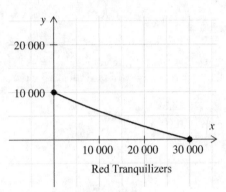

Red Tranquilizers

$y = \dfrac{900,000,000 - 30,000x}{x + 90,000}$

29. Fixed costs = \$40,000

Marginal cost = \$2.60/unit

a. $C(x) = 2.6x + 40,000$

b. Average cost per item:

$\overline{C}(x) = \dfrac{C(x)}{x} = \dfrac{2.6x + 40,000}{x}$

$= 2.6 + \dfrac{40,000}{x}$

c. As $|x|$ becomes large, $C(x)$ approaches 2.6. The horizontal asymptote is at $y = 2.6$. This means the average cost per item may get close to \$2.60, but never quite reach it.

31. $C(x) = .2x^3 - 25x^2 + 1531x + 25,000$

$$\overline{C}(x) = \frac{C(x)}{x} = .2x^2 - 25x + 1531 + \frac{25000}{x}$$

Graph the average cost function and use the minimum finder to find the point with the smallest y-coordinate.

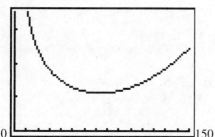

About 73.9 units should be produced to have the lowest possible average cost.

33. **a.** $g(x) = \dfrac{x^3 - 2}{x - 1}$; $-4 \le x \le 6$ and $-6 \le y \le 12$.

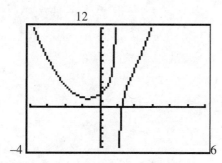

b. $g(x)$ and $y = x^2 + x + 1$

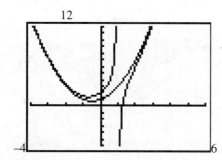

They appear almost identical because the parabola is an asymptote of the graph.

Chapter 3 Review Exercises

1. This rule does not define y as a function of x because the x-value of 2 corresponds to both 5 and –5.

2. This rule defines y as a function of x because each value of x determines one and only one value of y.

3. $y = \sqrt{x}$

   This rule defines y as a function of x because each value of x determines one and only one value of y.

4. $x = |y|$

   This rule does not define y as a function of x. A given value of x may define two values for y. For example, if $x = 4$, we have , so $y = 4$ and $y = -4$.

5. $x = y^2 + 1$

   This rule does not define y as a function of x. A given value of x may define two values for y. For example, if $x = 10$, we have $10 = y^2 + 1$, or $y^2 = 9$, so $y = \pm 3$.

6. $y = 5x - 2$

   This rule defines y as a function of x because each value of x determines one and only one value of y.

7. $f(x) = 4x - 1$

   a. $f(6) = 4(6) - 1 = 24 - 1$
   $= 23$

   b. $f(-2) = 4(-2) - 1 = -8 - 1$
   $= -9$

   c. $f(p) = 4p - 1$

   d. $f(r + 1) = 4(r + 1) - 1$
   $= 4r + 4 - 1 = 4r + 3$

8. $f(x) = 3 - 4x$

   a. $f(6) = 3 - 4(6) = -21$

   b. $f(-2) = 3 - 4(-2) = 3 + 8 = 11$

   c. $f(p) = 3 - 4p$

   d. $f(r + 1) = 3 - 4(r + 1)$
$$= -1 - 4r$$

9. $f(x) = -x^2 + 2x - 4$

   a. $f(6) = -6^2 + 2(6) - 4$
$$= -36 + 12 - 4$$
$$= -28$$

   b. $f(-2) = -(-2)^2 + 2(-2) - 4$
$$= -4 - 4 - 4$$
$$= -12$$

   c. $f(p) = -p^2 + 2p - 4$

   d. $f(r + 1)$
$$= -(r + 1)^2 + 2(r + 1) - 4$$
$$= -\left(r^2 + 2r + 1\right) + 2r + 2 - 4$$
$$= -r^2 - 2r - 1 + 2r + 2 - 4$$
$$= -r^2 - 3$$

10. $f(x) = 8 - x - x^2$

   a. $f(6) = 8 - 6 - 36 = -34$

   b. $f(-2) = 8 - (-2) - (-2)^2$
$$= 8 + 2 - 4$$
$$= 6$$

   c. $f(p) = 8 - p - p^2$

   d. $f(r + 1)$
$$= 8 - (r + 1) - (r + 1)^2$$
$$= 8 - r - 1 - \left(r^2 + 2r + 1\right)$$
$$= 8 - r - 1 - r^2 - 2r - 1$$
$$= -r^2 - 3r + 6$$

11. $f(x) = 5x - 3$ and $g(x) = -x^2 + 4x$

   a. $f(-2) = 5(-2) - 3$
$$= -10 - 3$$
$$= -13$$

   b. $g(3) = -3^2 + 4(3)$
$$= -9 + 12$$
$$= 3$$

   c. $g(-k) = -(-k)^2 + 4(-k)$
$$= -k^2 - 4k$$

   d. $g(3m) = -(3m)^2 + 4(3m)$
$$= -9m^2 + 12m$$

   e. $g(k - 5)$
$$= -(k - 5)^2 + 4(k - 5)$$
$$= -\left(k^2 - 10k + 25\right) + 4k - 20$$
$$= -k^2 + 10k - 25 + 4k - 20$$
$$= -k^2 + 14k - 45$$

   f. $f(3 - p) = 5(3 - p) - 3$
$$= 15 - 5p - 3$$
$$= 12 - 5p$$

12. $f(x) = x^2 + x + 1$

   a. $f(3) = 3^2 + 3 + 1 = 13$

   b. $f(1) = 1^2 + 1 + 1 = 3$

   c. $f(4) = 4^2 + 4 + 1 = 21$

   d. No, $f(3) + f(1) = 13 + 3 = 16$, while $f(3 + 1) = f(4) = 21$.

13. $f(x) = |x| - 3$

$$y = \begin{cases} x - 3 & \text{if} \quad x \geq 0 \\ -x - 3 & \text{if} \quad x < 0 \end{cases}$$

We graph the line $y = x - 3$ for $x \geq 0$. We graph the line $y = -x - 3$ for $x < 0$.

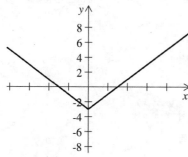

$f(x) = |x| - 3$

14. $f(x) = -|x| - 2$

$$y = \begin{cases} -x - 2 & \text{for} \quad x \geq 0 \\ -(-x) - 2 & \text{for} \quad x < 0 \end{cases}$$

or

$$y = \begin{cases} -x - 2 & \text{for} \quad x \geq 0 \\ x - 2 & \text{for} \quad x < 0 \end{cases}$$

We graph the line $y = -x - 2$ for $x \geq 0$. We graph the line $y = x - 2$ for $x < 0$.

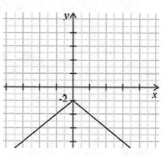

$f(x) = -|x| - 2$

15. $f(x) = -|x + 1| + 3$

$$y = \begin{cases} -(x + 1) + 3 & \text{if} \quad x + 1 \geq 0 \\ -[-(x + 1)] + 3 & \text{if} \quad x + 1 < 0 \end{cases}$$

or

$$y = \begin{cases} -x + 2 & \text{if} \quad x \geq -1 \\ x + 4 & \text{if} \quad x < -1 \end{cases}$$

We graph the line $y = -x + 2$ for $x \geq -1$. We graph the line $y = x + 4$ for $x < -1$.

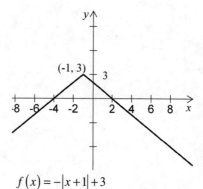

$f(x) = -|x + 1| + 3$

16. $f(x) = 2|x - 3| - 4$

$$y = \begin{cases} 2(x - 3) - 4 & \text{if} \quad x - 3 \geq 0 \\ -2(x - 3) - 4 & \text{if} \quad x - 3 < 0 \end{cases}$$

or

$$y = \begin{cases} 2x - 10 & \text{if} \quad x \geq 3 \\ -2x + 2 & \text{if} \quad x < 3 \end{cases}$$

We graph the line $y = 2x - 10$ for $x \geq 3$. We graph the line $y = -2x + 2$ for $x < 3$.

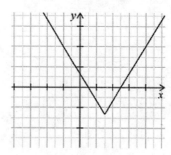

$f(x) = 2|x - 3| - 4$

17. $f(x) = [x - 3]$

For x in the interval $[0, 1)$, $[x - 3] = -3$.
For x in the interval $[1, 2)$, $[x - 3] = -2$.
For x in the interval $[2, 3)$, $[x - 3] = -1$.
Continue in this pattern.
The graph consists of a series of line segments. In each case the left endpoint is included, and the right endpoint is excluded.

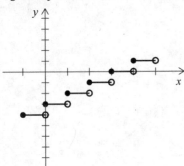

$f(x) = [x - 3]$

18. $f(x) = \left[\dfrac{1}{2}x - 2\right]$

$f(x)$ is a step function with breaks in the graph whenever x is an even integer.
If $x \in [-2, 0)$, $f(x) = -3$.
If $x \in [0, 2)$, $f(x) = -2$.
If $x \in [2, 4)$, $f(x) = -1$.
If $x \in [4, 6)$, $f(x) = 0$.
If $x \in [6, 8)$, $f(x) = 1$.

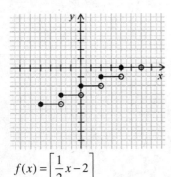

$f(x) = \left[\dfrac{1}{2}x - 2\right]$

19. $f(x) = \begin{cases} -4x + 2 & \text{if } x \le 1 \\ 3x - 5 & \text{if } x > 1 \end{cases}$

For $x \le 1$, graph the line $y = -4x + 2$ using the two points $(0, 2)$ and $(1, -2)$. For $x > 1$, graph the line $y = 3x - 5$ using the two points $(1, -2)$ and $(4, 7)$. Note that the two lines meet at their common endpoint $(1, -2)$.

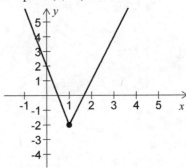

$f(x) = \begin{cases} -4x + 2 & \text{if } x \le 1 \\ 3x - 5 & \text{if } x > 1 \end{cases}$

20. $f(x) = \begin{cases} 3x + 1 & \text{if } x < 2 \\ -x + 4 & \text{if } x \ge 2 \end{cases}$

$x < 2$

x	0	1	2
y	1	4	7

$x \ge 2$

x	2	3	4
y	2	1	0

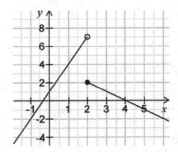

$f(x) = \begin{cases} 3x + 1 & \text{if } x < 2 \\ -x + 4 & \text{if } x \ge 2 \end{cases}$

21. $f(x) = \begin{cases} |x| & \text{if } x < 3 \\ 6-x & \text{if } x \geq 3 \end{cases}$

For $x < 3$, graph $y = |x|$ using points from the following table.

x	-3	-2	-1	0	1	2	2.9
y	3	2	1	0	1	2	2.9

For $x \geq 3$, graph the line $y = 6 - x$ using the two points (3, 3) and (6, 0).

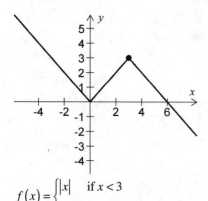

$f(x) = \begin{cases} |x| & \text{if } x < 3 \\ 6-x & \text{if } x \geq 3 \end{cases}$

22. $f(x) = \sqrt{x^2}$

x	-4	-3	-2	-1	0	1	2	3	4	5
f(x)	4	3	2	1	0	1	2	3	4	5

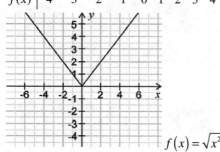

$f(x) = \sqrt{x^2}$

23. $g(x) = \dfrac{x^2}{8} - 3$

x	-8	-4	-2	0	2	4	8
g(x)	5	-1	$-2\frac{1}{2}$	-3	$-2\frac{1}{2}$	-1	5

23. Continued

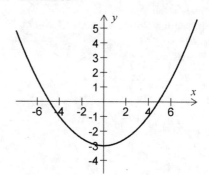

$g(x) = \dfrac{x^2}{8} - 3$

24. $h(x) = \sqrt{x} + 2$; x must be greater than or equal to 0.

x	0	1	4	9
h(x)	2	3	4	5

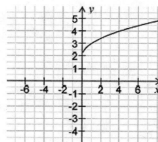

$h(x) = \sqrt{x} + 2$

25. a.

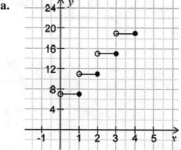

Hours

b. The domain is $(0, \infty)$.
The range is $\{7, 11, 15, 19, \ldots\}$.

c. If he can spend no more than \$15, he can use the polisher for at most 3 days.

26. a. The charge for 20 miles is
$45 + (20)(2) = 85$.
Therefore, $90 is enough for a 20-mile haul.

b.

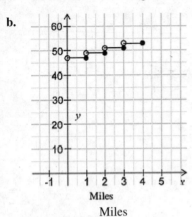

Miles

Miles

c. The domain is $(0, \infty)$
The range is $\{47, 49, 51, 53, \ldots\}$.

27.

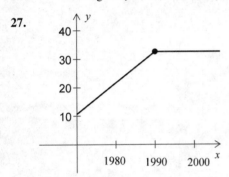

The graph suggests that the number of births to unmarried mothers appear to be leveling off.

28. a. Let (x_1, y_1) be $(70, 131.7)$ and (x_2, y_2) be $(100, 107.7)$.
Find the slope.
$$m = \frac{107.7 - 131.7}{100 - 70} = -.8$$
Use the point-slope form with $(70, 131.7)$.
$(y - 131.7) = -.8(x - 70)$
$f(x) = y = -.8x + 187.7$

b. Let $y = 100$.
$100 = -.8x + 187.7$

$x = 109.625$
In about the middle of 2009, the capita consumption will be 100 pounds.

29. Eight units cost $300; fixed cost is $60.

a. $C(x) = mx + b$
$b = 60$ and $C(8) = 300$
$C(x) = mx + 60$
$300 = m(8) + 60$
$8m = 240$
$m = 30$
$C(x) = 30x + 60$

b. The marginal cost is given by the slope, so the marginal cost is $30.

c. Average cost $= \dfrac{C(100)}{100}$

$= \dfrac{30(100) + 60}{100}$

$= \dfrac{3060}{100}$

$= 30.6$

The average cost per unit to produce 100 units is $30.60.

30. Fixed cost is $2000; 36 units cost $8480.

a. $C(x) = mx + 2000$
$C(36) = 8480 = 36m + 2000$
$$m = \frac{8480 - 2000}{36} = 180$$
$C(x) = 180x + 2000$

b. Since $m = 180$, the marginal cost is $180.

c. Average cost $= \dfrac{C(100)}{100}$

$= \dfrac{180(100) + 2000}{100}$

$= \dfrac{20,000}{100}$

$= 200$

The average cost per unit to produce 100 units is $200.

31. Twelve units cost $445; 50 units cost $1585.

    a. $C(x) = mx + b$
Use the two points (12, 445) and
(50, 1585).
$$m = \frac{1585 - 445}{50 - 12} = \frac{1140}{38} = 30$$
$$y - 1585 = 30(x - 50)$$
$$y - 1585 = 30x - 1500$$
$$y = 30x + 85$$
$$C(x) = 30x + 85$$

    b. The slope is 30, so the marginal cost is $30.

    c. Average cost $= \dfrac{C(100)}{100}$
$$= \frac{30(100) + 85}{100}$$
$$= \frac{3085}{100}$$
$$= 30.85$$

The average cost per unit to produce
100 units is $30.85.

32. Thirty units cost $1500; 120 units cost $5640. Use the points (30, 1500) and (120, 5640).

    a. $m = \dfrac{5640 - 1500}{120 - 30} = \dfrac{4140}{90} = 46$
$$C(x) = 46x + b$$
$$C(30) = 1500 = 46(30) + b = 1380 + b$$
$$b = 120$$
$$C(x) = 46x + 120$$

    b. The slope is 46, so the marginal cost is $46.

    c. Average cost $= \dfrac{C(100)}{100}$
$$= \frac{46(100) + 120}{100}$$
$$= \frac{4720}{100}$$
$$= 47.20$$

The average cost per item to produce
100 units is $47.20.

33. a. The fixed cost is $18,000.

    b. Revenue = (Price per item) $\times$ (Number of items)
$$R(x) = 28x$$

    c. Set the cost function equal to the revenue function and solve for x.
$$24x + 18,000 = 28x$$
$$-4x = -18,000$$
$$x = 4500 \text{ cartridges}$$

    d. Let $x = 4500$.
$$R(4500) = 28(4500) = \$126,000$$

34. Set $f(x)$ equal to $g(x)$ and solve for x.
$$-1.2x + 68 = 2.3x - 3.6$$
$$-3.5x = -71.6$$
$$x \approx 20$$
In 2000, the network and cable stations had the same viewing share.

35. $-.5q + 30.95 = .3q + 2.15$
$$-.8q = -28.8$$
$$q = 36 \text{ million subscribers}$$
$-.5(36) + 30.95 = \$12.95$ per month
The equilibrium quantity is 36 million subscribers at an equilibrium price of $12.95 per month.

36. $.0015q + 1 = -.0025q + 64.36$
$$.004q = 63.36$$
$$q = 15,840$$
$p = .0015(15,840) + 1 = 24.76$
The equilibrium quantity is 15,840 prescriptions at an equilibrium price of $24.76.

37. $f(x) = 3(x - 2)^2 + 6$
Here $a = 3$, $h = 2$, and $k = 6$. The vertex is (2, 6). The parabola opens upward since $a = 3 > 0$.

38. $f(x) = 2(x + 3)^2 - 5$
Here $a = 2$, $b = -3$, and $c = -5$. The vertex is $(-3, -5)$. The parabola opens upward since $a = 2 > 0$.

39. $g(x) = -4(x + 1)^2 + 8$
$a = -4$, $b = -1$, $c = 8$. The vertex is

$(-1, 8)$.
The parabola opens downward since $a = -4 < 0$.

40. $g(x) = -5(x-4)^2 - 6$

$a = -5$, $b = 4$, $c = -6$. The vertex is
$(4, -6)$. The parabola opens downward since $a = -5 < 0$.

41. $f(x) = x^2 - 4$

First, locate the vertex of the parabola and find the axis of the parabola.

$y = 1(x-0)^2 - 4$

The vertex is at the point with coordinates $(0, -4)$. The axis is the line with equation $x = 0$. Make a table of values to find points on one side of the axis and then use the axis of symmetry to find the corresponding points on the other side.

x	1	2	3
y	-3	0	5

Connect the points with a smooth curve.

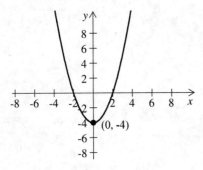

$f(x) = x^2 - 4$

42. $f(x) = 6 - x^2$

x	-2	-1	0	1	2
y	2	5	6	5	2

Vertex: $(0, 6)$
Axis: $x = 0$

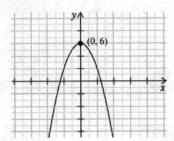

$f(x) = 6 - x^2$

43. $f(x) = x^2 + 2x - 3$

$\qquad = (x^2 + 2x + 1 - 1) - 3$

$\qquad = (x+1)^2 - 4$

x	-3	-2	-1	0	1
y	0	-3	-4	-3	0

Vertex: $(-1, -4)$
Axis: $x = -1$

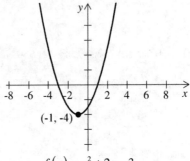

$f(x) = x^2 + 2x - 3$

44. $f(x) = -x^2 + 6x - 3$

$\qquad y = -x^2 + 6x - 3$

$\qquad = -1(x^2 - 6x) - 3$

$\qquad = -1(x^2 - 6x + 9) - 3 + 9$

$\qquad = -1(x-3)^2 + 6$

The vertex is at the point with coordinates $(3, 6)$. The axis is the line with equation $x = 3$.

x	4	5	6
y	5	2	-3

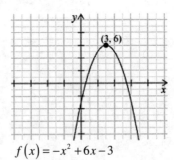

$f(x) = -x^2 + 6x - 3$

45. $f(x) = -x^2 - 4x + 1$

$\qquad = -\left(x^2 + 4x\right) + 1$

$\qquad = -\left(x^2 + 4x + 4 - 4\right) + 1$

$\qquad = -\left(x^2 + 4x + 4\right) + 4 + 1$

$\qquad = -(x+2)^2 + 5$

x	−4	−3	−2	−1	0
y	1	4	5	4	1

Vertex: (−2, 5)

Axis: $x = -2$

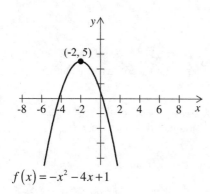

$f(x) = -x^2 - 4x + 1$

46. $f(x) = 4x^2 - 8x + 3$

$\qquad y = 4x^2 - 8x + 3$

$\qquad = \left(4x^2 - 8x\right) + 3$

$\qquad = 4\left(x^2 - 2x + 1\right) + 3 - 4$

$\qquad = 4(x-1)^2 - 1$

The vertex is at the point with coordinates (1, −1). The axis is the line with equation $x = 1$.

x	2	3	4
y	3	15	35

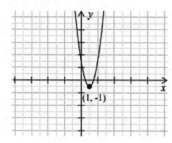

$f(x) = 4x^2 - 8x + 3$

47. $f(x) = 2x^2 + 4x - 3$

$\qquad = 2\left(x^2 + 2x\right) - 3$

$\qquad = 2\left(x^2 + 2x + 1 - 1\right) - 3$

$\qquad = 2(x+1)^2 - 5$

x	−3	−2	−1	0	1
y	3	−3	−5	−3	3

Vertex: (−1, −5)

Axis: $x = -1$

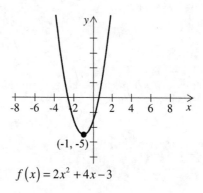

$f(x) = 2x^2 + 4x - 3$

48. $f(x) = -3x^2 - 12x - 8$

$\qquad y = -3x^2 - 12x - 8$

$\qquad y = -3\left(x^2 + 4x\right) - 8$

$\qquad y = -3\left(x^2 + 4x + 4\right) - 8 + 12$

$\qquad y = -3(x+2)^2 + 4$

The vertex is at the point with coordinates (−2, 4). The axis is the line with equation $x = -2$.

x	−1	0	1
y	1	−8	−23

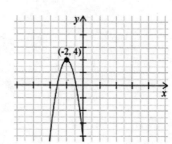

$f(x) = -3x^2 - 12x - 8$

49.
$$f(x) = x^2 + 6x - 2$$
$$= \left(x^2 + 6x + 9 - 9\right) - 2$$
$$= \left(x^2 + 6x + 9\right) - 9 - 2$$
$$= (x+3)^2 - 11$$

Vertex: $(-3, -11)$
Because $a = 1 > 0$, the parabola opens upward and the function has a minimum value. This is the y-value of the vertex, which is -11.

50.
$$f(x) = x^2 + 4x + 5$$
$$= \left(x^2 + 4x + 4\right) - 4 + 5$$
$$= (x+2)^2 + 1$$

Vertex: $(-2, 1)$
Because $a = 1 > 0$, the parabola opens upward and the function has a minimum value, which is 1.

51.
$$g(x) = -4x^2 + 8x + 3$$
$$= \left(-4x^2 + 8x\right) + 3$$
$$= -4\left(x^2 - 2x\right) + 3$$
$$= -4\left(x^2 - 2x + 1\right) + 3 + 4$$
$$= -4(x-1)^2 + 7$$

Vertex: $(1, 7)$
Because $a = -4 < 0$, the parabola opens downward and the function has a maximum value, which is 7.

52.
$$g(x) = -3x^2 - 6x + 3$$
$$= \left(-3x^2 - 6x\right) + 3$$
$$= -3\left(x^2 + 2x\right) + 3$$
$$= -3\left(x^2 + 2x + 1\right) + 3 + 3$$
$$= -3(x+1)^2 + 6$$

Vertex: $(-1, 6)$
The parabola opens downward and the function has a maximum value, which is 6.

53.
$$P = -4t^2 + 32t - 20$$
$$= -4(t^2 - 8t) - 20$$
$$= -4(t^2 - 8t + 16) + 4(16) - 20$$
$$= -4(t-4)^2 + 44$$

Vertex: $(4, 44)$
The parabola opens downward and the function has a maximum value of $P = 44$ when $t = 4$. So, the time of the largest profit is 4 months after she began.

54. $h = -16t^2 + 800t$

a. Let $h = 3200$ and solve for t.
$$3200 = -16t^2 + 800t$$
$$16t^2 - 800t + 3200 = 0$$
$$16\left(t^2 - 50t + 200\right) = 0$$
$$t^2 - 50t + 200 = 0$$
$$t = \frac{50 \pm \sqrt{(-50)^2 - 4(1)(200)}}{2(1)}$$
$$t = \frac{50 \pm \sqrt{1700}}{2}$$
$$t = \frac{50 + \sqrt{1700}}{2} \quad \text{or} \quad t = \frac{50 - \sqrt{1700}}{2}$$
$$t \approx 45.62 \quad \text{or} \quad t = 4.38$$
The first time it reaches a height of 3200 ft is about 4.38 sec.

b. Find the vertex of the parabola that is the graph of this function.
$$h = -16t^2 + 800t$$
$$= -16\left(t^2 - 50t\right)$$
$$= -16\left(t^2 - 50t + 625\right) + 10,000$$
$$= -16(t - 25)^2 + 10,000$$
The vertex is $(25, 10,000)$, so the maximum height is 10,000 ft.

55.
$$P = -x^2 + 250x - 15,000$$
$$= -(x^2 - 250x) - 15,000$$
$$= -(x^2 - 250x + 15,625) - 15,000 + 15,625$$
$$= -(x - 125)^2 + 625$$
Vertex: $(125, 625)$
The parabola opens downward and the function has a maximum value of $P = 625$ when $x = 125$. So, the complex produces the largest profit when it rents 125 units.

56. Let x represent the parallel redwood sides. Let y represent the parallel redwood and cement block sides.

$15x + 15x + 15y + 30y = 900$

$30x + 45y = 900$

$45y = 900 - 30x$

$y = 20 - \dfrac{2}{3}x$

Area $= x \cdot y = x\left(20 - \dfrac{2}{3}x\right)$

$A = -\dfrac{2}{3}x^2 + 20x$

$\quad = -\dfrac{2}{3}\left(x^2 - 30x\right)$

$\quad = -\dfrac{2}{3}\left(x^2 - 30x + 225 - 225\right)$

$\quad = -\dfrac{2}{3}\left(x^2 - 30x + 225\right) + \left(-\dfrac{2}{3}\right)(-225)$

$\quad = -\dfrac{2}{3}\left(x^2 - 30x + 225\right) + 150$

$\quad = -\dfrac{2}{3}(x - 15)^2 + 150$

The vertex of the parabola with this equation is (15, 150). The maximum area occurs when $x = 15$. If $x = 15$, $y = 20 - \dfrac{2}{3}(15) = 10$. The dimensions of the enclosure are 10 ft by 15 ft. The maximum possible area is 150 sq ft.

57. a. The data point for 2001 is (5, .38).

$f(x) = a(x - 5)^2 + .38$

$2 = a(25 - 5)^2 + .38$

$1.62 = a(20)^2$

$a = .00405$

$f(x) = .00405(x - 5)^2 + .38$

b. Let $x = 28$ to correspond to 2008.

$f(28) = .00405(28 - 5)^2 + .38$

$f(28) = 2.5225$

This is about $2.5225 trillion.
Let $x = 30$ to correspond to 2010.

$f(30) = .00405(30 - 5)^2 + .38$

$f(30) = 2.91125,$

about $2.91125 trillion.

58. a.

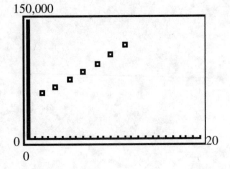

b. Using a graphing calculator,
$$f(x) = 143.494x^2 + 2653.9524x + 53194.2857$$

c. Let $x = 9$.
$$f(x) = 143.494(9)^2 + 2653.9524(9) + 53194.2857$$
$$\approx 88,703$$

Let $x = 15$.
$$f(x) = 143.494(15)^2 + 2653.9524(15) + 53194.2857$$
$$\approx 125,290$$

59. $f(x) = x^4 - 2$

First we find several ordered pairs.

x	−2	−1.5	−1	0	1	1.5	2
y	14	3.1	−1	−2	−1	3.1	14

Plot these ordered pairs and draw a smooth curve through them.

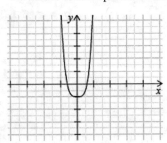

$$f(x) = x^4 - 2$$

60. $g(x) = x^3 - x$

$= x(x^2 - 1)$

$= x(x + 1)(x - 1)$

The x-intercepts are 0, –1, and 1.

Region	Test No.	Value of $f(x)$	Graph
$x < -1$	–2	–6	below x-axis
$-1 < x < 0$	$-\frac{1}{2}$	$\frac{3}{8}$	above x-axis
$0 < x < 1$	$\frac{1}{2}$	$-\frac{3}{8}$	below x-axis
$1 < x$	2	6	above x-axis

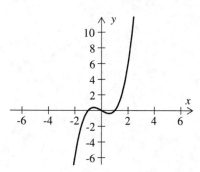

$g(x) = x^3 - x$

61. $f(x) = x(x - 2)(x + 3)$
The x-intercepts are 0, 2, and –3.

Region	Test No.	Value of $f(x)$	Graph
$x < -3$	–4	–24	below x-axis
$-3 < x < 0$	–2	8	above x-axis
$0 < x < 2$	1	–4	below x-axis
$2 < x$	3	18	above x-axis

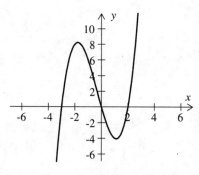

$f(x) = x(x - 2)(x + 3)$

62. $f(x) = (x - 1)(x + 2)(x - 3)$
The x-intercepts are 1, –2, and 3.

Region	Test No.	Value of $f(x)$	Graph
$x < -2$	–3	–24	below x-axis
$-2 < x < 1$	0	6	above x-axis
$1 < x < 3$	2	–4	below x-axis
$3 < x$	4	18	above x-axis

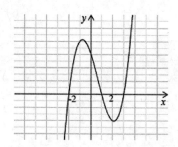

$f(x) = (x - 1)(x + 2)(x - 3)$

63. $f(x) = 3x(3x + 2)(x - 1)$

The x-intercepts are 0, $-\dfrac{2}{3}$, and 1.

Region	Test No.	Value of $f(x)$	Sign
$x < -\frac{2}{3}$	–1	–6	–
$-\frac{2}{3} < x < 0$	$-\frac{1}{3}$	$\frac{4}{3}$	+
$0 < x < 1$	$\frac{1}{2}$	$-\frac{21}{8}$	–
$1 < x$	2	48	+

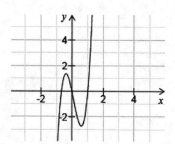

$f(x) = 3x(3x + 2)(x - 1)$

64. $f(x) = x^3 - 3x^2 - 4x$

$\qquad = x(x^2 - 3x - 4)$

$\qquad = x(x-4)(x+1)$

x-intercepts: $0, 4, -1$

Region	Test No.	Value of $f(x)$	Sign
$x < -1$	-2	-12	$-$
$-1 < x < 0$	$-\frac{1}{2}$	$\frac{9}{8}$	$+$
$0 < x < 4$	1	-6	$-$
$4 < x$	5	30	$+$

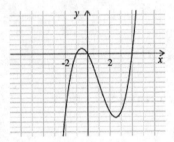

$$f(x) = x^3 - 3x^2 - 4x$$

65. $f(x) = x^4 - 5x^2 - 6$

$\qquad f(x) = (x^2 - 6)(x^2 + 1)$

First, find x-intercepts by setting $f(x) = 0$ and solving for x.

$$(x^2 - 6)(x^2 + 1) = 0$$

$$x^2 = 6$$

$$x = \pm\sqrt{6} \quad \text{(approximately } \pm 2.4\text{)}$$

These two numbers divide the x-axis into three regions. In each region the value of f is either positive or negative. Choose a point in each region as a test point and compute the function value.

65. Continued

x	-3	-1	0	1	3
$f(x)$	30	-10	-6	-10	30

Use these points and the x-intercepts to sketch the graph.

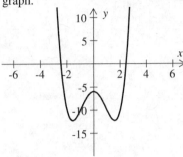

$$f(x) = x^4 - 5x^2 - 6$$

66. $f(x) = x^4 - 7x^2 - 8$

$\qquad = (x^2 - 8)(x^2 + 1)$

$\qquad = (x - \sqrt{8})(x + \sqrt{8})(x^2 + 1)$

x-intercepts: $-\sqrt{8}, \sqrt{8}$

(Note that $x^2 + 1 \neq 0$ for any x.)

Region	Test No.	Value of $f(x)$	Sign
$x < -\sqrt{8}$	-3	10	$+$
$-\sqrt{8} < x < \sqrt{8}$	0	-8	$-$
$\sqrt{8} < x$	3	10	$+$

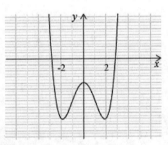

$$f(x) = x^4 - 7x^2 - 8$$

67. Demand equation:

$$p = -.000012q^3 - .00498q^2 + .1264q + 1508$$

Supply equation:

$$p = -.000001q^3 + .00097q^2 + 2q$$

Graph the supply and demand equations in first quadrant. Use the intersection finder to determine the equilibrium point.

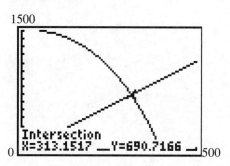

The equilibrium quantity is about 313,152 and the equilibrium price is about $690.72 per thousand.

68. $A(x) = -.000006x^4 + .0017x^3 + .03x^2 - 24x + 1110$

Use the average cost function and use the minimum finder to find point where $A(x)$ is smallest.

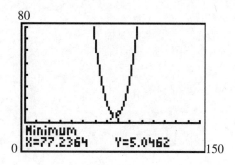

About 77,236 cans should be manufactured for an average cost of about $5.05 per can.

69. $C(x) = -.000006x^3 + .07x^2 + 2x + 1200$

a. $R(x) = 23x$

$$P(x) = R(x) - C(x)$$
$$= 23x + .000006x^3 - .07x^2 - 2x - 1200$$
$$= .000006x^3 - .07x^2 + 21x - 1200$$

69. Continued

    b. Graph the profit function and use root finder to find where $P(x)$ first crosses the
x-axis.

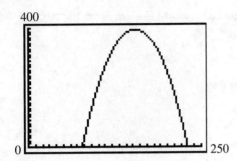

    The break-even point is about 76.54, or
77 racks.

    c. Use the root finder to find where $P(x)$ next crosses the x-axis.
At most about 230 racks can be made without losing money.

    d. Use the maximum finder to find the point where $P(x)$ is the largest.
About 153 racks should be made for a maximum profit of about $395.86.

70. a.

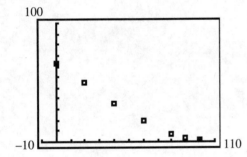

    b. Using a graphing calculator,
$$f(x) = .00000004638x^4 + .00003859x^3 - .002347x^2 - .90913x + 73.584$$

    c. Let $x = 25$.
$$f(25) = .00000004638(25)^4 + .00003859(25)^3 - .002347(25)^2 - .90913(25) + 73.584 \approx 50$$
Let $x = 35$.
$$f(35) = .00000004638(35)^4 + .00003859(35)^3 - .002347(35)^2 - .90913(35) + 73.584 \approx 40.6$$
Let $x = 50$.
$$f(50) = .00000004638(50)^4 + .00003859(50)^3 - .002347(50)^2 - .90913(50) + 73.584 \approx 27.4$$

    d. Answers vary.

71. $f(x) = \dfrac{1}{x-3}$

Find the vertical asymptote by setting the denominator equal to zero and solving for x.

$x - 3 = 0$

$\quad x = 3$

The line $x = 3$ is a vertical asymptote. The horizontal asymptote of the function in the form

$y = \dfrac{0x+1}{1x-3}$ is

$y = \dfrac{0}{1}$ or $y = 0$.

If $x = 0$, $y = -\dfrac{1}{3}$.

If $y = 0$, there is no solution for x.

x	0	1	2	4	5	6
y	$-\frac{1}{3}$	$-\frac{1}{2}$	-1	1	$\frac{1}{2}$	$\frac{1}{3}$

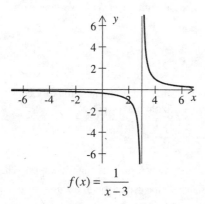

$f(x) = \dfrac{1}{x-3}$

72. $f(x) = \dfrac{-2}{x+4}$

Vertical asymptote: $x = -4$
Horizontal asymptote: $y = 0$

x	-8	-6	-5	$-\frac{9}{2}$	$-\frac{7}{2}$	-3	-2	0
y	$\frac{1}{2}$	1	2	4	-4	-2	-1	$-\frac{1}{2}$

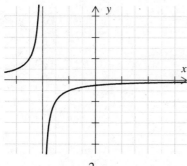

$f(x) = \dfrac{-2}{x+4}$

73. $f(x) = \dfrac{-3}{2x-4}$

Find the vertical asymptote by setting the denominator equal to zero and solving for x.

$2x - 4 = 0$

$\quad x = 2$

The line $x = 2$ is a vertical asymptote. The horizontal asymptote of the function in the form

$y = \dfrac{0x-3}{2x-4}$

is $y = \dfrac{0}{2}$ or $y = 0$.

If $x = 0$, $y = \dfrac{3}{4}$.

If $y = 0$, there is no solution for x.

x	-1	0	1	3	4	5
y	$\frac{1}{2}$	$\frac{3}{4}$	$\frac{3}{2}$	$-\frac{3}{2}$	$-\frac{3}{4}$	$-\frac{1}{2}$

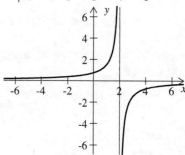

$f(x) = \dfrac{-3}{2x-4}$

74. $f(x) = \dfrac{5}{3x+7}$

Vertical asymptote: $x = -\dfrac{7}{3}$

Horizontal asymptote: $y = 0$

x	-6	-5	-4	-3	-2	-1	0	1
y	$-\frac{5}{11}$	$-\frac{5}{8}$	-1	$-\frac{5}{2}$	5	$\frac{5}{4}$	$\frac{5}{7}$	$\frac{1}{2}$

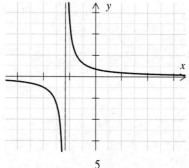

$f(x) = \dfrac{5}{3x+7}$

75. $g(x) = \dfrac{5x-2}{4x^2-4x-3}$

Find the vertical asymptotes by setting the denominator equal to zero and solving for x.

$4x^2 - 4x - 3 = 0$

$(2x-3)(2x+1) = 0$

$2x - 3 = 0 \quad$ or $\quad 2x + 1 = 0$

$\qquad 2x = 3 \quad$ or $\qquad 2x = -1$

$\qquad x = \dfrac{3}{2} \quad$ or $\qquad x = -\dfrac{1}{2}$

The lines $x = \dfrac{3}{2}$ and $x = -\dfrac{1}{2}$ are vertical asymptotes.

Divide the numerator and denominator by x^2 to find the horizontal asymptotes.

$$\frac{\frac{5x}{x^2} - \frac{2}{x^2}}{\frac{4x^2}{x^2} - \frac{4x}{x^2} - \frac{3}{x^2}} = \frac{\frac{5}{x} - \frac{2}{x^2}}{4 - \frac{4}{x} - \frac{3}{x^2}}$$

As $|x|$ gets very large, the numerator gets close to 0.

The function has a horizontal asymptote at $y = 0$.

Find the intercepts.

If $x = 0$, $y = \dfrac{-2}{-3} = \dfrac{2}{3}$. The y-intercept is $\dfrac{2}{3}$.

If $y = 0$,

$0 = \dfrac{5x-2}{4x^2-4x-3}$

$5x = 2$

$x = \dfrac{2}{5}$

The x-intercept is $\dfrac{2}{5}$.

x	-5	-3	-1	$-.3$	0	$\frac{2}{5}$
y	$-.23$	$-.38$	-1.4	2.4	$\frac{2}{3}$	0

x	1	$1\frac{1}{4}$	$1\frac{3}{4}$	2	4
y	-1	-2.4	3	1.6	$.4$

75. Continued

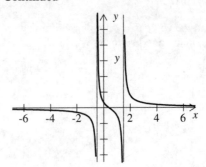

$$g(x) = \frac{5x-2}{4x^2-4x-3}$$

76. $g(x) = \dfrac{x^2}{x^2-1}$

Find the vertical asymptotes by setting the denominator equal to zero and solving for x.

$x^2 - 1 = 0$

$\quad x^2 = 1$

$x = 1$ or $x = -1$

The lines $x = 1$ and $x = -1$ are vertical asymptotes.

Divide the numerator and denominator by x^2 to find the horizontal asymptotes.

$$\frac{\frac{x^2}{x^2}}{\frac{x^2}{x^2} - \frac{1}{x^2}} = \frac{1}{1 - \frac{1}{x^2}}$$

As $|x|$ gets very large, the function approaches 1.

The function has a horizontal asymptote at $y = 1$.

Find the intercepts.

If $x = 0$, $y = 0$. The y-intercept is 0.

If $y = 0$,

$0 = \dfrac{x^2}{x^2-1}$

$x = 0$

The x-intercept is 0.

x	-3	-2	$-\frac{3}{2}$	$-\frac{2}{3}$	$-\frac{1}{2}$	0
y	$\frac{9}{8}$	$\frac{4}{3}$	$\frac{9}{5}$	$-\frac{4}{5}$	$-\frac{1}{3}$	0

x	$\frac{1}{2}$	$\frac{2}{3}$	$\frac{3}{2}$	2	3
y	$-\frac{1}{3}$	$-\frac{4}{5}$	$\frac{9}{5}$	$\frac{4}{3}$	$\frac{9}{8}$

76. Continued

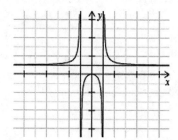

$$g(x) = \frac{x^2}{x^2 - 1}$$

77. $C(x) = \dfrac{650}{2x + 40}$

a. $C(10) = \dfrac{650}{2(10) + 40} \approx 10.83$

The average cost per carton to produce 10 cartons is about \$10.83.

b. $C(50) = \dfrac{650}{2(50) + 40} \approx 4.64$

The average cost per carton to produce 50 cartons is about \$4.64.

c. $C(70) = \dfrac{650}{2(70) + 40} \approx 3.61$

The average cost per carton to produce 70 cartons is about \$3.61.

d. $C(100) = \dfrac{650}{2(100) + 40} \approx 2.71$

The average cost per carton to produce 100 cartons is about \$2.71.

e. We see that $y = 0$ (the x-axis) is a horizontal asymptote.

x	0	10	30	50	70	100
y	16.25	10.83	6.50	4.64	3.61	2.71

Using these points and the horizontal asymptote, we graph the function.

77. Continued

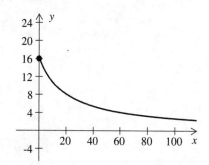

$$C(x) = \frac{650}{2x + 40}$$

78. $C(x) = \dfrac{400x + 400}{x + 4}$ and $R(x) = 100x$

$C(x) = \dfrac{400(x + 1)}{x + 4}$ and $R(x) = 100x$

a. In graphing the function $C(x)$ we see that $y = 400$ is a horizontal asymptote.

x	0	1	2	3	4	5
y	100	160	200	228.6	250	266.7

Using the asymptote and these points we graph the function $C(x)$. The graph of the function $R(x)$ is a line with a slope of 100 and a y-intercept of 0.

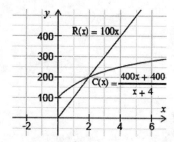

b. To find the break-even point, we use $C(x) = R(x)$ and solve for x.

$$\frac{400(x + 1)}{x + 4} = 100x$$

Divide both sides by 100.

$$\frac{4(x + 1)}{x + 4} = x$$

$$4x + 4 = x^2 + 4x$$

$$x^2 = 4$$

$$x = \pm 2$$

78. Continued

Reject −2 as a solution because the number of hundreds of units cannot be negative. $x = 2$ represents 200 units, so the break-even point is 200 units.

c. $P(1)$ represents a loss. From the graph we see that $C(1) > R(1)$. If the cost is greater than revenue, there is a loss.

d. $P(4)$ represents a profit. From the graph we see that $R(4) > C(4)$. If revenue is greater than cost, there is a profit.

79. Supply: $p = \dfrac{q^2}{4} + 25$

Demand: $p = \dfrac{500}{q}$

a.

q	0	5	10	15
Supply	25	31.25	50	81.25
Demand	No value	100	50	33.3

The equilibrium point is (10, 50).

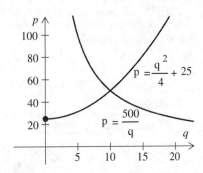

b. Supply exceeds demand if q is in the interval $(10, \infty)$.

c. Demand exceeds supply if q is in the interval $(0, 10)$.

80. $y = \dfrac{9.2x}{106 - x}$

a. If $x = 50$

$$y = \frac{9.2(50)}{106 - 50}$$

$y \approx 8.2$ thousand dollars

or about $8200

b. If $x = 98$

$$y = \frac{9.2(98)}{106 - 98}$$

$y = 112.7$ thousand dollars

or about $113,000

c. If $y = \$22,000$, we have

$$22 = \frac{9.2x}{106 - x}$$

$$22(106 - x) = 9.2x$$

$$2332 - 22x = 9.2x$$

$$2332 = 31.2x$$

$$x \approx 74.7$$

About 75% of the pollutant can be removed for $22,000.

Case 3 Architectural Arches

1. $f(x) = \dfrac{-k}{c^2}x^2 + k$ If $k = 15$, and $c = 8$, then

$$f(x) = \frac{-15}{64}x^2 + 15$$

2. $g(x) = \sqrt{225 - x^2}$; 30 ft because the radius at the base is 15ft.

3. $h(x) = \sqrt{64 - x^2} + 7$; The height is $(15 - 8)$ ft = 7 ft

4. It would fit through the semicircular and Norman arches, but not the parabolic arch. To allow it to fit through the parabolic arch, increase the height of the arch to 18 ft given by the function

$$f(x) = -\frac{18}{64}x^2 + 18.$$

Chapter 4: Exponential and Logarithmic Functions

Section 4.1 Exponential Functions

1. $f(x) = 6^x$

    This function is exponential, because the variable is in the exponent and the base is a positive constant.

3. $h(x) = 4x^2 - x + 5$

    This function is quadratic because it is a polynomial function of degree 2.

5. $f(x) = 675(1.055^x)$

    This function is exponential because the variable is in the exponent and the base is a positive constant other than 1.

7. $f(x) = .6^x$

    a. The equation describing this function has the format $f(x) = a^x$ with

        $0 < a < 1$, so the graph lies entirely above the x-axis and falls from left to right. It falls relatively steeply until it reaches the y-intercept 1 and then falls slowly, with the positive x-axis as a horizontal asymptote.

    b. $f(0) = .6^0 = 1; (0, 1)$

        $f(1) = .6^1 = .8; (1, .6)$

9. $h(x) = 2^{.5x}$

    a. The base of this exponential function is larger than 1, so the graph lies entirely above the x-axis and rises from left to right. The negative x-axis is a horizontal asymptote. The graph rises slowly until it reaches the y-intercept 1 and then rises quite steeply.

    b. $h(0) = 2^{.5(0)} = 2^0 = 1; (0, 1)$

        $h(1) = 2^{.5(1)} = 2^{.5}; (1, 2^{.5})$

11. $f(x) = e^{-x}$

    a. $e^{-x} = \left(\dfrac{1}{e}\right)^x$, so this exponential function is of the form $f(x) = a^x$ with $0 < a < 1$. The graph lies entirely above the x-axis and falls from left to right. It falls very steeply until it reaches the y-intercept 1 and then falls slowly, with the positive x-axis as a horizontal asymptote.

    b. $f(0) = e^{-0} = 1; (0, 1)$

        $f(1) = e^{-1} = \dfrac{1}{e}; \left(1, \dfrac{1}{e}\right)$

13. $f(x) = 3^x$

    Construct a table of ordered pairs, plot the points, and connect them to form a smooth curve.

x	-2	-1	0	1	2
y	$\frac{1}{9}$	$\frac{1}{3}$	1	3	9

    Note that this graph crosses the y-axis at 1 and it is always above the x-axis.

    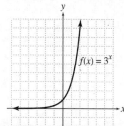

15. $f(x) = 2^{x/2}$

Construct a table of ordered pairs.

x	-4	-2	0	2	4
y	$\frac{1}{4}$	$\frac{1}{2}$	1	2	4

This graph crosses the y-axis at 1 and it is always above the x-axis.

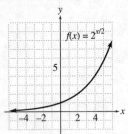

17. $f(x) = \left(\dfrac{1}{5}\right)^x$

x	-2	-1	0	1	2
y	25	5	1	$\frac{1}{5}$	$\frac{1}{25}$

This graph crosses the y-axis at 1 and it is always above the x-axis.

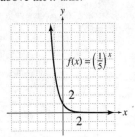

19. a. $f(x) = 2^x$

x	-2	-1	0	1	2
y	$\frac{1}{4}$	$\frac{1}{2}$	1	2	4

b. $g(x) = 2^{x+3}$

x	-2	-1	0	1
y	2	4	8	16

19. Continued

c. $h(x) = 2^{x-4}$

x	-1	0	1	2	3	4
y	$\frac{1}{32}$	$\frac{1}{16}$	$\frac{1}{8}$	$\frac{1}{4}$	$\frac{1}{2}$	1

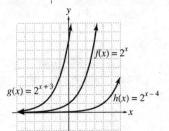

d. The graph of $y = 2^{x+c}$ is the graph of $f(x) = 2^x$ shifted c units to the left. The graph of $y = 2^{x-c}$ is the graph of $f(x) = 2^x$ shifted c units to the right.

21. $y = a^x$

$y = 2.3^x$ is A

23. $y = a^x$

$y = .75^x$ is C

25. $y = a^x$

$y = .31^x$ is E

27. a. $a > 1$

b. Domain: $(-\infty, \infty)$
Range: $(0, \infty)$

c.

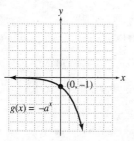

$g(x) = -a^x$

d. Domain: $(-\infty, \infty)$
Range: $(-\infty, 0)$

e.

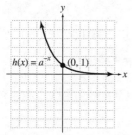

$h(x) = a^{-x}$

f. Domain: $(-\infty, \infty)$
Range: $(0, \infty)$

29. Since $f(x) = a^x$ and $f(3) = 27$,

$a^3 = 27$

$a^3 = 3^3$

$a = 3$

a. $f(x) = 3^x$

$f(1) = 3^1 = 3$

b. $f(x) = 3^x$

$f(-1) = 3^{-1} = \dfrac{1}{3}$

c. $f(x) = 3^x$

$f(2) = 3^2 = 9$

d. $f(x) = 3^x$

$f(0) = 3^0 = 1$

31. $f(x) = 2^{-x^2 + 2}$

x	-2	-1	0	1	2
y	$\frac{1}{4}$	2	4	2	$\frac{1}{4}$

For this graph, the x-axis is an asymptote on the left and on the right.

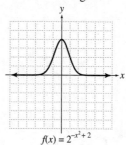

$f(x) = 2^{-x^2 + 2}$

33. $f(x) = x \cdot 2^x$

x	-4	-2	0	1	2
y	$-\frac{1}{4}$	$-\frac{1}{2}$	0	2	8

This graph passes through the origin, and it has the negative x-axis as a horizontal asymptote towards the left.

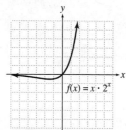

$f(x) = x \cdot 2^x$

35. $y = (1.06)^t$

a.

t	0	1	2	3	4	5
y	1	1.06	1.12	1.19	1.26	1.34

t	6	7	8	9	10
y	1.42	1.50	1.59	1.69	1.79

35. Continued

b.   Plot the eleven points in the table and draw a smooth curve connecting them. (Realize that negative values are not acceptable for t, which represents time.)

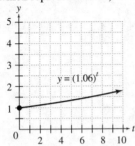

37. $y = (.97)^t$

a.   When $t = 10$, $y = (.97)^{10} \approx .74$.

Let x represent the unknown cost.

$.74x = 105,000$

$x = \dfrac{105,000}{.74}$

$x \approx 141,892$

The house will cost about \$141,892.

b.   When $t = 8$, $y = (.97)^8 \approx .78$.

Let x represent the unknown cost.

$.78x = 50$

$x = \dfrac{50}{.78}$

$x \approx 64.10$

The book will cost about \$64.10.

39. $f(x) = 15.76(1.1976^x)$ $(5 \le x \le 13)$

$x = 5$ corresponds to 1995.

a. $x = 2000 - 1995 + 5 = 10$

$f(10) = 15.76(1.1976^{10})$

≈ 95.648

There will be approximately 95,648,000 cell phone accounts in 2000.

b. $x = 2003 - 1995 + 5 = 13$

$f(13) = 15.76(1.1976^{13})$

≈ 164.289

There will be approximately 164,290,000 cell phone accounts in 2003.

c.   No, this model estimates about 580,490,000 cell phone accounts in 2010, when the U.S. population is projected to be about 309,000,000.

41. $W(x) = 2^{-x/24,360}$

a. $W(1000) = 2^{-1000/24,360} \approx .97$

After 1000 years, about .97 kg will be left.

b. $W(10,000) = 2^{-10,000/24,360} \approx .75$

After 10,000 years, about .75 kg will be left.

c. $W(15,000) = 2^{-15,000/24,360} \approx .65$

After 15,000 years, about .65 kg will be left.

d. $W(24,360) = 2^{-24,360/24,360} \approx .5$

It will take 24,360 years for the one kilogram to decay to half its original weight.

43. $P(t) = 4.834\left(1.01^{(t-1980)}\right)$

a. $t = 2005$

$P(2005) = 4.834\left(1.01^{(2005-1980)}\right)$

$= 4.834(1.01^{25})$

≈ 6.2

In 2005, the population will be about 6.2 billion.

b. $t = 2010$

$P(2010) = 4.834\left(1.01^{(2010-1980)}\right)$

$= 4.834(1.01^{30})$

≈ 6.5

In 2010, the population will be about 6.5 billion.

c. $t = 2030$

$P(2030) = 4.834\left(1.01^{(2030-1980)}\right)$

$= 4.834(1.01^{50})$

≈ 8.0

In 2030, the population will be about 8 billion.

d.   Answers vary.

45. $S = C(1-r)^n, C = \$54,000, n = 8, r = .12$

$S = 54,000(1-.12)^8$

$S = 54,000(.88)^8$

$S \approx \$19,420.26$

47.   Answer varies.
Possible answer:
When x is large, 2^{-x} is very close to 0 and 2^x is huge. Hence, $2^x + 2^{-x} \approx 2^x + 0 = 2^x$.

49. $g(x) = 7311e^{-.00823x}$

a. $x = 2000 - 1975 = 25$

$g(25) = 7311e^{-.00823(25)} \approx 5951$

$x = 2005 - 1975 = 30$

$g(30) = 7311e^{-.00823(30)} \approx 5711$

$x = 2007 - 1975 = 32$

$g(32) = 7311e^{-.00823(32)} \approx 5618$

b. $g(x) = 7311e^{-.00823x} < 5300$

$e^{-.00823x} < \dfrac{5300}{7311}$

Solving using a graphing calculator or a computer gives $x \approx 39.09$. So, in 2014 (1975 + 39.09) the number of hospitals will be less than 5300.

Section 4.2 Applications of Exponential Functions

1. $B(t) = 800(.9898^t)$

a. $B(6) = 800(.9898^6) \approx 752.27$
Your balance after 6 months is $752.27.

b. $B(12) = 800(.9898^{12}) \approx 707.39$
Your balance after 1 year is $707.39.

c. $B(60) = 800(.9898^{60}) \approx 432.45$
Your balance after 5 years is $732.45.

d. $B(96) = 800(.9898^{96}) \approx 298.98$
Your balance after 8 years is $298.98.

e. Answers vary. Your balance will never reach 0.

3. $f(x) = 4295.5e^{-.2294t}$ $(t \geq 4)$

a. $t = 2006 - 2004 + 4 = 6$
$f(6) = 4295.5e^{-.2294(6)} \approx 1084.55$
The average price of an LCD television in 2006 is about $1084.55.

b. $t = 2009 - 2004 + 4 = 9$
$f(9) = 4295.5e^{-.2294(9)} \approx 544.97$
The average price of an LCD television in 2009 is about $544.97.

5. a. Let $f(t)$ represent the Hispanic population in millions in the year t after 2000. The values of $f(t)$ at $t = 0$ and $t = 2050 - 2000 = 50$ are given, that is, $f(0) = 32.5$ and $f(50) = 98.2$. Solving the first of these equations for y_0 in $f(t) = y_0 b^t$:

$f(0) = 32.5$

$y_0 b^0 = 32.5$

$y_0 = 32.5$

The model has the form $f(t) = 32.5b^t$. Solving the second equation, $f(50) = 98.2$, for b:

$f(50) = 98.2$

$32.5b^{50} = 98.2$

$b^{50} = \dfrac{98.2}{32.5}$

$b = \left(\dfrac{98.2}{32.5}\right)^{1/50} \approx 1.022362$

The model is $f(t) = 32.5(1.022362)^t$.

b. $t = 2010 - 2000 = 10$
$f(10) = 32.5(1.022362)^{10} \approx 40.5$
On 2010 the population is established to be 40.5 million.
$t = 2035 - 2000 = 35$
$f(35) = 32.5(1.022362)^{35} \approx 70.5$
In 2035, the population is estimated to be 70.5 million.

c. By graphing and computing the intersection of

$y_1 = 32.5(1.022362)^t$ and

$y_2 = 54$

or by trial and error, the population should reach 54 million 23 years after 2000, or in 2023.

7. a. Let $f(t)$ be the amount of exports (in billions) t years after 1997. Solving $f(0) = 2$ for y_0:

$$f(0) = 2$$
$$y_0 b^0 = 2$$
$$y_0 = 2$$

The model has the form $f(t) = 2b^t$. Solving $f(7) = 12$ for b:

$$f(7) = 12$$
$$2b^7 = 12$$
$$b^7 = 6$$
$$b = 6^{1/7} \approx 1.2917$$

The model is $f(t) = 2(1.2917^t)$

b. $t = 2003 - 1997 = 6$

$$f(6) = 2(1.2917^6) \approx 9.289$$

In 2003, exports of vehicles and vehicle parts amounted to about $9,289,700,000.

$t = 2005 - 1999 = 8$

$$f(8) = 2(1.2917^8) \approx 15.5$$

In 2005, exports of vehicles and vehicle parts amounted to about $15,500,000,000.

9. a. Let $t = 0$ correspond to 1985. Using the data for 1985 and 2005 to find a function of the form $f(t) = y_0 b^t$, first solve for y_0:

$$f(0) = .928$$
$$y_0 b^0 = .928$$
$$y_0 = .928$$

The function has the form $f(t) = .928b^t$.

Solving for b:

$t = 2005 - 1985 = 20$

$$f(20) = .515$$
$$.928b^{20} = .515$$
$$b^{20} = \frac{.515}{.928}$$
$$b = \left(\frac{.515}{.928}\right)^{1/20} \approx .97098599$$

Thus the function becomes $f(t) = .928(.97098599^t)$. Using exponential regression on a graphing calculator, the function produced is $g(t) = .90383(.9709574^t)$.

b. $t = 2000 - 1985 = 15$

$$f(15) = .928(.97098599^{15}) \approx .597$$
$$g(15) = .90383(.9709574^{15}) \approx .581$$

In 2000, a dollar bought what 60 or 59 cents did in 1982-1984.

$t = 2006 - 1985 = 21$

$$f(21) = .928(.97098599^{21}) \approx .500$$
$$g(21) = .90383(.9709574^{21}) \approx .487$$

In 2006, a dollar buys what 50 or 49 cents did in 1982-1984.

$t = 2008 - 1985 = 23$

$$f(23) = .928(.97098599^{23}) \approx .471$$
$$g(23) = .90383(.9709574^{23}) \approx .459$$

In 2008, a dollar buys what 47 or 46 cents did in 1982-1984.

c. $f(t) = .928(.97098599^t)$

$$.928(.97098599) = .40$$
$$.97098599^t = \frac{.40}{.928}$$

Using a graphing calculator or PC, we solve this equation and obtain $t = 28.58$. Thus purchasing power will drop to 40 cents sometime during 2013 (1985 + 28.58). Using the regression model, we obtain 2012.

11. a. Let $f(t)$ be the death rate per 100,000 population in the year t after 1970.
The two-point model:
$f(0) = 492.7$

$y_0 b^0 = 492.7$

$y_0 = 492.7$

$t = 2002 - 1970 = 32$
$f(32) = 240.4$

$492.7 b^{32} = 240.4$

$b^{32} = \dfrac{240.4}{492.7}$

$b = \left(\dfrac{240.4}{492.7}\right)^{1/32} \approx .97783$

$f(t) = 492.7(.97783^t)$ where $t = 0$ corresponds to 1970.
Regression model by calculator:
$g(t) = 502.67(.977755^t)$

b. $t = 1995 - 1970 = 25$
$f(25) = 492.7(.97783^{25}) \approx 281.3$

$g(25) = 502.67(.977755^{25}) \approx 286.4$
The death rate in 1995 is 281.3 or 286.4 per 100,000.
$t = 2005 - 1970 = 35$
$f(35) = 492.7(.97783^{35}) \approx 224.8$

$g(35) = 502.67(.977755^{35}) \approx 228.7$
The death rate in 2005 is about 224.8 or 228.7 per 100,000.

c. By graphing the two-point model,
$y_1 = 492.7(.97783^t)$ and

$y_2 = 100$
and calculating the point of intersection, the death rate will be 100 when $t = 71.1$ or in 2041.
By graphing the regression model,
$y_1 = 502.67(.977755)^t$ and

$y_2 = 100$
and calculating the point of intersection, the death rate will be 100 when $t = 71.1$ or again in 2041.

13. $P(t) = 25 - 25e^{-3t}$

a. $P(1) = 25 - 25e^{-.3(1)} \approx 6$

b. $P(8) = 25 - 25e^{-.3(8)} \approx 23$

c. The maximum number of items that can be produced is 25.

15. $F(t) = T_0 + Cb^t$, where T_0 is the temperature of the constant environment. At $t = 0$, $F(0) = 100$.
So, $100 = T_0 + Cb^0$

$100 = -18 + Cb^0$

$100 = -18 + C(1)$

$C = 118$
The function then becomes $F(t) = -18 + 118b^t$.
At $t = 24$, $F(24) = 50$.
So $50 = -18 + 118b^{24}$

$68 = 118b^{24}$

$b^{24} = \dfrac{68}{118}$

$b = \left(\dfrac{68}{118}\right)^{1/24} \approx .977$

The function now becomes
$F(t) = -18 + 118(.977^t)$. At $t = 96$,

$F(96) = -18 + 118(.977^{96}) \approx -5.36$. So after 96 minutes, the temperature of the water will be about -5 degrees.

17. $y(t) = \dfrac{y_0 e^{kt}}{1 - y_0(1 - e^{kt})}$

a. Let $k = .1$ and $y_0 = .05$. Then
$$y(10) = \frac{.05 e^{.1(10)}}{1 - .05(1 - e^{.1(10)})}$$

$$= \frac{.05e}{1 - .05(1 - e)}$$

$$\approx .13.$$

b. Let $k = .2$ and $y_0 = .1$. Then
$$y(5) = \frac{.1 e^{.2(5)}}{1 - .1(1 - e^{.2(5)})}$$

$$= \frac{.1e}{1 - .1(1 - e)}$$

$$\approx .23.$$

19. a. $x = 2000 - 1970 = 30$

$$f(30) = \frac{74.22}{1 + 22.34e^{-.21(30)}} \approx 71.295$$

The function estimates 71.295 million subscribers in 2000.

$x = 2005 - 1970 = 35$

$$f(35) = \frac{74.22}{1 + 22.34e^{-.21(35)}} \approx 73.170$$

The function estimates 73.170 million subscribers in 2005.

b. To graph $f(x) = \dfrac{74.22}{1 + 22.34e^{-.21x}}$ enter this as y_1 and use $0 \le x \le 50$ and $0 \le y \le 100$.

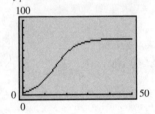

c. To determine the year in which the number of subscribers reached 74 million, graph y_1 from part b and $y_2 = 74$ on the same screen. On the CALC menu, select "intersect" and compute the point of intersection. The coordinates of this point are (42.5, 74) so the number of subscribers should reach 74 million when $t = 42$ or in 2012.

d. Viewing the graph of y_1 and $y_2 = 90$, the number of subscribers will appear to level off below 80 million in the foreseeable future.

21. a. $x = 2000 - 1990 = 10$

$$f(10) = \frac{95.8}{1 + 26.1e^{-.0512(10)}} \approx 5.7567$$

The function estimates the national debt to be $5.7567 trillion in 2000.

$x = 2005 - 1990 = 15$

$$f(15) = \frac{95.8}{1 + 26.1e^{-.0512(15)}} \approx 7.308$$

The function estimates the national debt to be $7.308 trillion in 2005.

$x = 2008 - 1990 = 18$

$$f(18) = \frac{95.8}{1 + 26.1e^{-.0512(18)}} \approx 8.4148$$

The function estimates the national debt to be $8.4148 trillion in 2008.

b. To graph $f(x) = \dfrac{95.8}{1 + 26.1e^{-.0512x}}$, enter this as y_1 and use $0 \le x \le 50$ and $0 \le y \le 35$.

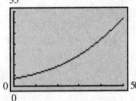

c. To determine the year when the debt will reach $10 trillion, graph y_1 from part b and $y_2 = 10$ on the same screen. On the CALC menu, select "intersect" and compute the point of intersection. The coordinates of this point are (21.7, 10) so the national debt should reach $10 trillion when $x = 21.7$ or in the latter part of 2011.

d. Viewing the model and using the graphing calculator's "TRACE" function, the debt does appear to level off after $t = 60$ or around 2020.

Section 4.3 Logarithmic Functions

1. $x = a^y$

3. It is missing the value that equals b^y.
If that value is x, it should read $y = \log_b x$.

5. $\log 100{,}000 = 5$ is equivalent to $10^5 = 100{,}000$.
(The base of the logarithm is understood to be 10.)

7. $\log_9 81 = 2$ is equivalent to $9^2 = 81$.

9. $10^{1.8751} = 75$
means
$\log 75 = 1.8751$.

11. $3^{-2} = \dfrac{1}{9}$ is equivalent to $\log_3\left(\dfrac{1}{9}\right) = -2$.

13. $\log 1000 = \log_{10} 10^3 = 3$

15. $\log_5 25 = \log_5 5^2 = 2$

17. $\log_4 64 = \log_4 4^3 = 3$

19. $\log_2 \dfrac{1}{4} = \log_2 2^{-2} = -2$

21. $\ln \sqrt{e} = \ln e^{1/2} = \dfrac{1}{2}$

23. $\ln e^{8.77} = 8.77$

25. $\log 53 \approx 1.724$

27. $\ln .452 \approx -.794$

29. $\log_a 1 = 0$ because $a^0 = 1$ for any valid base a.

31. $\log 4 + \log 8 - \log 2 = \log \dfrac{4(8)}{2}$
$\qquad\qquad\qquad\qquad\quad = \log 16$

33. $2 \ln 5 - \dfrac{1}{2} \ln 25 = \ln 5^2 - \ln 25^{1/2}$
$\qquad\qquad\qquad\quad = \ln 25 - \ln 5$
$\qquad\qquad\qquad\quad = \ln \dfrac{25}{5}$
$\qquad\qquad\qquad\quad = \ln 5$

35. $2 \log u + 3 \log w - 6 \log v$
$= \log u^2 + \log w^3 - \log v^6$
$= \log\left(\dfrac{u^2 w^3}{v^6}\right)$

37. $2 \ln(x+1) - \ln(x+2)$
$= \ln(x+1)^2 - \ln(x+2)$
$= \ln\left(\dfrac{(x+1)^2}{x+2}\right)$

39. $\ln \sqrt{6m^4 n^2} = \ln(6m^4 n^2)^{1/2}$
$\qquad\qquad\quad = \dfrac{1}{2} \ln 6m^4 n^2$
$\qquad\qquad\quad = \dfrac{1}{2}(\ln 6 + \ln m^4 + \ln n^2)$
$\qquad\qquad\quad = \dfrac{1}{2}(\ln 6 + 4 \ln m + 2 \ln n)$
$\qquad\qquad\quad = \dfrac{1}{2} \ln 6 + 2 \ln m + \ln n$

41. $\log \dfrac{\sqrt{xz}}{z^3} = \log \dfrac{(xz)^{1/2}}{z^3}$
$\qquad\qquad = \log(xz)^{1/2} - \log z^3$
$\qquad\qquad = \dfrac{1}{2} \log(xz) - 3 \log z$
$\qquad\qquad = \dfrac{1}{2} \log x + \dfrac{1}{2} \log z - 3 \log z$
$\qquad\qquad = \dfrac{1}{2} \log x - \dfrac{5}{2} \log z$

43. $\ln(x^2 y^5) = \ln x^2 + \ln y^5$
$\qquad\qquad = 2 \ln x + 5 \ln y$
$\qquad\qquad = 2u + 5v$

45. $\ln\left(\dfrac{x^3}{y^2}\right) = \ln x^3 - \ln y^2$
$\qquad\qquad = 3 \ln x - 2 \ln y$
$\qquad\qquad = 3u - 2v$

47. $\log_6 346 = \dfrac{\ln 346}{\ln 6}$
$\qquad\qquad \approx \dfrac{5.8464}{1.7918}$
$\qquad\qquad \approx 3.26296$

49. $\log_{35} 7646 = \dfrac{\ln 7646}{\ln 35}$

$\approx \dfrac{8.9419}{3.5553}$

≈ 2.5151

51. $\log (b + c) = \log b + \log c$
Consider the values $b = 1$ and $c = 2$.
Then $\log (b + c)$ becomes
$\log (1 + 2) = \log 3 \approx .4771$,
while $\log b + \log c$ becomes
$\log 1 + \log 2 \approx .3010$.
Many other choices for the b and c values would
demonstrate just as clearly that the statement $\log (b + c) = \log b + \log c$ is generally false.

53. $y = \ln (x + 2)$
We must have $x + 2 > 0$, so the domain is
$x > -2$.

x	y
-1.99	-4.6
-1.5	$-.7$
-1	0
0	$.7$
2	1.4
4	1.8

Connect these points with a smooth curve.

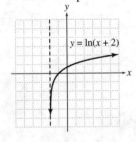

55. $y = \log (x - 3)$
We must have $x - 3 > 0$, so the domain is
$x > 3$.

x	y
3.01	-2
3.5	$-.30$
4	0
6	$.48$
8	$.70$

Connect these points with a smooth curve.

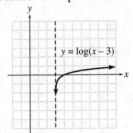

57. Answer varies.
Possible answer:

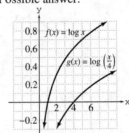

$\log \left(\dfrac{x}{4} \right) = \log x - \log 4$

$g(x)$ equals $\log 4$ subtracted from $f(x)$

59. $\ln 2.75 = 1.0116009$

$e^{1.0116009} = 2.75$

61. $D(r) = \dfrac{\ln 2}{\ln(1+r)}$

a. $D(4\%) = D(.04) = \dfrac{\ln 2}{\ln(1+.04)}$

≈ 17.67

It takes 17.67 years to double.

b. $D(8\%) = D(.08) = \dfrac{\ln 2}{\ln(1+.08)}$

≈ 9.01

It takes 9.01 years to double.

c. $D(18\%) = D(.18) = \dfrac{\ln 2}{\ln(1+.18)}$

≈ 4.19

It takes 4.19 years to double.

d. $D(36\%) = D(.36) = \dfrac{\ln 2}{\ln(1+.36)}$

≈ 2.25

It takes 2.25 years to double.

e. $17.67 \approx 18 = \dfrac{72}{4}$;

$9.01 \approx 9 = \dfrac{72}{8}$;

$4.19 \approx 4 = \dfrac{72}{18}$;

$2.25 \approx 2 = \dfrac{72}{36}$

The pattern is that it takes about 72/*k* years for money to double at *k*% interest.

63. a. In 1995, $x = 1995 - 1992 + 2 = 5$.

$f(5) = 5.03 + 10.24 \ln 5$

$= 5.03 + 10.24(1.6094)$

$= 5.03 + 16.4806$

$= 21.511$

In 1995, the number of McDonald's restaurants worldwide was about 21,511.

In 2000, $x = 2000 - 1992 + 2 = 10$.

$f(10) = 5.03 + 10.24 \ln 10$

$= 5.03 + 10.24(2.3026)$

$= 5.03 + 23.5785$

$= 28.608$

In 2000, the number of McDonald's restaurants worldwide was about 28,608.

In 2005, $x = 2005 - 1992 + 2 = 15$.

$f(15) = 5.03 + 10.24 \ln 15$

$= 5.03 + 10.24(2.7081)$

$= 5.03 + 27.7304$

$= 32.760$

In 2005, the number of McDonald's restaurants worldwide was about 32,760.

b.

x	5	10	15	20	25
y	21.5	28.6	32.7	35.7	37.9

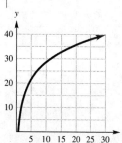

c. The number of restaurants is increasing at a slower rate as time goes on.

65. a. $x = 2007 - 2001 + 1 = 7$

$g(7) = 15.93 + 2.174 \ln 7$

≈ 20.16

By 2007, approximately 20.2% of the U.S. population will be age 60 or older.

$x = 2015 - 2001 + 1 = 15$

$g(15) = 15.93 + 2.174 \ln 15$

≈ 21.82

By 2015, approximately 21.8% of the U.S. population will be age 60 or older.

$x = 2030 - 2001 + 1 = 30$

$g(30) = 15.93 + 2.174 \ln 30$

≈ 23.32

By 2030, approximately 23.3% of the U.S. population will be age 60 or older.

$x = 2050 - 2001 + 1 = 50$

$g(50) = 15.93 + 2.174 \ln 50$

≈ 24.43

By 2050, approximately 24.4% of the U.S. population will be age 60 or older.

b.

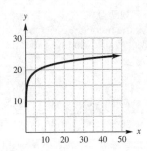

c. As time goes on, the percentage the U.S. population age 60 or over will level off just below 25%.

67. $n_1 = 2754$, $n_2 = 689$, $n_3 = 4428$, and $n_4 = 629$

$N = n_1 + n_2 + n_3 + n_4 = 8500$

So, the index of diversity is

$$H = \frac{N \log_2 N - \left[n_1 \log_2 n_1 + n_2 \log_2 n_2 + n_3 \log_2 n_3 + n_4 \log_2 n_4 \right]}{N}$$

$$= \frac{8500 \log_2 8500 - \left[2754 \log_2 2754 + 689 \log_2 689 + 4428 \log_2 4428 + 629 \log_2 629 \right]}{8500}$$

$$= \frac{8500 \dfrac{\ln 8500}{\ln 2} - \left[2754 \dfrac{\ln 2754}{\ln 2} + 689 \dfrac{\ln 689}{\ln 2} + 4428 \dfrac{\ln 4428}{\ln 2} + 629 \dfrac{\ln 629}{\ln 2} \right]}{8500}$$

≈ 1.5887

Section 4.4 Logarithmic and Exponential Equations

1. $\ln(3x+1) - \ln(5+x) = \ln 2$

$$\ln\frac{3x+1}{5+x} = \ln 2$$

$$\frac{3x+1}{5+x} = 2$$

$$3x+1 = 10+2x$$

$$x = 9$$

3. $\ln(x+1) = \ln(x-4)$

$$x+1 = x-4$$

$$1 = -4$$

Since this is not true for any value of x, there is no solution.

5. $2\ln(x-3) = \ln(x+5) + \ln 4$

$$\ln(x-3)^2 = \ln\left[4(x+5)\right]$$

$$(x-3)^2 = 4(x+5)$$

$$x^2 - 6x + 9 = 4x + 20$$

$$x^2 - 10x - 11 = 0$$

$$(x-11)(x+1) = 0$$

$$x = 11 \text{ or } x = -1$$

Since -1 is not in the domain of $\ln(x-3)$, the only solution is 11.

7. $\log_3(6x-2) = 2$

$$6x-2 = 3^2$$

$$6x-2 = 9$$

$$6x = 11$$

$$x = \frac{11}{6}$$

9. $\log x - \log(x+5) = -1$

$$\log\left(\frac{x}{x+5}\right) = -1$$

$$\frac{x}{x+5} = 10^{-1}$$

$$\frac{x}{x+5} = \frac{1}{10}$$

$$10x = x+5$$

$$9x = 5$$

$$x = \frac{5}{9}$$

11. $\log_3(y+2) = \log_3(y-7) + \log_3 4$

$$\log_3(y+2) = \log_3\left[4(y-7)\right]$$

$$y+2 = 4(y-7)$$

$$y+2 = 4y-28$$

$$30 = 3y$$

$$y = 10$$

13. $\ln(x+9) - \ln x = 1$

$$\ln\left(\frac{x+9}{x}\right) = 1$$

$$\frac{x+9}{x} = e^1$$

$$x+9 = ex$$

$$x - ex = -9$$

$$(1-e)x = -9$$

$$x = -\frac{-9}{e-1} \approx 5.2378$$

15. $\log x + \log(x-3) = 1$

$$\log\left[x(x-3)\right] = 1$$

$$x(x-3) = 10^1$$

$$x^2 - 3x = 10$$

$$x^2 - 3x - 10 = 0$$

$$(x-5)(x+2) = 0$$

$$x-5 = 0 \text{ or } x+2 = 0$$

$$x = 5 \text{ or } x = -2$$

$x = -2$ is not in the domain of $\log x$ nor that of $\log(x-3)$, so the only solution is $x = 5$.

17. $\log(3+b) = \log(4c-1)$

$$3+b = 4c-1$$

$$4+b = 4c$$

$$\frac{4+b}{4} = c$$

$$c = \frac{4+b}{4}$$

19. $2-b = \log(6c+5)$

$$6c+5 = 10^{2-b}$$

$$6c = 10^{2-b} - 5$$

$$c = \frac{10^{2-b} - 5}{6}$$

21. Answer varies.

23. $3^{x-1} = 9$

$3^{x-1} = 3^2$

$x - 1 = 2$

$x = 3$

25. $25^{-3x} = 3125$

$(5^2)^{-3x} = 5^5$

$5^{-6x} = 5^5$

$-6x = 5$

$x = -\dfrac{5}{6}$

27. $7^{-x} = 49^{x+3}$

$7^{-x} = (7^2)^{x+3}$

$7^{-x} = 7^{2(x+3)}$

$-x = 2(x+3)$

$-x = 2x+6$

$-3x = 6$

$x = -2$

29. $\left(\dfrac{3}{4}\right)^x = \dfrac{16}{9}$

$\left(\dfrac{3}{4}\right)^x = \left(\dfrac{4}{3}\right)^2$

$\left(\dfrac{3}{4}\right)^x = \left(\dfrac{3}{4}\right)^{-2}$

$x = -2$

31. $2^x = 5$

Take natural logarithms of both sides.

$\ln 2^x = \ln 5$

$x \ln 2 = \ln 5$

$x = \dfrac{\ln 5}{\ln 2} \approx \dfrac{1.6094}{.6931} \approx 2.3219$

33. $2^x = 3^{x-1}$

$\ln 2^x = \ln 3^{x-1}$

$x \ln 2 = (x-1)(\ln 3)$

$x \ln 2 = x \ln 3 - 1 \ln 3$

$x \ln 2 - x \ln 3 = -1 \ln 3$

$(\ln 2 - \ln 3)x = -1 \ln 3$

$x = \dfrac{-\ln 3}{\ln 2 - \ln 3}$

$x \approx 2.710$

35. $3^{1-2x} = 5^{x+5}$

$\ln 3^{1-2x} = \ln 5^{x+5}$

$(1-2x)(\ln 3) = (x+5)(\ln 5)$

$\ln 3 - 2x \ln 3 = x \ln 5 + 5 \ln 5$

$\ln 3 - 5 \ln 5 = x \ln 5 + 2x \ln 3$

$\ln 3 - 5 \ln 5 = (\ln 5 + 2 \ln 3)x$

$\dfrac{\ln 3 - 5 \ln 5}{\ln 5 + 2 \ln 3} = x$

$x \approx -1.825$

37. $e^{2x} = 7$

$\ln e^{2x} = \ln 7$

$2x = \ln 7$

$x = \dfrac{\ln 7}{2} \approx .973$

39. $2e^{5a+2} = 8$

$e^{5a+2} = 4$

$\ln e^{5a+2} = \ln 4$

$5a + 2 = \ln 4$

$5a = -2 + \ln 4$

$a = \dfrac{-2 + \ln 4}{5}$

$a \approx -.123$

41. $10^{4c-3} = d$

$\log 10^{4c-3} = \log d$

$4c - 3 = \log d$

$4c = \log d + 3$

$c = \dfrac{\log d + 3}{4}$

43. $e^{2c-1} = b$

$\ln e^{2c-1} = \ln b$

$2c - 1 = \ln b$

$2c = \ln b + 1$

$c = \dfrac{\ln b + 1}{2}$

45. $\log_5(r+2)+\log_5(r-2)=1$

$\log_5\left[(r+2)(r-2)\right]=1$

$(r+2)(r-2)=5^1$

$r^2-4=5$

$r^2=9$

$r=\pm 3$

Since -3 is not in the domain of $\log_5(r-2)$, 3 is the only solution.

47. $\log_3(a-3)=1+\log_3(a+1)$

$\log_3(a-3)-\log_3(a+1)=1$

$\log_3\dfrac{a-3}{a+1}=1$

$\dfrac{a-3}{a+1}=3^1$

$a-3=3a+3$

$-6=2a$

$a=-3$

Since -3 is not in the domain of $\log_3(a-3)$ nor in that of $\log_3(a+1)$, there is no solution.

49. $\log_2\sqrt{2y^2}-1=\dfrac{3}{2}$

$\log_2\sqrt{2y^2}=\dfrac{5}{2}$

$\sqrt{2y^2}=2^{5/2}$

$2y^2=2^5$

$2y^2=32$

$y^2=16$

$y=\pm 4$

51. $\log z=\sqrt{\log z}$

$(\log z)^2=\left(\sqrt{\log z}\right)^2$

$(\log z)^2=\log z$

$(\log z)^2-\log z=0$

$(\log z)(\log z-1)=0$

$\log z=0$ or $\log z-1=0$

$\log z=1$

$z=10^0$ · $z=10^1$

$z=1$ or $z=10$

53. $5^{-2x}=\dfrac{1}{25}$

$5^{-2x}=5^{-2}$

$-2x=-2$

$x=1$

55. $2^{|x|}=16$

$2^{|x|}=2^4$

$|x|=4$

$x=4$ or $x=-4$

57. $2^{x^2-1}=10$

$\ln 2^{x^2-1}=\ln 10$

$(x^2-1)\ln 2=\ln 10$

$x^2-1=\dfrac{\ln 10}{\ln 2}$

$x^2-1\approx\dfrac{2.3026}{.6931}$

$x^2-1\approx 3.3219$

$x^2=4.3219$

$x=\pm\sqrt{4.3219}$

$x\approx\pm 2.0789$

59. $2(e^x+1)=10$

$e^x+1=5$

$e^x=4$

$\ln e^x=\ln 4$

$x=\ln 4$

$x\approx 1.386$

61. Answer varies.
Possible answer:
Since $x^2\geq 0$ for every x, $x^2+1\geq 1$ for every x.
Hence $4^{x^2+1}\geq 4^1=4$ for every x.

63. a. In 1997, $x = 1997 - 1960 = 37$.

$$f(37) = 24.2\left(.9792^{37}\right)$$

$$= 24.2(.4595)$$

$$= 11.1188$$

In 1997, there were about 11 people per automobile.

In 2005, $x = 2005 - 1960 = 45$.

$$f(45) = 24.2\left(.9792^{45}\right)$$

$$= 24.2(.3883)$$

$$= 9.3979$$

In 2005, there were about 9 people per automobile.

b. To find x such that $f(x) = 7$, solve:

$$24.2\left(.9792^{x}\right) = 7$$

$$.9792^{x} = \frac{7}{24.2}$$

$$.9792^{x} = .2893$$

$$\log .9792^{x} = \log .2893$$

$$x \log .9792 = \log .2893$$

$$x = \frac{\log .2893}{\log .9792}$$

$$\approx 59.0071$$

If this model remains accurate, the number of people per automobile will reach 7 in 2019 (1960+59).

65.

$$y = y_0 (.90)^{t-1}$$

$$50 = 200(.90)^{t-1}$$

$$\frac{50}{200} = (.90)^{t-1}$$

$$\log \frac{50}{200} = \log(.90)^{t-1}$$

$$\log .25 = (t-1)\log(.90)$$

$$\frac{\log .25}{\log .90} = t - 1$$

$$\frac{\log .25}{\log .90} + 1 = t$$

$$14.2 \approx t$$

It will take about 14.2 hours.

67. $C(t) = 25e^{-.14t}$

a. $C(0) = 25e^{-.14(0)}$

$$= 25e^0$$

$$= 25$$

Initially, 25 g of cobalt was present.

b. We determine the half-life by finding a value of t such that

$$C(t) = \left(\frac{1}{2}\right)(25) = 12.5.$$

$$12.5 = 25e^{-.14t}$$

$$\frac{1}{2} = e^{-.14t}$$

$$\ln \frac{1}{2} = \ln e^{-.14t}$$

$$\ln \frac{1}{2} = -.14t$$

$$t = \frac{\ln \frac{1}{2}}{-.14}$$

$$= \frac{\ln .5}{-.14} \approx 4.95$$

The half-life of cobalt is about 4.95 years.

69. $y = y_0 \left(.5^{t/5730}\right)$

36% lost means 64% remains. Since $64\% = .64$, replace y with $.64y_0$ and solve for t.

$$.64y_0 = y_0 \left(.5^{t/5730}\right)$$

$$.64 = \left(.5^{t/5730}\right)$$

$$\ln .64 = \ln \left(.5^{t/5730}\right)$$

$$\ln .64 = \frac{t}{5730} \ln .5$$

$$t = \frac{5730 \ln .64}{\ln .5}$$

$$t \approx 3689.3$$

The ivory is about 3689 years old.

71. a. $R(i) = \log\left(\dfrac{i}{i_0}\right)$

$$R(i) = 6.6$$

$$\log\left(\frac{i}{i_0}\right) = 6.6$$

$$10^{6.6} = \frac{i}{i_0}$$

$$i = 10^{6.6} i_0$$

$$i \approx 3,981,071.7 i_0$$

71. Continued

b.
$$R(i) = 6.5$$
$$\log\left(\frac{i}{i_0}\right) = 6.5$$
$$10^{6.5} = \frac{i}{i_0}$$
$$i = 10^{6.5} i_0$$
$$i \approx 3{,}162{,}277.7 i_0$$

c. The July earthquake was
$$\frac{10^{6.6}}{10^{6.5}} = 10^{6.6-6.5} = 10^{.1} \approx 1.258$$
times stronger than the February earthquake.

73. $D(i) = 10 \cdot \log\left(\dfrac{i}{i_0}\right)$

a.
$$D(115 i_0) = 10\log\left(\frac{115 i_0}{i_0}\right)$$
$$= 10\log 115 \approx 21$$

b.
$$D(10^{10} i_0) = 10\log\left(\frac{10^{10} i_0}{i_0}\right)$$
$$= 10 \cdot (10\log 10)$$
$$= 10 \cdot (10 \cdot 1)$$
$$= 100$$

c.
$$D(31{,}600{,}000{,}000 i_0)$$
$$= 10\log\left(\frac{31{,}600{,}000{,}000 i_0}{i_0}\right)$$
$$= 10\log 31{,}600{,}000{,}000$$
$$\approx 105$$

d.
$$D(895{,}000{,}000{,}000 i_0)$$
$$= 10\log\left(\frac{895{,}000{,}000{,}000 i_0}{i_0}\right)$$
$$= 10\log(895{,}000{,}000{,}000)$$
$$\approx 120$$

73. Continued

e.
$$D(109{,}000{,}000{,}000{,}000 i_0)$$
$$= 10\log\left(\frac{109{,}000{,}000{,}000{,}000 i_0}{i_0}\right)$$
$$= 10\log(109{,}000{,}000{,}000{,}000)$$
$$\approx 140$$

75. a.
$$P(T) = 1 - e^{-.0034 - .0053T}$$
$$P(60) = 1 - e^{-.0034 - .0053(60)}$$
$$P(60) \approx .275$$
The reduction will be 27.5% when the tax is $60.

b.
$$P(T) = 1 - e^{-.0034 - .0053T}$$
$$.5 = 1 - e^{-.0034 - .0053T}$$
$$T = \frac{\ln(1 - .5) + .0034}{-.0053}$$
$$T = 130.14$$
The tax would be $130.14

Exercise 77 requires the use of a grapher.

77. a. To graph
$p = 86.3 \ln h - 680$,
enter this function as y_1 and use
$3000 \le x \le 8500$ and $0 \le y \le 120$.

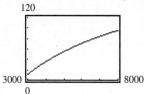

b. Graph $y = 50$, entering this function as y_2.
On the CALC menu use "intersect."

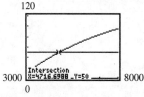

The point of intersection has approximate
coordinates (4716.70, 50). At about
4717 ft, 50% of the moisture is snow.

79. a. Let $f(x)$ be the logarithmic function that models this data. Using a graphing calculator, $f(x) = -4964.2 + 6284 \ln x$.

b. $x = 2002 - 1990 + 10 = 22$

$f(22) = -4964.2 + 6284 \ln 22$

$\approx 14,459.9$

This function predicts 14,460 transplants in 2002.

c. $f(x) = 16,500$

$-4964.2 + 6284 \ln x = 16,500$

$6284 \ln x = 21,464.2$

$\ln x = \dfrac{21,464.2}{6284}$

$x = e^{21464.2/6284}$

$x \approx 30.4$

If this model remains accurate, there will be 16,500 transplants in $1990 + 30 - 10 = 2010..$

Chapter 4 Review Exercises

1. $y = a^{x+2}$ is (c).

2. $y = a^x + 2$ is (a).

3. $y = -a^x + 2$ is (d).

4. $y = a^{-x} + 2$ is (b).

5. $0 < a < 1$

6. Domain of f: $(-\infty, \infty)$ All real numbers.

7. Range of f: $(0, \infty)$ All positive real numbers.

8. $f(0) = 1$

9. $f(x) = 4^x$

x	-2	-1	0	1	2
y	$\frac{1}{16}$	$\frac{1}{4}$	1	4	16

The negative x-axis is a horizontal asymptote for the graph.

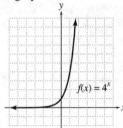

10. $g(x) = 4^{-x}$

x	-2	-1	0	1	2
y	16	4	1	$\frac{1}{4}$	$\frac{1}{16}$

The positive x-axis is a horizontal asymptote for the graph.

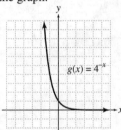

11. $f(x) = \ln x + 5$

x	.01	.1	1	2	4	6	8
y	.4	2.7	5	5.7	6.4	6.8	7.1

The negative y-axis is a vertical asymptote for the graph.

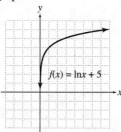

12. $g(x) = \log x - 3$

x	.01	.1	1	2	4	6	8
y	–5	–4	–3	–2.7	–2.4	–2.2	–2.1

The negative y-axis is a vertical asymptote for the graph.

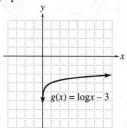

13. $f(x) = .789\left(1.638^x\right)$

a. $x = 2006 - 2000 = 6$

$f(6) = .789\left(1.638^6\right) \approx 15.239$

This model estimates that 15.24% of all U.S households will have a digital video recorder by the year 2006.

$x = 2008 - 2000 = 8$

$f(8) = .789\left(1.638^8\right) \approx 40.887$

This model estimates that 40.89% of all U.S households will have a digital video recorder by the year 2008.

b. To determine x for $f(x) = 100$, solve:

$.789\left(1.638^x\right) = 100$

$1.638^x = \dfrac{100}{.789}$

$\ln\left(1.638^x\right) = \ln\left(\dfrac{100}{.789}\right)$

$x\ln 1.638 = \ln\left(\dfrac{100}{.789}\right)$

$x = \dfrac{\ln\left(100 / .789\right)}{\ln 1.638} \approx 9.812$

If this model remains accurate, 100% of the U.S. households will have a digital video recorder when $x = 9.812$ or in 2009. No, the model seems unlikely to be accurate after 2008.

14. $p(t) = 250 - 120(2.8)^{-.5t}$

a. $p(2) = 250 - 120(2.8)^{-.5(2)}$

$p(2) \approx 207$

b. $p(4) = 250 - 120(2.8)^{-.5(4)}$

$p(4) \approx 235$

c. $p(10) = 250 - 120(2.8)^{-.5(10)}$

$p(10) = 249$

d.

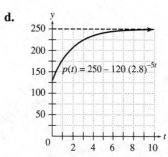

15. $10^{1.7404} = 55$ is equivalent to
$\log 55 = 1.7404$.

16. $4^5 = 1024$ is equivalent to $\log_4 1024 = 5$.

17. $e^{3.8067} = 45$ is equivalent to $\ln 45 = 3.8067$.

18. $7^{1/2} = \sqrt{7}$ is equivalent to $\log_7 \sqrt{7} = \dfrac{1}{2}$.

19. $\log 10,000 = 4$ is equivalent to $10^4 = 10,000$.

20. $\log 26.3 = 1.4200$ is equivalent to $10^{1.4200} = 26.3$.

21. $\ln 81.1 = 4.3957$ is equivalent to $e^{4.3957} = 81.1$.

22. $\log_2 4096 = 12$ is equivalent to $2^{12} = 4096$.

23. $\ln e^5 = 5$ because $\ln e^k = k$ for every real number k.

24. $\log \sqrt[3]{10} = \log_{10} 10^{1/3} = \dfrac{1}{3}$ because $\log_a a^y = y$ for every positive real number y.

25. $10^{\log 8.9} = 8.9$ because $a^{\log_a x} = x$ for every positive real number x.

26. $\ln e^{5t} = 5t$ because $\ln e^j = j$ for every real number j.

27. Let $x = \log_{36} 6$.
$$\log_{36} 6 = x$$
$$36^x = 6$$
$$x = \frac{1}{2}$$
Therefore, $\log_{36} 6 = \dfrac{1}{2}$.

28. $\log_8 32 = \log_8 8^{5/3} = \dfrac{5}{3}$

29. $\log 3x + \log 4x^4 = \log(3x \cdot 4x^4)$
$$= \log\left(12x^5\right)$$

30. $5\log u - 3\log u^4 = \log u^5 - \log\left(u^4\right)^3$
$$= \log u^5 - \log u^{12}$$
$$= \log\left(\frac{u^5}{u^{12}}\right)$$
$$= \log\left(\frac{1}{u^7}\right)$$

31. $3\log b - 2\log c = \log b^3 - \log c^2$
$$= \log\left(\frac{b^3}{c^2}\right)$$

32. $7\ln x - 3(\ln x^3 + 5\ln x)$
$$= 7\ln x - 3\ln x^3 - 15\ln x$$
$$= -3\ln x^3 - 8\ln x$$
$$= \ln\left(x^3\right)^{-3} + \ln x^{-8}$$
$$= \ln\left(x^{-9}\right) + \ln x^{-8}$$
$$= \ln\left(x^{-9} \cdot x^{-8}\right)$$
$$= \ln x^{-17}$$
$$= \ln\left(\frac{1}{x^{17}}\right)$$

33. $\ln(m+3) - \ln m = \ln 2$
$$\ln\frac{m+3}{m} = \ln 2$$
$$\frac{m+3}{m} = 2$$
$$m+3 = 2m$$
$$m = 3$$

34. $2\ln(y+1) = \ln(y^2-1) + \ln 5$
$$\ln(y+1)^2 = \ln\left[(y^2-1)(5)\right]$$
$$(y+1)^2 = 5(y^2-1)$$
$$y^2 + 2y + 1 = 5y^2 - 5$$
$$0 = 4y^2 - 2y - 6$$
$$0 = 2(2y^2 - y - 3)$$
$$0 = 2(2y-3)(y+1)$$
$$y = \frac{3}{2} \text{ or } y = -1$$

Since -1 is not in the domain of $\ln(y+1)$ nor in that of $\ln(y^2-1)$, the only solution is $\dfrac{3}{2}$.

35. $\log(m+2) = 1$

$\qquad m+2 = 10^1$

$\qquad m+2 = 10$

$\qquad\quad m = 8$

36. $\log x^2 = 2$

$\qquad x^2 = 10^2$

$\qquad x^2 = 100$

$\qquad x^2 = \pm\sqrt{100}$

$\qquad\; x = \pm 10$

37. $\log_2(3k-2) = 4$

$\qquad 3k-2 = 2^4$

$\qquad 3k-2 = 16$

$\qquad\quad 3k = 18$

$\qquad\quad\; k = 6$

38. $\log_5\left(\dfrac{5z}{z-2}\right) = 2$

$\qquad\quad 5^2 = \dfrac{5z}{z-2}$

$\qquad 25(z-2) = 5z$

$\qquad 25z - 50 = 5z$

$\qquad\quad 20z = 50$

$\qquad\qquad z = \dfrac{50}{20} = 2.5$

39. $\log x + \log(x+3) = 1$

$\qquad \log\left[x(x+3)\right] = 1$

$\qquad\quad x(x+3) = 10^1$

$\qquad\quad x^2 + 3x = 10$

$\qquad x^2 + 3x - 10 = 0$

$\qquad (x+5)(x-2) = 0$

$\qquad\qquad x = -5 \text{ or } x = 2$

Since −5 is not in the domain of $\log x$ nor in that of $\log(x+3)$, the only solution is 2.

40. $\log_2 r + \log_2(r-2) = 3$

$\qquad \log_2\left[r(r-2)\right] = 3$

$\qquad\quad r(r-2) = 2^3$

$\qquad r^2 - 2r - 8 = 0$

$\qquad (r-4)(r+2) = 0$

$\qquad\qquad r = 4 \text{ or } r = -2$

Since −2 is not in the domain of $\log_2 r$ nor in that of $\log_2(r-2)$, the only solution is 4.

41. $2^{3x} = \dfrac{1}{8}$

$\quad 2^{3x} = 2^{-3}$

$\quad 3x = -3$

$\quad\; x = -1$

42. $\left(\dfrac{9}{16}\right)^x = \dfrac{3}{4}$

$\left[\left(\dfrac{3}{4}\right)^2\right]^x = \left(\dfrac{3}{4}\right)^1$

$\left(\dfrac{3}{4}\right)^{2x} = \left(\dfrac{3}{4}\right)^1$

$\qquad 2x = 1$

$\qquad\; x = \dfrac{1}{2}$

43. $9^{2y-1} = 27^y$

$(3^2)^{2y-1} = (3^3)^y$

$\quad 3^{4y-2} = 3^{3y}$

$\quad 4y-2 = 3y$

$\qquad y = 2$

44. $\dfrac{1}{2} = \left(\dfrac{b}{4}\right)^{1/4}$

$\left(\dfrac{1}{2}\right)^4 = \left[\left(\dfrac{b}{4}\right)^{1/4}\right]^4$

$\dfrac{1}{16} = \dfrac{b}{4}$

$16b = 4$

$b = \dfrac{4}{16} = \dfrac{1}{4}$

45. $8^p = 19$

$\ln 8^p = \ln 19$

$p\ln 8 = \ln 19$

$p = \dfrac{\ln 19}{\ln 8}$

$p \approx \dfrac{2.9444}{2.0794}$

$p \approx 1.416$

46. $\quad 3^z = 11$

$\ln 3^z = \ln 11$

$z \ln 3 = \ln 11$

$z = \dfrac{\ln 11}{\ln 3}$

$z \approx 2.183$

47. $\quad 5 \cdot 2^{-m} = 35$

$2^{-m} = 7$

$\ln(2^{-m}) = \ln 7$

$-m \ln 2 = \ln 7$

$-m = \dfrac{\ln 7}{\ln 2}$

$-m = \dfrac{1.9459}{.6931}$

$m \approx -2.807$

48. $\quad 2 \cdot 15^{-k} = 18$

$15^{-k} = 9$

$\ln 15^{-k} = \ln 9$

$-k \ln 15 = \ln 9$

$k = -\dfrac{\ln 9}{\ln 15}$

$k \approx -.811$

49. $\quad e^{-5-2x} = 5$

$\ln e^{-5-2x} = \ln 5$

$-5 - 2x = \ln 5$

$-5 - 2x \approx 1.6094$

$-2x = 6.6094$

$x \approx -3.305$

50. $\quad e^{3x-1} = 12$

$\ln e^{3x-1} = \ln 12$

$3x - 1 = \ln 12$

$x = \dfrac{1 + \ln 12}{3}$

$x \approx 1.162$

51. $\qquad 6^{2-m} = 2^{3m+1}$

$\ln 6^{2-m} = \ln 2^{3m+1}$

$(2 - m) \ln 6 = (3m + 1) \ln 2$

$2 \ln 6 - m \ln 6 = 3m \ln 2 + \ln 2$

$2 \ln 6 - \ln 2 = 3m \ln 2 + m \ln 6$

$2 \ln 6 - \ln 2 = (3 \ln 2 + \ln 6)m$

$m = \dfrac{2 \ln 6 - \ln 2}{3 \ln 2 + \ln 6}$

$m \approx .747$

52. $\qquad 5^{3r-1} = 6^{2r+5}$

$\ln 5^{3r-1} = \ln 6^{2r+5}$

$3r \ln 5 - \ln 5 = 2r \ln 6 + 5 \ln 6$

$(3 \ln 5 - 2 \ln 6)r = 5 \ln 6 + \ln 5$

$r = \dfrac{5 \ln 6 + \ln 5}{3 \ln 5 - 2 \ln 6}$

≈ 8.490

53. $\quad (1 + .003)^k = 1.089$

$1.003^k = 1.089$

$\ln 1.003^k = \ln 1.089$

$k \ln 1.003 = \ln 1.089$

$k = \dfrac{\ln 1.089}{\ln 1.003}$

$k \approx 28.463$

54. $\quad (1 + .094)^z = 2.387$

$1.094^z = 2.387$

$\ln 1.094^z = \ln 2.387$

$z \ln 1.094 = \ln 2.387$

$z = \dfrac{\ln 2.387}{\ln 1.094}$

$z \approx 9.684$

55. $\quad G(t) = 15 + 2 \log t$

a. $\quad G(1) = 15 + 2 \log 1$

$G(1) = 15$

The GNP at 1 year is \$15 million.

b. $\quad G(2) = 15 + 2 \log 2$

$G(2) \approx 15.6$

The GNP at 2 years is \$15.6 million.

c. $\quad G(5) = 15 + 2 \log 5$

$G(5) \approx 16.4$

The GNP at 5 years is \$16.4 million.

56. $y = 2e^{.02t}$

  a. The population will triple, so $y = 3 \cdot 2$.
  The answer is B.

  b. The population will be 3 million, so
  $y = 3$.
  The answer is D.

  c. $t = 3$
  The answer is C.

  d. $t = \dfrac{4}{12} = \dfrac{1}{3}$
  The answer is A.

57. $A(t) = 10e^{-.00495t}$

  a. $A(0) = 10e^{-.00495(0)}$
  $= 10e^0$
  $= 10$
  The amount of polonium present initially was 10 g.

  b. We determine the half-life by finding a value
  of t such that $A(t) = \left(\dfrac{1}{2}\right)(10) = 5$.

  $5 = 10e^{-.00495(t)}$

  $\dfrac{1}{2} = e^{-.00495t}$

  $\ln \dfrac{1}{2} = \ln e^{-.00495t}$

  $\ln \dfrac{1}{2} = -.00495t$

  $t = \dfrac{\ln .5}{-.00495} \approx 140$

  The half-life of polonium is about
  140 days.

  c. Find t such that $A(t) = 3$.
  $10e^{-.00495(t)} = 3$

  $e^{-.00495(t)} = .3$

  $\ln e^{-.00495(t)} = \ln .3$

  $-.00495t = \ln .3$

  $t = \dfrac{\ln .3}{-.00495} \approx 243$

  It will take about 243 days for the polonium
  to decay to 3 g.

58. For earthquakes, increasing the ground motion by
a factor of 10^k increases the Richter magnitude
by k units.
In this problem, the second earthquake, with
ground motion $1000 = 10^3$ times greater than the
first earthquake, will measure
$4.6 + 3 = 7.6$ on the Richter scale.

59. $F(t) = T_0 + Ce^{-kt}$

  $T_0 = 50, F(0) = 50 + Ce^{-k(0)}, C = 250$

  $F(t) = 50 + 250e^{-kt}$

  $F(4) = 175 = 50 + 250e^{-k(4)}$

  $125 = 250e^{-k(4)}$

  $\dfrac{125}{250} = e^{-k(4)}$

  $\ln \dfrac{125}{250} = -k(4)$

  $k = .1733$

  $F(12) = 50 + 250e^{-.1733(12)}$

  $F(12) = 81.25$

  The temperature after 12 minutes is $81.25°$ C.

60. $F(t) = T_0 + Ce^{-kt}$

  $T_0 = 18$

  $F(0) = 3.4 = 18 + C$

  $C = -14.6$

  $F(30) = 18 - 14.6e^{-k(30)} = 7.2$

  $-14.6e^{-k(30)} = 7.2 - 18$

  $e^{-k(30)} = \dfrac{7.2 - 18}{-14.6}$

  $k = \dfrac{\ln 0.739726}{-30}$

  $k \approx 0.01$

  $F(t) = 18 - 14.6e^{-(0.01)t} = 10$

  $-14.6e^{-(0.01)t} = 10 - 18$

  $e^{-(0.01)t} = \dfrac{-8}{-14.6}$

  $t \approx \dfrac{\ln 0.5479}{-0.01}$

  $t \approx 60$

  It will thaw to $10°$C in 1 hour.

61. **a.** At $x = 0$, $f(0) = 272.6$ and at
$t = 2003 - 1995 = 8$, $f(8) = 17.2$. To find
$f(x) = a(b^x)$, first solve for a:

$$f(0) = 272.6$$

$$a(b^0) = 272.6$$

$$a = 272.6$$

Now solve for b:

$$f(8) = 17.2$$

$$272.6b^8 = 17.2$$

$$b^8 = \frac{17.2}{272.6}$$

$$b = \left(\frac{17.2}{272.6}\right)^{1/8} \approx .70795$$

The model becomes $f(x) = 272.6(.70795^x)$

b. Let $g(x)$ be the exponential function that
models this data. Using a graphing calculator,
$g(x) = 350.29(.7116^x)$

c. $x = 2004 - 1995 = 9$

$$f(9) = 272.6(.70795^9) \approx 12.177$$

$$g(9) = 350.29\left(.7116^9\right) \approx 16.389$$

About 12.177 or 16.389 million cassettes
were sold in 2004.

d. Since f and g are both decreasing functions,
we only need to find the x-value for which
$f(x) = 1.5$ (or $g(x) = 1.5$).

$$272.6(.70795^x) = 1.5$$

$$.70795^x = \frac{1.5}{272.6}$$

$$x \ln .70795 = \ln\left(\frac{1.5}{272.6}\right)$$

$$x = \frac{\ln(1.5/272.6)}{\ln .70795} \approx 15$$

Cassette sales will drop below 1.5 million in
$1995 + 15 = 2010$. A similar calculation
using g concludes again that sales should
drop below 1.5 million in 2011.

62. **a.** At $x = 0$, $f(0) = 1013$ and at $x = 10$,
$f(10) = 265$.
To find $f(x) = a(b^x)$, first solve for a:

$$f(0) = 1013$$

$$a(b^0) = 1013$$

$$a = 1013$$

Now solve for b:

$$f(10) = 265$$

$$1013b^{10} = 265$$

$$b^{10} = \frac{265}{1013}$$

$$b = \left(\frac{265}{1013}\right)^{1/10} \approx .8745$$

The model becomes $f(x) = 1013(.8745^x)$

b. Let $g(x)$ be the exponential function that
models this data. Using a graphing calculator,
$g(x) = 1035.52(.8747^x)$

c. $x = 1.5$

$$f(1.5) = 1013(.8745^{1.5}) \approx 828.4$$

$$g(1.5) = 1035.52(.8747^{1.5}) \approx 847.1$$

The models estimate the pressure at 1500m
to be 828.4 and 847.1 millibars respectively.
The models under and over estimate the
actual value of 846 millibars respectively.
$x = 11$

$$f(11) = 1013(.8745^{11}) \approx 231.7$$

$$g(11) = 1035.52(.8747^{11}) \approx 237.5$$

The models estimate the pressure at
11,000 m to be 231.7 and 237.5 millibars
respectively. Both models overestimate the
actual pressure of 227 millibars.

62. Continued

d. $f(x) = 500$

$1013(.8745^x) = 500$

$.8745^x = \dfrac{500}{1013}$

$x \ln .8745 = \ln \dfrac{500}{1013}$

$x = \dfrac{\ln \frac{500}{1013}}{\ln .8745} \approx 5.265$

$g(x) = 500$

$1035.52(.8747^x) = 500$

$.8747^x = \dfrac{500}{1035.52}$

$x \ln .8747 = \ln \dfrac{500}{1035.52}$

$x = \dfrac{\ln \frac{500}{1035.52}}{\ln .8747} \approx 5.438$

The pressure is 500 millibars at 5265 m under model f, and 5438 m under model g.

63. a. At $t = 1$, $f(1) = 27.8$ and at $t = 13$,

$f(13) = 70.9$. To find $f(x) = a + b \ln x$,

first solve for a:

$f(1) = 27.8$

$a + b \ln 1 = 27.8$

$a + 0 = 27.8$

$a = 27.8$

Now solve for b:

$f(13) = 70.9$

$27.8 + b \ln 13 = 70.9$

$b \ln 13 = 43.1$

$b = \dfrac{43.1}{\ln 13} \approx 16.80435$

The model becomes

$f(x) = 27.8 + 16.80345 \ln x$.

b. Let $g(x)$ be the logarithmic regression function that models the data. Using a graphing calculator,

$g(x) = 25.83 + 16.04 \ln x$.

c. $x = 2008 - 1992 + 1 = 17$

$f(17) = 27.8 + 16.80345 \ln 17 \approx 75.408$

$g(17) = 25.83 + 16.04 \ln 17 \approx 71.275$

There are expected to be about 75,408,000 (or 71,275,000) passengers in 2008.

63. Continued

d. Find the value of x for which $f(x) = 80$:

$f(x) = 80$

$27.8 + 16.80345 \ln x = 80$

$16.80345 \ln x = 52.2$

$\ln x = \dfrac{52.2}{16.80345}$

$x = e^{52.2/16.80345} \approx 22.3$

If this model remains accurate, the number of passengers will reach 80 million when $x = 1992 + 22 - 1$ or in 2013. A similar calculation using $g(x)$ concludes that 80 million passengers will be reached in 2020.

If this model remains accurate, the number of passengers will reach 80 million when x = 1992 + 29 − 1 or in 2020.

64. a.

x	1	16	19	23	25	30	33	34
y	2300	275,000	1,200,000	3,100,000	5,500,000	42,000,000	220,000,000	592,000,000

700,000,00

b. The number of transistors on computer chips is always positive and increases slowly until about 2003. In recent years the number of transistors on chips has increased sharply. An exponential function would fit this data best.

c. Let $f(x)$ be the exponential function that models this data. Using a graphing calculator,
$$f(x) = 1049.9(1.44^x)$$

d. 700,000,000

The function appears to fit all the data except for the last point.

e. $f(x) = 1,000,000,000$
$$1049.9(1.44^x) = 1,000,000,000$$
$$1.44^x = \frac{1,000,000,000}{1049.9}$$
$$x \ln 1.44 = \ln\left(\frac{1,000,000,000}{1049.9}\right)$$
$$x = \frac{\ln \frac{1,000,000,000}{1049.9}}{\ln 1.44}$$
$$x \approx 37.75$$

If the model remains accurate, the number of transistors will reach 1 billion in
$1971 + 37 - 1 = 2007$.

Case 4 Characteristics of the Monkeyface
          Prickleback

1. $L_t = L_x(1 - e^{-kt})$

    Let $L_x = 71.5$ and $k = .1$; then

    $L_t = 71.5(1 - e^{-.1t})$.

    When $t = 4$,

    $L_t = 71.5(1 - e^{-.1(4)})$.

    $= 71.5(1 - e^{-.4}) \approx 23.6$.

    When $t = 11$,

    $L_t = 71.5(1 - e^{-.1(11)})$.

    $= 71.5(1 - e^{-1.1}) \approx 47.7$.

    When $t = 17$,

    $L_t = 71.5(1 - e^{-.1(17)})$.

    $= 71.5(1 - e^{-1.7}) \approx 58.4$.

    Comparing these answers with the results in
    Figure 1, the estimates are a bit low.

2. $W = aL^b$

    Let $a = .01289$ and $b = 2.9$; then $W = .01289L^{2.9}$.

    When $L = 25$,

    $W = .01289(25)^{2.9} \approx 146.0$.

    When $L = 40$,

    $W = .01289(40)^{2.9} \approx 570.5$.

    When $L = 60$,

    $W = .01289(60)^{2.9} \approx 1848.8$.

    Yes, compared to the curve, these answers are
    reasonable estimates.

Chapter 5: Mathematics of Finance

Section 5.1 Simple Interest and Discount

Note: Exercises in this chapter have been completed with a calculator. To ensure as much accuracy as possible, rounded values have been avoided in intermediate steps. When rounded values have been necessary, several decimal places have been carried throughout the exercise; only the final answer has been rounded to 1 or 2 decimal places. In most cases, answers involving money have been rounded to the nearest cent. Students who use rounded intermediate values should expect their final answers to differ from the answers given here. Depending on the magnitude of the numbers used in the exercise, the difference could be a few pennies or several thousand dollars.

1. The factors are time and interest rate.

3. $2850 at 7% for 8 months

$$P = 2850, r = .07, \text{ and } t = \frac{8}{12}.$$

$$I = Prt$$

$$= 2850(.07)\left(\frac{8}{12}\right)$$

$$= 133.00$$

The simple interest is $133.00.

5. $3650 at 6.5% for 11 months

$$P = 3650, r = .065, t = \frac{11}{12}$$

$$I = Prt$$

$$= 3650(.065)\left(\frac{11}{12}\right)$$

$$= 217.48$$

The simple interest is $217.48.

7. $2830 at 8.9% for 125 days

$$P = 2830, r = .089, t = \frac{125}{365}$$

$$I = Prt$$

$$= 2830(.089)\left(\frac{125}{365}\right)$$

$$= 86.26$$

The simple interest is $86.26.

9. $5328 at 8%; loan made on August 16 is due December 30.
The duration of this loan is
$(31 - 16) + 30 + 31 + 30 + 30 = 136$ days.

$$P = 5328, r = .08, t = \frac{136}{365}$$

$$I = Prt$$

$$= 5328(.08)\left(\frac{136}{365}\right)$$

$$= 158.82$$

The simple interest is $158.82.

11. $12,000 at 9.5%; made on February 19 and due May 31
The duration of the loan is
$(28 - 19) + 31 + 30 + 31 = 101$ days.

$$P = 12,000, r = .095, t = \left(\frac{101}{365}\right)$$

$$I = Prt$$

$$= 12,000(.095)\left(\frac{101}{365}\right)$$

$$= 315.45$$

The simple interest is $315.45

13. $39,086 at 9.4%; made on September 12 and due July 30
The duration of the loan is 321 days.

$$P = 39,086, r = .094, t = \frac{321}{365}$$

$$I = Prt$$

$$= 39,086(.094)\left(\frac{321}{365}\right)$$

$$= 3231.18$$

The simple interest is $3231.18.

15. Answer varies.

17. $48,000 for 8 months; money earns 5%.

$$A = 48,000, t = \frac{8}{12}, \text{ and } r = .05.$$

$$P = \frac{A}{1 + rt}$$

$$= \frac{48,000}{1 + (.05)\left(\frac{8}{12}\right)}$$

$$= \frac{48,000}{1.0333}$$

$$= 46,451.61$$

The present value is $46,451.61.

19. $29,764 for 310 days; money earns 7.2%.

$A = 29,764, t = \dfrac{310}{365}, r = .072$

$P = \dfrac{A}{1+rt}$

$= \dfrac{29,764}{1+(.072)\left(\frac{310}{365}\right)}$

$= \dfrac{29,764}{1.0612}$

$= 28,048.80$

The present value is $28,048.80.

21. $9450; discount rate 10%; length of loan 7 months. Use the formula for proceeds of a discounted loan with

$A = 9450, r = .10$ and $t = \dfrac{7}{12}$

$P = A(1-rt)$

$= 9450\left[1-(.10)\left(\frac{7}{12}\right)\right]$

$= 8898.75$

The proceeds of the loan are $8898.75.

23. $50,900; discount rate 8.2%; length of loan 238 days

$A = 50,900, r = .082, t = \dfrac{238}{365}$

$P = A(1-rt)$

$= 50,900\left[1-(.082)\left(\frac{238}{365}\right)\right]$

$= 48,178.45$

The proceeds of the loan are $48,178.45.

25. $6200; discount rate 7%; length of loan 8 months

Discount $= 6200(.07)\left(\frac{8}{12}\right) \approx 289.33$

Proceeds $= 6200 - 289.33 = 5910.67$

$I = 289.33, P = 5910.67, t = \dfrac{8}{12}$

$I = Prt$

$289.33 = 5910.67(r)\left(\frac{8}{12}\right)$

$\dfrac{289.33}{5910.67\left(\frac{8}{12}\right)} = r$

$.0734 \approx r$

The actual interest rate is about 7.34%.

27. $58,000; discount rate 10.8%; length of loan 9 months

Discount $= 58,000(.108)\left(\frac{9}{12}\right) = 4698$

Proceeds $= 58,000 - 4698 = 53,302$

$I = 4698, P = 53,302, t = \dfrac{9}{12}$

$I = Prt$

$4698 = 53,302(r)\left(\frac{9}{12}\right)$

$\dfrac{4698}{53,302\left(\frac{9}{12}\right)} = r$

$.1175 \approx r$

The actual interest rate is about 11.75%

29. $P = 25,900, r = .084, t = \dfrac{11}{12}$

$A = P(1+rt)$

$= 25,900\left[1+(.084)\left(\frac{11}{12}\right)\right]$

$= 25,900(1.077)$

$= 27,894.30$

The amount Anne repaid was $27,894.30.

31. $P = 3000, r = .025, t = \dfrac{9}{12}$

$A = P(1+rt)$

$= 3000\left[1+(.025)\left(\frac{9}{12}\right)\right]$

$= 3056.25$

After 9 months, the amount will be $3056.25.

33. Interest $= 67,359.39 - 67,081.20 = 278.19$

$P = 67,081.20, t = \dfrac{1}{12}$

$I = Prt$

$r = \dfrac{I}{Pt}$

$= \dfrac{278.19}{(67,081.20)\left(\frac{1}{12}\right)}$

$= .050$

The interest rate was 5.0%.

35. Interest $= (24 - 22) + .50 = 2.50,$
$P = 22, t = 1$

$I = Prt$

$2.50 = 22(r)(1)$

$\dfrac{2.50}{22(1)} = r$

$.114 \approx r$

The simple interest rate is about 11.4%.

37. Want present value of $1769 in 4 months at 6.25% interest.

$$A = 1769, \; r = .0325, \; t = \frac{4}{12}$$

$$P = \frac{A}{1+rt}$$

$$= \frac{1769}{1+(.0325)\left(\dfrac{4}{12}\right)}$$

$$= 1750.04$$

The student should deposit $1750.04 today.

39. $A = 6000, \; r = .036, \; t = \dfrac{10}{12} = \dfrac{5}{6}$

$$P = \frac{A}{1+rt}$$

$$= \frac{6000}{1+.036\left(\frac{5}{6}\right)}$$

$$= 5825.24$$

Yee should deposit $5825.24.

41. $4200; discount rate 12.2%; length of loan 10 mo

$$\text{Discount} = 4200(.122)\left(\frac{10}{12}\right) = 427$$

$$\text{Proceeds} = 4200 - 427 = 3773$$

The net proceeds would be $3773, so

$$I = 427, \; P = 3773, \text{ and } t \approx \frac{10}{12} ..$$

$$I = Prt$$

$$427 = 3773(r)\left(\tfrac{10}{12}\right)$$

$$\frac{427}{3773\left(\frac{10}{12}\right)} = r$$

$$.1358 \approx r$$

The actual interest rate is about 13.58%.

43. The amount (with interest) that Shalia Johnson must pay the music shop is
$$A = P(1+rt)$$

$$= 7000\left[1+(.10)\left(\frac{7}{12}\right)\right]$$

$$A = \$7408.33.$$

The bank applies its discount rate to this total.

$$\text{Discount } = 7408.33(.105)\left(\frac{2}{12}\right)$$

$$= 129.65$$

The music shop will receive
$7408.33 − 129.65 = $7278.68
in cash from the bank. This amount will not be enough to pay the $7350 wholesaler's bill.

45. a. Unpaid charges = 457.80 − 87.50 = 370.30

$$r = .08, \; t = \frac{1}{12}$$

$$A = P(1+rt)$$

$$= 370.30\left[1+.08\left(\tfrac{1}{12}\right)\right]$$

$$= 372.77$$

b. Month 2:
372.77 − 87.50 = 285.27
Month 3:

$$285.27\left[1+.08\left(\frac{1}{12}\right)\right] - 87.50 = 199.67$$

Month 4:

$$199.67\left[1+.08\left(\frac{1}{12}\right)\right] - 87.50 = 113.50$$

Month 5:

$$113.50\left[1+.08\left(\frac{1}{12}\right)\right] - 87.50 = 26.76$$

Michael will owe

$$\$26.76\left[1+.08\left(\frac{1}{12}\right)\right] = \$26.94.$$

c. Month 2:
372.77 − 72.50 = 300.27
Month 3:

$$300.27\left[1+.08\left(\frac{1}{12}\right)\right] - 57.50 \approx 244.77$$

Month 4:

$$244.77\left[1+.08\left(\frac{1}{12}\right)\right] - 42.50 \approx 203.90$$

Month 5:

$$203.90\left[1+.08\left(\frac{1}{12}\right)\right] - 27.50 \approx 177.76$$

Month 6:

$$177.76\left[1+.08\left(\frac{1}{12}\right)\right] - 12.50 \approx 166.45$$

Michael cannot pay off the charges in 6 months. He will still owe $166.45.

Section 5.2 Compound Interest

1. r is the interest rate per year, while i is the interest rate per compounding period. t is the number of years, while n is the number of compounding periods.

3. The interest rate and number of compounding periods determine the amount of interest earned on a fixed principal.

5. Answer varies.

7. $1000 at 4% compounded annually for 8 yr

$P = 1000, i = \dfrac{4\%}{1} = .04$, and $n = 8(1) = 8$

$A = P(1+i)^n$

$\quad = 1000(1.04)^8$

$\quad \approx 1000(1.3686)$

$\quad = 1368.57$

The compound amount is $1368.57.

9. $470 at 8% compounded semiannually for 12 yr

$P = 470, i = \dfrac{8\%}{2} = .04, n = 12(2) = 24$

$A = P(1+i)^n$

$\quad = 470(1.04)^{24}$

$\quad \approx 470(2.5633)$

$\quad = 1204.75$

The compound amount is $1204.75.

11. $6500 at 5.5% compounded quarterly for 6 yr

$P = 6500, i = \dfrac{5.5\%}{4} = .01375, n = 6(4) = 24$

$A = P(1+i)^n$

$\quad = 6500(1.01375)^{24}$

$\quad \approx 6500(1.38784)$

$\quad \approx 9020.99$

The compound amount is $9020.99.

13. $26,000 at 6% compounded annually for 5 yr

$P = 26,000, i = .06, n = 5$

Find the compound amount and then amount of interest.

$A = P(1+i)^n$

$\quad = 26,000(1.06)^5$

$\quad \approx 34,793.87$

The compound amount is $34,793.87. The amount of interest earned is $34,793.87 − $26,000 = $8793.87.

15. $8000 at 4% compounded semiannually for 6.4 yr

$P = 8000, i = \dfrac{.04}{2} = .02, n = 6.4(2) = 12.8$

$A = P(1+i)^n$

$\quad = 8000(1.02)^{12.8}$

$\quad \approx 10,307.95$

The amount of interest earned is $10,307.95 − $8000 = $2307.95.

17. $5124.98 at 6.3% compounded quarterly for 5.2 yr

$P = 5124.98, i = \dfrac{.063}{4} = .01575,$

$n = 5.2(4) = 20.8$

$A = P(1+i)^n$

$\quad = 5124.98(1.01575)^{20.8}$

$\quad \approx 7093.46$

The amount of interest earned is $7093.46 − $5124.98 = $1968.48.

19. $P = 3000, A = 3907, n = 6$. Solve for i.

$P(1+i)^n = A$

$3000(1+i)^6 = 3907$

$(1+i)^6 = \dfrac{3907}{3000} = 1.3023$

$\sqrt[6]{(1+i)^6} = \sqrt[6]{1.3023}$

$1+i = \sqrt[6]{1.3023}$

$i = \sqrt[6]{1.3023} - 1$

$\quad \approx 0.0450$

The interest rate is about 4.50%

21. $P = 8500, A = 12{,}161, n = 7$. Solve for i.

$P(1+i)^n = A$

$8500(1+i)^7 = 12{,}161$

$(1+i)^7 = \dfrac{12{,}161}{8500} = 1.4307$

$\sqrt[7]{(1+i)^7} = \sqrt[7]{1.4307}$

$1+i = \sqrt[7]{1.4307}$

$i = \sqrt[7]{1.4307} - 1$

$\quad \approx 0.0525$

The interest rate is about 5.25%

23. $P = 25{,}000, r = .06, t = 2$
Substitute these values into the formula for continuous compound interest.

$A = Pe^{rt}$

$\quad = 25{,}000e^{(.06)2}$

$\quad = 25{,}000e^{.12}$

$\quad \approx 28{,}187.42$

The compound amount is \$28,187.42.

25. $P = 25{,}000, r = .06, t = 10$

$A = Pe^{rt}$

$\quad = 25{,}000e^{(.06)10}$

$\quad = 25{,}000e^{.6}$

$\quad \approx 45{,}552.97$

The compound amount is \$45,552.97.

27. $P = 25{,}000, r = .06, t = 20$

$A = Pe^{rt}$

$\quad = 25{,}000e^{(.06)20}$

$\quad = 25{,}000e^{1.2}$

$\quad \approx 83{,}002.92$

The compound amount is \$83,002.92.

29. 4% compounded semiannually
Use the formula for effective rate with $r = .04$ and $m = 2$.

$r_E = \left(1 + \dfrac{r}{m}\right)^m - 1$

$\quad = \left(1 + \dfrac{.04}{2}\right)^2 - 1$

$\quad = (1.02)^2 - 1$

$\quad = 1.0404 - 1$

$\quad = .0404$

The effective rate is 4.04%.

31. 6% compounded semiannually
$r = .06, m = 2$

$r_E = \left(1 + \dfrac{r}{m}\right)^m - 1$

$\quad = \left(1 + \dfrac{.06}{2}\right)^2 - 1$

$\quad = .06090$

The effective rate is 6.09%.

33. 5.2% compounded semiannually
$r = .052, m = 2$

$r_E = \left(1 + \dfrac{r}{m}\right)^m - 1$

$\quad = \left(1 + \dfrac{.052}{2}\right)^2 - 1$

$\quad = .05268$

The effective rate is 5.268%.

35. \$12,000 at 5% compounded annually for 6 yr
Use the present value formula for compound interest with
$A = 12{,}000, i = .05$, and $n = 6$.

$P = A(1 + i)^{-n}$

$\quad = 12{,}000(1.05)^{-6}$

$\quad \approx 8954.58$

The present value is \$8954.58.

37. \$4253.91 at 5.8% compounded semiannually for 4 yr

$A = 4253.91, i = \dfrac{.058}{2} = .029, n = 4(2) = 8.$

$P = A(1 + i)^{-n}$

$\quad = 4253.91(1.029)^{-8}$

$\quad \approx 3384.27$

The present value is \$3384.27.

39. \$17,230 at 4% compounded quarterly for 10 yr

$A = 17{,}230, i = \dfrac{.04}{4} = .01, n = 10(4) = 40$

$P = A(1 + i)^{-n}$

$\quad = 17{,}230(1.01)^{-40}$

$\quad \approx 11{,}572.58$

The present value is \$11,572,58.

41. $A = 1210, i = \dfrac{.08}{4} = .02, n = 5(4) = 20$

$P = A(1 + i)^{-n}$

$\quad = 1210(1.02)^{-20}$

$\quad \approx 814.30$

Since this amount is less than \$1000, "\$1000 now" is greater.

43. $P = 50{,}000, i = \dfrac{.12}{12} = .01, n = 4(12) = 48$

Find the compound amount and then the amount of interest.

$A = P(1+i)^n$

$\quad = 50{,}000(1.01)^{48}$

$\quad \approx \$80{,}611.30$

The business will pay interest in the amount of $\$80{,}611.30 - \$50{,}000 = \$30{,}611.30$.

45. $A = 2.9, i = \dfrac{.08}{12} \approx .0067, n = 5(12) = 60$

Use the formula for present value with compound interest.

$P = A(1+i)^{-n}$

$\quad \approx 2.9(1.0067)^{-60}$

$\quad \approx 1.946$

The company should invest about $\$1.946$ million now.

47. For Flagstar Bank, $r = 4.38\%, m = 4$.

$r_E = \left(1 + \dfrac{r}{m}\right)^m - 1$

$\quad = \left(1 + \dfrac{.0438}{4}\right)^4 - 1$

$\quad \approx .0445$

The effective rate for Flagstar Bank is about 4.45%.

For Principal Bank, $r = 4.37\%, m = 12$.

$r_E = \left(1 + \dfrac{r}{m}\right)^m - 1$

$\quad = \left(1 + \dfrac{.0437}{12}\right)^{12} - 1$

$\quad \approx .04459$

The effective rate for Principal Bank is about 4.46%.

Principal Bank pays a higher APY.

49.

$$r_E = \left(1 + \dfrac{r}{m}\right)^m - 1$$

$$r_E + 1 = \left(1 + \dfrac{r}{12}\right)^{12}$$

$$\sqrt[12]{r_E + 1} = 1 + \dfrac{r}{12}$$

$$-1 + \sqrt[12]{r_E + 1} = \dfrac{r}{12}$$

$$12\left(-1 + \sqrt[12]{r_E + 1}\right) = r$$

Term:	6 mo	1 yr	2 yr	3 yr	5 yr
APY(%):	3.25	3.75	4.00	4.25	4.75
Nominal Rates (%)	3.20	3.69	3.93	4.17	4.65

51. $P_1 = 5200, r_1 = .07, t_1 = \dfrac{10}{12} = \dfrac{5}{6}$

$A_1 = P_1(1 + r_1 t_1)$

$\quad = 5200\left[1 + .07\left(\tfrac{5}{6}\right)\right]$

$\quad \approx 5503.33$

$P_2 = 5503.33, i_2 = \dfrac{.063}{4} = .01575,$

$n_2 = 4(5) = 20$

$A_2 = P_2(1 + i_2)^{n_2}$

$\quad = 5503.33(1 + .01575)^{20}$

$\quad \approx 7522.50$

Vetere will have $7522.50 after 5 years.

53. a. $P = 16{,}000 - 30 = 15{,}970, i = \dfrac{.055}{12}, n = 12$

Use the formula for compound amount.

$$A = P(1+i)^n = 15{,}970\left(1 + \dfrac{.055}{12}\right)^{12}$$

After the annual charges,

$A - .0125A = A(1 - .0125)$

$\quad = A(.9875)$

$\quad \approx \$16{,}659.95$

53. Continued

    b. Note that the amount in part (a) can be written in terms of P.

$$A = P\left(1+\frac{.055}{12}\right)^{12}(.9875) \approx \$16,659.95$$

Therefore, the amount in the account after 7 years can be found by repeatedly multiplying P by the factors

$\left(1+\frac{.055}{12}\right)^{12}(.9875)$. So, by commutativity

of multiplication this amount is

$$A = P\left(1+\frac{.055}{12}\right)^{12(7)}(.9875)^7$$
$$\approx \$21,472.67.$$

55. $P = 1000$, $i = .06$, $n = 5$

$A = P(1+i)^n$

$\quad = 1000(1.06)^5$

$\quad \approx 1338.23$

Since this amount is greater than $1210, "$1000 now" is larger.

57. To find the number of years it will take $1 to inflate to $2, use the formula for compound amount with $A = 2$, $P = 1$, and $i = .04$.

$A = P(1+i)^n$

$2 = 1(1.04)^n$

$2 = (1.04)^n$

Take the log of both side and solve for n. Prices will double in about 17.7 yr.

59. To find the number of years it will take for a demand of 1 unit of electricity to increase to a demand of 2 units, use the formula for compound amount with $A = 2$, $P = 1$, and $i = .06$.

$A = P(1+i)^n$

$2 = 1(1.06)^n$

$2 = (1.06)^n$

Take the log of both side and solve for n. The electric utilities will need to double their generating capacity in about 11.9 yr.

61. $A = P(1+i)^n$

$$420,000,000 + 100 = 100\left(1+\frac{r}{1}\right)^{160}$$

$$\frac{420,000,100}{100} = (1+r)^{160}$$

$$\sqrt[160]{\frac{420,000,100}{100}} - 1 = r$$

$$0.1000 = r$$

The interest rate was 10%.

63. $P = 10,000$, $r = 0.05$, $t = 10$

    a. $A = P(1+i)^n$

$$= 10,000\left(1+\frac{.05}{1}\right)^{10(1)}$$

$$\approx 16,288.95$$

The future value is $16,288.95.

    b. $A = 10,000\left(1+\frac{.05}{4}\right)^{10(4)} \approx 16,436.19$

The future value is $16,436.19.

    c. $A = 10,000\left(1+\frac{.05}{12}\right)^{10(12)} \approx 16,470.09$

The future value is $16,470.09.

    d. $A = 10,000\left(1+\frac{.05}{365}\right)^{10(365)} = 16,486.65$

The future value is $16,486.65.

65. First consider the case of earning interest at a rate of k per annum compounded quarterly for all eight years and earning $2203.76 interest on the $1000 investment.

$$2203.76 = 1000\left(1+\frac{k}{4}\right)^{8(4)}$$

$$2.20376 = \left(1+\frac{k}{4}\right)^{32}$$

Use a calculator to raise both sides to the power of $\frac{1}{32}$.

$$1.025 = 1+\frac{k}{4}$$

$$.025 = \frac{k}{4}$$

$$.1 = k$$

Next consider the actual investments. The $1000 was invested for the first five years at a rate of j per annum compounded semiannually.

$$A = 1000\left(1+\frac{j}{2}\right)^{5(2)}$$

$$A = 1000\left(1+\frac{j}{2}\right)^{10}$$

This amount was then invested for the remaining three years at $k = .1$ per annum compounded quarterly for a final compound amount of $1990.76.

$$1990.76 = A\left(1+\frac{.1}{4}\right)^{3(4)}$$

$$1990.76 = A(1.025)^{12}$$

$$1480.24 \approx A$$

Recall that $A = 1000\left(1+\frac{j}{2}\right)^{10}$ and substitute this value into the above equation.

$$1480.24 = 1000\left(1+\frac{j}{2}\right)^{10}$$

$$1.48024 = \left(1+\frac{j}{2}\right)^{10}$$

Use a calculator to raise both sides to the power of $\frac{1}{10}$.

$$1.04 \approx 1+\frac{j}{2}$$

$$.04 = \frac{j}{2}$$

$$.08 = j$$

65. Continued

The ratio of k to j is

$$\frac{k}{j} = \frac{.1}{.08} = 1.25,$$

which is choice (a).

Section 5.3 Future Value of an Annuity and Sinking Funds

1. Answer varies.

3. a. $a_1 = 1276(.916)^1 \approx 1169$, and $r = .916$.

   b. $a_{10} = 1276(.916)^{10} \approx 531$ This means that a person who is 10 yr from retirement should have savings of 531% of his or her annual salary.

$$a_{20} = 1276(.916)^{20} \approx 221$$

This means that a person who is 20 yr from retirement should have savings of 221% of his or her annual salary.

5. $a = 4, r = 3$
The fourth term is
$$ar^{n-1} = 4(3)^{4-1}$$
$$= 4(3)^3$$
$$= 4(27)$$
$$= 108.$$

7. $a = 24, r = .5$
The fourth term is
$$ar^3 = 24(.5)^3$$
$$= 24(.125)$$
$$= 3$$

9. $a = 2000, r = 1.05$
The fourth term is
$$ar^3 = 2000(1.05)^3$$
$$= 2000(1.157625)$$
$$= 2315.25.$$

11. $a = 3, r = 2$
The sum of the first four terms of this geometric sequence is found as follows.
$$S_4 = \frac{3(2^4 - 1)}{2-1}$$
$$= \frac{3(15)}{1}$$
$$= 45$$

13. $a = 5, r = .2$

$$S_4 = \frac{a(r^4 - 1)}{r - 1}$$

$$= \frac{5[(.2)^4 - 1]}{.2 - 1}$$

$$= \frac{5(-.9984)}{-.8}$$

$$= 6.24$$

15. $a = 128, r = 1.1$

$$S_4 = \frac{a(r^4 - 1)}{r - 1}$$

$$= \frac{128[(1.1)^4 - 1]}{1.1 - 1}$$

$$= \frac{128(.4641)}{.1}$$

$$= 594.048$$

17. Here, $n = 12$, and $i = .05$.

$$S_{\overline{12}|.05} = \frac{(1 + .05)^{12} - 1}{.05}$$

$$\approx 15.91713$$

19. Here, $n = 16$ and $i = .04$.

$$s_{\overline{16}|.04} = \frac{(1 + .04)^{16} - 1}{.04}$$

$$\approx 21.82453$$

21. Here, $n = 40$ and $i = .01$.

$$s_{\overline{40}|.01} = \frac{(1 + .01)^{40} - 1}{.01}$$

$$\approx 48.88637$$

In Exercises 23–28, use the formula

$$S = R \cdot s_{\overline{n}|i} \text{ or } S = R\left[\frac{(1+i)^n - 1}{i}\right].$$

23. $R = 12,000, i = .062, n = 8$

$$S = 12,000\left[\frac{(1.062)^8 - 1}{.062}\right]$$

$$\approx 119,625.61$$

The future value is $119,625.61.

25. $R = 865, i = \frac{.06}{2} = .03, n = 10(2) = 20$

$$S = 865\left[\frac{(1.03)^{20} - 1}{.03}\right]$$

$$\approx 23,242.87$$

The future value is $23,242.87.

27. $R = 1200, i = \frac{.08}{4} = .02, n = 10(4) = 40$

$$S = 1200\left[\frac{(1.02)^{40} - 1}{.02}\right]$$

$$\approx 72,482.38$$

The future value is $72,482.38.

In Exercises 29–33, use the formula

$$S = R \cdot s_{\overline{n}|i} \text{ or } S = R\left[\frac{(1+i)^n - 1}{i}\right].$$

29. $S = 14,500, i = \frac{.04}{2} = .02, n = 8(2) = 16$

$$14,500 = R\left[\frac{(1.02)^{16} - 1}{.02}\right]$$

$$14,500 \approx R(18.639285)$$

$$R \approx 777.93$$

The periodic payment for this sinking fund is $777.93.

31. $S = 62,000, i = \frac{.06}{4} = .015, n = 6(4) = 24$

$$62,000 = R\left[\frac{(1.015)^{24} - 1}{.015}\right]$$

$$62,000 \approx R(28.633521)$$

$$R \approx 2165.29$$

The periodic payment for this sinking fund is $2165.29.

33. Answer varies.

35. $S = 65,000, i = \frac{.06}{2} = .03, n = \left(4\frac{1}{2}\right)(2) = 9$

$$65,000 = R\left[\frac{(1.03)^9 - 1}{.03}\right]$$

$$65,000 \approx R(10.15911)$$

$$R \approx 6398.20$$

The periodic payment is $6398.20.

37. $S = 25,000, i = \frac{.09}{4} = .0225, n = \left(3\frac{1}{2}\right)(4) = 14$

$$25,000 = R\left[\frac{(1.0225)^{14} - 1}{.0225}\right]$$

$$25,000 \approx R(16.24371)$$

$$R \approx 1539.06$$

The periodic payment is $1539.06.

39. $S = 9000$, $i = \dfrac{.07}{12} = .005833$, $n = \left(2\dfrac{1}{2}\right)(12) = 30$

$$9000 = R\left[\frac{(1.005833)^{30} - 1}{.005833}\right]$$

$9000 \approx R(32.6811)$

≈ 275.39

The periodic payment is $275.39.

In Exercises 41–48, use the formula

$$S = R \cdot s_{\overline{n+1}|i} - R \text{ or } S = R\left[\frac{(1+i)^{n+1} - 1}{i}\right] - R.$$

41. $R = 500$, $i = .05$, $n = 10$

$$S = 500\left[\frac{(1.05)^{11} - 1}{.05}\right] - 500$$

≈ 6603.39

The future value is $6603.39.

43. $R = 16,000$, $i = .047$, $n = 11$

$$S = 16,000\left[\frac{(1.047)^{12} - 1}{.047}\right] - 16,000$$

$\approx 234,295.32$

The future value is $234,295.32.

45. $R = 1000$, $i = \dfrac{.08}{2} = .04$, $n = 9(2) = 18$

$$S = 1000\left[\frac{(1.04)^{19} - 1}{.04}\right] - 1000$$

$\approx 26,671.23$

The future value is $26,671.23.

47. $R = 100$, $i = \dfrac{.09}{4} = .0225$, $n = 7(4) = 28$

$$S = 100\left[\frac{(1.0225)^{29} - 1}{.0225}\right] - 100$$

≈ 3928.88

The future value is $3928.88.

49. $R = 100$ $i = \dfrac{.048}{12}$, $n = 40(12) = 480$

Use the formula for future value of an ordinary annuity.

$S = R \cdot s_{\overline{n}|i}$

$$= 100\left[\frac{\left(1 + \frac{.048}{12}\right)^{480} - 1}{\frac{.048}{12}}\right]$$

$\approx 100(1448.7214)$

$\approx 144,872.14$

After 40 yr, the amount in the account will be $144,872.14

51. $R = 80$, $i = \dfrac{.075}{12}$, $n + 1 = 46$

Use the formula for future value of an annuity due.

$S = R \cdot s_{\overline{n+1}|i} - R$

$$= 80\left[\frac{\left(1 + \frac{.075}{12}\right)^{46} - 1}{\frac{.075}{12}}\right] - 80$$

≈ 4168.30

After 3 yr 9 mo, the amount in the account will be $4168.30.

53. $R = 12,000$, $i = \dfrac{.06}{1}$, $n = 9$

a. Use the formula for future value of an ordinary annuity.

$S = R \cdot s_{\overline{n}|i} - R$

$S = R \cdot s_{\overline{n}|i}$

$$= 12,000\left[\frac{(1 + .06)^9 - 1}{.06}\right]$$

$\approx 137,895.79$

She will have $137,895.79 on deposit after 9 yr.

b. $i = \dfrac{.05}{1}$

$$S = 12,000\left[\frac{(1 + .05)^9 - 1}{.05}\right] \approx 132,318.77$$

She will have $132,318.77 on deposit after 9 yr.

c. $137,895.79 - 132,318.77 = 5577.02$

She would lose $5577.02.

55. From ages 50 to 60, we have an ordinary annuity with $R = 1200$, $i = \dfrac{.07}{4} = .0175$, $n = 10(4) = 40$.

Use the formula for the future value of an ordinary annuity.

$$S = \left[\frac{(1+i)^n - 1}{i}\right]$$

$$= 1200\left[\frac{(1.0175)^{40} - 1}{.0175}\right]$$

$$= 68,680.96$$

At age 60, the value of the retirement account is $68,680.96. This amount now earns 9% interest compounded monthly for 5 yr. Use the formula for compound amount with

$P = 68,680.96$, $i = \dfrac{.09}{12} = .0075$, and

$n = 5(12) = 60$ to find the value of this amount after 5 yr.

$$A = P(1+i)^n$$

$$= 68,680.96(1.0075)^{60}$$

$$= 107,532.48$$

The value of the amount she withdraws from the retirement account will be $107,532.48 when she reaches 65. The deposits of $300 at the end of each month into the mutual fund form another ordinary annuity. Use the formula for the future value of an ordinary annuity with

$R = 300$, $i = \dfrac{.09}{12} = .0075$, and $n = 5(12) = 60$.

$$S = R\left[\frac{(1+i)^n - 1}{i}\right]$$

$$= 300\left[\frac{(1.0075)^{60} - 1}{.0075}\right]$$

$$= 22,627.24$$

The value of this annuity after 5 yr is $22,627.24. The total amount in the mutual fund account when the woman reaches age 65 will be $107,532.48 + $22,627.24 = $130,159.72.

57. For the first 12 yr, we have an annuity due. To find the amount in this account after 12 yr, use the formula for the future value of an annuity due with $R = 10,000$, $i = .05$, and $n = 12$.

$$S = R\left[\frac{(1+i)^{n+1} - 1}{i}\right] - R$$

$$= 10,000\left[\frac{(1.05)^{13} - 1}{.05}\right] - 10,000$$

$$\approx 167,129.83$$

This amount, $167,129.83, now earns 6% interest compounded semiannually for another 9 yr, but no new deposits are made. Use the formula for compound amount with

$P = 167,129.83$, $i = \dfrac{.06}{2} = .03$, and

$n = 9(2) = 18$.

$$A = P(1+i)^n$$

$$= 167,129.83(1.03)^{18}$$

$$= 284,527.35.$$

The final amount on deposit after 21 yr is $284,527.35.

59. $S = 10,000$, $n = 8(4) = 32$

a. $i = \dfrac{.05}{4} = .0125$

$$S = R \cdot s_{\overline{n}|i}$$

$$10,000 = R\left[\frac{(1.0125)^{32} - 1}{.0125}\right]$$

$$10,000 \approx R(39.05044)$$

$$R = \$256.08$$

He should deposit $256.08 at the end of each quarter.

b. $i = \dfrac{.058}{4} = .0145$

$$S = R \cdot s_{\overline{n}|i}$$

$$10,000 = R\left[\frac{(1.0145)^{32} - 1}{.0145}\right]$$

$$10,000 \approx R(40.35398)$$

$$R = \$247.81$$

He should deposit $247.81 quarterly.

61. $S = 18{,}000$, $i = \dfrac{.061}{4} = .01525$, $n = 6(4) = 24$

$$S = R \cdot s_{\overline{n}|i}$$

$$18{,}000 = R\left[\frac{(1.01525)^{24} - 1}{.01525}\right]$$

$18{,}000 \approx R(28.71981)$

$R = \$626.75$

She should deposit $626.75 at the end of each quarter.

63.
$$S = R\left[\frac{(1+i)^n - 1}{i}\right]$$

$$330{,}000 = 250\left[\frac{(1+\frac{x}{12})^{12(30)} - 1}{\frac{x}{12}}\right]$$

Graph each side of this equation. The intersection is (.07397, 330,000). Alternatively, use the built-in financial solver of a TI-83. The screen is shown below.

```
N=360.000
I%=7.397
PV=0.000
PMT=-250.000
FV=330000.000
P/Y=12.000
C/Y=12.000
PMT:ERC BEGIN
```

Exercise 65 requires the use of a grapher.

65. a. $P = 60{,}000$, $r = .08$, $t = \dfrac{1}{4}$

$I = \mathrm{Pr}\,t$

$\quad = 60{,}000(.08)\left(\dfrac{1}{4}\right)$

$I = \$1200$

Each quarterly interest payment is $1200.

b. $S = 60{,}000$, $i = \dfrac{.06}{2} = .03$,

$n = 7(2) = 14$

$$S = R \cdot s_{\overline{n}|i}$$

$$60{,}000 = R\left[\frac{(1.03)^{14} - 1}{.03}\right]$$

$60{,}000 \approx R(17.08632)$

$R = \$3511.58$

The amount of each payment is $3511.58.

c. Use the amount of each deposit as calculated in part (b). The interest earned is calculated as follows:

$\left(\begin{array}{c}\text{previous}\\ \text{total}\end{array}\right)(.03)$.

Each total is calculated as follows:

$\left(\begin{array}{c}\text{previous}\\ \text{total}\end{array}\right) + \left(\begin{array}{c}\text{amount of}\\ \text{deposit}\end{array}\right) + \left(\begin{array}{c}\text{interest}\\ \text{earned}\end{array}\right)$.

See the table below. Notice that the last payment in a sinking fund table may differ slightly from the others because of earlier rounding.

Payment Number	Amount of Deposit	Interest Earned	Total
1	$3511.58	$0.00	$3511.58
2	3511.58	105.35	7128.51
3	3511.58	213.86	10,853.9
4	3511.58	325.62	14,691.1
5	3511.58	440.73	18,643.4
6	3511.58	559.30	22,714.3

Section 5.4 Present Value of an Annuity and Amortization

1. $a_{\overline{n}|i} = \dfrac{1-(1+i)^{-n}}{i}$

This corresponds to choice (c).

In Exercises 3–8, use the formula

$$a_{\overline{n}|i} = \dfrac{1-(1+i)^{-n}}{i}.$$

3. $n = 15, i = .06$

$$a_{\overline{15}|.06} = \dfrac{1-(1.05)^{-15}}{.05}$$
$$\approx 10.37966$$

5. $n = 18, i = .04$

$$a_{\overline{18}|.04} = \dfrac{1-(1.04)^{-18}}{.04}$$
$$\approx 12.65930$$

7. $n = 16, i = .015$

$$a_{\overline{16}|.015} = \dfrac{1-(1.015)^{-16}}{.015}$$
$$\approx 14.13126$$

9. Answer varies.

11. $R = 1400, i = .06, n = 8$

$$P = 1400\left[\dfrac{1-(1.06)^{-8}}{.06}\right] \approx 8693.71$$

The present value is $8693.71.

13. $R = 50,000, i = \dfrac{.05}{4} = .0125, n = 10(4) = 40$

$$P = 50,000\left[\dfrac{1-(1.0125)^{-40}}{.0125}\right] \approx 1,566,346.66$$

The present value is $1,566,346.66.

15. $R = 18,579, i = \dfrac{.074}{2} = .037, n = 8(2) = 16$

$$P = 18,579\left[\dfrac{1-(1.037)^{-16}}{.037}\right] \approx 221,358.80$$

The present value is $221,358.80.

In Exercises 16–18, use the formula

$$P = R \cdot a_{\overline{n}|i} \text{ or } P = R\left[\dfrac{1-(1+i)^{-n}}{i}\right].$$

17. $R = 10,000, i = .04, n = 15$

The lump sum is the same as the present value of the annuity.

$$P = 10,000\left[\dfrac{1-(1.04)^{-15}}{.04}\right]$$
$$\approx 111,183.87$$

The required lump sum is $111,183.87.

19. Answer varies.

21. $P = 41,000, i = \dfrac{.09}{2} = .045, n = 10$

$$R = 41,000\left[\dfrac{.045}{1-(1.045)^{-10}}\right] \approx 5181.53$$

Semiannual payments of $5181.53 are required to amortize this loan.

23. $P = 140,000, i = \dfrac{.12}{4} = .03, n = 15$

$$R = 140,000\left[\dfrac{.03}{1-(1.03)^{-15}}\right] \approx 11,727.32$$

Quarterly payments of $11,727.32 are required to amortize this loan.

25. $P = 5500, i = \dfrac{.095}{12}, n = 24$

$$R = 5500\left[\dfrac{\frac{.095}{12}}{1-\left(1+\frac{.095}{12}\right)^{-24}}\right] \approx 252.53$$

Monthly payments of $252.53 are required to amortize this loan.

27. $P = 170,892, i = \dfrac{.0711}{12}, n = 30(12)$

$$R = 170,892\left[\dfrac{\frac{.0711}{12}}{1-\left(1+\frac{.0711}{12}\right)^{-30(12)}}\right] \approx 1149.60$$

Monthly payments of $1149.60 are required to amortize this loan.

29. $P = 96,511, i = \dfrac{.0857}{12}, n = 25(12)$

$$R = 96,511\left[\dfrac{\frac{.0857}{12}}{1-\left(1+\frac{.0857}{12}\right)^{-25(12)}}\right] \approx 781.69$$

Monthly payments of $781.69 are required to amortize this loan.

31. Locate the entry of the table that is in the row labeled "Payment Number 10" and in the column labeled "Portion to Principal," to observe that $86.24 of the tenth payment is used to reduce the debt.

33. To find out how much interest is paid in the last 5 months of the loan, add the last five entries in the column labeled "Interest for Period."

$4.31
3.47
2.61
1.75
+ .88

$13.02

The amount of interest paid in the last 5 months of the loan is $13.02.

35. Find the future value of an annuity with

$$R = 4000, i = \frac{.06}{2} = .03, \text{ and } n = 10(2) = 20.$$

$$S = R\left[\frac{(1+i)^n - 1}{i}\right]$$

$$= 4000\left[\frac{(1.03)^{20} - 1}{.03}\right]$$

$$\approx 107,481.50$$

Now find the present value of $107,481.50 at 8% compounded quarterly for 10 years.

$$A = 107,481.50, i = \frac{.08}{4} = .02, n = 10(4) = 40$$

$$A = P(1+i)^n$$

$$107,481.50 = P(1.02)^{40}$$

$$P = \frac{107,481.50}{(1.02)^{40}} \approx 48,677.34$$

The required lump sum is $48,677.34.

37. $P = 6000, i = \dfrac{.12}{12} = .01, n = 4(12) = 48$

a. $R = \dfrac{P}{a_{\overline{n}|i}}$

$$= \frac{6000}{\left[\frac{1-(1.01)^{-48}}{.01}\right]}$$

$$\approx \frac{6000}{37.97396}$$

$$\approx 158.00$$

The amount of each monthly payment is $158.00.

b. Kushida will pay a total of
48($158) = $7584.
The price of the car is $6000, so the total amount of interest he will pay is
$7584 − $6000 = $1584.

39. $R = 1,133,334, \quad i = \dfrac{.04215}{1} = .04215$

Present Value of an annuity of $1,133,334 is

$$P = R\left[\frac{1-(1+i)^{-n}}{i}\right]$$

$$= 1,133,334\left[\frac{1-(1+.04215)^{-29}}{.04215}\right]$$

$$\approx 18,767,598.64$$

Cash value = $18,767,598.64 + $1,133,334

$$= \$19,900,932.64$$

41. a. $P = 54,000,000$ and so each payment would be $\dfrac{\$54,000,000}{30} = \$1,800,000$

b. $R = 1,800,000 \quad i = \dfrac{.048}{1} = .048$

Present Value of an annuity of $1,800,000 is

$$P = R\left[\frac{1-(1+i)^{-n}}{i}\right]$$

$$= 1,800,000\left[\frac{1-(1+.048)^{-29}}{.048}\right]$$

$$\approx 27,871,601.18$$

Cash value = $27,871,601.18 + $1,800,000

$$= \$29,671,601.18$$

43. $R = \dfrac{35,000}{\left[\dfrac{1-(1+\frac{.0743}{12})^{-120}}{\frac{.0743}{12}}\right]}$

$R \approx 414.18$

Total payments = $10(12)(414.18) = 49,701.60$

Interest = $49,701.60 - 35,000 = 14,701.60$

The monthly payments are \$414.18 and the total interest paid is \$14,701.60.

45. a. $P = 14,000 + 7200 - 1200 = 20,000$

$i = \dfrac{.12}{2} = .06, n = 5(2) = 10$

$R = \dfrac{20,000}{\left[\dfrac{1-(1.06)^{-10}}{.06}\right]}$

$R \approx 2717.36$

Each payment is \$2717.36.

b. There were 2 payments left at the time she decided to pay off the loan.

47. a. $P = 212,000 - .2(212,000) = 169,600$

$i = \dfrac{.072}{12} = .006, n = 30(12) = 360$

$R = \dfrac{169,600}{\left[\dfrac{1-(1.006)^{-360}}{.006}\right]}$

$R \approx 1151.22$

Their monthly payments are \$1151.22.

b. There are $360 - 96 = 264$ payments left to be made. They can be thought of as an annuity consisting of 264 payments of \$1151.22 at .072% interest

$\left(\dfrac{.072}{12} = .006\right)$ per period.

$1151.22\left[\dfrac{1-(1.006)^{-264}}{.006}\right]$

$= 152,320.58$

Their loan balance is \$152,320.58.

c. They will have made $6(12) = 72$ payments. The loan balance at the beginning of the seventh year will be:

$1151.22\left[\dfrac{1-(1.006)^{-288}}{.006}\right]$

$= 157,609.90$

The loan balance at the end of the seventh year will be:

$1151.22\left[\dfrac{1-(1.006)^{-276}}{.006}\right] = 155,060.12$

So they pay

\$157,609.90 − \$155,060.12 = \$2549.78 interest during the 7th year.

47. Continued

d. $1151.22 + 150 = \dfrac{169,600}{\left[\dfrac{1-(1.006)^{-12x}}{.006}\right]}$

Solve by graphing $y_1 = 1301.22$ and

$y_2 = \dfrac{169,600}{\left[\dfrac{1-(1.006)^{-12x}}{.006}\right]}$. It will take 21.22 years to

pay off the loan.

49. Amount needed for retirement:

$3500 = \dfrac{P}{\left[\dfrac{1-(1+\frac{.105}{12})^{-120}}{\frac{.105}{12}}\right]}$

$P = 259,384.15$

Monthly contribution to acquire \$259,384.15:

$259,384.15 = R\left[\dfrac{(1.00875)^{240} - 1}{.00875}\right]$

$R = 320.03$

She must make monthly payments of \$320.03

51. Balance of loan after 12 years of payments:

Monthly payment:

$R = \dfrac{160,000}{\left[\dfrac{1-(1+\frac{.098}{12})^{-30(12)}}{\frac{.098}{12}}\right]}$

$R \approx 1380.53$

since 216 payments need to be made after they have paid 144 payments, they can be thought of as an annuity consisting of 216 payments of \$1380.53 at .81667% interest

$\left(\dfrac{.098}{12} = .0081667\right)$.

$1380.53\left[\dfrac{1-(1.0081667)^{-216}}{.0081667}\right] = 139,868.73$

Monthly payments on refinancing \$139,868.73:

$R = \dfrac{139,868.73}{\left[\dfrac{1-(1+\frac{.072}{12})^{-25(12)}}{\frac{.072}{12}}\right]}$

$R \approx 1006.48$

Balance 5 years after the refinance:

They have made 60 payments and have 240 remaining.

$1006.48\left[\dfrac{1-(1.006)^{-240}}{.006}\right] = 127,831.45$

Their balance is \$127,831.45.

53. $P = 72,000,\ i = \dfrac{.06}{2} = .03,\ n = 9$

First find the amount of each semiannual payment.

$$R = \frac{P}{a_{\overline{n}|i}} = \frac{72,000}{\left[\frac{1-(1.03)^{-9}}{.03}\right]} \approx 9247.24$$

The amount of each semiannual payment is $9247.24.

The interest for each period is 3% of the principal at the end of each previous period.

The portion to principal for each period is the difference between the amount of the payment and the interest for the period.

The principal at the end of each new period is obtained by subtracting the new portion to principal from the principal at the end of the previous period.

Payment Number	Amount of Payment	Interest for Period	Portion to Principal	Principal at End of Period
0	------	------	------	$72,000.00
1	$9247.24	$2160.00	$7087.24	64,912.76
2	9247.24	1947.38	7299.85	57,612.91
3	9247.24	1728.39	7518.85	50,094.06
4	9247.24	1502.82	7744.42	42,349.64

55. $P = 20,000,\ i = \dfrac{.07}{2} = .035,\ n = 5(2) = 10$

First find the amount of each semiannual payment.

$$R = \frac{P}{a_{\overline{n}|i}} = \frac{20,000}{\left[\frac{1-(1.035)^{-10}}{.035}\right]} \approx 2404.83.$$

The amount of each semiannual payment is $2404.83.

The interest for the first period is

$(.035)($20,000) = $700.$

The portion to principal for the first period is

$2404.83 – $700 = $1704.83.$

The principal at the end of the first period is

$20,000 – $1704.83 = $18,295.17$

Repeat these steps to construct three more rows of the table, which will look as follows.

Payment Number	Amount of Payment	Interest for Period	Portion to Principal	Principal at End of Period
0	------	------	------	$20,000.00
1	$2404.83	$700.00	$1704.83	18,295.17
2	2404.83	640.33	1764.50	16,530.67
3	2404.83	578.57	1826.26	14,704.42
4	2404.83	514.65	1890.18	12,814.24

Chapter 5 Review Exercises

1. $P = 4902, r = .065, t = \dfrac{11}{12}$

$I = Prt$

$\quad = 4902(.065)\left(\dfrac{11}{12}\right)$

$\quad \approx 292.08$

The simple interest is \$292.08.

2. $P = 42,368, r = .0922, t = \dfrac{5}{12}$

$I = Prt$

$\quad = (42,368)(.0922)\left(\dfrac{5}{12}\right)$

$\quad \approx 1627.64$

The simple interest is \$1627.64.

3. $P = 3478, r = .074, t = \dfrac{88}{365}$

$I = Prt$

$\quad = 3478(.074)\left(\dfrac{88}{365}\right)$

$\quad \approx 62.05$

The simple interest is \$62.05.

4. $P = 2390, r = .087, t = \dfrac{[(31-3)+30+28]}{365} = \dfrac{86}{365}$

$I = Prt$

$\quad = (2390)(.087)\left(\dfrac{86}{365}\right)$

$\quad \approx 48.99$

The simple interest is \$48.99.

5. Answer varies.

6. $A = 459.57, r = .055, t = \dfrac{7}{12}$

$P = \dfrac{A}{1+rt}$

$\quad = \dfrac{459.57}{1+.055\left(\frac{7}{12}\right)}$

$\quad \approx 445.28$

The present value is \$445.28.

7. $A = 80{,}612, r = .0677, t = \dfrac{128}{365}$

$P = \dfrac{A}{1+rt}$

$\quad = \dfrac{80{,}612}{1+(.0677)\left(\frac{128}{365}\right)}$

$\quad \approx 78{,}742.54$

The present value is \$78,742.54.

8. Answer varies.

9. $A = 802.34, r = .082, t = \dfrac{11}{12}$

$P = A(1-rt)$

$\quad = 802.34\left[1-(.082)\left(\dfrac{11}{12}\right)\right]$

$\quad \approx 802.34(.924833)$

$\quad \approx 742.03$

The proceeds of this discounted loan are \$742.03.

10. $A = 12{,}000, r = .0709, t = \dfrac{145}{365}$

$P = A(1-rt)$

$\quad = 12{,}000\left[1-(.0709)\left(\dfrac{145}{365}\right)\right]$

$\quad \approx 12{,}000(.97183)$

$\quad \approx 11{,}662.01$

The proceeds of this loan are \$11,662.01.

11. Answer varies.
Possible answer:
Compound interest produces more interest because interest is paid on the previously earned interest as well as on the original principal.

12. $P = 2800, i = .06, n = 12$

$A = P(1+i)^n$

$\quad = 2800(1.06)^{12}$

$\quad \approx 5634.15$

The compound amount is \$5634.15. The amount of interest earned is
\$5634.15 − \$2800 = \$2834.15.

13. $P = 57,809.34$, $i = \dfrac{.04}{4} = .01$, $n = 6(4) = 24$

$A = P(1+i)^n$

$= 57,809.34(1.01)^{24}$

$\approx 73,402.52$

The compound amount is $73,402.52.
The amount of interest earned is
$73,402.52 – $57,809.34 = $15,593.18.

14. $P = 12,903.45$, $i = \dfrac{.0637}{4} = .015925$, $n = 29$

$A = P(1+i)^n$

$= 12,903.45(1.015925)^{29}$

$\approx \$20,402.98$

The compound amount is $20,402.98. The
amount of interest earned is
$20,402.98 – $12,903.45 = $7499.53.

15. $P = 4677.23$, $i = \dfrac{.0457}{12}$, $n = 32$

$A = P(1+i)^n$

$= 4677.23\left(1+\dfrac{.0457}{12}\right)^{32}$

≈ 5282.19

The compound amount is $5282.19. The amount
of interest earned is
$5282.19 – $4677.23 = $604.96.

16. $A = 42,000$, $i = \dfrac{.12}{12} = .01$, $n = 7(12) = 84$

$P = \dfrac{A}{(1+i)^n}$

$= \dfrac{42,000}{(1.01)^{84}}$

$\approx 18,207.65$

The present value is $18,207.65.

17. $A = 17,650$, $i = \dfrac{.08}{4} = .02$, $n = 4(4) = 16$

$P = \dfrac{A}{(1+i)^n}$

$= \dfrac{17,650}{(1.02)^{16}}$

$\approx 12,857.07$

The present value is $12,857.07.

18. $A = 1347.89$, $i = \dfrac{.062}{2} = .031$,

$n = (3.5)(2) = 7$

$P = \dfrac{A}{(1+i)^n}$

$= \dfrac{1347.89}{(1.031)^7}$

≈ 1088.54

The present value is $1088.54.

19. $A = 2388.90$, $i = \dfrac{.0575}{12}$, $n = 44$

$P = \dfrac{A}{(1+i)^n}$

$= \dfrac{2388.90}{\left(1+\frac{.0575}{12}\right)^{44}}$

≈ 1935.77

The present value is $1935.77.

20. $a = 5$, $r = 3$
The first five terms of the geometric sequence are
$5, 5(3), 5(3)^2, 5(3)^3, 5(3)^4,$
or
5, 15, 45, 135, 405.

21. $a = 16$, $r = \dfrac{1}{2}$

The first four terms of the geometric sequence are

$16, 16\left(\dfrac{1}{2}\right), 16\left(\dfrac{1}{2}\right)^2, 16\left(\dfrac{1}{2}\right)^3,$

or
16, 8, 4, 2.

22. $a = -3$, $r = 2$
The sixth term of the geometric sequence is
$ar^{n-1} = -3(2)^{6-1}$

$= -3(2)^5$

$= -3(32)$

$= -96.$

23. $a = -2$, $r = -2$
The fifth term of the geometric sequence is
$ar^{n-1} = -2(-2)^{5-1}$

$= -2(-2)^4$

$= -2(16)$

$= -32.$

24. $a = -3, r = 3$

The sum of the first four terms of the geometric sequence is found as follows.

$$S_n = \frac{a(r^n - 1)}{r - 1}$$

$$S_4 = \frac{-3[(-3)^4 - 1]}{3 - 1}$$

$$= \frac{-3(80)}{2}$$

$$= -120$$

25. $a = 8000, r = -\dfrac{1}{2}$

$$S_n = \frac{a(r^n - 1)}{r - 1}$$

$$S_5 = \frac{8000\left[\left(-\frac{1}{2}\right)^5 - 1\right]}{-\frac{1}{2} - 1}$$

$$= \frac{8000\left(-\frac{1}{32} - 1\right)}{-\frac{3}{2}}$$

$$= \frac{8000\left(-\frac{33}{32}\right)}{-\frac{3}{2}}$$

$$= \frac{-8250}{-\frac{3}{2}} = 5500$$

26. $s_{\overline{30}|.02} = \dfrac{(1.02)^{30} - 1}{.02}$

$$\approx 40.56807921$$

27. Answer varies.

28. $R = 1288, i = .07, n = 14$

$$S = R \cdot s_{\overline{n}|i}$$

$$= 1288\left[\frac{(1.07)^{14} - 1}{.07}\right]$$

$$\approx 1288(22.550488)$$

$$\approx 29,045.03$$

The future value of this ordinary annuity is $29,045.03.

29. $R = 4000, i = \dfrac{.06}{4} = .015, n = 8(4) = 32$

$$S = R \cdot s_{\overline{n}|i}$$

$$= 4000\left[\frac{(1.015)^{32} - 1}{.015}\right]$$

$$\approx 4000(40.688288)$$

$$\approx 162,753.15$$

The future value of this ordinary annuity is $162,753.15.

30. $R = 233, i = \dfrac{.06}{12} = .005, n = 4(12) = 48$

$$S = R \cdot s_{\overline{n}|i}$$

$$= 233\left[\frac{(1.005)^{48} - 1}{.005}\right]$$

$$\approx 233(54.097832)$$

$$\approx 12,604.79$$

The future value of this ordinary annuity is $12,604.79.

31. $R = 672, i = \dfrac{.05}{4} = .0125, n = 7(4) = 28$

Because deposits are made at the beginning of each time period, this is the annuity due.

$$S = R \cdot s_{\overline{n+1}|i} - R$$

$$= 672\left[\frac{(1.0125)^{29} - 1}{.0125}\right] - 672$$

$$\approx 672(34.695377) - 672$$

$$\approx 22,643.29$$

The future value of this annuity due is $22,643.29.

32. $R = 11,900, i = \dfrac{.07}{12} = .005833, n = 13$

$$S = R \cdot s_{\overline{n+1}|i} - R$$

$$= 11,900\left[\frac{(1.005833)^{14} - 1}{.005833}\right] - 11,900$$

$$\approx 11,900(14.5433887) - 11,900$$

$$\approx 161,166.33$$

The future value of this annuity due is $161,166.33.

33. Answer varies.

Possible answer:

The purpose of a sinking fund is to receive funds deposited regularly for some goal.

34. $S = 6500, i = .05, n = 6$

$$S = R \cdot s_{\overline{n}|i}$$

$$R = \frac{S}{s_{\overline{n}|i}}$$

$$= \frac{6500}{\left[\frac{(1.05)^6 - 1}{.05}\right]}$$

$$\approx \frac{6500}{6.80191}$$

$$\approx 955.61$$

The amount of each payment into this sinking fund is $955.61.

35. $S = 57,000, i = \frac{.06}{2} = .03, n = \left(8\frac{1}{2}\right)(2) = 17$

$$R = \frac{S}{s_{\overline{n}|i}}$$

$$= \frac{57,000}{\left[\frac{(1.03)^{17} - 1}{.03}\right]}$$

$$\approx \frac{57,000}{21.76159}$$

$$\approx 2619.29$$

The amount of each payment is $2619.29.

36. $S = 233,188, i = \frac{.057}{4} = .01425,$

$$n = \left(7\frac{3}{4}\right)(4) = 31$$

$$R = \frac{S}{s_{\overline{n}|i}}$$

$$= \frac{233,188}{\left[\frac{(1.01425)^{31} - 1}{.01425}\right]}$$

$$\approx \frac{233,188}{38.63753}$$

$$\approx 6035.27$$

The amount of each payment is $6035.27.

37. $S = 56,788, i = \frac{.0612}{12} = .0051,$

$$n = \left(4\frac{1}{2}\right)(12) = 54$$

$$R = \frac{S}{s_{\overline{n}|i}}$$

$$= \frac{56,788}{\left[\frac{(1.0051)^{54} - 1}{.0051}\right]}$$

$$\approx \frac{56,788}{61.98743}$$

$$\approx 916.12$$

The amount of each payment is $916.12.

38. $R = 850, i = .05, n = 4$

$$P = R \cdot a_{\overline{n}|i}$$

$$= 850\left[\frac{1 - (1.05)^{-4}}{.05}\right]$$

$$\approx 850(3.54595)$$

$$\approx 3014.06$$

The present value of this ordinary annuity is $3014.06.

39. $R = 1500, i = \frac{.08}{4} = .02, n = 7(4) = 28$

$$P = R \cdot a_{\overline{n}|i}$$

$$= 1500\left[\frac{1 - (1.02)^{-28}}{.02}\right]$$

$$\approx 1500(21.28127)$$

$$\approx 31,921.91$$

The present value is $31,921.91.

40. $R = 4210, i = \frac{.056}{2} = .028, n = 8(2) = 16$

$$P = R \cdot a_{\overline{n}|i}$$

$$= 4210\left[\frac{1 - (1.028)^{-16}}{.028}\right]$$

$$\approx 4210(12.75533)$$

$$\approx \$53,699.94$$

The present value is $53,699.94.

41. $R = 877.34$, $i = \dfrac{.064}{12} \approx .0053$, $n = 17$

$P = R \cdot a_{\overline{n}|i}$

$= 877.34 \left[\dfrac{1 + (.064)^{-17}}{\dfrac{.064}{12}} \right]$

$\approx 14{,}222.42$

The present value is $14,222.42

42. Answers vary.
For example, home loans and car loans are commonly amortized.

43. $P = 32{,}000$, $i = \dfrac{.084}{4} = .021$, $n = 10$

$P = R \cdot a_{\overline{n}|i}$

$R = \dfrac{P}{a_{\overline{n}|i}}$

$= \dfrac{32{,}000}{\left[\dfrac{1 - (1.021)^{-10}}{.021} \right]}$

$\approx \dfrac{32{,}000}{8.93577}$

≈ 3581.11

Quarterly payments of $3581.11 are necessary to amortize this loan.

44. $P = 5607$, $i = \dfrac{.076}{12} \approx .006\overline{3}$, $n = 32$

$R = \dfrac{P}{a_{\overline{n}|i}}$

$\approx \dfrac{5607}{\left[\dfrac{1 - (1.0063)^{-32}}{.0063} \right]}$

$\approx \dfrac{5607}{28.89858}$

≈ 194.13

Monthly payments of $194.13 are needed to amortize this loan.

45. $P = 56{,}890$, $i = \dfrac{.0674}{12} = .005617$,

$n = 25(12) = 300$

$R = \dfrac{P}{a_{\overline{n}|i}}$

$= \dfrac{56{,}890}{\left[\dfrac{1 - \left(1 + \dfrac{.0674}{12}\right)^{-300}}{\dfrac{.0674}{12}} \right]}$

≈ 392.70

The monthly payment for this mortgage is $392.70.

46. $P = 77{,}110$, $i = \dfrac{.0845}{12} = .007041\overline{6}$,

$n = 30(12) = 360$

$R = \dfrac{P}{a_{\overline{n}|i}}$

$\approx \dfrac{77{,}110}{\left[\dfrac{1 - \left(1 + \dfrac{.0845}{12}\right)^{-360}}{\dfrac{.0845}{12}} \right]}$

≈ 590.18

The monthly payment for this mortgage is $590.18.

47. Locate the entry of the table that is in the row labeled "Payment Number 5" and in the column labeled "Interest for Period," to observe that $896.06 of the fifth payment is interest.

48. Locate the entry of the table that is in the row labeled "Payment Number 6" and in the column labeled "Portion to Principal," to observe that $127.48 of the sixth payment is used to reduce the debt.

49. To find out how much interest is paid in the first 3 months of the loan, add the first three nonzero entries in the column labeled "Interest for Period."

$899.58
898.71
$\underline{+897.83}$
$2696.12

50. To find out how much the debt has been reduced by the end of the first 6 months, add the first six nonzero entries in the column labeled "Portion to Principal."

$123.06

123.93

124.80

125.69

126.58

$\underline{127.48}$

$751.54

51. $P = 9812, r = .12, t = \dfrac{7}{12}$

Use the formula for simple discount.

$P = A(1 - rt)$

$A = \dfrac{P}{1 - rt}$

$\quad = \dfrac{9812}{1 - (.12)\left(\frac{7}{12}\right)}$

$\quad = \dfrac{9812}{.93} \approx 10,550.54$

The amount of the loan is $10,550.54.

52. Find the maturity value of the loan, the amount Tom Wilson must pay his mother.

$A = P(1 + rt)$

$\quad = 5800\left[1 + .06\left(\dfrac{10}{12}\right)\right]$

$\quad = 5800(1.05)$

$\quad = 6090.00$

The bank will apply its discount rate to the 6090.00, to obtain amount of discount

$\quad = 6090(.1045)\left(\dfrac{3}{12}\right)$

$\quad \approx 159.10$

Tom's mother will receive from the bank $6090.00 − $159.10 = $5930.90, which isn't enough because it falls short of the $6000 needed for the new furniture

Alternatively, we can apply the equation $P = A(1 - rt)$

$P = A(1 - rt)$

$P = 6090(1 - .1045(\dfrac{3}{12}))$

$\quad = 5930.90$

53. $P = 15,000, \ i = \dfrac{.06}{2} = .03, \ n = 7.5(2) = 15$

$I = P(1 + i)^n - P$

$\quad = 15,000(1 + .03)^{15} - 15,000$

$\quad \approx 8369.51$

54. Starting at age 23:

$P = 500, \ i = \dfrac{.05}{4} = .0125,$

$n = (65 - 23)4 = 42(4) = 168$

$A = P(1 + i)^n$

$\quad = 500(1.0125)^{168}$

$\quad \approx 4030.28$

Starting at age 40:

$t = (65 - 40)4 = 25(4) = 100$

$A = 500(1.0125)^{100} \approx 1731.70$

$4030.28 - 1731.70 = 2298.58$

He will have $2298.58 more if he invests now.

55. a. $A = 75,302, \ P = 10,000, \ t = 10$

$A = P(1 + r)^t$

$\dfrac{A}{P} = (1 + r)^t$

$\left(\dfrac{A}{P}\right)^{1/t} = 1 + r$

$r = \left(\dfrac{A}{P}\right)^{1/t} - 1$

$\quad = (7.5302)^{1/10} - 1$

$\quad \approx .223716$

The annual rate of return was approximately 22.3716%.

b. $A = 75,302, \ P = 60,000, \ t = 4$

$r = \left(\dfrac{A}{P}\right)^{1/t} - 1$

$\quad = (1.25503)^{1/4} - 1$

$\quad \approx .058434$

The annual rate of return was approximately 5.8434%.

56. $r_F = \left(1+\dfrac{r}{m}\right)^m - 1 \quad r_E = \left(1+\dfrac{r}{m}\right)^m - 1$

Frontenac Bank:

$$r_F = \left(1+\dfrac{.0394}{4}\right)^4 - 1 \approx .04$$

$$r_E = \left(1+\dfrac{.0394}{4}\right)^4 - 1 \approx .04$$

The effective rate is about 4.00%.
E*-TRADE Bank:

$$r_E = \left(1+\dfrac{.0393}{365}\right)^{365} - 1 \approx .0401$$

The effective rate is about 4.01%.
E*-TRADE Bank pays a higher effective rate.

57. $A = 2$, $P = 1$, $t = 2005 - 1982 = 23$

$$A = Pe^{rt}$$

$$\dfrac{A}{P} = e^{rt}$$

$$\ln\dfrac{A}{P} = \ln e^{rt}$$

$$\ln\dfrac{A}{P} = rt$$

$$\dfrac{1}{t}\ln\dfrac{A}{P} = r$$

$$\dfrac{1}{23}\ln\dfrac{2}{1} = r$$

$$.03014 \approx r$$

The inflation rate was approximately 3.014%.

58. $A = 194$, $P = 172$, $t = 2005 - 2000 = 5$

$$A = Pe^{rt}$$

$$\dfrac{A}{P} = e^{rt}$$

$$\ln\dfrac{A}{P} = \ln e^{rt}$$

$$\ln\dfrac{A}{P} = rt$$

$$\dfrac{1}{t}\ln\dfrac{A}{P} = r$$

$$\dfrac{1}{5}\ln\dfrac{194}{172} = r$$

$$.02407 \approx r$$

The inflation rate was approximately 2.407%.

59. $R = 3200$, $i = \dfrac{.068}{2} = .034$, $n = (3.5)(2) = 7$

Use the formula for the future value of an ordinary annuity.

$$S = R \cdot s_{\overline{n}|i}$$

$$= 3200\left[\dfrac{(1.034)^7 - 1}{.034}\right]$$

$$\approx 3200(7.75586)$$

$$\approx 24{,}818.76$$

The final amount in the account will be $24,818.76.
The interest earned will be
$24,818.76 - 7(\$3200) = \$2418.76$.

60. $S = 52{,}000$, $i = \dfrac{.075}{12} = .00625$,

$n = 20(12) = 240$

Use the formula for future value of an ordinary annuity to find the periodic payment.

$$S = R \cdot s_{\overline{n}|i}$$

$$R = \dfrac{S}{s_{\overline{n}|i}}$$

$$= \dfrac{52{,}000}{\left[\dfrac{(1.00625)^{240} - 1}{.00625}\right]}$$

$$\approx \dfrac{52{,}000}{553.73073}$$

$$\approx 93.91$$

The firm should invest $93.91 monthly.

61. $S = 150{,}000$, $i = \dfrac{.0525}{4} = .013125$

$n = 79(4) = 316$

Use the formula for future value of an annuity.

$$R = S\left[\dfrac{i}{(1+i)^n - 1}\right]$$

$$\approx 150{,}000\left[\dfrac{.013125}{(1.013125)^{316} - 1}\right]$$

$$\approx 32.49$$

She would have to put $32.49 into her savings account every three months.

62. Age 55:

$$P = 6000, \quad i = \frac{.08}{1} = .08, \quad n = 20(1) = 20$$

$$R = 6000, \quad i = \frac{.08}{1} = .08, \quad n = 20(1) = 20$$

$$S = 6000\left[\frac{(1.08)^{20} - 1}{.08}\right] \approx 274,571.79$$

Age 65:

$$P = 12,000, \quad i = \frac{.08}{1} = .08, \quad n = 10(1) = 10$$

$$R = 12,000, \quad i = \frac{.08}{1} = .08, \quad n = 10(1) = 10$$

$$S = 12,000\left[\frac{(1.08)^{10} - 1}{.08}\right] \approx 173,838.75$$

If the pension starts at age 55, the total is $274,571.79 whereas, if the pension starts at age 65, the total is $173,838.75.

63. $A = 7500, \quad i = \dfrac{.10}{2} = .05, \quad n = 3(2) = 6$

$$A = P(1 + i)^n$$

$$P = \frac{A}{(1 + i)^n}$$

$$= \frac{7500}{(1.05)^6}$$

$$\approx 5596.62$$

The required lump sum is $5596.62.

64. $P = 15,000, \quad i = \dfrac{.072}{12} = .006, \quad n = 36$

This is an ordinary annuity.

$$P = R \cdot a_{\overline{n}|i}$$

$$R = \frac{P}{a_{\overline{n}|i}}$$

$$= \frac{15,000}{\left[\frac{1-(1.006)^{-36}}{.006}\right]}$$

$$\approx \frac{15,000}{32.29075}$$

$$\approx 464.53$$

The amount of each payment will be $464.53.

65. $P = 40,000, \quad i = \dfrac{.09}{2} = .045, \quad n = 8(2) = 16$

This is an ordinary annuity.

$$R = \frac{P}{a_{\overline{n}|i}}$$

$$= \frac{40,000}{\left[\frac{1-(1.045)^{-16}}{.045}\right]}$$

$$\approx \frac{40,000}{11.23402}$$

$$\approx \$3560.61$$

The amount of each payment will be $3560.61.

66. Amount of loan
= $91,000 – $20,000
= $71,000

a. Use the formula for amortization payments with $P = 71,000, \quad i = \dfrac{.09}{12} = .0075,$ and $n = 30(12) = 360.$

$$R = \frac{Pi}{1-(1+i)^{-n}}$$

$$= \frac{71,000(.0075)}{1-(1.0075)^{-360}}$$

$$\approx \frac{532.50}{.932114}$$

$$\approx 571.28$$

The monthly payment for this mortgage is $571.28.

b. To find the amount of the first payment that goes to interest, use $I = Prt$ with $P = 71,000, \quad i = .0075,$ and $t = 1.$

$$I = (71,000)(.0075)(1)$$

$$= 532.50$$

Of the first payment, $532.50 is interest.

66. Continued

c. Using method 1, since 180 payments were made, there are 180 remaining payments. The present value is

$$571.28\left[\frac{1-(1.0075)^{-180}}{.0075}\right] \approx 56,324.44,$$

so the remaining balance is $56,324.44.
Using method 2, since 180 payments were already made, we have

$$571.28\left[\frac{1-(1.0075)^{-180}}{.0075}\right] \approx 56,324.44.$$

They still owe
$71,000 - $56,324.44 = $14,675.56.
Furthermore, they owe the interest on this amount for 180 months, for a total remaining balance of
$(14,675.56)(1.0075)^{180} = $56,325.43.

d. Closing costs
= $3700 + (.025)($136,000)
= $3700 + $3400
= $7100

e. Amount of money received
= Selling price – Closing costs –
    Current mortgage balance
Using method 1, the amount received is
$136,000 – $7100 – $56,324.44
= $72,575.56.
Using method 2, the amount received is
$136,000 – $7100 – $56,325.43
= $72,574.57.

67. $P = 10,000, i = .05, n = 7$

$A = P(1+i)^n$

$= 10,000(1.05)^7$

$\approx $14,071.00$

At the end of 7 yr, the value of the death benefit will be $14,071.00. This balance of $14,071.00 is to be paid out in 120 equal monthly payments, with $i = \dfrac{.03}{12} = .0025$. The full balance will be paid out, so we use the amortization formula. Let X be the amount of each payment.

$$X = \frac{14,071.00}{a_{\overline{120}|.0025}}$$

$$= \frac{14,071.00}{\left[\frac{1-(1.0025)^{-120}}{.0025}\right]}$$

$$\approx \frac{14,071.00}{103.56175}$$

$$\approx $135.87$$

This corresponds to choice (d).

68. Total amount after 16 deposits:

$$R = 500, i = \frac{.09}{4} = .0225, n = 16$$

$$S = 500\left[\frac{(1.0225)^{16}-1}{.0225}\right] = 9502.70$$

Total amount after next 52 quarters:
$A = 9502.70(1.+.0225)^{52} = 30.223.13$
Total of 32 $750 deposits:

$$R = 750, i = \frac{.09}{4} = .0225, n = 32$$

$$S = 750\left[\frac{(1.0225)^{32}-1}{.0225}\right]$$

$$\approx 34.603.43$$

Gene's account balance is
$30,223.13 + $34,603.43 = $64,826.56.

69. Amount needed for retirement:

$$55,000 = \frac{x}{\left[\frac{1-(1.09)^{-20}}{.09}\right]}$$

$$x = 502,070.01$$

Annual contribution to acquire $502,070.01:

$$502,070.01 = R\left[\frac{(1.09)^{25}-1}{.09}\right]$$

$$R \approx 5927.56$$

She will have to make annual deposits of $5927.56.

70. a.

$$r_E = \left(1 + \frac{r}{m}\right)^m - 1$$

$$r_E + 1 = \left(1 + \frac{r}{m}\right)^m$$

$$\log(r_E + 1) = m\log\left(1 + \frac{r}{m}\right)$$

$$\frac{\log(r_E + 1)}{m} = \log\left(1 + \frac{r}{m}\right)$$

$$10^{\frac{\log(r_E+1)}{m}} = 1 + \frac{r}{m}$$

$$m\left[10^{\frac{\log(r_E+1)}{m}} - 1\right] = r$$

$$12\left(10^{\frac{\log 1.1}{12}} - 1\right) = r$$

$$.09569 \approx r$$

The annual interest rate is about 9.569%.

b. $P = 140,000$, $i = \dfrac{.06625}{12}$, $n = 30(12) = 360$

Use the formula for amortization payments.

$$R = P\left[\frac{i}{1 - (1+i)^{-n}}\right]$$

$$= 140,000\left[\frac{\frac{.06625}{12}}{1 - \left(1 + \frac{.06625}{12}\right)^{-360}}\right]$$

$$\approx 896.44$$

Her payment is $896.44.

c. $R = 1200 - 896.44 = 303.56$, $i = \dfrac{.09569}{12}$,

$n = 30(12) = 360$

Use the formula for the future value of an annuity.

$$S = R\left[\frac{(1+i)^n - 1}{i}\right]$$

$$= 303.56\left[\frac{\left(1 + \frac{.09569}{12}\right)^{360} - 1}{\frac{.09569}{12}}\right]$$

$$\approx 626,200.88$$

She will have $626,200.88 in the fund.

d. $R = 140,000\left[\dfrac{\frac{.0625}{12}}{1 - \left(1 + \frac{.0625}{12}\right)^{-12(15)}}\right]$

$$\approx 1200.39$$

His payment is $1200.39.

70. Continued

e. $S = 1200\left[\dfrac{\left(1 + \frac{.09569}{12}\right)^{15(12)} - 1}{\frac{.09569}{12}}\right]$

$$\approx 478,134.14$$

He will have $478,134.14 in the fund.

f. Sue is ahead by
626,200.88 − 478,134.14 = 148,066.74

g. Answer varies.

Case 5 Time, Money, and Polynomials

1.

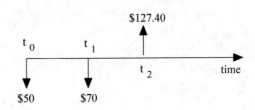

$50(1+i)^2 + 70(1+i) = 127.40$
Let $x = 1 + i$.
$$50x^2 + 70x = 127.40$$

$50x^2 + 70x - 127.40 = 0$
Use the quadratic formula to solve for x.

$$x = \frac{-70 \pm \sqrt{70^2 - 4(50)(-127.40)}}{2(50)}$$

We get $x \approx 1.043$ and $x \approx -2.443$. Since $x = 1 + i$, the two values for i are $.043 = 4.3\%$ and $-3.443 = -344.3\%$, and we reject the negative value in this case. The YTM is 4.3%.

2. a.

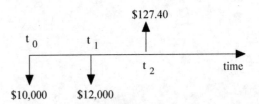

b. The amount in the fund at the end of the second year was
$[10,000(1.05) + 12,000](1.045)$
$= \$23,512.50$.

2. Continued

c. $10,000(1+i)^2 + 12,000(1+i)$

$= 23,512.50$

$10,000x^2 + 12,000x - 23,512.50 = 0$

The solutions are

$x \approx 1.047$ and $x \approx -2.247$,

which correspond to

$i = .047 = 4.7\%$ and

$i = -3.247 = -324.7\%$

We reject the negative value in this case. The YTM is 4.7% (which lies between the yearly rates of 5% and 4.5%, as might have been expected).

3. a.

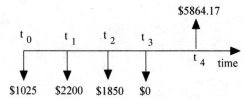

b. $1025(1+i)^4 + 2200(1+i)^3$

$+ 1850(1+i)^2 = 5864.17$

$f(x) = 1025x^4 + 2200x^3$

$+ 1850x^2 - 5864.17 = 0$

We expect that $0 < i < 1$, so that $1 < x < 2$. Calculate that

$f(1) = -789.17$ and

$f(2) = 35,535.83$.

This change in sign guarantees that there is a solution to $f(x) = 0$ between 1 and 2.

Next find $f(1.1), f(1.2), f(1.3)$ and so on, and look for another change in sign.

$f(1.0) = -789.17$ and

$f(1.1) = 803.2325$

Now, subdivide the interval between 1.0 and 1.1.

$f(1.05) \approx -31.88$ and

$f(1.06) \approx 128.76$

Subdivide again, and find

$f(1.052) \approx .00$, so $x = 1.052$ is the approximate solution to the polynomial equation, and this corresponds to

$i = \text{YTM} = .052 = 5.2\%$

4. a.

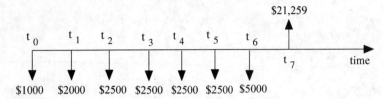

b. $1000(1+i)^7 + 2000(1+i)^6 + 2500(1+i)^5 + 2500(1+i)^4$

$\quad +2500(1+i)^3 + 2500(1+i)^2 + 5000(1+i) = 21,259$

$\quad f(x) = 1000x^7 + 2000x^6 + 2500x^5 + 2500x^4 + 2500x^3 + 2500x^2$

$\quad +5000x - 21,259 = 0$

The following solutions presuppose the use of a TI-83 graphing calculator. Similar results can be obtained from other graphing calculators.

c. Answer varies.

Possible answer: Let $x = (1+i)$.

We graph the function

$f(x) = 1000(1+x)^7 + 2000(1+x)^6 + 2500(1+x)^5 + 2500(1+x)^4$

$\quad +2500(1+x)^3 + 2500(1+x)^2 + 5000(1+x) - 21,259$

by entering it as y_1 with

$.0505 \le x \le .0507$ and $-1 \le y \le 1$.

We see the graph intersects the x-axis between these two values, so 5.05% < YTM < 5.07.

d. On the CALC menu, use zero to find the x-intercept of this graph. We read $x = .05059847$, so the YTM is 5.06%.

5. a.

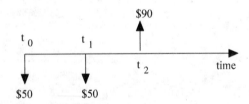

Let $x = (1+i)$.

$50(1+i)^2 + 50(1+i) = 90$

$f(x) = 50x^2 + 50x - 90 = 0$

The solutions are

$x \approx .932$ and $x \approx -1.932$,

which correspond to

$1 + i = .932 \quad$ and $\quad 1 + i = -1.932$

$\quad i = -.068 \quad$ and $\quad\quad i = -2.932$.

b. Consider $A = 50(1+i)^2$.

When $i = -.068$, $A \approx \$43.43$, which is reasonable because a negative interest rate should indicate a loss in value. When $i = -2.932$, $A \approx \$186.63$, which is not reasonable.

Chapter 6: Systems of Linear Equations and Matrices

Section 6.1 Systems of Linear Equations

1. Is $(-1, 3)$ a solution of
$$2x + y = 1$$
$$-3x + 2y = 9$$
Check $(-1, 3)$ in equation (1).
$2(-1) + 3 = 1$?
$\quad -2 + 3 = 1$? True
Check $(-1, 3)$ in equation (2).
$-3(-1) + 2(3) = 9$?
$$3 + 6 = 9 ? \quad \text{True}$$
The ordered pair $(-1, 3)$ is a solution of this system of equations.

3. $3x - y = 1$ (1)
$\quad x + 2y = -9$ (2)
Solve for y from equation (1) to get $y = 3x - 1$.
Then substitute $y = 3x - 1$
into (2) $x + 2(3x - 1) = -9$. Solve for x.

$$x + 2(3x - 1) = -9$$
$$x + 6x - 2 = -9$$
$$7x - 2 = -9$$
$$7x = -9 + 2$$
$$7x = -7$$
$$x = -1$$

Substitute $x = -1$ into
$\quad y = 3x - 1$ to
$\qquad$ solve for y.
$\quad y = 3(-1) - 1$
$\quad y = -4$.
$\qquad$ The solution set is
$\qquad (-1, -4)$

5. $3x - 2y = 4$ (1)
$\quad 2x + y = -1$ (2)
Solve for y from equation 2 to get
$\quad y = -1 - 2x$.

Substitute $y = -1 - 2x$ into equation (1)
$$3x - 2(-1 - 2x) = 4$$
$$3x + 2 + 4x = 4$$
$$7x + 2 = 4$$
$$7x = 4 - 2$$
$$7x = 2$$
$$x = \frac{2}{7}$$

Substitute $x = \frac{2}{7}$ into $y = -1 - 2x$
to solve for y.

$$y = -1 - 2\left(\frac{2}{7}\right)$$
$$y = -1 - \frac{4}{7}$$
$$y = -\frac{11}{7}$$

The solution set is $\left(\frac{2}{7}, -\frac{11}{7}\right)$.

Substitute -2 for x in equation (1) to find y.
$\quad 3x + 2y = -4$ (1)
$\quad 3(-2) + 2y = -4$
$\quad\quad -6 + 2y = -4$
$\quad\quad\quad 2y = 2$
$\quad\quad\quad\quad y = 1$
The solution is $(-2, 1)$.

7. $r + s = 0$ (1)
$\quad r - s = 5$ (2)
Solve for r from (1) to get $r = -s$. Substitute
$r = -s$ into (2) to get

$$-s - s = 5$$
$$-2s = 5$$
$$s = -\frac{5}{2}$$

7.   Continued

Substitute

$$s = -\frac{5}{2} \text{ into } r = -s$$

to get $r = -(-\frac{5}{2})$

$$r = \frac{5}{2}$$

The solution of the system is $\left(\frac{5}{2}, -\frac{5}{2}\right)$.

9. $x - 2y = 5$ (1)

$2x + y = 3$ (2)

Multiply equation (2) by 2 to get

$x - 2y = 5$ (1)

$4x + 2y = 6$

Add the two equations to get

$5x = 11$

$$x = \frac{11}{5}$$

Substitute $\frac{11}{5}$ for x in equation (1) to get

$$\frac{11}{5} - 2y = 5$$

$$-2y = 5 - \frac{11}{5}$$

$$-2y = \frac{14}{5}$$

$$y = -\frac{7}{5}.$$

The solution of the system is $\left(\frac{11}{5}, -\frac{7}{5}\right)$.

11. $2x - 2y = 12$ (1)

$-2x + 3y = 10$ (2)

Add the two equations to get $y = 22$.
Substitute 22 for y in equation (1) to get

$2x - 2(22) = 12$

$2x - 44 = 12$

$2x = 56$

$x = 28$

The solution of the system is (28, 22).

13. $x + 3y = -1$ (1)

$2x - y = 5$ (2)

Multiply equation (2) by 3 to get

$x + 3y = -1$ (1)

$6x - 3y = 15$ (2)

Add the two equations to get

$7x = 14$

$x = 2.$

Substitute 2 for x in equation (1) to get

$2 + 3y = -1$

$3y = -3$

$y = -1.$

The solution of the system is $(2, -1)$.

15. $2x + 3y = 15$ (1)

$8x + 12y = 40$ (2)

Multiply equation (1) by –4 and add the result to
equation (1).

$-8x - 12y = -60$

$\underline{8x + 12y = 40 \quad (2)}$

$0 = -20$

The statement "$0 = -20$" is false, so the system is
inconsistent. There is no solution.

17. $2x - 8y = 2$ (1)

$3x - 12y = 3$ (2)

Multiply equation (1) by 3 and equation (2) by –2
to get

$6x - 24y = 6$ (3)

$-6x + 24y = -6.$ (4)

Add the two equations.

$6x - 24y = 6$ (3)

$\underline{-6x + 24y = -6 \quad (4)}$

$0 = 0$

The system is dependent, so there is an infinite
number of solutions. To solve the system in terms
of one of the parameters, say y, solve equation (1)
for x in terms of y.

$2x - 8y = 2$ (1)

$2x = 8y + 2$

$x = 4y + 1$

The solution to the system is the infinite set of
pairs of the form

$(4y + 1, y)$ for any real number y.
Note: the system could also be solved in terms of
x. The ordered pairs would all have the form:

$$\left(x, \frac{1}{4}x - \frac{1}{4}\right).$$

19. $3x + 2y = 5$ *(1)*

$6x + 4y = 8$ *(2)*

Multiply equation (1) by –2 to get

$-6x - 4y = -10$ *(3)*

$6x + 4y = 8$ *(2)*

Add the two equations.

$-6x - 4y = -10$ *(3)*

$\underline{6x + 4y = 8 \quad (2)}$

$0 = -2$

The statement "$0 = -2$" is false, so the system is inconsistent. There is no solution.

21. $4x - 3y = -1$

$3y = 4x + 1$

$y = \dfrac{4x + 1}{3}$

$y = \dfrac{4}{3}x + \dfrac{1}{3}$

$\text{slope} = \dfrac{4}{3}$

$y - \text{int ercept} = \dfrac{1}{3}$

The slopes and y-intercepts of these two equations match graphical solution (a).

23. $\dfrac{x}{5} + 3y = 31$ (1)

$2x - \dfrac{y}{5} = 8$ (2)

Multiply both equations (1) and (2) by 5 to clear the fractions.

$x + 15y = 155$ *(3)*

$10x - y = 40$ *(4)*

Multiply equation (4) by 15 and add the result to equation (3). Substitute 5 for x in equation (1) to find y.

$150x - 15y = 600$

$\underline{x + 15y = 155}$

$151x \qquad = 755$

$x = 5$

$\dfrac{x}{5} + 3y = 31$ *(1)*

$\dfrac{5}{5} + 3y = 31$

$1 + 3y = 31$

$3y = 30$

$y = 10$

The solution is (5, 10).

25. a. Multiply the second equation by –1 and then add the equations:

Rate of increase (Red Army) $= 200,000 - .5r - .3b = 0$

Rate of increase (Blue Army) $= \underline{-350,000 + .5r + .7b = 0}$

$-150,000 \qquad + .4b = 0$

$.4b = 150,000$

$b = 375,000$

By substituting $b = 375,000$ back into one of the rate of increase equations, $r = 175,000$.

b. Answers vary.

27. $\begin{aligned} -1490x + 2y &= 40,586 \quad (1) \\ -1686x + 3y &= 30,210 \quad (2) \end{aligned}$

Multiply (1) by 3 and multiply (2) by -2 to get

$-4470x + 6y = 121758$

$\underline{3372x - 6y = -60420}$

$-1098x = 61338$

$x = -55.86$

Because x is negative, the income

level will never be the same.

29. $30x + 70y = 16000 \quad (1)$

$1.50(30x) + 3(30y) = 345000 \quad (2)$

Simplify equation (2)

$30x + 70y = 16000 \quad (1)$

$45x + 210y = 34500 \quad (2)$

Multiply equation (1) by -3 and add

$-90x - 210y = -48000$

$\underline{45x + 210y = 34500}$

$-45x = -13500$

$x = 300$

Substitute $x = 300$ into (1) and solve for y

$30(300) + 70y = 16000$

$9000 + 70y = 16000$

$70y = 7000$

$y = 100$

300 of Boeing and 100 of GE.

31. $2x + 3y = 34 \quad (1)$

$25x + 30y = 365 \quad (2)$

Multiply (1) by -10, add and then solve for x.

$-20x - 30y = -340$

$\underline{25x + 30y = 365}$

$5x = 25$

$x = 5$ skirts

$y = 8$ blouses

33. $\begin{aligned} x \quad\quad - 3z &= 2 \\ 2x - 4y + 5z &= 1 \\ 5x - 8y + 7z &= 6 \\ 3x - 4y + 2z &= 3 \end{aligned}$

35. $\begin{aligned} 3x \quad + z + 2w + 18v &= 0 \\ -4x + y \quad - w - 24v &= 0 \\ 7x - y + z + 3w + 42v &= 0 \\ 4x \quad + z + 2w + 24v &= 0 \end{aligned}$

37. $\begin{aligned} x + y + 2z + 3w &= 1 \\ -y + z - 2w &= -1 \\ 3x + y + 4z + 5w &= 2 \end{aligned}$

39. $\begin{aligned} x + 12y - 3z + 4w &= 10 \\ 2y + 3z \quad + w &= 4 \\ -z &= -7 \\ 6y - 2z - 3w &= 0 \end{aligned}$

41. $\begin{aligned} x + 3y - 4z + 2w &= 1 \quad (1) \\ y + z - w &= 4 \quad (2) \\ 2z + 2w &= -6 \quad (3) \\ 3w &= 9 \quad (4) \end{aligned}$

Divide (4) by 3. Substitute $w = 3$ into (3) and solve for z.

$x + 3y - 4z + 2w = 1$

$y + z - w = 4$

$2z + 2(3) = -6 \Rightarrow 2z = -12 \Rightarrow z = -6$

$w = 3$

Substitute $w = 3$ and $z = -6$ into (2) to solve for y.

$x + 3y - 4z + 2w = 1$

$y + (-6) - 3 = 4$

$y = 13$

Substitute $y = 13$, $w = 3$, and $z = -6$ into (1) to solve for x.

$x + 3(13) - 4(-6) + 2(3) = 1$

$x + 39 + 24 + 6 = 1$

$x = -68$

The solution set is $(-68, 13, -6, 3)$.

43. $2x + 2y - 4z + w = -5$ (1)

$3y + 4z - w = 0$ (2)

$2z - 7w = -6$ (3)

$5w = 15$ (4)

Divide (4) by 5 and substitute $w = 3$ into (3) to solve for z.

$2x + 2y - 4z + w = -5$

$3y + 4z - w = 0$

$2z - 7(3) = -6 \Rightarrow 2z - 21 = -6 \Rightarrow z = \dfrac{15}{2}$

$w = 3$

$2x + 2y - 4z + w = -5$

$3y + 4z - w = 0$

$2z - 7(3) = -6 \Rightarrow 2z - 21 = -6 \Rightarrow z = \dfrac{15}{2}$

Substitute $w = 3$ and $z = 15/2$ into (2) to solve for y.

$2x + 2y - 4z + w = -5$

$3y + 4(\dfrac{15}{2}) - (3) = 0 \Rightarrow 3y + 30 = 3$

$3y = -27$

$y = -9$

$2x + 2(-9) - 4\left(\dfrac{15}{2}\right) + (3) = -5$

$2x - 18 - 30 + 3 = -5$

$2x - 45 = -5$

$2x = 40$

$x = 20$

The solution set is $\left(20, -9, \dfrac{15}{2}, 3\right)$.

45. The augmented matrix for the system
$2x + y + z = 3$

$3x - 4y + 2z = -5$

$x + y + z = 2$

is $\begin{bmatrix} 2 & 1 & 1 & 3 \\ 3 & -4 & 2 & -5 \\ 1 & 1 & 1 & 2 \end{bmatrix}$.

47. The system of equations for the augmented matrix
$\begin{bmatrix} 2 & 3 & 8 & 20 \\ 1 & 4 & 6 & 12 \\ 0 & 3 & 5 & 10 \end{bmatrix}$ is

$2x + 3y + 8z = 20$

$x + 4y + 6z = 12$

$3y + 5z = 10.$

49. $\begin{bmatrix} 1 & 2 & 3 & -1 \\ 6 & 5 & 4 & 6 \\ 2 & 0 & 7 & -4 \end{bmatrix}$

Interchange R_2 and R_3 to obtain

$\begin{bmatrix} 1 & 2 & 3 & -1 \\ 2 & 0 & 7 & -4 \\ 6 & 5 & 4 & 6 \end{bmatrix}$.

51. $\begin{bmatrix} -4 & -3 & 1 & -1 & 2 \\ 8 & 2 & 5 & 0 & 6 \\ 0 & -2 & 9 & 4 & 5 \end{bmatrix}$

Replace R_2 with $2R_1 + R_2$ to obtain

$\begin{bmatrix} -4 & -3 & 1 & -1 & 2 \\ 0 & -4 & 7 & -2 & 10 \\ 0 & -2 & 9 & 4 & 5 \end{bmatrix}$.

53. $x + 2y \quad = 0$ (1)

$y - z = 2$ (2)

$x + y + z = -2$ (3)

Multiply equation (3) by -1 and add to equation (1).

$x + 2y \quad = 0$

$\underline{-x - y - z = 2}$

$y - z = 2$

The system now becomes

$x + 2y \quad = 0$ (1)

$y - z = 2$ (2)

$y - z = 2$ (4)

Multiply equation (4) by -1 and add to equation (2).

$y - z = 2$

$\underline{-y + z = -2}$

$0 = 0$ (5)

Since equation (5) is true, the system is dependent.

55. $x + 2y + 4z = 6$ (1)

$y + z = 1$ (2)

$x + 3y + 5z = 10$ (3)

Multiply equation (3) by -1 and add the result to equation (1).

$x + 2y + 4z = 6$

$\underline{-x - 3y - 5z = -10}$

$-y - z = -4$

The system now becomes

$x + 2y + 4z = 6$ (1)

$y + z = 1$ (2)

$-y - z = -4.$ (4)

Add equation (2) to equation (4).

$y + z = 1$ (2)

$\underline{-y - z = -4}$ (4)

$0 = -3$ (5)

Since equation (5) is false, the system is inconsistent.

57. $a - 3b - 2c = -3$ (1)

$3a + 2b - c = 12$ (2)

$-a - b + 4c = 3$ (3)

Multiply equation (1) by -3 and add the result to equation (2). Also, add equation (1) to equation (3).

The new system is

$a - 3b - 2c = -3$ (1)

$11b + 5c = 21$ (4)

$-4b + 2c = 0.$ (5)

Multiply equation (4) by $\dfrac{1}{11}$. This gives

$a - 3b - 2c = -3$ (1)

$b + \dfrac{5}{11}c = \dfrac{21}{11}$ (6)

$-4b + 2c = 0.$ (5)

Multiply equation (6) by 4 and add the result to equation (5). This gives

$a - 3b - 2c = -3$ (1)

$b + \dfrac{5}{11}c = \dfrac{21}{11}$ (6)

$\dfrac{42}{11}c = \dfrac{84}{11}.$ (7)

Multiply equation (7) by $\dfrac{11}{42}$ to get

$a - 3b - 2c = -3$ (1)

$b + \dfrac{5}{11}c = \dfrac{21}{11}$ (6)

$c = 2.$ (8)

Substitute 2 for c in equation (6) to get

$b + \dfrac{10}{11} = \dfrac{21}{11},$ or $b = \dfrac{11}{11} = 1.$

Substitute 2 for c and 1 for b in equation (1) to get $a = 4$. The system is independent.

59. The augmented matrix is

$$\begin{bmatrix} -1 & 3 & 2 & | & 0 \\ 2 & -1 & -1 & | & 3 \\ 1 & 2 & 3 & | & 0 \end{bmatrix}$$

Now use the matrix method to solve the system:

$$\begin{bmatrix} 1 & -3 & -2 & | & 0 \\ 2 & -1 & -1 & | & 3 \\ 1 & 2 & 3 & | & 0 \end{bmatrix} \quad -1R_1$$

$$\begin{bmatrix} 1 & -3 & -2 & | & 0 \\ 0 & 5 & 3 & | & 3 \\ 1 & 2 & 3 & | & 0 \end{bmatrix} \quad -2R_1 + R_2$$

$$\begin{bmatrix} 1 & -3 & -2 & | & 0 \\ 0 & 5 & 3 & | & 3 \\ 0 & -5 & -5 & | & 0 \end{bmatrix} \quad R_1 - R_2$$

$$\begin{bmatrix} 1 & -3 & -2 & | & 0 \\ 0 & 5 & 3 & | & 3 \\ 0 & 0 & -2 & | & 3 \end{bmatrix} \quad R_2 + R_3$$

$$\begin{bmatrix} 1 & -3 & -2 & | & 0 \\ 0 & 5 & 3 & | & 3 \\ 0 & 0 & 1 & | & -\dfrac{3}{2} \end{bmatrix} \quad -\tfrac{1}{2}R_3$$

The system is now in row echelon form, and

$$z = -\frac{3}{2}.$$

Using back substitution, we find that

$$5y + 3(-3/2) = 3$$

$$5y = \frac{15}{2}$$

$$y = \frac{3}{2}$$

Solve for x as follows:

$$-x + 3\left(\frac{3}{2}\right) + 2\left(-\frac{3}{2}\right) = 0$$

$$-x + \frac{9}{2} - 3 = 0$$

$$-x + \frac{3}{2} = 0$$

$$x = \frac{3}{2}$$

The solution set is $\left(\dfrac{3}{2}, \dfrac{3}{2}, -\dfrac{3}{2}\right)$.

61. The augmented matrix is

$$\begin{bmatrix} 1 & -2 & 4 & | & 6 \\ 1 & 1 & 13 & | & 6 \\ -2 & 6 & -1 & | & -10 \end{bmatrix}$$

Now use the matrix method to solve the system:

$$\begin{bmatrix} 1 & -2 & 4 & | & 6 \\ 0 & -3 & -9 & | & 0 \\ -2 & 6 & -z & | & -10 \end{bmatrix} \quad R_1 - R_2$$

$$\begin{bmatrix} 1 & -2 & 4 & | & 6 \\ 0 & 1 & 3 & | & 0 \\ 0 & 2 & 7 & | & 2 \end{bmatrix} \quad \begin{array}{l} -\tfrac{1}{3}R_2 \\ 2R_1 + R_3 \end{array}$$

$$\begin{bmatrix} 1 & -2 & 4 & | & 6 \\ 0 & 1 & 3 & | & 0 \\ 0 & 0 & 1 & | & 2 \end{bmatrix} \quad -2R_2 + R_3$$

The system is now in row echelon form, and

$$z = 2$$

$$y + 3(2) = 0$$

$$y = -6$$

$$x + -2(-6) + 4(2) = 6$$

$$x + 20 = 6$$

$$x = -14$$

The solution set for the system is

$$(-14, -6, 2)$$

63. The augmented matrix is

$$\begin{bmatrix} 1 & 1 & 1 & | & 200 \\ 1 & -2 & 0 & | & 0 \\ 2 & 3 & 5 & | & 600 \\ 2 & -1 & 1 & | & 200 \end{bmatrix}$$

Use the matrix method to solve the system:

$$\begin{bmatrix} 1 & 1 & 1 & | & 200 \\ 0 & 3 & 1 & | & 200 \\ 0 & 4 & 4 & | & 400 \\ 2 & -1 & 1 & | & 200 \end{bmatrix} \begin{matrix} \\ R_1 - R_2 \\ R_3 - R_4 \\ \\ \end{matrix}$$

$$\begin{bmatrix} 1 & 1 & 1 & | & 200 \\ 0 & 3 & 1 & | & 200 \\ 0 & 4 & 4 & | & 400 \\ 0 & 3 & 1 & | & 200 \end{bmatrix} \begin{matrix} \\ \\ \\ 2R_1 - R_4 \end{matrix}$$

$$\begin{bmatrix} 1 & 1 & 1 & | & 200 \\ 0 & 3 & 1 & | & 200 \\ 0 & 1 & 1 & | & 100 \\ 0 & 0 & 0 & | & 0 \end{bmatrix} \begin{matrix} \\ \\ \frac{1}{4}R_3 \\ R_2 - R_4 \end{matrix}$$

$$\begin{bmatrix} 1 & 1 & 1 & | & 200 \\ 0 & 3 & 1 & | & 200 \\ 0 & 0 & 2 & | & 100 \\ 0 & 0 & 0 & | & 0 \end{bmatrix} \begin{matrix} \\ \\ 3R_3 - R_2 \\ \\ \end{matrix}$$

$$\begin{bmatrix} 1 & 1 & 1 & | & 200 \\ 0 & 3 & 1 & | & 200 \\ 0 & 0 & 1 & | & 50 \\ 0 & 0 & 0 & | & 0 \end{bmatrix} \begin{matrix} \\ \\ \frac{1}{2}R_3 \\ \\ \end{matrix}$$

The system is now in row echelon form, and $z = 50$.

Use back-substitution to solve for x and y.

$3y + 1(50) = 200$

$3y = 150$

$y = 50$

$x + 1(50) + 1(50) = 200$

$x + 100 = 200$

$x = 100$

The solution to the system is

$(100, 50, 50)$

65. $x + y + z = 2$

   $2x + y - z = 5$

   $x - y + z = -2$

The augmented matrix is

$$\begin{bmatrix} 1 & 1 & 1 & | & 2 \\ 2 & 1 & -1 & | & 5 \\ 1 & -1 & 1 & | & -2 \end{bmatrix}$$

Now use the matrix method to solve the system:

$$\begin{bmatrix} 1 & 1 & 1 & | & 2 \\ 0 & -1 & -3 & | & 1 \\ 1 & -1 & 1 & | & -2 \end{bmatrix} -2R_1 + R_2$$

$$\begin{bmatrix} 1 & 1 & 1 & | & 2 \\ 0 & -1 & -3 & | & 1 \\ 0 & -2 & 0 & | & -4 \end{bmatrix} -R_1 + R_3$$

$$\begin{bmatrix} 1 & 1 & 1 & | & 2 \\ 0 & -1 & -3 & | & 1 \\ 0 & 0 & 6 & | & -6 \end{bmatrix} -2R_2 + R_3$$

$$\begin{bmatrix} 1 & 1 & 1 & | & 2 \\ 0 & 1 & 3 & | & -1 \\ 0 & 0 & 1 & | & -1 \end{bmatrix} \begin{matrix} \\ -R_2 \\ \frac{1}{6}R_3 \end{matrix}$$

The system is now in row echelon form, and $z = -1$. Using back substitution, we find that $y + 3(-1) = -1$ or $y = 2$,

and $x + 2 + (-1) = 2$ or $x = 1$. The solution is $(1, 2, -1)$.

67. $x + 3y + 4z = 14$
$2x - 3y + 2z = 10$
$3x - y + z = 9$
$4x + 2y + 5z = 23$

The augmented matrix is

$$\begin{bmatrix} 1 & 3 & 4 & | & 14 \\ 2 & -3 & 2 & | & 10 \\ 3 & -1 & 1 & | & 9 \\ 4 & 2 & 5 & | & 23 \end{bmatrix}$$

Now use the matrix method to solve the system:

$$\begin{bmatrix} 1 & 3 & 4 & | & 14 \\ 0 & -9 & -6 & | & -18 \\ 0 & -10 & -11 & | & -33 \\ 0 & -10 & -11 & | & -33 \end{bmatrix} \begin{matrix} \\ -2R_1 + R_2 \\ -3R_1 + R_3 \\ -4R_1 + R_4 \end{matrix}$$

$$\begin{bmatrix} 1 & 3 & 4 & | & 14 \\ 0 & 1 & \frac{2}{3} & | & 2 \\ 0 & -10 & -11 & | & -33 \\ 0 & 0 & 0 & | & 0 \end{bmatrix} \begin{matrix} \\ -\frac{1}{9}R_2 \\ \\ -R_3 + R_4 \end{matrix}$$

$$\begin{bmatrix} 1 & 3 & 4 & | & 14 \\ 0 & 1 & \frac{2}{3} & | & 2 \\ 0 & 0 & -\frac{13}{3} & | & -13 \\ 0 & 0 & 0 & | & 0 \end{bmatrix} \begin{matrix} \\ \\ 10R_2 + R_3 \\ \\ \end{matrix}$$

$$\begin{bmatrix} 1 & 3 & 4 & | & 14 \\ 0 & 1 & \frac{2}{3} & | & 2 \\ 0 & 0 & 1 & | & 3 \\ 0 & 0 & 0 & | & 0 \end{bmatrix} \begin{matrix} \\ \\ -\frac{3}{13}R_3 \\ \\ \end{matrix}$$

The system is now in row echelon form, and $z = 3$.
Using back substitution, we find $y = 0$ and $x = 2$.
The solution is

$(2, 0, 3)$.

69. $x + 2y + 3z = 8$
$3x - y + 2z = 5$
$-2x - 4y - 6z = 5$

The augmented matrix is

$$\begin{bmatrix} 1 & 2 & 3 & | & 8 \\ 3 & -1 & 2 & | & 5 \\ -2 & -4 & -6 & | & 5 \end{bmatrix}$$

Now use the matrix method to solve the system:

$$\begin{bmatrix} 1 & 2 & 3 & | & 8 \\ 3 & -1 & 2 & | & 5 \\ 0 & 0 & 0 & | & 21 \end{bmatrix} \begin{matrix} \\ \\ 2R_1 + R_3 \end{matrix}$$

The last row is equivalent to the equation
$0 = 21$, which is false. Therefore the system is
inconsistent and has no solution.

71. $2x - 4y + z = -4$
$x + 2y - z = 0$
$-x + y + z = 6$
$2x - y + z = 2$

The augmented matrix is

$$\begin{bmatrix} 2 & -4 & 1 & | & -4 \\ 1 & 2 & -1 & | & 0 \\ -1 & 1 & 1 & | & 6 \\ 2 & -1 & 1 & | & 2 \end{bmatrix}$$

Now use the matrix method to solve the system:

$$\begin{bmatrix} 1 & 2 & -1 & | & 0 \\ 2 & -4 & 1 & | & -4 \\ -1 & 1 & 1 & | & 6 \\ 2 & -1 & 1 & | & 2 \end{bmatrix} \text{Interchange } R_1 \text{ and } R_2$$

$$\begin{bmatrix} 1 & 2 & -1 & | & 0 \\ 0 & -8 & 3 & | & -4 \\ 0 & 3 & 0 & | & 6 \\ 0 & 3 & 0 & | & 6 \end{bmatrix} \begin{matrix} \\ -2R_1 + R_2 \\ R_1 + R_3 \\ -R_2 + R_4 \end{matrix}$$

$$\begin{bmatrix} 1 & 2 & -1 & | & 0 \\ 0 & 1 & -\frac{3}{8} & | & \frac{1}{2} \\ 0 & 1 & 0 & | & 2 \\ 0 & 0 & 0 & | & 0 \end{bmatrix} \begin{matrix} \\ -\frac{1}{8}R_2 \\ \frac{1}{3}R_3 \\ -R_3 + R_4 \end{matrix}$$

From the last row, we have $y = 2$. Using back
substitution we find $z = 4$ and $x = 0$. The solution is
$(0, 2, 4)$.

73. $5x + 3y + 4z = 19$

$\qquad 3x - y + z = -4$

The augmented matrix is

$$\begin{bmatrix} 5 & 3 & 4 & | & 19 \\ 3 & -1 & 1 & | & -4 \end{bmatrix}$$

Now use the matrix method to solve the system:

$$\begin{bmatrix} 14 & 0 & 7 & | & 7 \\ 3 & -1 & 1 & | & -4 \end{bmatrix} 3R_2 + R_1$$

$$\begin{bmatrix} 1 & 0 & \frac{1}{2} & | & \frac{1}{2} \\ 3 & -1 & 1 & | & -4 \end{bmatrix} \frac{1}{14}R_1$$

Since there are only two equations, the system has an infinite number of solutions. Solve the first equation for x in terms of the parameter z.

$$x + \frac{1}{2}z = \frac{1}{2}$$

$$x = \frac{1}{2} - \frac{1}{2}z$$

$$= \frac{1-z}{2}$$

Now substitute $\dfrac{1-z}{2}$ for x in the second equation and solve for y in terms of the parameter z.

$$3\left(\frac{1-z}{2}\right) - y + z = -4$$

$$3\left(\frac{1-z}{2}\right) + z + 4 = y$$

$$\frac{3(1-z) + 2x + 8}{2} = y$$

$$\frac{3 - 3z + 2z + 8}{2} = y$$

$$\frac{11 - z}{2} = y$$

The solution is $\left(\dfrac{1-z}{2}, \dfrac{11-z}{2}, z\right)$ for any real number z.

75. $11x + 10y + 9z = 5$

$\qquad x + 2y + 3z = 1$

$\qquad 3x + 2y + z = 1$

The augmented matrix is

$$\begin{bmatrix} 11 & 10 & 9 & | & 5 \\ 1 & 2 & 3 & | & 1 \\ 3 & 2 & 1 & | & 1 \end{bmatrix}$$

Now use the matrix method to solve the system:

$$\begin{bmatrix} 1 & 2 & 3 & | & 1 \\ 11 & 10 & 9 & | & 5 \\ 3 & 2 & 1 & | & 1 \end{bmatrix} \text{Interchange } R_1 + R_2$$

$$\begin{bmatrix} 1 & 2 & 3 & | & 1 \\ 0 & -12 & -24 & | & -6 \\ 0 & -4 & -8 & | & -2 \end{bmatrix} \begin{matrix} \\ -11R_1 + R_2 \\ -3R_1 + R_3 \end{matrix}$$

$$\begin{bmatrix} 1 & 2 & 3 & | & 1 \\ 0 & 1 & 2 & | & \frac{1}{2} \\ 0 & 1 & 2 & | & \frac{1}{2} \end{bmatrix} \begin{matrix} \\ -\frac{1}{12}R_2 \\ -\frac{1}{4}R_3 \end{matrix}$$

$$\begin{bmatrix} 1 & 2 & 3 & | & 1 \\ 0 & 1 & 2 & | & \frac{1}{2} \\ 0 & 0 & 0 & | & 0 \end{bmatrix} -R_2 + R_3$$

The last row gives $0 = 0$, which is true, so the system is dependent. Use the second equation to solve for y in terms of the parameters z.

$$y + 2z = \frac{1}{2}$$

$$y = \frac{1}{2} - 2z$$

Substitute $\dfrac{1}{2} - 2z$ for y in the first equation.

$$x + 2\left(\frac{1}{2} - 2z\right) + 3z = 1$$

$$x + 1 - 4z + 3z = 1$$

$$x = z$$

The solution is $\left(z, \dfrac{1}{2} - 2z, z\right)$ for any real number z.

77. a. To find the equation of the line through (1, 2) and (3, 4), first find the slope.

$$m = \frac{4-2}{3-1} = \frac{2}{2} = 1$$

Use the point-slope form with $(x_1, y_1) = (1, 2)$.

$$y - y_1 = m(x - x_1)$$
$$y - 2 = 1(x - 1)$$
$$y - 2 = x - 1$$
$$y = x + 1$$

An alternate solution can be written using a system of equations. Using the form $y = mx + b$, the line through (1, 2) and (3, 4) would yield the system of equations.

$$2 = m(1) + b \quad (1)$$
$$4 = m(3) + b. \quad (2)$$

This simplifies to

$$m + b = 2 \quad (1)$$
$$3m + b = 4. \quad (2)$$

This system can be solved by multiplying equation (1) by –1 and adding the result to equation (2).

$$-m - b = -2$$
$$\underline{3m + b = 4 \quad (2)}$$
$$2m = 2$$
$$m = 1$$

Substitute 1 for m in equation (1).

$$m + b = 2 \quad (1)$$
$$1 + b = 2$$
$$b = 1$$

So, the equation is $y = x + 1$.

b. To find the equation of the line with slope $m = 3$ through (–1, 1), use the form

$y = mx + b$. Since $m = 3$, the equation becomes $y = 3x + b$.

Substitute –1 for x and 1 for y.

$$1 = 3(-1) + b$$
$$1 = -3 + b$$
$$4 = b$$

The equation is $y = 3x + 4$.

c. To find a point on both lines, solve the system

$$y = x + 1 \quad (1)$$
$$y = 3x + 4 \quad (2)$$

Multiply equation (1) by –1 and add the result to equation (2).

$$-y = -x - 1$$
$$\underline{y = 3x + 4 \quad (2)}$$
$$0 = 2x + 3$$
$$-2x = 3$$
$$x = -\frac{3}{2}$$

Substitute $-\dfrac{3}{2}$ for x in equation (1).

$$y = x + 1 \quad (1)$$
$$y = -\frac{3}{2} + 1$$
$$y = -\frac{1}{2}$$

The point $\left(-\dfrac{3}{2}, -\dfrac{1}{2}\right)$ is on both lines.

79.

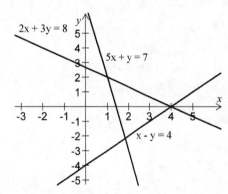

The three lines do not have one common

intersection point.

81. Let $x =$ investment in B account

And $2x =$ investment in AAA

Then $30,000 - 3x =$ investment in A.

Total interest is $2000.

Sum of the interests of the three is 2000.

$0.10(x) + 0.06(30,000 - 3x) + 0.05(2x) = 2000.00$
$0.10x + 1800 - 0.18x + 0.10x = 2000.00$
$0.02x + 1800 = 2000$
$0.02x = 2000 - 1800$
$0.02x = 200$
$x = 10,000$

$10,000 was invested in the B account.

No money was invested in the A account.

$20,000 was invested in the AAA account.

83. Let $x =$ number of pounds of corn chips

$y =$ number of pounds of nuts

$z =$ number of pounds of pretzels

The system of linear equations that corresponds to the problem is written below. Use substitution to solve.

$$\begin{cases} z + y + z = 100 & (1) \\ 2x + 6y + 3z = 4(100) & (2) \\ x = 3z & (3) \end{cases}$$

Substitute $x = 3z$ into (1) and (2)

$3z + y + z = 100$
$2(3z) + 6y + 3z = 400$
$x = 3z$

$4z + y = 100 \quad (1)$
$9z + 6y = 400 \quad (2)$

Solve for y from (1)

$y = 100 - 4z$
$9z + 6y = 400$

Substitute $y = 100 - 4z$ into (2)

83. Continued

$9z + 6(100 - 4z) = 400$
$9z + 600 - 24z = 400$
$-15z + 600 = 400$
$-15z = -200$
$z = 13\frac{1}{3}$ lbs (weight of pretzels)

Substitute $z = 13\frac{1}{3}$ into $y = 100 - 4z$

$y = 100 - 4\left(13\frac{1}{3}\right)$
$y = 100 - 4\left(\frac{40}{3}\right)$
$y = 100 - \frac{160}{3}$
$y = \frac{300}{3} - \frac{160}{3}$
$y = 46\frac{2}{3}$ lbs (weight of nuts)

Because $x = 3z$,

$x = 3\left(13\frac{1}{3}\right)$
$x = 3\left(\frac{40}{3}\right)$
$x = 40$ lbs (weight of corn chips)

85. Let $x =$ the number of box seats

$y =$ the number of grandstand seats

$z =$ the number of bleacher seats

The system of linear equations that describes the problem is written below.

$x + y + z = 7000 \quad (1)$

$6x + 4y + 2z = 26,400 \quad (2)$

$x = \frac{1}{3}z \quad (3)$

Use substitution to solve. First substitute $x = \frac{1}{3}z$ into (1) and (2) to get

85. Continued

$$\frac{1}{3}z + y + z = 7000 \qquad (4)$$

$$6\left(\frac{1}{3}z\right) + 4y + 2z = 26,400 \quad (5)$$

Multiply (4) and (5) by 3 to clear the fractions. Then combine like terms.

$$4z + 3y = 21000 \quad (6)$$
$$12z + 12y = 79200 \quad (7)$$

Multiply (6) by (-3) and add to (7).

$$-12z - 9y = -21000$$
$$\underline{12z + 12y = 79200}$$
$$3y = 16200$$
$$y = 5400$$

Substitute $y = 5400$ into (6) to solve for z

$$4z + 3(5400) = 21000$$
$$4z + 16200 = 21000$$
$$4z = 4800$$
$$z = 1200$$

Substitute $z = 1200$ into (3).

$$x = \frac{1}{3}(1200) = 400$$

So the answer is 400 box seats 5400　grandstand seats 1200 bleacher seats.

87 a. The attenuation value for beam 3 is $b + c$.

b. The system of equations is
$$a + b = .8$$
$$a + c = .55$$
$$b + c = .65$$
The augmented matrix is
$$\begin{bmatrix} 1 & 1 & 0 & | & .8 \\ 1 & 0 & 1 & | & .55 \\ 0 & 1 & 1 & | & .65 \end{bmatrix}$$
In row echelon form
$$\begin{bmatrix} 1 & 1 & 0 & | & .8 \\ 0 & 1 & -1 & | & .25 \\ 0 & 0 & 1 & | & .2 \end{bmatrix}$$
So $c = .2$. Back substitution yields $a = .35$, $b = .45$. Therefore, A is tumorous, B is bone, and C is healthy tissue.

89. a. For intersection C, x_2 cars enter on 11th street and x_3 on N street. The number of cars entering C must equal the number leaving, so that
$$x_2 + x_3 = 300 + 400$$
$$x_2 + x_3 = 700$$
For intersection D, x_3 cars enter on N street and x_4 cars enter on 10th street. The figure shows that 200 cars leave D on N street and 400 on 10th street.
$$x_3 + x_4 = 200 + 400$$
$$x_3 + x_4 = 600$$

b. The system of equations is
$$x_1 + x_4 = 1000$$
$$x_1 + x_2 = 1100$$
$$x_2 + x_3 = 700$$
$$x_3 + x_4 = 600$$
Solve each equation in terms of x_4.
$$x_1 = 1000 - x_4$$
$$x_3 = 600 - x_4$$
$$x_2 = 1100 - (1000 - x_4) = 100 + x_4$$
The solution is $(1000 - x_4, 100 + x_4, 600 - x_4, x_4)$.

Section 6.2 The Gauss-Jordan Method

1. $\begin{bmatrix} 1 & 0 & 0 & 0 & \frac{3}{2} \\ 0 & 1 & 0 & 0 & 7 \\ 0 & 0 & 1 & 0 & -3 \\ 0 & 0 & 0 & 1 & 0 \end{bmatrix}$

The solution of the system is $\left(\frac{3}{2}, 7, -3, 0\right)$.

3. $\begin{bmatrix} 1 & 0 & 0 & 1 & 2 \\ 0 & 1 & 0 & 2 & -3 \\ 0 & 0 & 1 & 0 & 5 \\ 0 & 0 & 0 & 0 & 0 \end{bmatrix}$

The linear system associated with this matrix is
$x + w = 2$ (1)
$y + 2w = -3$ (2)
$\quad z = 5$ (3)

To put the solution in terms of w, solve for x in equation (1) and y in equation (2).
The solution is $(2 - w, -3 - 2w, 5, w)$.

5. $x + 2y + z = 5$
$2x + y - 3z = -2$
$3x + y + 4z = -5$
The matrix is
$\begin{bmatrix} 1 & 2 & 1 & 5 \\ 2 & 1 & -3 & -2 \\ 3 & 1 & 4 & -5 \end{bmatrix}$

$\begin{bmatrix} 1 & 2 & 1 & 5 \\ 0 & -3 & -5 & -12 \\ 0 & -5 & 1 & -20 \end{bmatrix} \begin{matrix} \\ -2R_1 + R_2 \\ -3R_1 + R_2 \end{matrix}$

$\begin{bmatrix} 1 & 2 & 1 & 5 \\ 0 & 1 & \frac{5}{3} & 4 \\ 0 & -5 & 1 & -20 \end{bmatrix} -\frac{1}{3}R_2$

$\begin{bmatrix} 1 & 0 & -\frac{7}{3} & -3 \\ 0 & 1 & \frac{5}{3} & 4 \\ 0 & 0 & \frac{28}{3} & 0 \end{bmatrix} \begin{matrix} -2R_2 + R_1 \\ \\ 5R_2 + R_3 \end{matrix}$

$\begin{bmatrix} 1 & 0 & -\frac{7}{3} & -3 \\ 0 & 1 & \frac{5}{3} & 4 \\ 0 & 0 & 1 & 0 \end{bmatrix} \frac{3}{28}R_3$

$\begin{bmatrix} 1 & 0 & 0 & -3 \\ 0 & 1 & 0 & 4 \\ 0 & 0 & 1 & 0 \end{bmatrix} \begin{matrix} \frac{7}{3}R_3 + R_1 \\ -\frac{3}{5}R_3 + R_2 \end{matrix}$

The solution is $(-3, 4, 0)$.

7. $x + 3y - 6z = 7$
$2x - y + 2z = 0$
$x + y + 2z = -1$
The matrix is
$\begin{bmatrix} 1 & 3 & -6 & 7 \\ 2 & -1 & 2 & 0 \\ 1 & 1 & 2 & -1 \end{bmatrix}$

$\begin{bmatrix} 1 & 3 & -6 & 7 \\ 0 & -7 & 14 & -14 \\ 0 & -2 & 8 & -8 \end{bmatrix} \begin{matrix} \\ -2R_1 + R_2 \\ -1R_1 + R_3 \end{matrix}$

$\begin{bmatrix} 1 & 3 & -6 & 7 \\ 0 & 1 & -2 & 2 \\ 0 & -2 & 8 & -8 \end{bmatrix} -\frac{1}{7}R_2$

$\begin{bmatrix} 1 & 0 & 0 & 1 \\ 0 & 1 & -2 & 2 \\ 0 & 0 & 4 & -4 \end{bmatrix} \begin{matrix} -3R_2 + R_1 \\ \\ 2R_2 + R_3 \end{matrix}$

$\begin{bmatrix} 1 & 0 & 0 & 1 \\ 0 & 1 & -2 & 2 \\ 0 & 0 & 1 & -1 \end{bmatrix} \frac{1}{4}R_3$

$\begin{bmatrix} 1 & 0 & 0 & 1 \\ 0 & 1 & 0 & 0 \\ 0 & 0 & 1 & -1 \end{bmatrix} 2R_3 + R_2$

The solution is $(1, 0, -1)$.

9. $x - 2y + 4z = 6$
$x + y + 13z = 6$
$-2x + 6y - z = -10$

The augmented matrix is

$\begin{bmatrix} 1 & -2 & 4 & 6 \\ 1 & 1 & 13 & 6 \\ -2 & 6 & -1 & -10 \end{bmatrix}$

$\begin{bmatrix} 1 & -2 & 4 & 6 \\ 0 & 3 & 9 & 0 \\ 0 & 2 & 7 & 2 \end{bmatrix} \begin{matrix} \\ -R_1 + R_2 \\ 2R_1 + R_3 \end{matrix}$

$\begin{bmatrix} 1 & -2 & 4 & 6 \\ 0 & 3 & 9 & 0 \\ 0 & 0 & 3 & 6 \end{bmatrix} \begin{matrix} \\ \\ -2R_1 + 3R_3 \end{matrix}$

$\begin{bmatrix} 1 & -2 & 4 & 6 \\ 0 & 1 & 3 & 0 \\ 0 & 0 & 1 & 2 \end{bmatrix} \begin{matrix} \\ \frac{1}{3}R_2 \\ \frac{1}{3}R_3 \end{matrix}$

9. Continued

$$\begin{bmatrix} 1 & 0 & 10 & | & 6 \\ 0 & 1 & 3 & | & 0 \\ 0 & 0 & 1 & | & 2 \end{bmatrix} \quad R_1 + 2R_2$$

$$\begin{bmatrix} 1 & 0 & 0 & | & -14 \\ 0 & 1 & 0 & | & 0 \\ 0 & 0 & 1 & | & 2 \end{bmatrix} \begin{matrix} R_1 + -10R_3 \\ R_2 + -3R_3 \end{matrix}$$

The solution is $(-14, -6, 2)$.

11.
$$3x + 5y - z = 0$$
$$4x - y + 2z = 1$$
$$-6x - 10y + 2z = 0$$

The matrix is

$$\begin{bmatrix} 3 & 5 & -1 & | & 0 \\ 4 & -1 & 2 & | & 1 \\ -6 & -10 & 2 & | & 0 \end{bmatrix}$$

$$\begin{bmatrix} 1 & \frac{5}{3} & -\frac{1}{3} & | & 0 \\ 4 & -1 & 2 & | & 1 \\ 1 & \frac{5}{3} & -\frac{1}{3} & | & 0 \end{bmatrix} \begin{matrix} \frac{1}{3}R_1 \\ \\ -\frac{1}{6}R_3 \end{matrix}$$

$$\begin{bmatrix} 1 & \frac{5}{3} & -\frac{1}{3} & | & 0 \\ 0 & -\frac{23}{3} & \frac{10}{3} & | & 1 \\ 0 & 0 & 0 & | & 0 \end{bmatrix} \begin{matrix} \\ -4R_1 + R_2 \\ -1R_1 + R_3 \end{matrix}$$

$$\begin{bmatrix} 1 & \frac{5}{3} & -\frac{1}{3} & | & 0 \\ 0 & 1 & -\frac{10}{23} & | & -\frac{3}{23} \\ 0 & 0 & 0 & | & 0 \end{bmatrix} -\frac{3}{23}R_2$$

$$\begin{bmatrix} 1 & 0 & \frac{9}{23} & | & \frac{5}{23} \\ 0 & 1 & -\frac{10}{23} & | & -\frac{3}{23} \\ 0 & 0 & 0 & | & 0 \end{bmatrix} -\frac{5}{3}R_2 + R_1$$

The row of zeros indicates a dependent system. Solve the first two equations respectively for x and y in terms of the parameter z to obtain

$$x = \frac{-9z + 5}{23}$$

$$y = \frac{10z - 3}{23}.$$

There is an infinite number of solutions of the form

$$\left(\frac{-9z + 5}{23}, \frac{10z - 3}{23}, z \right)$$

where z can be any real number.

13.
$$x + y - z = 6$$
$$2x - y + z = -9$$
$$x - 2y + 3z = 1$$

The matrix is

$$\begin{bmatrix} 1 & 1 & -1 & | & 6 \\ 2 & -1 & 1 & | & -9 \\ 1 & -2 & 3 & | & 1 \end{bmatrix}$$

$$\begin{bmatrix} 1 & 1 & -1 & | & 6 \\ 0 & -3 & 3 & | & -21 \\ 0 & -3 & 4 & | & -5 \end{bmatrix} \begin{matrix} \\ -2R_1 + R_2 \\ -1R_1 + R_3 \end{matrix}$$

$$\begin{bmatrix} 1 & 1 & -1 & | & 6 \\ 0 & 1 & -1 & | & 7 \\ 0 & -3 & 4 & | & -5 \end{bmatrix} -\frac{1}{3}R_2$$

$$\begin{bmatrix} 1 & 0 & 0 & | & -1 \\ 0 & 1 & -1 & | & 7 \\ 0 & 0 & 1 & | & 16 \end{bmatrix} \begin{matrix} -1R_2 + R_1 \\ \\ 3R_2 + R_3 \end{matrix}$$

$$\begin{bmatrix} 1 & 0 & 0 & | & -1 \\ 0 & 1 & 0 & | & 23 \\ 0 & 0 & 1 & | & 16 \end{bmatrix} R_3 + R_2$$

The solution is $(-1, 23, 16)$.

15. $x + y + z = 1$

$$x - 2y + 2z = 4$$
$$2x - y + 3z = 5$$

The augmented matrix is

$$\begin{bmatrix} 1 & 1 & 1 & | & 1 \\ 1 & -2 & 2 & | & 4 \\ 2 & -1 & 3 & | & 5 \end{bmatrix}$$

$$\begin{bmatrix} 1 & 1 & 1 & | & 1 \\ 0 & -3 & 1 & | & 3 \\ 0 & 3 & -1 & | & -3 \end{bmatrix} \begin{matrix} \\ -R_1 + R_2 \\ -2R_2 + R_1 \end{matrix}$$

$$\begin{bmatrix} 1 & 1 & 1 & | & 1 \\ 0 & -3 & 1 & | & 3 \\ 0 & 0 & 0 & | & 0 \end{bmatrix} R_2 + R_3$$

Since the last row is all zeros, the system has no unique solution. Rows 1 and 2 can be written as the system

15. Continued

$$x + y + z = 1 \quad (1)$$
$$-3y + z = 3 \quad (2)$$

Let z be arbitrary, so

$$y = -1 + \frac{1}{3}z$$

Substitute $y = -1 + \frac{1}{3}z$ in (1)

$$x + \frac{1}{3}z + z = 1$$

$$x = 2 - \frac{4}{3}z$$

The solution is $(2 - \frac{4}{3}z, -1 + \frac{1}{3}z, z)$ for any

real number z.

17.
$$x - 2y + z = 5$$
$$2x + y - z = 2$$
$$-2x + 4y - 2z = 2$$

The matrix is

$$\begin{bmatrix} 1 & -2 & 1 & | & 5 \\ 2 & 1 & -1 & | & 2 \\ -2 & 4 & -2 & | & 2 \end{bmatrix}$$

$$\begin{bmatrix} 1 & -2 & 1 & | & 5 \\ 0 & 5 & -3 & | & -8 \\ 0 & 0 & 0 & | & 12 \end{bmatrix} \begin{matrix} \\ -2R_1 + R_2 \\ 2R_1 + R_3 \end{matrix}$$

The third row represents the equation
$0 = 12$, which is false. The system is inconsistent
and has no solution.

19.
$$-8x - 9y = 11$$
$$24x + 34y = 2$$
$$16x + 11y = -57$$

The matrix is

$$\begin{bmatrix} -8 & -9 & | & 11 \\ 24 & 34 & | & 2 \\ 16 & 11 & | & -57 \end{bmatrix}$$

$$\begin{bmatrix} -8 & -9 & | & 11 \\ 0 & 7 & | & 35 \\ 0 & -7 & | & -35 \end{bmatrix} \begin{matrix} \\ 3R_1 + R_2 \\ 2R_1 + R_2 \end{matrix}$$

$$\begin{bmatrix} 1 & \frac{9}{8} & | & -\frac{11}{8} \\ 0 & 1 & | & 5 \\ 0 & 1 & | & 5 \end{bmatrix} \begin{matrix} -\frac{1}{8}R_1 \\ \frac{1}{7}R_2 \\ -\frac{1}{7}R_3 \end{matrix}$$

$$\begin{bmatrix} 1 & 0 & | & -7 \\ 0 & 1 & | & 5 \\ 0 & 0 & | & 0 \end{bmatrix} \begin{matrix} -\frac{9}{8}R_2 + R_1 \\ \\ -1R_2 + R_3 \end{matrix}$$

The solution is $(-7, 5)$.

21.
$$x + 2y = 3$$
$$2x + 3y = 4$$
$$3x + 4y = 5$$
$$4x + 5y = 6$$

The matrix is

$$\begin{bmatrix} 1 & 2 & | & 3 \\ 2 & 3 & | & 4 \\ 3 & 4 & | & 5 \\ 4 & 5 & | & 6 \end{bmatrix}$$

Multiply row 1 by –2 and add to row 2; multiply
row 1 by –3 and add to row 3; multiply row 1
by –4 and add to row 4 to obtain the following.

$$\begin{bmatrix} 1 & 2 & | & 3 \\ 0 & -1 & | & -2 \\ 0 & -2 & | & -4 \\ 0 & -3 & | & -6 \end{bmatrix}$$

Next, multiply row 2 by –2 to get

$$\begin{bmatrix} 1 & 2 & | & 3 \\ 0 & 1 & | & 2 \\ 0 & -2 & | & -4 \\ 0 & -3 & | & -6 \end{bmatrix}$$

21. Continued

Finally, multiply row 2 by –2 and add to row 1; multiply row 2 by 2 and add to row 3; multiply row 2 by 3 and add to row 4 to obtain

$$\begin{bmatrix} 1 & 0 & -1 \\ 0 & 1 & 2 \\ 0 & 0 & 0 \\ 0 & 0 & 0 \end{bmatrix}$$

The solution is $(-1, 2)$.

23. $x + y - z = -20$

$2x - y + z = 11$

The matrix is

$$\begin{bmatrix} 1 & 1 & -1 & -20 \\ 2 & -1 & 1 & 11 \end{bmatrix}$$

$$\begin{bmatrix} 1 & 1 & -1 & -20 \\ 0 & -3 & 3 & 51 \end{bmatrix} -2R_1 + R_2$$

$$\begin{bmatrix} 1 & 1 & -1 & -20 \\ 0 & 1 & -1 & -17 \end{bmatrix} -\tfrac{1}{3} R_2$$

$$\begin{bmatrix} 1 & 0 & 0 & -3 \\ 0 & 1 & -1 & -17 \end{bmatrix} -R_3 + R_1$$

This last matrix is the augmented matrix for the system

$x = -3$ *(1)*

$y - z = -17.$ *(2)*

Solve equation (2) for y in terms of z.

$y = z - 17$. The solution is all ordered triples of the form $(-3, z - 17, z)$ for any real number z.

25. $2x + y + 3z - 2w = -6$

$4x + 3y + z - w = -2$

$x + y + z + w = -5$

$-2x - 2y + 2z + 2w = -10$

The matrix is

$$\begin{bmatrix} 2 & 1 & 3 & -2 & -6 \\ 4 & 3 & 1 & -1 & -2 \\ 1 & 1 & 1 & 1 & -5 \\ -2 & -2 & 2 & 2 & -10 \end{bmatrix}$$

Interchange R_1 and R_3.

$$\begin{bmatrix} 1 & 1 & 1 & 1 & -5 \\ 4 & 3 & 1 & -1 & -2 \\ 2 & 1 & 3 & -2 & -6 \\ -2 & -2 & 2 & 2 & -10 \end{bmatrix}$$

$$\begin{bmatrix} 1 & 1 & 1 & 1 & -5 \\ 0 & -1 & -3 & -5 & 18 \\ 0 & -1 & 1 & -4 & 4 \\ 0 & 0 & 4 & 4 & -20 \end{bmatrix} \begin{matrix} \\ -4R_1 + R_2 \\ -2R_1 + R_3 \\ 2R_1 + R_4 \end{matrix}$$

$$\begin{bmatrix} 1 & 1 & 1 & 1 & -5 \\ 0 & 1 & 3 & 5 & -18 \\ 0 & -1 & 1 & -4 & 4 \\ 0 & 0 & 4 & 4 & -20 \end{bmatrix} -1R_2$$

$$\begin{bmatrix} 1 & 0 & -2 & -4 & 13 \\ 0 & 1 & 3 & 5 & -18 \\ 0 & 0 & 4 & 1 & -14 \\ 0 & 0 & 4 & 4 & -20 \end{bmatrix} \begin{matrix} -1R_2 + R_1 \\ \\ R_2 + R_3 \\ \end{matrix}$$

$$\begin{bmatrix} 1 & 0 & -2 & -4 & 13 \\ 0 & 1 & 3 & 5 & -18 \\ 0 & 0 & 1 & \tfrac{1}{4} & -\tfrac{7}{2} \\ 0 & 0 & 4 & 4 & -20 \end{bmatrix} \tfrac{1}{4} R_3$$

25. Continued

$$\begin{bmatrix} 1 & 0 & 0 & -\frac{7}{2} & | & 6 \\ 0 & 1 & 0 & \frac{17}{4} & | & -\frac{15}{2} \\ 0 & 0 & 1 & \frac{1}{4} & | & -\frac{7}{2} \\ 0 & 0 & 0 & 3 & | & -6 \end{bmatrix} \begin{matrix} 2R_3 + R_1 \\ -3R_3 + R_2 \\ \\ -4R_3 + R_4 \end{matrix}$$

$$\begin{bmatrix} 1 & 0 & 0 & -\frac{7}{2} & | & 6 \\ 0 & 1 & 0 & \frac{17}{4} & | & -\frac{15}{2} \\ 0 & 0 & 1 & \frac{1}{4} & | & -\frac{7}{2} \\ 0 & 0 & 0 & 1 & | & -2 \end{bmatrix} \frac{1}{3}R_4$$

$$\begin{bmatrix} 1 & 0 & 0 & 0 & | & -1 \\ 0 & 1 & 0 & 0 & | & 1 \\ 0 & 0 & 1 & 0 & | & -3 \\ 0 & 0 & 0 & 1 & | & -2 \end{bmatrix} \begin{matrix} \frac{7}{2}R_4 + R_1 \\ -\frac{17}{4}R_4 + R_2 \\ -\frac{1}{4}R_4 + R_3 \\ \\ \end{matrix}$$

The solution is $(-1, 1, -3, -2)$.

27.
$$x + 2y - z = 3$$
$$3x + y + w = 4$$
$$2x - y + z + w = 2$$

The matrix is

$$\begin{bmatrix} 1 & 2 & -1 & 0 & | & 3 \\ 3 & 1 & 0 & 1 & | & 4 \\ 2 & -1 & 1 & 1 & | & 2 \end{bmatrix}$$

$$\begin{bmatrix} 1 & 2 & -1 & 0 & | & 3 \\ 0 & -5 & 3 & 1 & | & -5 \\ 0 & -5 & 3 & 1 & | & -4 \end{bmatrix} \begin{matrix} \\ -3R_1 + R_2 \\ -2R_1 + R_3 \end{matrix}$$

$$\begin{bmatrix} 1 & 2 & -1 & 0 & | & 3 \\ 0 & -5 & 3 & 1 & | & -5 \\ 0 & 0 & 0 & 0 & | & 1 \end{bmatrix} -1R_2 + R_3$$

Since the last row yields a false statement, there is no solution.

29. $\dfrac{3}{x} - \dfrac{1}{y} + \dfrac{4}{z} = -13$

$\dfrac{1}{x} + \dfrac{2}{y} - \dfrac{1}{z} = 12$

$\dfrac{4}{x} - \dfrac{1}{y} + \dfrac{3}{z} = -7$

Let $u = \dfrac{1}{x}, v = \dfrac{1}{y}, w = \dfrac{1}{z}$. Then the system becomes

$$3u - v + 4w = -13$$
$$u + 2v - w = 12$$
$$4u + v + 3w = -7$$

29. Continued

The matrix is

$$\begin{bmatrix} 3 & -1 & 4 & | & -13 \\ 1 & 2 & -1 & | & 12 \\ 4 & -1 & 3 & | & -7 \end{bmatrix}$$

Interchange rows 1 and 2:

$$\begin{bmatrix} 1 & 2 & -1 & | & 12 \\ 3 & -1 & 4 & | & 13 \\ 4 & -1 & 3 & | & -7 \end{bmatrix}$$

Multiply row 1 by -3 and add to row 2; multiply row 1 by -4 and add to row 3 to obtain

$$\begin{bmatrix} 1 & 2 & -1 & | & 12 \\ 0 & -7 & 7 & | & -49 \\ 0 & -9 & 7 & | & -55 \end{bmatrix} -\frac{1}{7}R_2 \to R_2$$

$$\begin{bmatrix} 1 & 2 & -1 & | & 12 \\ 0 & 1 & -1 & | & 7 \\ 0 & -9 & 7 & | & -55 \end{bmatrix}$$

Multiply row 2 by -2 and add to row 2; multiply row 2 by 9 and add to row 3 to obtain

$$\begin{bmatrix} 1 & 0 & 1 & | & -2 \\ 0 & 1 & -1 & | & 7 \\ 0 & 0 & -2 & | & 8 \end{bmatrix}$$

Multiply row 3 by $-\dfrac{1}{2}$. The new matrix is

$$\begin{bmatrix} 1 & 0 & 1 & | & -2 \\ 0 & 1 & -1 & | & 7 \\ 0 & 0 & 1 & | & -4 \end{bmatrix}$$

Multiply row 3 by -1 and add to row 1; multiply row 3 by 1 and add to row 2 to obtain

$$\begin{bmatrix} 1 & 0 & 0 & | & 2 \\ 0 & 1 & 0 & | & 3 \\ 0 & 0 & 1 & | & -4 \end{bmatrix}$$

So $u = 2$, $v = 3$ and $w = -4$. Since $u = \dfrac{1}{x}$, $v = \dfrac{1}{y}$, $w = \dfrac{1}{2}$, the solution is $\left(\dfrac{1}{2}, \dfrac{1}{3}, -\dfrac{1}{4} \right)$.

31. Let x = the number of hours for the Garcia firm;

y = the number of hours for the Wong firm.

The system of equations is

$10x + 20y = 500$

$30x + 10y = 750$

$5x + 10y = 250.$

The matrix is

$$\begin{bmatrix} 10 & 20 & | & 500 \\ 30 & 10 & | & 750 \\ 5 & 10 & | & 250 \end{bmatrix}$$

$$\begin{bmatrix} 1 & 2 & | & 50 \\ 30 & 10 & | & 750 \\ 5 & 10 & | & 250 \end{bmatrix} \frac{1}{10}R_1$$

$$\begin{bmatrix} 1 & 2 & | & 50 \\ 0 & -50 & | & -750 \\ 0 & 0 & | & 0 \end{bmatrix} \begin{matrix} \\ -30R_1 + R_2 \\ -5R_1 + R_3 \end{matrix}$$

$$\begin{bmatrix} 1 & 2 & | & 50 \\ 0 & 1 & | & 15 \\ 0 & 0 & | & 0 \end{bmatrix} -\frac{1}{50}R_2$$

$$\begin{bmatrix} 1 & 0 & | & 20 \\ 0 & 1 & | & 15 \\ 0 & 0 & | & 0 \end{bmatrix} -2R_2 + R_1$$

The Garcia firm should be hired for 20 hr, and the Wong firm should be hired for 15 hr.

33. Let x = the number of pounds of pretzels, y = the number of pounds of dried fruit, and z = the number of pounds of nuts.

The system of equations is

$x + y + z = 140$

$3x + 4y + 8z = 6(140)$

$x = 2y.$

The system simplifies to

$x + y + z = 140$

$3x + 4y + 8z = 840$

$x - 2y = 0.$

The matrix is

$$\begin{bmatrix} 1 & 1 & 1 & | & 140 \\ 3 & 4 & 8 & | & 840 \\ 1 & -2 & 0 & | & 0 \end{bmatrix}$$

$$\begin{bmatrix} 1 & 1 & 1 & | & 140 \\ 0 & 1 & 5 & | & 420 \\ 0 & -3 & -1 & | & -140 \end{bmatrix} \begin{matrix} \\ -3R_1 + R_2 \\ -R_1 + R_3 \end{matrix}$$

$$\begin{bmatrix} 1 & 0 & -4 & | & -280 \\ 0 & 1 & 5 & | & 420 \\ 0 & 0 & 14 & | & 1120 \end{bmatrix} \begin{matrix} -R_2 + R_1 \\ \\ 3R_2 + R_3 \end{matrix}$$

$$\begin{bmatrix} 1 & 0 & -4 & | & -280 \\ 0 & 1 & 5 & | & 420 \\ 0 & 0 & 1 & | & 80 \end{bmatrix} \frac{1}{14}R_3$$

$$\begin{bmatrix} 1 & 0 & 0 & | & 40 \\ 0 & 1 & 0 & | & 20 \\ 0 & 0 & 1 & | & 80 \end{bmatrix} \begin{matrix} 4R_3 + R_1 \\ -5R_3 + R_2 \\ \end{matrix}$$

Use 40 lb of pretzels, 20 lb of dried fruit, and 80 lb of nuts.

35. a. The two equations are

$$a(25)^2 + b(25) = 625a + 25b = 61.7$$

$$a(35)^2 + b(35) = 1225a + 35b = 106$$

The matrix is

$$\begin{bmatrix} 625 & 25 & | & 61.7 \\ 1225 & 35 & | & 106 \end{bmatrix}$$

$$\begin{bmatrix} 1 & \frac{1}{25} & | & \frac{617}{6250} \\ 1225 & 35 & | & 106 \end{bmatrix} \frac{1}{625}R_1$$

$$\begin{bmatrix} 1 & \frac{1}{25} & | & \frac{617}{6250} \\ 0 & -14 & | & \frac{-3733}{250} \end{bmatrix} -1225R_1 + R_2$$

$$\begin{bmatrix} 1 & \frac{1}{25} & | & \frac{617}{6250} \\ 0 & 1 & | & \frac{3733}{3500} \end{bmatrix} -\frac{1}{14}R_2$$

$$\begin{bmatrix} 1 & 0 & | & \frac{981}{17,500} \\ 0 & 1 & | & \frac{3733}{3500} \end{bmatrix} -\frac{1}{25}R_2 + R_1$$

Therefore, $a = \dfrac{981}{17,500} \approx .056057$ and

$b = \dfrac{3733}{3500} \approx 1.06657$.

b. Let $x = 55$

$$y = .056057(55)^2 + 1.06657(55)$$

$$y \approx 228 \text{ ft}$$

37. Let x = amount in mutual fund

$\quad y$ = amount in corporate bonds

$\quad z$ = amount in fast food franchise

$$x + y + z = x + 2x + z = 3x + z = 70,000$$

$$.02x + .1(2x) + .06z = .22x + .06z = 4800$$

The matrix is

$$\begin{bmatrix} 3 & 1 & | & 70,000 \\ .22 & .06 & | & 4800 \end{bmatrix}$$

$$\begin{bmatrix} 3 & 1 & | & 70,000 \\ 22 & 6 & | & 480,000 \end{bmatrix}$$

$$\begin{bmatrix} 1 & \frac{1}{3} & | & \frac{70,000}{3} \\ 22 & 6 & | & 480,000 \end{bmatrix} \frac{1}{3}R_1$$

$$\begin{bmatrix} 1 & \frac{1}{3} & | & \frac{70,000}{3} \\ 0 & -\frac{4}{3} & | & -\frac{100,000}{3} \end{bmatrix} -22R_1 + R_2$$

$$\begin{bmatrix} 1 & \frac{1}{3} & | & \frac{70,000}{3} \\ 0 & 1 & | & 25,000 \end{bmatrix} -\frac{3}{4}R_2$$

$$\begin{bmatrix} 1 & 0 & | & 15,000 \\ 0 & 1 & | & 25,000 \end{bmatrix} -\frac{1}{3}R_2 + R_1$$

Therefore, she should invest $15,000 in the mutual fund, $30,000 in bonds, and $25,000 in the food franchise.

39. Let x = number of cups of Roasted Chicken Rotini
y = number of cups of Hearty Chicken
z = number of cups of Chunky Chicken Noodle

The system of equations is
$100x + 130y + 130z = 2030$
$970x + 480y + 880z = 11,900$
$6x + 8x + 8z = 124$

After simplifying the equations, the matrix is

$$\begin{bmatrix} 10 & 13 & 13 & 203 \\ 97 & 48 & 88 & 1190 \\ 3 & 4 & 4 & 62 \end{bmatrix}$$

$$\begin{bmatrix} 1 & \frac{13}{10} & \frac{13}{10} & \frac{203}{10} \\ 97 & 48 & 88 & 1190 \\ 3 & 4 & 4 & 62 \end{bmatrix} \frac{1}{10}R_1$$

$$\begin{bmatrix} 1 & \frac{13}{10} & \frac{13}{10} & \frac{203}{10} \\ 0 & -78.1 & -38.1 & -779.1 \\ 0 & .1 & .1 & 1.1 \end{bmatrix} \begin{matrix} \\ -97R_1 + R_3 \\ -3R_1 + R_3 \end{matrix}$$

$$\begin{bmatrix} 1 & \frac{13}{10} & \frac{13}{10} & \frac{203}{10} \\ 0 & -78.1 & -38.1 & -779.1 \\ 0 & 1 & 1 & 11 \end{bmatrix} \begin{matrix} \\ \\ 10R_3 \end{matrix}$$

Interchange R_3 and R_2.

$$\begin{bmatrix} 1 & \frac{13}{10} & \frac{13}{10} & \frac{203}{10} \\ 0 & 1 & 1 & 11 \\ 0 & -78.1 & -38.1 & -779.1 \end{bmatrix}$$

$$\begin{bmatrix} 1 & \frac{13}{10} & \frac{13}{10} & \frac{203}{10} \\ 0 & 1 & 1 & 11 \\ 0 & 0 & 40 & 80 \end{bmatrix} 78.1R_2 + R_3$$

$$\begin{bmatrix} 1 & \frac{13}{10} & \frac{13}{10} & \frac{203}{10} \\ 0 & 1 & 1 & 11 \\ 0 & 0 & 1 & 2 \end{bmatrix} \frac{1}{40}R_3$$

$$\begin{bmatrix} 1 & 0 & 0 & 6 \\ 0 & 1 & 1 & 11 \\ 0 & 0 & 1 & 2 \end{bmatrix} -\frac{13}{10}R_2 + R_1$$

$$\begin{bmatrix} 1 & 0 & 0 & 6 \\ 0 & 1 & 0 & 9 \\ 0 & 0 & 1 & 2 \end{bmatrix} -R_3 + R_2$$

Therefore, use 6 cups of Roasted Chicken Rotini, 9 cups of Hearty Chicken, and 2 cups of Chunky Chicken Noodle.

Serving size = $\dfrac{6+9+2}{10} = 1.7$ cups

41. Let x = the amount invested in AAA bonds;
y = the amount invested in A bonds;
z = the amount invested in B bonds.

a. Solve the system
$x + y + z = 25,000$ (1)
$.06x + .07y + .1z = 1810$ (2)
$x = 2z.$ (3)

Equation (3) should be rewritten so that the system becomes
$x + y + z = 25,000$ (1)
$.06x + .07y + .1z = 1810$ (2)
$x - 2z = 0.$ (3)

The matrix is

$$\begin{bmatrix} 1 & 1 & 1 & 25,000 \\ .06 & .07 & -1 & 1810 \\ 1 & 0 & -2 & 0 \end{bmatrix}$$

$$\begin{bmatrix} 1 & 1 & 1 & 25,000 \\ 6 & 7 & 10 & 181,000 \\ 1 & 0 & -2 & 0 \end{bmatrix} 100R_2$$

$$\begin{bmatrix} 1 & 1 & 1 & 25,000 \\ 0 & 1 & 4 & 31,000 \\ 0 & -1 & -3 & -25,000 \end{bmatrix} \begin{matrix} \\ -6R_1 + R_2 \\ -R_1 + R_3 \end{matrix}$$

$$\begin{bmatrix} 1 & 0 & -3 & -6000 \\ 0 & 1 & 4 & 31,000 \\ 0 & 0 & 1 & 6000 \end{bmatrix} \begin{matrix} -R_2 + R_1 \\ \\ R_2 + R_3 \end{matrix}$$

$$\begin{bmatrix} 1 & 0 & 0 & 12,000 \\ 0 & 1 & 0 & 7000 \\ 0 & 0 & 1 & 6000 \end{bmatrix} \begin{matrix} 3R_3 + R_1 \\ -4R_3 + R_2 \\ \end{matrix}$$

The client should invest $12,000 in AAA bonds at 6%, $7000 in A bonds at 7%, and $6000 in B bonds at 10%.

41. Continued

b. The new system is
$$x + y + z = 30,000$$
$$.06x + .07y + .1z = 2150$$
$$x - 2z = 0.$$

The matrix is
$$\begin{bmatrix} 1 & 1 & 1 & | & 30,000 \\ .06 & .07 & .1 & | & 2150 \\ 1 & 0 & -2 & | & 0 \end{bmatrix}$$

$$\begin{bmatrix} 1 & 1 & 1 & | & 30,000 \\ 6 & 7 & 10 & | & 215,000 \\ 1 & 0 & -2 & | & 0 \end{bmatrix} 100R_2$$

$$\begin{bmatrix} 1 & 1 & 1 & | & 30,000 \\ 0 & 1 & 4 & | & 35,000 \\ 0 & -1 & -3 & | & -30,000 \end{bmatrix} \begin{matrix} \\ -6R_1 + R_2 \\ -R_1 + R_3 \end{matrix}$$

$$\begin{bmatrix} 1 & 0 & -3 & | & -5000 \\ 0 & 1 & 4 & | & 35,000 \\ 0 & 0 & 1 & | & 5000 \end{bmatrix} \begin{matrix} -R_2 + R_1 \\ \\ R_2 + R_3 \end{matrix}$$

$$\begin{bmatrix} 1 & 0 & 0 & | & 10,000 \\ 0 & 1 & 0 & | & 15,000 \\ 0 & 0 & 1 & | & 5000 \end{bmatrix} \begin{matrix} 3R_3 + R_1 \\ -4R_3 + R_2 \\ \\ \end{matrix}$$

The client should invest $10,000 in AAA bonds at 6%, $15,000 in A bonds at 7%, and $5000 in B bonds at 10%.

c. The new system is
$$x + y + z = 40,000$$
$$.06x + .07y + .1z = 2900$$
$$x - 2z = 0.$$

The matrix is
$$\begin{bmatrix} 1 & 1 & 1 & | & 40,000 \\ .06 & .07 & .1 & | & 2900 \\ 1 & 0 & -2 & | & 0 \end{bmatrix}$$

$$\begin{bmatrix} 1 & 1 & 1 & | & 40,000 \\ 6 & 7 & 10 & | & 290,000 \\ 1 & 0 & -2 & | & 0 \end{bmatrix} 100R_2$$

$$\begin{bmatrix} 1 & 1 & 1 & | & 40,000 \\ 0 & 1 & 4 & | & 50,000 \\ 0 & -1 & -3 & | & -40,000 \end{bmatrix} \begin{matrix} \\ -6R_1 + R_2 \\ -R_1 + R_3 \end{matrix}$$

$$\begin{bmatrix} 1 & 0 & -3 & | & -10,000 \\ 0 & 1 & 4 & | & 50,000 \\ 0 & 0 & 1 & | & 10,000 \end{bmatrix} \begin{matrix} -R_2 + R_1 \\ \\ R_2 + R_3 \end{matrix}$$

41. Continued

$$\begin{bmatrix} 1 & 0 & 0 & | & 20,000 \\ 0 & 1 & 0 & | & 10,000 \\ 0 & 0 & 1 & | & 10,000 \end{bmatrix} \begin{matrix} 3R_3 + R_1 \\ -4R_3 + R_2 \\ \\ \end{matrix}$$

The client should invest $20,000 in AAA bonds at 6%, $10,000 in A bonds at 7%, and $10,000 in B bonds at 10%.

43. a. An equation $y = ax^2 + bx + c$ is sought.
When $x = 6$, $y = 2.80$. Therefore
$36a + 6b + c = 2.80$.
When $x = 8$, $y = 2.48$. Therefore,
$64a + 8b + c = 2.48$.
When $x = 10$, $y = 2.24$. Therefore,
$100a + 10b + c = 2.24$.
Thus, the system of equations is
$$36a + 6b + c = 2.80 \quad (1)$$
$$64a + 8b + c = 2.48 \quad (2)$$
$$100a + 10b + c = 2.24. \quad (3)$$
Solve the system by elimination. Subtract equation (1) from equation (2) to obtain
$28a + 2b = -.32$,
and subtract equation (2) from equation (3) to obtain
$36a + 2b = -.24$.
Now the system is
$$28a + 2b = -.32 \quad (4)$$
$$36a + 2b = -.24 \quad (5)$$
Subtract equation (5) from equation (4).
$$-8a = -.08$$
Thus, $a = .01$. Substituting into
$28a + 2b = -.32$ gives
$$28(.01) + 2b = -.32$$
$$2b = -.60$$
$$b = -.30.$$
Substituting into equation (1) gives
$$36(.01) + 6(-.30) + c = 2.80$$
$$.36 - 1.80 + c = 2.80$$
$$-1.44 + c = 2.80$$
$$c = 4.24.$$
Thus, the equation is
$$y = .01x^2 - .3x + 4.24.$$

43. Continued

b. Write $y = .01x^2 - .3x + 4.24$ in the form
$y = a(x-h)^2 + k$.
$y = .01(x^2 - 30x) + 4.24$
$y = .01(x^2 - 30x + 225) + 4.24 - 2.25$
$y = .01(x-15)^2 + 1.99$
The minimum value of y is 1.99, occurring when $x = 15$. Thus, 15 platters should be fired at one time to minimize the fuel cost. The minimum fuel cost is $1.99.

45. a. $f(x) = ax^2 + bx + c$
$11 = a(3)^2 + b(3) + c$
$20 = a(28)^2 + b(28) + c$
$30 = a(44)^2 + b(44) + c$
The three equations reduce to:
$$9a + 3b + c = 11 \quad (1)$$
$$784a + 28b + c = 20 \quad (2)$$
$$1936a + 44b + c = 30 \quad (3)$$
Subtract (1) from (2) and (3) to get
$$775a + 25b = 9 \quad (4)$$
$$1927a + 41b = 19 \quad (5)$$
Multiply equation (4) by -41, and equation (5) by 25 and then add to obtain

$$-31775a - 1025b = -369$$
$$\underline{48175a + 1025b = 475}$$
$$1640a = 106$$
$$a = \frac{106}{1640}$$
$$= .006463$$

Substitute .006463 for a in equation (4) and solve for b.
$775(.006463) + 25b = 9$
$$b = .159634$$

Substitute .006463 for a and .159634 for b in equation (1) and solve for c.
$9(.006463) + 3(.159634) + c = 11$
$$c = 10.4629$$
The quadratic function is
$f(x) = .006463x^2 + .159634x + 10.4629$

45. Continued

b. Let $x = 9$. Then the GDP estimate in 2009 is

$f(9) = .006463(9)^2 + .15963(9) + 10.4629$
$= 12.4231$
This is about $12.4 trillion.

Let $x = 15$. Then the GDP estimate in 20015
is $\begin{aligned} f(15) &= .006463(15)^2 + .15963(15) \\ &\quad + 10.4629 \end{aligned}$

$= 14.3115$ This is about $14.3 trillion.

c. Use the quadratic formula to solve

$.006463x^2 + .159634x + 10.4629 = 25$ where $a = .006463$, $b = .159634$, and $c = -14.5371$ to get $x = 36.6505$. The GDP will reach $25 trillion in 2036.

47. a. $C = aS^2 + bS + c$
$33 = a(320)^2 + b(320) + c$
$40 = a(600)^2 + b(600) + c$
$50 = a(1283)^2 + b(1283) + c$
Simplify the equations
$c + 320b + 102,400a = 33$
$c + 600b + 360,000a = 40$
$c + 1283b + 1,646,089a = 50$
The matrix is
$$\begin{bmatrix} 1 & 320 & 102,400 & 33 \\ 1 & 600 & 360,000 & 40 \\ 1 & 1283 & 1,646,089 & 50 \end{bmatrix}$$
$$\begin{bmatrix} 1 & 320 & 102,400 & 33 \\ 0 & 280 & 257,600 & 7 \\ 0 & 963 & 1,543,689 & 17 \end{bmatrix} \begin{matrix} \\ -R_1 + R_2 \\ -R_1 + R_3 \end{matrix}$$

$$\begin{bmatrix} 1 & 320 & 102,400 & 33 \\ 0 & 280 & 257,600 & 7 \\ 0 & 0 & 657,729 & -7.075 \end{bmatrix} \begin{matrix} \\ \\ -\frac{963}{280}R_2 + R_3 \end{matrix}$$

$$\begin{bmatrix} 1 & 320 & 102,400 & 33 \\ 0 & 1 & 920 & 0.025 \\ 0 & 0 & 1 & -.00001 \end{bmatrix} \begin{matrix} \\ \frac{1}{280}R_2 \\ \frac{1}{657,729}R_3 \end{matrix}$$

. **47.Continued**

$$\begin{bmatrix} 1 & 0 & -192,000 \\ 0 & 1 & 0 \\ 0 & 0 & 1 \end{bmatrix}\begin{array}{c} 25 \\ 0.034896 \\ -.00001 \end{array}\begin{array}{l} -320R_2+R_1 \\ -920R_3+R_2 \end{array}$$

$$\begin{bmatrix} 1 & 0 & 0 \\ 0 & 1 & 0 \\ 0 & 0 & 1 \end{bmatrix}\begin{array}{c} 22.9 \\ .034896 \\ -.00001 \end{array}192,000R_3+R_1$$

The relationship is expressed as
$C = aS^2 + bS + c$
$C = -.0000108S^2 + .034896S + 22.9$

b. $45 = -.0000108S^2 + .034896S + 22.9$
Plot $y_1 = 45$ and
$y_2 = -.0000108x^2 + .034896x + 22.9$ They
intersect at (864.7, 45). The top speed is
approximately 865 knots.

Section 6.3 Basic Matrix Operations

1. $\begin{bmatrix} 7 & -8 & 4 \\ 0 & 13 & 9 \end{bmatrix}$ is a 2×3 matrix.

Its additive inverse is $\begin{bmatrix} -7 & 8 & -4 \\ 0 & -13 & -9 \end{bmatrix}$.

3. $\begin{bmatrix} -3 & 0 & 11 \\ 1 & \frac{1}{4} & -7 \\ 5 & -3 & 9 \end{bmatrix}$ is a 3×3 square matrix.

Its additive inverse is $\begin{bmatrix} 3 & 0 & -11 \\ -1 & -\frac{1}{4} & 7 \\ -5 & 3 & -9 \end{bmatrix}$.

5. $\begin{bmatrix} 7 \\ 11 \end{bmatrix}$ is a 2×1 column matrix.

Its additive inverse is $\begin{bmatrix} -7 \\ -11 \end{bmatrix}$.

7. If $A + B = A$, then B must be a zero matrix. Because A is a 5×3 matrix and only matrices of the same size can be added, B must also be 5×3. Therefore, B is a 5×3 zero matrix.

9. $\begin{bmatrix} 1 & 2 & 5 & -1 \\ 3 & 0 & 2 & -4 \end{bmatrix}+\begin{bmatrix} 8 & 12 & -5 & 5 \\ -2 & -1 & 0 & 0 \end{bmatrix}$

$=\begin{bmatrix} 1+8 & 2+12 & 5+(-5) & -1+5 \\ 3+(-2) & 0+(-1) & 2+0 & -4+0 \end{bmatrix}$

$=\begin{bmatrix} 9 & 14 & 0 & 4 \\ 1 & -1 & 2 & -4 \end{bmatrix}$

11. $\begin{bmatrix} -1 & 5 & 9 \\ 2 & 2 & 3 \end{bmatrix}+\begin{bmatrix} 4 & 8 & -7 \\ 1 & -1 & 5 \end{bmatrix}$

$=\begin{bmatrix} -1+4 & 5+8 & 9+(-7) \\ 2+1 & 2+(-1) & 3+5 \end{bmatrix}$

$=\begin{bmatrix} 3 & 13 & 2 \\ 3 & 1 & 8 \end{bmatrix}$

13. $\begin{bmatrix} 4 & -2 & 5 \\ 3 & 7 & 0 \end{bmatrix}-\begin{bmatrix} 1 & 5 & -2 \\ -3 & 3 & 8 \end{bmatrix}$

$=\begin{bmatrix} 4-1 & -2-5 & 5-(-2) \\ 3-(-3) & 7-3 & 0-8 \end{bmatrix}$

$=\begin{bmatrix} 3 & -7 & 7 \\ 6 & 4 & -8 \end{bmatrix}$

15. $2A = 2\begin{bmatrix} -2 & 0 \\ 5 & 3 \end{bmatrix}=\begin{bmatrix} -4 & 0 \\ 10 & 6 \end{bmatrix}$

17. $-4B = -4\begin{bmatrix} 0 & 2 \\ 4 & -6 \end{bmatrix}=\begin{bmatrix} 0 & -8 \\ -16 & 24 \end{bmatrix}$

19. $-4A+5B = -4\begin{bmatrix} -2 & 0 \\ 5 & 3 \end{bmatrix}+5\begin{bmatrix} 0 & 2 \\ 4 & -6 \end{bmatrix}$

$=\begin{bmatrix} 8 & 0 \\ -20 & -12 \end{bmatrix}+\begin{bmatrix} 0 & 10 \\ 20 & -30 \end{bmatrix}$

$=\begin{bmatrix} 8 & 10 \\ 0 & -42 \end{bmatrix}$

21. $A = \begin{bmatrix} 1 & -2 \\ 4 & 3 \end{bmatrix}, B = \begin{bmatrix} 2 & -1 \\ 0 & 5 \end{bmatrix}$

$2A + 3B = 2\begin{bmatrix} 1 & -2 \\ 4 & 3 \end{bmatrix} + 3\begin{bmatrix} 2 & -1 \\ 0 & 5 \end{bmatrix}$

$= \begin{bmatrix} 2 & -4 \\ 8 & 6 \end{bmatrix} + \begin{bmatrix} 6 & -3 \\ 0 & 15 \end{bmatrix}$

$= \begin{bmatrix} 8 & -7 \\ 8 & 21 \end{bmatrix}$

If $2X = 2A + 3B$, then

$2X = \begin{bmatrix} 8 & -7 \\ 8 & 21 \end{bmatrix}$

and $X = \begin{bmatrix} 4 & -\frac{7}{2} \\ 4 & \frac{21}{2} \end{bmatrix}$.

23.
$X + T = \begin{bmatrix} x & y \\ z & w \end{bmatrix} + \begin{bmatrix} r & s \\ t & u \end{bmatrix}$

$= \begin{bmatrix} x+r & y+s \\ z+t & w+u \end{bmatrix},$

which is another 2×2 matrix.

25. Show that $X + (T + P) = (X + T) + P$. On the left-hand side, the sum of $T + P$ is obtained first, and then $X + (T + P)$. This gives the matrix

$\begin{bmatrix} x+(r+m) & y+(s+n) \\ z+(t+p) & w+(u+q) \end{bmatrix}.$

For the right-hand side, first the sum

$X + T$ is obtained, and then $(X + T) + P$. This gives the matrix

$\begin{bmatrix} (x+r)+m & (y+s)+n \\ (z+t)+p & (w+u)+q \end{bmatrix}.$

Comparing corresponding elements shows that they are equal by the associative property of addition of real numbers. Thus, $X + (T + P) = (X + T) + P$.

27. Show that $P + O = P$.

$P + O = \begin{bmatrix} m & n \\ p & q \end{bmatrix} + \begin{bmatrix} 0 & 0 \\ 0 & 0 \end{bmatrix}$

$= \begin{bmatrix} m+0 & n+0 \\ p+0 & q+0 \end{bmatrix}$

$= \begin{bmatrix} m & n \\ p & q \end{bmatrix}$

$= P$

29. Several possible answers, including:

	basketball	hockey	football	baseball
percent of no shows	16	16	20	18
lost revenue per fan ($)	18.20	18.25	19	15.40
lost annual revenue (millions $)	22.7	35.8	51.9	96.3

31.

	1998	2000	2005
heart	4121	4143	3140
lung	3171	3614	3601
liver	12,070	15,539	17,376
kidney	38,270	45,273	62,130

33. Here is a possible answer.

	1995	2000	2001	2002
Ages 15-24	2.9	2.6	2.5	2.5
Ages 45-54	111	94.2	92.9	90.7
Ages 65-74	799.9	666.6	635.1	615.9

35. a. A matrix for the death rate of male drivers is

$$A = \begin{bmatrix} 2.61 & 4.39 & 6.29 & 9.08 \\ 1.63 & 2.77 & 4.61 & 6.92 \\ .92 & .75 & .62 & .54 \end{bmatrix}$$

b. A matrix for the death rate of female drivers is

$$B = \begin{bmatrix} 1.38 & 1.72 & 1.94 & 3.31 \\ 1.26 & 1.48 & 2.82 & 2.28 \\ .41 & .33 & .27 & .40 \end{bmatrix}$$

c. Subtract matrix B from matrix A to see the difference between the death rate of males and females.

$$\begin{bmatrix} 1.23 & 2.67 & 4.35 & 5.77 \\ .37 & 1.29 & 1.79 & 4.64 \\ .51 & .42 & .35 & .14 \end{bmatrix}$$

Section 6.4 Matrix Products and Inverses

1. AB is 2×2. A has 2 rows, B has 2 columns.
BA is 2×2. B has 2 rows, A has 2 columns.

3. AB is 3×3. A has 3 rows, B has 3 columns.
BA is 5×5. B has 3 columns, A has 3 rows.

5. AB does not exist because the number of columns of A is not the same as the number of rows of B.
BA is 3×2. B has 3 rows, A has 2 columns.

7. Columns; rows

9. $\begin{bmatrix} 1 & 2 \\ 3 & 4 \end{bmatrix}\begin{bmatrix} -1 \\ 3 \end{bmatrix} = \begin{bmatrix} 1(-1)+2(3) \\ 3(-1)+4(3) \end{bmatrix} = \begin{bmatrix} 5 \\ 9 \end{bmatrix}$

11. $\begin{bmatrix} 2 & 2 & -1 \\ 3 & 0 & 1 \end{bmatrix}\begin{bmatrix} 0 & 2 \\ -1 & 5 \\ 0 & 2 \end{bmatrix} = \begin{bmatrix} (2)(0)+(2)(-1)+(-1)(0) & (2)(2)+(2)(5)+(-1)(2) \\ (3)(0)+(0)(-1)+(1)(0) & (3)(2)+(0)(4)+(1)(2) \end{bmatrix} = \begin{bmatrix} -2 & 12 \\ 0 & 8 \end{bmatrix}$

13. $\begin{bmatrix} -4 & 1 \\ 2 & -3 \end{bmatrix}\begin{bmatrix} 1 & 0 \\ 0 & 1 \end{bmatrix} = \begin{bmatrix} (-4)(1)+(1)(0) & (-4)(0)+(1)(1) \\ (2)(1)+(-3)(0) & (2)(0)+(-3)(1) \end{bmatrix} = \begin{bmatrix} -4 & 1 \\ 2 & -3 \end{bmatrix}$

15. $\begin{bmatrix} 1 & 0 & 0 \\ 0 & 1 & 0 \\ 0 & 0 & 1 \end{bmatrix}\begin{bmatrix} 3 & -5 & 7 \\ -2 & 1 & 6 \\ 0 & -3 & 4 \end{bmatrix}$

$= \begin{bmatrix} (1)(3)+(0)(-2)+(0)(0) & (1)(-5)+(0)(1)+(0)(-3) & (1)(7)+(0)(6)+(0)(4) \\ (0)(3)+(1)(-2)+(0)(0) & (0)(-5)+(1)(1)+(0)(-3) & (0)(7)+(1)(6)+(0)(4) \\ (0)(3)+(0)(-2)+(1)(0) & (0)(-5)+(0)(1)+(1)(-3) & (0)(7)+(0)(6)+(1)(4) \end{bmatrix}$

$= \begin{bmatrix} 3 & -5 & 7 \\ -2 & 1 & 6 \\ 0 & -3 & 4 \end{bmatrix}$

17. $\begin{bmatrix} 1 & 2 & 3 \\ 4 & 5 & 6 \\ 7 & 8 & 9 \end{bmatrix}\begin{bmatrix} -1 & 5 \\ 7 & 0 \\ 1 & 3 \end{bmatrix} = \begin{bmatrix} (1)(-1)+(2)(7)+(3)(1) & (1)(5)+(2)(0)+(3)(3) \\ (4)(-1)+(5)(7)+(6)(1) & (4)(5)+(5)(0)+(6)(3) \\ (7)(-1)+(8)(7)+(9)(1) & (7)(5)+(8)(0)+(9)(3) \end{bmatrix} = \begin{bmatrix} 16 & 14 \\ 37 & 38 \\ 58 & 62 \end{bmatrix}$

19. $AB = \begin{bmatrix} -3 & -9 \\ 2 & 6 \end{bmatrix}\begin{bmatrix} 4 & 6 \\ 2 & 3 \end{bmatrix} = \begin{bmatrix} -30 & -45 \\ 20 & 30 \end{bmatrix}$

$BA = \begin{bmatrix} 4 & 6 \\ 2 & 3 \end{bmatrix}\begin{bmatrix} -3 & -9 \\ 2 & 6 \end{bmatrix} = \begin{bmatrix} 0 & 0 \\ 0 & 0 \end{bmatrix}$

Since $\begin{bmatrix} -30 & -45 \\ 20 & 30 \end{bmatrix} \neq \begin{bmatrix} 0 & 0 \\ 0 & 0 \end{bmatrix}$, $AB \neq BA$.

Therefore, matrix multiplication is not commutative.

21.

$A + B = \begin{bmatrix} -3 & -9 \\ 2 & 6 \end{bmatrix} + \begin{bmatrix} 4 & 6 \\ 2 & 3 \end{bmatrix} = \begin{bmatrix} 1 & -3 \\ 4 & 9 \end{bmatrix}$

$A - B = \begin{bmatrix} -3 & -9 \\ 2 & 6 \end{bmatrix} - \begin{bmatrix} 4 & 6 \\ 2 & 3 \end{bmatrix} = \begin{bmatrix} -7 & -15 \\ 0 & 3 \end{bmatrix}$

$(A + B)(A - B) = \begin{bmatrix} 1 & -3 \\ 4 & 9 \end{bmatrix}\begin{bmatrix} -7 & -15 \\ 0 & 3 \end{bmatrix} = \begin{bmatrix} -7 & -24 \\ -28 & -33 \end{bmatrix}$

$A^2 = \begin{bmatrix} -3 & -9 \\ 2 & 6 \end{bmatrix}\begin{bmatrix} -3 & -9 \\ 2 & 6 \end{bmatrix} = \begin{bmatrix} -9 & -27 \\ 6 & 18 \end{bmatrix}$

$B^2 = \begin{bmatrix} 4 & 6 \\ 2 & 3 \end{bmatrix}\begin{bmatrix} 4 & 6 \\ 2 & 3 \end{bmatrix} = \begin{bmatrix} 28 & 42 \\ 14 & 21 \end{bmatrix}$

$A^2 - B^2 = \begin{bmatrix} -9 & -27 \\ 6 & 18 \end{bmatrix} - \begin{bmatrix} 28 & 42 \\ 14 & 21 \end{bmatrix} = \begin{bmatrix} -37 & -69 \\ -8 & -3 \end{bmatrix}$

Since $\begin{bmatrix} -7 & -24 \\ -28 & -33 \end{bmatrix} \neq \begin{bmatrix} -37 & -69 \\ -8 & -3 \end{bmatrix}$, $(A + B)(A - B) \neq A^2 - B^2$.

23. Verify that $(PX)T = P(XT)$.

$$(PX)T = \left(\begin{bmatrix} m & n \\ p & q \end{bmatrix}\begin{bmatrix} x & y \\ z & w \end{bmatrix}\right)\begin{bmatrix} r & s \\ t & u \end{bmatrix} = \begin{bmatrix} mx+nz & my+nw \\ px+qz & py+qw \end{bmatrix}\begin{bmatrix} r & s \\ t & u \end{bmatrix}$$

$$= \begin{bmatrix} (mx+nz)r+(my+nw)t & (mx+nz)s+(my+nw)u \\ (px+qz)r+(py+qw)t & (px+qz)s+(py+qw)u \end{bmatrix}$$

$$= \begin{bmatrix} mxr+nzr+myt+nwt & mxs+nzs+myu+nwu \\ pxr+qzr+pyt+qwt & pxs+qzs+pyu+qwu \end{bmatrix} \begin{matrix} \text{Distributive property} \\ \text{for real numbers} \end{matrix}$$

$$P(XT) = \begin{bmatrix} m & n \\ p & q \end{bmatrix}\left(\begin{bmatrix} x & y \\ z & w \end{bmatrix}\begin{bmatrix} r & s \\ t & u \end{bmatrix}\right) = \begin{bmatrix} m & n \\ p & q \end{bmatrix}\begin{bmatrix} xr+yt & xs+yu \\ zr+wt & zs+wu \end{bmatrix}$$

$$= \begin{bmatrix} m(xr+yt)+n(zr+wt) & m(xs+yu)+n(zs+wu) \\ p(xr+yt)+q(zr+wt) & p(xs+yu)+q(zs+wu) \end{bmatrix}$$

$$= \begin{bmatrix} mxr+myt+nzr+nwt & mxs+myu+nzs+nwu \\ pxr+pyt+qzr+qwt & pxs+pyu+qzs+qwu \end{bmatrix} \begin{matrix} \text{Distributive property} \\ \text{for real numbers} \end{matrix}$$

$$= \begin{bmatrix} mxr+nzr+myt+nwt & mxs+nzs+myu+nwu \\ pxr+qzr+pyt+qwt & pxs+qzs+pyu+nwu \end{bmatrix} \begin{matrix} \text{Commutative property} \\ \text{for real numbers} \end{matrix}$$

Thus, $(PX)T = P(XT)$.

25. Verify $k(X + T) = kX + kT$ for any real number k.

$$k(X+T) = k\begin{bmatrix} x & y \\ z & w \end{bmatrix} + \begin{bmatrix} r & s \\ t & u \end{bmatrix}$$

$$= k\begin{bmatrix} x+r & y+s \\ z+t & w+u \end{bmatrix}$$

$$= \begin{bmatrix} k(x+r) & k(y+s) \\ k(z+t) & k(w+u) \end{bmatrix}$$

$$= \begin{bmatrix} kx+kr & ky+ks \\ kz+kt & kw+ku \end{bmatrix} \begin{matrix} \text{Distributive property} \\ \text{for real numbers} \end{matrix}$$

$$= \begin{bmatrix} kx & ky \\ kz & kw \end{bmatrix} + \begin{bmatrix} kr & ks \\ kt & ku \end{bmatrix}$$

$$= k\begin{bmatrix} x & y \\ z & w \end{bmatrix} + k\begin{bmatrix} r & s \\ t & u \end{bmatrix} = kX + kT$$

27. $\begin{bmatrix} 5 & 2 \\ 3 & -1 \end{bmatrix}\begin{bmatrix} -1 & 2 \\ 3 & -4 \end{bmatrix} = \begin{bmatrix} 1 & 2 \\ -6 & 10 \end{bmatrix} \neq I$,

so the given matrices are not inverses of each other.

29. $\begin{bmatrix} 3 & -1 \\ -4 & 2 \end{bmatrix}\begin{bmatrix} 1 & \frac{1}{2} \\ 2 & \frac{3}{2} \end{bmatrix} = \begin{bmatrix} 1 & 0 \\ 0 & 1 \end{bmatrix} = I$

$\begin{bmatrix} 1 & \frac{1}{2} \\ 2 & \frac{3}{2} \end{bmatrix}\begin{bmatrix} 3 & -1 \\ -4 & 2 \end{bmatrix} = \begin{bmatrix} 1 & 0 \\ 0 & 1 \end{bmatrix} = I$

Therefore, the given matrices are inverses of each other.

31. $\begin{bmatrix} 1 & 1 & 1 \\ 2 & 3 & 0 \\ 1 & 2 & 1 \end{bmatrix}\begin{bmatrix} 1.5 & .5 & -1.5 \\ -1 & 0 & 1 \\ .5 & -.5 & .5 \end{bmatrix}$

$= \begin{bmatrix} 1 & 0 & 0 \\ 0 & 1 & 0 \\ 0 & 0 & 1 \end{bmatrix} = I$,

The given matrices are inverses of each other.

33. To find the inverse of $\begin{bmatrix} 2 & -3 \\ -1 & 2 \end{bmatrix}$, write the augmented matrix $\begin{bmatrix} A \mid I \end{bmatrix}$.

$\begin{bmatrix} 2 & -3 & | & 1 & 0 \\ -1 & 2 & | & 0 & 1 \end{bmatrix}$.

$\begin{bmatrix} 1 & -2 & | & 1 & 1 \\ -1 & 2 & | & 0 & 1 \end{bmatrix} R_1 + R_2$

$\begin{bmatrix} 1 & -1 & | & 1 & 1 \\ 0 & 1 & | & 1 & 2 \end{bmatrix} R_1 + R_2$

$\begin{bmatrix} 1 & 0 & | & 2 & 3 \\ 0 & 1 & | & 1 & 2 \end{bmatrix} R_1 + R_2$

The inverse is $\begin{bmatrix} 2 & 3 \\ 1 & 2 \end{bmatrix}$.

35. To find the inverse of $\begin{bmatrix} 2 & 4 \\ 3 & 6 \end{bmatrix}$, write the augmented matrix $\begin{bmatrix} A \mid I \end{bmatrix}$.

$\begin{bmatrix} 2 & 4 & | & 1 & 0 \\ 3 & 6 & | & 0 & 1 \end{bmatrix}$

$\begin{bmatrix} 1 & 2 & | & .5 & 0 \\ 3 & 6 & | & 0 & 1 \end{bmatrix} \frac{1}{2} R_1$

$\begin{bmatrix} 1 & 2 & | & .5 & 0 \\ 0 & 0 & | & -1.5 & 1 \end{bmatrix} -3R_1 + R_2$

Since there is no way to continue the desired transformation, the given matrix has no inverse.

37. To find the inverse of $\begin{bmatrix} 2 & 6 \\ 1 & 4 \end{bmatrix}$, write the augmented matrix $\begin{bmatrix} A \mid I \end{bmatrix}$.

$\begin{bmatrix} 2 & 6 & | & 1 & 0 \\ 1 & 4 & | & 0 & 1 \end{bmatrix}$

Interchange R_1 and R_2.

$\begin{bmatrix} 1 & 4 & | & 0 & 1 \\ 2 & 6 & | & 1 & 0 \end{bmatrix}$

$\begin{bmatrix} 1 & 4 & | & 0 & 1 \\ 0 & -2 & | & 1 & -2 \end{bmatrix} -2R_1 + R_2$

$\begin{bmatrix} 1 & 4 & | & 0 & 1 \\ 0 & 1 & | & -\frac{1}{2} & 1 \end{bmatrix} -\frac{1}{2} R_2$

$\begin{bmatrix} 1 & 0 & | & 2 & -3 \\ 0 & 1 & | & -\frac{1}{2} & 1 \end{bmatrix} -4R_2 + R_1$

The inverse of the given matrix is

$\begin{bmatrix} 2 & -3 \\ -\frac{1}{2} & 1 \end{bmatrix}$.

39. To find the inverse of $\begin{bmatrix} 1 & -1 & 0 \\ -1 & 2 & 3 \\ 1 & 0 & 2 \end{bmatrix}$, write the augmented matrix $\begin{bmatrix} A \mid I \end{bmatrix}$.

$\begin{bmatrix} 1 & -1 & 0 & | & 1 & 0 & 0 \\ -1 & 2 & 3 & | & 0 & 1 & 0 \\ 1 & 0 & 2 & | & 0 & 0 & 1 \end{bmatrix}$

$\begin{bmatrix} 1 & -1 & 0 & | & 1 & 0 & 0 \\ 0 & 1 & 3 & | & 1 & 1 & 0 \\ 1 & 0 & 2 & | & 0 & 0 & 1 \end{bmatrix} R_1 + R_2$

$\begin{bmatrix} 1 & -1 & 0 & | & 1 & 0 & 0 \\ 0 & 1 & 3 & | & 1 & 1 & 0 \\ 0 & -1 & -2 & | & 1 & 0 & -1 \end{bmatrix} R_1 - R_2$

$\begin{bmatrix} 1 & -1 & 0 & | & 1 & 0 & 0 \\ 0 & 1 & 3 & | & 1 & 1 & 0 \\ 0 & 0 & 1 & | & 2 & 1 & -1 \end{bmatrix} R_2 + R_3$

$\begin{bmatrix} 1 & 0 & 3 & | & 2 & 1 & 0 \\ 0 & 1 & 3 & | & 1 & 1 & 0 \\ 0 & 0 & 1 & | & 2 & 1 & -1 \end{bmatrix} R_1 + R_2$

$\begin{bmatrix} 1 & 0 & 0 & | & -4 & -2 & 3 \\ 0 & 1 & 3 & | & 1 & 1 & 0 \\ 0 & 0 & 1 & | & 2 & 1 & -1 \end{bmatrix} R_1 - 3R_3$

39. Continued

$$\begin{bmatrix} 1 & 0 & 0 & | & -4 & -2 & 3 \\ 0 & 1 & 0 & | & -5 & -2 & 3 \\ 0 & 0 & 1 & | & 2 & 1 & -1 \end{bmatrix} R_2 - 3R_3$$

The inverse of the given matrix is

$$\begin{bmatrix} -4 & -2 & 3 \\ -5 & -2 & 3 \\ 2 & 1 & -1 \end{bmatrix}.$$

41. To find the inverse of $\begin{bmatrix} 1 & 4 & 3 \\ 1 & -3 & -2 \\ 2 & 5 & 4 \end{bmatrix}$,

write the augmented matrix $[A \mid I]$.

$$\begin{bmatrix} 1 & 4 & 3 & | & 1 & 0 & 0 \\ 1 & -3 & -2 & | & 0 & 1 & 0 \\ 2 & 5 & 4 & | & 0 & 0 & 1 \end{bmatrix}$$

$$\begin{bmatrix} 1 & 4 & 3 & | & 1 & 0 & 0 \\ 0 & -7 & -5 & | & -1 & 1 & 0 \\ 0 & -3 & -2 & | & -2 & 0 & 1 \end{bmatrix} \begin{matrix} -1R_1 + R_2 \\ -2R_1 + R_3 \end{matrix}$$

$$\begin{bmatrix} 1 & 4 & 3 & | & 1 & 0 & 0 \\ 0 & 1 & \frac{5}{7} & | & \frac{1}{7} & -\frac{1}{7} & 0 \\ 0 & -3 & -2 & | & -2 & 0 & 1 \end{bmatrix} -\frac{1}{7}R_2$$

$$\begin{bmatrix} 1 & 0 & \frac{1}{7} & | & \frac{3}{7} & \frac{4}{7} & 0 \\ 0 & 1 & \frac{5}{7} & | & \frac{1}{7} & -\frac{1}{7} & 0 \\ 0 & 0 & \frac{1}{7} & | & -\frac{11}{7} & -\frac{3}{7} & 1 \end{bmatrix} \begin{matrix} -4R_2 + R_1 \\ \\ 3R_2 + R_3 \end{matrix}$$

$$\begin{bmatrix} 1 & 0 & \frac{1}{7} & | & \frac{3}{7} & \frac{4}{7} & 0 \\ 0 & 1 & \frac{5}{7} & | & \frac{1}{7} & -\frac{1}{7} & 0 \\ 0 & 0 & 1 & | & -11 & -3 & 7 \end{bmatrix} 7R_3$$

$$\begin{bmatrix} 1 & 0 & 0 & | & 2 & 1 & -1 \\ 0 & 1 & 0 & | & 8 & 2 & -5 \\ 0 & 0 & 1 & | & -11 & -3 & 7 \end{bmatrix} \begin{matrix} -\frac{1}{7}R_3 + R_1 \\ -\frac{5}{7}R_3 + R_2 \end{matrix}$$

The inverse of the given matrix is

$$\begin{bmatrix} 2 & 1 & -1 \\ 8 & 2 & -5 \\ -11 & -3 & 7 \end{bmatrix}.$$

43. To find the inverse of $\begin{bmatrix} 1 & -3 & 4 \\ 2 & -5 & 7 \\ 0 & -1 & 1 \end{bmatrix}$,

write the augmented matrix $[A \mid I]$.

$$\begin{bmatrix} 1 & -3 & 4 & | & 1 & 0 & 0 \\ 2 & -5 & 7 & | & 0 & 1 & 0 \\ 0 & -1 & 1 & | & 0 & 0 & 1 \end{bmatrix}$$

$$\begin{bmatrix} 1 & -3 & 4 & | & 1 & 0 & 0 \\ 0 & 1 & -1 & | & -2 & 1 & 0 \\ 0 & -1 & 1 & | & 0 & 0 & 1 \end{bmatrix} -2R_1 + R_2$$

$$\begin{bmatrix} 1 & -3 & 4 & | & 1 & 0 & 0 \\ 0 & 1 & -1 & | & -2 & 1 & 0 \\ 0 & 0 & 0 & | & -2 & 1 & 1 \end{bmatrix} R_2 + R_3$$

Since there is no way to continue the desired transformation, the given matrix has no inverse.

45. Use a graphing calculator to find the inverse of

$$\begin{bmatrix} 2 & 4 & 6 \\ -1 & -4 & -3 \\ 0 & 1 & -1 \end{bmatrix}.$$

The inverse of the given matrix is

$$\begin{bmatrix} 3 & -5 & 8 \\ -\frac{1}{2} & 1 & -1 \\ -\frac{1}{2} & 1 & -2 \end{bmatrix}.$$

47. Use a graphing calculator to find the inverse of

$$\begin{bmatrix} 1 & -2 & 3 & 0 \\ 0 & 1 & -1 & 1 \\ -2 & 2 & -2 & 4 \\ 0 & 2 & -3 & 1 \end{bmatrix}.$$

The inverse of the given matrix is

$$\begin{bmatrix} \frac{1}{2} & \frac{1}{2} & -\frac{1}{4} & \frac{1}{2} \\ -1 & 4 & -\frac{1}{2} & -2 \\ -\frac{1}{2} & \frac{5}{2} & -\frac{1}{4} & -\frac{3}{2} \\ \frac{1}{2} & -\frac{1}{2} & \frac{1}{4} & \frac{1}{2} \end{bmatrix}.$$

49. a. $R = \begin{bmatrix} .024 & .008 \\ .025 & .007 \\ .015 & .009 \\ .011 & .011 \end{bmatrix}$

b. $P = \begin{bmatrix} 1996 & 286 & 226 & 460 \\ 2440 & 365 & 252 & 484 \\ 2906 & 455 & 277 & 499 \\ 3683 & 519 & 310 & 729 \\ 4723 & 697 & 364 & 702 \end{bmatrix}$

c.

$PR = \begin{bmatrix} 1996 & 286 & 226 & 460 \\ 2440 & 365 & 252 & 484 \\ 2906 & 455 & 277 & 499 \\ 3683 & 519 & 310 & 729 \\ 4723 & 697 & 364 & 702 \end{bmatrix} \begin{bmatrix} .024 & .008 \\ .025 & .007 \\ .015 & .009 \\ .011 & .011 \end{bmatrix}$

$= \begin{bmatrix} 1996(.024) + 286(.025) + 226(.015) + 460(.011) & 1996(.008) + 286(.007) + 226(.009) + 460(.011) \\ 2440(.024) + 365(.025) + 252(.015) + 484(.011) & 2440(.008) + 365(.007) + 252(.009) + 484(.011) \\ 2906(.024) + 455(.025) + 277(.015) + 499(.011) & 2906(.008) + 455(.007) + 277(.009) + 499(.011) \\ 3683(.024) + 519(.025) + 310(.015) + 729(.011) & 3683(.008) + 519(.007) + 310(.009) + 729(.011) \\ 4723(.024) + 697(.025) + 364(.015) + 702(.011) & 4723(.008) + 697(.007) + 364(.009) + 702(.011) \end{bmatrix}$

$= \begin{bmatrix} 63.504 & 25.064 \\ 76.789 & 29.667 \\ 90.763 & 34.415 \\ 114.036 & 43.906 \\ 143.959 & 53.661 \end{bmatrix}$

d. The rows represent the years 1970, 1980, 1990, 2000, 2025. Column 1 gives the total births in those years, column 2 the total deaths.

e. The total number of births in 2000 was 114,036,000.
The total number of deaths projected for 2025 is 53,661,000.

51. a. $\begin{pmatrix} 278.1 & 31.6 & 37.4 & 126.8 \\ 300.1 & 34.3 & 41.1 & 127.3 \end{pmatrix}$

b. $\begin{pmatrix} .01425 & .00865 \\ .01145 & .00775 \\ .0175 & .00755 \\ .00945 & .00925 \end{pmatrix}$

c. $\begin{pmatrix} 6.177505 & 4.105735 \\ 6.591395 & 4.349520 \end{pmatrix}$

d. The rows correspond to years. The entries in each column give the total number of births and deaths, respectively, in the four countries, taken together.

e. 6,177,505

53. a. $A = \begin{bmatrix} 128.4 & 73.0 \\ 140.8 & 73.5 \\ 158.7 & 73.4 \\ 182.1 & 73.2 \end{bmatrix}$

b. $B = \begin{bmatrix} 47.37 & 48.4 & 49.91 & 50.64 \\ 35.33 & 36.47 & 37.64 & 38.79 \end{bmatrix}$

c. $AB = \begin{bmatrix} 128.4 & 73.0 \\ 140.8 & 73.5 \\ 158.7 & 73.4 \\ 182.1 & 73.2 \end{bmatrix} \begin{bmatrix} 47.37 & 48.4 & 49.91 & 50.64 \\ 35.33 & 36.47 & 37.64 & 38.79 \end{bmatrix}$

$= \begin{bmatrix} 8661.40 & 8876.87 & 9156.16 & 9333.85 \\ 9266.45 & 9495.27 & 9793.87 & 9981.18 \\ 10,110.84 & 10,357.98 & 10,683.49 & 10,883.75 \\ 11,212.23 & 11,483.24 & 11,843.86 & 12,060.97 \end{bmatrix}$

d. $BA =$
$\begin{bmatrix} 47.37 & 48.4 & 49.91 & 50.64 \\ 35.33 & 36.47 & 37.64 & 38.79 \end{bmatrix}$
$\begin{bmatrix} 128.4 & 73.0 \\ 140.8 & 73.5 \\ 158.7 & 73.4 \\ 182.1 & 73.2 \end{bmatrix}$

$= \begin{bmatrix} 30,039.29 & 14,385.65 \\ 22,708.48 & 10,861.84 \end{bmatrix}$

e. All the dollar figures are in millions. In AB, row 1, column 1 is the combined monthly cost of both cell phones and basic cable for all subscribers in 2001; row 2, column 2, is the same total for 2002; row 3, column 3, is the same total for 2003; row 4, column 4, is the same total for 2004. The remaining entries are meaningless for this problem. In BA, row 1, column 1, is the total monthly cost for all cell phone subscribers over the four-year period; row 2, column 2, is the total monthly cost for all basic cable subscribers over the four-year period. The remaining entries are meaningless for this problem.

55. a. Let matrix P contain the amount of products needed.

$P = \begin{array}{c} \\ \text{Dept. 1} \\ \text{Dept. 2} \\ \text{Dept. 3} \\ \text{Dept. 4} \end{array} \begin{bmatrix} \text{Paper} & \text{Tape} & \begin{array}{c}\text{Print}\\\text{Rib.}\end{array} & \begin{array}{c}\text{Memo}\\\text{Pads}\end{array} & \text{Pens} \\ 10 & 4 & 3 & 5 & 6 \\ 7 & 2 & 2 & 3 & 8 \\ 4 & 5 & 1 & 0 & 10 \\ 0 & 3 & 4 & 5 & 5 \end{bmatrix}$

Let matrix C contain the cost from each supplier.

$C = \begin{array}{c} \\ \text{Paper} \\ \text{Tape} \\ \text{Print. Rib.} \\ \text{Memo Pads} \\ \text{Pens} \end{array} \begin{bmatrix} A & B \\ 2 & 3 \\ 1 & 1 \\ 4 & 3 \\ 3 & 3 \\ 1 & 2 \end{bmatrix}$

To find the total departmental cost from each supplier, multiply P times C.

$PC = \begin{bmatrix} 10 & 4 & 3 & 5 & 6 \\ 7 & 2 & 2 & 3 & 8 \\ 4 & 5 & 1 & 0 & 10 \\ 0 & 3 & 4 & 5 & 5 \end{bmatrix} \begin{bmatrix} 2 & 3 \\ 1 & 1 \\ 4 & 3 \\ 3 & 3 \\ 1 & 2 \end{bmatrix}$

$= \begin{array}{c} \\ \text{Dept. 1} \\ \text{Dept. 2} \\ \text{Dept. 3} \\ \text{Dept. 4} \end{array} \begin{bmatrix} A & B \\ 57 & 70 \\ 41 & 54 \\ 27 & 40 \\ 39 & 40 \end{bmatrix}$

55. Continued

b. The total cost from each supplier would be found by adding each column. The total cost from Supplier A is \$164; the total cost from Supplier B is \$204. The company should buy from Supplier A.

d. The combined cost of producing 1 kg of Mucho Mocha in Managua is found in row 3, column 2. This cost is \$1.60.

e. Let matrix S contain the amount of products in the special shipment.

$$S = \begin{matrix} CC \\ MM \\ AD \end{matrix} \begin{bmatrix} 100 \\ 200 \\ 500 \end{bmatrix}$$

The total cost for producing this shipment at each location would be found by multiplying the matrix in part (c) by S.

$$\begin{bmatrix} 2.1 & 2.1 & 1.8 \\ 1.6 & 1.7 & 1.5 \\ 1.3 & 1.6 & 1.5 \end{bmatrix} \begin{bmatrix} 100 \\ 200 \\ 500 \end{bmatrix} = \begin{matrix} SD \\ MC \\ M \end{matrix} \begin{bmatrix} 1530 \\ 1250 \\ 1200 \end{bmatrix}$$

The smallest cost is \$1200, in Managua

Section 6.5 Applications of Matrices

1. The solution to the matrix equation $AX = B$ is $X = A^{-1}B$.

$$A = \begin{bmatrix} 1 & -1 \\ 5 & 6 \end{bmatrix}, B = \begin{bmatrix} -4 \\ 2 \end{bmatrix}$$

Use a graphing calculator or row operations on $[A \mid I]$ to find A^{-1}. Then,

$$X = A^{-1}B$$
$$= \begin{bmatrix} .455 & -.091 \\ -.455 & -.091 \end{bmatrix} \begin{bmatrix} -4 \\ 2 \end{bmatrix}$$
$$= \begin{bmatrix} -2 \\ 2 \end{bmatrix}.$$

3. The solution to the matrix equation $AX = B$ is $X = A^{-1}B$.

$$A = \begin{bmatrix} 3 & 1 \\ 4 & 2 \end{bmatrix}, B = \begin{bmatrix} 3 & 5 \\ 4 & 6 \end{bmatrix}.$$

Use a graphing calculator or row operations on $[A \mid I]$ to find that

$$A^{-1} = \begin{bmatrix} 1 & -\frac{1}{2} \\ -2 & \frac{3}{2} \end{bmatrix}.$$

Then
$$X = A^{-1}B$$
$$= \begin{bmatrix} 1 & -\frac{1}{2} \\ -2 & \frac{3}{2} \end{bmatrix} \begin{bmatrix} 3 & 4 \\ 4 & 6 \end{bmatrix}$$
$$= \begin{bmatrix} 1 & 2 \\ 0 & -1 \end{bmatrix}.$$

5. The solution to the matrix equation $AX = B$ is $X = A^{-1}B$.

$$A = \begin{bmatrix} 2 & 1 & 0 \\ -4 & -1 & 3 \\ 3 & 1 & -2 \end{bmatrix}, B = \begin{bmatrix} 2 \\ 7 \\ 4 \end{bmatrix}.$$

Use a graphing calculator or row operations on $[A \mid I]$ to find that

$$A^{-1} = \begin{bmatrix} 1 & -2 & -3 \\ -1 & 4 & 6 \\ 1 & -1 & -2 \end{bmatrix}.$$

Then,
$$X = A^{-1}B$$
$$= \begin{bmatrix} 1 & -2 & -3 \\ -1 & 4 & 6 \\ 1 & -1 & -2 \end{bmatrix} \begin{bmatrix} 2 \\ 7 \\ 4 \end{bmatrix}$$
$$= \begin{bmatrix} -24 \\ 50 \\ -13 \end{bmatrix}.$$

7.
$$x + 2y + 3z = 10$$
$$2x + 3y + 2z = 6$$
$$-x - 2y - 4z = -1$$

has coefficient matrix

$$A = \begin{bmatrix} 1 & 2 & 3 \\ 2 & 3 & 2 \\ -1 & -2 & -4 \end{bmatrix}.$$

Find A^{-1}:

$$[A \mid I] = \begin{bmatrix} 1 & 2 & 3 & 1 & 0 & 0 \\ 2 & 3 & 2 & 0 & 1 & 0 \\ -1 & -2 & -4 & 0 & 0 & 1 \end{bmatrix}$$

$$\begin{bmatrix} 1 & 2 & 3 & 1 & 0 & 0 \\ 0 & -1 & -4 & -2 & 1 & 0 \\ 0 & 0 & -1 & 1 & 0 & 1 \end{bmatrix} \begin{matrix} \\ -2R_1 + R_2 \\ R_1 + R_3 \end{matrix}$$

$$\begin{bmatrix} 1 & 0 & -5 & -3 & 2 & 0 \\ 0 & 1 & 4 & 2 & -1 & 0 \\ 0 & 0 & 1 & -1 & 0 & -1 \end{bmatrix} \begin{matrix} 2R_2 + R_1 \\ -1R_2 \\ -1R_3 \end{matrix}$$

$$\begin{bmatrix} 1 & 0 & 0 & -8 & 2 & -5 \\ 0 & 1 & 0 & 6 & -1 & 4 \\ 0 & 0 & 1 & -1 & 0 & -1 \end{bmatrix} \begin{matrix} 5R_3 + R_1 \\ -4R_3 + R_2 \\ \end{matrix}$$

$$A^{-1} = \begin{bmatrix} -8 & 2 & -5 \\ 6 & -1 & 4 \\ -1 & 0 & -1 \end{bmatrix}$$

$$X = A^{-1}B = \begin{bmatrix} -8 & 2 & -5 \\ 6 & -1 & 4 \\ -1 & 0 & -1 \end{bmatrix} \begin{bmatrix} 10 \\ 6 \\ -1 \end{bmatrix} = \begin{bmatrix} -63 \\ 50 \\ -9 \end{bmatrix}$$

The solution is $(-63, 50, -9)$.

9.
$$x + 4y + 3z = -12$$
$$x - 3y - 2z = 0$$
$$2x + 5y + 4z = 7$$

has the coefficient matrix

$$A = \begin{bmatrix} 1 & 4 & 3 \\ 1 & -3 & -2 \\ 2 & 5 & 4 \end{bmatrix}.$$

From Exercise 41 of Section 6.4,

$$A^{-1} = \begin{bmatrix} 2 & 1 & -1 \\ 8 & 2 & -5 \\ -11 & -3 & 7 \end{bmatrix}.$$

$$X = A^{-1}B$$

$$= \begin{bmatrix} 2 & 1 & -1 \\ 8 & 2 & -5 \\ -11 & -3 & 7 \end{bmatrix} \begin{bmatrix} -12 \\ 0 \\ 7 \end{bmatrix}$$

$$= \begin{bmatrix} -31 \\ -131 \\ 181 \end{bmatrix}.$$

The solution is $(-31, -131, 181)$.

11.
$$2x + 4y + 6z = 4$$
$$x + 4y + 2z = 8$$
$$y - z = -4$$

has coefficient matrix

$$A = \begin{bmatrix} 2 & 4 & 6 \\ 1 & 4 & 2 \\ 0 & 1 & -1 \end{bmatrix}.$$

From Exercise 45 of Section 6.4,

$$A^{-1} = \begin{bmatrix} 3 & -5 & 8 \\ -\dfrac{1}{2} & -1 & -1 \\ -\dfrac{1}{2} & -\dfrac{1}{2} & -2 \end{bmatrix}.$$

$$X = A^{-1}B = A^{-1} = \begin{bmatrix} 3 & -5 & 8 \\ -\dfrac{1}{2} & -1 & -1 \\ -\dfrac{1}{2} & -\dfrac{1}{2} & -2 \end{bmatrix} \begin{bmatrix} 4 \\ 8 \\ -4 \end{bmatrix}$$

$$= \begin{bmatrix} -60 \\ 10 \\ 14 \end{bmatrix}$$

The solution is $(-60, 10, 14)$.

13.
$$\begin{aligned} x - 2z \quad\quad &= 4 \\ -2x + y + 2z + 2w &= -8 \\ 3x - y - 2z - 3w &= 12 \\ y + 4z + w &= -4 \end{aligned}$$

The coefficient matrix is
$$A = \begin{bmatrix} 1 & 0 & -2 & 0 \\ -2 & 1 & 2 & 2 \\ 3 & -1 & -2 & -3 \\ 0 & 1 & 4 & 1 \end{bmatrix}.$$

From Exercise 47 in Section 6.4,
$$A^{-1} = \begin{bmatrix} \frac{1}{2} & -1 & -\frac{1}{2} & \frac{1}{2} \\ \frac{1}{2} & 4 & \frac{5}{2} & -\frac{1}{2} \\ -\frac{1}{4} & -\frac{1}{2} & -\frac{1}{4} & \frac{1}{4} \\ \frac{1}{2} & -2 & -\frac{3}{2} & \frac{1}{2} \end{bmatrix}$$

$$X = A^{-1} = \begin{bmatrix} \frac{1}{2} & -1 & -\frac{1}{2} & \frac{1}{2} \\ \frac{1}{2} & 4 & \frac{5}{2} & -\frac{1}{2} \\ -\frac{1}{4} & -\frac{1}{2} & -\frac{1}{4} & \frac{1}{4} \\ \frac{1}{2} & -2 & -\frac{3}{2} & \frac{1}{2} \end{bmatrix} \begin{bmatrix} 4 \\ -8 \\ 12 \\ -4 \end{bmatrix}$$

$$= \begin{bmatrix} 2 \\ 2 \\ -1 \\ -2 \end{bmatrix}.$$

The solution is $(2, 2, -1, -2)$.

15. Since $N = X - MX$,
$$\begin{aligned} N &= IX - MX \\ N &= (I - M)X \\ (I - M)^{-1}N &= (I - M)^{-1}(I - M)X \\ (I - M)^{-1}N &= IX. \end{aligned}$$

Thus, $X = (I - M)^{-1}N$.
$$I - M = \begin{bmatrix} 1 & 0 \\ 0 & 1 \end{bmatrix} - \begin{bmatrix} 0 & 1 \\ -2 & 1 \end{bmatrix} = \begin{bmatrix} 1 & -1 \\ 2 & 0 \end{bmatrix}$$

If $I - M = \begin{bmatrix} 1 & -1 \\ 2 & 0 \end{bmatrix}$,

$$(I - M)^{-1} = \begin{bmatrix} 0 & \frac{1}{2} \\ -1 & \frac{1}{2} \end{bmatrix}.$$

15 Continued

Since $X = (I - M)^{-1}N$
$$X = \begin{bmatrix} 0 & \frac{1}{2} \\ -1 & \frac{1}{2} \end{bmatrix} \begin{bmatrix} 8 \\ -12 \end{bmatrix}$$

$$= \begin{bmatrix} -6 \\ -14 \end{bmatrix}.$$

17. Let x = the number of buffets;
y = the number of chairs;
z = the number of tables.
The information can be summarized in a table.

	Construction hours	Finishing hours
Buffets	$30x$	$10x$
Chairs	$10y$	$10y$
Tables	$10z$	$30z$
Total	350	150

The system of equations is

$$30x + 10y + 10z = 350 \quad (1)$$
$$10x + 10y + 30z = 150 \quad (2)$$

To solve in terms of z, multiply equation (1) by –1 and add to equation (2).
$$\begin{aligned} -30x - 10y - 10z &= -350 \\ \underline{10x + 10y + 30z} &= \underline{150} \\ -20x \quad\quad + 20z &= -200 \end{aligned}$$

Solve equation the above equation for x in terms of z.
$$\begin{aligned} -20x &= -20z - 200 \\ x &= z + 10 \end{aligned}$$

Substitute $z + 10$ for x in equation (1) and solve for y.
$$\begin{aligned} 30(z + 10) + 10y + 10z &= 350 \\ 30z + 300 + 10y + 10z &= 350 \\ 10y + 40z + 300 &= 350 \\ 10y &= -40z + 50 \\ y &= -4z + 5 \end{aligned}$$

Since y cannot be negative and x, y, and z must all be whole numbers, z must be 0 or 1.
Each week, the factory should produce either 10 buffets, 5 chairs, and 0 tables or 11 buffets, 1 chair, and 1 table.

19. The following solution presupposes the use of a TI-82 graphing calculator. Similar results can be obtained from other graphing calculators.

Let x = number of bacterium 1;

y = number of bacterium 2;

z = number of bacterium 3.

Food

	I	II	III
Bacterium 1	1.3	1.3	2.3
Bacterium 2	1.1	2.4	3.7
Bacterium 3	8.1	2.9	5.1
Totals	16,000	28,000	44,000

The data in the table produce the system of equations

$1.3x + 1.1y + 8.1z = 16,000$

$1.3x + 2.4y + 2.9z = 28,000$

$2.3x + 3.7y + 5.1z = 44,000.$

We store the following matrix as matrix A:

$$\begin{bmatrix} 1.3 & 1.1 & 8.1 & | & 16,000 \\ 1.3 & 2.4 & 2.9 & | & 28,000 \\ 2.3 & 3.7 & 5.1 & | & 44,000 \end{bmatrix}.$$

Using row operations, we transform this to obtain

$$\begin{bmatrix} 1 & 0 & 0 & | & 2339.74359 \\ 0 & 1 & 0 & | & 10,128.20513 \\ 0 & 0 & 1 & | & 224.3589744 \end{bmatrix}$$

Thus, 2340 of the first species, 10,128 of the second species, and 224 of the third species can be maintained.

21. Let x = the wholesale price of jeans

y = the wholesale price of jackets

z = the wholesale price of sweaters

w = the wholesale price of shirts

The system is

$3000x + 3000y + 2200z + 4200w = \$507,650$

$2700x + 2500y + 2100z + 4300w = \$459,075$

$5000x + 2000y + 1400z + 7500w = \$541,225$

$7000x + 1800y + 600z + 8000w = \$571,500$

Store the following matrix as matrix A.

$$\begin{bmatrix} 3000 & 3000 & 2200 & 4200 & | & 507,650 \\ 2700 & 2500 & 2100 & 4300 & | & 459,075 \\ 5000 & 2000 & 1400 & 7500 & | & 541,225 \\ 7000 & 1800 & 600 & 8000 & | & 571,500 \end{bmatrix}$$

Using row operations, transform this matrix to obtain

$$\begin{bmatrix} 1 & 0 & 0 & 0 & | & 34.5 \\ 0 & 1 & 0 & 0 & | & 72 \\ 0 & 0 & 1 & 0 & | & 44 \\ 0 & 0 & 0 & 1 & | & 21.75 \end{bmatrix}$$

Therefore, the wholesale price for jeans is \$34.50, jacket is \$72, sweater is \$44, and a shirt is \$21.75.

23. $A = \begin{bmatrix} \frac{1}{2} & \frac{2}{5} \\ \frac{1}{4} & \frac{1}{5} \end{bmatrix}$, $D = \begin{bmatrix} 2 \\ 4 \end{bmatrix}$

To find the production matrix, first calculate $I - A$.

$$I - A = \begin{bmatrix} 1 & 0 \\ 0 & 1 \end{bmatrix} - \begin{bmatrix} \frac{1}{2} & \frac{2}{5} \\ \frac{1}{4} & \frac{1}{5} \end{bmatrix}$$

$$= \begin{bmatrix} \frac{1}{2} & -\frac{2}{5} \\ -\frac{1}{4} & \frac{4}{5} \end{bmatrix}$$

Using row operations, find the inverse of $I - A$.

$$(I - A)^{-1} = \begin{bmatrix} \frac{8}{3} & \frac{4}{3} \\ \frac{5}{6} & \frac{5}{3} \end{bmatrix}$$

Since $X = (I - A)^{-1}D$,

$$X = \begin{bmatrix} \frac{8}{3} & \frac{4}{3} \\ \frac{5}{6} & \frac{5}{3} \end{bmatrix} \begin{bmatrix} 2 \\ 4 \end{bmatrix}$$

$$= \begin{bmatrix} \frac{32}{3} \\ \frac{25}{3} \end{bmatrix}.$$

25. $A = \begin{bmatrix} .25 & .08 \\ .33 & .11 \end{bmatrix}$

$D = \begin{bmatrix} 690 \\ 920 \end{bmatrix}$

$(I-A)^{-1} = \begin{bmatrix} 1.388 & 0.1248 \\ 0.5147 & 1.1698 \end{bmatrix}$

$X = (I-A)^{-1} D$

$X = \begin{bmatrix} 1.388 & 0.1248 \\ 0.5147 & 1.1698 \end{bmatrix} \begin{bmatrix} 690 \\ 920 \end{bmatrix}$

$X = \begin{bmatrix} 1073 \\ 1431 \end{bmatrix}$

Produce 1073 metric tons of wheat, and 1431 metric tons of oil.

27. First, write the input-output matrix A.

$A = \begin{bmatrix} .4 & .6 \\ .5 & .25 \end{bmatrix} = \begin{bmatrix} \frac{2}{5} & \frac{3}{5} \\ \frac{1}{2} & \frac{1}{4} \end{bmatrix}$

$I-A = \begin{bmatrix} 1 & 0 \\ 0 & 1 \end{bmatrix} - \begin{bmatrix} \frac{2}{5} & \frac{3}{5} \\ \frac{1}{2} & \frac{1}{4} \end{bmatrix}$

$= \begin{bmatrix} \frac{3}{5} & -\frac{3}{5} \\ -\frac{1}{2} & \frac{3}{4} \end{bmatrix}$

Form $\begin{bmatrix} A-I \mid I \end{bmatrix}$.

$\begin{bmatrix} \frac{3}{5} & -\frac{3}{5} & 1 & 0 \\ -\frac{1}{2} & \frac{3}{4} & 0 & 1 \end{bmatrix}$

$\begin{bmatrix} 1 & -1 & \frac{5}{3} & 0 \\ -\frac{1}{2} & \frac{3}{4} & 0 & 1 \end{bmatrix} \frac{5}{3} R_1$

$\begin{bmatrix} 1 & -1 & \frac{5}{3} & 0 \\ 0 & \frac{1}{4} & \frac{5}{6} & 1 \end{bmatrix} \frac{1}{2} R_1 + R_2$

$\begin{bmatrix} 1 & -1 & 5 & 0 \\ 0 & 1 & \frac{10}{3} & 4 \end{bmatrix} 4R_2$

$\begin{bmatrix} 1 & 0 & 5 & 4 \\ 0 & 1 & \frac{10}{3} & 4 \end{bmatrix} R_2 + R_1$

Thus, $(I-A)^{-1} = \begin{bmatrix} 5 & 4 \\ \frac{10}{3} & 4 \end{bmatrix}$

Since $D = \begin{bmatrix} 15 \text{ million} \\ 12 \text{ million} \end{bmatrix}$ and $X = \left(I-A \right)^{-1} D$,

$X = \begin{bmatrix} 5 & 4 \\ \frac{10}{3} & 4 \end{bmatrix} \begin{bmatrix} 15 \text{ million} \\ 12 \text{ million} \end{bmatrix}$

$= \begin{bmatrix} 123 \text{ million} \\ 98 \text{ million} \end{bmatrix}.$

The output should be $123 million of electricity and $98 million of gas.

29. $A = \begin{bmatrix} \frac{1}{4} & \frac{1}{6} \\ \frac{1}{2} & 0 \end{bmatrix}$

$I-A = \begin{bmatrix} \frac{3}{4} & -\frac{1}{6} \\ -\frac{1}{2} & 1 \end{bmatrix}$

$(I-A)^{-1} = \begin{bmatrix} \frac{3}{2} & \frac{1}{4} \\ \frac{3}{4} & \frac{9}{8} \end{bmatrix}$

a. $X = \begin{bmatrix} \frac{3}{2} & \frac{1}{4} \\ \frac{3}{4} & \frac{9}{8} \end{bmatrix} \begin{bmatrix} 1 \\ 1 \end{bmatrix} = \begin{bmatrix} \frac{7}{4} \\ \frac{15}{8} \end{bmatrix}$

Thus, $\frac{7}{4}$ bushes of yams and $\frac{15}{8} \approx 2$ pigs should be produced.

b. $X = \begin{bmatrix} \frac{3}{2} & \frac{1}{4} \\ \frac{3}{4} & \frac{9}{8} \end{bmatrix} \begin{bmatrix} 100 \\ 70 \end{bmatrix} = \begin{bmatrix} 167.5 \\ 153.75 \end{bmatrix}$

Thus, 167.5 bushels of yams and $153.75 \approx 154$ pigs should be produced.

31. The input-output matrix, A, and the matrix $I - A$, are

   a.   .40 unit of agriculture, .12 unit of manufacturing, and 3.60 units of households.

   b.

$$A = \begin{bmatrix} .25 & .40 & .133 \\ .14 & .12 & .100 \\ .80 & 3.60 & .133 \end{bmatrix} \text{ and } I - A = \begin{bmatrix} .75 & -.40 & -.133 \\ -.14 & .88 & -.100 \\ -.80 & -3.60 & .867 \end{bmatrix}$$

Next, calculate $(I - A)^{-1}$.

$$(I - A)^{-1} \approx \begin{bmatrix} 6.61 & 13.53 & 2.57 \\ 3.3 & 8.91 & 1.53 \\ 19.8 & 49.5 & 9.9 \end{bmatrix}$$

$$X = (I - A)^{-1} D \approx \begin{bmatrix} 6.61 & 13.53 & 2.57 \\ 3.3 & 8.91 & 1.53 \\ 19.8 & 49.5 & 9.9 \end{bmatrix} \begin{bmatrix} 35 \\ 38 \\ 40 \end{bmatrix} \approx \begin{bmatrix} 848 \\ 516 \\ 2970 \end{bmatrix}$$

Therefore, 848 units of agriculture, 516 units of manufacturing, and 2970 units of households need to be produced.

   c.   The number of units of agriculture is about 813.

33. a. 017 unit of manufacturing and .216 unit of energy.

b. The input-output matrix A, and the matrix I–A, are

$$A = \begin{bmatrix} .293 & 0 & 0 \\ .014 & .207 & .017 \\ .044 & .010 & .216 \end{bmatrix} \text{ and } X = \begin{bmatrix} 175,000 \\ 22,000 \\ 12,000 \end{bmatrix}$$

Next, calculate $D = X - AX$.

$$D = \begin{bmatrix} 175,000 \\ 22,000 \\ 12,000 \end{bmatrix} \begin{bmatrix} .293 & 0 & 0 \\ .014 & .207 & .017 \\ .044 & .010 & .216 \end{bmatrix} \begin{bmatrix} 175,000 \\ 22,000 \\ 12,000 \end{bmatrix}$$

$$D \approx \begin{bmatrix} 123,725 \\ 14,792 \\ 1,488 \end{bmatrix}$$

Therefore, about 123,725,000 pounds of agriculture, 14,792,000 pounds of manufacturing, and 1,488,000 pounds of energy should be produced.

c. $X = (I - A)^{-1} D \approx \begin{bmatrix} 1.41 & 0 & 0 \\ .03 & 1.26 & .03 \\ .08 & .02 & 1.28 \end{bmatrix} \begin{bmatrix} 138,213 \\ 17,597 \\ 1786 \end{bmatrix} \approx \begin{bmatrix} 195,492 \\ 25,933 \\ 13,580 \end{bmatrix}$

Therefore, about 195,492,000 pounds of agriculture, 25,933,000 pounds of manufacturing, and 13,580,000 pounds of energy should be produced.

35. $A = \begin{bmatrix} .1045 & .0428 & .0029 & .0031 \\ .0826 & .1087 & .0584 & .0321 \\ .0867 & .1019 & .2032 & .3555 \\ .6253 & .3448 & .6106 & .0798 \end{bmatrix}, D = \begin{bmatrix} 450 \\ 300 \\ 125 \\ 100 \end{bmatrix}$

$$(I - A)^{-1} = \begin{bmatrix} 1.133 & .062 & .019 & .013 \\ .202 & 1.19 & .171 & .108 \\ .748 & .536 & 1.87 & .742 \\ 1.343 & .845 & 1.315 & 1.63 \end{bmatrix}$$

$$X = (I - A)^{-1} D$$

$$X = \begin{bmatrix} 1.133 & .062 & .019 & .013 \\ .202 & 1.19 & .171 & .108 \\ .748 & .536 & 1.87 & .742 \\ 1.343 & .845 & 1.315 & 1.63 \end{bmatrix} \begin{bmatrix} 450 \\ 300 \\ 125 \\ 100 \end{bmatrix} = \begin{bmatrix} 532 \\ 480 \\ 805 \\ 1185 \end{bmatrix}$$

This means $532 million of natural resources, $481 million of manufacturing, $805 million of trade and services, and $1185 million of personal consumption.

37. a. The input-output matrix A, and the matrix $I - A$, are

$$A = \begin{bmatrix} .2 & .1 & .1 \\ .1 & .1 & 0 \\ .5 & .6 & .7 \end{bmatrix} \text{ and } I - A = \begin{bmatrix} .8 & -.1 & -.1 \\ -.1 & .9 & 0 \\ -.5 & -.6 & .3 \end{bmatrix}$$

Next, calculate $(I - A)^{-1}$.

$$(I - A)^{-1} \approx \begin{bmatrix} 1.67 & .56 & .56 \\ .19 & 1.17 & .06 \\ 3.15 & 3.27 & 4.38 \end{bmatrix}$$

b. These multipliers imply that if the demand for one community's output increases by $1 then the output in the other community will increase by the amount in the row and column of that matrix. For example, if the demand for Hermitage's output increases by $1, then output from Sharon will increase by $.56, Farrell by $.06, and Hermitage by $4.38.

39. The message *Head for the hills* broken into groups of 2 letters would have the following matrices.

$$\begin{bmatrix} 8 \\ 5 \end{bmatrix}, \begin{bmatrix} 1 \\ 4 \end{bmatrix}, \begin{bmatrix} 27 \\ 6 \end{bmatrix}, \begin{bmatrix} 15 \\ 18 \end{bmatrix}, \begin{bmatrix} 27 \\ 20 \end{bmatrix},$$

$$\begin{bmatrix} 8 \\ 5 \end{bmatrix}, \begin{bmatrix} 27 \\ 8 \end{bmatrix}, \begin{bmatrix} 9 \\ 12 \end{bmatrix}, \begin{bmatrix} 12 \\ 19 \end{bmatrix}.$$

Multiply $M = \begin{bmatrix} 1 & 3 \\ 2 & 7 \end{bmatrix}$ by each matrix in the code to get

$$\begin{bmatrix} 23 \\ 51 \end{bmatrix}, \begin{bmatrix} 13 \\ 30 \end{bmatrix}, \begin{bmatrix} 45 \\ 96 \end{bmatrix}, \begin{bmatrix} 69 \\ 156 \end{bmatrix}, \begin{bmatrix} 87 \\ 194 \end{bmatrix},$$

$$\begin{bmatrix} 23 \\ 51 \end{bmatrix}, \begin{bmatrix} 51 \\ 110 \end{bmatrix}, \begin{bmatrix} 45 \\ 102 \end{bmatrix}, \begin{bmatrix} 69 \\ 157 \end{bmatrix}.$$

41.

$$A = \begin{bmatrix} 0 & 1 & 2 & 2 \\ 1 & 0 & 1 & 0 \\ 2 & 1 & 0 & 1 \\ 2 & 0 & 1 & 0 \end{bmatrix}$$

$$A^2 = \begin{bmatrix} 9 & 2 & 3 & 2 \\ 2 & 2 & 2 & 3 \\ 3 & 2 & 6 & 4 \\ 2 & 3 & 4 & 5 \end{bmatrix}$$

a. The number of ways to travel from city 1 to city 3 by passing through exactly one city is the entry in row 1, column 3 of A^2, which is 3.

b. The number of ways to travel from city 2 to city 4 by passing through exactly one city is the entry in row 2, column 4 of A^2, which is 3.

c. The number of ways to travel from city 1 to city 3 by passing through at most one city is the sum of the entries in row 1, column 3 of A and A^2, which is 2 + 3 or 5.

d. The number of ways to travel from city 2 to city 4 by passing through at most one city is the sum of the entries in row 2, column 4 of A and A^2, which is 0 + 3 or 3.

43. a.

$$\begin{array}{cc} & \begin{array}{ccc} 1 & 2 & 3 \end{array} \\ B = \begin{array}{c} 1 \\ 2 \\ 3 \end{array} & \begin{bmatrix} 0 & 2 & 3 \\ 2 & 0 & 4 \\ 3 & 4 & 0 \end{bmatrix} \end{array}$$

b.

$$B^2 = \begin{bmatrix} 0 & 2 & 3 \\ 2 & 0 & 4 \\ 3 & 4 & 0 \end{bmatrix} \begin{bmatrix} 0 & 2 & 3 \\ 2 & 0 & 4 \\ 3 & 4 & 0 \end{bmatrix}$$

$$= \begin{bmatrix} 13 & 12 & 8 \\ 12 & 20 & 6 \\ 8 & 6 & 25 \end{bmatrix}$$

c. Use the entry in row 1, column 2 of B^2, which is 12.

d. Use the sum of the entries in row 1, column 2 of B and B^2, which is 2 + 12 or 14.

45. a.

$$\begin{array}{cc} & \begin{array}{cccc} d & r & c & m \end{array} \\ C = \begin{array}{c} d \\ r \\ c \\ m \end{array} & \begin{bmatrix} 0 & 1 & 1 & 1 \\ 0 & 0 & 0 & 1 \\ 0 & 1 & 0 & 1 \\ 0 & 0 & 0 & 0 \end{bmatrix} \end{array}$$

b.

$$C^2 = \begin{bmatrix} 0 & 1 & 1 & 1 \\ 0 & 0 & 0 & 1 \\ 0 & 1 & 0 & 1 \\ 0 & 0 & 0 & 0 \end{bmatrix} \begin{bmatrix} 0 & 1 & 1 & 1 \\ 0 & 0 & 0 & 1 \\ 0 & 1 & 0 & 1 \\ 0 & 0 & 0 & 0 \end{bmatrix}$$

$$= \begin{bmatrix} 0 & 1 & 0 & 2 \\ 0 & 0 & 0 & 0 \\ 0 & 0 & 0 & 1 \\ 0 & 0 & 0 & 0 \end{bmatrix}$$

C^2 gives the number of food sources once removed from the feeder. Thus, since dogs eat rats and rats eat mice, mice are an indirect as well as a direct food source of dogs

Chapter 6 Review Exercises

1. $-5x - 3y = -3$ (1)
  $2x + y = 4$ (2)
Multiply equation (2) by 3 and add to equation (1).
$-5x - 3y = -3$
$\underline{6x + 3y = 12}$
$x = 9$
Substitute 9 for x in equation (2) to solve for y.
$2(9) + y = 4$
$18 + y = 4$
$y = 14$
The solution is $(9, -14)$.

2. $3x - y = 8$ (1)
  $2x + 3y = 6$ (2)
Multiply equation (1) by 3 and add to equation (2).
$9x - 3y = 24$
$\underline{2x + 3y = 6}$
$11x = 30$
$x = \dfrac{30}{11}$
Substitute $\dfrac{30}{11}$ for x in equation (1) to solve for y.
$3\left(\dfrac{30}{11}\right) - y = 8$
$-y = 8 - \dfrac{90}{11}$
$-y = -\dfrac{2}{11}$
$y = \dfrac{2}{11}$
The solution is $\left(\dfrac{30}{11}, \dfrac{2}{11}\right)$.

3. $4x - 5y = 6$ (1)
  $3x - 3y = 9$ (2)
Multiply equation (1) by 3 and equation (2) by –4 and add the results.
$12x - 15y = 18$
$\underline{-12x + 12y = -36}$
$ -3y = -18$
$y = 6$
Substitute 6 for y in equation (1) to solve for x.
$4x - 5(6) = 6$
$4x - 30 = 6$
$4x = 36$
$x = 9$
The solution is $(9, 6)$.

4. $\dfrac{1}{4}x - \dfrac{1}{3}y = -\dfrac{1}{4}$ (1)
  $\dfrac{1}{10}x + \dfrac{2}{5}y = \dfrac{2}{5}$ (2)
Multiply equation (1) by 12 and equation (2) by 10 and add the results.
$3x - 4y = -3$
$\underline{x + 4y = 4}$
$4x = 1$
$x = \dfrac{1}{4}$
Substitute $\dfrac{1}{4}$ for x in equation (4) to solve for y.
$\dfrac{1}{4} + 4y = 4$
$4y = 4 - \dfrac{1}{4}$
$4y = \dfrac{15}{4}$
$y = \dfrac{15}{16}$
The solution is $\left(\dfrac{1}{4}, \dfrac{15}{16}\right)$.

5.
$$x - 2y = 1 \quad (1)$$
$$4x + 4y = 2 \quad (2)$$
$$10x + 8y = 4 \quad (3)$$

Multiply equation (1) by 2 and add to equation (2).

$$2x - 4y = 2$$
$$\underline{4x + 4y = 2}$$
$$6x \quad\quad = 4$$
$$x = \frac{2}{3}$$

Substitute $\frac{2}{3}$ for x in equation (1) to solve for y.

$$\frac{2}{3} - 2y = 1$$
$$-2y = 1 - \frac{2}{3}$$
$$-2y = \frac{1}{3}$$
$$y = -\frac{1}{6}$$

The ordered pair $\left(\frac{2}{3}, -\frac{1}{6}\right)$ satisfies equations (1) and (2), but not equation (3). The system is inconsistent; there is no solution.

6.
$$x + y - 4z = 0 \quad (1)$$
$$2x + y - 3z = 2 \quad (2)$$

Multiply equation (1) by -1 and add to equation (2).

$$-x - y + 4z = 0 \quad (3)$$
$$\underline{2x + y - 3z = 2} \quad (2)$$
$$x \quad\quad + z = 2 \quad (4)$$

Solve equation (4) for x in terms of z.

$$x + z = 2$$
$$x = 2 - z$$

Substitute $2 - z$ for x in equation (1) and solve for y in terms of z.

$$2 - z + y - 4z = 0$$
$$2 + y - 5z = 0$$
$$y = 5z - 2$$

The system is dependent, and the solution is all ordered triples of the form
$(2 - z, 5z - 2, z)$ for any real number z.

7.
$$3x + y - z = 3 \quad (1)$$
$$x \quad + 2z = 6 \quad (2)$$
$$-3x - y + 2z = 9 \quad (3)$$

Add equations (1) and (3).

$$3x + y - z = 3$$
$$\underline{-3x - y + 2z = 9}$$
$$z = 12$$

Substitute 12 for z in equation (2) and solve for x.

$$x + 2(12) = 6$$
$$x + 24 = 6$$
$$x = -18$$

Substitute -18 for x and 12 for z in equation (1) and solve for y.

$$3(-18) + y - 12 = 3$$
$$-54 + y - 12 = 3$$
$$y - 66 = 3$$
$$y = 69$$

The solution is $(-18, 69, 12)$.

8.
$$4x - y - 2z = 4 \quad (1)$$
$$x - y - \frac{1}{2}z = 1 \quad (2)$$
$$2x - y - z = 8 \quad (3)$$

Multiply equation (2) by -4 and add to equation (1).

$$4x - y - 2z = 4 \quad (1)$$
$$\underline{-4x + 4y + 2z = -4} \quad (4)$$
$$3y = 0$$
$$y = 0$$

Substitute 0 for y in equations (1) and (3).

$$4x - 2z = 4 \quad (5)$$
$$2x - z = 8 \quad (6)$$

Multiply equation (6) by -2 and add to equation (5).

$$4x - 2z = 4 \quad (5)$$
$$\underline{-4x + 2z = -16} \quad (7)$$
$$0 = -12 \quad (8)$$

Since equation (8) is false, this system is inconsistent; there is no solution.

9. Let x = the number of standard paper clips (in 1000s);

y = the number of extra large paper clips (in 1000s).

We will solve the following system.

$\frac{1}{4}x+\frac{1}{3}y=4$ (1)

$\frac{1}{2}x+\frac{1}{3}y=6$ (2)

Multiply equation (1) by 12 and equation (2) by –6 and add the results.

$3x+4y=48$ (3)

$\underline{-3x-2y=-36}$ (4)

$2y=12$

$y=6$

Substitute 6 for y in equation (3) and solve for x.

$3x+4(6)=48$

$3x+24=48$

$3x=24$

$x=8$

The manufacturer can make 8000 standard and 6000 extra large paper clips.

10. Let x = the number of shares of the first stock;

y = the number of shares of the second stock.

We will solve the following system.

$32x+23y=10{,}100$ (1)

$1.2x+1.4y=540$ (2)

Multiply equation (1) by 1.2 and equation (2) by –32 and add the results.

$38.4x+27.6y=12{,}120$ (3)

$\underline{-38.4x-44.8y=-17{,}280}$ (4)

$-17.2y=-5160$

$y=300$

Substitute 300 for y in equation (2) and solve for x.

$1.2x+1.4(300)=540$

$1.2x+420=540$

$1.2x=120$

$x=100$

She should buy 100 shares of the first stock and 300 shares of the second stock.

11. Let x = the amount invested in the first fund;

y = the amount invested in the second fund.

We will solve the system

$.08x+.02y=780$ (1)

$.1x+.01y=810$ (2)

Multiply equation (2) by –2 and add to equation (1).

$.08x+.02y=780$ (1)

$\underline{-.2x-.02y=-1620}$ (3)

$-.12x=-840$

$x=7000$

Substitute 7000 for x in equation (1) and solve for y.

$.08(7000)+.02y=780$

$560+.02y=780$

$.02y=220$

$y=11{,}000$

Joyce has $7000 invested in the first fund and $11,000 in the second.

12. Let x = the number of one dollar bills;
y = the number of five dollar bills;
z = the number of ten dollar bills.
From the given information, we have the system

$$
\begin{aligned}
x + y + z &= 35 \quad (1)\\
x + 5y + 10z &= 144 \quad (2)\\
z &= y + 2. \quad (3)
\end{aligned}
$$

Substitute $y + 2$ for z in equation (1) to get

$$
\begin{aligned}
x + y + y + 2 &= 35\\
x + 2y + 2 &= 35\\
x + 2y &= 33. \quad (4)
\end{aligned}
$$

Substitute $y + 2$ for z in equation (2) to get

$$
\begin{aligned}
x + 5y + 10(y + 2) &= 144\\
x + 5y + 10y + 20 &= 144\\
x + 15y &= 124. \quad (5)
\end{aligned}
$$

Multiply equation (4) by –1 and add to equation (5).

$$
\begin{aligned}
-x - 2y &= -33\\
\underline{x + 15y} &= \underline{124}\\
13y &= 91\\
y &= 7
\end{aligned}
$$

Substitute 7 for y in equation (4) and solve for x.

$$
\begin{aligned}
x + 2(7) &= 33\\
x + 14 &= 33\\
x &= 19
\end{aligned}
$$

Substitute 7 for y in equation (3) and solve for z.

$$
\begin{aligned}
z &= 7 + 2\\
z &= 9
\end{aligned}
$$

The solution is 19 ones, 7 fives, and 9 tens.

13. Organize the information into a table.

	I	II	III
Food	1	200	150
Shelter	2	0	200
Counseling	0	100	100

Let x = the number of clients form source I;
y = the number of clients from source II;
z = the number of clients from source III.

13. Continued

The system is

$$
\begin{aligned}
100x + 200y + 150z &= 50,000 \quad (1)\\
250x \qquad\ \ + 200z &= 32,500 \quad (2)\\
100y + 100z &= 25,000. \quad (3)
\end{aligned}
$$

Multiply equation (1) by $\dfrac{1}{100}$.

$$
\begin{aligned}
x + 2y + 1.5z &= 500 \quad (4)\\
250x \qquad\ \ + 200z &= 32,500 \quad (2)\\
100y + 100z &= 25,000 \quad (3)
\end{aligned}
$$

Multiply equation (4) by –250 and add it to equation (2).

$$
\begin{aligned}
x + 2y + 1.5z &= 500 \quad (4)\\
-500y - 175z &= -92,500 \quad (5)\\
100y + 100z &= 25,000 \quad (3)
\end{aligned}
$$

Multiply equation (5) by $-\dfrac{1}{500}$.

$$
\begin{aligned}
x + 2y + 1.5z &= 500 \quad (4)\\
y + .35z &= 185 \quad (6)\\
100y + 100z &= 25,000 \quad (3)
\end{aligned}
$$

Multiply equation (6) by –100 and add it to equation (3).

$$
\begin{aligned}
x + 2y + 1.5z &= 500 \quad (4)\\
y + .35z &= 185 \quad (6)\\
65z &= 6500 \quad (7)
\end{aligned}
$$

From equation (7) $z = 100$. Substituting into equation (6) gives

$$
\begin{aligned}
y + .35(100) &= 185\\
y &= 150.
\end{aligned}
$$

Substituting into equation (4) gives

$$
\begin{aligned}
x + 2(150) + 1.5(100) &= 500\\
x + 300 + 150 &= 500\\
x &= 50.
\end{aligned}
$$

Thus, the agency can serve 50 clients from source I, 150 from source II, and 100 from source III.

14. Let x = the number of blankets;
y = the number of rugs;
z = the number of skirts.
Solve the following system.

$$24x + 30y + 12z = 306 \quad (1)$$
$$4x + 5y + 3z = 59 \quad (2)$$
$$15x + 18y + 9z = 201 \quad (3)$$

Simplify the system by dividing equation (1) by 6 and equation (3) by 3.

$$4x + 5y + 2z = 51 \quad (4)$$
$$4x + 5y + 3z = 59 \quad (2)$$
$$5x + 6y + 3z = 67 \quad (5)$$

Multiply equation (4) by -1 and add to equation (2).

$$-4x - 5y - 2z = -51$$
$$\underline{4x + 5y + 3z = 59}$$
$$z = 8$$

Substitute 8 for z in equation (4).

$$4x + 5y + 2(8) = 51$$
$$4x + 5y = 35 \quad (5)$$

Substitute 8 for z in equation (5).

$$5x + 6y + 3(8) = 67$$
$$5x + 6y = 43 \quad (6)$$

Multiply equation (6) by 5 and equation (7) by -4 and add the results.

$$20x + 25y = 175$$
$$\underline{-20x - 24y = -172}$$
$$y = 3$$

Substitute 3 for y in equation (6) and solve for x.

$$4x + 5(3) = 35$$
$$4x = 20$$
$$x = 5$$

They can make 5 blankets, 3 rugs, and 8 skirts.

15.
$$x + z = -3$$
$$y - z = 6$$
$$2x + 3z = 5$$

The augmented matrix is

$$\begin{bmatrix} 1 & 0 & 1 & | & -3 \\ 0 & 1 & -1 & | & 6 \\ 2 & 0 & 3 & | & 5 \end{bmatrix}$$

$$\begin{bmatrix} 1 & 0 & 1 & | & -3 \\ 0 & 1 & -1 & | & 6 \\ 0 & 0 & 1 & | & 11 \end{bmatrix} \begin{matrix} \\ \\ -2R_1 + R_3 \end{matrix}$$

$$\begin{bmatrix} 1 & 0 & 0 & | & -14 \\ 0 & 1 & 0 & | & 17 \\ 0 & 0 & 1 & | & 11 \end{bmatrix} \begin{matrix} -1R_3 + R_1 \\ R_3 + R_2 \\ \\ \end{matrix}$$

The solution is $(-14, 17, 11)$.

16.
$$3x + 2y + 4z = -1$$
$$-2x + y - 2z = 6$$
$$3x + 3y + 6z = 3$$

The augmented matrix is

$$\begin{bmatrix} 3 & 2 & 4 & | & -1 \\ -2 & 1 & -2 & | & 6 \\ 3 & 3 & 6 & | & 3 \end{bmatrix}$$

$$\begin{bmatrix} 1 & \dfrac{2}{3} & \dfrac{4}{3} & | & -\dfrac{1}{3} \\ -2 & 1 & -2 & | & 6 \\ 3 & 3 & 6 & | & 3 \end{bmatrix} \dfrac{1}{3}R_1$$

$$\begin{bmatrix} 1 & \dfrac{2}{3} & \dfrac{4}{3} & | & -\dfrac{1}{3} \\ 0 & \dfrac{7}{3} & \dfrac{2}{3} & | & \dfrac{16}{3} \\ 0 & 1 & 2 & | & 4 \end{bmatrix} \begin{matrix} \\ 2R_1 + R_2 \\ -3R_1 + R_3 \end{matrix}$$

$$\begin{bmatrix} 1 & \dfrac{2}{3} & \dfrac{4}{3} & | & -\dfrac{1}{3} \\ 0 & 1 & \dfrac{2}{7} & | & \dfrac{16}{7} \\ 0 & 1 & 2 & | & 4 \end{bmatrix} \begin{matrix} \\ \dfrac{3}{7}R_2 \\ \\ \end{matrix}$$

16. Continued

$$\begin{bmatrix} 1 & 0 & \frac{8}{7} & -\frac{13}{7} \\ 0 & 1 & \frac{2}{7} & \frac{16}{7} \\ 0 & 0 & \frac{12}{7} & \frac{12}{7} \end{bmatrix} \begin{matrix} -\frac{2}{3}R_2 + R_1 \\ \\ -1R_2 + R_3 \end{matrix}$$

$$\begin{bmatrix} 1 & 0 & \frac{8}{7} & -\frac{13}{7} \\ 0 & 1 & \frac{2}{7} & \frac{16}{7} \\ 0 & 0 & 1 & 1 \end{bmatrix} \begin{matrix} \\ \\ \frac{7}{12}R_3 \end{matrix}$$

$$\begin{bmatrix} 1 & 0 & 0 & -3 \\ 0 & 1 & 0 & 2 \\ 0 & 0 & 1 & 1 \end{bmatrix} \begin{matrix} -\frac{8}{7}R_3 + R_1 \\ -\frac{2}{7}R_3 + R_2 \\ \end{matrix}$$

The solution is $(-3, 2, 1)$.

17.
$$5x - 8y + z = 1$$
$$3x - 2y + 4z = 3$$
$$10x - 16y + 2z = 3$$

The augmented matrix is

$$\begin{bmatrix} 5 & -8 & 1 & 1 \\ 3 & -2 & 4 & 3 \\ 10 & -16 & 2 & 3 \end{bmatrix}$$

$$\begin{bmatrix} 1 & -\frac{8}{5} & \frac{1}{5} & \frac{1}{5} \\ 3 & -2 & 4 & 3 \\ 10 & -16 & 2 & 3 \end{bmatrix} \frac{1}{5}R_1$$

$$\begin{bmatrix} 1 & -\frac{8}{5} & \frac{1}{5} & \frac{1}{5} \\ 0 & \frac{14}{5} & \frac{17}{5} & \frac{12}{5} \\ 0 & 0 & 0 & 1 \end{bmatrix} \begin{matrix} \\ -3R_1 + R_2 \\ -10R_1 + R_3 \end{matrix}$$

Since row 3 has all zeros except for the last entry, there is no solution.

18.
$$x - 2y + 3z = 4$$
$$2x + y - 4z = 3$$
$$-3z + 4y - z = -2$$

The augmented matrix is

$$\begin{bmatrix} 1 & -2 & 3 & 4 \\ 2 & 1 & -4 & 3 \\ -3 & 4 & -1 & -2 \end{bmatrix}$$

$$\begin{bmatrix} 1 & -2 & 3 & 4 \\ 0 & 5 & -10 & -5 \\ 0 & -2 & 8 & 10 \end{bmatrix} \begin{matrix} \\ -2R_1 + R_2 \\ 3R_1 + R_3 \end{matrix}$$

$$\begin{bmatrix} 1 & -2 & 3 & 4 \\ 0 & 1 & -2 & -1 \\ 0 & -2 & 8 & 10 \end{bmatrix} \begin{matrix} \\ \frac{1}{5}R_2 \\ \end{matrix}$$

$$\begin{bmatrix} 1 & 0 & -1 & 2 \\ 0 & 1 & -2 & -1 \\ 0 & 0 & 4 & 8 \end{bmatrix} \begin{matrix} 2R_2 + R_1 \\ \\ 2R_2 + R_3 \end{matrix}$$

$$\begin{bmatrix} 1 & 0 & -1 & 2 \\ 0 & 1 & -2 & -1 \\ 0 & 0 & 1 & 2 \end{bmatrix} \begin{matrix} \\ \\ \frac{1}{4}R_3 \end{matrix}$$

$$\begin{bmatrix} 1 & 0 & 0 & 4 \\ 0 & 1 & 0 & 3 \\ 0 & 0 & 1 & 2 \end{bmatrix} \begin{matrix} R_3 + R_1 \\ 2R_3 + R_2 \\ \end{matrix}$$

The solution is $(4, 3, 2)$.

19. $3x + 2y - 6z = 3$

$x + y + 2z = 2$

$2x + 2y + 5z = 0$

The augmented matrix is

$$\begin{bmatrix} 3 & 2 & -6 & | & 3 \\ 1 & 1 & 2 & | & 2 \\ 2 & 2 & 5 & | & 0 \end{bmatrix}$$

Interchange R_1 and R_2.

$$\begin{bmatrix} 1 & 1 & 2 & | & 2 \\ 3 & 2 & -6 & | & 3 \\ 2 & 2 & 5 & | & 0 \end{bmatrix}$$

$$\begin{bmatrix} 1 & 1 & 2 & | & 2 \\ 0 & -12 & 0 & | & -3 \\ 0 & 0 & 1 & | & -4 \end{bmatrix} \begin{matrix} \\ -3R_1 + R_2 \\ -2R_1 + R_3 \end{matrix}$$

$$\begin{bmatrix} 1 & 1 & 2 & | & 2 \\ 0 & 1 & 12 & | & 3 \\ 0 & 0 & 1 & | & -4 \end{bmatrix} -1R_2$$

$$\begin{bmatrix} 1 & 0 & -10 & | & -1 \\ 0 & 1 & 12 & | & 3 \\ 0 & 0 & 1 & | & -4 \end{bmatrix} -1R_2 + R_1$$

$$\begin{bmatrix} 1 & 0 & 0 & | & -41 \\ 0 & 1 & 0 & | & 51 \\ 0 & 0 & 1 & | & -4 \end{bmatrix} \begin{matrix} 10R_3 + R_1 \\ -12R_3 + R_2 \end{matrix}$$

The solution is $(-41, 51, -4)$.

20. Let $x = $ the number of chairs;

$y = $ number tables;

$z = $ number of chests.

	Construction	Painting	Packing
Chair	2	1	2
Table	4	3	3
Chest	8	6	4
Totals	2000	1400	1300

$2x + 4y + 8z = 2000$

$x + 3y + 6z = 1400$

$2x + 3y + 4z = 1300$

20. Continued

Divide the first equation by 2 and write the augmented matrix.

$$\begin{bmatrix} 1 & 2 & 4 & | & 1000 \\ 1 & 3 & 6 & | & 1400 \\ 2 & 3 & 4 & | & 1300 \end{bmatrix}$$

$$\begin{bmatrix} 1 & 2 & 4 & | & 1000 \\ 0 & 1 & 2 & | & 400 \\ 0 & -1 & -4 & | & -700 \end{bmatrix} \begin{matrix} \\ -1R_1 + R_2 \\ -2R_1 + R_3 \end{matrix}$$

$$\begin{bmatrix} 1 & 0 & 0 & | & 200 \\ 0 & 1 & 2 & | & 400 \\ 0 & 0 & -2 & | & -300 \end{bmatrix} \begin{matrix} -2R_2 + R_1 \\ \\ R_2 + R_3 \end{matrix}$$

$$\begin{bmatrix} 1 & 0 & 0 & | & 200 \\ 0 & 1 & 0 & | & 100 \\ 0 & 0 & -2 & | & -300 \end{bmatrix} R_3 + R_2$$

$$\begin{bmatrix} 1 & 0 & 0 & | & 200 \\ 0 & 1 & 0 & | & 100 \\ 0 & 0 & 1 & | & 150 \end{bmatrix} -\frac{1}{2}R_3$$

The factory can produce 200 chairs, 100 tables, and 150 chests.

21. $\begin{bmatrix} 2 & 3 \\ 5 & 9 \end{bmatrix}$

The matrix is 2×2. Since the matrix is 2×2, it is square.

22. $\begin{bmatrix} 2 & -1 \\ 4 & 6 \\ 5 & 7 \end{bmatrix}$

The size of this matrix is 3×2.

23. $\begin{bmatrix} 12 & 4 & -8 & -1 \end{bmatrix}$

The matrix is 1×4. The matrix is a row matrix.

24. $\begin{bmatrix} -7 & 5 & 6 & 4 \\ 3 & 2 & -1 & 2 \\ -1 & 12 & 8 & -1 \end{bmatrix}$

This is a 3×4 square matrix.

25. $\begin{bmatrix} 6 & 8 & 10 \\ 5 & 3 & -2 \end{bmatrix}$

This matrix is 2×3.

26. $\begin{bmatrix} -9 \\ 15 \\ 4 \end{bmatrix}$

This matrix is 3×1. It is a column matrix.

27. $\begin{bmatrix} 8 & 8 & 8 \\ 10 & 5 & 9 \\ 7 & 10 & 7 \\ 8 & 9 & 7 \end{bmatrix}$

28. $\begin{bmatrix} 5 & 7 & 2532 & 52\frac{3}{8} & -\frac{1}{4} \\ 3 & 9 & 1464 & 56 & \frac{1}{8} \\ 2.50 & 5 & 4974 & 41 & -1\frac{1}{2} \\ 1.36 & 10 & 1754 & 18 & \frac{1}{2} \end{bmatrix}$

29. $B = \begin{bmatrix} 1 & 3 & -2 \\ 2 & 3 & 0 \\ 0 & 1 & 5 \end{bmatrix}$

$-B = \begin{bmatrix} -1 & -3 & 2 \\ -2 & -3 & 0 \\ 0 & -1 & -5 \end{bmatrix}$

30. $D = \begin{bmatrix} 6 \\ 1 \\ 0 \end{bmatrix}$

$-D = \begin{bmatrix} -6 \\ -1 \\ 0 \end{bmatrix}$

31. $A = \begin{bmatrix} 4 & 10 \\ -2 & -3 \\ 6 & 9 \end{bmatrix}, C = \begin{bmatrix} 5 & 0 \\ -1 & 3 \\ 4 & 7 \end{bmatrix}$

$3A - 2C$

$= \begin{bmatrix} 12 & 30 \\ -6 & -9 \\ 18 & 27 \end{bmatrix} - \begin{bmatrix} 10 & 0 \\ -2 & 6 \\ 8 & 14 \end{bmatrix}$

$= \begin{bmatrix} 2 & 30 \\ -4 & -15 \\ 10 & 13 \end{bmatrix}$

32. $F = \begin{bmatrix} -1 & 2 \\ 8 & 7 \end{bmatrix}, G = \begin{bmatrix} 2 & 5 \\ 1 & 6 \end{bmatrix}$

$F + 3G$

$= \begin{bmatrix} -1 & 2 \\ 8 & 7 \end{bmatrix} + \begin{bmatrix} 6 & 15 \\ 3 & 18 \end{bmatrix}$

$= \begin{bmatrix} 5 & 17 \\ 11 & 25 \end{bmatrix}$

33. $B = \begin{bmatrix} 1 & 3 & -2 \\ 2 & 3 & 0 \\ 0 & 1 & 5 \end{bmatrix} C = \begin{bmatrix} 5 & 0 \\ -1 & 3 \\ 4 & 7 \end{bmatrix}$

$2B - 5C$ is not defined, since the matrices have different sizes.

34. $G = \begin{bmatrix} 2 & 5 \\ 1 & 6 \end{bmatrix}, F = \begin{bmatrix} -1 & 2 \\ 8 & 7 \end{bmatrix}$

$G - 2F$

$= \begin{bmatrix} 2 & 5 \\ 1 & 6 \end{bmatrix} - \begin{bmatrix} -2 & 8 \\ 6 & 14 \end{bmatrix} = \begin{bmatrix} 4 & -3 \\ -5 & -8 \end{bmatrix}$

35. Using the data from Exercise 28, the first day matrix is the following.

$$\begin{array}{c} \\ \text{ATT} \\ \text{GE} \\ \text{GO} \\ \text{S} \end{array} \begin{array}{cc} \text{Sales} & \text{Price} \\ & \text{change} \\ \begin{bmatrix} 2532 & -\frac{1}{4} \\ 1464 & \frac{1}{8} \\ 4974 & -\frac{3}{2} \\ 1754 & \frac{1}{2} \end{bmatrix} \end{array}$$

The next day matrix is the following.

$\begin{bmatrix} 2310 & -\frac{1}{4} \\ 1258 & -\frac{1}{4} \\ 5061 & \frac{1}{2} \\ 1812 & \frac{1}{2} \end{bmatrix}$

The total sales and price changes for the two days are given by the sum

$\begin{bmatrix} 2532 & -\frac{1}{4} \\ 1464 & \frac{1}{8} \\ 4974 & -\frac{3}{2} \\ 1754 & \frac{1}{2} \end{bmatrix} + \begin{bmatrix} 2310 & -\frac{1}{4} \\ 1258 & -\frac{1}{4} \\ 5061 & \frac{1}{2} \\ 1812 & \frac{1}{2} \end{bmatrix}$

$= \begin{bmatrix} 4842 & -\frac{1}{2} \\ 2722 & -\frac{1}{8} \\ 10,035 & -1 \\ 3566 & 1 \end{bmatrix}$

36. a. First shipment:

Tulsa   New Orleans

$$\begin{array}{c} \text{Chicago} \\ \text{Dallas} \\ \text{Atlanta} \end{array} \begin{bmatrix} 110{,}000 & 85{,}000 \\ 73{,}000 & 108{,}000 \\ 95{,}000 & 69{,}000 \end{bmatrix}$$

Second shipment:

$$\begin{bmatrix} 58{,}000 & 40{,}000 \\ 33{,}000 & 52{,}000 \\ 80{,}000 & 30{,}000 \end{bmatrix}$$

b. Add the 2 matrices in part (a).

$$\begin{bmatrix} 168{,}000 & 125{,}000 \\ 106{,}000 & 160{,}000 \\ 175{,}000 & 99{,}000 \end{bmatrix}$$

37. $A = \begin{bmatrix} 4 & 10 \\ -2 & -3 \\ 6 & 9 \end{bmatrix},\ G = \begin{bmatrix} 2 & 5 \\ 1 & 6 \end{bmatrix}$

$$AG = \begin{bmatrix} (4)(2)+(10)(1) & (4)(5)+(10)(6) \\ (-2)(2)+(-3)(1) & (-2)(5)+(-3)(6) \\ (6)(2)+(9)(1) & (6)(5)+(9)(6) \end{bmatrix}$$

$$= \begin{bmatrix} 18 & 80 \\ -7 & -28 \\ 21 & 84 \end{bmatrix}$$

38. $E = \begin{bmatrix} 1 & 3 & -4 \end{bmatrix},\ B = \begin{bmatrix} 1 & 3 & -2 \\ 2 & 3 & 0 \\ 0 & 1 & 5 \end{bmatrix}$

$$EB = \begin{bmatrix} 7 & 8 & -22 \end{bmatrix}$$

39. $G = \begin{bmatrix} 2 & 5 \\ 1 & 6 \end{bmatrix},\ F = \begin{bmatrix} -1 & 2 \\ 8 & 7 \end{bmatrix}$

$$GF = \begin{bmatrix} (2)(-1)+(5)(8) & (2)(2)+(5)(7) \\ (1)(-1)+(6)(8) & (1)(2)+(6)(7) \end{bmatrix}$$

$$= \begin{bmatrix} 38 & 39 \\ 47 & 44 \end{bmatrix}$$

40. $C = \begin{bmatrix} 5 & 0 \\ -1 & 3 \\ 4 & 7 \end{bmatrix},\ A = \begin{bmatrix} 4 & 10 \\ -2 & -3 \\ 6 & 9 \end{bmatrix}$

CA is not defined since the number of columns in C is not equal to the number of rows in A.

41. $A = \begin{bmatrix} 4 & 10 \\ -2 & -3 \\ 6 & 9 \end{bmatrix},\ G = \begin{bmatrix} 2 & 5 \\ 1 & 6 \end{bmatrix},\ F = \begin{bmatrix} -1 & 2 \\ 8 & 7 \end{bmatrix}$

From Exercise 39, $GF = \begin{bmatrix} 38 & 39 \\ 47 & 44 \end{bmatrix}$.

Then, $AGF = A(GF)$

$$= \begin{bmatrix} 4 & 10 \\ -2 & -3 \\ 6 & 9 \end{bmatrix}\begin{bmatrix} 38 & 39 \\ 47 & 44 \end{bmatrix}$$

$$= \begin{bmatrix} 622 & 596 \\ -217 & -210 \\ 651 & 630 \end{bmatrix}$$

42. The matrix is

$$\begin{bmatrix} 3.54 & 1.41 \\ 1.53 & 1.57 \\ .34 & .29 \\ 7.53 & 6.21 \end{bmatrix}\cdot\begin{bmatrix} 4 \\ 4 \end{bmatrix} = \begin{bmatrix} 3.54(4)+1.41(4) \\ 1.53(4)+1.57(4) \\ .34(4)+.29(4) \\ 7.53(4)+6.21(4) \end{bmatrix}$$

$$= \begin{bmatrix} 19.8 \\ 12.4 \\ 2.52 \\ 54.96 \end{bmatrix}$$

Therefore, there are about 20 head and face injuries, about 12 concussions, about 3 neck injuries, and about 55 other injuries.

43. a.

Cutting   Shaping

$$\begin{array}{c} \text{Standard} \\ \text{Extra Large} \end{array} \begin{bmatrix} \frac{1}{4} & \frac{1}{2} \\ \frac{1}{3} & \frac{1}{3} \end{bmatrix}$$

b. $\begin{bmatrix} 48 & 66 \end{bmatrix}\begin{bmatrix} \frac{1}{4} & \frac{1}{2} \\ \frac{1}{3} & \frac{1}{3} \end{bmatrix} = \begin{bmatrix} 34 & 46 \end{bmatrix}$

The cutting machine will operate for 34 hr and the shaping machine for 46 hr.

44. a.

	Cost Per Share	Earnings Per Share
Stock 1	32	1.20
Stock 2	23	1.49
Stock 3	54	2.10

b.

$$\text{Number of shares} \begin{bmatrix} 50 & 20 & 15 \end{bmatrix} \begin{matrix} \text{Stock} \\ 1 \quad 2 \quad 3 \end{matrix}$$

c.

$$\begin{bmatrix} 50 & 20 & 15 \end{bmatrix} \begin{bmatrix} 32 & 1.20 \\ 23 & 1.49 \\ 54 & 2.10 \end{bmatrix} = \begin{bmatrix} 2870 & 121.30 \end{bmatrix}$$

Total cost = $2870.

Total dividend = $121.30

45. There are many correct answers. Here is one example.

$$A = \begin{bmatrix} 3 & 0 \\ 2 & 1 \end{bmatrix} ; \text{ let } B = \begin{bmatrix} 1 & 2 \\ 3 & 4 \end{bmatrix}.$$

$$AB = \begin{bmatrix} 3 & 6 \\ 5 & 8 \end{bmatrix}$$

$$BA = \begin{bmatrix} 7 & 2 \\ 17 & 4 \end{bmatrix}$$

Thus, AB and BA are both defined, and $AB \neq BA$.

46. No. $A = 4I$, so $AB = BA = 4B$.

47. Let $A = \begin{bmatrix} -4 & 4 \\ 0 & 5 \end{bmatrix}$.

$$[A \mid I] = \begin{bmatrix} -4 & 4 & | & 1 & 0 \\ 0 & 5 & | & 0 & 1 \end{bmatrix}$$

$$= \begin{bmatrix} 1 & -1 & | & -\frac{1}{4} & 0 \\ 0 & 1 & | & 0 & \frac{1}{5} \end{bmatrix} \begin{matrix} -\frac{1}{4}R_1 \\ \frac{1}{5}R_2 \end{matrix}$$

$$= \begin{bmatrix} 1 & 0 & | & -\frac{1}{4} & \frac{1}{5} \\ 0 & 1 & | & 0 & \frac{1}{5} \end{bmatrix} R_2 + R_1$$

$$A^{-1} = \begin{bmatrix} -\frac{1}{4} & \frac{1}{5} \\ 0 & \frac{1}{5} \end{bmatrix}$$

48. Let $A = \begin{bmatrix} 3 & -1 \\ -5 & 2 \end{bmatrix}$.

$$[A \mid I] = \begin{bmatrix} 3 & -1 & | & 1 & 0 \\ -5 & 2 & | & 0 & 1 \end{bmatrix}$$

$$\begin{bmatrix} 1 & -\frac{1}{3} & | & \frac{1}{3} & 0 \\ 0 & \frac{1}{3} & | & \frac{5}{3} & 1 \end{bmatrix} \begin{matrix} \frac{1}{3}R_1 \\ 5R_1 + R_2 \end{matrix}$$

$$\begin{bmatrix} 1 & 0 & | & 2 & 1 \\ 0 & 1 & | & 5 & 3 \end{bmatrix} \begin{matrix} \frac{1}{3}R_2 + R_1 \\ 3R_2 \end{matrix}$$

$$A^{-1} = \begin{bmatrix} 2 & 1 \\ 5 & 3 \end{bmatrix}$$

49. Let $A = \begin{bmatrix} 6 & 4 \\ 3 & 2 \end{bmatrix}$.

$$[A \mid I] = \begin{bmatrix} 6 & 4 & | & 1 & 0 \\ 3 & 2 & | & 0 & 1 \end{bmatrix}$$

$$\begin{bmatrix} 1 & \frac{2}{3} & | & \frac{1}{6} & 0 \\ 3 & 2 & | & 0 & 1 \end{bmatrix} \frac{1}{6}R_1$$

$$\begin{bmatrix} 1 & \frac{2}{3} & | & \frac{1}{6} & 0 \\ 0 & 0 & | & -\frac{1}{2} & 1 \end{bmatrix} -3R_1 + R_2$$

The second row can never become $\begin{bmatrix} 0 & 1 \end{bmatrix}$, so A has no inverse.

50. Let $A = \begin{bmatrix} 3 & 0 \\ -1 & 4 \end{bmatrix}$.

$$\begin{bmatrix} 3 & 0 & | & 1 & 0 \\ -1 & 4 & | & 0 & 1 \end{bmatrix}$$

Multiply row 1 by $\dfrac{1}{3}$ and continue.

$$\begin{bmatrix} 1 & 0 & | & \frac{1}{3} & 0 \\ 0 & 4 & | & \frac{1}{3} & 1 \end{bmatrix} R_1 + R_2$$

$$\begin{bmatrix} 1 & 0 & | & \frac{1}{3} & 0 \\ 0 & 1 & | & \frac{1}{12} & \frac{1}{4} \end{bmatrix} \frac{1}{4}R_2$$

$$A^{-1} = \begin{bmatrix} \frac{1}{3} & 0 \\ \frac{1}{12} & \frac{1}{4} \end{bmatrix}$$

51. Let $A = \begin{bmatrix} 2 & 0 & 6 \\ 1 & -1 & 0 \\ 0 & 1 & -3 \end{bmatrix}$

$[A \mid I] = \begin{bmatrix} 2 & 0 & 6 & 1 & 0 & 0 \\ 1 & -1 & 0 & 0 & 1 & 0 \\ 0 & 1 & -3 & 0 & 0 & 1 \end{bmatrix}$

Interchange rows.

$\begin{bmatrix} 1 & -1 & 0 & 0 & 1 & 0 \\ 0 & 1 & -3 & 0 & 0 & 1 \\ 2 & 0 & 6 & 1 & 0 & 0 \end{bmatrix}$

$\begin{bmatrix} 1 & -1 & 0 & 0 & 1 & 0 \\ 0 & 1 & -3 & 0 & 0 & 1 \\ 0 & 2 & 6 & 1 & -2 & 0 \end{bmatrix} -2R_1 + R_3$

$\begin{bmatrix} 1 & 0 & -3 & 0 & 1 & 1 \\ 0 & 1 & -3 & 0 & 0 & 1 \\ 0 & 0 & 12 & 1 & -2 & -2 \end{bmatrix} \begin{matrix} R_2 + R_1 \\ \\ -2R_2 + R_3 \end{matrix}$

$\begin{bmatrix} 1 & 0 & -3 & 0 & 1 & 1 \\ 0 & 1 & -3 & 0 & 0 & 1 \\ 0 & 0 & 1 & \frac{1}{12} & -\frac{1}{6} & -\frac{1}{6} \end{bmatrix} \frac{1}{12}R_3$

$[I \mid B] = \begin{bmatrix} 1 & 0 & 0 & \frac{1}{4} & \frac{1}{2} & \frac{1}{2} \\ 0 & 1 & 0 & \frac{1}{4} & -\frac{1}{2} & \frac{1}{2} \\ 0 & 0 & 1 & \frac{1}{12} & -\frac{1}{6} & -\frac{1}{6} \end{bmatrix} \begin{matrix} 3R_3 + R_1 \\ 3R_3 + R_2 \\ \\ \end{matrix}$

$A^{-1} = \begin{bmatrix} \frac{1}{4} & \frac{1}{2} & \frac{1}{2} \\ \frac{1}{4} & -\frac{1}{2} & \frac{1}{2} \\ \frac{1}{12} & -\frac{1}{6} & -\frac{1}{6} \end{bmatrix}$

52. Let $A = \begin{bmatrix} 2 & -1 & 0 \\ 1 & 0 & 2 \\ 1 & -4 & 0 \end{bmatrix}$

$\begin{bmatrix} 2 & -1 & 0 & 1 & 0 & 0 \\ 1 & 0 & 2 & 0 & 1 & 0 \\ 1 & -4 & 0 & 0 & 0 & 1 \end{bmatrix}$

Interchange rows 1 and 2.

$\begin{bmatrix} 1 & 0 & 2 & 0 & 1 & 0 \\ 2 & -1 & 0 & 1 & 0 & 0 \\ 1 & -4 & 0 & 0 & 0 & 1 \end{bmatrix}$

$\begin{bmatrix} 1 & 0 & 2 & 0 & 1 & 0 \\ 0 & -1 & -4 & 1 & -2 & 0 \\ 0 & -4 & -2 & 0 & -1 & 1 \end{bmatrix} \begin{matrix} \\ -2R_1 + R_2 \\ -1R_1 + R_3 \end{matrix}$

Multiply row 2 by -1 and continue.

$\begin{bmatrix} 1 & 0 & 1 & 0 & 1 & 0 \\ 0 & 1 & 4 & -1 & 2 & 0 \\ 0 & 0 & 14 & -4 & 7 & 1 \end{bmatrix} 4R_2 + R_3$

Multiply row 3 by $\dfrac{1}{14}$ and continue.

$\begin{bmatrix} 1 & 0 & 0 & \frac{4}{7} & 0 & -\frac{1}{7} \\ 0 & 1 & 0 & \frac{1}{7} & 0 & -\frac{2}{7} \\ 0 & 0 & 1 & -\frac{2}{7} & \frac{1}{2} & \frac{1}{14} \end{bmatrix} \begin{matrix} -1R_3 + R_1 \\ -4R_3 + R_2 \\ \\ \end{matrix}$

$A^{-1} = \begin{bmatrix} \frac{4}{7} & 0 & -\frac{1}{7} \\ \frac{1}{7} & 0 & -\frac{2}{7} \\ -\frac{2}{7} & \frac{1}{2} & \frac{1}{14} \end{bmatrix}$

53. Let $A = \begin{bmatrix} 2 & 3 & 5 \\ -2 & -3 & -5 \\ 1 & 4 & 2 \end{bmatrix}$.

$[A \mid I] = \begin{bmatrix} 2 & 3 & 5 & 1 & 0 & 0 \\ -2 & -3 & -5 & 0 & 1 & 0 \\ 1 & 4 & 2 & 0 & 0 & 1 \end{bmatrix}$

$\begin{bmatrix} 1 & 4 & 2 & 0 & 0 & 1 \\ -2 & -3 & -5 & 0 & 1 & 0 \\ 2 & 3 & 5 & 1 & 0 & 0 \end{bmatrix}$ Interchange R_1 and R_3

$\begin{bmatrix} 1 & 4 & 2 & 0 & 0 & 1 \\ -2 & -3 & -5 & 0 & 1 & 0 \\ 0 & 0 & 0 & 1 & 1 & 0 \end{bmatrix} R_2 + R_3$

The third row can never become $\begin{bmatrix} 0 & 0 & 1 \end{bmatrix}$, so A does not have an inverse.

54. Let $A = \begin{bmatrix} 1 & 3 & 6 \\ 4 & 0 & 9 \\ 5 & 15 & 30 \end{bmatrix}$

$\begin{bmatrix} 1 & 3 & 6 & | & 1 & 0 & 0 \\ 4 & 0 & 9 & | & 0 & 1 & 0 \\ 5 & 15 & 30 & | & 0 & 0 & 1 \end{bmatrix}$

$\begin{bmatrix} 1 & 3 & 6 & | & 1 & 0 & 0 \\ 0 & -12 & -15 & | & -4 & 1 & 0 \\ 0 & 0 & 0 & | & -5 & 0 & 1 \end{bmatrix} \begin{matrix} \\ -4R_1 + R_2 \\ -5R_1 + R_3 \end{matrix}$

The last row is all zeros, so no inverse exists.

55. Let $A = \begin{bmatrix} 1 & 3 & -2 & -1 \\ 0 & 1 & 1 & 2 \\ -1 & -1 & 1 & -1 \\ 1 & -1 & -3 & -2 \end{bmatrix}$.

Use a graphing calculator to find A^{-1}.

$A^{-1} = \begin{bmatrix} -\frac{2}{3} & -\frac{17}{3} & -\frac{14}{3} & -3 \\ \frac{1}{3} & \frac{1}{3} & \frac{1}{3} & 0 \\ -\frac{1}{3} & -\frac{10}{3} & -\frac{7}{3} & -2 \\ 0 & 2 & 1 & 1 \end{bmatrix}$

56. Let $A = \begin{bmatrix} 3 & 2 & 0 & -1 \\ 2 & 0 & 1 & 2 \\ 1 & 2 & -1 & 0 \\ 2 & -1 & 1 & 1 \end{bmatrix}$.

Use a graphing calculator to find A^{-1}.

$A^{-1} = \begin{bmatrix} 0 & -\frac{1}{3} & \frac{1}{3} & \frac{2}{3} \\ \frac{1}{3} & \frac{2}{3} & -\frac{1}{3} & -1 \\ \frac{2}{3} & 1 & -\frac{4}{3} & -\frac{4}{3} \\ -\frac{1}{3} & \frac{1}{3} & \frac{1}{3} & \frac{10}{3} \end{bmatrix}$

57. $F = \begin{bmatrix} -1 & 2 \\ 8 & 7 \end{bmatrix}$

To find F^{-1}, form the augmented matrix $[F \mid I]$.

$\begin{bmatrix} -1 & 2 & | & 1 & 0 \\ 8 & 7 & | & 0 & 1 \end{bmatrix}$

$\begin{bmatrix} 1 & -2 & | & -1 & 0 \\ 8 & 7 & | & 0 & 1 \end{bmatrix} -1R_1$

$\begin{bmatrix} 1 & -2 & | & -1 & 0 \\ 0 & 23 & | & 8 & 1 \end{bmatrix} -3R_1 + R_2$

$\begin{bmatrix} 1 & -2 & | & -1 & 0 \\ 0 & 1 & | & \frac{8}{23} & \frac{1}{23} \end{bmatrix} \frac{1}{23}R_2$

$\begin{bmatrix} 1 & 0 & | & -\frac{7}{23} & \frac{2}{23} \\ 0 & 1 & | & \frac{8}{23} & \frac{1}{23} \end{bmatrix} 2R_2 + R_1$

$F^{-1} = \begin{bmatrix} -\frac{7}{23} & \frac{2}{23} \\ \frac{8}{23} & \frac{1}{23} \end{bmatrix}$

58. $G = \begin{bmatrix} 2 & 5 \\ 1 & 6 \end{bmatrix}$

To find G^{-1}, form the augmented matrix $[G \mid I]$.

$\begin{bmatrix} 2 & 5 & | & 1 & 0 \\ 1 & 6 & | & 0 & 1 \end{bmatrix}$

Interchange R_1 and R_2.

$\begin{bmatrix} 1 & 6 & | & 0 & 1 \\ 2 & 5 & | & 1 & 0 \end{bmatrix}$

$\begin{bmatrix} 1 & 6 & | & 0 & 1 \\ 0 & -7 & | & 1 & -2 \end{bmatrix} -2R_1 + R_2$

$\begin{bmatrix} 1 & 6 & | & 0 & 1 \\ 0 & 1 & | & -\frac{1}{7} & \frac{2}{7} \end{bmatrix} -\frac{1}{7}R_2$

$\begin{bmatrix} 1 & 0 & | & \frac{6}{7} & -\frac{5}{7} \\ 0 & 1 & | & -\frac{1}{7} & \frac{2}{7} \end{bmatrix} -6R_2 + R_1$

$G^{-1} = \begin{bmatrix} \frac{6}{7} & -\frac{5}{7} \\ -\frac{1}{7} & \frac{2}{7} \end{bmatrix}$

59. $G = \begin{bmatrix} 2 & 5 \\ 1 & 6 \end{bmatrix}, F = \begin{bmatrix} -1 & 2 \\ 8 & 7 \end{bmatrix}$

$G - F = \begin{bmatrix} 3 & 3 \\ -7 & -1 \end{bmatrix}$

To find $(G - F)^{-1}$, form the augmented matrix $[G - F \mid I]$.

$\begin{bmatrix} 3 & 3 & | & 1 & 0 \\ -7 & -1 & | & 0 & 1 \end{bmatrix}$

$\begin{bmatrix} 1 & 1 & | & \frac{1}{3} & 0 \\ -7 & -1 & | & 0 & 1 \end{bmatrix} \frac{1}{3} R_1$

$\begin{bmatrix} 1 & 1 & | & \frac{1}{3} & 0 \\ 0 & 6 & | & \frac{7}{3} & 1 \end{bmatrix} 7 R_1 + R_2$

$\begin{bmatrix} 1 & 1 & | & \frac{1}{3} & 0 \\ 0 & 1 & | & \frac{7}{18} & \frac{1}{6} \end{bmatrix} \frac{1}{6} R_2$

$\begin{bmatrix} 1 & 0 & | & -\frac{1}{18} & -\frac{1}{6} \\ 0 & 1 & | & \frac{7}{18} & \frac{1}{6} \end{bmatrix} -1 R_2 + R_1$

$(G - F)^{-1} = \begin{bmatrix} -\frac{1}{18} & -\frac{1}{6} \\ \frac{7}{18} & \frac{1}{6} \end{bmatrix}$

60. $F = \begin{bmatrix} -1 & 2 \\ 8 & 7 \end{bmatrix}, G = \begin{bmatrix} 2 & 5 \\ 1 & 6 \end{bmatrix}$

$F + G = \begin{bmatrix} 1 & 7 \\ 9 & 13 \end{bmatrix}$

To find $(F + G)^{-1}$, form the augmented matrix $[F + G \mid I]$.

$\begin{bmatrix} 1 & 7 & | & 1 & 0 \\ 9 & 13 & | & 0 & 1 \end{bmatrix}$

$\begin{bmatrix} 1 & 7 & | & 1 & 0 \\ 0 & -50 & | & -9 & 1 \end{bmatrix} -9 R_1 + R_2$

$\begin{bmatrix} 1 & 7 & | & 1 & 0 \\ 0 & 1 & | & \frac{9}{50} & -\frac{1}{50} \end{bmatrix} -\frac{1}{50} R_2$

$\begin{bmatrix} 1 & 0 & | & -\frac{13}{50} & \frac{7}{50} \\ 0 & 1 & | & \frac{9}{50} & -\frac{1}{50} \end{bmatrix} -50 R_2 + R_1$

$(F + G)^{-1} = \begin{bmatrix} -\frac{13}{50} & \frac{7}{50} \\ \frac{9}{50} & -\frac{1}{50} \end{bmatrix}$

61. $B = \begin{bmatrix} 1 & 3 & -2 \\ 2 & 3 & 0 \\ 0 & 1 & 5 \end{bmatrix}$

To find B^{-1} form the augmented matrix $[B \mid I]$.

$\begin{bmatrix} 1 & 3 & -2 & | & 1 & 0 & 0 \\ 2 & 3 & 0 & | & 0 & 1 & 0 \\ 0 & 1 & 5 & | & 0 & 0 & 1 \end{bmatrix}$

$\begin{bmatrix} 1 & 3 & -2 & | & 1 & 0 & 0 \\ 0 & -3 & 4 & | & -2 & 1 & 0 \\ 0 & 1 & 5 & | & 0 & 0 & 1 \end{bmatrix} -2 R_1 + R_2$

$\begin{bmatrix} 1 & 3 & -2 & | & 1 & 0 & 0 \\ 0 & 1 & -\frac{4}{3} & | & \frac{2}{3} & -\frac{1}{3} & 0 \\ 0 & 1 & 5 & | & 0 & 0 & 1 \end{bmatrix} -\frac{1}{3} R_2$

$\begin{bmatrix} 1 & 0 & 2 & | & -1 & 1 & 0 \\ 0 & 1 & -\frac{4}{3} & | & \frac{2}{3} & -\frac{1}{3} & 0 \\ 0 & 0 & \frac{19}{3} & | & -\frac{2}{3} & \frac{1}{3} & 1 \end{bmatrix} \begin{array}{l} -3 R_2 + R_1 \\ \\ -1 R_2 + R_3 \end{array}$

$\begin{bmatrix} 1 & 0 & 2 & | & -1 & 1 & 0 \\ 0 & 1 & -\frac{4}{3} & | & \frac{2}{3} & -\frac{1}{3} & 0 \\ 0 & 0 & 1 & | & -\frac{2}{19} & \frac{1}{19} & \frac{3}{19} \end{bmatrix} \frac{3}{19} R_3$

$\begin{bmatrix} 1 & 0 & 0 & | & -\frac{15}{19} & \frac{17}{19} & -\frac{6}{19} \\ 0 & 1 & 0 & | & \frac{10}{19} & -\frac{5}{19} & \frac{4}{19} \\ 0 & 0 & 1 & | & -\frac{2}{19} & \frac{1}{19} & \frac{3}{19} \end{bmatrix} \begin{array}{l} -2 R_3 + R_2 \\ \frac{4}{3} R_3 + R_2 \end{array}$

$B^{-1} = \begin{bmatrix} -\frac{15}{19} & \frac{17}{19} & -\frac{6}{19} \\ \frac{10}{19} & -\frac{5}{19} & \frac{4}{19} \\ -\frac{2}{19} & \frac{1}{19} & \frac{3}{19} \end{bmatrix}$

62. Answer varies.

63. $A = \begin{bmatrix} 2 & 4 \\ -1 & -3 \end{bmatrix}, B = \begin{bmatrix} 8 \\ 3 \end{bmatrix}$

If $AX = B$, then $X = A^{-1}B$.

If $A = \begin{bmatrix} 2 & 4 \\ -1 & -3 \end{bmatrix}$, then $A^{-1} = \begin{bmatrix} \frac{3}{2} & 2 \\ -\frac{1}{2} & -1 \end{bmatrix}$.

Thus, $X = \begin{bmatrix} \frac{3}{2} & 2 \\ -\frac{1}{2} & -1 \end{bmatrix} \begin{bmatrix} 8 \\ 3 \end{bmatrix} = \begin{bmatrix} 18 \\ -7 \end{bmatrix}$.

64. $A = \begin{bmatrix} 1 & 3 \\ -2 & 4 \end{bmatrix}$, $B = \begin{bmatrix} 9 \\ 6 \end{bmatrix}$

$A^{-1} = \begin{bmatrix} \frac{2}{5} & -\frac{3}{10} \\ \frac{1}{5} & \frac{1}{10} \end{bmatrix}$

Then $X = A^{-1}B$

$\begin{bmatrix} \frac{2}{5} & -\frac{3}{10} \\ \frac{1}{5} & \frac{1}{10} \end{bmatrix} \begin{bmatrix} 9 \\ 6 \end{bmatrix} = \begin{bmatrix} \frac{9}{5} \\ \frac{12}{5} \end{bmatrix}$.

65. $A = \begin{bmatrix} 1 & 0 & 2 \\ -1 & 1 & 0 \\ 3 & 0 & 4 \end{bmatrix}$, $B = \begin{bmatrix} 8 \\ 4 \\ -6 \end{bmatrix}$

If $AX = B$, then $X = A^{-1}B$.

If $A = \begin{bmatrix} 1 & 0 & 2 \\ -1 & 1 & 0 \\ 3 & 0 & 4 \end{bmatrix}$,

then $A^{-1} = \begin{bmatrix} -2 & 0 & 1 \\ -2 & 1 & 1 \\ \frac{3}{2} & 0 & -\frac{1}{2} \end{bmatrix}$.

Thus,

$X = \begin{bmatrix} -2 & 0 & 1 \\ -2 & 1 & 1 \\ \frac{3}{2} & 0 & -\frac{1}{2} \end{bmatrix} \begin{bmatrix} 8 \\ 4 \\ -6 \end{bmatrix} = \begin{bmatrix} -22 \\ -18 \\ 15 \end{bmatrix}$.

66. $A = \begin{bmatrix} 2 & 4 & 0 \\ 1 & -2 & 0 \\ 0 & 0 & 3 \end{bmatrix}$, $B = \begin{bmatrix} 72 \\ -24 \\ 48 \end{bmatrix}$

$A^{-1} = \begin{bmatrix} \frac{1}{4} & \frac{1}{2} & 0 \\ \frac{1}{8} & -\frac{1}{4} & 0 \\ 0 & 0 & \frac{1}{3} \end{bmatrix}$

$X = A^{-1}B = \begin{bmatrix} \frac{1}{4} & \frac{1}{2} & 0 \\ \frac{1}{8} & -\frac{1}{4} & 0 \\ 0 & 0 & \frac{1}{3} \end{bmatrix} \begin{bmatrix} 72 \\ -24 \\ 48 \end{bmatrix} = \begin{bmatrix} 6 \\ 15 \\ 16 \end{bmatrix}$.

67. $x + y = 2$
$2x + 3y = 8$

The system as a matrix equation is

$\begin{bmatrix} 1 & 1 \\ 2 & 3 \end{bmatrix} \begin{bmatrix} x \\ y \end{bmatrix} = \begin{bmatrix} 2 \\ 8 \end{bmatrix}$.

Let $A = \begin{bmatrix} 1 & 1 \\ 2 & 3 \end{bmatrix}$.

$A^{-1} = \begin{bmatrix} 3 & -1 \\ -2 & 1 \end{bmatrix}$

Thus,

$\begin{bmatrix} x \\ y \end{bmatrix} = \begin{bmatrix} 3 & -1 \\ -2 & 1 \end{bmatrix} \begin{bmatrix} 2 \\ 8 \end{bmatrix} = \begin{bmatrix} -2 \\ 4 \end{bmatrix}$.

The solution is (–2, 4).

68. $5x - 3y = -2$
$2x + 7y = -9$

Let $A = \begin{bmatrix} 5 & -3 \\ 2 & 7 \end{bmatrix}$ and $B = \begin{bmatrix} -2 \\ -9 \end{bmatrix}$.

$A^{-1} = \begin{bmatrix} \frac{7}{41} & \frac{3}{41} \\ -\frac{2}{41} & \frac{5}{41} \end{bmatrix}$

$X = A^{-1}B = \begin{bmatrix} -1 \\ -1 \end{bmatrix}$

The solution is (–1, –1).

69. $2x + y = 10$
$3x - 2y = 8$

The system as a matrix equation is

$\begin{bmatrix} 2 & 1 \\ 3 & -2 \end{bmatrix} \begin{bmatrix} x \\ y \end{bmatrix} = \begin{bmatrix} 10 \\ 8 \end{bmatrix}$.

Let $A = \begin{bmatrix} 2 & 1 \\ 3 & -2 \end{bmatrix}$.

$A^{-1} = \begin{bmatrix} \frac{2}{7} & \frac{1}{7} \\ \frac{3}{7} & -\frac{2}{7} \end{bmatrix}$

Thus,

$X = \begin{bmatrix} \frac{2}{7} & \frac{1}{7} \\ \frac{3}{7} & -\frac{2}{7} \end{bmatrix} \begin{bmatrix} 10 \\ 8 \end{bmatrix} = \begin{bmatrix} 4 \\ 2 \end{bmatrix}$.

The solution is (4, 2).

70. $x - 2y = 7$

$3x + y = 7$

$A = \begin{bmatrix} 1 & -2 \\ 3 & 1 \end{bmatrix}, B = \begin{bmatrix} 7 \\ 7 \end{bmatrix}$

$A^{-1} = \begin{bmatrix} \frac{1}{7} & \frac{2}{7} \\ -\frac{3}{7} & \frac{1}{7} \end{bmatrix}$

$X = A^{-1}B = \begin{bmatrix} 3 \\ -2 \end{bmatrix}$

The solution is $(3, -2)$.

71. $x + y + z = 1$

$2x - y = -2$

$ 3y + z = 2$

The system as a matrix equation is

$\begin{bmatrix} 1 & 1 & 1 \\ 2 & -1 & 0 \\ 0 & 3 & 1 \end{bmatrix} \begin{bmatrix} x \\ y \\ z \end{bmatrix} = \begin{bmatrix} 1 \\ -2 \\ 2 \end{bmatrix}$.

Let $A = \begin{bmatrix} 1 & 1 & 1 \\ 2 & -1 & 0 \\ 0 & 3 & 1 \end{bmatrix}$.

$A^{-1} = \begin{bmatrix} -\frac{1}{3} & \frac{2}{3} & \frac{1}{3} \\ -\frac{2}{3} & \frac{1}{3} & \frac{2}{3} \\ 2 & -1 & -1 \end{bmatrix}$

Thus,

$X = \begin{bmatrix} -\frac{1}{3} & \frac{2}{3} & \frac{1}{3} \\ -\frac{2}{3} & \frac{1}{3} & \frac{2}{3} \\ 2 & -1 & -1 \end{bmatrix} \begin{bmatrix} 1 \\ -2 \\ 2 \end{bmatrix} = \begin{bmatrix} -1 \\ 0 \\ 2 \end{bmatrix}$.

The solution is $(-1, 0, 2)$.

72. $x = -3$

$ y + z = 6$

$2x - 3z = -9$

$A = \begin{bmatrix} 1 & 0 & 0 \\ 0 & 1 & 1 \\ 2 & 0 & -3 \end{bmatrix}, B = \begin{bmatrix} -3 \\ 6 \\ -9 \end{bmatrix}$

$A^{-1} = \begin{bmatrix} 1 & 0 & 0 \\ -\frac{2}{3} & 1 & \frac{1}{3} \\ \frac{2}{3} & 0 & -\frac{1}{3} \end{bmatrix}$

$X = A^{-1}B = \begin{bmatrix} -3 \\ 5 \\ 1 \end{bmatrix}$

The solution is $(-3, 5, 1)$.

73. $3x - 2y + 4z = 4$

$4x + y - 5z = 2$

$-6x + 4y - 8z = -2$

The system as a matrix equation is

$\begin{bmatrix} 3 & -2 & 4 \\ 4 & 1 & -5 \\ -6 & 4 & -8 \end{bmatrix} \begin{bmatrix} x \\ y \\ z \end{bmatrix} = \begin{bmatrix} 4 \\ 2 \\ -2 \end{bmatrix}$.

Let $A = \begin{bmatrix} 3 & -2 & 4 \\ 4 & 1 & -5 \\ -6 & 4 & -8 \end{bmatrix}$.

Since row 3 is -2 times row 1, the matrix will have no inverse, and the system cannot be solved by this method. Another method should be used to complete the solution. Use the elimination method. Multiply equation (1) by 2 and add the result to equation (3).

$6x - 4y + 8z = 8$

$\underline{-6x + 4y - 8z = -2}$

$ 0 = 6$

This false result indicates that the system has no solution.

74. $x + 2y = -1$

$ 3y - z = -5$

$x + 2y - z = -3$

$A = \begin{bmatrix} 1 & 2 & 0 \\ 0 & 3 & -1 \\ 1 & 2 & -1 \end{bmatrix}, B = \begin{bmatrix} -1 \\ -5 \\ -3 \end{bmatrix}$

$A^{-1} = \begin{bmatrix} \frac{1}{3} & -\frac{2}{3} & \frac{2}{3} \\ \frac{1}{3} & \frac{1}{3} & -\frac{1}{3} \\ 1 & 0 & -1 \end{bmatrix}$

$X = A^{-1}B = \begin{bmatrix} 1 \\ -1 \\ 2 \end{bmatrix}$

The solution is $(1, -1, 2)$.

75. Let x = the amount of the 9% wine;
y = the amount of the 14% wine.
We have the system

$$x + y = 40$$
$$.09x + .14y = .12(40).$$

This system can be simplified to

$$x + y = 40 \quad (1)$$
$$.09x + .14y = 4.8. \quad (2)$$

Multiply equation (1) by $-.09$ and add to equation (2).

$$-.09x - .09y = -3.6$$
$$\underline{.09x + .14y = 4.8}$$
$$.05y = 1.2$$
$$y = 24$$

Substitute 24 for y in equation (1) to get $x = 16$. The wine maker should mix 16 liters of the 9% wine and 24 liters of the 14% wine.

76. Let x = number of grams of 12 carat gold
y = number of grams of 22 carat gold.
The system is

$$x + \phantom{\frac{12}{24}}y = 25 \quad (1)$$
$$\frac{12}{24}x + \frac{22}{24}y = \frac{15}{24} \quad (2)$$

Multiply equation (2) by 24.

$$x + y = 25 \quad (1)$$
$$12x + 22y = 375 \quad (3)$$

Multiply equation (1) by -12 and add the result to equation (3).

$$-12x - 12y = -300$$
$$\underline{12x + 22y = 375}$$
$$10y = 75$$
$$y = 7.5$$

Substitute 7.5 for y in equation (1)

$$x + 7.5 = 25$$
$$x = 17.5$$

The merchant should mix 17.5 grams of 12 carat gold and 7.5 grams of 22 carat gold.

77. Let x = the number of liters of 40% solution;
y = the number of liters of 60% solution.

$$x + y = 40$$
$$.4x + .6y = .45(40) = 18$$

Multiply the second equation by 5.

$$x + y = 40$$
$$2x + 3y = 90$$

Use the Gauss-Jordan method.

$$\begin{bmatrix} 1 & 1 & | & 40 \\ 2 & 3 & | & 90 \end{bmatrix}$$

$$\begin{bmatrix} 1 & 1 & | & 40 \\ 0 & 1 & | & 10 \end{bmatrix} -2R_1 + R_2$$

$$\begin{bmatrix} 1 & 0 & | & 30 \\ 0 & 1 & | & 10 \end{bmatrix} -1R_3 + R_1$$

The chemist should use 30 liters of the 40% solution and 10 liters of the 60% solution.

78. Let x = the number of pounds of tea worth \$4.60 a pound;
y = the number of pounds of tea worth \$6.50 a pound.
The system is

$$x + y = 10 \quad (1)$$
$$4.60x + 6.50y = 5.74(10). \quad (2)$$

Multiply equation (1) by -4.60 and add the result to equation (2).

$$-4.60x - 4.60y = -46.0$$
$$\underline{4.60x + 6.50y = 57.4}$$
$$1.90y = 11.4$$
$$y = 6$$

Substitute 6 for y in equation (1) and solve for x.

$$x + 6 = 10$$
$$y = 4$$

Thus, 4 pounds of the tea worth \$4.60 a pound should be used.

79. Let x = the number of bowls;
  y = the number of plates.
We will solve the system

$$3x + 2y = 480 \quad (1)$$
$$.25x + .2y = 44 \quad (2)$$

Multiply equation (2) by -10 and add to equation (1).

$$3x + 2y = 480$$
$$-2.5x - 2y = -440$$
$$.5x = 40$$
$$x = 80$$

Substitute 80 for x in equation (1) and solve for y.

$$3(80) + 2y = 480$$
$$2y = 240$$
$$y = 120$$

The factory can produce 80 bowls and 120 plates.

80. Let x = the speed of the boat;
  y = the speed of the current.
The system is

$$3(x + y) = 57 \quad (1)$$
$$5(x - y) = 55. \quad (2)$$

Dividing equation (1) by 3 and equation (2) by 5 gives

$$x + y = 19 \quad (3)$$
$$x - y = 11. \quad (4)$$

Adding the above two equations gives

$$2x = 30$$
$$x = 15.$$

Substituting into (3) gives

$$15 + y = 19$$
$$y = 4.$$

The speed of the boat is 15 km/hr, and the speed of the current is 4 km/hr.

81. Let x = the amount invested at 8%;
  y = the amount invested at $8\frac{1}{2}$%;
  z = the amount invested at 11%.
The system is

$$x + y + z = 50{,}000$$
$$.08x + .085y + .11z = 4436.25$$
$$.11z = .08x + 80.$$

Write the system in the proper form.

$$x + y + z = 50{,}000$$
$$.08x + .085y + .11z = 4436.25$$
$$-.08x + .11z = 80$$

Write the augmented matrix of the system.

$$\begin{bmatrix} 1 & 1 & 1 & | & 50{,}000 \\ .08 & .085 & .11 & | & 4436.25 \\ -.08 & 0 & .11 & | & 80 \end{bmatrix}$$

$$\begin{bmatrix} 1 & 1 & 1 & | & 50{,}000 \\ 0 & .005 & .03 & | & 436.25 \\ 0 & .08 & .19 & | & 4080 \end{bmatrix} \begin{matrix} \\ -.08R_1 + R_2 \\ .08R_1 + R_3 \end{matrix}$$

$$\begin{bmatrix} 1 & 1 & 1 & | & 50{,}000 \\ 0 & 1 & 6 & | & 87{,}250 \\ 0 & .08 & .19 & | & 4080 \end{bmatrix} \begin{matrix} \\ \frac{1}{.005}R_2 \\ \\ \end{matrix}$$

$$\begin{bmatrix} 1 & 0 & -5 & | & -37{,}250 \\ 0 & 1 & 6 & | & 87{,}250 \\ 0 & 0 & -.29 & | & -2900 \end{bmatrix} \begin{matrix} -1R_2 + R_1 \\ \\ -.08R_2 + R_3 \end{matrix}$$

$$\begin{bmatrix} 1 & 0 & -5 & | & -37{,}250 \\ 0 & 1 & 6 & | & 87{,}250 \\ 0 & 0 & 1 & | & 10{,}000 \end{bmatrix} \begin{matrix} \\ \\ -\frac{1}{.29}R_3 \end{matrix}$$

$$\begin{bmatrix} 1 & 0 & 0 & | & 12{,}750 \\ 0 & 1 & 0 & | & 27{,}250 \\ 0 & 0 & 1 & | & 10{,}000 \end{bmatrix} \begin{matrix} 5R_3 + R_1 \\ -6R_3 + R_2 \\ \\ \end{matrix}$$

Thus, $x = 12{,}750$, $y = 27{,}250$, and $z = 10{,}000$. Ms. Tham invested \$12,750 at 8%, \$27,250 at $8\frac{1}{2}$%, and \$10,000 at 11%.

82. Let x = the number of children;

y = the number of teenagers;

z = the number of adults.

$$x + y + z = 570$$
$$2x + 3y + 5z = 1950$$
$$y = \frac{3}{4}x$$

Simplify the third equation and write the augmented matrix.

$$\begin{bmatrix} 1 & 1 & 1 & | & 570 \\ 2 & 3 & 5 & | & 1950 \\ 3 & -4 & 0 & | & 0 \end{bmatrix}$$

$$\begin{bmatrix} 1 & 1 & 1 & | & 570 \\ 0 & 1 & 3 & | & 810 \\ 0 & -7 & -3 & | & -1710 \end{bmatrix} \begin{matrix} \\ -2R_1 + R_2 \\ -3R_1 + R_3 \end{matrix}$$

$$\begin{bmatrix} 1 & 0 & -2 & | & -240 \\ 0 & 1 & 3 & | & 810 \\ 0 & 0 & 18 & | & 3960 \end{bmatrix} \begin{matrix} -1R_2 + R_1 \\ \\ 7R_2 + R_3 \end{matrix}$$

$$\begin{bmatrix} 1 & 0 & -2 & | & -240 \\ 0 & 1 & 3 & | & 810 \\ 0 & 0 & 1 & | & 220 \end{bmatrix} \begin{matrix} \\ \\ \frac{1}{18}R_3 \end{matrix}$$

$$\begin{bmatrix} 1 & 0 & 0 & | & 200 \\ 0 & 1 & 0 & | & 150 \\ 0 & 0 & 1 & | & 220 \end{bmatrix} \begin{matrix} 2R_3 + R_1 \\ -3R_3 + R_2 \end{matrix}$$

There were 200 children, 150 teenagers, and 220 adults at the concert.

83. $A = \begin{bmatrix} 0 & \frac{1}{4} \\ \frac{1}{2} & 0 \end{bmatrix}$, $D = \begin{bmatrix} 2100 \\ 1400 \end{bmatrix}$

a. $I - A = \begin{bmatrix} 1 & 0 \\ 0 & 1 \end{bmatrix} - \begin{bmatrix} 0 & \frac{1}{4} \\ \frac{1}{2} & 0 \end{bmatrix} = \begin{bmatrix} 1 & -\frac{1}{4} \\ -\frac{1}{2} & 1 \end{bmatrix}$

b. $\begin{bmatrix} 1 & -\frac{1}{4} & | & 1 & 0 \\ -\frac{1}{2} & 1 & | & 0 & 1 \end{bmatrix}$

$\begin{bmatrix} 1 & -\frac{1}{4} & | & 1 & 0 \\ 1 & \frac{7}{8} & | & \frac{1}{2} & 1 \end{bmatrix} \frac{1}{2}R_1 + R_2$

$\begin{bmatrix} 1 & -\frac{1}{4} & | & 1 & 0 \\ 0 & 1 & | & \frac{4}{7} & \frac{8}{7} \end{bmatrix} \frac{8}{7}R_2$

$\begin{bmatrix} 1 & 0 & | & \frac{8}{7} & \frac{2}{7} \\ 0 & 1 & | & \frac{4}{7} & \frac{8}{7} \end{bmatrix} \frac{1}{4}R_2 + R_1$

Thus,

$(I - A)^{-1} = \begin{bmatrix} \frac{8}{7} & \frac{2}{7} \\ \frac{4}{7} & \frac{8}{7} \end{bmatrix}$.

83. Continued

c. $X = (I - A)^{-1}D$

$X = \begin{bmatrix} \frac{8}{7} & \frac{2}{7} \\ \frac{4}{7} & \frac{8}{7} \end{bmatrix} \begin{bmatrix} 2100 \\ 1400 \end{bmatrix} = \begin{bmatrix} 2800 \\ 2800 \end{bmatrix}$

84. a. The input-output matrix is

$$\begin{array}{cc} & c \quad g \end{array}$$
$$\begin{array}{c} c \\ g \end{array} \begin{bmatrix} 0 & \frac{1}{2} \\ \frac{2}{3} & 0 \end{bmatrix} = A.$$

b. $I - A = \begin{bmatrix} 1 & -\frac{1}{2} \\ -\frac{2}{3} & 1 \end{bmatrix}$, $D = \begin{bmatrix} 400 \\ 800 \end{bmatrix}$

$(I - A)^{-1} = \begin{bmatrix} \frac{3}{2} & \frac{3}{4} \\ 1 & \frac{3}{2} \end{bmatrix}$

$X = (I - A)^{-1}D$

$= \begin{bmatrix} 1200 \\ 1600 \end{bmatrix}$

The production required is 1200 units of cheese and 1600 units of goats.

85. Write the input-output matrix.

$$\begin{array}{cc} & A \quad M \end{array}$$
$$A = \begin{array}{c} A \\ M \end{array} \begin{bmatrix} .10 & .70 \\ .40 & .20 \end{bmatrix} = \begin{bmatrix} \frac{1}{10} & \frac{7}{10} \\ \frac{4}{10} & \frac{2}{10} \end{bmatrix}$$

$I - A = \begin{bmatrix} 1 & 0 \\ 0 & 1 \end{bmatrix} - \begin{bmatrix} \frac{1}{10} & \frac{7}{10} \\ \frac{4}{10} & \frac{2}{10} \end{bmatrix} = \begin{bmatrix} \frac{9}{10} & -\frac{7}{10} \\ -\frac{4}{10} & \frac{8}{10} \end{bmatrix}$

Find $(I - A)^{-1}$.

$\begin{bmatrix} \frac{9}{10} & -\frac{7}{10} & | & 1 & 0 \\ -\frac{4}{10} & \frac{8}{10} & | & 0 & 1 \end{bmatrix}$

$\begin{bmatrix} 1 & -\frac{7}{9} & | & \frac{10}{9} & 0 \\ -\frac{4}{10} & \frac{8}{10} & | & 0 & 1 \end{bmatrix} \frac{10}{9}R_1$

$\begin{bmatrix} 1 & -\frac{7}{9} & | & \frac{10}{9} & 0 \\ 0 & \frac{44}{90} & | & \frac{4}{9} & 1 \end{bmatrix} \frac{4}{10}R_1 + R_2$

$\begin{bmatrix} 1 & -\frac{7}{9} & | & \frac{10}{9} & 0 \\ 0 & 1 & | & \frac{10}{11} & \frac{90}{44} \end{bmatrix} \frac{90}{44}R_2$

$\begin{bmatrix} 1 & 0 & | & \frac{180}{99} & \frac{70}{44} \\ 0 & 1 & | & \frac{10}{11} & \frac{90}{44} \end{bmatrix} \frac{7}{9}R_2 + R_1$

$(I - A)^{-1} = \begin{bmatrix} \frac{180}{99} & \frac{70}{44} \\ \frac{10}{11} & \frac{90}{44} \end{bmatrix}$

85. Continued

$$X = (I - A)^{-1} D$$

$$X = \begin{bmatrix} \frac{180}{99} & \frac{70}{44} \\ \frac{10}{11} & \frac{90}{44} \end{bmatrix} \begin{bmatrix} 60,000 \\ 20,000 \end{bmatrix}$$

$$X = \begin{bmatrix} 140,909 \\ 95,455 \end{bmatrix} \text{ Rounded}$$

The agriculture industry should produce $140,909 while the manufacturing industry should produce $95,455.

86. a. 9 unit agriculture; .4 unit services; .02 unit mining; .9 unit manufacturing.

b. The input-output matrix, A, and the matrix, $I - A$, are

$$A = \begin{bmatrix} .02 & .9 & 0 & .001 \\ 0 & .4 & 0 & .06 \\ .01 & .02 & .06 & .07 \\ .25 & .9 & .9 & .4 \end{bmatrix}$$

$$I - A = \begin{bmatrix} .980 & -.900 & 0 & -.001 \\ 0 & .600 & 0 & -.060 \\ -.010 & -.020 & .940 & -.070 \\ -.250 & -.900 & -.900 & .600 \end{bmatrix}$$

Next, calculate $(I - A)^{-1}$.

$$(I - A)^{-1} \approx \begin{bmatrix} 1.079 & 1.960 & .2132 & .223 \\ .064 & 2.129 & .230 & .240 \\ .060 & .4107 & 1.242 & .186 \\ .635 & 4.627 & 2.296 & 2.398 \end{bmatrix}$$

$$D = \begin{bmatrix} 760 \\ 1600 \\ 1000 \\ 2000 \end{bmatrix}$$

$$X = (I - A)^{-1} D \approx \begin{bmatrix} 4615 \\ 4165 \\ 2317 \\ 14,979 \end{bmatrix}$$

This is about 4615 units of agriculture, 4165 units of services, 2317 units of mining, and 14,979 units of manufacturing.

c. About 12,979 units.

87. a. .4 unit agriculture; .09 unit construction, .4 unit energy, .1 unit manufacturing, .9 unit transportation.

b. The input-output matrix, A and the matrix, $I - A$, are

$$A = \begin{bmatrix} .18 & .017 & .4 & .005 & 0 \\ .14 & .018 & .09 & .001 & 0 \\ .9 & 0 & .4 & .06 & .002 \\ .19 & .16 & .1 & .008 & .5 \\ .14 & .25 & .9 & .4 & .12 \end{bmatrix}$$

$$(I-A)^{-1} \approx \begin{bmatrix} 7.744 & .294 & 5.745 & .509 & .302 \\ 2.319 & 1.111 & 1.890 & .167 & .099 \\ 13.100 & .543 & 11.546 & 1.006 & .598 \\ 14.119 & .976 & 12.002 & 2.357 & 1.367 \\ 21.707 & 1.361 & 18.715 & 2.229 & 2.445 \end{bmatrix} \qquad X = \begin{bmatrix} 28,067 \\ 9383 \\ 51,372 \\ 61,364 \\ 90,403 \end{bmatrix}$$

Solve $X = (I - A)^{-1} D$ for D.

$$D = \begin{bmatrix} 1999.809 \\ 599.882 \\ 1700.254 \\ 3700.378 \\ 2499.110 \end{bmatrix}$$

This represents 2000 units of agriculture, 600 units of construction, 1700 units of energy, 3700 units of manufacturing, and 2500 units of transportation.

c. Multiply matrix $(I - A)^{-1}$ by matrix D where $\qquad D = \begin{bmatrix} 2400 \\ 850 \\ 1400 \\ 3200 \\ 1800 \end{bmatrix}$

$$(I-A)^{-1} D \approx \begin{bmatrix} 29,049 \\ 9869 \\ 52,362 \\ 61,520 \\ 90,987 \end{bmatrix}$$

This represents 29,049 units of agriculture, 9869 units of construction, 52,520 units of energy, 61,520 units of manufacturing, and 90,987 units of transportation.

88. Let $M = \begin{array}{c} \\ A \\ B \\ C \\ D \end{array} \begin{array}{c} A\ B\ C\ D \\ \begin{bmatrix} 0 & 1 & 0 & 1 \\ 1 & 0 & 0 & 1 \\ 0 & 0 & 0 & 1 \\ 1 & 1 & 1 & 0 \end{bmatrix} \end{array}$

a. M^2 gives the number of one-step flights between cities.

$$M^2 = \begin{bmatrix} 0 & 1 & 0 & 1 \\ 1 & 0 & 0 & 1 \\ 0 & 0 & 0 & 1 \\ 1 & 1 & 1 & 0 \end{bmatrix} \begin{bmatrix} 0 & 1 & 0 & 1 \\ 1 & 0 & 0 & 1 \\ 0 & 0 & 0 & 1 \\ 1 & 1 & 1 & 0 \end{bmatrix}$$

$$= \begin{bmatrix} 2 & 1 & 1 & 1 \\ 1 & 2 & 1 & 1 \\ 1 & 1 & 1 & 0 \\ 1 & 1 & 0 & 3 \end{bmatrix}$$

$m_{13}^2 = 1$ so there is 1 one-stop flight between cities A and C.

b. The number of direct or one-stop flights from B to C is
$m_{23}^2 + m_{23} = 1 + 0 = 1$.

c. The number of two-stop flights is given by
$M^3 = M^2 \cdot M$

$$= \begin{bmatrix} 2 & 3 & 1 & 4 \\ 3 & 2 & 1 & 4 \\ 1 & 1 & 0 & 3 \\ 4 & 4 & 3 & 2 \end{bmatrix}$$

89. a. The message *leave now* broken into groups of 2 letters would have the following matrices:

$$\begin{bmatrix} 12 \\ 5 \end{bmatrix}, \begin{bmatrix} 1 \\ 22 \end{bmatrix}, \begin{bmatrix} 5 \\ 27 \end{bmatrix}, \begin{bmatrix} 14 \\ 15 \end{bmatrix}, \begin{bmatrix} 23 \\ 27 \end{bmatrix}$$

Multiply $M = \begin{bmatrix} 2 & 6 \\ 1 & 4 \end{bmatrix}$ by each matrix in the code to get

$$\begin{bmatrix} 54 \\ 32 \end{bmatrix}, \begin{bmatrix} 134 \\ 89 \end{bmatrix}, \begin{bmatrix} 172 \\ 113 \end{bmatrix}, \begin{bmatrix} 118 \\ 74 \end{bmatrix}, \begin{bmatrix} 208 \\ 131 \end{bmatrix}$$

b. To decode the message, use
$M^{-1} = \begin{bmatrix} 2 & -3 \\ -\frac{1}{2} & 1 \end{bmatrix}$.Case 6 Matrix Operations and Airline Route Maps

Case 6 Matrix Operations and Airline Route Maps

1. As the matrix A^2 indicates, the only city *not* reachable from New Bedford in a two-flight sequence is Provincetown, since this is the only city whose column has a zero in the row for New Bedford. Thus, the cities reachable by a two-flight sequence from New Bedford are Boston, Hyannis, Martha's Vineyard, Nantucket, New Bedford, and Providence.

   In the matrix A^3, there are no zero entries in the row for New Bedford, so all Cape Air cities may be reached by a three-flight sequence.

2. Hyannis—Boston—Nantucket—New Bedford
   Hyannis—Boston—Martha's Vineyard—New Bedford
   Hyannis—Nantucket—Martha's Vineyard—New Bedford
   Hyannis—Martha's Vineyard—Nantucket—New Bedford

3. The trips between two different cities for which the matrix entries remain zero until A^3 are trips between Provincetown and Providence, and Provincetown and New Bedford. These trips each take three flights.

4. There are 14 vertices, so the adjacency matrix is a 14 3 14 matrix. If the vertices in the Big Sky Airlines graph respectively correspond to Spokane (S), Kalispell (K), Missoula (M), Helena (H), Great Falls (GF), Billings (B), Lewistown (L), Havre (HV), Glasgow (G), Miles City (MC), Sidney (SY), Wolf (W), Glendine (GD), and Bismarck (BK), then the adjacency matrix for Big Sky Airlines is

$$A = \begin{matrix} & \text{S} & \text{K} & \text{M} & \text{H} & \text{GF} & \text{B} & \text{L} & \text{HV} & \text{G} & \text{MC} & \text{SY} & \text{W} & \text{GD} & \text{BK} \\ \left[\begin{array}{cccccccccccccc} 0 & 1 & 0 & 0 & 1 & 0 & 0 & 0 & 0 & 0 & 0 & 0 & 0 & 0 \\ 1 & 0 & 1 & 1 & 0 & 0 & 0 & 0 & 0 & 0 & 0 & 0 & 0 & 0 \\ 0 & 1 & 0 & 1 & 0 & 1 & 0 & 0 & 0 & 0 & 0 & 0 & 0 & 0 \\ 0 & 1 & 1 & 0 & 0 & 1 & 0 & 0 & 0 & 0 & 0 & 0 & 0 & 0 \\ 1 & 0 & 0 & 0 & 0 & 1 & 0 & 0 & 0 & 0 & 0 & 0 & 0 & 0 \\ 0 & 0 & 1 & 1 & 1 & 0 & 1 & 0 & 1 & 1 & 1 & 1 & 0 & 0 \\ 0 & 0 & 0 & 0 & 0 & 1 & 0 & 1 & 0 & 0 & 0 & 0 & 0 & 0 \\ 0 & 0 & 0 & 0 & 0 & 0 & 1 & 0 & 0 & 0 & 0 & 0 & 0 & 0 \\ 0 & 0 & 0 & 0 & 0 & 1 & 0 & 0 & 0 & 0 & 0 & 1 & 0 & 0 \\ 0 & 0 & 0 & 0 & 0 & 1 & 0 & 0 & 0 & 0 & 0 & 0 & 1 & 0 \\ 0 & 0 & 0 & 0 & 0 & 1 & 0 & 0 & 0 & 0 & 0 & 0 & 1 & 1 \\ 0 & 0 & 0 & 0 & 0 & 1 & 0 & 0 & 1 & 0 & 0 & 0 & 0 & 0 \\ 0 & 0 & 0 & 0 & 0 & 0 & 0 & 0 & 0 & 1 & 1 & 0 & 0 & 0 \\ 0 & 0 & 0 & 0 & 0 & 0 & 0 & 0 & 0 & 0 & 1 & 0 & 0 & 0 \end{array}\right] & \begin{matrix} \text{S} \\ \text{K} \\ \text{M} \\ \text{H} \\ \text{GF} \\ \text{B} \\ \text{L} \\ \text{HV} \\ \text{G} \\ \text{MC} \\ \text{SY} \\ \text{W} \\ \text{GD} \\ \text{BK} \end{matrix} \end{matrix}.$$

5. To find the cities that may be reached by a three-flight sequence from Helena, find all the columns which have a nonzero element in row 4 of matrix $S = A + A^2 + A^3$.

Since row 4 of $S_3 = A + A^2 + A^3$ is

$$\begin{matrix} \text{S} & \text{K} & \text{M} & \text{H} & \text{GF} & \text{B} & \text{L} & \text{HV} & \text{G} & \text{MC} & \text{SY} & \text{W} & \text{GD} & \text{BK} \\ [3 & 8 & 8 & 7 & 3 & 13 & 2 & 1 & 3 & 2 & 2 & 3 & 2 & 1] \end{matrix} \text{H},$$

all Big Sky cities may be reached by a three-flight sequence from Helena. The cities that need at least three flights to get from them to Helena are the cities represented by columns with zeros in row 4 of $S_2 = A + A^2$.

Since row 4 of $S_2 = A + A^2$ is

$$\begin{matrix} \text{S} & \text{K} & \text{M} & \text{H} & \text{GF} & \text{B} & \text{L} & \text{HV} & \text{G} & \text{MC} & \text{SY} & \text{W} & \text{GD} & \text{BK} \\ [1 & 2 & 3 & 3 & 1 & 2 & 1 & 0 & 1 & 1 & 1 & 1 & 0 & 0] \end{matrix} \text{H},$$

at least three flights must be used to get to Helena from Havre, Glendine, and Bismarck.

6. From Exercise 5 we know that some trips take 3 flights. By looking for zeros in S_3, we can determine if any trips need more than 3 flights.

$$
S_3 =
\begin{array}{c}
\begin{array}{cccccccccccccc}
\text{S} & \text{K} & \text{M} & \text{H} & \text{GF} & \text{B} & \text{L} & \text{HV} & \text{G} & \text{MC} & \text{SY} & \text{W} & \text{GD} & \text{BK}
\end{array} \\
\left[
\begin{array}{cccccccccccccc}
2 & 5 & 3 & 3 & 4 & 3 & 1 & 0 & 1 & 1 & 1 & 1 & 0 & 0 \\
5 & 5 & 8 & 8 & 3 & 5 & 2 & 0 & 2 & 2 & 2 & 2 & 0 & 0 \\
3 & 8 & 7 & 8 & 3 & 13 & 2 & 1 & 3 & 2 & 2 & 3 & 2 & 1 \\
3 & 8 & 8 & 7 & 3 & 13 & 2 & 1 & 3 & 2 & 2 & 3 & 2 & 1 \\
4 & 3 & 3 & 3 & 2 & 10 & 1 & 1 & 2 & 1 & 1 & 2 & 2 & 1 \\
3 & 5 & 13 & 13 & 10 & 12 & 10 & 1 & 11 & 11 & 12 & 11 & 2 & 1 \\
1 & 2 & 2 & 2 & 1 & 10 & 2 & 3 & 2 & 1 & 1 & 2 & 2 & 1 \\
0 & 0 & 1 & 1 & 1 & 1 & 3 & 1 & 1 & 1 & 1 & 1 & 0 & 0 \\
1 & 2 & 3 & 3 & 2 & 11 & 2 & 1 & 4 & 2 & 2 & 5 & 2 & 1 \\
1 & 2 & 2 & 2 & 1 & 11 & 1 & 1 & 2 & 2 & 2 & 2 & 5 & 2 \\
1 & 2 & 2 & 2 & 1 & 12 & 1 & 1 & 2 & 2 & 3 & 2 & 6 & 4 \\
1 & 2 & 3 & 3 & 2 & 11 & 2 & 1 & 5 & 2 & 2 & 4 & 2 & 1 \\
0 & 0 & 2 & 2 & 2 & 2 & 2 & 0 & 2 & 5 & 6 & 2 & 2 & 1 \\
0 & 0 & 1 & 1 & 1 & 1 & 1 & 0 & 1 & 2 & 4 & 1 & 1 & 1
\end{array}
\right]
\begin{array}{l}
\text{S} \\ \text{K} \\ \text{M} \\ \text{H} \\ \text{GF} \\ \text{B} \\ \text{L} \\ \text{HV} \\ \text{G} \\ \text{MC} \\ \text{SY} \\ \text{W} \\ \text{GD} \\ \text{BK}
\end{array}
\end{array}
$$

From S_3 we see that trips between Spokane and Havre, Spokane and Glendine, Spokane and Bismarck, Kalispell and Havre, Kalispell and Glendine, Kalispell and Bismarck, Havre and Glendine, and Havre and Bismarck require more than 3 flights. By looking at $S_4 = A + A^2 + A^3 + A^4$, we find that all entries have positive values; thus, these trips can be made in 4 flights.

Chapter 7: Linear Programming

Section 7.1 Graphing Linear Inequalities in Two Variables

1. $y \geq -x - 2$

   Graph of $y = -x - 2$ is a solid line. The intercepts are (−2, 0) and (0, −2). Use the origin as a test point. Since $0 \geq -0 - 2$ is true, the origin will be included in the region, so shade the half-plane above the line. This matches choice F.

   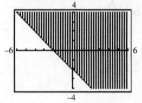

3. $y \leq x + 2$

   Graph $y = x + 2$ as a solid line. The intercepts are (−2, 0) and (0, 2). Use the origin as a test point. Since $0 \leq 0 + 2$ is true, the origin will be included in the region, so shade the half-plane below the line. This matches choice A.

   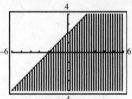

5. $6x + 4y \geq -12$

   Graph $6x + 4y = -12$ is as a solid line. The intercepts are (−2, 0) and (0, −3). Use the origin as a test point. Since $6(0) + 4(0) \geq -12$ is true, the origin will be included in the region, so shade the half-plane above the line. This matches choice E.

   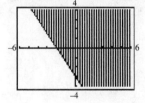

7. $y < 5 - 2x$

   Graph $y = 5 - 2x$ as a dashed line. The intercepts are (0, 5) and $\left(\frac{5}{2}, 0\right)$. Since the O inequality symbol is <, shade the half-plane below the line.

   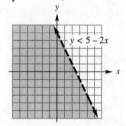

9. $3x - 2y \geq 18$

   Graph $3x - 2y = 18$ as a solid line. The intercepts are (0, −9) and (6, 0). Use the origin as a test point. Since $0 - 0 \geq 18$ is false, the origin will not be included in the region, so shade the half-plane below the line.

   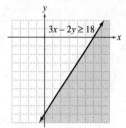

11. $2x - y \leq 4$

   Graph $2x - y = 4$ as a solid line. The intercepts are (0, −4) and (2, 0). Use the origin as a test point. Since $2(0) - 0 \leq 4$ is true, the origin will be included in the region, so shade the half-plane above the line.

   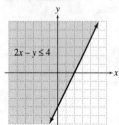

13. $y \leq -4$

Graph $y = -4$ as a solid line. $y = -4$ is a horizontal line crossing the y-axis at $(0, -4)$. Since the inequality symbol is $\leq$, shade the half-plane below the line.

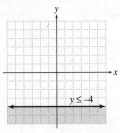

15. $3x - 2y \geq 18$

Graph the line $3x - 2y = 18$ as a solid line. The intercepts are $(0, -9)$ and $(6, 0)$. Use the origin as a test point. Since $3(0) - 2(0) \geq 18$ is false, the origin will not be included in the region, so shade the half-plane below the line.

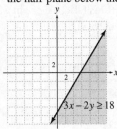

17. $3x + 4y > 12$

Graph $3x + 4y = 12$ as a dashed line. The intercepts are $(0, 3)$ and $(4, 0)$. Use the origin as a test point. Since $3(0) + 4(0) > 12$ is false, the origin will not be included in the region, so shade the half-plane above the line.

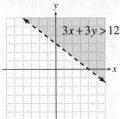

19. $2x - 4y < 3$

Graph $2x - 4y = 3$ as a dashed line. The intercepts are $\left(\dfrac{3}{2}, 0\right)$ and $\left(0, -\dfrac{3}{4}\right)$. Use the origin as a test point. $2(0) - 4(0) < 3$ is true, so the region above the line, which includes the origin, is the correct region to shade.

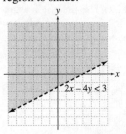

21. $x \leq 5y$

Graph $x = 5y$ as a solid line. Since this line contains the origin, some point other than $(0, 0)$ must be used as a test point. The point $(1, 2)$ gives $1 \leq 5(2)$ or $1 \leq 10$, a true sentence. Shade the side of the line containing $(1, 2)$, that is, the side above the line.

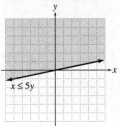

23. $-3x < y$

Graph $y = -3x$ as a dashed line. Since this line contains the origin, use some point other than $(0, 0)$ as a test point. $(1, 1)$ used as a test point gives $-3 < 1$, a true sentence. Shade the region containing $(1, 1)$, which is the region above the line.

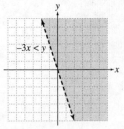

25. $y < x$

Graph $y = x$ as a dashed line. Since this line contains the origin, choose a point other than (0, 0) as a test point. (2, 3) gives $3 < 2$, which is false. Shade the region that does not contain the test point, that is, the region below the line.

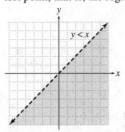

27. Answers may vary.

Possible answer: When the inequality is $<$ or $>$, the line is dashed. When the inequality is $\leq$ or $\geq$, the line is solid.

29. $y \geq 3x - 6$

$y \geq -x + 1$

Graph $y \geq 3x - 6$ as the region on or above the solid line $y = 3x - 6$. Graph $y \geq -x + 1$ as the region on or above the solid line $y = -x + 1$. The feasible region is the overlap of the two half-planes.

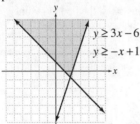

31. $2x + y \leq 5$

$x + 2y \leq 5$

Graph $2x + y \leq 5$ as the region on or below the solid line $2x + y = 5$, which has intercepts (0, 5) and $\left(\dfrac{5}{2}, 0\right)$. Graph $x + 2y \leq 5$ as the region on or below the solid line $x + 2y = 5$, which has intercepts $\left(0, \dfrac{5}{2}\right)$ and (5, 0). The feasible region is the overlap of the two half-planes.

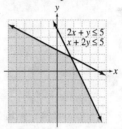

33. $2x + y > 8$

$4x - y < 3$

Graph $2x + y > 8$ as the region above the dashed line $2x + y = 8$, which has intercepts (0, 8) and (4, 0). Graph $4x - y < 3$ as the region above the dashed line $4x - y = 3$, which has intercepts (0, –3) and $\left(\dfrac{3}{4}, 0\right)$.

The overlap of these two regions is the feasible region.

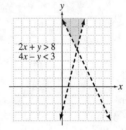

35. $2x - y < 1$

$3x + y < 6$

Graph $2x - y < 1$ as the region above the dashed line $2x - y = 1$. Graph $3x + y < 6$ as the region below the dashed line $3x + y < 6$. Shade the overlapping part of these two regions to show the feasible region.

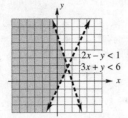

37. $-x - y < 5$

$2x - y < 4$

Graph $-x - y < 5$ as the region above the dashed line $-x - y = 5$. Graph $2x - y < 4$ as the region above the dashed line $2x - y = 4$. Shade the overlapping part of these two regions to show the feasible region.

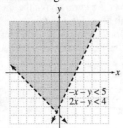

39. $3x + y \geq 6$

$x + 2y \geq 7$

$x \geq 0$

$y \geq 0$

The inequalities $x \geq 0$ and $y \geq 0$ restrict the feasible region to the first quadrant. Graph $3x + y \geq 6$ as the region on or above the solid line $3x + y = 6$. Graph $x + 2y \geq 7$ as the region on or above the solid line $x + 2y = 7$. The feasible region is the overlap of these half-planes.

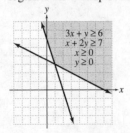

41. $-2 < x < 3$

$-1 \leq y \leq 5$

$2x + y < 6$

The graph of $-2 < x < 3$ is the region between the vertical line $x = -2$ and $x = 3$, but not including the lines themselves. The graph of $-1 \leq y \leq 5$ is the region between the horizontal lines $y = -1$ and $y = 5$, including the lines. The graph of $2x + y < 6$ is the region below the dashed line $2x + y = 6$. Shade the region common to all three graphs to show the feasible region.

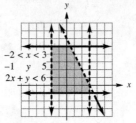

43. $2y - x \geq -5$

$y \leq 3 + x$

$x \geq 0$

$y \geq 0$

The graph of $2y - x \geq -5$ consists of the boundary line $2y - x = 5$ and the region above it. The graph of $y \leq 3 + x$ consists of the boundary line $y = 3 + x$ and the region below it. The inequalities $x \geq 0$ and $y \geq 0$ restrict the feasible region to the first quadrant. Shade the region common to all of these graphs to show the feasible region.

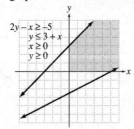

45. $3x + 4y > 12$
$2x - 3y < 6$
$0 \le y \le 2$
$x \ge 0$

$3x + 4y > 12$ is the set of points above the dashed line $3x + 4y = 12$. $2x - 3y < 6$ is the set of points above the dashed line $2x - 3y = 6$. $0 \le y \le 2$ is the rectangular strip of points lying on or between the horizontal lines $y = 0$ and $y = 2$. $x \ge 0$ consists of all the points on and to the right of the y-axis. The feasible region is the triangular region satisfying all the inequalities.

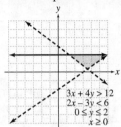

47. The shaded region lies between the horizontal lines $y = 0$ and $y = 4$, so this inequality is $0 \le y \le 4$. The shaded region also lies to the right of the vertical line $x = 0$, so this inequality is $x \ge 0$. The shaded region also lies below the line containing the points $(3, 4)$ and $(6, 0)$ which has the equation $4x + 3y = 24$. Using the origin as a test point, we determine that $4(0) + 3(0) \le 24$ is true, so the inequality must be $\le$. So the third inequality is $4x + 3y \le 24$. So, the inequalities are
$x \ge 0$
$0 \le y \le 4$
$4x + 3y \le 24$

49. The feasible region is the interior of the rectangle with vertices $(2, 3)$, $(2, -1)$, $(7, 3)$, and $(7, -1)$. The x-values range from 2 to 7, not including 2 or 7. The y-values range from -1 to 3, not including -1 or 3. Thus, the system of inequalities is
$2 < x < 7$
$-1 < y < 3.$

51. a.

	Number	Hours Spinning	Hours Dyeing	Hours Weaving
Shawls	x	1	1	1
Afghans	y	2	1	4
Maximum number of hours available		8	6	14

b. $x + 2y \le 8$ Spinning inequality
$x + y \le 6$ Dyeing inequality
$x + 4y \le 14$ Weaving inequality
$x \ge 0$ Ensures a non-negative number of each
$y \ge 0$

c. Graph the solid lines $x + 2y = 8$, $x + y = 6$, $x + 4y = 14$, $x = 0$, and $y = 0$, and shade the appropriate half-planes to get the feasible region. See the graph in the answer section of the textbook.

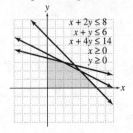

53. The system is

$$x \geq 3000$$
$$y \geq 5000$$
$$x + y \leq 10,000.$$

The first inequality gives the region to the right of the vertical line $x = 3000$, including the points on the line. The second inequality gives the region above the horizontal line $y = 5000$, including the points on the line. The third inequality gives the region below the line $x + y = 10,000$, including the points on the line. See the graph in the answer section of your textbook.

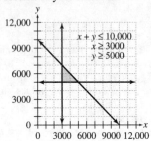

55.

	Number	Emissions	Cost
Type 1	x	.5 lb	\$.16
Type 2	y	.3 lb	\$.20
Maximum		1.8 lb	\$.8

The manufacturer produces at least 3.2 million barrels annually, so
$$x + y \geq 3.2.$$
The cost is not to exceed \$.8 million, so $.16x + .20y \leq .8$. Total emissions must not exceed 1.8 million lb, so $.5x + .3y \leq 1.8$.
We obtain

$$x + \quad y \geq 3.2$$
$$.16x + .20y \leq \;.8$$
$$.5x + \quad .3y \leq 1.8$$
$$x \geq 0$$
$$y \geq 0.$$

Using the above system, graph solid lines and shade appropriate half-planes to get the feasible region bounded by the y axis on the left, lines $.16x + .20y = .8$ and $.5x + .3y = 1.8$ above, and the line
$x + y = 3.2$ below. See the graph in the answer section of the textbook.

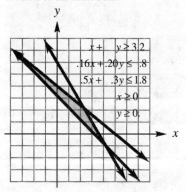

Section 7.2 Linear Programming: The Graphical Method

1. Make a table indicating the value of the objective function $z = 3x + 5y$ at each corner point.

Corner Point	Value of $z = 3x + 5y$
(1, 1)	$3(1) + 5(1) = 8$ Minimum
(2, 7)	$3(2) + 5(7) = 41$
(5, 10)	$3(5) + 5(10) = 65$ Maximum
(6, 3)	$3(6) + 5(3) = 33$

The maximum value of 65 occurs at (5, 10). The minimum value of 8 occurs as (1, 1).

3.

Corner Point	Value of $z = .75x + .40y$
(0, 0)	0 Minimum
(0, 12)	4.8
(4, 8)	6.2
(7, 3)	6.45
(9, 0)	6.75 Maximum

The maximum is 6.75 at (9, 0); the minimum is 0 at (0, 0).

5. a.

Corner Point	Value of $z = x + 5y$
(0, 8)	40
(3, 4)	23
$\left(\frac{13}{2}, 2\right)$	16.5
(12, 0)	12 Minimum

The minimum is 12 at (12, 0). There is no maximum because the feasible region is unbounded.

b.

Corner Point	Value of $z = 2x + 3y$
(0, 8)	24
(3, 4)	18 Minimum
$\left(\frac{13}{2}, 2\right)$	19
(12, 0)	24

The minimum is 18 at (3, 4). There is no maximum because the feasible region is unbounded.

5. **Continued**

c.

Corner Point	Value of $z = 2x + 4y$
(0, 8)	32
(3, 4)	22
$\left(\frac{13}{2}, 2\right)$	21 Minimum
(12, 0)	24

The minimum is 21 at $\left(\dfrac{13}{2}, 2\right)$. There is no maximum.

d.

Corner Point	Value of $z = 4x + y$
(0, 8)	8 Minimum
(3, 4)	16
$\left(\frac{13}{2}, 2\right)$	28
(12, 0)	48

The minimum is 8 at (0, 8). There is no maximum

7. Maximize $z = 2x + 5y$
   subject to: $2x + 3y \le 6$

   $\qquad\qquad\quad 4x + y \le 6$

   $\qquad\qquad\quad x \ge 0,\ y \ge 0.$

Graph the feasible region, and identify the corner points.

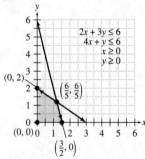

The graph shows the feasible region is bounded. The corner points

are $(0, 0), (0, 2), \left(\dfrac{3}{2}, 0\right), \left(\dfrac{6}{5}, \dfrac{6}{5}\right)$, which is the

intersection of $2x + 3y = 6$ and
$4x + y = 6$. Use the corner points to find the maximum value of the objective function.

7. Continued

Corner Point	Value of $z = 2x + 5y$
$(0, 0)$	0
$(0, 2)$	10 Maximum
$\left(\frac{6}{5}, \frac{6}{5}\right)$	$\frac{42}{5}$
$\left(\frac{3}{2}, 0\right)$	3

The maximum value of $z = 2x + 5y$ is 10 at $x = 0$, $y = 2$ at $(0, 2)$.

9. Minimize $z = 2x + y$
  subject to: $\quad 3x - y \geq 12$
  $\qquad\qquad\quad x + y \leq 15$
  $\qquad\qquad\quad x \geq 2, \ y \geq 3$.

Graph the feasible region, and identify the corner points

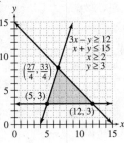

The feasible region is bounded with corner points $(5, 3)$, $\left(\frac{27}{4}, \frac{33}{4}\right)$, $(12, 3)$.

Corner Point	Value of $z = 2x + y$
$(5, 3)$	13 Minimum
$\left(\frac{27}{4}, \frac{33}{4}\right)$	$\frac{87}{4}$
$(12, 3)$	27

The minimum is 13 at $(5, 3)$.

11. Maximize $z = 2x + 4y$
  subject to: $\quad x - y \leq 10$
  $\qquad\qquad\quad 5x + 3y \leq 75$
  $\qquad\qquad\quad x \geq 0, \ y \geq 0$.

Graph the feasible region, and identify the corner points.

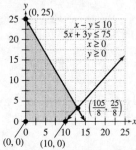

This region is bounded, with corner points $(0, 0)$, $(0, 25)$, $\left(\frac{105}{8}, \frac{25}{8}\right)$, $(10, 0)$.

Corner Point	Value of $z = 2x + 4y$
$(0, 0)$	0
$(0, 25)$	100 Maximum
$\left(\frac{105}{8}, \frac{25}{8}\right)$	$\frac{155}{4}$
$(10, 0)$	20

The maximum is 100 at $x = 0$, $y = 25$ or $(0, 25)$.

13. $\quad 3x + 2y \geq 6$
  $\quad\ \ x + 2y \geq 4$
  $\ x \geq 0, \ y \geq 0$

Graph the feasible region and identify the corner points.

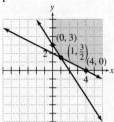

13. Continued

Corner Point	Value of $z = 3x + 4y$
$(0, 3)$	12
$\left(1, \dfrac{3}{2}\right)$	9 Minimum
$(4, 0)$	12

The minimum value of 9 occurs at $\left(1, \dfrac{3}{2}\right)$.

There is no maximum value because the feasible region is unbounded.

15. $x + y \le 6$

$-x + y \le 2$

$2x - y \le 8$

Graph the feasible region, and identify the corner points.

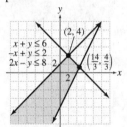

Corner Point	Value of $z = 3x + 4y$
$(2, 4)$	22 Maximum
$\left(\dfrac{14}{3}, \dfrac{4}{3}\right)$	$\dfrac{58}{3}$

The maximum value of z is 22 at $(2, 4)$. There is no minimum because the feasible region is unbounded.

17. a. $x + y \le 20$

$x + 3y \le 24$

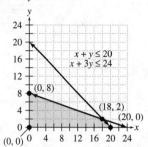

Corner Point	Value of $z = 10x + 12y$
$(0, 0)$	0
$(0, 8)$	96
$(18, 2)$	204 Maximum
$(20, 0)$	200

The maximum value of 204 occurs when $x = 18$ and $y = 2$, or at $(18, 2)$.

b. $3x + y \le 15$

$x + 2y \le 18$

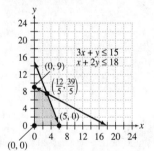

17. Continued

Corner Point	Value of $z = 10x + 12y$
$(0, 0)$	0
$(0, 9)$	108
$\left(\dfrac{12}{5}, \dfrac{39}{5}\right)$	$\dfrac{588}{5}$
$(5, 0)$	50

The maximum value of $\dfrac{588}{5}$ or $117\dfrac{3}{5}$

occurs when

$x = \dfrac{12}{5}$ and $y = \dfrac{39}{5}$, or at $\left(\dfrac{12}{5}, \dfrac{39}{5}\right)$.

c. $x + 2y \geq 10$

$2x + y \geq 12$

$x - y \leq 8$

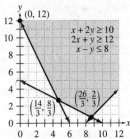

The feasible region is unbounded, so there is no maximum.

19. Answer varies. Sample response: The constraints do not describe a feasible region, i.e. there does not exist a point that satisfies all five constraints.

Section 7.3 Applications of Linear Programming

1. Let $x =$ the number of canoes and
  $y =$ the number of rowboats.
The inequality is $8x + 5y \leq 110$.

3. Let $x =$ the number of radio spots and
  $y =$ the number of TV ads.
Since each radio spot costs \$250 and each TV ad cost \$750, the inequality is
$250x + 750y \leq 9500$.

5. Let $x =$ the number of chain saws
   and
  $y =$ the number of wood chippers.
Assembling x chain saws at 4 hours each takes $4x$ hours while assembling y wood chippers at 6 hours each takes $6y$ hours. There are only 48 available hours, the first constraint is
$4x + 6y \leq 48$.
The number of chain saws and wood chippers assembled cannot be negative, so $x \geq 0$, $y \geq 0$. Each chain saw produces a profit of \$150 and each wood chipper, \$220. If z represents total profit, then $z = 150x + 220y$ is the objective function to be maximized.
Maximize $z = 150x + 220y$
subject to: $4x + 6y \leq 48$

$x \geq 0,\ y \geq 0$

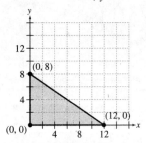

Corner Point	$z = 150x + 220y$
$(0, 0)$	$150(0) + 220(0) = 0$
$(12, 0)$	$150(12) + 220(0) = 1800$ (maximum)
$(0, 8)$	$150(0) + 220(8) = 1760$

12 chain saws and no wood chippers should be assembled for maximum profit.

7. Let x = the number of pounds of deluxe coffee
and
y = the number of pounds of regular coffee.
The mixture of deluxe and regular coffee needs to be at least 50 pounds.
$x + y \geq 50$
The mixture must have at least 10 pounds of deluxe coffee.
$x \geq 10$
The pounds of coffee cannot be negative, so
$x \geq 0, y \geq 0$.
At $6 per pound of deluxe coffee and $5 per pound of regular coffee, the total cost, z, is
$z = 6x + 5y$
which is the objective function to be minimized.
Minimize: $z = 6x + 5y$
subject to: $x + y \geq 50$
$$x \geq 10$$
$$x \geq 0, y \geq 0$$

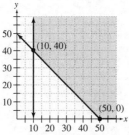

Corner Point	$z = 6x + 5y$
(50, 0)	$6(50) + 5(0) = 300$
(10, 40)	$6(10) + 5(40) = 260$ (minimum)

The mixture should contain 10 pounds of deluxe and 40 pounds of regular coffee to minimize cost.

9. Let x = the number of units of Policy A to purchase and
y = the number of units of Policy B to purchase.

a. Minimize: $z = 50x + 40y$
subject to: $10x + 15y \geq 300$
$$180x + 120y \geq 3000$$
$$x \geq 0$$
$$y \geq 0.$$

9. Continued

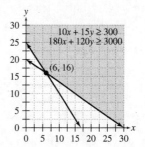

$50(6) + 40(16) = 940$

The minimum premium cost is from purchasing 6 units of Policy A and 16 units of Policy B and is $940.

b. Minimize: $z = 25x + 40y$
The constraints and graph remain unchanged. Now the minimum premium cost is from purchasing 30 units of
Policy A and no units of Policy B and is $750.

11. Let x = the number of servings of A
and
y = the number of servings of B.
Minimize $z = .25x + .40y$
subject to: $3x + 2y \geq 15$
$$2x + 4y \geq 15$$
$$x \geq 0, y \geq 0$$

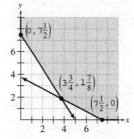

Corner Point	$z = .25x + .40y$
$\left(0, 7\frac{1}{2}\right)$	$.25(0) + .40\left(7\frac{1}{2}\right) = 3$
$\left(3\frac{3}{4}, 1\frac{7}{8}\right)$	$.25\left(3\frac{3}{4}\right) + .40\left(1\frac{7}{8}\right) = 1.69$ (minimum)
$\left(7\frac{1}{2}, 0\right)$	$.25\left(7\frac{1}{2}\right) + .40(0) = 1.88$

He can satisfy the requirements with $3\frac{3}{4}$ servings of A and $1\frac{7}{8}$ servings of B for a minimum cost of $1.69.

13. Let x = the number of Type 1 bolts and
y = the number of Type 2 bolts.

Maximize $z = .10x + .12y$

subject to: $.1x + .1y \leq 240$

$.1x + .4y \leq 720$

$.1x + .02y \leq 160$

$x \geq 0$

$y \geq 0$

Graph the feasible region, and label the corner points. Use the corner points to find the maximum value of the objective function.

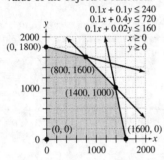

Corner Point	Value of $z = .10x + .12y$
$(0, 0)$	0
$(0, 1800)$	216
$(1600, 0)$	160
$(800, 1600)$	272 Maximum
$(1400, 1000)$	260

Manufacture 800 Type 1 bolts and 1600 Type 2 bolts for a maximum revenue of $272/day.

15. Summarizing the data provided,

	Energy	Protein	Hides	Availability	Cost(hrs)
plant	30	10	0	25	30
animal	20	25	1	25	15
Requirements	360	300	8		

Let x = the number of plants to gather

and

y = the number of animals to gather

Translate the table into a linear programming problem.

Minimize $z = 30x + 15y$

subject to: $30x + 20y \geq 360$

$10x + 25y \geq 300$

$y \geq 8$

$x \leq 25$

$y \leq 25$

$x \geq 0,\ y \geq 0$

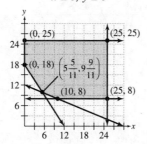

Corner Point	$z = 30x + 15y$
$(0, 25)$	$30(0) + 15(25) = 375$
$(0, 18)$	$30(0) + 15(18) = 270$
	minimum
$\left(5\frac{5}{11},\ 9\frac{9}{11}\right)$	$30\left(5\frac{5}{11}\right) + 15\left(9\frac{9}{11}\right) = 310\frac{10}{11}$
$(10, 8)$	$30(10) + 15(8) = 420$
$(25, 8)$	$30(25) + 15(8) = 870$
$(25, 25)$	$30(25) + 15(25) = 1125$

The population should gather no plants and 18 animals for a minimum 270 hours.

17. Let x = the number of cards from warehouse I to San Jose. Then $350 - x$ is the number of cards from warehouse II to San Jose. Let y = the number of cards from warehouse I to Memphis. Then $250 - y$ is the number of cards from warehouse II to Memphis. The constraints are represented by the inequalities

$$x \geq 0, \ y \geq 0$$
$$x \leq 350$$
$$y \leq 250$$
$$x + y \leq 500$$

Since

$$(350 - x) + (250 - y) \leq 290,$$
$$x + y \geq 310.$$

Minimize

$$.25x + .23(350 - x) + .22y + .21(250 - y)$$
$$= 133 + .02x + .01y.$$

The graph and corner points are shown below.

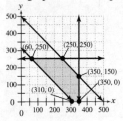

Corner Point	Value of $z = 133 + .02x + .01y$
(60, 250)	136.7 Minimum
(250, 250)	140.5
(350, 150)	141.5
(350, 0)	140
(310, 0)	139.2

The minimum cost is $136.70. From warehouse I ship 60 boxes to San Jose and 250 boxes to Memphis. From warehouse II ship 290 boxes to San Jose and none to Memphis.

19. Let x = the amount invested in bonds and
y = the amount invested in mutual funds. (Both are in millions of dollars.)
The amount of annual interest is $.04x + .06y$.
Maximize $z = .04x + .06y$
subject to:
$$x \geq 20$$
$$y \geq 6$$
$$300x + 100y \leq 8400$$
$$x + y \leq 50$$
$$x \geq 0, \ y \geq 0$$

Graph the feasible region, and label the corner points.

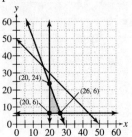

Corner Point	Value of $z = .04x + .06y$
(20, 24)	2.24 Maximum
(26, 6)	14.
(20, 6)	1.16

He should invest $20 million in bonds and $24 million in mutual funds for maximum annual interest of $4 million.

21. Let x = the number of humanities courses and y = the number of science courses.

Maximize $\quad z = 3(5y) + 2\left(\dfrac{1}{2}\right)(4x)$

$\qquad +3\left(\dfrac{1}{4}\right)(4x) + 4\left(\dfrac{1}{4}\right)(4x)$

subject to: $\qquad x \geq 4$

$\qquad 4 \leq y \leq 12$

$\qquad 4x + 5y \leq 92.$

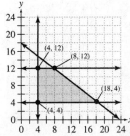

The objective function simplifies to $z = 11x + 15y.$

Corner Point	Value of $z = 11x + 15y$
(4, 4)	104
(4, 12)	224
(8, 12)	268 Maximum
(18, 4)	258

She should take 8 humanities and 12 science courses.

23. 1 Zeta + 2 Beta must not exceed 1000; thus (b) is the correct answer.

25. \$4 Zeta + \$5.25 Beta equals the total contribution margin; (c) is the correct answer.

Section 7.4 The Simplex Method: Maximization

1. Maximize $\quad z = 32x_1 + 9x_2$

subject to: $\quad 4x_1 + 2x_2 \leq 20$

$\qquad 5x_1 + \ x_2 \leq 50$

$\qquad 2x_1 + 3x_2 \leq 25$

$\qquad x_1 \geq 0,\ x_2 \geq 0.$

a. There are 3 constraints, so 3 slack variables are needed.

b. Use x_3, x_4, and x_5 for the slack variables.

c. $\quad 4x_1 + 2x_2 + x_3 \qquad\qquad = 20$

$\qquad 5x_1 + \ x_2 \qquad + x_4 \qquad = 50$

$\qquad 2x_1 + 3x_2 \qquad\qquad + x_5 = 25$

3. Maximize $\quad z = 8x_1 + 3x_2 + x_3$

subject to: $\qquad 3x_1 - x_2 + \ 4x_3 \leq 95$

$\qquad\qquad 7x_1 + 6x_2 + \ 8x_3 \leq 118$

$\qquad\qquad 4x_1 + 5x_2 + 10x_3 \leq 220$

$\qquad\qquad x_1 \geq 0,\ x_2 \geq 0,\ x_3 \geq 0.$

a. There are 3 constraints, so 3 slack variables are needed.

b. Use x_4, x_5, and x_6 for the slack variables.

c. $\quad 3x_1 - \ x_2 + 4x_3 + x_4 \qquad\qquad = 95$

$\qquad 7x_1 + 6x_2 + 8x_3 \qquad + x_5 \qquad = 118$

$\qquad 4x_1 + 5x_2 + 10x_3 \qquad\qquad + x_6 = 220$

5. Maximize $\quad z = 5x_1 + x_2$

subject to: $\quad 2x_1 + 5x_2 \leq 6$

$\qquad 4x_1 + \ x_2 \leq 6$

$\qquad 5x_1 + 3x_2 \leq 15$

$\qquad x_1 \geq 0,\ x_2 \geq 0.$

Since there are 3 constraints, 3 slack variables are needed: x_3, x_4, and $x_5.$

The constraints are now

$2x_1 + 5x_2 + x_3 \qquad\qquad = 6$

$4x_1 + \ x_2 \qquad + x_4 \qquad = 6$

$5x_1 + 3x_2 \qquad\qquad + x_5 = 15.$

5. The initial simplex tableau is

$$x_1 \quad x_2 \quad x_3 \quad x_4 \quad x_5 \quad z$$

$$\begin{bmatrix} 2 & 5 & 1 & 0 & 0 & 0 & | & 6 \\ 4 & 1 & 0 & 1 & 0 & 0 & | & 6 \\ 5 & 3 & 0 & 0 & 1 & 0 & | & 15 \\ \hline -5 & -1 & 0 & 0 & 0 & 1 & | & 0 \end{bmatrix}.$$

7. Maximize $\quad z = x_1 + 5x_2 + 10x_3$

subject to: $\quad x_1 + 2x_2 + 3x_3 \le 10$

$$2x_1 + x_2 + x_3 \le 8$$

$$3x_1 \qquad + 4x_3 \le 6$$

$$x_1 \ge 0, \ x_2 \ge 0, \ x_3 \ge 0.$$

Since there are 3 constraints, 3 slack variables are needed.: $x_4, \ x_5, \ x_6.$

The constraints are now

$$x_1 + 2x_2 + 3x_3 + x_4 \qquad\qquad = 10$$

$$2x_1 + x_2 + x_3 \qquad + x_5 \qquad = 8$$

$$3x_1 \qquad + 4x_3 \qquad\qquad + x_6 = 6$$

The initial tableau is

$$x_1 \quad x_2 \quad x_3 \quad x_4 \quad x_5 \quad x_6 \quad z$$

$$\begin{bmatrix} 1 & 2 & 3 & 1 & 0 & 0 & 0 & | & 10 \\ 2 & 1 & 1 & 0 & 1 & 0 & 0 & | & 8 \\ 3 & 0 & 4 & 0 & 0 & 1 & 0 & | & 6 \\ \hline -1 & -5 & -10 & 0 & 0 & 0 & 1 & | & 0 \end{bmatrix}.$$

9.

$$x_1 \quad x_2 \quad x_3 \quad x_4 \quad x_5 \quad z$$

$$\begin{bmatrix} 2 & 2 & 0 & 3 & 1 & 0 & | & 15 \\ 3 & 4 & 1 & 6 & 0 & 0 & | & 20 \\ \hline -2 & -3 & 0 & 1 & 0 & 1 & | & 10 \end{bmatrix}$$

The most negative indicator is –3, so the pivot column is column 2. Since $\dfrac{20}{4}$ is the smaller quotient, the pivot row is row two. The pivot is the 4 in row two, column two.

11.

$$x_1 \quad x_2 \quad x_3 \quad x_4 \quad x_5 \quad x_6 \quad z$$

$$\begin{bmatrix} 6 & 2 & 1 & 3 & 0 & 0 & 0 & | & 8 \\ 0 & 2 & 0 & 1 & 0 & 1 & 0 & | & 7 \\ 6 & 1 & 0 & 3 & 1 & 0 & 0 & | & 6 \\ \hline -3 & -2 & 0 & 2 & 0 & 0 & 1 & | & 12 \end{bmatrix}$$

The most negative indicator is –3, so the pivot column is column one. Since $\dfrac{6}{6} = 1$ is the smallest quotient, the pivot row is row three. Thus, the pivot is the 6 in row three, column one.

13.

$$x_1 \quad x_2 \quad x_3 \quad x_4 \quad x_5 \quad z$$

$$\begin{bmatrix} 1 & 2 & 4 & 1 & 0 & 0 & | & 56 \\ 2 & \underline{2} & 1 & 0 & 1 & 0 & | & 40 \\ \hline -1 & -3 & -2 & 0 & 0 & 1 & | & 0 \end{bmatrix}$$

Start by multiplying each entry of row 2 by $\dfrac{1}{2}$ in order to change the pivot to 1.

$$x_1 \quad x_2 \quad x_3 \quad x_4 \quad x_5 \quad z$$

$$\begin{bmatrix} 1 & 2 & 4 & 1 & 0 & 0 & | & 56 \\ 1 & 1 & \frac{1}{2} & 0 & \frac{1}{2} & 0 & | & 20 \\ \hline -1 & -3 & -2 & 0 & 0 & 1 & | & 0 \end{bmatrix} \frac{1}{2} R_2$$

Now use row operations to change the entry in row one column two and the indicator –3 to 0.

$$x_1 \quad x_2 \quad x_3 \quad x_4 \quad x_5 \quad z$$

$$\begin{bmatrix} -1 & 0 & 3 & 1 & -1 & 0 & | & 16 \\ 1 & 1 & \frac{1}{2} & 0 & \frac{1}{2} & 0 & | & 20 \\ \hline 2 & 0 & -\frac{1}{2} & 0 & \frac{3}{2} & 1 & | & 60 \end{bmatrix} \begin{matrix} -2R_2 + R_1 \\ \\ 3R_2 + R_3 \end{matrix}$$

15.

$$x_1 \quad x_2 \quad x_3 \quad x_4 \quad x_5 \quad x_6 \quad z$$

$$\begin{bmatrix} 1 & 1 & 1 & 1 & 0 & 0 & 0 & | & 60 \\ 3 & 1 & \underline{2} & 0 & 1 & 0 & 0 & | & 100 \\ 1 & 2 & 3 & 0 & 0 & 1 & 0 & | & 200 \\ \hline -1 & -1 & -2 & 0 & 0 & 0 & 1 & | & 0 \end{bmatrix}$$

$$x_1 \quad x_2 \quad x_3 \quad x_4 \quad x_5 \quad x_6 \quad z$$

$$\begin{bmatrix} 1 & 1 & 1 & 1 & 0 & 0 & 0 & | & 60 \\ \frac{3}{2} & \frac{1}{2} & 1 & 0 & \frac{1}{2} & 0 & 0 & | & 50 \\ 1 & 2 & 3 & 0 & 0 & 1 & 0 & | & 200 \\ \hline -1 & -1 & -2 & 0 & 0 & 0 & 1 & | & 0 \end{bmatrix} \frac{1}{2} R_2$$

$$x_1 \quad x_2 \quad x_3 \quad x_4 \quad x_5 \quad x_6 \quad z$$

$$\begin{bmatrix} -\frac{1}{2} & \frac{1}{2} & 0 & 1 & -\frac{1}{2} & 0 & 0 & | & 10 \\ \frac{3}{2} & \frac{1}{2} & 1 & 0 & \frac{1}{2} & 0 & 0 & | & 50 \\ -\frac{7}{2} & \frac{1}{2} & 0 & 0 & -\frac{3}{2} & 1 & 0 & | & 50 \\ \hline 2 & 0 & 0 & 0 & 1 & 0 & 1 & | & 100 \end{bmatrix} \begin{matrix} -1R_2 + R_1 \\ \\ -3R_2 + R_3 \\ 2R_2 + R_4 \end{matrix}$$

17.

$$\begin{array}{cccccc} x_1 & x_2 & x_3 & x_4 & x_5 & z \end{array}$$

$$\begin{bmatrix} 3 & 2 & 0 & -3 & 1 & 0 & | & 29 \\ 4 & 0 & 1 & -2 & 0 & 0 & | & 16 \\ \hline -5 & 0 & 0 & -1 & 0 & 1 & | & 11 \end{bmatrix}$$

a. The basic variables are x_3 and x_5. x_1, x_2, and x_4 are nonbasic.

b. The basic feasible solution is $x_1 = 0$, $x_2 = 0$, $x_3 = 16$, $x_4 = 0$, $x_5 = 29$, $z = 11$

c. Because there are still negative indicators, this solution is not the maximum.

19.

$$\begin{array}{ccccccc} x_1 & x_2 & x_3 & x_4 & x_5 & x_6 & z \end{array}$$

$$\begin{bmatrix} 1 & 0 & 2 & \frac{1}{2} & 0 & \frac{1}{3} & 0 & | & 6 \\ 0 & 1 & -1 & 5 & 0 & -1 & 13 & | & 13 \\ 0 & 0 & 1 & \frac{3}{2} & 1 & -\frac{1}{3} & 0 & | & 21 \\ \hline 0 & 0 & 2 & \frac{1}{2} & 0 & 3 & 1 & | & 18 \end{bmatrix}$$

a. The basic variables are x_1, x_2, and x_5. x_3, x_4, and x_6 are nonbasic.

b. The basic feasible solution is $x_1 = 6$, $x_2 = 13$, $x_3 = 0$, $x_4 = 0$, $x_5 = 21$, $x_6 = 0$, $z = 18$.

c. Because there are no negative indicators, this solution is the maximum.

21. Maximize $z = x_1 + 3x_2$

subject to: $x_1 + x_2 \le 10$

$5x_1 + 2x_2 \le 20$

$x_1 + 2x_2 \le 36$

$x_1 \ge 0, x_2 \ge 0.$

Using slack variables, x_3, x_4, and x_5, the constraints become:

$x_1 + \quad x_2 + x_3 \qquad = 10$

$5x_1 + 2x_2 \qquad + x_4 \qquad = 20$

$x_1 + 2x_2 \qquad\qquad + x_5 = 36.$

21. Continued

The initial simplex tableau is

$$\begin{array}{cccccc} x_1 & x_2 & x_3 & x_4 & x_5 & z \end{array}$$

$$\begin{bmatrix} 1 & \underline{1} & 1 & 0 & 0 & 0 & | & 10 \\ 5 & 2 & 0 & 1 & 0 & 0 & | & 20 \\ 1 & 2 & 0 & 0 & 1 & 0 & | & 36 \\ \hline -1 & -3 & 0 & 0 & 0 & 1 & | & 0 \end{bmatrix}.$$

Pivot the 1 in row one, column two.

$$\begin{array}{cccccc} x_1 & x_2 & x_3 & x_4 & x_5 & z \end{array}$$

$$\begin{bmatrix} 1 & 1 & 1 & 0 & 0 & 0 & | & 10 \\ 3 & 0 & -2 & 1 & 0 & 0 & | & 0 \\ -1 & 0 & -2 & 0 & 1 & 0 & | & 16 \\ \hline 2 & 0 & 3 & 0 & 0 & 1 & | & 30 \end{bmatrix} \begin{array}{l} \\ -2R_1 + R_2 \\ -2R_1 + R_3 \\ 3R_1 + R_4 \end{array}$$

Since there are no negative indicators, this matrix is the final tableau. The maximum is 30 when $x_1 = 0$, $x_2 = 10$, $x_3 = 0$, $x_4 = 0$, and $x_5 = 16$.

23. Maximize $z = 2x_1 + x_2$

subject to: $x_1 + 3x_2 \le 12$

$2x_1 + \quad x_2 \le 10$

$x_1 + \quad x_2 \le 4$

$x_1 \ge 0, x_2 \ge 0.$

Using slack variables x_3, x_4, and x_5 the constraints become:

$x_1 + 3x_2 + x_3 \qquad\qquad = 12$

$2x_1 + \quad x_2 \qquad + x_4 \qquad = 10$

$x_1 + \quad x_2 \qquad\qquad + x_5 = 4.$

The initial simplex tableau is

$$\begin{array}{cccccc} x_1 & x_2 & x_3 & x_4 & x_5 & z \end{array}$$

$$\begin{bmatrix} 1 & 3 & 1 & 0 & 0 & 0 & | & 12 \\ 2 & 1 & 0 & 1 & 0 & 0 & | & 10 \\ \underline{1} & 1 & 0 & 0 & 1 & 0 & | & 4 \\ \hline -2 & -1 & 0 & 0 & 0 & 1 & | & 0 \end{bmatrix}.$$

Pivot on the 1 in row three, column one.

$$\begin{array}{cccccc} x_1 & x_2 & x_3 & x_4 & x_5 & z \end{array}$$

$$\begin{bmatrix} 0 & 2 & 1 & 0 & -1 & 0 & | & 8 \\ 0 & -1 & 0 & 1 & -2 & 0 & | & 2 \\ 1 & 1 & 0 & 0 & 1 & 0 & | & 4 \\ \hline 0 & 1 & 0 & 0 & 2 & 1 & | & 8 \end{bmatrix} \begin{array}{l} -1R_3 + R_1 \\ -2R_3 + R_2 \\ \\ 2R_3 + R_4 \end{array}$$

Since there are no negative indicators, this matrix is the final tableau.

The maximum is 8 when $x_1 = 4$, $x_2 = 0$, $x_3 = 8$, $x_4 = 2$, $x_5 = 0$.

25. Maximize $\quad z = 5x_1 + 4x_2 + x_3$

subject to: $\quad -2x_1 + x_2 + 2x_3 \le 3$

$$x_1 - x_2 + x_3 \le 1$$

$$x_1 \ge 0, \ x_2 \ge 0, \ x_3 \ge 0.$$

Using the slack variables x_4 and x_5, the constraints become:

$$-2x_1 + x_2 + 2x_3 + x_4 \quad\quad = 3$$

$$x_1 - x_2 + x_3 \quad\quad + x_5 = 1.$$

The initial simplex tableau is

$$
\begin{array}{cccccc}
x_1 & x_2 & x_3 & x_4 & x_5 & z
\end{array}
$$

$$
\left[\begin{array}{cccccc|c}
-2 & 1 & 2 & 1 & 0 & 0 & 3 \\
\underline{1} & -1 & 1 & 0 & 1 & 0 & 1 \\
\hline
-5 & -4 & -1 & 0 & 0 & 1 & 0
\end{array}\right].
$$

Pivot on the 1 in row two, column one.

$$
\begin{array}{cccccc}
x_1 & x_2 & x_3 & x_4 & x_5 & z
\end{array}
$$

$$
\left[\begin{array}{cccccc|c}
0 & -1 & 4 & 1 & 2 & 0 & 5 \\
1 & -1 & 1 & 0 & 1 & 0 & 1 \\
\hline
0 & -9 & 4 & 0 & 5 & 1 & 5
\end{array}\right]
\begin{array}{l}
2R_2 + R_1 \\
\\
5R_2 + R_3
\end{array}
$$

There is a negative indicator in column two, but all entries in that column are negative, so there is no place to continue pivoting.
Therefore, there is no maximum.

27. Maximize $\quad z = 2x_1 + x_2 + x_3$

subject to: $\quad x_1 - 3x_2 + x_3 \le 3$

$$x_1 - 2x_2 + 2x_3 \le 12$$

$$x_1 \ge 0, \ x_2 \ge 0, \ x_3 \ge 0.$$

Using slack variables x_4 and x_5, the constraints become:

$$x_1 - 3x_2 + x_3 + x_4 \quad\quad = 3$$

$$x_1 - 2x_2 + 2x_3 \quad\quad + x_5 = 12.$$

The initial simplex tableau is

$$
\begin{array}{cccccc}
x_1 & x_2 & x_3 & x_4 & x_5 & z
\end{array}
$$

$$
\left[\begin{array}{cccccc|c}
\underline{1} & -3 & 1 & 1 & 0 & 0 & 3 \\
1 & -2 & 2 & 0 & 1 & 0 & 12 \\
\hline
-2 & -1 & -1 & 0 & 0 & 1 & 0
\end{array}\right].
$$

Pivot one the 1 in row one, column one.

$$
\begin{array}{cccccc}
x_1 & x_2 & x_3 & x_4 & x_5 & z
\end{array}
$$

$$
\left[\begin{array}{cccccc|c}
1 & -3 & 1 & 1 & 0 & 0 & 3 \\
0 & \underline{1} & 1 & -1 & 1 & 0 & 9 \\
\hline
0 & -7 & 1 & 2 & 0 & 1 & 6
\end{array}\right]
\begin{array}{l}
\\
-R_1 + R_2 \\
2R_1 + R_3
\end{array}
$$

Pivot on the 1 in row two, column two.

27. Continued

$$
\begin{array}{cccccc}
x_1 & x_2 & x_3 & x_4 & x_5 & z
\end{array}
$$

$$
\left[\begin{array}{cccccc|c}
1 & 0 & 4 & -2 & 3 & 0 & 30 \\
0 & 1 & 1 & -1 & 1 & 0 & 9 \\
\hline
0 & 0 & 8 & -5 & 7 & 1 & 69
\end{array}\right]
\begin{array}{l}
3R_2 + R_1 \\
\\
7R_2 + R_3
\end{array}
$$

The only negative indicator is in column four, which has all negative entries, so there is no place to continue pivoting. Therefore, there is no maximum.

29. Maximize $\quad z = 2x_1 + 2x_2 - 4x_3$

Subject to: $\quad 3x_1 + 3x_2 - 6x_3 \le 51$

$$5x_1 + 5x_2 + 10x_3 \le 99$$

$$x_1 \ge 0, \ x_2 \ge 0, \ x_3 \ge 0.$$

Using slack variables x_4 and x_5, the constraints become:

$$3x_1 + 3x_2 - 6x_3 + x_4 \quad\quad = 51$$

$$5x_1 + 5x_2 + 10x_3 \quad\quad + x_5 = 99.$$

The initial simplex tableau is

$$
\begin{array}{cccccc}
x_1 & x_2 & x_3 & x_4 & x_5 & z
\end{array}
$$

$$
\left[\begin{array}{cccccc|c}
\underline{3} & 3 & -6 & 1 & 0 & 0 & 51 \\
5 & 5 & 10 & 0 & 1 & 0 & 99 \\
\hline
-2 & -2 & 4 & 0 & 0 & 1 & 0
\end{array}\right].
$$

Pivot on the 3 in row one, column one.

$$
\begin{array}{cccccc}
x_1 & x_2 & x_3 & x_4 & x_5 & z
\end{array}
$$

$$
\left[\begin{array}{cccccc|c}
1 & 1 & -2 & \frac{1}{3} & 0 & 0 & 17 \\
5 & 5 & 10 & 0 & 1 & 0 & 99 \\
\hline
-2 & -2 & 4 & 0 & 0 & 1 & 0
\end{array}\right]
\begin{array}{l}
\frac{1}{3}R_1 \\
\\
\\
\end{array}
$$

$$
\begin{array}{cccccc}
x_1 & x_2 & x_3 & x_4 & x_5 & z
\end{array}
$$

$$
\left[\begin{array}{cccccc|c}
1 & 1 & -2 & \frac{1}{3} & 0 & 0 & 17 \\
0 & 0 & 20 & -\frac{5}{3} & 1 & 0 & 14 \\
\hline
0 & 0 & 0 & \frac{2}{3} & 0 & 1 & 34
\end{array}\right]
\begin{array}{l}
\\
-5R_1 + R_2 \\
2R_1 + R_3
\end{array}
$$

The maximum is 34 when

$x_1 = 17, \ x_2 = 0, \ x_3 = 0, \ x_4 = 0, \ x_5 = 14$

or when $x_1 = 0, \ x_2 = 17, \ x_3 = 0, \ x_4 = 0,$

and $x_5 = 14$.

31. Maximize $z = 300x_1 + 200x_2 + 100x_3$
subject to: $x_1 + x_2 + x_3 \le 100$
$2x_1 + 3x_2 + 4x_3 \le 320$
$2x_1 + x_2 + x_3 \le 160$
$x_1 \ge 0, x_2 \ge 0, x_3 \ge 0.$

The initial simplex tableau is

$$\begin{array}{ccccccc|c}
x_1 & x_2 & x_3 & x_4 & x_5 & x_6 & z & \\
1 & 1 & 1 & 1 & 0 & 0 & 0 & 100 \\
2 & 3 & 4 & 0 & 1 & 0 & 0 & 320 \\
\underline{2} & 1 & 1 & 0 & 0 & 1 & 0 & 160 \\
\hline
-300 & -200 & -100 & 0 & 0 & 0 & 1 & 0
\end{array}$$

Pivot on the 2 in row three, column one.

$$\begin{array}{ccccccc|cl}
x_1 & x_2 & x_3 & x_4 & x_5 & x_6 & z & \\
1 & 1 & 1 & 1 & 0 & 0 & 0 & 100 \\
2 & 3 & 4 & 0 & 1 & 0 & 0 & 320 \\
1 & \frac{1}{2} & \frac{1}{2} & 0 & 0 & \frac{1}{2} & 0 & 80 & \frac{1}{2}R_3 \\
\hline
-300 & -200 & -100 & 0 & 0 & 0 & 1 & 0
\end{array}$$

$$\begin{array}{ccccccc|cl}
x_1 & x_2 & x_3 & x_4 & x_5 & x_6 & z & \\
0 & \frac{1}{2} & \frac{1}{2} & 1 & 0 & -\frac{1}{2} & 0 & 20 & -1R_3+R_1 \\
0 & 2 & 3 & 0 & 1 & -1 & 0 & 160 & -2R_3+R_2 \\
1 & \frac{1}{2} & \frac{1}{2} & 0 & 0 & \frac{1}{2} & 0 & 80 & \\
\hline
0 & -50 & 50 & 0 & 0 & 150 & 1 & 24{,}000 & 300R_3+R_4
\end{array}$$

Pivot on the $\dfrac{1}{2}$ in row one, column two.

$$\begin{array}{ccccccc|cl}
x_1 & x_2 & x_3 & x_4 & x_5 & x_6 & z & \\
0 & 1 & 1 & 2 & 0 & -1 & 0 & 40 & 2R_1 \\
0 & 2 & 3 & 0 & 1 & -1 & 0 & 160 \\
1 & \frac{1}{2} & \frac{1}{2} & 0 & 0 & \frac{1}{2} & 0 & 80 \\
\hline
0 & -50 & 50 & 0 & 0 & 150 & 1 & 24{,}000
\end{array}$$

$$\begin{array}{ccccccc|cl}
x_1 & x_2 & x_3 & x_4 & x_5 & x_6 & z & \\
0 & 1 & 1 & 2 & 0 & -1 & 0 & 40 & \\
0 & 0 & 1 & -4 & 1 & 1 & 0 & 80 & -2R_1+R_2 \\
1 & 0 & 0 & -1 & 0 & 1 & 0 & 60 & -\frac{1}{2}R_1+R_3 \\
\hline
0 & 0 & 100 & 100 & 0 & 100 & 1 & 26{,}000 & 50R_1+R_4
\end{array}$$

The maximum value is 26,000 when
$x_1 = 60, x_2 = 40, x_3 = 0, x_4 = 0, x_5 = 80,$ and
$x_6 = 0.$

33. Maximize $z = 4x_1 - 3x_2 + 2x_3$
subject to: $2x_1 - x_2 + 8x_3 \le 40$
$4x_1 - 5x_2 + 6x_3 \le 60$
$2x_1 - 2x_2 + 6x_3 \le 24$
$x_1 \ge 0, x_2 \ge 0, x_3 \ge 0.$

Note: The third constraint simplifies to
$x_1 - x_2 + 3x_3 \le 12.$

The initial simplex tableau is

$$\begin{array}{ccccccc|c}
x_1 & x_2 & x_3 & x_4 & x_5 & x_6 & z & \\
2 & -1 & 8 & 1 & 0 & 0 & 0 & 40 \\
4 & -5 & 6 & 0 & 1 & 0 & 0 & 60 \\
\underline{1} & -1 & 3 & 0 & 0 & 1 & 0 & 12 \\
\hline
-4 & 3 & -2 & 0 & 0 & 0 & 1 & 0
\end{array}$$

Pivot on the 1 in row three, column one.

$$\begin{array}{ccccccc|cl}
x_1 & x_2 & x_3 & x_4 & x_5 & x_6 & z & \\
0 & \underline{1} & 2 & 1 & 0 & -2 & 0 & 16 & -2R_3+R_1 \\
0 & -1 & -6 & 0 & 1 & -4 & 0 & 12 & -4R_3+R_2 \\
1 & -1 & 3 & 0 & 0 & 1 & 0 & 12 & \\
\hline
0 & -1 & 10 & 0 & 0 & 4 & 1 & 48 & 4R_3+R_4
\end{array}$$

Pivot on the 1 in row one, column two.

$$\begin{array}{ccccccc|cl}
x_1 & x_2 & x_3 & x_4 & x_5 & x_6 & z & \\
0 & 1 & 2 & 1 & 0 & -2 & 0 & 16 & \\
0 & 0 & -4 & 1 & 1 & -6 & 0 & 28 & R_1+R_2 \\
1 & 0 & 5 & 1 & 0 & -1 & 0 & 28 & R_1+R_3 \\
\hline
0 & 0 & 12 & 1 & 0 & 2 & 1 & 64 & R_1+R_4
\end{array}$$

The maximum is 64 when
$x_1 = 28, x_2 = 16, x_3 = 0, x_4 = 0, x_5 = 28,$ and
$x_6 = 0.$

35. Maximize $\quad z = x_1 + 2x_2 + x_3 + 5x_4$

subject to: $\qquad x_1 + 2x_2 + x_3 + x_4 \le 50$

$\qquad\qquad 3x_1 + x_2 + 2x_3 + x_4 \le 100$

$\qquad x_1 \ge 0,\, x_2 \ge 0,\, x_3 \ge 0,\, x_4 \ge 0.$

The initial simplex tableau is

$$
\begin{array}{ccccccc}
x_1 & x_2 & x_3 & x_4 & x_5 & x_6 & z
\end{array}
$$

$$
\left[\begin{array}{ccccccc|c}
1 & 2 & 1 & \underline{1} & 1 & 0 & 0 & 50 \\
3 & 1 & 2 & 1 & 0 & 1 & 0 & 100 \\
\hline
-1 & -2 & -1 & -5 & 0 & 0 & 1 & 0
\end{array}\right].
$$

The pivot is the 1 in row one, column four.

$$
\begin{array}{ccccccc}
x_1 & x_2 & x_3 & x_4 & x_5 & x_6 & z
\end{array}
$$

$$
\left[\begin{array}{ccccccc|c}
1 & 2 & 1 & 1 & 1 & 0 & 0 & 50 \\
2 & -1 & 1 & 0 & -1 & 1 & 0 & 50 \\
\hline
4 & 8 & 4 & 0 & 5 & 0 & 1 & 250
\end{array}\right]
\begin{array}{l}
\\
-1R_1 + R_2 \\
5R_1 + R_3
\end{array}
$$

The maximum is 250 when

$x_1 = 0,\, x_2 = 0,\, x_3 = 0,\, x_4 = 50,\, x_5 = 0,\,$ and

$x_6 = 50.$

37.
$$
\begin{array}{cccccc}
x_1 & x_2 & x_3 & x_4 & x_5 & z
\end{array}
$$

$$
\left[\begin{array}{cccccc|c}
1 & 1 & 1 & 1 & 0 & 0 & 12 \\
2 & 1 & 2 & 0 & 1 & 0 & 30 \\
\hline
-2 & -2 & -1 & 0 & 0 & 1 & 0
\end{array}\right]
$$

a. Pivot on the 1 in row one, column one.

$$
\begin{array}{cccccc}
x_1 & x_2 & x_3 & x_4 & x_5 & z
\end{array}
$$

$$
\left[\begin{array}{cccccc|c}
1 & 1 & 1 & 1 & 0 & 0 & 12 \\
0 & -1 & 0 & -2 & 1 & 0 & 6 \\
\hline
0 & 0 & 1 & 2 & 0 & 1 & 24
\end{array}\right]
\begin{array}{l}
\\
-2R_1 + R_2 \\
2R_1 + R_3
\end{array}
$$

The maximum is 24 when

$x_1 = 12,\, x_2 = 0,\, x_3 = 0,\, x_4 = 0,\, x_5 = 6.$

b. Pivot on the 1 in row one, column two.

$$
\begin{array}{cccccc}
x_1 & x_2 & x_3 & x_4 & x_5 & z
\end{array}
$$

$$
\left[\begin{array}{cccccc|c}
1 & 1 & 1 & 1 & 0 & 0 & 12 \\
1 & 0 & 1 & -1 & 1 & 0 & 18 \\
\hline
0 & 0 & 1 & 2 & 0 & 1 & 24
\end{array}\right]
\begin{array}{l}
\\
-1R_1 + R_2 \\
2R_1 + R_3
\end{array}
$$

The maximum is 24 when

$x_1 = 0,\, x_2 = 12,\, x_3 = 0,\, x_4 = 0,\, x_5 = 18.$

c. This problem has a unique maximum value of z, which is 24, but it occurs at two different basic feasible solutions.

Section 7.5 Maximization Applications

1. Let x_1 = the number of Siamese cats and

$\qquad x_2$ = the number of Persian cats.

The problem is to maximize $z = 12x_1 + 10x_2$ subject to:

$\qquad 2x_1 + x_2 \le 90$

$\qquad x_1 + 2x_2 \le 80$

$\qquad x_1 + x_2 \le 50$

$\qquad x_1 \ge 0,\, x_2 \ge 0.$

There are three constraints to be changed into equalities, so introduce three slack variables, $x_3,\, x_4$ and x_5. The problem can now be restated as:

Find $x_1 \ge 0,\, x_2 \ge 0,\, x_3 \ge 0,\, x_4 \ge 0,\, x_5 \ge 0,\,$ such that

$\qquad 2x_1 + x_2 + x_3 \qquad\qquad = 90$

$\qquad x_1 + 2x_2 \qquad + x_4 \qquad = 80$

$\qquad x_1 + x_2 \qquad\qquad + x_5 = 50$

and $z = 12x_1 + 10x_2$ is maximized.

The initial simplex tableau is

$$
\begin{array}{ccccc}
x_1 & x_2 & x_3 & x_4 & x_5
\end{array}
$$

$$
\left[\begin{array}{ccccc|c}
2 & 1 & 1 & 0 & 0 & 90 \\
1 & 2 & 0 & 1 & 0 & 80 \\
1 & 1 & 0 & 0 & 1 & 50 \\
\hline
-12 & -10 & 0 & 0 & 0 & 0
\end{array}\right].
$$

3. Let x_1 = the number of kg of P,

$\qquad x_2$ = the number of kg of Q,

$\qquad x_3$ = the number of kg of R, and

$\qquad x_4$ = the number of kg of S

The constraints are

$\qquad .375x_3 + .625x_4 \le 500$

$\qquad .75x_2 + .50x_3 + .375x_4 \le 600$

$\qquad x_1 + .25x_2 + .125x_3 \qquad \le 300$

$\qquad x_1 \ge 0,\, x_2 \ge 0,\, x_3 \ge 0,\, x_4 \ge 0.$

(Notice the food contents are given in *percent* of nutrient per kilogram.) The objective function to maximize is $z = 90x_1 + 70x_2 + 60x_3 + 50x_4$.

The initial tableau is

$$
\begin{array}{ccccccc}
x_1 & x_2 & x_3 & x_4 & x_5 & x_6 & x_7
\end{array}
$$

$$
\left[\begin{array}{ccccccc|c}
0 & 0 & .375 & .625 & 1 & 0 & 0 & 500 \\
0 & .75 & .5 & .375 & 0 & 1 & 0 & 600 \\
1 & .25 & .125 & 0 & 0 & 0 & 1 & 300 \\
\hline
-90 & -70 & -60 & -50 & 0 & 0 & 0 & 0
\end{array}\right]
$$

5. a. The information is contained in the table.

	Aluminum	Steel	Profit
1 Speed	12	20	$8
3 Speed	21	30	$12
10 Speed	16	40	$24
Amount Available	42,000	91,800	

Let x_1 = number of 1-speed bikes,

x_2 = number of 3-speed bikes, and

x_3 = number of 10-speed bikes.

The problem is to maximize

$z = 8x_1 + 12x_2 + 24x_3$

subject to:

$12x_1 + 21x_2 + 16x_3 \le 42,000$

$20x_1 + 30x_2 + 40x_3 \le 91,800$

$x_1 \ge 0, \ x_2 \ge 0, \ x_3 \ge 0.$

The initial simplex tableau is

$$\begin{array}{ccccc} x_1 & x_2 & x_3 & x_4 & x_5 \end{array}$$

$$\left[\begin{array}{ccccc|c} 12 & 21 & 16 & 1 & 0 & 42,000 \\ 20 & 30 & 40 & 0 & 1 & 91,800 \\ \hline -8 & -12 & -24 & 0 & 0 & 0 \end{array}\right].$$

Pivot on the 40 in row two, column three.

$$\begin{array}{ccccc} x_1 & x_2 & x_3 & x_4 & x_5 \end{array}$$

$$\left[\begin{array}{ccccc|cl} 12 & 21 & 16 & 1 & 0 & 42,000 & \\ \frac{1}{2} & \frac{3}{4} & 1 & 0 & \frac{1}{40} & 2295 & \frac{1}{40}R_2 \\ \hline -8 & -12 & -24 & 0 & 0 & 0 & \end{array}\right]$$

$$\begin{array}{ccccc} x_1 & x_2 & x_3 & x_4 & x_5 \end{array}$$

$$\left[\begin{array}{ccccc|cl} 4 & 9 & 0 & 1 & -\frac{2}{5} & 5280 & -16R_2 + R_1 \\ \frac{1}{2} & \frac{3}{4} & 1 & 0 & \frac{1}{40} & 2295 & \frac{1}{40}R_2 \\ \hline 4 & 6 & 0 & 0 & \frac{3}{5} & 55,080 & 24R_2 + R_3 \end{array}\right]$$

From the final tableau, the maximum is 55,080 when $x_3 = 2295$, $x_4 = 5280$, and $x_1 = x_2 = x_5 = 0$. Thus, the manufacturer should make no 1-speed or 3-speed bicycles, and should make 2295 10-speed bicycles for a maximum profit of $55,080.

b. In the optimal solution, $x_4 = 5280$ and $x_5 = 0$. $x_4 = 5280$ means 5280 units of aluminum should be left unused. $x_5 = 0$ means all of the steel is used.

7. a. Let x_1 = the number of minutes allotted to the senator,

x_2 = the number of minutes allotted to the congresswoman and

x_3 = the number of minutes allotted to the governor.

Maximize $z = 40x_1 + 60x_2 + 50x_3$

Subject to: $x_1 + x_2 + x_3 \le 27$

$x_1 \quad\quad\quad \ge 2x_3$

$x_1 \quad + x_3 \ge 2x_2$

$x_1 \ge 0, \ x_2 \ge 0, \ x_3 \ge 0$

Rewrite problem in standard maximum form.

Maximize $z = 40x_1 + 60x_2 + 50x_3$

$x_1 + x_2 + x_3 \le 27$

$-x_1 \quad\quad + 2x_3 \le 0$

$-x_1 + 2x_2 - x_3 \le 0$

$x_1 \ge 0, \ x_2 \ge 0, \ x_3 \ge 0$

The initial simplex tableau is

$$\begin{array}{cccccc} x_1 & x_2 & x_3 & x_4 & x_5 & x_6 \end{array}$$

$$\left[\begin{array}{cccccc|c} 1 & 1 & 1 & 1 & 0 & 0 & 27 \\ -1 & 0 & 2 & 0 & 1 & 0 & 0 \\ -1 & \underline{2} & -1 & 0 & 0 & 1 & 0 \\ \hline -40 & -60 & -50 & 0 & 0 & 0 & 0 \end{array}\right]$$

Pivot on the 2 in row three, column two.

Multiply row three by $\frac{1}{2}$ and complete the pivot to get

$$\begin{array}{cccccc} x_1 & x_2 & x_3 & x_4 & x_5 & x_6 \end{array}$$

$$\left[\begin{array}{cccccc|cl} \frac{3}{2} & 0 & \frac{3}{2} & 1 & 0 & -\frac{1}{2} & 27 & -R_3 + R_1 \\ -1 & 0 & 2 & 0 & 1 & 0 & 0 & \\ -\frac{1}{2} & 1 & -\frac{1}{2} & 0 & 0 & \frac{1}{2} & 0 & \frac{1}{2}R_3 \\ \hline -70 & 0 & -80 & 0 & 0 & 30 & 0 & 60R_3 + R_4 \end{array}\right]$$

Pivot on the 2 in row two, column three.

Multiply row 2 by $\frac{1}{2}$ and complete the pivot to get

$$\begin{array}{cccccc} x_1 & x_2 & x_3 & x_4 & x_5 & x_6 \end{array}$$

$$\left[\begin{array}{cccccc|cl} \frac{9}{4} & 0 & 0 & 1 & -\frac{3}{4} & -\frac{1}{2} & 27 & -\frac{3}{2}R_2 + R_1 \\ -\frac{1}{2} & 0 & 1 & 0 & \frac{1}{2} & 0 & 0 & \frac{1}{2}R_2 \\ -\frac{3}{4} & 1 & 0 & 0 & \frac{1}{4} & \frac{1}{2} & 0 & \frac{1}{2}R_2 + R_3 \\ \hline -110 & 0 & 0 & 0 & 40 & 30 & 0 & 80R_2 + R_4 \end{array}\right]$$

Pivot on the $\frac{9}{4}$ in row one, column one.

7. Continued

$$
\begin{array}{ccccccc}
x_1 & x_2 & x_3 & x_4 & x_5 & x_6 & \\
\end{array}
$$

$$
\left[\begin{array}{cccccc|c}
1 & 0 & 0 & \frac{4}{9} & -\frac{1}{3} & -\frac{2}{9} & 12 \\
0 & 0 & 1 & \frac{2}{9} & \frac{1}{3} & -\frac{1}{9} & 6 \\
0 & 1 & 0 & \frac{1}{3} & 0 & \frac{1}{3} & 9 \\
\hline
0 & 0 & 0 & \frac{440}{9} & \frac{10}{3} & \frac{50}{9} & 1320
\end{array}\right]
\begin{array}{l}
\frac{4}{9}R_1 \\
\frac{1}{2}R_1 + R_2 \\
\frac{3}{4}R_1 + R_3 \\
110R_1 + R_4
\end{array}
$$

The senator should get 12 minutes of airtime, the congresswoman 9 minutes and the governor 6 minutes for a maximum of 1,320,000 viewers.

b. In the optimal solution, $x_4 = 0$, $x_5 = 0$ and $x_6 = 0$. $x_4 = 0$ means that all of the 27 minutes will be used by the politicians, i.e. there is no slack. $x_5 = 0$ means that the senator has exactly twice as much time as the governor. $x_6 = 0$ means that the senator and governor have exactly twice as much time as the congresswoman.

9. Let x_1, x_2, and x_3 = the number of newspapers, radio, and TV ads respectively.

Maximize $\quad z = 2000x_1 + 1200x_2 + 10{,}000x_3$

subject to:

$$400x_1 + 200x_2 + 1200x_3 \le 8000$$

$$x_1 \le 20,\ x_2 \le 30,\ x_3 \le 6$$

$$x_1 \ge 0,\ x_2 \ge 0,\ x_3 \ge 0.$$

The initial simplex tableau is

$$
\begin{array}{ccccccc}
x_1 & x_2 & x_3 & x_4 & x_5 & x_6 & x_7 \\
\end{array}
$$

$$
\left[\begin{array}{ccccccc|c}
400 & 200 & 1200 & 1 & 0 & 0 & 0 & 8000 \\
1 & 0 & 0 & 0 & 1 & 0 & 0 & 20 \\
0 & 1 & 0 & 0 & 0 & 1 & 0 & 30 \\
0 & 0 & 1 & 0 & 0 & 0 & 1 & 6 \\
\hline
-2000 & -1200 & -10{,}000 & 0 & 0 & 0 & 0 & 0
\end{array}\right]
$$

Pivot on the 1 in row four, column three.

$$
\begin{array}{ccccccc}
x_1 & x_2 & x_3 & x_4 & x_5 & x_6 & x_7 \\
\end{array}
$$

$$
\left[\begin{array}{ccccccc|c}
400 & 200 & 0 & 1 & 0 & 0 & -1200 & 800 \\
1 & 0 & 0 & 0 & 1 & 0 & 0 & 20 \\
0 & 1 & 0 & 0 & 0 & 1 & 0 & 30 \\
0 & 0 & 1 & 0 & 0 & 0 & 1 & 6 \\
\hline
-2000 & -1200 & 0 & 0 & 0 & 0 & 10{,}000 & 60{,}000
\end{array}\right]
\begin{array}{l}
-1200R_4 + R_1 \\
\\
\\
\\
10{,}000R_4 + R_5
\end{array}
$$

Pivot on the 400 in row one, column one.

9. Continued

$$
\begin{array}{ccccccc}
x_1 & x_2 & x_3 & x_4 & x_5 & x_6 & x_7 \\
\end{array}
$$

$$
\left[
\begin{array}{ccccccc|c}
1 & \frac{1}{2} & 0 & \frac{1}{400} & 0 & 0 & -3 & 2 \\
1 & 0 & 0 & 0 & 1 & 0 & 0 & 20 \\
0 & 1 & 0 & 0 & 0 & 1 & 0 & 30 \\
0 & 0 & 1 & 0 & 0 & 0 & 1 & 6 \\
\hline
-2000 & -1200 & 0 & 0 & 0 & 0 & 10{,}000 & 60{,}000 \\
\end{array}
\right]
\begin{array}{l}
\frac{1}{400}R_1 \\
\\
\\
\\
\\
\end{array}
$$

$$
\begin{array}{ccccccc}
x_1 & x_2 & x_3 & x_4 & x_5 & x_6 & x_7 \\
\end{array}
$$

$$
\left[
\begin{array}{ccccccc|c}
1 & \frac{1}{2} & 0 & \frac{1}{400} & 0 & 0 & -3 & 2 \\
0 & -\frac{1}{2} & 0 & -\frac{1}{400} & 1 & 0 & 3 & 18 \\
0 & 1 & 0 & 0 & 0 & 1 & 0 & 30 \\
0 & 0 & 1 & 0 & 0 & 0 & 1 & 6 \\
\hline
0 & -200 & 0 & 5 & 0 & 0 & 4000 & 64{,}000 \\
\end{array}
\right]
\begin{array}{l}
\\
-1R_1 + R_2 \\
\\
\\
2000R_1 + R_5 \\
\end{array}
$$

Pivot on the $\dfrac{1}{2}$ in row one, column two.

$$
\begin{array}{ccccccc}
x_1 & x_2 & x_3 & x_4 & x_5 & x_6 & x_7 \\
\end{array}
$$

$$
\left[
\begin{array}{ccccccc|c}
2 & 1 & 0 & \frac{1}{200} & 0 & 0 & -6 & 4 \\
0 & -\frac{1}{2} & 0 & -\frac{1}{400} & 1 & 0 & 3 & 18 \\
0 & 1 & 0 & 0 & 0 & 1 & 0 & 30 \\
0 & 0 & 1 & 0 & 0 & 0 & 1 & 6 \\
\hline
0 & -200 & 0 & 5 & 0 & 0 & 4000 & 64{,}000 \\
\end{array}
\right]
\begin{array}{l}
2R_1 \\
\\
\\
\\
\\
\end{array}
$$

$$
\begin{array}{ccccccc}
x_1 & x_2 & x_3 & x_4 & x_5 & x_6 & x_7 \\
\end{array}
$$

$$
\left[
\begin{array}{ccccccc|c}
2 & 1 & 0 & \frac{1}{200} & 0 & 0 & -6 & 4 \\
1 & 0 & 0 & 0 & 1 & 0 & 0 & 20 \\
-2 & 0 & 0 & -\frac{1}{200} & 0 & 1 & 6 & 26 \\
0 & 0 & 1 & 0 & 0 & 0 & 1 & 6 \\
\hline
400 & 0 & 0 & 6 & 0 & 0 & 2800 & 64{,}800 \\
\end{array}
\right]
\begin{array}{l}
\\
\frac{1}{2}R_1 + R_2 \\
-1R_1 + R_3 \\
\\
200R_1 + R_5 \\
\end{array}
$$

No newspaper ads, 4 radio ads, and 6 TV ads will give a maximum exposure of 64,800 people.

11. a. Let x_1 = the number of fund-raising parties,

x_2 = the number of mailings, and

x_3 = the number of dinner parties.

Maximize $z = 200,000x_1 + 100,000x_2 + 600,000x_3$

Subject to: $\quad x_1 \quad + x_2 \quad + x_3 \le 25$

$\quad\quad 3000x_1 + 1000x_2 + 12,000x_3 \le 102,000$

$\quad\quad x_1 \ge 0,\ x_2 \ge 0,\ x_3 \ge 0$

The initial simplex tableau is

$$
\begin{array}{ccccc}
x_1 & x_2 & x_3 & x_4 & x_5 \\
\end{array}
$$

$$
\left[\begin{array}{ccccc|c}
1 & 1 & 1 & 1 & 0 & 25 \\
3000 & 1000 & \underline{12,000} & 0 & 1 & 102,000 \\
\hline
-200,000 & -100,000 & -600,000 & 0 & 0 & 0
\end{array}\right]
$$

Pivot on the 12,000, row two, column three.

$$
\begin{array}{ccccc}
x_1 & x_2 & x_3 & x_4 & x_5 \\
\end{array}
$$

$$
\left[\begin{array}{ccccc|c}
\frac{3}{4} & \frac{11}{12} & 0 & 1 & -\frac{1}{12,000} & \frac{33}{2} \\
\frac{1}{4} & \frac{1}{12} & 1 & 0 & \frac{1}{12,000} & \frac{17}{2} \\
\hline
-50,000 & -50,000 & 0 & 0 & 50 & 5,100,000
\end{array}\right]
\begin{array}{l}
-R_2 + R_1 \\
\frac{1}{12,000}R_2 \\
600,000R_2 + R_3
\end{array}
$$

Pivot on $\dfrac{11}{12}$ in row one, column two.

$$
\begin{array}{ccccc}
x_1 & x_2 & x_3 & x_4 & x_5 \\
\end{array}
$$

$$
\left[\begin{array}{ccccc|c}
\frac{9}{11} & 1 & 0 & \frac{12}{11} & -\frac{1}{11,000} & 18 \\
\frac{2}{11} & 0 & 1 & -\frac{1}{11} & \frac{1}{11,000} & 7 \\
\hline
-\frac{100,000}{11} & 0 & 0 & \frac{600,000}{11} & \frac{500}{11} & 6,000,000
\end{array}\right]
\begin{array}{l}
\frac{12}{11}R_1 \\
-\frac{1}{12}R_1 + R_2 \\
50,000R_1 + R_3
\end{array}
$$

Pivot on $\dfrac{9}{11}$ in row one, column one.

$$
\begin{array}{ccccc}
x_1 & x_2 & x_3 & x_4 & x_5 \\
\end{array}
$$

$$
\left[\begin{array}{ccccc|c}
1 & \frac{11}{9} & 0 & \frac{4}{3} & -\frac{1}{9000} & 22 \\
0 & -\frac{2}{9} & 1 & -\frac{1}{3} & \frac{1}{9000} & 3 \\
\hline
0 & \frac{100,000}{9} & 0 & \frac{200,000}{3} & \frac{400}{9} & 6,200,000
\end{array}\right]
\begin{array}{l}
\frac{11}{9}R_1 \\
-\frac{2}{11}R_1 + R_2 \\
\frac{100,000}{11}R_1 + R_3
\end{array}
$$

The party should plan 22 fund raising parties and 3 dinner parties and no mailings to raise a maximum of $6,200,000.

b. Answer will vary.

13. Using the data given, the problem should be stated as follows:

Maximize $\quad 5x_1 + 4x_2 + 3x_3$

subject to:

$2x_1 + 3x_2 + \ x_3 \leq 400$

$4x_1 + 2x_2 + 3x_3 \leq 600$

$x_1 \geq 0, \ x_2 \geq 0, \ x_3 \geq 0.$

when x_1 = the number of type A lamps,

$\qquad x_2$ = the number of type B lamps, and

$\qquad x_3$ = the number of type C lamps.

a. (3) $\quad$ 5, 4, 3

b. (4) $\quad$ 400, 600

c. (3) $\quad 2x_1 + 3x_2 + 1x_3 \leq 400$

15. From Exercise 1, the initial simplex tableau is

$$
\begin{array}{ccccc}
x_1 & x_2 & x_3 & x_4 & x_5 \\
\end{array}
$$

$$
\left[\begin{array}{ccccc|c}
2 & 1 & 1 & 0 & 0 & 90 \\
1 & 2 & 0 & 1 & 0 & 80 \\
1 & 1 & 0 & 0 & 1 & 50 \\
\hline
-12 & -10 & 0 & 0 & 0 & 0
\end{array}\right].
$$

Pivot on the 2 in row one, column one.

$$
\begin{array}{ccccc}
x_1 & x_2 & x_3 & x_4 & x_5 \\
\end{array}
$$

$$
\left[\begin{array}{ccccc|c}
1 & \frac{1}{2} & \frac{1}{2} & 0 & 0 & 45 \\
1 & 2 & 0 & 1 & 0 & 80 \\
1 & 1 & 0 & 0 & 1 & 50 \\
\hline
-12 & -10 & 0 & 0 & 0 & 0
\end{array}\right]
\begin{array}{l}
\frac{1}{2}R_1 \\ \\ \\ \\
\end{array}
$$

$$
\begin{array}{ccccc}
x_1 & x_2 & x_3 & x_4 & x_5 \\
\end{array}
$$

$$
\left[\begin{array}{ccccc|c}
1 & \frac{1}{2} & \frac{1}{2} & 0 & 0 & 45 \\
0 & \frac{3}{2} & -\frac{1}{2} & 1 & 0 & 35 \\
0 & \frac{1}{2} & -\frac{1}{2} & 0 & 1 & 5 \\
\hline
0 & -4 & 6 & 0 & 0 & 540
\end{array}\right]
\begin{array}{l}
\\
-1R_1 + R_2 \\
-1R_1 + R_3 \\
12R_1 + R_4
\end{array}
$$

Pivot on the $\dfrac{1}{2}$ in row three, column two.

$$
\begin{array}{ccccc}
x_1 & x_2 & x_3 & x_4 & x_5 \\
\end{array}
$$

$$
\left[\begin{array}{ccccc|c}
1 & \frac{1}{2} & \frac{1}{2} & 0 & 0 & 45 \\
0 & \frac{3}{2} & -\frac{1}{2} & 1 & 0 & 35 \\
0 & 1 & -1 & 0 & 2 & 10 \\
\hline
0 & -4 & 6 & 0 & 0 & 540
\end{array}\right]
\begin{array}{l}
\\ \\
2R_3 \\ \\
\end{array}
$$

$$
\begin{array}{ccccc}
x_1 & x_2 & x_3 & x_4 & x_5 \\
\end{array}
$$

$$
\left[\begin{array}{ccccc|c}
1 & 0 & 1 & 0 & -1 & 40 \\
0 & 0 & 1 & 1 & -3 & 20 \\
0 & 1 & -1 & 0 & 2 & 10 \\
\hline
0 & 0 & 2 & 0 & 8 & 580
\end{array}\right]
\begin{array}{l}
-\frac{1}{2}R_3 + R_1 \\
-\frac{3}{2}R_3 + R_2 \\ \\
4R_3 + R_4
\end{array}
$$

The breeder should raise 40 Siamese and 10 Persian cats for a maximum gross income of $580.

17. The initial simplex tableau is

$$
\begin{array}{ccccccc}
x_1 & x_2 & x_3 & x_4 & x_5 & x_6 & x_7
\end{array}
$$

$$
\left[\begin{array}{ccccccc|c}
0 & 0 & .375 & .625 & 1 & 0 & 0 & 500 \\
0 & .75 & .5 & .375 & 0 & 1 & 0 & 600 \\
1 & .25 & .125 & 0 & 0 & 0 & 1 & 300 \\
\hline
-90 & -70 & -60 & -50 & 0 & 0 & 0 & 0
\end{array}\right]
$$

Use a computer program for the simplex method to produce the solution:
Maximum total growth value is 87,454.55 when 163.6 kg of food P, none of food Q, 1090.9 kg of food R, and 145.5 kg of food S are used.

Section 7.6 The Simplex Method: Duality and Minimization

1. The transpose of

$$
\begin{bmatrix}
3 & -4 & 5 \\
1 & 10 & 7 \\
0 & 3 & 6
\end{bmatrix}
$$

is

$$
\begin{bmatrix}
3 & 1 & 0 \\
-4 & 10 & 3 \\
5 & 7 & 6
\end{bmatrix}.
$$

3. The transpose of

$$
\begin{bmatrix}
3 & 0 & 14 & -5 & 3 \\
4 & 17 & 8 & -6 & 1
\end{bmatrix}
$$

is

$$
\begin{bmatrix}
3 & 4 \\
0 & 17 \\
14 & 8 \\
-5 & -6 \\
3 & 1
\end{bmatrix}.
$$

5. Minimize $\quad w = 3y_1 + 5y_2$

subject to: $\quad 3y_1 + y_2 \geq 4$

$\qquad\qquad\quad -y_1 + 2y_2 \geq 6$

$\qquad\qquad\quad y_1 \geq 0, \ y_2 \geq 0.$

The augmented matrix is

$$
\left[\begin{array}{cc|c}
3 & 1 & 4 \\
-1 & 2 & 6 \\
\hline
3 & 5 & 0
\end{array}\right].
$$

The transpose of this matrix is

$$
\left[\begin{array}{cc|c}
3 & -1 & 3 \\
1 & 2 & 5 \\
\hline
4 & 6 & 0
\end{array}\right].
$$

The entries in this second matrix can be used to write the following dual maximization problem:

Maximize $\quad z = 4x_1 + 6x_2$

subject to: $\quad 3x_1 - 2x_2 \leq 3$

$\qquad\qquad\quad x_1 + 2x_2 \leq 5$

$\qquad\qquad\quad x_1 \geq 0, \ x_2 \geq 0.$

7. Minimize $\quad w = 2y_1 + 8y_2$

subject to: $\quad y_1 + 7y_2 \geq 18$

$\qquad\qquad\quad 4y_1 + y_2 \geq 15$

$\qquad\qquad\quad 5y_1 + 3y_2 \geq 20$

$\qquad\qquad\quad y_1 \geq 0, \ y_2 \geq 0.$

The augmented matrix is

$$
\left[\begin{array}{cc|c}
1 & 7 & 18 \\
4 & 1 & 15 \\
5 & 3 & 20 \\
\hline
2 & 8 & 0
\end{array}\right].
$$

The transpose of this matrix is

$$
\left[\begin{array}{ccc|c}
1 & 4 & 5 & 2 \\
7 & 1 & 3 & 8 \\
\hline
18 & 15 & 20 & 0
\end{array}\right].
$$

The entries of this second matrix can be used to write the following dual maximization problem:

Maximize $\quad z = 18x_1 + 15x_2 + 20x_3$

subject to: $\qquad x_1 + 4x_2 + 5x_3 \leq 2$

$\qquad\qquad\quad 7x_1 + x_2 + 3x_3 \leq 8$

$\qquad\qquad\quad x_1 \geq 0, \ x_2 \geq 0, \ x_3 \geq 0.$

9. Minimize $w = y_1 + 2y_2 + 6y_3$

subject to: $3y_1 + 4y_2 + 6y_3 \geq 8$

$y_1 + 5y_2 + 2y_3 \geq 12$

$y_1 \geq 0,\ y_2 \geq 0,\ y_3 \geq 0.$

The augmented matrix is

$$\begin{bmatrix} 3 & 4 & 6 & 8 \\ 1 & 5 & 2 & 12 \\ \hline 1 & 2 & 6 & 0 \end{bmatrix}.$$

The transpose of this matrix is

$$\begin{bmatrix} 3 & 1 & 1 \\ 4 & 5 & 2 \\ 6 & 2 & 6 \\ \hline 8 & 12 & 0 \end{bmatrix}.$$

The entries of this second matrix can be used to write the following dual maximization problem:

Maximize $z = 8x_1 + 12x_2$

subject to: $3x_1 + x_2 \leq 1$

$4x_1 + 5x_2 \leq 2$

$6x_1 + 2x_2 \leq 6$

$x_1 \geq 0,\ x_2 \geq 0.$

11. Maximize $w = 8y_1 + 9y_2 + 3y_3$

subject to: $y_1 + y_2 + y_3 \geq 5$

$y_1 + y_2 \qquad \geq 4$

$2y_1 + y_2 + 3y_3 \geq 15$

$y_1 \geq 0,\ y_2 \geq 0,\ y_3 \geq 0.$

The augmented matrix is

$$\begin{bmatrix} 1 & 1 & 1 & 5 \\ 1 & 1 & 0 & 4 \\ 2 & 1 & 3 & 15 \\ \hline 8 & 9 & 3 & 0 \end{bmatrix}.$$

The transpose of this matrix is

$$\begin{bmatrix} 1 & 1 & 2 & 8 \\ 1 & 1 & 1 & 9 \\ 1 & 0 & 3 & 3 \\ \hline 5 & 4 & 15 & 0 \end{bmatrix}.$$

The entries of this second matrix can be used to write the following dual maximization problem:

Maximize $z = 5x_1 + 4x_2 + 15x_3$

subject to: $x_1 + x_2 + 2x_3 \leq 8$

$x_1 + x_2 + x_3 \leq 9$

$x_1 \qquad + 3x_3 \leq 3$

$x_1 \geq 0,\ x_2 \geq 0,\ x_3 \geq 0.$

13. Minimize $w = 2y_1 + y_2 + 3y_3$

subject to: $y_1 + y_2 + y_3 \geq 100$

$2y_1 + y_2 \qquad \geq 50$

$y_1 \geq 0,\ y_2 \geq 0,\ y_3 \geq 0.$

The augmented matrix is

$$\begin{bmatrix} 1 & 1 & 1 & 100 \\ 2 & 1 & 0 & 50 \\ \hline 2 & 1 & 3 & 0 \end{bmatrix}.$$

The transpose is

$$\begin{bmatrix} 1 & 2 & 2 \\ 1 & 1 & 1 \\ 1 & 0 & 3 \\ \hline 100 & 50 & 0 \end{bmatrix}.$$

The dual maximization problem is:

Maximize $z = 100x_1 + 50x_2$

subject to: $x_1 + 2x_2 \leq 2$

$x_1 + x_2 \leq 1$

$x_1 \qquad \leq 3$

$x_1 \geq 0,\ x_2 \geq 0.$

The initial simplex tableau is

$\quad x_1 \quad x_2 \ \ x_3 \ x_4 \ x_5$

$$\begin{bmatrix} 1 & 2 & 1 & 0 & 0 & 2 \\ 1 & 1 & 0 & 1 & 0 & 1 \\ 1 & 0 & 0 & 0 & 1 & 3 \\ \hline -100 & -50 & 0 & 0 & 0 & 0 \end{bmatrix}.$$

Pivot on the 1 in row two, column one.

$\quad x_1 \ \ x_2 \ x_3 \quad x_4 \ \ x_5$

$$\begin{bmatrix} 0 & 1 & 1 & -1 & 0 & 1 \\ 1 & 1 & 0 & 1 & 0 & 1 \\ 0 & -1 & 0 & -1 & 1 & 2 \\ \hline 0 & 50 & 0 & 100 & 0 & 100 \end{bmatrix} \begin{matrix} -1R_2 + R_1 \\ \\ -R_2 + R_3 \\ 100R_2 + R_4 \end{matrix}$$

The solution to the minimization problem is found in the bottom row of the final matrix in the entries corresponding to the slack variables.

The minimum is 100 when

$y_1 = 0,\ y_2 = 100,\ \text{and}\ y_3 = 0.$

15. Minimize $w = 3y_1 + y_2 + 4y_3$

subject to: $2y_1 + y_2 + y_3 \geq 6$

$y_1 + 2y_2 + y_3 \geq 8$

$2y_1 + y_2 + 2y_3 \geq 12$

$y_1 \geq 0,\ y_2 \geq 0,\ y_3 \geq 0$

The augmented matrix is

$$\begin{bmatrix} 2 & 1 & 1 & 6 \\ 1 & 2 & 1 & 8 \\ 2 & 1 & 2 & 12 \\ \hline 3 & 1 & 4 & 0 \end{bmatrix}.$$

The transpose is

$$\begin{bmatrix} 2 & 1 & 2 & 3 \\ 1 & 2 & 1 & 1 \\ 1 & 1 & 2 & 4 \\ \hline 6 & 8 & 12 & 0 \end{bmatrix}.$$

The dual maximization problem is

Maximize $z = 6x_1 + 8x_2 + 12x_3$

subject to: $2x_1 + x_2 + 2x_3 \leq 3$

$x_1 + 2x_2 + x_3 \leq 1$

$x_1 + x_2 + 2x_3 \leq 4$

$x_1 \geq 0,\ x_2 \geq 0,\ x_3 \geq 0.$

The initial tableau is

$\quad x_1 \ \ x_2 \quad x_3 \ x_4 \ x_5 \ x_6$

$$\begin{bmatrix} 2 & 1 & 2 & 1 & 0 & 0 & 3 \\ 1 & 2 & 1 & 0 & 1 & 0 & 1 \\ 1 & 1 & 2 & 0 & 0 & 1 & 4 \\ \hline -6 & -8 & -12 & 0 & 0 & 0 & 0 \end{bmatrix}.$$

Pivot on the 1 in row two, column three.

$\quad x_1 \ \ x_2 \ x_3 \ x_4 \ \ x_5 \ x_6$

$$\begin{bmatrix} 0 & -3 & 0 & 1 & -2 & 0 & 1 \\ 1 & 2 & 1 & 0 & 1 & 0 & 1 \\ -1 & -3 & 0 & 0 & -2 & 1 & 2 \\ \hline 6 & 16 & 0 & 0 & 12 & 0 & 12 \end{bmatrix} \begin{matrix} -2R_2 + R_1 \\ \\ -2R_2 + R_3 \\ 12R_2 + R_4 \end{matrix}$$

The minimum is 12 when $y_1 = 0,\ y_2 = 12,\ y_3 = 0.$

17. Minimize $\quad w = 6y_1 + 4y_2 + 2y_3$

subject to: $\quad 2y_1 + 2y_2 + \ y_3 \geq 2$

$\qquad\qquad\quad y_1 + 3y_2 + 2y_3 \geq 3$

$\qquad\qquad\quad y_1 + \ y_2 + 2y_3 \geq 4$

$\qquad y_1 \geq 0, \ y_2 \geq 0, \ y_3 \geq 0.$

The augmented matrix is

$$\begin{bmatrix} 2 & 2 & 1 & 2 \\ 1 & 3 & 2 & 3 \\ 1 & 1 & 2 & 4 \\ \hline 6 & 4 & 2 & 0 \end{bmatrix}.$$

The transpose is

$$\begin{bmatrix} 2 & 1 & 1 & 6 \\ 2 & 3 & 1 & 4 \\ 1 & 2 & 2 & 2 \\ \hline 2 & 3 & 4 & 0 \end{bmatrix}.$$

The dual maximization problem is

Maximize $\quad z = 2x_1 + 3x_2 + 4x_3$

subject to: $\quad 2x_1 + \ x_2 + \ x_3 \leq 6$

$\qquad\qquad\quad 2x_1 + 3x_2 + \ x_3 \leq 4$

$\qquad\qquad\quad x_1 + 2x_2 + 2x_3 \leq 2$

$\qquad x_1 \geq 0, \ x_2 \geq 0, \ x_3 \geq 0.$

The initial simplex tableau is

$x_1 \quad x_2 \quad x_3 \ x_4 \ x_5 \ x_6$

$$\begin{bmatrix} 2 & 1 & 1 & 1 & 0 & 0 & 6 \\ 2 & 3 & 1 & 0 & 1 & 0 & 4 \\ 1 & 2 & \underline{2} & 0 & 0 & 1 & 2 \\ \hline -2 & -3 & -4 & 0 & 0 & 0 & 0 \end{bmatrix}.$$

Pivot on the 2 in row three, column three.

$x_1 \quad x_2 \quad x_3 \ x_4 \ x_5 \ x_6$

$$\begin{bmatrix} 2 & 1 & 1 & 1 & 0 & 0 & 6 \\ 2 & 3 & 1 & 0 & 1 & 0 & 4 \\ \frac{1}{2} & 1 & 1 & 0 & 0 & \frac{1}{2} & 1 \\ \hline -2 & -3 & -4 & 0 & 0 & 0 & 0 \end{bmatrix} \frac{1}{2}R_3$$

$x_1 \ x_2 \ x_3 \ x_4 \ x_5 \quad x_6$

$$\begin{bmatrix} \frac{3}{2} & 0 & 0 & 1 & 0 & -\frac{1}{2} & 5 \\ \frac{3}{2} & 2 & 0 & 0 & 1 & -\frac{1}{2} & 3 \\ \frac{1}{2} & 1 & 1 & 0 & 0 & \frac{1}{2} & 1 \\ \hline 0 & 1 & 0 & 0 & 0 & 2 & 4 \end{bmatrix} \begin{matrix} -R_3+R_1 \\ -R_3+R_2 \\ \\ 4R_3+R_4 \end{matrix}$$

The minimum is 4 when $y_1 = 0$, $y_2 = 0$, and $y_3 = 2$.

19. Minimize $\quad w = 20y_1 + 12y_2 + 40y_3$

subject to: $\qquad y_1 + y_2 + 5y_3 \geq 20$

$\qquad\qquad\quad 2y_1 + y_2 + \ y_3 \geq 30$

$\qquad\quad y_1 \geq 0, \ y_2 \geq 0, \ y_3 \geq 0.$

The augmented matrix is

$$\begin{bmatrix} 1 & 1 & 5 & 20 \\ 2 & 1 & 1 & 30 \\ \hline 20 & 12 & 40 & 0 \end{bmatrix}.$$

The transpose is

$$\begin{bmatrix} 1 & 2 & 20 \\ 1 & 1 & 12 \\ 5 & 1 & 40 \\ \hline 20 & 30 & 0 \end{bmatrix}.$$

The dual maximization problem is

Maximize $\quad z = 20x_1 + 30x_2$

subject to: $\quad x_1 + 2x_2 \leq 20$

$\qquad\qquad\quad x_1 + \ x_2 \leq 12$

$\qquad\qquad\quad 5x_1 + \ x_2 \leq 40$

$\qquad\quad x_1 \geq 0, \ x_2 \geq 0.$

The initial simplex tableau is

$x_1 \quad x_2 \ x_3 \ x_4 \ x_5$

$$\begin{bmatrix} 1 & \underline{2} & 1 & 0 & 0 & 20 \\ 1 & 1 & 0 & 1 & 0 & 12 \\ 5 & 1 & 0 & 0 & 1 & 40 \\ \hline -20 & -30 & 0 & 0 & 0 & 0 \end{bmatrix}.$$

Pivot on the 2 in row one, column two.

$x_1 \quad x_2 \ x_3 \ x_4 \ x_5$

$$\begin{bmatrix} \frac{1}{2} & 1 & \frac{1}{2} & 0 & 0 & 10 \\ 1 & 1 & 0 & 1 & 0 & 12 \\ 5 & 1 & 0 & 0 & 1 & 40 \\ \hline -20 & -30 & 0 & 0 & 0 & 0 \end{bmatrix} \frac{1}{2}R_1$$

$x_1 \ x_2 \quad x_3 \ x_4 \ x_5$

$$\begin{bmatrix} \frac{1}{2} & 1 & \frac{1}{2} & 0 & 0 & 10 \\ \frac{1}{2} & 0 & -\frac{1}{2} & 1 & 0 & 2 \\ \frac{9}{2} & 0 & -\frac{1}{2} & 0 & 1 & 30 \\ \hline -5 & 0 & 15 & 0 & 0 & 300 \end{bmatrix} \begin{matrix} \\ -R_1+R_2 \\ -R_1+R_3 \\ 30R_1+R_4 \end{matrix}$$

Pivot on the $\dfrac{1}{2}$ in row two, column one.

19. Continued

$$x_1 \quad x_2 \quad x_3 \quad x_4 \quad x_5$$

$$\begin{bmatrix} \frac{1}{2} & 1 & \frac{1}{2} & 0 & 0 & 10 \\ 1 & 0 & -1 & 2 & 0 & 4 \\ \frac{9}{2} & 0 & -\frac{1}{2} & 0 & 1 & 30 \\ \hline -5 & 0 & 15 & 0 & 0 & 300 \end{bmatrix} 2R_2$$

$$x_1 \quad x_2 \quad x_3 \quad x_4 \quad x_5$$

$$\begin{bmatrix} 0 & 1 & 1 & -1 & 0 & 8 \\ 1 & 0 & -1 & 2 & 0 & 4 \\ 0 & 0 & 4 & -9 & 1 & 12 \\ \hline 0 & 0 & 10 & 10 & 0 & 320 \end{bmatrix} \begin{matrix} -\frac{1}{2}R_2 + R_1 \\ \\ -\frac{9}{2}R_2 + R_3 \\ 5R_2 + R_4 \end{matrix}$$

The minimum is 320 when $y_1 = 10$, $y_2 = 10$, and $y_3 = 0$.

21. Minimize $\quad w = 4y_1 + 2y_2 + y_3$

subject to: $\quad y_1 + y_2 + \ y_3 \geq 4$

$$3y_1 + y_2 + 3y_3 \geq 6$$

$$y_1 + y_2 + 3y_3 \geq 5$$

$$y_1 \geq 0, \ y_2 \geq 0, \ y_3 \geq 0.$$

The augmented matrix is

$$\begin{bmatrix} 1 & 1 & 1 & 4 \\ 3 & 1 & 3 & 6 \\ 1 & 1 & 3 & 5 \\ \hline 4 & 2 & 1 & 0 \end{bmatrix}.$$

The transpose is

$$\begin{bmatrix} 1 & 3 & 1 & 4 \\ 1 & 1 & 1 & 2 \\ 1 & 3 & 3 & 1 \\ \hline 4 & 6 & 5 & 0 \end{bmatrix}.$$

The dual maximization problem is

Maximize $\quad z = 4x_1 + 6x_2 + 5x_3$

subject to: $\quad x_1 + 3x_2 + \ x_3 \leq 4$

$$x_1 + \ x_2 + \ x_3 \leq 2$$

$$x_1 + 3x_2 + 3x_3 \leq 1$$

$$x_1 \geq 0, \ x_2 > 0, \ x_3 \geq 0.$$

21. Continued

The initial simplex tableau is

$$x_1 \quad x_2 \quad x_3 \ x_4 \ x_5 \ x_6$$

$$\begin{bmatrix} 1 & 3 & 1 & 1 & 0 & 0 & 4 \\ 1 & 1 & 1 & 0 & 1 & 0 & 2 \\ 1 & 3 & 3 & 0 & 0 & 1 & 1 \\ \hline -4 & -6 & -5 & 0 & 0 & 0 & 0 \end{bmatrix}.$$

Pivot on the 3 in row three, column two.

$$x_1 \quad x_2 \quad x_3 \ x_4 \ x_5 \ x_6$$

$$\begin{bmatrix} 1 & 3 & 1 & 1 & 0 & 0 & 4 \\ 1 & 1 & 1 & 0 & 1 & 0 & 2 \\ \frac{1}{3} & 1 & 1 & 0 & 0 & \frac{1}{3} & \frac{1}{3} \\ \hline -4 & -6 & -5 & 0 & 0 & 0 & 0 \end{bmatrix} \frac{1}{3}R_3$$

$$x_1 \ x_2 \ x_3 \ x_4 \ x_5 \quad x_6$$

$$\begin{bmatrix} 0 & 0 & -2 & 1 & 0 & -1 & 3 \\ \frac{2}{3} & 0 & 0 & 0 & 1 & -\frac{1}{3} & \frac{5}{3} \\ \frac{1}{3} & 1 & 1 & 0 & 0 & \frac{1}{3} & \frac{1}{3} \\ \hline -2 & 0 & 1 & 0 & 0 & 2 & 2 \end{bmatrix} \begin{matrix} -3R_3 + R_1 \\ -1R_3 + R_2 \\ \\ 6R_3 + R_4 \end{matrix}$$

Pivot on the $\dfrac{1}{3}$ in row three, column one.

$$x_1 \ x_2 \ x_3 \ x_4 \ x_5 \quad x_6$$

$$\begin{bmatrix} 0 & 0 & -2 & 1 & 0 & -1 & 3 \\ \frac{2}{3} & 0 & 0 & 0 & 1 & -\frac{1}{3} & \frac{5}{3} \\ 1 & 3 & 3 & 0 & 0 & 1 & 1 \\ \hline -2 & 0 & 1 & 0 & 0 & 2 & 2 \end{bmatrix} 3R_3$$

$$x_1 \ x_2 \ x_3 \ x_4 \ x_5 \quad x_6$$

$$\begin{bmatrix} 0 & 0 & -2 & 1 & 0 & -1 & 3 \\ 0 & -2 & -2 & 0 & 1 & -1 & 1 \\ 1 & 3 & 3 & 0 & 0 & 1 & 1 \\ \hline 0 & 6 & 7 & 0 & 0 & 4 & 4 \end{bmatrix} \begin{matrix} \\ -\frac{2}{3}R_3 + R_2 \\ \\ 2R_3 + R_4 \end{matrix}$$

The minimum is 4 when $y_1 = 0$, $y_2 = 0$, and $y_3 = 4$.

23. Let y_1 = the amount of product A;

y_2 = the amount of product B.

The given information can be expressed as the following maximization problem:

Minimize $\quad w = 24y_1 + 40y_2$

subject to: $\quad 4y_1 + 2y_2 \geq 20$

$\qquad\qquad 2y_1 + 5y_2 \geq 18$

$\qquad\qquad y_1 \geq 0,\ y_2 \geq 0.$

The augmented matrix is

$$\begin{bmatrix} 4 & 2 & 20 \\ 2 & 5 & 18 \\ \hline 24 & 40 & 0 \end{bmatrix}.$$

The transpose is

$$\begin{bmatrix} 4 & 2 & 24 \\ 2 & 5 & 40 \\ \hline 20 & 18 & 0 \end{bmatrix}.$$

The dual maximization problem is

Maximize $\quad z = 20x_1 + 18x_2$

subject to: $\quad 4x_1 + 2x_2 \leq 24$

$\qquad\qquad 2x_1 + 5x_2 \leq 40$

$\qquad\qquad x_1 \geq 0,\ x_2 \geq 0.$

The initial simplex tableau is

$$\begin{array}{cccc} x_1 & x_2 & x_3 & x_4 \end{array}$$
$$\begin{bmatrix} \underline{4} & 2 & 1 & 0 & 24 \\ 2 & 5 & 0 & 1 & 40 \\ \hline -20 & -18 & 0 & 0 & 0 \end{bmatrix}.$$

Pivot on the 4 in row one, column one.

$$\begin{array}{ccccc} x_1 & x_2 & x_3 & x_4 & x_5 \end{array}$$
$$\begin{bmatrix} 1 & \frac{1}{2} & \frac{1}{4} & 0 & 6 \\ 2 & 5 & 0 & 1 & 40 \\ \hline -20 & -18 & 0 & 0 & 0 \end{bmatrix} \frac{1}{4}R_1$$

$$\begin{array}{cccc} x_1 & x_2 & x_3 & x_4 \end{array}$$
$$\begin{bmatrix} 1 & \frac{1}{2} & \frac{1}{4} & 0 & 6 \\ 0 & \underline{4} & -\frac{1}{2} & 1 & 28 \\ \hline 0 & -8 & 5 & 0 & 120 \end{bmatrix} \begin{array}{l} -2R_1 + R_2 \\ 20R_1 + R_3 \end{array}$$

Pivot on the 4 in row two, column two.

$$\begin{array}{ccccc} x_1 & x_2 & x_3 & x_4 & x_5 \end{array}$$
$$\begin{bmatrix} 1 & \frac{1}{2} & \frac{1}{4} & 0 & 6 \\ 0 & \underline{1} & -\frac{1}{8} & \frac{1}{4} & 7 \\ \hline 0 & -8 & 5 & 0 & 120 \end{bmatrix} \frac{1}{4}R_2$$

23. Continued

$$\begin{array}{cccc} x_1 & x_2 & x_3 & x_4 \end{array}$$
$$\begin{bmatrix} 1 & 0 & \frac{5}{16} & -\frac{1}{8} & \frac{5}{2} \\ 0 & 1 & -\frac{1}{8} & \frac{1}{4} & 7 \\ \hline 0 & 0 & 4 & 2 & 176 \end{bmatrix} \begin{array}{l} -\frac{1}{2}R_2 + R_1 \\ \\ 8R_2 + R_3 \end{array}$$

Glenn should use 4 servings of product A and 2 servings of product B for a minimum cost of $1.76.

25. Let y_1 = the number of additional units of regular beer to produce

y_2 = the number of additional units of light beer to produce

The sale of 12 units of regular beer and 10 units of light beer already generates

$12 \cdot 100{,}000 + 10 \cdot 300{,}000 = \$4{,}200{,}000$ in revenue. Additional units need only to generate $7{,}000{,}000 - 4{,}200{,}000 = \$2{,}800{,}000$ in revenue.

Minimize $\quad w = 36{,}000y_1 + 48{,}000y_2$

subject to: $\quad 100{,}000y_1 + 300{,}000y_2 \geq 2{,}800{,}000$

$\qquad\qquad\quad y_1 \qquad\quad + y_2 \geq 20$

$\qquad\qquad\quad y_1 \geq 0,\ y_2 \geq 0$

The augmented matrix is

$$\begin{bmatrix} 100{,}000 & 300{,}000 & 2{,}800{,}000 \\ 1 & 1 & 20 \\ \hline 36{,}000 & 48{,}000 & 0 \end{bmatrix}$$

The transpose is

$$\begin{bmatrix} 100{,}000 & 1 & 36{,}000 \\ 300{,}000 & 1 & 48{,}000 \\ \hline 2{,}800{,}000 & 20 & 0 \end{bmatrix}$$

The dual maximization problem is

Maximize $\quad z = 2{,}800{,}000x_1 + 20x_2$

subject to: $\quad 100{,}000x_1 + x_2 \leq 36{,}000$

$\qquad\qquad\quad 300{,}000x_1 + x_2 \leq 48{,}000$

$x_1 \geq 0,\ x_2 \geq 0$

The initial simplex tableau is

$$\begin{array}{cccc} x_1 & x_2 & x_3 & x_4 \end{array}$$
$$\begin{bmatrix} 100{,}000 & 1 & 1 & 0 & 36{,}000 \\ 300{,}000 & 1 & 0 & 1 & 48{,}000 \\ \hline -2{,}800{,}000 & -20 & 0 & 0 & 0 \end{bmatrix}$$

25. Continued

Pivot on the 300,000 in row two, column one.

$$
\begin{array}{cccc}
x_1 & x_2 & x_3 & x_4
\end{array}
$$

$$
\left[\begin{array}{cccc|c}
0 & \frac{2}{3} & 1 & -\frac{1}{3} & 20,000 \\
1 & \frac{1}{300,000} & 0 & \frac{1}{300,000} & \frac{16}{100} \\
0 & -\frac{32}{3} & 0 & \frac{28}{3} & 448,000
\end{array}\right]
\begin{array}{l}
-100,000R_2 + R_1 \\
\frac{1}{300,000}R_2 \\
2,800,000R_2 + R_3
\end{array}
$$

Pivot on the $\dfrac{2}{3}$ in row one, column two.

$$
\left[\begin{array}{cccc|c}
0 & 1 & \frac{3}{2} & -\frac{1}{2} & 30,000 \\
1 & 0 & -\frac{1}{200,000} & \frac{1}{200,000} & \frac{6}{100} \\
0 & 0 & 16 & 4 & 768,000
\end{array}\right]
\begin{array}{l}
\frac{3}{2}R_1 \\
-\frac{1}{300,000}R_1 + R_2 \\
\frac{32}{3}R_1 + R_3
\end{array}
$$

The total cost is the cost of the original 12 and 10 units of beer plus the cost of the additional units of beer
$768,000 + 12 \cdot 36,000 + 10 \cdot 48,000 = \$1,680,000$
The brewery should make 16 additional units of regular beer and 4 additional units of light beer for a total of 28 units of regular beer and 14 units of light beer at a minimum cost of $1,680,000.

27. The linear program (P)
Minimize $z = x_1 + 2x_2$
$$-2x_1 + x_2 \geq 1$$
$$x_1 - 2x_2 \geq 1$$
$$x_1 \geq 0, \ x_2 \geq 0$$

has no feasible solution. Graphing the constraints of (P) shows that the constraints do not form a feasible space. To determine the dual of (P), (D), we have the augmented matrix

$$
\left[\begin{array}{cc|c}
-2 & 1 & 1 \\
1 & -2 & 1 \\
\hline
1 & 2 & 0
\end{array}\right]
$$

the transpose

$$
\left[\begin{array}{cc|c}
-2 & 1 & 1 \\
1 & -2 & 2 \\
\hline
1 & 1 & 0
\end{array}\right]
$$

and the dual maximization problem
Maximize $w = y_1 + y_2$
subject to: $-2y_1 + y_2 \leq 1$
$$y_1 - 2y_2 \leq 2$$
$$y_1 \geq 0, \ y_2 \geq 0$$

27. Continued

Graphing the constraints of (D) does show a feasible space but the objective function is unbounded.

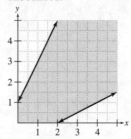

The answer is (a).

29. Let x_1 = the number of toy bears and
$\quad\quad x_2$ = the number of monkeys.
Maximize $\quad z = x_1 + 1.5x_2$
subject to: $\quad\quad x_1 + 2x_2 \leq 200$
$$4x_1 + 3x_2 \leq 600$$
$$x_2 \leq 90$$
$$x_1 \geq 0, \ x_2 \geq 0.$$

a. The augmented matrix is
$$
\left[\begin{array}{cc|c}
1 & 2 & 200 \\
4 & 3 & 600 \\
0 & 1 & 90 \\
\hline
1 & 1.5 & 0
\end{array}\right].
$$
The transpose is
$$
\left[\begin{array}{ccc|c}
1 & 4 & 0 & 1 \\
2 & 3 & 1 & 1.5 \\
\hline
200 & 600 & 90 & 0
\end{array}\right].
$$
The dual minimization problem is
Minimize $\quad w = 200y_1 + 600y_2 + 90y_3$
subject to: $\quad\quad y_1 + 4y_2 \quad\quad \geq 1$
$$2y_1 + 3y_2 + y_3 \geq 1.5$$
$$y_1 \geq 0, \ y_2 \geq 0, \ y_3 \geq 0.$$

b. The solution to the dual minimization problem is found in the last row of the final simplex. tableau of the original problem. The minimum is 180 when $y_1 = .6$, $y_2 = .1$, and
$y_3 = 0$.

29. Continued

c. The shadow cost for felt is in row four, column three of the final matrix, $.60$/unit. So, an increase of 10 squares of felt would increase the profit by $6. The profit would be $186.

d. The shadow cost for stuffing is $.10$/unit; the shadow cost for trim is 0/unit. So, a decrease of 10 oz of stuffing and 10 ft of trim would decrease the profit by $1. The profit would be $179.

Section 7.7 The Simplex Method: Nonstandard Problems

1. Maximize $z = 5x_1 + 2x_2 - x_3$

subject to
$$2x_1 + 3x_2 + 5x_3 \geq 8$$
$$4x_1 - x_2 + 3x_3 \leq 7$$
$$x_1 \geq 0,\ x_2 \geq 0,\ x_3 \geq 0.$$

a. Insert a surplus variable in the first constraint and slack variables in the last two constraints. The problem then becomes:

Maximize $z = 5x_1 + 2x_2 - x_3$

subject to:
$$2x_1 + 3x_2 + 5x_3 - x_4 \qquad = 8$$
$$4x_1 - x_2 + 3x_3 \qquad + x_5 = 7$$
$$x_1 \geq 0,\ x_2 \geq 0,\ x_3 \geq 0,\ x_4 \geq 0,\ x_5 \geq 0.$$

b. The initial simplex tableau is

$$
\begin{array}{ccccc}
x_1 & x_2 & x_3 & x_4 & x_5 \\
\end{array}
$$
$$
\left[\begin{array}{ccccc|c}
2 & 3 & 5 & -1 & 0 & 8 \\
4 & -1 & 3 & 0 & 1 & 7 \\
\hline
-5 & -2 & 1 & 0 & 0 & 0 \\
\end{array} \right].
$$

3. Maximize $z = 2x_1 - 3x_2 + 4x_3$

subject to:
$$x_1 + x_2 + x_3 \leq 100$$
$$x_1 + x_2 + x_3 \geq 75$$
$$x_1 + x_2 \qquad \geq 27$$
$$x_1 \geq 0,\ x_2 \geq 0,\ x_3 \geq 0.$$

a. Insert a slack variable in the first constraints and surplus variables in the last two constraints. The problem then becomes:

Maximize $z = 2x_1 - 3x_2 + 4x_3$

subject to:
$$x_1 + x_2 + x_3 + x_4 \qquad\qquad = 100$$
$$x_1 + x_2 + x_3 \qquad - x_5 \qquad = 75$$
$$x_1 + x_2 \qquad\qquad\qquad - x_6 = 27$$
$$x_1 \geq 0,\ x_2 \geq 0,\ x_3 \geq 0, x_4 \geq 0,\ x_5 \geq 0,\ x_6 \geq 0.$$

b. The initial simplex tableau is

$$
\begin{array}{cccccc}
x_1 & x_2 & x_3 & x_4 & x_5 & x_6 \\
\end{array}
$$
$$
\left[\begin{array}{cccccc|c}
1 & 1 & 1 & 1 & 0 & 0 & 100 \\
1 & 1 & 1 & 0 & -1 & 0 & 75 \\
1 & 1 & 0 & 0 & 0 & -1 & 27 \\
\hline
-2 & 3 & -4 & 0 & 0 & 0 & 0 \\
\end{array} \right].
$$

5. Minimize $w = 2y_1 + 5y_2 - 3y_3$

subject to:
$$y_1 + 2y_2 + 3y_3 \geq 115$$
$$2y_1 + y_2 + y_3 \leq 200$$
$$y_1 + y_3 \geq 50$$
$$y_1 \geq 0,\ y_2 \geq 0,\ y_3 \geq 0.$$

Rewrite the objective function to get a maximization problem.

Maximize $z = -2y_1 - 5y_2 + 3y_3$

subject to:
$$y_1 + 2y_2 + 3y_3 \geq 115$$
$$2y_1 + y_2 + y_3 \leq 200$$
$$y_1 + y_3 \geq 50$$
$$y_1 \geq 0,\ y_2 \geq 0,\ y_3 \geq 0.$$

Insert surplus variables in the first and third constraints and a slack variable in the second to get the initial simplex tableau.

$$
\begin{array}{cccccc}
y_1 & y_2 & y_3 & y_4 & y_5 & y_6 \\
\end{array}
$$
$$
\left[\begin{array}{cccccc|c}
1 & 2 & 3 & -1 & 0 & 0 & 115 \\
2 & 1 & 1 & 0 & 1 & 0 & 200 \\
1 & 0 & 1 & 0 & 0 & -1 & 50 \\
\hline
2 & 5 & -3 & 0 & 0 & 0 & 0 \\
\end{array} \right]
$$

7. Minimize $\quad w = y_1 - 4y_2 + 2y_3$

subject to: $\quad -7y_1 + 6y_2 - 8y_3 \leq -18$

$$4y_1 + 5y_2 + 10y_3 \geq 20$$

$$y_1 \geq 0,\ y_2 \geq 0,\ y_3 \geq 0.$$

Rewrite the objective function and multiply the first constraint by -1 to get a maximization problem.

Maximize $\quad z = -y_1 + 4y_2 - 2y_3$

subject to: $\quad 7y_1 - 6y_2 + 8y_3 \geq 18$

$$4y_1 + 5y_2 + 10y_3 \geq 20$$

$$y_1 \geq 0,\ y_2 \geq 0,\ y_3 \geq 0.$$

Insert surplus variables in both constraints to get the initial simplex tableau.

$$
\begin{array}{ccccc}
y_1 & y_2 & y_3 & y_4 & y_5
\end{array}
$$

$$
\left[
\begin{array}{ccccc|c}
7 & -6 & 8 & -1 & 0 & 18 \\
4 & 5 & 10 & 0 & -1 & 20 \\
\hline
1 & -4 & 2 & 0 & 0 & 0
\end{array}
\right]
$$

9. Maximize $\quad z = 12x_1 + 10x_2$

subject to: $\quad x_1 + 2x_2 \geq 24$

$$x_1 + x_2 \leq 40$$

$$x_1 \geq 0,\ x_2 \geq 0.$$

The initial simplex tableau is

$$
\begin{array}{cccc}
x_1 & x_2 & x_3 & x_4
\end{array}
$$

$$
\left[
\begin{array}{cccc|c}
1 & 2 & -1 & 0 & 24 \\
1 & 1 & 0 & 1 & 40 \\
\hline
-12 & -10 & 0 & 0 & 0
\end{array}
\right]
$$

For Stage I pivoting, pivot on the 1 in row one, column one.

$$
\begin{array}{cccc}
x_1 & x_2 & x_3 & x_4
\end{array}
$$

$$
\left[
\begin{array}{cccc|c}
1 & 2 & -1 & 0 & 24 \\
0 & -1 & 1 & 1 & 16 \\
\hline
0 & 14 & -12 & 0 & 288
\end{array}
\right]
\begin{array}{l}
-R_1 + R_2 \\
12R_2 + R_3
\end{array}
$$

For Stage II pivoting, pivot on the 1 in row two, column three.

$$
\begin{array}{cccc}
x_1 & x_2 & x_3 & x_4
\end{array}
$$

$$
\left[
\begin{array}{cccc|c}
1 & 1 & 0 & 1 & 40 \\
0 & -1 & 1 & 1 & 16 \\
\hline
0 & 2 & 0 & 12 & 480
\end{array}
\right]
\begin{array}{l}
R_2 + R_1 \\
12R_1 + R_3
\end{array}
$$

The maximum is 480 when $x_1 = 40$ and $x_2 = 0$.

11. Find $x_1 \geq 0,\ x_2 \geq 0$, and $x_3 \geq 0$ such that

$$x_1 + x_2 + 2x_3 \leq 38$$

$$2x_1 + x_2 + x_3 \geq 24$$

and

$$z = 3x_1 + 2x_2 + 2x_3 \text{ is maximized.}$$

The initial simplex tableau is

$$
\begin{array}{ccccc}
x_1 & x_2 & x_3 & x_4 & x_5
\end{array}
$$

$$
\left[
\begin{array}{ccccc|c}
1 & 1 & 2 & 1 & 0 & 38 \\
2 & 1 & 1 & 0 & -1 & 24 \\
\hline
-3 & -2 & -2 & 0 & 0 & 0
\end{array}
\right]
$$

For Stage I pivoting, pivot on the 2 in row two, column one.

$$
\begin{array}{ccccc}
x_1 & x_2 & x_3 & x_4 & x_5
\end{array}
$$

$$
\left[
\begin{array}{ccccc|c}
1 & 1 & 2 & 1 & 0 & 38 \\
1 & \frac{1}{2} & \frac{1}{2} & 0 & -\frac{1}{2} & 12 \\
\hline
-3 & -2 & -2 & 0 & 0 & 0
\end{array}
\right]
\frac{1}{2}R_2
$$

$$
\begin{array}{ccccc}
x_1 & x_2 & x_3 & x_4 & x_5
\end{array}
$$

$$
\left[
\begin{array}{ccccc|c}
0 & \frac{1}{2} & \frac{3}{2} & 1 & \frac{1}{2} & 26 \\
1 & \frac{1}{2} & \frac{1}{2} & 0 & -\frac{1}{2} & 12 \\
\hline
0 & -\frac{1}{2} & -\frac{1}{2} & 0 & -\frac{3}{2} & 36
\end{array}
\right]
\begin{array}{l}
-1R_2 + R_1 \\
\\
3R_2 + R_3
\end{array}
$$

For Stage II pivoting, pivot on the $\dfrac{1}{2}$ in row one, column five.

$$
\begin{array}{ccccc}
x_1 & x_2 & x_3 & x_4 & x_5
\end{array}
$$

$$
\left[
\begin{array}{ccccc|c}
0 & 1 & 3 & 2 & 1 & 52 \\
1 & \frac{1}{2} & \frac{1}{2} & 0 & -\frac{1}{2} & 12 \\
\hline
0 & -\frac{1}{2} & -\frac{1}{2} & 0 & -\frac{3}{2} & 36
\end{array}
\right]
2R_1
$$

$$
\begin{array}{ccccc}
x_1 & x_2 & x_3 & x_4 & x_5
\end{array}
$$

$$
\left[
\begin{array}{ccccc|c}
0 & 1 & 3 & 2 & 1 & 52 \\
1 & 1 & 2 & 1 & 0 & 38 \\
\hline
0 & 1 & 4 & 3 & 0 & 114
\end{array}
\right]
\begin{array}{l}
\frac{1}{2}R_1 + R_2 \\
\frac{3}{2}R_1 + R_3
\end{array}
$$

The maximum is 114 when $x_1 = 38$, $x_2 = 0$, and $x_3 = 0$.

13. Find $x_1 \geq 0$ and $x_2 \geq 0$ such that

$$x_1 + 2x_2 \leq 18$$
$$x_1 + 3x_2 \geq 12$$
$$2x_1 + 2x_2 \leq 30$$

and

$z = 5x_1 + 10x_2$ is maximized.

The initial simplex tableau is

$$\begin{array}{ccccc} x_1 & x_2 & x_3 & x_4 & x_5 \end{array}$$

$$\left[\begin{array}{ccccc|c} 1 & 2 & 1 & 0 & 0 & 18 \\ 1 & 3 & 0 & -1 & 0 & 12 \\ 2 & 2 & 0 & 0 & 1 & 30 \\ \hline -5 & -10 & 0 & 0 & 0 & 0 \end{array}\right].$$

For Stage I pivoting, we can pivot on the 1 in row two, column one or the 3 in row two, column two. Choosing the 1 in row two, column one as the pivot, we proceed as follows.

$$\begin{array}{ccccc} x_1 & x_2 & x_3 & x_4 & x_5 \end{array}$$

$$\left[\begin{array}{ccccc|c} 0 & -1 & 1 & 1 & 0 & 6 \\ 1 & 3 & 0 & -1 & 0 & 12 \\ 0 & -4 & 0 & 2 & 1 & 6 \\ \hline 0 & 5 & 0 & -5 & 0 & 60 \end{array}\right]\begin{array}{l} -R_2 + R_1 \\ \\ -2R_2 + R_3 \\ 5R_2 + R_4 \end{array}$$

For Stage II, pivot on the 2 in row three, column four.

$$\begin{array}{ccccc} x_1 & x_2 & x_3 & x_4 & x_5 \end{array}$$

$$\left[\begin{array}{ccccc|c} 0 & -1 & 1 & 1 & 0 & 6 \\ 1 & 3 & 0 & -1 & 0 & 12 \\ 0 & -2 & 0 & 1 & \frac{1}{2} & 3 \\ \hline 0 & 5 & 0 & -5 & 0 & 60 \end{array}\right]\begin{array}{l} \\ \\ \frac{1}{2}R_3 \\ \\ \end{array}$$

$$\begin{array}{ccccc} x_1 & x_2 & x_3 & x_4 & x_5 \end{array}$$

$$\left[\begin{array}{ccccc|c} 0 & 1 & 1 & 0 & -\frac{1}{2} & 3 \\ 1 & 1 & 0 & 0 & \frac{1}{2} & 15 \\ 0 & -2 & 0 & 1 & \frac{1}{2} & 3 \\ \hline 0 & -5 & 0 & 0 & \frac{5}{2} & 75 \end{array}\right]\begin{array}{l} -R_3 + R_1 \\ R_3 + R_2 \\ \\ 5R_3 + R_4 \end{array}$$

Now pivot on the 1 in row one, column two.

$$\begin{array}{ccccc} x_1 & x_2 & x_3 & x_4 & x_5 \end{array}$$

$$\left[\begin{array}{ccccc|c} 0 & 1 & 1 & 0 & -\frac{1}{2} & 3 \\ 1 & 0 & -1 & 0 & 1 & 12 \\ 0 & 0 & 2 & 1 & -\frac{1}{2} & 9 \\ \hline 0 & 0 & 5 & 0 & 0 & 90 \end{array}\right]\begin{array}{l} \\ -R_1 + R_2 \\ 2R_1 + R_3 \\ 5R_1 + R_4 \end{array}$$

The above tableau gives the solution:

The maximum is 90 when $x_1 = 12$ and $x_2 = 3$.

13. Continued

However, in Stage I, we could also choose the 3 in row two, column two as the pivot and proceed as follows.

$$\begin{array}{ccccc} x_1 & x_2 & x_3 & x_4 & x_5 \end{array}$$

$$\left[\begin{array}{ccccc|c} 1 & 2 & 1 & 0 & 0 & 18 \\ \frac{1}{3} & 1 & 0 & -\frac{1}{3} & 0 & 4 \\ 2 & 2 & 0 & 0 & 1 & 30 \\ \hline -5 & -10 & 0 & 0 & 0 & 0 \end{array}\right]\begin{array}{l} \\ \frac{1}{3}R_2 \\ \\ \\ \end{array}$$

$$\begin{array}{ccccc} x_1 & x_2 & x_3 & x_4 & x_5 \end{array}$$

$$\left[\begin{array}{ccccc|c} \frac{1}{3} & 0 & 1 & \frac{2}{3} & 0 & 10 \\ \frac{1}{3} & 1 & 0 & -\frac{1}{3} & 0 & 4 \\ \frac{4}{3} & 0 & 0 & \frac{2}{3} & 1 & 22 \\ \hline -\frac{5}{3} & 0 & 0 & -\frac{10}{3} & 0 & 40 \end{array}\right]\begin{array}{l} -2R_2 + R_1 \\ \\ -2R_2 + R_3 \\ 10R_2 + R_4 \end{array}$$

For Stage II, pivot on the $\dfrac{2}{3}$ in row one, column four.

$$\begin{array}{ccccc} x_1 & x_2 & x_3 & x_4 & x_5 \end{array}$$

$$\left[\begin{array}{ccccc|c} \frac{1}{2} & 0 & \frac{3}{2} & 1 & 0 & 15 \\ \frac{1}{3} & 1 & 0 & -\frac{1}{3} & 0 & 4 \\ \frac{4}{3} & 0 & 0 & \frac{2}{3} & 1 & 22 \\ \hline -\frac{5}{3} & 0 & 0 & -\frac{10}{3} & 0 & 40 \end{array}\right]\begin{array}{l} \frac{3}{2}R_1 \\ \\ \\ \\ \end{array}$$

$$\begin{array}{ccccc} x_1 & x_2 & x_3 & x_4 & x_5 \end{array}$$

$$\left[\begin{array}{ccccc|c} \frac{1}{2} & 0 & \frac{3}{2} & 1 & 0 & 15 \\ \frac{1}{2} & 1 & \frac{1}{2} & 0 & 0 & 9 \\ 1 & 0 & -1 & 0 & 1 & 12 \\ \hline 0 & 0 & 5 & 0 & 0 & 90 \end{array}\right]\begin{array}{l} \\ \frac{1}{3}R_1 + R_2 \\ -\frac{2}{3}R_1 + R_3 \\ \frac{10}{3}R_1 + R_4 \end{array}$$

This tableau gives the solution:

The maximum is 90 when $x_1 = 0$ and $x_2 = 9$.

Thus, the maximum is 90 when $x_1 = 12$ and $x_2 = 3$ or when $x_1 = 0$ and $x_2 = 9$.

15. Minimize $w = 3y_1 + 2y_2$

subject to: $2y_1 + 3y_2 \geq 60$

$y_1 + 4y_2 \geq 40$

$y_1 \geq 0,\ y_2 \geq 0.$

The initial simplex tableau is

$$
\begin{array}{cccc}
y_1 & y_2 & y_3 & y_4 \\
\end{array}
$$

$$
\left[
\begin{array}{cccc|c}
2 & 3 & -1 & 0 & 60 \\
1 & 4 & 0 & -1 & 40 \\
\hline
3 & 2 & 0 & 0 & 0
\end{array}
\right].
$$

For Stage I, pivot on the 2 in row one, column one.

$$
\begin{array}{cccc}
y_1 & y_2 & y_3 & y_4 \\
\end{array}
$$

$$
\left[
\begin{array}{cccc|c}
1 & \frac{3}{2} & -\frac{1}{2} & 0 & 30 \\
0 & \frac{5}{2} & \frac{1}{2} & -1 & 10 \\
\hline
0 & -\frac{5}{2} & \frac{3}{2} & 0 & -90
\end{array}
\right]
\begin{array}{l}
\frac{1}{2}R_1 \\
-R_1 + R_2 \\
-3R_1 + R_3
\end{array}
$$

To continue Stage I, pivot on the $\dfrac{5}{2}$ in row two, column two.

$$
\begin{array}{cccc}
y_1 & y_2 & y_3 & y_4 \\
\end{array}
$$

$$
\left[
\begin{array}{cccc|c}
1 & 0 & -\frac{4}{5} & \frac{3}{5} & 24 \\
0 & 1 & \frac{1}{5} & -\frac{2}{5} & 4 \\
\hline
0 & 0 & 2 & -1 & -80
\end{array}
\right]
\begin{array}{l}
-\frac{3}{5}R_2 + R_1 \\
\frac{2}{5}R_2 \\
\frac{5}{2}R_2 + R_3
\end{array}
$$

For Stage II, pivot on the $\dfrac{3}{5}$ in row one, column four.

$$
\begin{array}{cccc}
y_1 & y_2 & y_3 & y_4 \\
\end{array}
$$

$$
\left[
\begin{array}{cccc|c}
\frac{5}{3} & 0 & -\frac{4}{3} & 1 & 40 \\
\frac{2}{3} & 1 & -\frac{1}{3} & 0 & 20 \\
\hline
\frac{5}{3} & 0 & \frac{2}{3} & 0 & -40
\end{array}
\right]
\begin{array}{l}
\frac{5}{3}R_1 \\
\frac{2}{5}R_1 + R_2 \\
R_1 + R_3
\end{array}
$$

The minimum is 40 when $y_1 = 0$ and $y_2 = 20.$

17. Maximize $z = 3x_1 + 2x_2$

subject to: $x_1 + x_2 = 50$

$4x_1 + 2x_2 \geq 120$

$5x_1 + 2x_2 \leq 200$

$x_1 \geq 0,\ x_2 \geq 0$

The initial simplex tableau is

$$
\begin{array}{cccccc}
x_1 & x_2 & x_3 & x_4 & x_5 & x_6 \\
\end{array}
$$

$$
\left[
\begin{array}{cccccc|c}
1 & 1 & -1 & 0 & 0 & 0 & 50 \\
1 & 1 & 0 & 1 & 0 & 0 & 50 \\
4 & 2 & 0 & 0 & -1 & 0 & 120 \\
5 & 2 & 0 & 0 & 0 & 1 & 200 \\
\hline
-3 & -2 & 0 & 0 & 0 & 0 & 0
\end{array}
\right]
$$

For Stage I, pivot on the 4 in row three, column one.

$$
\begin{array}{cccccc}
x_1 & x_2 & x_3 & x_4 & x_5 & x_6 \\
\end{array}
$$

$$
\left[
\begin{array}{cccccc|c}
0 & \frac{1}{2} & -1 & 0 & \frac{1}{4} & 0 & 20 \\
0 & \frac{1}{2} & 0 & 1 & \frac{1}{4} & 0 & 20 \\
1 & \frac{1}{2} & 0 & 0 & -\frac{1}{4} & 0 & 30 \\
0 & -\frac{1}{2} & 0 & 0 & \frac{5}{4} & 1 & 50 \\
\hline
0 & -\frac{1}{2} & 0 & 0 & -\frac{3}{4} & 0 & 90
\end{array}
\right]
\begin{array}{l}
-R_3 + R_1 \\
-R_3 + R_2 \\
\frac{1}{4}R_3 \\
-5R_3 + R_4 \\
3R_3 + R_5
\end{array}
$$

To continue Stage I, pivot on the $\dfrac{1}{2}$ in row one column two.

$$
\begin{array}{cccccc}
x_1 & x_2 & x_3 & x_4 & x_5 & x_6 \\
\end{array}
$$

$$
\left[
\begin{array}{cccccc|c}
0 & 1 & -2 & 0 & \frac{1}{2} & 0 & 40 \\
0 & 0 & 1 & 1 & 0 & 0 & 0 \\
1 & 0 & 1 & 0 & -\frac{1}{2} & 0 & 10 \\
0 & 0 & -1 & 0 & \frac{3}{2} & 1 & 70 \\
\hline
0 & 0 & -1 & 0 & -\frac{1}{2} & 0 & 110
\end{array}
\right]
\begin{array}{l}
2R_1 \\
-\frac{1}{2}R_1 + R_2 \\
-\frac{1}{2}R_1 + R_3 \\
\frac{1}{2}R_1 + R_4 \\
\frac{1}{2}R_1 + R_5
\end{array}
$$

For Stage II, pivot on the 1 in row two, column three.

$$
\begin{array}{cccccc}
x_1 & x_2 & x_3 & x_4 & x_5 & x_6 \\
\end{array}
$$

$$
\left[
\begin{array}{cccccc|c}
0 & 1 & 0 & 2 & \frac{1}{2} & 0 & 40 \\
0 & 0 & 1 & 1 & 0 & 0 & 0 \\
1 & 0 & 0 & -1 & -\frac{1}{2} & 0 & 10 \\
0 & 0 & 0 & 1 & \frac{3}{2} & 1 & 70 \\
\hline
0 & 0 & 0 & 1 & -\frac{1}{2} & 0 & 110
\end{array}
\right]
\begin{array}{l}
2R_2 + R_1 \\
\\
-R_2 + R_3 \\
R_2 + R_4 \\
R_2 + R_5
\end{array}
$$

Pivot on the $\dfrac{3}{2}$ in row four, column five.

17. Continued

$$\begin{array}{cccccc} x_1 & x_2 & x_3 & x_4 & x_5 & x_6 \end{array}$$

$$\left[\begin{array}{cccccc|c} 0 & 1 & 0 & \frac{5}{3} & 0 & -\frac{1}{3} & \frac{50}{3} \\ 0 & 0 & 1 & 1 & 0 & 0 & 0 \\ 1 & 0 & 0 & -\frac{2}{3} & 0 & \frac{1}{3} & \frac{100}{3} \\ 0 & 0 & 0 & \frac{2}{3} & 1 & \frac{2}{3} & \frac{140}{3} \\ \hline 0 & 0 & 0 & \frac{4}{3} & 0 & \frac{1}{3} & \frac{400}{3} \end{array}\right] \begin{array}{l} -\frac{1}{2}R_4 + R_1 \\ \\ \frac{1}{2}R_4 + R_3 \\ \frac{2}{3}R_4 \\ \frac{1}{2}R_4 + R_5 \end{array}$$

The maximum is $133\frac{1}{3}$ when $x_1 = 33\frac{1}{3}$ and

$x_2 = 16\frac{2}{3}$.

19. Minimize $w = 32y_1 + 40y_2$

Maximize $z = -w = -32y_1 - 40y_2$.

subject to: $20y_1 + 10y_2 = 200$

$\qquad\qquad 25y_1 + 40y_2 \le 500$

$\qquad\qquad 18y_1 + 24y_2 \ge 300$

$\qquad\qquad y_1 \ge 0, \ y_2 \ge 0$

The initial simplex tableau is

$$\begin{array}{cccccc} y_1 & y_2 & y_3 & y_4 & y_5 & y_6 \end{array}$$

$$\left[\begin{array}{cccccc|c} \underline{20} & 10 & -1 & 0 & 0 & 0 & 200 \\ 20 & 10 & 0 & 1 & 0 & 0 & 200 \\ 25 & 40 & 0 & 0 & 1 & 0 & 500 \\ 18 & 24 & 0 & 0 & 0 & -1 & 300 \\ \hline 32 & 40 & 0 & 0 & 0 & 0 & 0 \end{array}\right]$$

For Stage I, pivot on the 20 in row one, column one.

$$\begin{array}{cccccc} y_1 & y_2 & y_3 & y_4 & y_5 & y_6 \end{array}$$

$$\left[\begin{array}{cccccc|c} 1 & \frac{1}{2} & -\frac{1}{20} & 0 & 0 & 0 & 10 \\ 0 & 0 & 1 & 1 & 0 & 0 & 0 \\ 0 & \frac{55}{2} & \frac{5}{4} & 0 & 1 & 0 & 250 \\ 0 & \underline{15} & \frac{9}{10} & 0 & 0 & -1 & 120 \\ \hline 0 & 24 & \frac{8}{5} & 0 & 0 & 0 & -320 \end{array}\right] \begin{array}{l} \frac{1}{20}R_1 \\ -20R_1 + R_2 \\ -25R_1 + R_3 \\ -18R_1 + R_4 \\ -32R_1 + R_5 \end{array}$$

Pivot on the 15 in row four, column two.

$$\begin{array}{cccccc} y_1 & y_2 & y_3 & y_4 & y_5 & y_6 \end{array}$$

$$\left[\begin{array}{cccccc|c} 1 & 0 & -\frac{4}{50} & 0 & 0 & \frac{1}{30} & 6 \\ 0 & 0 & 1 & 1 & 0 & 0 & 0 \\ 0 & 0 & -\frac{2}{5} & 0 & 1 & \frac{11}{6} & 30 \\ 0 & 1 & \frac{3}{50} & 0 & 0 & -\frac{1}{15} & 8 \\ \hline 0 & 0 & \frac{4}{25} & 0 & 0 & \frac{8}{5} & -512 \end{array}\right] \begin{array}{l} -\frac{1}{2}R_4 + R_1 \\ \\ -\frac{55}{2}R_4 + R_3 \\ \frac{1}{15}R_4 \\ -24R_4 + R_5 \end{array}$$

The program is optimal after Stage I.

The minimum is 512 when $y_1 = 6$ and $y_2 = 8$.

21. Let y_1 = the amount of ingredient I per barrel of gasoline;

$\qquad y_2$ = the amount of ingredient II per barrel of gasoline;

$\qquad y_3$ = the amount of ingredient III per barrel of gasoline.

Minimize $\quad w = .30y_1 + .09y_2 + .27y_3$

Maximize $z = -w = -.30y_1 - .09y_2 - .27y_3$

subject to:

$\qquad\qquad y_1 + y_2 + y_3 \ge 10$

$\qquad\qquad y_1 + y_2 + y_3 \le 15$

$\qquad\qquad y_1 - \frac{1}{4}y_2 \qquad \ge 0 \quad \left(\text{Since } y_1 \ge \frac{1}{4}y_2\right)$

$\qquad\qquad -y_1 \qquad + y_3 \ge 0 \quad (\text{Since } y_3 \ge y_1)$

$y_1 \ge 0, \ y_2 \ge 0, \ y_3 \ge 0$

The initial simplex tableau is

$$\begin{array}{ccccccc} y_1 & y_2 & y_3 & y_4 & y_5 & y_6 & y_7 \end{array}$$

$$\left[\begin{array}{ccccccc|c} 1 & 1 & 1 & -1 & 0 & 0 & 0 & 10 \\ 1 & 1 & 1 & 0 & 1 & 0 & 0 & 15 \\ 1 & -\frac{1}{4} & 0 & 0 & 0 & -1 & 0 & 0 \\ -1 & 0 & 1 & 0 & 0 & 0 & -1 & 0 \\ \hline .30 & .09 & .27 & 0 & 0 & 0 & 0 & 0 \end{array}\right].$$

23. Let y_1 = the number of computers from W_1 to D_1
    y_2 = the number of computers from W_1 to D_2;
    y_3 = the number of computers from W_2 to D_1;
    y_4 = the number of computers from W_2 to D_2.

Minimize $w = 14y_1 + 22y_2 + 12y_3 + 10y_4$

Maximize $z = -w = -14y_1 - 22y_2 - 12y_3 - 10y_4$

subject to: $y_1 \quad + y_3 \qquad \geq 32$
                        $y_2 \qquad + y_4 \geq 20$
                        $y_1 + y_2 \qquad\qquad \leq 25$
                        $y_3 + y_4 \leq 30$
                        $y_1 \geq 0, \ y_2 \geq 0, y_3 \geq 0, \ y_4 \geq 0.$

The initial simplex tableau is

y_1	y_2	y_3	y_4	y_5	y_6	y_7	y_8	
1	0	1	0	-1	0	0	0	32
0	1	0	1	0	-1	0	0	20
1	1	0	0	0	0	1	0	25
0	0	1	1	0	0	0	1	30
14	22	12	10	0	0	0	0	0

25. Let x_1 = the number of barrels of oil supplied by S_1 to D_1.

Let x_2 = the number of barrels of oil supplied by S_2 to D_1.

Let x_3 = the number of barrels of oil supplied by S_1 to D_2.

Let x_4 = the number of barrels of oil supplied by S_2 to D_2.

Minimize $w = 30x_1 + 25x_2 + 20x_3 + 22x_4$

subject to: $x_1 + x_2 \qquad\qquad \geq 3000$
                $x_3 + x_4 \geq 5000$
                $x_1 \qquad + x_3 \qquad \leq 5000$
                $x_2 \qquad + x_4 \leq 5000$
                $2x_1 + 5x_2 + 6x_3 + 4x_4 \leq 40,000$

The initial simplex tableau is

x_1	x_2	x_3	x_4	x_5	x_6	x_7	x_8	x_9	
1	1	0	0	-1	0	0	0	0	3000
0	0	1	1	0	-1	0	0	0	5000
1	0	1	0	0	0	1	0	0	5000
0	1	0	1	0	0	0	1	0	5000
2	5	6	4	0	0	0	0	1	40,000
30	25	20	22	0	0	0	0	0	0

25. Continued

For Stage I, pivot on the 1 in row one, column one.

x_1	x_2	x_3	x_4	x_5	x_6	x_7	x_8	x_9	
1	1	0	0	−1	0	0	0	0	3000
0	0	1	1	0	−1	0	0	0	5000
0	−1	1	0	1	0	1	0	0	2000
0	1	0	1	0	0	0	1	0	5000
0	3	6	4	2	0	0	0	1	34,000
0	−5	20	22	30	0	0	0	0	−90,000

To continue Stage I, pivot on the 1 in row three, column three.

x_1	x_2	x_3	x_4	x_5	x_6	x_7	x_8	x_9	
1	1	0	0	−1	0	0	0	0	3000
0	1	0	1	−1	−1	−1	0	0	3000
0	−1	1	0	1	0	1	0	0	2000
0	1	0	1	0	0	0	1	0	5000
0	9	0	4	−4	0	−6	0	1	22,000
0	15	0	22	10	0	−20	0	0	−130,000

To continue Stage I, pivot on the 1 in row two, column four.

x_1	x_2	x_3	x_4	x_5	x_6	x_7	x_8	x_9	
1	1	0	0	−1	0	0	0	0	3000
0	1	0	1	−1	−1	−1	0	0	3000
0	−1	1	0	1	0	1	0	0	2000
0	0	0	0	1	1	1	1	0	2000
0	5	0	0	0	4	−2	0	1	10,000
0	−7	0	0	32	22	2	0	0	−196,000

We have completed Stage I. For Stage II, pivot on the 5 in row five, column two.

x_1	x_2	x_3	x_4	x_5	x_6	x_7	x_8	x_9	
1	0	0	0	−1	$-\frac{4}{5}$	$\frac{2}{5}$	0	$-\frac{1}{5}$	1000
0	0	0	1	−1	$-\frac{9}{5}$	$-\frac{3}{5}$	0	$-\frac{1}{5}$	1000
0	0	1	0	1	$\frac{4}{5}$	$\frac{3}{5}$	0	$\frac{1}{5}$	4000
0	0	0	0	1	1	1	1	0	2000
0	1	0	0	0	$\frac{4}{5}$	$-\frac{2}{5}$	0	$\frac{1}{5}$	2000
0	0	0	0	32	$\frac{138}{5}$	$-\frac{4}{5}$	0	$\frac{7}{5}$	−182,000

Finally, pivot on the 1 in row four, column seven.

x_1	x_2	x_3	x_4	x_5	x_6	x_7	x_8	x_9	
1	0	0	0	$-\frac{7}{5}$	$-\frac{6}{5}$	0	$-\frac{2}{5}$	$-\frac{1}{5}$	200
0	0	0	1	$-\frac{2}{5}$	$-\frac{6}{5}$	0	$\frac{3}{5}$	$-\frac{1}{5}$	2200
0	0	1	0	$\frac{2}{5}$	$\frac{1}{5}$	0	$-\frac{3}{5}$	$\frac{1}{5}$	2800
0	0	0	0	1	1	1	1	0	2000
0	1	0	0	$\frac{2}{5}$	$\frac{6}{5}$	0	$\frac{2}{5}$	$\frac{1}{5}$	2800
0	0	0	0	$\frac{164}{5}$	$\frac{142}{5}$	0	$\frac{4}{5}$	$\frac{7}{5}$	−180,400

This gives $x_1 = 200$, $x_2 = 2800$, $x_3 = 2800$, and $x_4 = 2200$ for a minimum cost of $180,400.

27. Let y_1 = the amount of bluegrass seed

y_2 = the amount of rye seed

y_3 = the amount of Bermuda seed

The problem is to minimize:

$w = .12y_1 + .15y_2 + .05y_3$

subject to:

$$y_1 \geq .2(y_1 + y_2 + y_3)$$

$$y_3 \leq \frac{2}{3}y_2$$

$$y_1 + y_2 + y_3 \geq 5000$$

Or,

maximize: $z = -w = -.12y_1 - .15y_2 - .05y_3$

subject to:

$$.8y_1 - .2y_2 - .2y_3 \geq 0$$

$$2y_2 - 3y_3 \geq 0$$

$$y_1 + y_2 + y_3 \geq 5000$$

Adding surplus variables y_4, y_5, and y_6, the initial tableau is

$$
\begin{array}{cccccc}
y_1 & y_2 & y_3 & y_4 & y_5 & y_6 \\
\end{array}
$$
$$
\left[
\begin{array}{cccccc|c}
.8 & -.2 & -.2 & -1 & 0 & 0 & 0 \\
0 & 2 & -3 & 0 & -1 & 0 & 0 \\
1 & 1 & 1 & 0 & 0 & -1 & 5000 \\
\hline
.12 & .15 & .05 & 0 & 0 & 0 & 0
\end{array}
\right]
$$

Since y_4 through y_6 are negative, this does not have a feasible solution. For Stage I, use the .8 in row one, column one.

$$
\begin{array}{cccccc}
y_1 & y_2 & y_3 & y_4 & y_5 & y_6 \\
\end{array}
$$
$$
\left[
\begin{array}{cccccc|c}
1 & -.25 & -.25 & -1.25 & 0 & 0 & 0 \\
0 & 2 & -3 & 0 & -1 & 0 & 0 \\
0 & 1.25 & 1.25 & 1.25 & 0 & -1 & 5000 \\
\hline
0 & .18 & .08 & .15 & 0 & 0 & 0
\end{array}
\right]
$$

Continue Stage 1 by pivoting on the 2 in row two, column two.

$$
\begin{array}{cccccc}
y_1 & y_2 & y_3 & y_4 & y_5 & y_6 \\
\end{array}
$$
$$
\left[
\begin{array}{cccccc|c}
1 & 0 & -.625 & -1.25 & -.125 & 0 & 0 \\
0 & 1 & -1.5 & 0 & -.5 & 0 & 0 \\
0 & 0 & 3.125 & 1.25 & .625 & -1 & 5000 \\
\hline
0 & 0 & .35 & .15 & .09 & 0 & 0
\end{array}
\right]
$$

To finish Stage 1, pivot on the 3.125 in row three, column three.

27. Continued

$$
\begin{array}{cccccc}
y_1 & y_2 & y_3 & y_4 & y & _5 & y_6 \\
\end{array}
$$
$$
\left[
\begin{array}{cccccc|c}
1 & 0 & 0 & -1 & 0 & -.2 & 1000 \\
0 & 1 & 0 & .6 & -.2 & -.48 & 2400 \\
0 & 0 & 1 & .4 & .2 & -.32 & 1600 \\
\hline
0 & 0 & 0 & .01 & .02 & .112 & -560
\end{array}
\right]
$$

Stage II pivoting is not necessary. The solution is $y_1 = 1000$, $y_2 = 2400$, $y_3 = 1600$ and $w = 560$. In other words, Topgrade Turf should use 1000 lbs. of bluegrass seed, 2400 lbs. of rye seed, and 1600 lbs. of Bermuda seed for a minimum cost of \$560.

29. Let x_1 = amount allotted to commercial loans (in millions)

x_2 = amount allotted to home loans (in millions)

Maximize $z = .10x_1 + .12x_2$

subject to: $x_1 + x_2 \leq 25$

$$2x_1 + 3x_2 \leq 72$$

$$-4x_1 + x_2 \geq 0$$

$$x_1 + x_2 \geq 10$$

$$x_1 \geq 0, \; x_2 \geq 0$$

The initial simplex tableau is

$$
\begin{array}{cccccc}
x_1 & x_2 & x_3 & x_4 & x_5 & x_6 \\
\end{array}
$$
$$
\left[
\begin{array}{cccccc|c}
1 & 1 & 1 & 0 & 0 & 0 & 25 \\
2 & 3 & 0 & 1 & 0 & 0 & 72 \\
-4 & 1 & 0 & 0 & -1 & 0 & 0 \\
1 & 1 & 0 & 0 & 0 & -1 & 10 \\
\hline
-.10 & -.12 & 0 & 0 & 0 & 0 & 0
\end{array}
\right]
$$

For Stage I, pivot on the 1 in row three, column two.

$$
\begin{array}{cccccc}
x_1 & x_2 & x_3 & x_4 & x_5 & x_6 \\
\end{array}
$$
$$
\left[
\begin{array}{cccccc|c}
5 & 0 & 1 & 0 & 1 & 0 & 25 \\
14 & 0 & 0 & 1 & 3 & 0 & 72 \\
-4 & 1 & 0 & 0 & -1 & 0 & 0 \\
5 & 0 & 0 & 0 & 1 & -1 & 10 \\
\hline
-\frac{29}{50} & 0 & 0 & 0 & -\frac{3}{25} & 0 & 0
\end{array}
\right]
$$

To continue Stage I, pivot on the 5 in row four, column one.

29. Continued

$$\begin{array}{cccccc} x_1 & x_2 & x_3 & x_4 & x_5 & x_6 \end{array}$$

$$\left[\begin{array}{cccccc|c} 0 & 0 & 1 & 0 & 0 & 1 & 15 \\ 0 & 0 & 0 & 1 & \frac{1}{5} & \frac{14}{5} & 44 \\ 0 & 1 & 0 & 0 & -\frac{1}{5} & -\frac{4}{5} & 8 \\ 1 & 0 & 0 & 0 & \frac{1}{5} & -\frac{1}{5} & 2 \\ \hline 0 & 0 & 0 & 0 & -\frac{1}{250} & -\frac{29}{250} & \frac{29}{25} \end{array}\right]$$

For Stage II, pivot on the 1 in row one, column six.

$$\begin{array}{cccccc} x_1 & x_2 & x_3 & x_4 & x_5 & x_6 \end{array}$$

$$\left[\begin{array}{cccccc|c} 0 & 0 & 1 & 0 & 0 & 1 & 15 \\ 0 & 0 & -\frac{14}{5} & 1 & \frac{1}{5} & 0 & 2 \\ 0 & 1 & \frac{4}{5} & 0 & -\frac{1}{5} & 0 & 20 \\ 1 & 0 & \frac{1}{5} & 0 & \frac{1}{5} & 0 & 5 \\ \hline 0 & 0 & \frac{29}{250} & 0 & -\frac{1}{250} & 0 & \frac{29}{10} \end{array}\right]$$

To continue stage II, pivot on the $\dfrac{4}{5}$ in row two, column five.

$$\begin{array}{cccccc} x_1 & x_2 & x_3 & x_4 & x_5 & x_6 \end{array}$$

$$\left[\begin{array}{cccccc|c} 0 & 0 & 1 & 0 & 0 & 1 & 15 \\ 0 & 0 & -14 & 5 & 1 & 0 & 10 \\ 0 & 1 & -2 & 1 & 0 & 0 & 22 \\ 1 & 0 & 3 & -1 & 0 & 0 & 3 \\ \hline 0 & 0 & \frac{3}{50} & \frac{1}{50} & 0 & 0 & \frac{147}{50} \end{array}\right]$$

This gives $x_1 = 3$, $x_2 = 22$, and the maximum is

$\dfrac{147}{50} = 2.94$. Allot $3,000,000 for commercial loans and $22,000,000 for home loans for a maximum interest income of $2,940,000.

31. Let y_1 = the amount of regular beer and

$\quad\quad y_2$ = the amount of light beer.

Minimize $\quad 36,000y_1 + 48,000y_2$

subject to: $\quad\quad\quad y_1 \geq 12$

$\quad\quad\quad\quad\quad\quad\quad y_2 \geq 10$

$\quad\quad\quad\quad -2y_1 + y_2 \leq 0$ (Since $y_2 \leq 2y_1$)

$\quad\quad\quad\quad\quad y_1 + y_2 \geq 42.$

(Since the company already produces at least $12 + 10 = 22$ units and can produce at least 20 additional units: $y_1 + y_2 \geq 22 + 20 = 42$.)

The initial simplex tableau is

$$\begin{array}{cccccc} y_1 & y_2 & y_3 & y_4 & y_5 & y_6 \end{array}$$

$$\left[\begin{array}{cccccc|c} \underline{1} & 0 & -1 & 0 & 0 & 0 & 12 \\ 0 & 1 & 0 & -1 & 0 & 0 & 10 \\ -2 & 1 & 0 & 0 & 1 & 0 & 0 \\ 1 & 1 & 0 & 0 & 0 & -1 & 42 \\ \hline 36 & 48 & 0 & 0 & 0 & 0 & 0 \end{array}\right].$$

For Stage I, pivot on the 1 in row one, column one.

$$\begin{array}{cccccc} y_1 & y_2 & y_3 & y_4 & y_5 & y_6 \end{array}$$

$$\left[\begin{array}{cccccc|c} 1 & 0 & -1 & 0 & 0 & 0 & 12 \\ 0 & \underline{1} & 0 & -1 & 0 & 0 & 10 \\ 0 & 1 & -2 & 0 & 1 & 0 & 24 \\ 0 & 1 & 1 & 0 & 0 & -1 & 30 \\ \hline 0 & 48 & 36 & 0 & 0 & 0 & -432 \end{array}\right]\begin{array}{l} \\ \\ 2R_1 + R_3 \\ -R_1 + R_4 \\ -36R_1 + R_5 \end{array}$$

To continue Stage I, pivot on the 1 in row two, column two.

$$\begin{array}{cccccc} y_1 & y_2 & y_3 & y_4 & y_5 & y_6 \end{array}$$

$$\left[\begin{array}{cccccc|c} 1 & 0 & -1 & 0 & 0 & 0 & 12 \\ 0 & 1 & 0 & -1 & 0 & 0 & 10 \\ 0 & 0 & -2 & 1 & 1 & 0 & 14 \\ 0 & 0 & 1 & \underline{1} & 0 & -1 & 20 \\ \hline 0 & 0 & 36 & 48 & 0 & 0 & -912 \end{array}\right]\begin{array}{l} \\ \\ -R_2 + R_3 \\ -R_2 + R_4 \\ -48R_2 + R_5 \end{array}$$

Now pivot on the 1 in row four, column four.

$$\begin{array}{cccccc} y_1 & y_2 & y_3 & y_4 & y_5 & y_6 \end{array}$$

$$\left[\begin{array}{cccccc|c} 1 & 0 & -1 & 0 & 0 & 0 & 12 \\ 0 & 1 & 1 & 0 & 0 & -1 & 30 \\ 0 & 0 & \underline{-3} & 0 & 1 & 1 & -6 \\ 0 & 0 & 1 & 1 & 0 & -1 & 20 \\ \hline 0 & 0 & -12 & 0 & 0 & 48 & -1872 \end{array}\right]\begin{array}{l} \\ R_4 + R_2 \\ -R_4 + R_3 \\ \\ -48R_4 + R_5 \end{array}$$

31. Continued

For Stage II, pivot on the -3 in row three, column three.

$$
\begin{array}{cccccc}
y_1 & y_2 & y_3 & y_4 & y_5 & y_6
\end{array}
$$

$$
\left[
\begin{array}{cccccc|c}
1 & 0 & 0 & 0 & -\frac{1}{3} & -\frac{1}{3} & 14 \\
0 & 1 & 0 & 0 & \frac{1}{3} & -\frac{2}{3} & 28 \\
0 & 0 & 1 & 0 & -\frac{1}{3} & -\frac{1}{3} & 2 \\
0 & 0 & 0 & 1 & \frac{1}{3} & -\frac{2}{3} & 18 \\
\hline
0 & 0 & 0 & 0 & -4 & 44 & -1848
\end{array}
\right]
\begin{array}{l}
R_3 + R_1 \\
-R_3 + R_2 \\
-\frac{1}{3}R_3 \\
-R_3 + R_4 \\
12R_3 + R_5
\end{array}
$$

Finally, pivot on the $\dfrac{1}{3}$ in row four, column five.

$$
\begin{array}{cccccc}
y_1 & y_2 & y_3 & y_4 & y_5 & y_6
\end{array}
$$

$$
\left[
\begin{array}{cccccc|c}
1 & 0 & 0 & 1 & 0 & -1 & 32 \\
0 & 1 & 0 & -1 & 0 & 0 & 0 \\
0 & 0 & 1 & 1 & 0 & -1 & 20 \\
0 & 0 & 0 & 3 & 1 & -2 & 54 \\
\hline
0 & 0 & 0 & 12 & 0 & 36 & -1632
\end{array}
\right]
\begin{array}{l}
\frac{1}{3}R_4 + R_1 \\
-\frac{1}{3}R_4 + R_2 \\
\frac{1}{3}R_4 + R_3 \\
3R_4 \\
4R_4 + R_5
\end{array}
$$

Make 32 units of regular beer and 10 units of light beer for a minimum cost of $1,632,000.

33. From Exercise 21, we know the initial simplex tableau is

$$
\begin{array}{ccccccc}
y_1 & y_2 & y_3 & y_4 & y_5 & y_6 & y_7
\end{array}
$$

$$
\left[
\begin{array}{ccccccc|c}
1 & 1 & 1 & -1 & 0 & 0 & 0 & 10 \\
1 & 1 & 1 & 0 & 1 & 0 & 0 & 15 \\
1 & -\frac{1}{4} & 0 & 0 & 0 & -1 & 0 & 0 \\
-1 & 0 & 1 & 0 & 0 & 0 & -1 & 0 \\
\hline
.30 & .09 & .27 & 0 & 0 & 0 & 0 & 0
\end{array}
\right]
$$

Using technology we find the optimal solution is

$$
y_1 = 1\frac{2}{3}, \ y_2 = 6\frac{2}{3}, \ \text{and} \ y_3 = 1\frac{2}{3}.
$$

For each barrel of gasoline, the mixture should

contain $1\dfrac{2}{3}$ ounces of ingredient I, $6\dfrac{2}{3}$ ounces of

ingredient II, and $1\dfrac{2}{3}$ ounces of ingredient III at a

minimum cost of $1.55 per barrel.

35. From Exercise 23, we know the initial simplex tableau is

$$
\begin{array}{cccccccc}
y_1 & y_2 & y_3 & y_4 & y_5 & y_6 & y_7 & y_8
\end{array}
$$

$$
\left[
\begin{array}{cccccccc|c}
1 & 0 & 1 & 0 & -1 & 0 & 0 & 0 & 32 \\
0 & 1 & 0 & 1 & 0 & -1 & 0 & 0 & 20 \\
1 & 1 & 0 & 0 & 0 & 0 & 1 & 0 & 25 \\
0 & 0 & 1 & 1 & 0 & 0 & 0 & 1 & 30 \\
\hline
14 & 22 & 12 & 10 & 0 & 0 & 0 & 0 & 0
\end{array}
\right]
$$

Using technology we find the optimal solution is $y_1 = 22, \ y_2 = 0, \ y_3 = 10, \ \text{and} \ y_4 = 20.$

The manufacturer should send 22 computers from W_1 to D_1, 0 from W_1 to D_2, 10 from W_2 to D_1, and 20 computers from W_2 to D_2 for a minimum cost of $628.

Chapter 7 Review Exercises

1. $y \le 3x + 2$

 Graph the solid boundary line $y = 3x + 2$ which goes through the points (0, 2) and (2, 8). The inequality is $\le$, so shade the half-plane below the line.

 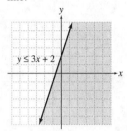

2. $2x - y \ge 6$

 Graph the solid boundary line $2x - y = 6$, which goes through the points (0, –6) and (3, 0). Testing the origin gives the statement $2(0) - 0 \ge 6$, which is false, so shade the half-plane not containing the origin, which is the region below the line.

 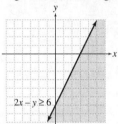

3. $3x + 4y \ge 12$

 Graph the solid boundary line $3x + 4y = 12$, which goes through (0, 3) and (4, 0). Testing the origin gives $3(0) + 4(0) \ge 12$, which is false, so shade the half-plane not containing the origin, which is the region above the line.

 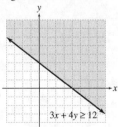

4. $y \le 4$

 Graph the solid boundary line $y = 4$, which is the horizontal line crossing the y-axis at (0, 4). Shade the half-plane below the line.

 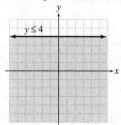

5. $x + y \le 6$

 $2x - y \ge 3$

 $x + y \le 6$ is the region on or below the line $x + y = 6$; $2x - y \ge 3$ is the region on or below the line $2x - y = 3$. The system of inequalities must meet both conditions so we shade the overlap of the two half-planes.

 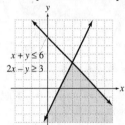

6. $4x + y \ge 8$

 $2x - 3y \le 6$

 Graph $4x + y = 8$ as a solid line using (2, 0) and (0, 8), and $2x - 3y = 6$ as a solid line using (3, 0) and (0, –2). The test point (0, 0) gives $0 + 0 \ge 8$, which is false and $0 - 0 \le 6$, which is true. Shade all points above $4x + y = 8$ and above $2x - 3y = 6$. The solution region is to the right of $4x + y = 8$ and above $2x - 3y = 6$.

 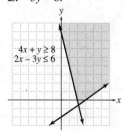

7. $2 \leq x \leq 5$

$1 \leq y \leq 7$

$x - y \leq 3$

The graph of $2 \leq x \leq 5$ is the region lying on or between the two vertical lines $x = 2$ and $x = 5$. The graph of $1 \leq y \leq 7$ is the region lying on or between the two horizontal lines $y = 1$ and $y = 7$. The graphs of $x - y \leq 3$ is the region lying on or below the line $x - y = 3$. Shade the region that is common to all three graphs.

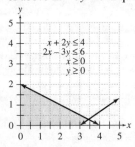

8. $x + 2y \leq 4$

$2x - 3y \leq 6$

$x \geq 0$

$y \geq 0$

Graph $x + 2y = 4$ and $2x - 3y = 6$ as solid lines. $x \geq 0$ and $y \geq 0$ restrict the region to quadrant I. Use $(0, 0)$ as a test point to get
$0 + 0 \leq 4$ and $0 + 0 \leq 6$, which are true. The region is all points on or below $x + 2y = 4$, and on or above $2x - 3y = 6$ in quadrant I.

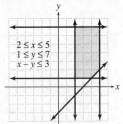

9. Let $x =$ the number of batches of cakes and $y =$ the number of batches of cookies. Then we have the following inequalities:

$2x + \frac{3}{2}y \leq 15$ Oven time

$3x + \frac{2}{3}y \leq 13$ Decorating

$x \geq 0$

$y \geq 0$.

The solution of this system of inequalities is the graph of the feasible region.

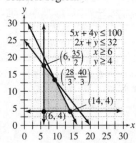

10. Let $x =$ the number of units of special and $y =$ the number of basic
Then we have the following inequalities:

$5x + 4y \leq 100$

$2x + y \leq 32$

$x \geq 6$

$y \geq 4$

The solution of this system is the graph of the feasible region.

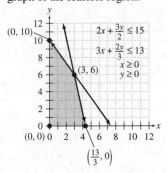

11.

Corner Point	Value of $z = 5x + 4y$
$(1, 6)$	11
$(6, 7)$	37
$(7, 3)$	38 Maximum
$\left(1, 2\frac{1}{2}\right)$	7.5 Minimum
$(2, 1)$	11

The maximum value of 38 occurs at $(7, 3)$ the minimum value of 7.5 occurs at $\left(1, 2\frac{1}{2}\right)$

12.

Corner Point	Value of $z = 5x + y$
$(0, 8)$	8 Minimum
$(8, 8)$	48 Maximum
$(5, 2)$	27
$(2, 0)$	10

The maximum value is 48 at $(8, 8)$; the minimum value is 8 at $(0, 8)$.

13. Maximize $\quad z = 2x + 4y$

subject to: $\quad 2x + 7y \le 14$

$\quad\quad\quad\quad 2x + 3y \le 10$

$\quad\quad\quad\quad x \ge 0, \ y \ge 0.$

Graph the feasible region.

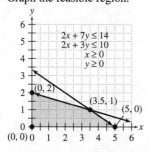

Corner Point	Value of $z = 2x + 4y$
$(0, 0)$	0
$(0, 2)$	8
$(3.5, 1)$	11 Maximum
$(5, 0)$	10

The maximum is 11 when $x = 3.5$, $y = 1$.

14. Find $x \ge 0$ and $y \ge 0$ such that

$8x + 9y \ge 72$

$6x + 8y \ge 72$

and $w = 10x + 2y$ is minimized. Graph the feasible region.

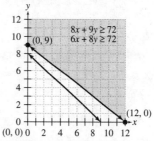

Corner Point	Value of $z = 10x + 2y$
$(0, 9)$	18 Minimum
$(12, 0)$	120

The minimum is 18 when $x = 0$, $y = 9$.

15. Find $x \ge 0$ and $y \ge 0$ such that

$x + y \le 50$

$2x + y \ge 20$

$x + 2y \ge 30$

and $w = 3x + 7y$ is minimized. Graph the feasible region.

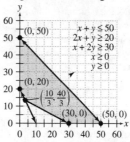

Corner Point	Value of $w = 3x + 7y$
$(0, 20)$	140
$(0, 50)$	350
$(50, 0)$	150
$(30, 0)$	90 Minimum
$\left(\frac{10}{3}, \frac{40}{3}\right)$	$\frac{310}{3} = 103\frac{1}{3}$

The minimum is 90 when $x = 30$, $y = 0$.

16. Maximize $z = -5x + 2y$

subject to: $3x + 2y \leq 12$

$5x + y \geq 5$

$x \geq 0, y \geq 0.$

Graph the feasible region.

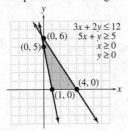

Corner Point	Value of $z = -5x + 2y$
(0, 5)	10
(0, 6)	12 Maximum
(4, 0)	–20
(1, 0)	–5

The maximum is 12 when $x = 0, y = 6.$

17. From the graph for Exercise 9, the corner points

are (0, 10), (3, 6), $\left(\dfrac{13}{3}, 0\right)$, and

(0, 0). Since x was the number of batches of cakes and y the number of batches of cookies, the revenue function is

$z = 30x + 20y.$ Evaluate this objective function at each corner point.

Corner Point	Value of $z = 30x + 20y$
(0, 10)	200
(3, 6)	210 Maximum
$\left(\frac{13}{3}, 0\right)$	130
(0, 0)	0

The bakery should make 3 batches of cakes and 6 batches of cookies to produce a maximum profit of $210.

18. The revenue function is $z = 20x + 18y.$ Refer to the graph in Exercise 10, and evaluate at the corner points.

Corner Point	Value of $z = 20x + 18y$
(6, 4)	192
$\left(6, \dfrac{35}{2}\right)$	435 Maximum
$\left(\dfrac{28}{3}, \dfrac{40}{3}\right)$	$\dfrac{1280}{3}$
(14, 4)	352

A maximum revenue of $435 is obtained by making 6 units of special pizza and

$\dfrac{35}{2}$ $\left(\text{or } 17\dfrac{1}{2}\right)$ units of basic pizza.

19. a. Let x_1 = the number of item A;

x_2 = the number of item B;

x_3 = the number of item C.

b. The objective function is

$z = 4x_1 + 3x_2 + 3x_3.$

c. The constraints are

$2x_1 + 3x_2 + 6x_3 \leq 1200$

$x_1 + 2x_2 + 2x_3 \leq 800$

$2x_1 + 2x_2 + 4x_3 \leq 500$

$x_1 \geq 0, x_2 \geq 0, x_3 \geq 0.$

20. a. Let x_1 = the amount invested in oil leases,

x_2 = the amount in bonds, and

x_3 = the amount invested in stock.

b. We want to maximize the objective function

$z = .15x_1 + .09x_2 + .05x_3.$

c. $x_1 + x_2 + x_3 \leq 50,000$

$x_1 + x_2 \quad\ \leq 15,000$

$x_1 \quad\ + x_3 \leq 25,000$

$x_1 \geq 0, x_2 \geq 0, x_3 \geq 0$

21. a. Let x_1 = the number of gallons of Fruity wine,

x_2 = the number of gallons of Crystal wine to be made.

b. The objective function is $z = 12x_1 + 15x_2$.

c. The ingredients available are the limitations. The constraints are

$2x_1 + x_2 \leq 110$

$2x_1 + 3x_2 \leq 125$

$2x_1 + x_2 \leq 90$

$x_1 \geq 0,\ x_2 \geq 0.$

22. a. Let x_1 = the number of 5-gallon bags,

x_2 = the number of 10-gallon bags,

x_3 = the number of 20-gallon bags.

b. The objective function is $z = x_1 + .9x_2 + .95x_3$.

c. $x_1 + 1.1x_2 + 1.5x_3 \leq 8$

$x_1 + 1.2x_2 + 1.3x_3 \leq 8$

$2x_1 + 3x_2 + 4x_3 \leq 8$

$x_1 \geq 0,\ x_2 \geq 0,\ x_3 \geq 0$

23. It is necessary to use the simplex method when there are more than two variables.

24. Problem with constraints involving "$\leq$" or "$=$" can be solved using slack variables, while those involving "$\geq$" or "$=$" can be solved using surplus variables.

25. Any standard minimization problem can be solved using the method of duals.

26. a. Maximize $z = 6x_1 + 7x_2 + 5x_3$

subject to: $4x_1 + 2x_2 + 3x_3 \leq 9$

$5x_1 + 4x_2 + x_3 \leq 10$

$x_1 \qquad\qquad\ \geq 0$

$x_2 \qquad\ \geq 0$

$x_3 \geq 0$

b. The first constraint would become

$4x_1 + 2x_2 + 3x_3 \geq 9$

c. $z = 227$ when $x_1 = 0,\ x_2 = 0,\ x_3 = 0$

d. Minimize: $w = 9y_1 + 10y_2$

subject to: $4y_1 + 5y_2 \geq 6$

$2y_1 + 4y_2 \geq 7$

$3y_1 + y_2 \geq 5$

$y_1 \qquad\ \geq 0$

$y_2 \geq 0$

e. The minimum value is $w = 227$, when $y_1 = 13$, and $y_2 = 11$.

27. Maximize $z = 2x_1 + 9x_2$

subject to: $3x_1 + 5x_2 \leq 47$

$x_1 + x_2 \leq 25$

$5x_1 + 2x_2 \leq 35$

$2x_1 + x_2 \leq 30$

$x_1 \geq 0,\ x_2 \geq 0.$

a. $3x_1 + 5x_2 + x_3 \qquad\qquad\qquad = 47$

$x_1 + x_2 \qquad + x_4 \qquad\qquad = 25$

$5x_1 + 2x_2 \qquad\qquad + x_5 \qquad = 35$

$2x_1 + x_2 \qquad\qquad\qquad + x_6 = 30$

b. $\begin{array}{cccccc} x_1 & x_2 & x_3 & x_4 & x_5 & x_6 \end{array}$

$$\left[\begin{array}{cccccc|c} 3 & 5 & 1 & 0 & 0 & 0 & 47 \\ 1 & 1 & 0 & 1 & 0 & 0 & 25 \\ 5 & 2 & 0 & 0 & 1 & 0 & 35 \\ 2 & 1 & 0 & 0 & 0 & 1 & 30 \\ \hline -2 & -9 & 0 & 0 & 0 & 0 & 0 \end{array}\right]$$

28. Maximize $z = 15x_1 + 12x_2$
subject to:
$$2x_1 + 5x_2 \le 50$$
$$x_1 + 3x_2 \le 25$$
$$4x_1 + x_2 \le 18$$
$$x_1 + x_2 \le 12$$
$$x_1 \ge 0, \; x_2 \ge 0.$$

a.
$$2x_1 + 5x_2 + x_3 \qquad = 50$$
$$x_1 + 3x_2 \qquad + x_4 \qquad = 25$$
$$4x_1 + x_2 \qquad + x_5 \qquad = 18$$
$$x_1 + x_2 \qquad + x_6 = 12$$

b.

$$
\begin{array}{cccccc}
x_1 & x_2 & x_3 & x_4 & x_5 & x_6
\end{array}
$$

$$
\left[
\begin{array}{cccccc|c}
2 & 5 & 1 & 0 & 0 & 0 & 50 \\
1 & 3 & 0 & 1 & 0 & 0 & 25 \\
4 & 1 & 0 & 0 & 1 & 0 & 18 \\
1 & 1 & 0 & 0 & 0 & 1 & 12 \\
\hline
-15 & -12 & 0 & 0 & 0 & 0 & 0
\end{array}
\right]
$$

29. Maximize $z = 4x_1 + 6x_2 + 3x_3$
subject to:
$$x_1 + x_2 + x_3 \le 100$$
$$2x_1 + 3x_2 \qquad \le 500$$
$$x_1 \qquad + 2x_3 \le 350$$
$$x_1 \ge 0, \; x_2 \ge 0, \; x_3 \ge 0.$$

a.
$$x_1 + x_2 + x_3 + x_4 \qquad = 100$$
$$2x_1 + 3x_2 \qquad + x_5 \qquad = 500$$
$$x_1 \qquad + 2x_3 \qquad + x_6 = 350$$

b.

$$
\begin{array}{cccccc}
x_1 & x_2 & x_3 & x_4 & x_5 & x_6
\end{array}
$$

$$
\left[
\begin{array}{cccccc|c}
1 & 1 & 1 & 1 & 0 & 0 & 100 \\
2 & 3 & 0 & 0 & 1 & 0 & 500 \\
1 & 0 & 2 & 0 & 0 & 1 & 350 \\
\hline
-5 & -6 & -3 & 0 & 0 & 0 & 0
\end{array}
\right]
$$

30. Maximize $z = x_1 + 8x_2 + 2x_3$
subject to:
$$x_1 + x_2 + x_3 \le 90$$
$$2x_1 + 5x_2 + x_3 \le 120$$
$$x_1 + 3x_2 \qquad \le 80$$
$$x_1 \ge 0, \; x_2 \ge 0, \; x_3 \ge 0.$$

a.
$$x_1 + x_2 + x_3 + x_4 \qquad = 90$$
$$2x_1 + 5x_2 + x_3 \qquad + x_5 \qquad = 120$$
$$x_1 + 3x_2 \qquad + x_6 = 80$$

b.

$$
\begin{array}{cccccc}
x_1 & x_2 & x_3 & x_4 & x_5 & x_6
\end{array}
$$

$$
\left[
\begin{array}{cccccc|c}
1 & 1 & 1 & 1 & 0 & 0 & 90 \\
2 & 5 & 1 & 0 & 1 & 0 & 120 \\
1 & 3 & 0 & 0 & 0 & 1 & 80 \\
\hline
-1 & -8 & -2 & 0 & 0 & 0 & 0
\end{array}
\right]
$$

31.

$$
\begin{array}{ccccc}
x_1 & x_2 & x_3 & x_4 & x_5
\end{array}
$$

$$
\left[
\begin{array}{ccccc|c}
1 & 2 & 3 & 1 & 0 & 28 \\
2 & 4 & 8 & 0 & 1 & 32 \\
\hline
-5 & -2 & -3 & 0 & 0 & 0
\end{array}
\right]
$$

Pivot on the 2 in row two, column one.

$$
\begin{array}{ccccc}
x_1 & x_2 & x_3 & x_4 & x_5
\end{array}
$$

$$
\left[
\begin{array}{ccccc|c}
1 & 2 & 3 & 1 & 0 & 28 \\
1 & 2 & 4 & 0 & \frac{1}{2} & 16 \\
\hline
-5 & -2 & -3 & 0 & 0 & 0
\end{array}
\right]
\begin{array}{l}
\\ \frac{1}{2}R_2 \\ \\
\end{array}
$$

$$
\begin{array}{ccccc}
x_1 & x_2 & x_3 & x_4 & x_5
\end{array}
$$

$$
\left[
\begin{array}{ccccc|c}
0 & 0 & -1 & 1 & -\frac{1}{2} & 12 \\
1 & 2 & 4 & 0 & \frac{1}{2} & 16 \\
\hline
0 & 8 & 17 & 0 & \frac{5}{2} & 80
\end{array}
\right]
\begin{array}{l}
-R_2 + R_1 \\ \\ 5R_2 + R_3
\end{array}
$$

The maximum is 80 when
$x_1 = 16$, $x_2 = 0$, $x_3 = 0$, $x_4 = 12$, $x_5 = 0$.

32.

$$
\begin{array}{cccc}
x_1 & x_2 & x_3 & x_4
\end{array}
$$

$$
\left[\begin{array}{cccc|c}
2 & 1 & 1 & 0 & 10 \\
9 & \underline{3} & 0 & 1 & 15 \\
\hline
-2 & -3 & 0 & 0 & 0
\end{array}\right]
$$

Pivot on the 3 in row two, column two.

$$
\begin{array}{cccc}
x_1 & x_2 & x_3 & x_4
\end{array}
$$

$$
\left[\begin{array}{cccc|c}
2 & 1 & 1 & 0 & 10 \\
3 & 1 & 0 & \frac{1}{3} & 5 \\
\hline
-2 & -3 & 0 & 0 & 0
\end{array}\right]\frac{1}{3}R_2
$$

$$
\begin{array}{cccc}
x_1 & x_2 & x_3 & x_4
\end{array}
$$

$$
\left[\begin{array}{cccc|c}
-1 & 0 & 1 & -\frac{1}{3} & 5 \\
3 & 1 & 0 & \frac{1}{3} & 5 \\
\hline
7 & 0 & 0 & 1 & 15
\end{array}\right]\begin{array}{l}-R_2+R_1\\ \\3R_2+R_3\end{array}
$$

The maximum is 15 when $x_1 = 0$, $x_2 = 5$, $x_3 = 5$, and $x_4 = 0$.

33.

$$
\begin{array}{cccccc}
x_1 & x_2 & x_3 & x_4 & x_5 & x_6
\end{array}
$$

$$
\left[\begin{array}{cccccc|c}
1 & 2 & 2 & 1 & 0 & 0 & 50 \\
\underline{4} & 24 & 0 & 0 & 1 & 0 & 20 \\
1 & 0 & 2 & 0 & 0 & 1 & 15 \\
\hline
-5 & -3 & -2 & 0 & 0 & 0 & 0
\end{array}\right]
$$

Pivot on the 4 in row two, column one.

$$
\begin{array}{cccccc}
x_1 & x_2 & x_3 & x_4 & x_5 & x_6
\end{array}
$$

$$
\left[\begin{array}{cccccc|c}
1 & 2 & 2 & 1 & 0 & 0 & 50 \\
1 & 6 & 0 & 0 & \frac{1}{4} & 0 & 5 \\
1 & 0 & 2 & 0 & 0 & 1 & 15 \\
\hline
-5 & -3 & -2 & 0 & 0 & 0 & 0
\end{array}\right]\frac{1}{4}R_2
$$

$$
\begin{array}{cccccc}
x_1 & x_2 & x_3 & x_4 & x_5 & x_6
\end{array}
$$

$$
\left[\begin{array}{cccccc|c}
0 & -4 & 2 & 1 & -\frac{1}{4} & 0 & 45 \\
1 & 6 & 0 & 0 & \frac{1}{4} & 0 & 5 \\
0 & -6 & \underline{2} & 0 & -\frac{1}{4} & 1 & 10 \\
\hline
0 & 27 & -2 & 0 & \frac{5}{4} & 0 & 25
\end{array}\right]\begin{array}{l}-R_2+R_1\\ \\-R_2+R_3\\5R_2+R_4\end{array}
$$

Pivot on the 2 in row three, column three.

$$
\begin{array}{cccccc}
x_1 & x_2 & x_3 & x_4 & x_5 & x_6
\end{array}
$$

$$
\left[\begin{array}{cccccc|c}
0 & -4 & 2 & 1 & -\frac{1}{4} & 0 & 45 \\
1 & 6 & 0 & 0 & \frac{1}{4} & 0 & 5 \\
0 & -3 & 1 & 0 & -\frac{1}{8} & \frac{1}{2} & 5 \\
\hline
0 & 27 & -2 & 0 & \frac{5}{4} & 0 & 25
\end{array}\right]\frac{1}{2}R_3
$$

33. Continued

$$
\begin{array}{cccccc}
x_1 & x_2 & x_3 & x_4 & x_5 & x_6
\end{array}
$$

$$
\left[\begin{array}{cccccc|c}
0 & 2 & 0 & 1 & 0 & -1 & 35 \\
1 & 6 & 0 & 0 & \frac{1}{4} & 0 & 5 \\
0 & -3 & 1 & 0 & -\frac{1}{8} & \frac{1}{2} & 5 \\
\hline
0 & 21 & 0 & 0 & 1 & 1 & 35
\end{array}\right]\begin{array}{l}-2R_3+R_1\\ \\ \\2R_3+R_4\end{array}
$$

The maximum is 35 when $x_1 = 5$, $x_2 = 0$, $x_3 = 5$, $x_4 = 35$, $x_5 = 0$, and $x_6 = 0$.

34.

$$
\begin{array}{ccccc}
x_1 & x_2 & x_3 & x_4 & x_5
\end{array}
$$

$$
\left[\begin{array}{ccccc|c}
1 & -2 & 1 & 0 & 0 & 38 \\
1 & -1 & 0 & 1 & 0 & 12 \\
2 & \underline{1} & 0 & 0 & 1 & 30 \\
\hline
-1 & -2 & 0 & 0 & 0 & 0
\end{array}\right]
$$

Pivot on the 1 in row three, column two.

$$
\begin{array}{ccccc}
x_1 & x_2 & x_3 & x_4 & x_5
\end{array}
$$

$$
\left[\begin{array}{ccccc|c}
5 & 0 & 1 & 0 & 2 & 98 \\
3 & 0 & 0 & 1 & 1 & 42 \\
2 & 1 & 0 & 0 & 1 & 30 \\
\hline
3 & 0 & 0 & 0 & 2 & 60
\end{array}\right]\begin{array}{l}2R_3+R_1\\R_3+R_2\\ \\2R_3+R_4\end{array}
$$

The maximum is 60 when $x_1 = 0$, $x_2 = 30$, $x_3 = 98$, $x_4 = 42$, and $x_5 = 0$.

35. Minimize $\quad w = 18y_1 + 10y_2$

subject to: $\quad y_1 + y_2 \geq 17$

$\qquad\qquad 5y_1 + 8y_2 \geq 42$

$\qquad\qquad y_1 \geq 0,\ y_2 \geq 0.$

Rewrite the objective function:

Maximize $\quad z = -18y_1 - 10y_2$

with the same constraints.

36. Minimize $\quad w = 12y_1 + 20y_2 - 8y_3$

subject to: $\quad y_1 + y_2 + 2y_3 \geq 48$

$\qquad\qquad y_1 + y_2 \qquad\ \geq 12$

$\qquad\qquad\qquad\qquad y_3 \geq 10$

$\qquad\qquad 3y_1 \qquad + y_3 \geq 30$

$\qquad\qquad y_1 \geq 0,\ y_2 \geq 0,\ y_3 \geq 0.$

Rewrite the objective function:

Maximize $\quad z = -12y_1 - 20y_2 + 8y_3$

with the same constraints.

37. Minimize $\quad w = 6y_1 - 3y_2 + 4y_3$

subject to: $\quad 2y_1 + y_2 + y_3 \geq 112$

$\qquad\qquad\quad y_1 + y_2 + y_3 \geq 80$

$\qquad\qquad\quad y_1 + y_2 \qquad \geq 45$

$\qquad\quad y_1 \geq 0,\ y_2 \geq 0,\ y_3 \geq 0.$

Rewrite the objective function:

Maximize $\quad z = -6y_1 + 3y_2 - 4y_3$

with the same constraints.

38. Maximize $\quad z = 2x_1 + 4x_2$

subject to: $\quad 3x_1 + 2x_2 \leq 12$

$\qquad\qquad\quad 5x_1 + x_2 \geq 5$

$\qquad\qquad\quad x_1 \geq 0,\ x_2 \geq 0.$

The initial simplex tableau is

$$\begin{array}{cccc} x_1 & x_2 & x_3 & x_4 \end{array}$$
$$\left[\begin{array}{cccc|c} 3 & 2 & 1 & 0 & 12 \\ 5 & \underline{1} & 0 & -1 & 5 \\ \hline -2 & -4 & 0 & 0 & 0 \end{array}\right].$$

For Stage I, pivot on the underlined 1.

$$\begin{array}{cccc} x_1 & x_2 & x_3 & x_4 \end{array}$$
$$\left[\begin{array}{cccc|c} -7 & 0 & 1 & \underline{2} & 2 \\ 5 & 1 & 0 & -1 & 5 \\ \hline 18 & 0 & 0 & -4 & 20 \end{array}\right]\begin{array}{l} -2R_2 + R_1 \\ \\ 4R_2 + R_3 \end{array}$$

For Stage II, pivot on the underlined 2.

$$\begin{array}{cccc} x_1 & x_2 & x_3 & x_4 \end{array}$$
$$\left[\begin{array}{cccc|c} -\frac{7}{2} & 0 & \frac{1}{2} & 1 & 1 \\ 5 & 1 & 0 & -1 & 5 \\ \hline 18 & 0 & 0 & -4 & 20 \end{array}\right]\frac{1}{2}R_1$$

$$\begin{array}{cccc} x_1 & x_2 & x_3 & x_4 \end{array}$$
$$\left[\begin{array}{cccc|c} -\frac{7}{2} & 0 & \frac{1}{2} & 1 & 1 \\ \frac{3}{2} & 1 & \frac{1}{2} & 0 & 6 \\ \hline 4 & 0 & 2 & 0 & 24 \end{array}\right]\begin{array}{l} \\ R_1 + R_2 \\ 4R_1 + R_3 \end{array}$$

The maximum is 24 when $x_1 = 0$ and $x_2 = 6$.

39. Minimize $\quad w = 4y_1 - 8y_2$

subject to: $\quad y_1 + y_2 \leq 50$

$\qquad\qquad\quad 2y_1 - 4y_2 \geq 20$

$\qquad\qquad\quad y_1 - y_2 \leq 22$

$\qquad\qquad\quad y_1 \geq 0,\ y_2 \geq 0.$

The initial simplex tableau is

$$\begin{array}{ccccc} y_1 & y_2 & y_3 & y_4 & y_5 \end{array}$$
$$\left[\begin{array}{ccccc|c} 1 & 1 & 1 & 0 & 0 & 50 \\ \underline{2} & -4 & 0 & -1 & 0 & 20 \\ 1 & -1 & 0 & 0 & 1 & 22 \\ \hline 4 & -8 & 0 & 0 & 0 & 0 \end{array}\right].$$

For Stage I, pivot on the 2 in row two, column one.

$$\begin{array}{ccccc} y_1 & y_2 & y_3 & y_4 & y_5 \end{array}$$
$$\left[\begin{array}{ccccc|c} 1 & 1 & 1 & 0 & 0 & 50 \\ 1 & -2 & 0 & -\frac{1}{2} & 0 & 10 \\ 1 & -1 & 0 & 0 & 1 & 22 \\ \hline 4 & -8 & 0 & 0 & 0 & 0 \end{array}\right]\frac{1}{2}R_2$$

$$\begin{array}{ccccc} y_1 & y_2 & y_3 & y_4 & y_5 \end{array}$$
$$\left[\begin{array}{ccccc|c} 0 & 3 & 1 & \frac{1}{2} & 0 & 40 \\ 1 & -2 & 0 & -\frac{1}{2} & 0 & 10 \\ 0 & 1 & 0 & \frac{1}{2} & 1 & 12 \\ \hline 0 & 0 & 0 & 2 & 0 & -40 \end{array}\right]\begin{array}{l} -R_2 + R_1 \\ \\ -R_2 + R_3 \\ -4R_2 + R_4 \end{array}$$

The minimum is 40 when $y_1 = 10$ and $y_2 = 0$.

40. If $w = -z$

$$\left[\begin{array}{cccccc|c} 0 & 1 & 0 & 2 & 5 & 0 & 17 \\ 0 & 0 & 1 & 3 & 1 & 1 & 25 \\ 1 & 0 & 0 & 4 & 2 & \frac{1}{2} & 8 \\ \hline 0 & 0 & 0 & 2 & 5 & 0 & -427 \end{array}\right]$$

indicates a minimum value of 427 at $(8, 17, 25, 0, 0, 0)$.

41.
$$\left[\begin{array}{ccccccc|c} 0 & 0 & 2 & 1 & 0 & 6 & 6 & 92 \\ 1 & 0 & 3 & 0 & 0 & 0 & 2 & 47 \\ 0 & 1 & 0 & 0 & 0 & 1 & 0 & 68 \\ 0 & 0 & 4 & 0 & 1 & 0 & 3 & 35 \\ \hline 0 & 0 & 5 & 0 & 0 & 2 & 9 & -1957 \end{array}\right]$$

The minimum value is 1957 at $(47, 68, 0, 92, 35, 0, 0)$.

42. Using the method of duals,

$$\begin{bmatrix} 1 & 0 & 0 & 3 & 1 & 2 & | & 12 \\ 0 & 0 & 1 & 4 & 5 & 3 & | & 5 \\ 0 & 1 & 0 & -2 & 7 & -6 & | & 8 \\ \hline 0 & 0 & 0 & 5 & 7 & 3 & | & 172 \end{bmatrix}$$

indicates a minimum value of 172 at (5, 7, 3, 0, 0, 0).

43. Using the method of duals,

$$\begin{bmatrix} 0 & 0 & 1 & 6 & 3 & 1 & | & 2 \\ 1 & 0 & 0 & 4 & -2 & 2 & | & 8 \\ 0 & 1 & 0 & 10 & 7 & 0 & | & 12 \\ \hline 0 & 0 & 0 & 9 & 5 & 8 & | & 62 \end{bmatrix}$$

indicates a minimum of 62 at (9, 5, 8, 0, 0, 0).

44. Using the method of duals,

$$\begin{bmatrix} 1 & 0 & 7 & -1 & | & 100 \\ 0 & 1 & 1 & 3 & | & 27 \\ \hline 0 & 0 & 7 & 2 & | & 640 \end{bmatrix}$$

indicates a minimum of 640 at (7, 2, 0, 0).

45. From Exercise 19, the initial simplex tableau is

$x_1 \quad x_2 \quad x_3 \quad x_4 \quad x_5 \quad x_6$

$$\begin{bmatrix} 2 & 3 & 6 & 1 & 0 & 0 & | & 1200 \\ 1 & 2 & 2 & 0 & 1 & 0 & | & 800 \\ \underline{2} & 2 & 4 & 0 & 0 & 1 & | & 500 \\ \hline -4 & -3 & -3 & 0 & 0 & 0 & | & 0 \end{bmatrix}.$$

Pivot on the 2 in row three, column one.

$x_1 \quad x_2 \quad x_3 \quad x_4 \quad x_5 \quad x_6$

$$\begin{bmatrix} 2 & 3 & 6 & 1 & 0 & 0 & | & 1200 \\ 1 & 2 & 2 & 0 & 1 & 0 & | & 800 \\ 1 & 1 & 2 & 0 & 0 & \frac{1}{2} & | & 250 \\ \hline -4 & -3 & -3 & 0 & 0 & 0 & | & 0 \end{bmatrix} \frac{1}{2}R_3$$

$x_1 \; x_2 \; x_3 \; x_4 \; x_5 \quad x_6$

$$\begin{bmatrix} 0 & 1 & 2 & 1 & 0 & -1 & | & 700 \\ 0 & 1 & 0 & 0 & 1 & -\frac{1}{2} & | & 550 \\ 1 & 1 & 2 & 0 & 0 & \frac{1}{2} & | & 250 \\ \hline 0 & 1 & 5 & 0 & 0 & 2 & | & 1000 \end{bmatrix} \begin{matrix} -2R_3 + R_1 \\ -1R_3 + R_2 \\ \\ 4R_3 + R_4 \end{matrix}$$

Roberta should get 250 units of item A, none of item B, and none of item C for a minimum profit of $1000.

46. From Exercise 20, we have the initial simplex tableau (after multiplying the fourth row by 100 to clear the decimals):

$x_1 \quad x_2 \quad x_3 \; x_4 \; x_5 \; x_6$

$$\begin{bmatrix} 1 & 1 & 1 & 1 & 0 & 0 & | & 50,000 \\ \underline{1} & 1 & 0 & 0 & 1 & 0 & | & 15,000 \\ 1 & 0 & 1 & 0 & 0 & 1 & | & 25,000 \\ \hline -15 & -9 & -5 & 0 & 0 & 0 & | & 0 \end{bmatrix}.$$

$x_1 \quad x_2 \quad x_3 \; x_4 \quad x_5 \; x_6$

$$\begin{bmatrix} 0 & 0 & 1 & 1 & -1 & 0 & | & 35,000 \\ 1 & 1 & 0 & 0 & 1 & 0 & | & 15,000 \\ 0 & -1 & \underline{1} & 0 & -1 & 1 & | & 10,000 \\ \hline 0 & 6 & -5 & 0 & 15 & 0 & | & 225,000 \end{bmatrix} \begin{matrix} -1R_2 + R_1 \\ \\ -1R_2 + R_3 \\ 15R_2 + R_4 \end{matrix}$$

$x_1 \quad x_2 \; x_3 \; x_4 \quad x_5 \quad x_6$

$$\begin{bmatrix} 0 & 1 & 0 & 1 & 0 & -1 & | & 25,000 \\ 1 & 1 & 0 & 0 & 1 & 0 & | & 15,000 \\ 0 & -1 & 1 & 0 & -1 & 1 & | & 10,000 \\ \hline 0 & 1 & 0 & 0 & 10 & 5 & | & 275,000 \end{bmatrix} \begin{matrix} -1R_3 + R_1 \\ \\ \\ 5R_3 + R_4 \end{matrix}$$

Since $x_1 = 15,000$ and $x_3 = 10,000$, he should invest $15,000 in oil leases and $10,000 in stock for a maximum return of .01(275,000) = $2750.

47. From Exercise 21, we have the initial simplex tableau:

$x_1 \quad x_2 \; x_3 \; x_4 \; x_5$

$$\begin{bmatrix} 2 & 1 & 1 & 0 & 0 & | & 110 \\ 2 & \underline{3} & 0 & 1 & 0 & | & 125 \\ 2 & 1 & 0 & 0 & 1 & | & 90 \\ \hline -12 & -15 & 0 & 0 & 0 & | & 0 \end{bmatrix}$$

$x_1 \qquad x_2 \; x_3 \; x_4 \; x_5$

$$\begin{bmatrix} 2 & 1 & 1 & 0 & 0 & | & 110 \\ \frac{2}{3} & \underline{1} & 0 & \frac{1}{3} & 0 & | & \frac{125}{3} \\ 2 & 1 & 0 & 0 & 1 & | & 90 \\ \hline -12 & -15 & 0 & 0 & 0 & | & 0 \end{bmatrix} \frac{1}{3}R_2$$

$x_1 \quad x_2 \; x_3 \quad x_4 \; x_5$

$$\begin{bmatrix} \frac{4}{3} & 0 & 1 & -\frac{1}{3} & 0 & | & \frac{205}{3} \\ \frac{2}{3} & 1 & 0 & \frac{1}{3} & 0 & | & \frac{125}{3} \\ \frac{4}{3} & 0 & 0 & -\frac{1}{3} & 1 & | & \frac{145}{3} \\ \hline -2 & 0 & 0 & 5 & 0 & | & 625 \end{bmatrix} \begin{matrix} -R_2 + R_1 \\ \\ -R_2 + R_3 \\ 15R_2 + R_4 \end{matrix}$$

47. Continued

$$x_1 \quad x_2 \quad x_3 \quad x_4 \quad x_5$$

$$\begin{bmatrix} \frac{4}{3} & 0 & 1 & -\frac{1}{3} & 0 & \frac{205}{3} \\ \frac{2}{3} & 1 & 0 & \frac{1}{3} & 0 & \frac{125}{3} \\ 1 & 0 & 0 & -\frac{1}{4} & \frac{3}{4} & \frac{145}{4} \\ \hline -2 & 0 & 0 & 5 & 0 & 625 \end{bmatrix} \frac{4}{3} R_3$$

$$\begin{bmatrix} 0 & 0 & 1 & 0 & -1 & 20 \\ 0 & 1 & 0 & \frac{1}{2} & -\frac{1}{2} & \frac{35}{2} \\ 1 & 0 & 0 & -\frac{1}{4} & \frac{3}{4} & \frac{145}{4} \\ \hline 0 & 0 & 0 & \frac{9}{2} & \frac{3}{2} & \frac{1395}{2} \end{bmatrix} \begin{matrix} -\frac{4}{3}R_3 + R_1 \\ -\frac{2}{3}R_3 + R_2 \\ \\ 2R_3 + R_4 \end{matrix}$$

The winery should make 17.5 gallons of Crystal wine and 36.25 gallons of Fruity wine for a maximum profit of $697.50.

48. From Exercise 22, we have the initial tableau:

$$x_1 \quad x_2 \quad x_3 \quad x_4 \ x_5 \ x_6$$

$$\begin{bmatrix} 1 & 1.1 & 1.5 & 1 & 0 & 0 & 8 \\ 1 & 1.2 & 1.3 & 0 & 1 & 0 & 8 \\ 2 & 3 & 4 & 0 & 0 & 1 & 8 \\ \hline -1 & -.9 & -.95 & 0 & 0 & 0 & 0 \end{bmatrix}.$$

$$x_1 \quad x_2 \quad x_3 \quad x_4 \ x_5 \ x_6$$

$$\begin{bmatrix} 1 & 1.1 & 1.5 & 1 & 0 & 0 & 8 \\ 1 & 1.2 & 1.3 & 0 & 1 & 0 & 8 \\ 1 & 1.5 & 2 & 0 & 0 & .5 & 4 \\ \hline -1 & -.9 & -.95 & 0 & 0 & 0 & 0 \end{bmatrix} \frac{1}{2} R_3$$

$$x_1 \quad x_2 \quad x_3 \ x_4 \ x_5 \quad x_6$$

$$\begin{bmatrix} 0 & -.4 & -.5 & 1 & 0 & -.5 & 4 \\ 0 & -.3 & -.7 & 0 & 1 & -.5 & 4 \\ 1 & 1.5 & 2 & 0 & 0 & .5 & 4 \\ \hline 0 & .6 & 1.05 & 0 & 0 & .5 & 4 \end{bmatrix} \begin{matrix} -1R_3 + R_1 \\ -1R_3 + R_2 \\ \\ R_3 + R_4 \end{matrix}$$

The final tableau gives $x_1 = 4$, $x_2 = 0$, $x_3 = 0$, and $z = 4$. Therefore, 4 units of 5-gallon bags (and none of the others) should be made for a maximum profit of $4 per unit.

49. Let y_1 = the number of cases of corn

$\quad y_2$ = the number of cases of beans

$\quad y_3$ = the number of cases of carrots

49. Continued

Minimize $w = 10y_1 + 15y_2 + 25y_3$

Maximize $z = -w = -10y_1 - 15y_2 - 25y_3$

subject to: $y_1 + y_2 + y_3 \geq 1000$

$\qquad\qquad y_1 - 2y_2 \qquad \geq 0$

$\qquad\qquad\qquad\qquad y_3 \geq 340$

$\qquad y_1 \geq 0, \ y_2 \geq 0, \ y_3 \geq 0$

The initial simplex tableau is

$$y_1 \quad y_2 \quad y_3 \quad y_4 \quad y_5 \quad y_6$$

$$\begin{bmatrix} 1 & 1 & 1 & -1 & 0 & 0 & 1000 \\ 1 & -2 & 0 & 0 & -1 & 0 & 0 \\ 0 & 0 & 1 & 0 & 0 & -1 & 340 \\ \hline 10 & 15 & 25 & 0 & 0 & 0 & 0 \end{bmatrix}$$

For Stage I, pivot on the 1 in row two, column one.

$$y_1 \quad y_2 \quad y_3 \ y_4 \quad y_5 \ y_6$$

$$\begin{bmatrix} 0 & 3 & 1 & -1 & 1 & 0 & 1000 \\ 1 & -2 & 0 & 0 & -1 & 0 & 0 \\ 0 & 0 & 1 & 0 & 0 & -1 & 340 \\ \hline 0 & 35 & 25 & 0 & 10 & 0 & 0 \end{bmatrix} \begin{matrix} -R_2 + R_1 \\ \\ \\ -10R_2 + R_4 \end{matrix}$$

Continuing Stage I, pivot on the 3 in row one, column two.

$$y_1 \ y_2 \ y_3 \quad y_4 \quad y_5 \ y_6$$

$$\begin{bmatrix} 0 & 1 & \frac{1}{3} & -\frac{1}{3} & \frac{1}{3} & 0 & \frac{1000}{3} \\ 1 & 0 & \frac{2}{3} & -\frac{2}{3} & -\frac{1}{3} & 0 & \frac{2000}{3} \\ 0 & 0 & 1 & 0 & 0 & -1 & 340 \\ \hline 0 & 0 & \frac{40}{3} & \frac{35}{3} & -\frac{5}{3} & 0 & -\frac{35,000}{3} \end{bmatrix} \begin{matrix} \frac{1}{3}R_1 \\ 2R_1 + R_2 \\ \\ -35R_1 + R_4 \end{matrix}$$

Pivot on the 1 in row three, column three.

$$y_1 \ y_2 \ y_3 \ y_4 \quad y_5 \quad y_6$$

$$\begin{bmatrix} 0 & 1 & 0 & -\frac{1}{3} & \frac{1}{3} & \frac{1}{3} & 220 \\ 1 & 0 & 0 & -\frac{2}{3} & -\frac{1}{3} & \frac{2}{3} & 440 \\ 0 & 0 & 1 & 0 & 0 & -1 & 340 \\ \hline 0 & 0 & 0 & \frac{35}{3} & -\frac{5}{3} & \frac{40}{3} & -16,200 \end{bmatrix} \begin{matrix} -\frac{1}{3}R_3 + R_1 \\ -\frac{2}{3}R_3 + R_2 \\ \\ -\frac{40}{3}R_3 + R_4 \end{matrix}$$

For Stage II, pivot on the $\dfrac{1}{3}$ in row one, column five.

$$y_1 \ y_2 \ y_3 \ y_4 \ y_5 \ y_6$$

$$\begin{bmatrix} 0 & 3 & 0 & -1 & 1 & 1 & 660 \\ 1 & 1 & 0 & -1 & 0 & 1 & 660 \\ 0 & 0 & 1 & 0 & 0 & -1 & 340 \\ \hline 0 & 5 & 0 & 10 & 0 & 15 & -15,100 \end{bmatrix} \begin{matrix} 3R_1 \\ \frac{1}{3}R_1 + R_2 \\ \\ \frac{5}{3}R_1 + R_4 \end{matrix}$$

The Cauchy Canners should produce 660 cases of corn, no cases of beans, and 340 cases of carrots for a minimum cost of $15,100.

50. First put the data into a table.

	Lumber	Concrete	Advertising	Total Spent
Atlantic	1000	3000	2000	$3000
Pacific	2000	3000	3000	$4000
Minimum Use	8000	18000	15000	

Let y_1 = the number of Atlantic boats and
y_2 = the number of pacific boats.
The problem is to minimize
$w = 3000y_1 + 4000y_2$
subject to:

$$1000y_1 + 2000y_2 \geq 8000$$
$$3000y_1 + 3000y_2 \geq 18,000$$
$$2000y_1 + 3000y_2 \geq 15,000$$
$$y_1 \geq 0, \ y_2 \geq 0.$$

The matrix for this problem is

$$\begin{bmatrix} 1000 & 2000 & 8000 \\ 3000 & 3000 & 18,000 \\ 2000 & 3000 & 15,000 \\ \hline 3000 & 4000 & 0 \end{bmatrix}.$$

The transpose of this matrix is

$$\begin{bmatrix} 1000 & 3000 & 2000 & 3000 \\ 2000 & 3000 & 3000 & 4000 \\ \hline 8000 & 18,000 & 15,000 & 0 \end{bmatrix}.$$

The dual problem is as follows.
Minimize
$z = 8000x_1 + 18,000x_2 + 15,000x_3$
subject to:
$$1000x_1 + 3000x_2 + 2000x_3 \leq 3000$$
$$2000x_1 + 3000x_2 + 3000x_3 \leq 4000$$
$$x_1 \geq 0, \ x_2 \geq 0, \ x_3 \geq 0.$$

The initial simplex tableau is

$$\begin{array}{ccccc} x_1 & x_2 & x_3 & x_4 & x_5 \end{array}$$

$$\begin{bmatrix} 1000 & \underline{3000} & 2000 & 1 & 0 & 3000 \\ 2000 & 3000 & 3000 & 0 & 1 & 4000 \\ \hline -8000 & -18,000 & -15,000 & 0 & 0 & 0 \end{bmatrix}.$$

Pivot on the 3000 in row one, column two.

$$\begin{array}{ccccc} x_1 & x_2 & x_3 & x_4 & x_5 \end{array} \qquad\qquad \begin{array}{ccccc} x_1 & x_2 & x_3 & x_4 & x_5 \end{array}$$

$$\begin{bmatrix} \frac{1}{3} & 1 & \frac{2}{3} & \frac{1}{3000} & 0 & 1 \\ 2000 & 3000 & 3000 & 0 & 1 & 4000 \\ \hline -8000 & -18,000 & -15,000 & 0 & 0 & 0 \end{bmatrix} \begin{array}{l} \frac{1}{3000}R_1 \end{array} \begin{bmatrix} \frac{1}{3} & 1 & \frac{2}{3} & \frac{1}{3000} & 0 & 1 \\ 1000 & 0 & \underline{1000} & -1 & 1 & 1000 \\ \hline -2000 & 0 & -3000 & 6 & 0 & 18,000 \end{bmatrix} \begin{array}{l} \\ -3000R_1 + R_2 \\ 18,000R_1 + R_3 \end{array}$$

Pivot on the 1000 in row two, column three.

50. Continued

$$
\begin{array}{ccccc}
x_1 & x_2 & x_3 & x_4 & x_5
\end{array}
$$

$$
\left[
\begin{array}{ccccc|c}
\frac{1}{3} & 1 & \frac{2}{3} & \frac{1}{3000} & 0 & 1 \\
1 & 0 & 1 & -\frac{1}{1000} & \frac{1}{1000} & 1 \\
\hline
-2000 & 0 & -3000 & 6 & 0 & 18{,}000
\end{array}
\right]
\begin{array}{l} \\ \frac{1}{1000}R_2 \\ \\ \end{array}
$$

$$
\begin{array}{ccccc}
x_1 & x_2 & x_3 & x_4 & x_5
\end{array}
$$

$$
\left[
\begin{array}{ccccc|c}
-\frac{1}{3} & 1 & 0 & \frac{1}{1000} & -\frac{1}{1500} & \frac{1}{3} \\
1 & 0 & 1 & -\frac{1}{1000} & \frac{1}{1000} & 1 \\
\hline
1000 & 0 & 0 & 3 & 3 & 21{,}000.
\end{array}
\right]
\begin{array}{l} -\frac{2}{3}R_2 + R_1 \\ \\ 3000R_2 + R_3 \end{array}
$$

The contractor should build 3 atlantic and 3 pacific models for a minimum cost of $21,000.

51. Let y_1 = kg of whole tomatoes   Minimize $w = 4y_1 + 3.25y_2$

y_2 = kg of tomato sauce

subject to: $y_1 + y_2 \le 3{,}000{,}000$

$\qquad\qquad\qquad y_1 \qquad\quad \ge 800{,}000$

$\qquad\qquad\qquad\qquad y_2 \ge 80{,}000$

$$\frac{6}{60}y_1 + \frac{3}{60}y_2 \ge 110{,}000$$

$\qquad\qquad y_1 \ge 0,\ y_2 \ge 0$

The initial simplex tableau is

$$
\begin{array}{cccccc}
y_1 & y_2 & y_3 & y_4 & y_5 & y_6
\end{array}
$$

$$
\left[
\begin{array}{cccccc|c}
1 & 1 & 1 & 0 & 0 & 0 & 3{,}000{,}000 \\
1 & 0 & 0 & -1 & 0 & 0 & 800{,}000 \\
0 & 1 & 0 & 0 & -1 & 0 & 80{,}000 \\
.1 & .05 & 0 & 0 & 0 & -1 & 110{,}000 \\
\hline
4 & 3.25 & 0 & 0 & 0 & 0 & 0
\end{array}
\right]
$$

For Stage I, pivot on the 1 in row two, column one.

$$
\begin{array}{cccccc}
y_1 & y_2 & y_3 & y_4 & y_5 & y_6
\end{array}
$$

$$
\left[
\begin{array}{cccccc|c}
0 & 1 & 1 & 1 & 0 & 0 & 2{,}200{,}000 \\
1 & 0 & 0 & -1 & 0 & 0 & 800{,}000 \\
0 & 1 & 0 & 0 & -1 & 0 & 80{,}000 \\
0 & .05 & 0 & .1 & 0 & -1 & 30{,}000 \\
\hline
0 & 3.25 & 0 & 4 & 0 & 0 & -3{,}200{,}000
\end{array}
\right]
\begin{array}{l} -R_2 + R_1 \\ \\ \\ -.1R_2 + R_4 \\ -4R_2 + R_5 \end{array}
$$

Pivot on the 1 in row three, column two.

$$
\begin{array}{cccccc}
y_1 & y_2 & y_3 & y_4 & y_5 & y_6
\end{array}
$$

$$
\left[
\begin{array}{cccccc|c}
0 & 0 & 1 & 1 & 1 & 0 & 2{,}120{,}000 \\
1 & 0 & 0 & -1 & 0 & 0 & 800{,}000 \\
0 & 1 & 0 & 0 & -1 & 0 & 80{,}000 \\
0 & 0 & 0 & .1 & .05 & -1 & 26{,}000 \\
\hline
0 & 0 & 0 & 4 & 3.25 & 0 & -3{,}460{,}000
\end{array}
\right]
\begin{array}{l} -R_3 + R_1 \\ \\ \\ -.05R_3 + R_4 \\ -3.25R_3 + R_5 \end{array}
$$

Pivot on the 1 in row four, column four.

51. Continued

$$\left[\begin{array}{cccccc|c}
0 & 0 & 1 & 0 & .5 & 10 & 1{,}860{,}000 \\
1 & 0 & 0 & 0 & .5 & -10 & 1{,}060{,}000 \\
0 & 1 & 0 & 0 & -1 & 0 & 80{,}000 \\
0 & 0 & 0 & 1 & .5 & -10 & 260{,}000 \\
\hline
0 & 0 & 0 & 0 & 1.25 & 40 & -4{,}500{,}000
\end{array}\right]
\begin{array}{l}
-R_4 + R_1 \\
R_4 + R_2 \\
\\
10R_4 \\
-4R_4 + R_5
\end{array}$$

This program is optimal after Stage I.

1,060,000 kg of tomatoes should be used for canned whole tomatoes and 80,000 kg of tomatoes should be used for sauce at a minimum cost of $4,500,000.

52. Let y_1 = the number of runs of type I and

y_2 = the number of runs of type II.

Minimize $w = 15{,}000y_1 + 6000y_2$

Maximize $z = -w = -15{,}000y_1 - 6000y_2$

subject to: $3000y_1 + 3000y_2 = 18{,}000$

$\qquad\qquad 2000y_1 + 1000y_2 = 7000$

$\qquad\qquad 2000y_1 + 3000y_2 \geq 14{,}000$

$\qquad\qquad\quad y_1 \geq 0,\ y_2 \geq 0$

The initial simplex tableau is

$$\begin{array}{ccccccc}
y_1 & y_2 & y_3 & y_4 & y_5 & y_6 & y_7
\end{array}$$
$$\left[\begin{array}{ccccccc|c}
3000 & 3000 & -1 & 0 & 0 & 0 & 0 & 18{,}000 \\
3000 & 3000 & 0 & 1 & 0 & 0 & 0 & 18{,}000 \\
\underline{2000} & 1000 & 0 & 0 & -1 & 0 & 0 & 7000 \\
2000 & 1000 & 0 & 0 & 0 & 1 & 0 & 7000 \\
2000 & 3000 & 0 & 0 & 0 & 0 & -1 & 14{,}000 \\
\hline
15{,}000 & 6000 & 0 & 0 & 0 & 0 & 0 & 0
\end{array}\right]$$

For Stage I, pivot on the 2000 in row three, column one.

$$\begin{array}{ccccccc}
y_1 & y_2 & y_3 & y_4 & y_5 & y_6 & y_7
\end{array}$$
$$\left[\begin{array}{ccccccc|c}
0 & 1500 & -1 & 0 & 1.5 & 0 & 0 & 7500 \\
0 & 1500 & 0 & 1 & 1.5 & 0 & 0 & 7500 \\
1 & .5 & 0 & 0 & -.0005 & 0 & 0 & 3.5 \\
0 & 0 & 0 & 0 & 1 & 1 & 0 & 0 \\
0 & \underline{2000} & 0 & 0 & 1 & 0 & -1 & 7000 \\
\hline
0 & -1500 & 0 & 0 & 7.5 & 0 & 0 & -52{,}500
\end{array}\right]
\begin{array}{l}
-3000R_3 + R_1 \\
-3000R_3 + R_2 \\
.0005R_3 \\
-2000R_3 + R_4 \\
-2000R_3 + R_5 \\
-15{,}000R_3 + R_6
\end{array}$$

Pivot on the 2000 in row five, column two.

$$\begin{array}{ccccccc}
y_1 & y_2 & y_3 & y_4 & y_5 & y_6 & y_7
\end{array}$$
$$\left[\begin{array}{ccccccc|c}
0 & 0 & -1 & 0 & .75 & 0 & \underline{.75} & 2250 \\
0 & 0 & 0 & 1 & .75 & 0 & .75 & 2250 \\
1 & 0 & 0 & 0 & -.00075 & 0 & .00025 & 1.75 \\
0 & 0 & 0 & 0 & 1 & 1 & 0 & 0 \\
0 & 1 & 0 & 0 & .0005 & 0 & -.0005 & 3.5 \\
\hline
0 & 0 & 0 & 0 & 8.25 & 0 & -.75 & -47{,}250
\end{array}\right]
\begin{array}{l}
-1500R_5 + R_1 \\
-1500R_5 + R_2 \\
-.5R_5 + R_3 \\
\\
.0005R_5 \\
1500R_5 + R_6
\end{array}$$

52. Continued

Pivot on the .75 in row one column seven.

$y_1 \ y_2 \ y_3 \ y_4 \ y_5 \ y_6 \ y_7$

$$
\begin{bmatrix}
0 & 0 & -\frac{4}{3} & 0 & 1 & 0 & 1 & 3000 \\
0 & 0 & 1 & 1 & 0 & 0 & 0 & 0 \\
1 & 0 & \frac{1}{3000} & 0 & -.001 & 0 & 0 & 1 \\
0 & 0 & 0 & 0 & 1 & 1 & 0 & 0 \\
0 & 1 & -\frac{1}{1500} & 0 & .001 & 0 & 0 & 5 \\
\hline
0 & 0 & -1 & 0 & 9 & 0 & 0 & -45,000
\end{bmatrix}
\begin{matrix}
\frac{4}{3}R_1 \\
-.75R_1 + R_2 \\
-.00025R_1 + R_3 \\
\\
.0005R_1 + R_5 \\
.75R_1 + R_6
\end{matrix}
$$

For Stage II, pivot on the 1 in row two, column three.

$y_1 \ y_2 \ y_3 \ y_4 \qquad y_5 \ y_6 \ y_7$

$$
\begin{bmatrix}
0 & 0 & 0 & \frac{4}{3} & 1 & 0 & 1 & 3000 \\
0 & 0 & 1 & 1 & 0 & 0 & 0 & 0 \\
1 & 0 & 0 & -\frac{1}{3000} & -.001 & 0 & 0 & 1 \\
0 & 0 & 0 & 0 & 1 & 1 & 0 & 0 \\
0 & 1 & 0 & \frac{1}{1500} & .001 & 0 & 0 & 5 \\
\hline
0 & 0 & 0 & 1 & 9 & 0 & 0 & -45,000
\end{bmatrix}
\begin{matrix}
\frac{4}{3}R_2 + R_1 \\
\\
-\frac{1}{3000}R_2 + R_3 \\
\\
\frac{1}{1500}R_2 + R_5 \\
R_2 + R_6
\end{matrix}
$$

The company should produce 1 run of type I and 5 runs of type II for a minimum cost of $45,000.

Case 7 Cooking with Linear Programming

1. Let x_1 = the number of 100 gram units of feta cheese

   x_2 = the number of 100 gram units of lettuce

   x_3 = the number of 100 gram units of salad dressing

   x_4 = the number of 100 gram units of tomato

Maximize $z = 4.09x_1 + 2.37x_2 + 2.5x_3 + 4.64x_4$

subject to:
$$263x_1 + \ 14x_2 + 448.8x_3 + 21x_4 < 260$$
$$492.5x_1 + 36x_2 \qquad\quad + 5x_4 > 210$$
$$10.33x_1 + 1.62x_2 \qquad\quad +.85x_4 > 6$$
$$x_1 + \ x_2 \qquad +x_3 + \ x_4 < 4$$
$$x_3 \qquad\quad \geq .3125$$
$$x_1 \geq 0, \ x_2 \geq 0, \ x_3 \geq 0, \ x_4 \geq 0$$

Using a computer with linear programming software, the optimal solution is:

$x_1 = .243037, \ x_2 = 2.35749, \ x_3 = .3125, \ x_4 = 1.08698$

Converting to kitchen units gives approximately $\frac{1}{6}$ cup feta cheese $4\frac{1}{4}$ cups of lettuce, $\frac{1}{8}$ cup of salad dressing

and $\frac{7}{8}$ of a tomato for a salad with a maximum of about 12.41 g of carbohydrates.

2. Let y_1 = the number of 100 gram units of beef

y_2 = the number of 100 gram units of oil

y_3 = the number of 100 gram units of onion

y_4 = the number of 100 gram units of soy sauce

Minimize $215y_1 + 884y_2 + 38y_3 + 60y_4$

subject to:
$$y_2 + 8.63y_3 + 5.57y_4 < 10$$
$$26y_1 \quad + 1.16y_3 + 10.51y_4 > 50$$
$$6.4y_3 \quad > 3.5$$
$$y_2 \quad \geq .045$$
$$y_1 \geq 0,\ y_2 \geq 0,\ y_3 \geq 0,\ y_4 \geq 0$$

Using a computer with linear programming software, the optimal solution is

$y_1 = 1.51873$, $y_2 = .045$, $y_3 = .546875$, $y_4 = .939941$

Converting to kitchen units gives approximately $5\frac{1}{3}$ ounces of beef, $\frac{1}{3}$ tablespoon of oil, $\frac{1}{2}$ an onion, and

$5\frac{1}{4}$ tablespoons of soy sauce for a stirfry with a minimum 443.48 calories.

Chapter 8: Sets and Probability

Section 8.1 Sets

1. $3 \in \{2, 5, 7, 9, 10\}$
This statement is false. The number 3 is not an element of the set.

3. $9 \notin \{2, 1, 5, 8\}$
The statement is true. 9 is not an element of the set.

5. $\{2, 5, 8, 9\} = \{2, 5, 9, 8\}$
The statement is true. The sets contain exactly the same elements, so they are equal. The ordering of the elements in a set is unimportant.

7. {all whole numbers greater than 7 and less than 10} = $\{8, 9\}$
The statement is true. 8 and 9 are the only such numbers.

9. $\{x \mid x$ is an odd integer, $6 \leq x \leq 18\}$
$= \{7, 9, 11, 15, 17\}$
The statement is false. The number 13 should be included.

11. Answers vary.
Possible answer:
$\{0\}$ is a set containing 1 element, namely zero.
$\{\varnothing\}$ is a set containing 1 element, namely, a mathematical symbol representing the empty set.

13. Since every element of A is also an element of U, $A \subseteq U$.

15. $A \nsubseteq E$ since A contains elements that do not belong to E, namely –3, 0, 3.

17. $\varnothing \subseteq A$ since the empty set is a subset of every set.

19. $D \subseteq B$, since every element of D is also an element of B.

21. $\{A, B, C\}$ contains 3 elements. Therefore, it has $2^3 = 8$ subsets.

23. $\{x \mid x$ is an integer strictly between 0 and 8\}
$= \{1, 2, 3, 4, 5, 6, 7\}$. This set contains 6 elements. Therefore it has $2^7 = 128$ subsets.

25. $\{x \mid x$ is an integer less than or equal to 0 or greater than or equal to 8\}

27. Answers vary.

29. $\{8, 11, 15\} \cap \{8, 11, 19, 20\} = \{8, 11\}$

31. $\{6, 12, 14, 16\} \cap \{6, 14, 19\} = \{6, 14\}$
$\{6, 14\}$ is the set of all elements belonging to both of the first two sets, so it is the intersection of those sets.

33. $\{3, 5, 9, 10\} \cup \varnothing = \{3, 5, 9, 10\}$

35. $\{1, 2, 4\} \cup \{1, 2\} = \{1, 2, 4\}$
The answer set $\{1, 2, 4\}$ consists of all elements belonging to the first set, to the second set, or to both sets, and therefore it is the union of the first two sets.

37. $X \cap Y = \{3, 5\}$ since only these elements are contained in both.

39. $X' = \{1, 7, 9, 11, 15\}$ since these are the elements that are in U but not in X.

41. $X' \cap Y' = \{1, 7, 9, 11, 15\} \cap \{1, 2, 4, 6, 11, 15\} = \{1, 11, 15\}$

43. $X \cup (Y \cap Z)$

$X = \{2, 3, 4, 5, 6\}$ and $Y \cap Z = \varnothing$.
Hence, the union of these two sets is
$X \cup (Y \cap Z) = \{2, 3, 4, 5, 6\}.$.

45. M' is the set of all students in this school not taking this course.

47. $N \cap P$ is the set of all students in this school taking both accounting and zoology.

49. A pair of sets is disjoint if the two sets have no elements in common. The pairs of these sets that are disjoint are C and D, A and E, C and E, and D and E.

51. C' is the set of all stocks with a zero or positive price change; $C' = \{\text{Allstate}\}$

53. $(A \cup B)'$ is the set of all stocks with a high price less than or equal to \$50 and a last price either less than or equal to \$20 or greater than or equal to \$45; $(A \cup B)' = \{\text{Goodyear}\}$.

55. $M \cap E$ is the set of all male employed applicants.

57. $M' \cup S'$ is the set of all female or married applicants.

59. $\{\text{Magazines, Internet}\}$

61. $F = \{\text{Comcast, Time Warner}\}$

63. $H = \{\text{Comcast, Time Warner, Cox Communications, Charter, Adelphia, Cablevision}\}$

65. $H \cup F = \{\text{Comcast Cable, Time Warner, Cox, Charter, Adelphia}\} \cup \{\text{ Comcast Cable, Time Warner }\}$
$\qquad = \{\text{Comcast, Time Warner}\}$

67. $U = \{s, d, c, g, i, m, h\}$
$O = \{i, m, h, g\}$
$O' = \{s, d, c\}$

69. $N = \{s, d, c, g\}$
$O = \{i, m, h, g\}$
$N \cap O = \{g\}$

Section 8.2 Applications of Venn Diagrams

1. $A \cap B'$
Shade the region inside B that is outside A.

$A \cap B'$

3. $B' \cup A'$
Shade the region outside B and the region outside of region A.

$B' \cup A'$

5. $B' \cup (A \cap B')$
First shade the common region that is in A and outside B. Then shade the rest of the region outside B.

$B' \cup (A \cap B')$

7. $U' = \varnothing$
There are no elements outside the universal set. Therefore, there is no region to be shaded.

9. Three sets divide the universal set into at most <u>8</u> regions.

11. $(A \cap C') \cup B$
First find $A \cap C'$, the region in A *and* not in C.

$A \cap C'$

For the union, we want the region in $(A \cap C')$ *or* in B, or both.

$(A \cap C') \cup B$

13. $A' \cap (B \cap C)$

First find $B \cap C$, the region in B *and* in C.

$B \cap C$

Now find A', the region not in A.

A'

For the intersection, we want the region in A' *and* in $(B \cap C)$.

$A' \cap (B \cap C)$

15. $(A \cap B') \cup C$

First find $A \cap B'$, the region in A *and* not in B.

$A \cap B'$

For the union, we want the region in $(A \cap B')$ *or* in C, or both.

$(A \cap B') \cup C$

17. Percentage of children under age 18 who lived with their mother only:
100% − 68.4% − 4.6% − 4.1% = 22.9%

19. Let W be the set of women, C the set of those who speak Cantonese, and F the set of those who lit firecrackers. We are given the following information.

$$n(W) = 120$$
$$n(C) = 150$$
$$n(F) = 170$$
$$n(W' \cap C) = 108$$
$$n(W' \cap F') = 100$$
$$n(W \cap C' \cap F) = 18$$
$$n(W' \cap C' \cap F') = 78$$
$$n(W \cap C \cap F) = 30$$

Note that
$n(W' \cap C \cap F')$
$= n(W' \cap F') - n(W' \cap C' \cap F')$
$= 100 - 78 = 22$.
Furthermore,
$n(W' \cap C \cap F)$
$= (W' \cap C) - n(W' \cap C \cap F')$
$= 108 - 22 = 86$.
We now have
$n(W \cap C \cap F')$
$= n(C) - n(W' \cap C \cap F)$
$\quad - n(W \cap C \cap F) - n(W' \cap C \cap F')$
$= 150 - 86 - 30 - 22 = 12$.
With all of the overlaps of W, C and F determined, we can now compute
$n(W \cap C' \cap F') = 60$ and
$n(W' \cap C' \cap F) = 36$.

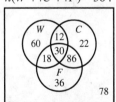

a. Adding up the disjoint components, we find the total attendance to be
60 + 12 + 18 + 30 + 22 + 86 + 36 + 78
= 342.

b. $n(C') = 342 - n(C)$
$\qquad = 342 - 150 = 192$

c. $n(W \cap F') = 60 + 12 = 72$

d. $n(W' \cap C \cap F) = 86$

21. Start with the innermost region: 6 with all three characteristics. Then work outward.

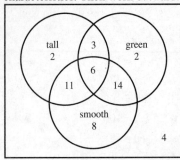

a. $4 + 6 + 3 + 14 + 11 + 2 + 8 + 2 = 50$ plants.

b. The number is inside the tall circle outside the green and smooth circles, or 2 plants.

c. The number is inside both the green and smooth circles but outside the tall circle, or 14 plants.

23. In the diagram, the regions are labeled (a) to (g). First, the number for each region is found by starting with innermost region (d). 15 had all three; then proceed to work outward as follows:
Region (a):
Since 25 had A, $25 - (2 + 15 + 1) = 7$.
Region (b):
Since 17 had A and B, $17 - 15 = 2$.
Region (c):
Since 27 had B, $27 - (2 + 15 + 7) = 3$.
Region (d):
15 had all three.
Region (e):
Since 22 had B and Rh, $22 - 15 = 7$.
Region (f):
Since 30 had Rh, $30 - (1 + 15 + 7) = 7$.
Region (g):
12 had none.
Region with no label:
Since 16 had A and Rh, $16 - 15 = 1$.

a. $7 + 2 + 3 + 15 + 7 + 7 + 12 + 1 = 54$ patients were represented.

b. $7 + 3 + 7 = 17$ patients had exactly one antigen.

c. $2 + 1 + 7 = 10$ patients had exactly two antigens.

d. Since a person having only the Rh antigen has type O-positive blood, 7 had type O-positive blood.

23. Continued

e. Since a person having A, B, and Rh antigens is AB-positive, 15 had AB-positive blood.

f. Since a person having only the B antigen is B-negative, 3 had B-negative blood.

g. Since a person having neither A, B, nor Rh antigens is O-negative, 12 had O-negative blood.

h. Since a person having A and Rh antigens is A-positive, 1 had A-positive blood.

25. a. $n(A \cap M) = 100$.

b. $n[C \cap (F \cup M)] = n(C \cap F) + n(C \cap M)$
$$= 52 + 36$$
$$= 88$$

c. $n(D \cup F) = n(D) + n(F) - n(D \cap F)$
$$= 18 + 288 - 11$$
$$= 295$$

d. $n(B' \cap E')$
$$= 100 + 128 + 36 + 52 + 7 + 11 = 334$$

27. a. $n(A \cap E) = 1{,}348$ thousand

b. $n(E \cup B) = n(E) + n(B) - n(E \cap B)$
$$= 23{,}201 + 213{,}219 - 21{,}853$$
$$= 214{,}567 \text{ thousand}$$

c. $n[D \cup (B \cap E)] = n(D) + n(B \cap E)$
$$= 143{,}571 + 21{,}853$$
$$= 165{,}424 \text{ thousand}$$

d. $n[E' \cap (A \cup C)] = n(C) + n(A \cap D)$
$$= 72{,}131 + 11{,}947$$
$$= 84{,}078 \text{ thousand}$$

e. $n(C' \cup A)$
$$= n(A \cup B) - n(B \cap C)$$
$$= 238{,}903 - 59{,}742$$
$$= 179{,}161 \text{ thousand}$$

f. $n(D' \cap A')$
$$= n(B) - n(B \cap D)$$
$$= 213{,}219 - 131{,}624$$
$$= 81{,}595 \text{ thousand}$$

29. Answers vary.

31. $n(A \cup B) = n(A) + n(B) - n(A \cap B)$
$$30 = 12 + 27 - n(A \cap B)$$
$$30 = 39 - n(A \cap B)$$
$$n(A \cap B) = 9$$

33. $n(A \cup B) = n(A) + n(B) - n(A \cap B)$
$$35 = 13 + n(B) - 5$$
$$35 = 8 + n(B)$$
$$n(B) = 27$$

35. $n(A) = 26$
$$n(B) = 10$$
$$n(A \cup B) = 30$$
$$n(A') = 17$$
This gives
$$n(A \cup B) = n(A) + n(B) - n(A \cap B)$$
$$30 = 26 + 10 - n(A \cap B)$$
$$n(A \cap B) = 26 + 10 - 30$$
$$= 6$$
$n(A) = 26$ and $n(A \cap B) = 6$, so
$$n(A \cap B') = 20$$
$n(B) = 10$ and $n(A \cap B) = 6$, so
$$n(B \cap A') = 4$$
Since $n(A')=17$, 4 of which are
accounted for in $B \cap A'$, 13 remain
in $A' \cap B'$.

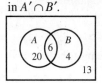

37. $n(A') = 28$
$$n(B) = 25$$
$$n(A' \cup B') = 45$$
$$n(A \cap B) = 12$$
$n(B) = 25$ and $n(A \cap B) = 12$, so
$$n(B \cap A') = 13$$
Since $n(A')=28$, of which 13 are
accounted for, 15 are in $A' \cap B'$.
$$n(A' \cup B') = n(A') + n(B') - n(A' \cap B')$$
$$45 = 28 + n(B') - 15$$
$$45 = 13 + n(B')$$
$$32 = n(B')$$
15 are in $A' \cap B'$, so the rest are in
$A \cap B'$, and $n(A \cap B')=17$.

39. $n(A) = 54$
$$n(A \cap B) = 22$$
$$n(A \cup B) = 85$$
$$n(A \cap B \cap C) = 4$$
$$n(A \cap C) = 15$$
$$n(B \cap C) = 16$$
$$n(C) = 44$$
$$n(B') = 63$$
Start with $A \cap B \cap C$. Now $n(A \cap C) = 15$, of
which 4 are in $A \cap B \cap C$, so
$n(A \cap B' \cap C) = 11$. $n(B \cap C) = 16$, of which 4
are in $A \cap B \cap C$, so $n(B \cap C \cap A') = 12$. $n(C) = 44$, so 17 are in $C \cap A' \cap B'$. $n(A \cap B) = 22$, so
18 are in $A \cap B \cap C'$. $n(A) = 54$, so $54 - 11 - 18$
$- 4 = 21$ are in $A \cap B' \cap C'$.
$$n(A \cup B) = n(A) + n(B) - n(A \cap B)$$
$$85 = 54 + n(B) - 22$$
$$53 = n(B)$$
This leaves 19 in $B \cap A' \cap C'$. $n(B') = 63$, of
which $21 + 11 + 17 = 49$ are accounted for,
leaving 14 in $A' \cap B' \cap C'$.

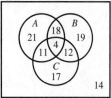

41. $(A \cap B)'$ is the complement of the intersection of A and B; hence it contains all elements not in $A \cap B$.

$(A \cap B)'$

$A' \cup B'$ is the union of the complements of A and B; hence it contains any element that is either not in A or not in B.

A'

B'

Note that $(A \cap B)' = A' \cup B'$, as claimed.

$A' \cup B'$

43. $A \cup (B \cap C)$ contains all points in A and the points where B and C overlap.

$A \cup (B \cap C)$

$(A \cup B) \cap (A \cup C)$ contains the points where $A \cup B$ and $A \cup C$ overlap.

$A \cup B$

$A \cup C$

Note that $A \cup (B \cap C) = (A \cup B) \cap (A \cup C)$, as claimed.

45. The complement of A intersect B equals the union of the complement of A and the complement of B.

47. A union (B intersect C) equals (A union B) intersect (A union C).

Section 8.3 Introduction to Probability

1. Answers vary.
Possible answer:
A coin or die is fair if the probability of any result is the same, that is, all results are equally likely.

3. There are 12 months in a year. The sample space is the set
{January, February, March, ..., December}.

5. There are 80 points on the test. The sample space is the set {0, 1, 2, ..., 80}.

7. There are only 2 choices the management can make. The sample space is the set
{go ahead, cancel}.

9. An event is a subset of outcomes from a sample space.

11. To have equally likely outcomes, each marble must be represented in the sample space. Let y, w, and b represent the three colors. Use subscripts to distinguish between marbles of the same color.

$S = \{y_1, y_2, y_3, w_1, w_2, w_3, w_4, b_1, b_2, b_3,$
$\quad b_4, b_5, b_6, b_7, b_8\}$

a. A yellow marble is drawn.
$\{y_1, y_2, y_3\}$

b. A blue marble is drawn.
$\{b_1, b_2, b_3, b_4, b_5, b_6, b_7, b_8\}$

c. A white marble is drawn.
$\{w_1, w_2, w_3, w_4\}$

d. A black marble is drawn. There are no black marbles, so this event is the empty set, Δ.

13. Let H = student guesses true, and T = student guesses false.

$S = \{$HHH, HHT, HTH, HTT, THH, THT, TTH, TTT$\}$

a. true twice and false once $\{$HHT, HTH, THH$\}$

b. all three false $\{$TTT$\}$

c. true once and false twice $\{$HTT, THT, TTH$\}$

15. There are 3 possibilities for the first choice and 3 for the second choice. The sample space is $\{$(flat, light beige), (flat, dark beige), (flat, black), (2 inch, light beige), (2 inch, dark beige), (2 inch, black), (3 inch, light beige), (3 inch, dark beige), (3 inch, black)$\}$

a. The shoe has a heel and is black: $\{$(2 inch, black), (3 inch, black)$\}$

b. The shoe has no heel and is beige: $\{$(flat, light beige), (flat, dark beige)$\}$

c. The shoe has a heel and is beige: $\{$(2 inch, light beige), (2 inch, dark beige), (3 inch, light beige), (3 inch, dark beige)$\}$.

17. $S = \{1, 2, 3, 4, 5, 6\}$
Let E be the event of "getting a 5."
$E = \{5\}$
$P(E) = \dfrac{n(E)}{n(S)} = \dfrac{1}{6}$

19. $S = \{1, 2, 3, 4, 5, 6\}$
Let E be the event "getting a number greater than 4."
$E = \{5, 6\}$
$P(E) = \dfrac{n(E)}{n(S)} = \dfrac{2}{6} = \dfrac{1}{3}$

21. $S = \{1, 2, 3, 4, 5, 6\}$
Let E be the event "getting a multiple of 3." So $E = \{3, 6\}$
Since S contains 6 elements, $P(E) = \dfrac{2}{6} = \dfrac{1}{3}$.

23. $P(8) = \dfrac{\text{number of ways to get a 8}}{\text{total number of cards}}$
$= \dfrac{4}{52} = \dfrac{1}{13}$

25. $P(\text{red } 10) = \dfrac{\text{number of ways to get a red 10}}{\text{total number of cards}}$
$= \dfrac{2}{52} = \dfrac{1}{26}$

27. $P(8 \text{ of diamonds})$
$= \dfrac{\text{number of ways to get a 8 of diamonds}}{\text{total number of cards}}$
$= \dfrac{1}{52}$

29. $P(\text{red 6 or black 9}) = \dfrac{4}{52} = \dfrac{1}{13}$

31. The table lists religious service attendance as percentages. The probability that a randomly selected respondent would attend a religious service

a. Nearly every week is 0.065.

b. Once a year is 0.139.

c. Every week is 0.165.

d. Less than every week is 0.75.

e. Every week or more is 0.242.

33. a. G' = The person is not overweight.

b. $F \cap G$ = The person has a family history of heart disease and is overweight.

c. $E \cup G'$ = The person smokes or is not overweight.

35. The total population (in thousands) for 2000 is 275,400. The total projected population (in thousands) for 2025 is 338,300. The probability that a randomly selected person is

    a. Hispanic in 2000 is $\dfrac{32,500}{275,400} \approx .118$.

    b. Hispanic in 2025 is $\dfrac{56,900}{338,300} \approx .168$.

    c. African-American in 2000 is $\dfrac{33,500}{275,400} \approx .122$.

    d. African-American in 2025 is
$\dfrac{44,700}{338,300} \approx .132$.

Section 8.4 Basic Concepts of Probability

1. Answers vary.

3. Wearing a hat and wearing glasses are not disjoint, because it is possible to wear both at the same time.

5. Being a doctor and being under 5 years old are disjoint, since it is impossible to be under 5 years old and be a doctor.

7. Being a female and being a pilot are not disjoint, since there are many female pilots.

For Exercises 9–13, count outcomes by referring to Figure 8.20 in the text.

9.   **a.** $P(\text{sum is } 8) = \dfrac{5}{36}$

    b. $P(\text{sum is } 9) = \dfrac{4}{36} = \dfrac{1}{9}$

    c. $P(\text{sum is } 10) = \dfrac{3}{36} = \dfrac{1}{12}$

    d. $P(\text{sum is } 13) = \dfrac{0}{36} = 0$

11.   **a.** $P(\text{not more than } 5) = \dfrac{10}{36} = \dfrac{5}{18}$

    b. $P(\text{not less than } 8) = \dfrac{15}{36} = \dfrac{5}{12}$

    c. $P(\text{between 3 and 7 (exclusive)}) = \dfrac{12}{36} = \dfrac{1}{3}$

13. The shoes come in two shades of beige (light and dark) and black, so $P(\text{shoes are black}) = \dfrac{1}{3}$.

15. **a.** $P(\text{a 4 or a queen}) = \dfrac{4}{52} + \dfrac{4}{52} = \dfrac{8}{52} = \dfrac{2}{13}$

    b. $P(\text{a 3 or a spade}) = \dfrac{4}{52} + \dfrac{13}{52} - \dfrac{1}{52}$

$$= \dfrac{16}{52}$$
$$= \dfrac{4}{13}$$

    c. $P(\text{a black card or a 9}) = \dfrac{26}{52} + \dfrac{4}{52} - \dfrac{2}{52}$

$$= \dfrac{28}{52}$$
$$= \dfrac{7}{13}$$

17. **a.** $P(\text{less than a 4}) = \dfrac{12}{52} = \dfrac{3}{13}$

    b. $P(\text{a club or a 4}) = \dfrac{13}{52} + \dfrac{4}{52} - \dfrac{1}{52}$

$$= \dfrac{16}{52}$$
$$= \dfrac{4}{13}$$

    c. $P(\text{a red card or an king}) = \dfrac{26}{52} + \dfrac{4}{52} - \dfrac{2}{52}$

$$= \dfrac{28}{52}$$
$$= \dfrac{7}{13}$$

    d. $P(\text{a club or a queen}) = \dfrac{13}{52} + \dfrac{4}{52} - \dfrac{1}{52}$

$$= \dfrac{16}{52}$$
$$= \dfrac{4}{13}$$

17. Continued

e. $P(\text{black card or face card}) = \dfrac{26}{52} + \dfrac{12}{52} - \dfrac{6}{52}$

$= \dfrac{32}{52}$

$= \dfrac{8}{13}$

19. a. $P(\text{an aunt or a cousin}) = \dfrac{3}{10} + \dfrac{2}{10} = \dfrac{5}{10} = \dfrac{1}{2}$

b. $P(\text{a male or a uncle}) = \dfrac{3}{10} + \dfrac{2}{10} - \dfrac{2}{10} = \dfrac{3}{10}$

c. $P(\text{a female or a cousin}) = \dfrac{4}{10} + \dfrac{2}{10} - \dfrac{1}{10}$

$= \dfrac{5}{10}$

$= \dfrac{1}{2}$

For Exercise 21, sample space is
$S = \{(\text{Connie, Casey}), (\text{Connie, Lindsey}),$
$(\text{Connie, Jackie}), (\text{Connie, Taisa}),$
$(\text{Connie, Lisa}), (\text{Casey, Connie}),$
$(\text{Casey, Lindsey}), (\text{Casey, Jackie}),$
$(\text{Casey, Taisa}), (\text{Casey, Lisa}),$
$(\text{Lindsey, Connie}), (\text{Lindsey, Casey}),$
$(\text{Lindsey, Jackie}), (\text{Lindsey, Taisa}),$
$(\text{Lindsey, Lisa}), (\text{Jackie, Connie}),$
$(\text{Jackie, Casey}), (\text{Jackie, Lindsey}),$
$(\text{Jackie, Taisa}), (\text{Jackie, Lisa}),$
$(\text{Taisa, Connie}), (\text{Taisa, Casey}),$
$(\text{Taisa, Lindsey}), (\text{Taisa, Jackie}),$
$(\text{Taisa, Lisa}), (\text{Lisa, Connie}),$
$(\text{Lisa, Casey}), (\text{Lisa, Lindsey}),$
$(\text{Lisa, Jackie}), (\text{Lisa, Taisa})\}$

The sample space has 30 elements.

21. $P(\text{first begins with "L" and second with "J"})$

$= \dfrac{2}{30} = \dfrac{1}{15}$

23.

a. $P(Z' \cap Y') = P((Z \cup Y)') = .42$

b. $P(Z' \cup Y') = 1 - P((Z \cap Y))$

$= 1 - .12$

$= .88$

c. $P(Z' \cup Y) = P(Z') + P(Y) - P(Z' \cap Y)$

$= (.42 + .18) + .3 - .18$

$= .72$

d. $P(Z \cap Y') = .28$

25. When rolling a die, there are 6 equally likely outcomes. The sample space is
$S = \{1, 2, 3, 4, 5, 6\}$.

Let E be the event "2 is rolled." Then $P(E) = \dfrac{1}{6}$

and $P(E') = \dfrac{5}{6}$. The odds in favor of rolling a 5

are $\dfrac{P(E)}{P(E')} = \dfrac{\frac{1}{6}}{\frac{5}{6}} = \dfrac{1}{5}$, written 1 to 5.

27. Let E be the event " 2, 3, 5, or 6 is rolled." Then
$P(E) = \dfrac{4}{6} = \dfrac{2}{3}$ and $P(E') = \dfrac{1}{3}$. The odds in favor

of rolling 1, 2, 3, or 4 are $\dfrac{P(E)}{P(E')} = \dfrac{\frac{2}{3}}{\frac{1}{3}} = 2$, written

2 to 1.

29. There are 3 yellow, 4 white, and 8 blue marbles.

a. Yellow: There are 3 ways to win and 12 ways to lose. The odds are 3 to 12 or 1 to 4.

b. Blue: There are 8 ways to win and 7 ways to lose; the odds are 8 to 7.

c. White: There are 4 ways to win and 11 ways to lose; the odds are 4 to 11.

31. Let E be the event "rolling a 7 or 11." Then
$P(E) = \dfrac{8}{36}$ and $P(E') = \dfrac{28}{36}$. The odds in favor of

rolling a 7 or 11 are $\dfrac{\frac{8}{36}}{\frac{28}{36}} = \dfrac{8}{28} = \dfrac{2}{7}$, written 2 to 7.

33. Answers vary.
Possible answer:
To correct this statement, one could say either
"odds in favor of a direct hit are very low" or
"odds against a direct hit are very high."

35. not relative frequency

37. relative frequency

39. relative frequency

41. not relative frequency

43. Answers vary.

45. This experiment is possible, since all probabilities are non-negative and
$.92 + .03 + 0 + .02 + .03 = 1$

47. This experiment is not possible, since the sum of the probabilities is $\dfrac{13}{12}$, which is greater than 1.

49. This experiment is not possible, since a probability cannot be negative.

51. Using a graphing calculator,

   a. $P(\text{the sum is 9 or more}) \approx .2778$

   b. $P(\text{the sum is less than 7}) \approx .4167$

   The probabilities compare very well to the results in Exercise 10.

53. Using a graphing calculator,

   a. $P(\text{exactly 4 heads}) \approx .15625$

   b. $P(\text{2 heads and 3 tails}) \approx .3125$

55. **a.** $P(\$35,000 \text{ or more})$
$= .151 + .083 + .110 + .140$
$= .484$

   b. $P(\text{less than } \$75,000) = 1 - P(\$75,000 \text{ or more}) = 1 - (.110 + .140) = 1 - .250 = .75$

   c. $P(\$25,000 \text{ to } \$74,999) = .123 + .151 + .083 = .357$

   d. $P(\$34,999 \text{ or less}) = .161 + .132 + .123 = .416$

57. **a.** $P(\text{less than } \$25) = .07 + .18 = .25$

   b. $P(\text{more than } \$24.99)$
$= .21 + .16 + .11 + .09 + .07 + .08 + .03$
$= .75$

   c. $P(\$50 \text{ to } \$199.99) = .16 + .11 + .09 = .36$

59. In fraction form, $.74 = \dfrac{37}{50}$. Therefore, the odds

against the company making a profit are $\dfrac{50-37}{37}$,

or 13 to 37.

61.

  a. $P(C') = .456 + .505 = .961$

  b. $P(M) = .035 + .456 = .491$

  c. $P(M') = .004 + .505 = .509$

  d. $P(M' \cap C') = 1 - (.004 + .035 + .456) = .505$

  e. $P(C \cap M') = .004$ (inside C and outside M)

  f. $P(C \cup M') = .004 + .035 + .505 = .544$

63. **a.** Since red is no longer dominant, RW or WR results in pink.
$P(\text{red}) = P(RR) = \dfrac{1}{4}$

   b. Pink is produced by RW or WR, so
$P(\text{pink}) = \dfrac{2}{4} = \dfrac{1}{2}$.

   c. $P(\text{white}) = P(WW) = \dfrac{1}{4}$

65.

	Employed (E)	Not Employed (U)
Male (M)	73,100	3339
Female (F)	62,464	3015

a. P(male and not employed)

$= \dfrac{3339}{141,918} \approx .024$

b. P(female and not employed)

$= \dfrac{3015}{141,918} \approx .021$

c. P(not employed)

$= \dfrac{3339+3015}{141,918} \approx .045$

d. P(female and employed)

$= \dfrac{62,464}{141,918} \approx .440$

67. a. Divide each number by total of 199,850.

	A	B	C	D
O	.053	.058	.039	.005
E	.310	.273	.215	.047

b. P(woman enlisted in the army)

$= P(E \cap A) = .310$

c. P(woman is officer in navy or marines)

$= P(O \cap C) + P(O \cap D)$

$= .039 + .005 = .044$

d. $P(A \cup B)$

$= P(A) + P(B) - P(A \cap B)$

$= .363 + .331 - 0 = .694$

e. $P(E \cup (C \cup D)) = P(E) + P(C \cup D)$

$- P(E \cap (C \cup D))$

$= .845 + .306 - (.215 + .047)$

$= .845 + .306 - .262 = .889$

Section 8.5 Conditional Probability and Independent Events

1. Roll a fair die.

$P(3 \mid \text{odd}) = \dfrac{P(3 \cap \text{odd})}{P(\text{odd})} = \dfrac{\frac{1}{6}}{\frac{1}{2}} = \dfrac{1}{3}$

3. $P(\text{odd} \mid 3) = \dfrac{P(\text{odd} \cap 3)}{P(3)} = \dfrac{\frac{1}{6}}{\frac{1}{6}} = 1$

5. There are 6 doubles, 1 of which has a sum of 6.

$P(\text{sum of } 6 \mid \text{double}) = \dfrac{1}{6}$

7. Since the first card is a heart, there are 51 cards remaining, 12 of them hearts.

$P(\text{second is heart} \mid \text{first is heart}) = \dfrac{12}{51} = \dfrac{4}{17}$

9. $P(\text{jack and } 10) = \dfrac{8 \cdot 4}{52 \cdot 51} \approx .012$

There are 8 possibilities for the first card (4 jacks and 4 tens), but for the second card there are only 4 (the 4 tens if a jack was picked or the 4 jacks if a 10 was picked).

11. Answers vary.

13. Answers vary.

15. No. Knowledge that a college is a religiously affiliated school does affect the probability that four semesters of theology would be required to graduate from that college. So, the events are dependent.

17. Yes. Knowledge that Tom Cruise's next movie grosses over $200 million does not give any information about the occurrence or nonoccurrence of the event that the Republicans have a majority in Congress in 2008. So, these events are independent.

19. No, for a two-child family, the knowledge that each child is the same sex influences the probability of the event that there is at most one male. Eliminating the "one of each" possiblity lowers the probability of "at most one male" from .75 to .5. However, the events are independent for a three-child family because the first event does not influence the probability of the second event. The probability of at most one male is .5 whether male-female mixes are allowed or not.

In Exercise 21, let S be the event: cabinet made by Sitlington; C be the event: cabinet made by Capek; Y be the event: cabinet is satisfactory; and N be the event: cabinet is unsatisfactory. Construct a probability tree.

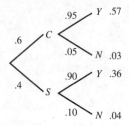

21. $P(C\mid N) = \dfrac{P(C \cap N)}{P(N)}$

$= \dfrac{.03}{.03 + .04}$

$= \dfrac{.03}{.07}$

$= \dfrac{3}{7}$

$\approx .43$

23. a.

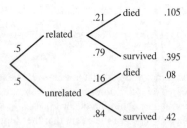

$P(\text{a child survives}) = .395 + .42$

$\qquad\qquad\qquad\qquad = .815$

b. Since .79 of children from a consanguineous marriage survived by age 10,

$.5(.79) = .395$

of the children from a consanguineous marriage survive.

25. a. $P(\text{via heterosexual contact} \mid \text{male})$

$= \dfrac{15,413}{88,866} \approx .17$

b. $P(\text{via intravenous drug use} \mid \text{female})$

$= \dfrac{6723}{34,592} \approx .19$

c. $P(\text{female})$

$= \dfrac{34,592}{123,458} \approx .28$

d. $P(\text{female} \mid \text{via heterosexual contact})$

$= \dfrac{26,882}{42,295} \approx .64$

27. $P(A \cup C') = \dfrac{2584}{5180} \approx ..499$

29. $P(F \cap B') = \dfrac{0}{5180} = 0$

31. $P(\text{rain} \mid \text{rain forecast})$

$= \dfrac{66}{222} \approx .30$

33. Answers vary.

35. $P(C) = .049$ (directly from the chart)

37. $P(M \cup C) = P(M) + P(C) - P(M \cap C)$

$\qquad\qquad = .527 + .049 - .042$

$\qquad\qquad = .534$

39. $P(M' \mid C) = \dfrac{P(M' \cap C)}{P(C)} = \dfrac{.007}{.049}$

$\qquad\qquad = \dfrac{1}{7} \approx .143$

41. $P(M') = .473$ Since $P(M') \neq P(M' \mid C)$,

$P(M' \mid C) = .007$

M' and C are dpendent.

43. Complete the probability tree

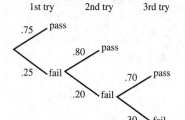

P(fails 1st and 2nd test)
$= P$(fails 1st) $\cdot P$(fails 2nd | fails 1st)
$= (.25)(.20) = .05$

45. P(requires at least 2 tries)
$= P$(does not pass on 1st try)
$= .25$

47.

$$P\left(\text{in 2002 read 51 or more books}\right) = \frac{8}{100} = .08$$

49. $P\left(\text{in 2005 read between 1 and 10 books}\right)$

$$= \frac{38+14}{100} = \frac{52}{100} = .52$$

51.

$$P\left(\text{vehicle registered in 2002}\right) = \frac{229,621}{879,833}$$

$$\approx .26$$

53. $P\left(\text{vehicle registered 1999 | vehicle a bus}\right)$

$$= \frac{729}{2986} \approx .24$$

55. Once the singing group has a hit, subsequent records are also hits. The only way to have exactly one hit in their first three records is for the first two not to be hits while the third is a hit.
P(one hit in first three records)
$= (.68)(.84)(.08)$
$\approx .0457$

57. P(have computer service)
$= 1 - P$(no computer service)
$= 1 - (.003)(.005)$
$= 1 - .000015$
$= .999985$
Answers vary. It is fairly realistic to assume independence because the chance of a failure of one computer does not usually depend on the failure of another, so long as the cause of failure does not lie with something the two systems have in common, like a power source.

59. **a.** The probability of success with one component is
$$1 - .03 = .97;$$
with 2, $1 - (.03)^2 = .9991;$
with 3, $1 - (.03)^3 = .999973;$
with 4, $1 - (.03)^4 = .99999919.$
Therefore, 4 (the original and 3 backups) will do the job.

   b. Answers vary. It is probably reasonable to assume independence here so long as the cause of failure is some internal defect and the components are from different manufacture lots.

61. Let A be the event "student studies" and B be the event "student gets a good grade." We are told that
$P(A) = .6$, $P(B) = .7$, and $P(A \cap B) = .52.$
$P(A) \cdot P(B) = (.6)(.7) = .42$
Since $P(A) \cdot P(B)$ is not equal to $P(A \cap B)$,
A and B are not independent. Rather, they are dependent events.

Section 8.6 Bayes' Formula

For Exercise 1:

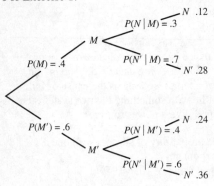

1. $P(M \mid N) = \dfrac{.12}{.12 + .24} = \dfrac{.12}{.36} = \dfrac{1}{3}$

For Exercises 3 and 5:

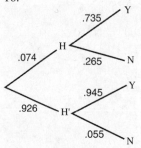

3. $P(R_1 \mid Q) = \dfrac{.02}{.02 + .18 + .21} = \dfrac{.02}{.41} = \dfrac{2}{41} \approx .0488$

5. $P(R_3 \mid Q)$

$= \dfrac{.21}{.02 + .18 + .21} = \dfrac{.21}{.41} = \dfrac{21}{41} \approx .5122$

For Exercise 7:

jar 1 $\underline{P(\text{white} \mid \text{jar 1}) = 1/3}$ white 1/6

$P(\text{jar 1}) = 1/2$

$P(\text{jar 2}) = 1/3$ jar 2 $\underline{P(\text{white} \mid \text{jar 2}) = 2/3}$ white 2/9

$P(\text{jar 3}) = 1/6$

jar 3 $\underline{P(\text{white} \mid \text{jar 3}) = 1/2}$ white 1/12

7. $P(\text{jar 2} \mid \text{white})$

$= \dfrac{\dfrac{2}{9}}{\dfrac{1}{6} + \dfrac{2}{9} + \dfrac{1}{12}} = \dfrac{8}{17} \approx .4706$

For Exercises 9 and 10:
Let H mean "Hispanic" and Y mean "living in U.S. at age 16."

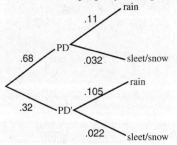

9. Use Bayes' formula.
$P(\text{H} \mid \text{live outside U.S. at age 16})$

$= \dfrac{(.074)(.265)}{(.074)(.265) + (.926)(.055)} = \dfrac{.01961}{.07054} \approx .278$

For Exercise 11:
Let PD mean "property damage only."

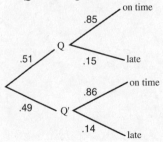

11. Use Bayes' formula.
$P(\text{PD} \mid \text{occurred during snow/sleet})$

$= \dfrac{(.68)(.032)}{(.68)(.032) + (.32)(.022)} = \dfrac{.02176}{.02880} \approx .7556$

For Exercise 13:
Let Q mean "Quantas Airline."

on time
.85

Q
.51   .15   late

on time
.86

.49 Q'

.14   late

13. $P(Q \mid \text{late}) = \dfrac{(.51)(.15)}{(.51)(.15)+(.49)(.14)}$

$= \dfrac{.0765}{.1451} \approx .5272$

For Exercise 15:
Let C mean "college graduate."

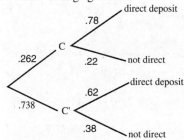

15. $P(C \mid \text{direct deposit})$

$= \dfrac{(.262)(.78)}{(.262)(.78)+(.738)(.62)} = \dfrac{.20436}{.66192} \approx .3087$

For Exercise 17:
Let H mean "household is headed by someone age 65 or older."

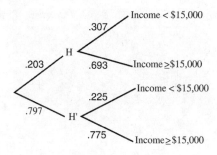

17. $P(H \mid \text{income} < \$15{,}000)$

$= \dfrac{(.203)(.307)}{(.203)(.307)+(.797)(.225)} = \dfrac{.0623}{.2416}$

$\approx .2579$

19. Let L be the event "the object was shipped by land," A be the event "the object was shipped by air," S be the event "the object was shipped by sea," and E be the event "an error occurred."
$P(L \mid E)$

$= \dfrac{P(L) \cdot P(E \mid L)}{P(L) \cdot P(E \mid L)+P(A) \cdot P(E \mid A)+P(S) \cdot P(E \mid S)}$

$= \dfrac{(.50)(.02)}{(.50)(.02)+(.40)(.04)+(.10)(.14)}$

$= \dfrac{.0100}{.0400}$

$= .25$

The correct response is (c).

For Exercise 21:

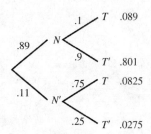

21. $P(N' \mid T) = \dfrac{.0825}{.089+.0825} = \dfrac{.0825}{.1715} \approx .481$

For Exercise 23:
Let Y mean "American was 18 years old or younger."

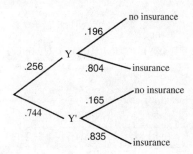

23. $P(Y' \mid \text{has health insurance}) =$

$\dfrac{(.744)(.835)}{(.744)(.835)+(.256)(.804)} = \dfrac{.6212}{.8271}$

$\approx .7511$

25. a. $P(D^+ | T^+) = \dfrac{.02(.54)}{.02(.54) + .98(.06)} \approx .155$

b. $P(D^- | T^-) = \dfrac{.98(.94)}{.98(.94) + .02(.46)} \approx .990$

c. $P(T^+ \cap D^-) = P(T^+ | D^-) \cdot P(D^-)$

$= (.06)(.98)$

$= .0588$

$= 5.88\%$

5.88% of 1000 exams is about 59 false positives.

27. Let FS mean "lives in fraternity/sorority house," D mean "lives in dormitory," and C mean "lives off campus."

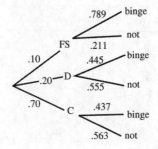

a. $P(\text{binge}) = .1(.789) + .2(.445) + .7(.437)$

$= .4738$

b. $P(FS | \text{binge}) = \dfrac{.1(.789)}{.4738}$

$= .1665$

29. $P(\text{In South} | \text{Uses Central AC})$

$= \dfrac{(.352)(.694)}{(.194)(.221) + (.237)(.513) + (.352)(.694) + (.215)(.269)}$

$\approx .524$

31. $P(55\text{–}74 | \text{lives alone})$

$= \dfrac{.212(.191)}{(.293)(.088) + (.412)(.107) + (.212)(.191) + (.083)(.393)}$

$= \dfrac{.040492}{.142979}$

$\approx .2832$

33. $P(\text{not living alone})$

$= 1 - \left[(.293)(.088) + (.412)(.107) + (.212)(.191) + (.083)(.393)\right]$

$= .857$

35. $P(75 \text{ or over} | \text{not living alone})$

$= \dfrac{(.083)(1 - .393)}{(.293)(1 - .088) + (.412)(1 - .107) + (.212)(1 - .191) + (.083)(1 - .393)}$

$= \dfrac{.050381}{.857021}$

$\approx .059$

Chapter 8 Review Exercises

1. $9 \in \{8, 4, -3, -9, 6\}$
Because 9 is not an element of the given set, the statement is false.

2. $4 \in \{3, 9, 7\}$
Because 4 is not an element of the given set, the statement is false.

3. $2 \notin \{0, 1, 2, 3, 4\}$
Because 2 is an element of the given set, the statement is false.

4. $0 \notin \{0, 1, 2, 3, 4\}$
Because 0 is an element of the given set, the statement is false.

5. $\{3, 4, 5\} \subseteq \{2, 3, 4, 5, 6\}$
The statement is true because every member of the first set is in the second set.

6. $(1, 2, 5, 8) \subseteq \{1, 2, 5, 10, 11\}$
This statement is false because 8 is an element of the first set but not of the second.

7. $\{1, 5, 9\} \subset \{1, 5, 6, 9, 10\}$
This statement is true because every member of the first set is a member of the second set and the second set contains at least one element not in the first set.

8. $0 \subseteq \emptyset \triangle$
This statement is false because the empty set has no subsets except itself. Also, 0 is an element, not a set, so it cannot be a subset.

9. $\{x \mid x$ is a national holiday$\}$ = {New Year's Day, Martin Luther King's Birthday, Presidents' Day, Memorial Day, Independence Day, Labor Day, Columbus Day, Veterans' Day, Thanksgiving, Christmas}

10. $\{x \mid x$ is an integer, $-3 \le x < 1\}$
$= \{-3, -2, -1, 0\}$

11. {all counting numbers less than 5}
$= \{1, 2, 3, 4\}$

12. $\{x \mid x$ is a leap year between 1989 and 2006$\}$
$= \{1992, 1996, 2000, 2004\}$

13. M' contains all the elements of U not in M.
$M' = \{B_1, B_2, B_3, B_6, B_{12}\}$

14. N' contains all the elements of U not in N.
$N' = \{B_3, B_6, B_{12}, D\}$

15. $M \cap N$ contains all the elements that are common to M and N.
$M \cap N = \{A, C, E\}$

16. $M \cup N$ contains all the elements in either M or N or both.
$M \cup N = \{A, B_1, B_2, C, D, E\}$

17. $M \cup N'$ contains all the elements in either M or not in N, or both.
$M \cup N' = \{A, B_3, B_6, B_{12}, C, D, E\}$

18. $M' \cap N$ contains all the elements in N and not in M.
$M' \cap N = \{B_1, B_2\}$

19. $A \cap C$ is the set of all students who are majoring in business and have brown eyes.

20. $B \cap D$ is the set of all students in the class have a GPA less than 3.0 and are younger than 25.

21. $A \cup D$ is the set of all students in the class are majoring in business or are younger than 25.

22. $A' \cap D$ is the set of all students in the class who are not majoring in business are younger than 25..

23. $B' \cap C'$ is the set of all students in the class who have a GPA greater than or equal to 3.0 and who do not have brown eyes.

24. $B \cup A'$

Shade the region inside B as well as all of the region outside of A.

$B \cup A'$

25. $A' \cap B$

Shade all the region inside B that is also outside of A.

26. $A' \cap (B' \cap C)$

First choose the regions that are inside C and outside of B. From those regions, then shade the region outside of A.

27. $(A \cup B)' \cap C$

First choose the regions that are inside either A or B. Then choose all the regions outside those you have chosen. Now shade from those regions the region in C.

For Exercises 28–30, use the following Venn diagram. Let F represent the set of officers, M the set of minorities, and W the set of women.

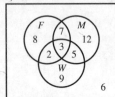

28. Add the numbers in all the regions.
$8 + 7 + 12 + 2 + 3 + 5 + 9 + 6 = 52$
A total of 52 people were interviewed.

29. Look for the region outside the officer circle, inside the minority circle, and inside the woman circle. There were 5 enlisted minority women.

30. Look for the region outside the woman circle, inside the minority circle, and inside the officer circle. There were 7 male minority officers.

31. The sample space for rolling a die is $\{1, 2, 3, 4, 5, 6\}$.

32. The sample space for drawing a card from a deck containing only 4 aces is
{ace of hearts, ace of diamonds, ace of spades, ace of clubs}.

33. The sample space for choosing a color and then a number is {(red, 10), (red, 20), (red, 30), (blue, 10), (blue, 20), (blue, 30), (green, 10), (green, 20), (green, 30)}.

34. The sample space for choosing one of 5 discs and then a color ball is {(2, blue), (2, yellow), (4,blue), (4, yellow), (6, blue), (6, yellow), (8, blue), (8, yellow), (10, blue), (10, yellow)}.

35. Event F, the ball is blue.
$F = \{(2, \text{blue}), (4, \text{blue}), (6, \text{blue}), (8, \text{blue}), (10, \text{blue})\}$

36. Event E, the disc shows a number greater than 5,
$E = \{(6, \text{blue}), (6, \text{yellow}), (8, \text{blue}), (8, \text{yellow}), (10, \text{blue}), (10, \text{yellow})\}$

37. No, the outcomes in this sample space are not equally likely because there are more yellow balls than blue balls.

38. "A customer buys neither" is written $E' \cap F'$.
(This event can also be written as $(E \cup F)'$.)

39. "A customer buys at least one" is written $E \cup F$.

40. Answers vary.
Possible answer:
This answer must be incorrect because any probability must be between 0 and 1 inclusive.

41. Answers vary.
Possible answer:
Disjoint sets are sets which have no elements in common, for example, the set of all females and the set of all people who have been President of the United States.

42. Answers vary.
Possible answer:
Two events are mutually exclusive if their intersection is the empty set. An example is the events "rolling a die and getting a 3" and "rolling a die and getting an even number."

43. Answers vary.
Possible answer:
Disjoint sets and mutually exclusive events both have no elements in common, that is, their intersection is the empty set. Mutually exclusive events are in fact disjoint subsets of the sample space.

44. There are 2 red queens (the queen of hearts and the queen of diamonds) in a deck of 52 cards, so

$$P(\text{red queen}) = \frac{2}{52} = \frac{1}{26}.$$

45. There are 12 face cards (jack, queen, and king of each suit) in a deck of 52 cards, so

$$P(\text{face card}) = \frac{12}{52} = \frac{3}{13}.$$

46. Let R be the event "red card is drawn" and F be the event "face card is drawn." There are 26 red cards and 12 face cards, of which 6 are red. Use the union rule of probability.

$$
\begin{aligned}
P(R \cup F) &= P(R) + P(F) - P(R \cap F) \\
&= \frac{26}{52} + \frac{12}{52} - \frac{6}{52} \\
&= \frac{32}{52} \\
&= \frac{8}{13}
\end{aligned}
$$

47.
$$
\begin{aligned}
P(\text{black} \mid 10) &= \frac{P(\text{black} \cap 10)}{P(10)} \\
&= \frac{\frac{2}{52}}{\frac{4}{52}} \\
&= \frac{2}{4} \\
&= \frac{1}{2}
\end{aligned}
$$

48. There are 12 face cards, of which 4 are jacks, so

$$P(\text{jack} \mid \text{face card}) = \frac{4}{12} = \frac{1}{3}.$$

49. There are 4 jacks, all of which are face cards, so

$$P(\text{face card} \mid \text{king}) = \frac{4}{4} = 1.$$

50. There are 13 clubs and 39 non-clubs. The odds in favor of drawing a club are

$$\frac{13}{39} = \frac{1}{3}, \text{ written 1 to 3.}$$

51. There are 2 red queens and 50 other cards. The odds in favor of drawing a red queen are

$$\frac{2}{50} = \frac{1}{25}, \text{ written 1 to 25.}$$

52. There are six black face cards, four 9's, and 42 cards that are neither black face cards nor 9's. The odds in favor of drawing a black face card or a 7 are

$$\frac{10}{42} = \frac{5}{21}, \text{ written 5 to 21.}$$

53. $P(\text{no more than 3 defective})$
$= .31 + .25 + .18 + .12 = .86$

54. $P\begin{pmatrix} \text{at least} \\ \text{3 defective} \end{pmatrix} = P(3) + P(4) + P(5)$
$$= .12 + .08 + .06$$
$$= .26$$

55.

		2nd Parent	
		N_2	T_2
1st	N_1	$N_1 N_2$	$N_1 T_2$
Parent	T_1	$T_1 N_2$	$T_1 T_2$

56. $P(\text{child has disease}) = P(T_1 T_2)$
$$= \frac{1}{4}$$

57. There are 4 possible combinations, but only 2 have a normal cell combined with a trait cell $(N_1 T_2, T_1 N_2)$.

$$P(\text{child is carrier}) = \frac{2}{4} = \frac{1}{2}$$

58. $P(\text{child is neither carrier nor has disease})$
$$= P(N_1 N_2) = \frac{1}{4}$$

59. There are 36 possibilities, with 5 having a sum of 8:
$(4, 4), (3, 5), (5, 3), (2, 6)$ and $(6, 2)$.

$$P(8) = \frac{5}{36} \approx .139$$

60. P(no more than 4)
$$= P(4) + P(3) + P(2)$$
$$= \frac{3}{36} + \frac{2}{36} + \frac{1}{36}$$
$$= \frac{6}{36}$$
$$= \frac{1}{6}$$
$$\approx .167$$

61. P(at least 9)
$$= P(2) + P(3) + P(4) + P(5)$$
$$= \frac{1}{36} + \frac{2}{36} + \frac{3}{36} + \frac{4}{36}$$
$$= \frac{10}{36}$$
$$= \frac{5}{18}$$
$$\approx .278$$

62. P(odd and greater than 8)
$$= P(9) + P(11)$$
$$= \frac{4}{36} + \frac{2}{36}$$
$$= \frac{6}{36}$$
$$= \frac{1}{6}$$
$$\approx .167$$

63. A roll less than 4 means 3 or 2. There are 2 ways to get 3 and 1 way to get 2. Hence,
$$P\left(2 \,|\, \text{less than 4}\right) = \frac{1}{3}.$$

64. $P(7 \,|\, \text{at least one is a 4}) = \dfrac{2}{11} \approx .182$, since there are 11 possibilities with at least one of the dice being a 4 {(4, 1), (1, 4), (4, 2), (2, 4), (4, 3), (3, 4), (4, 4), (5, 4), (4, 5), (6, 4), (4, 6)} with only (4, 3) and (3, 4) having a sum of 7.

For Exercises 65–68, draw a Venn diagram and use the given information to fill in the probabilities for each of the regions.

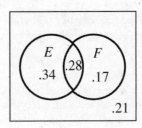

65. $P(E \cup F) = .34 + .28 + .17 = .79$

66. $P(E \cap F') = .34$

67. $P(E' \cup F) = .28 + .17 + .21 = .66$

68. $P(E' \cap F') = 1 - (.34 + .28 + .17) = .21$

69. Draw a probability tree.

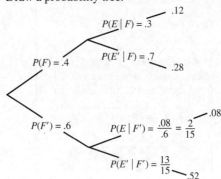

a. $P(E' \,|\, F) = .7$

b. $P(E \,|\, F') = \dfrac{2}{15} \approx .1333$

70. Answers vary.

71. Answers vary.

Use the following probability tree for Exercises 72–75.

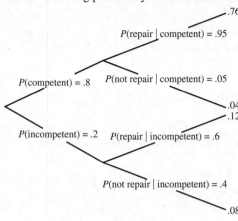

72. $P(\text{competent} \mid \text{repaired})$

$= \dfrac{.76}{.76 + .12} = \dfrac{.76}{.88} = \dfrac{19}{22}$

73. $P(\text{incompetent} \mid \text{repaired})$

$= \dfrac{.12}{.76 + .12} = \dfrac{12}{88} = \dfrac{3}{22}$

74. $P(\text{competent} \mid \text{not repaired})$

$= \dfrac{.04}{.04 + .08} = \dfrac{.04}{.12} = \dfrac{1}{3}$

75. $P(\text{incompetent} \mid \text{not repaired})$

$= \dfrac{.08}{.04 + .08} = \dfrac{8}{12} = \dfrac{2}{3}$

76. Let D mean defective.

$P(D) = .17(.04) + .39(.02) + .35(.07) + .09(.03)$

$\qquad = .0418$

a. $P(4 \mid D) = \dfrac{P(4 \cap D)}{P(D)}$

$\qquad = \dfrac{.09(.03)}{.0418}$

$\qquad = \dfrac{.0027}{.0418}$

$\qquad \approx .0646$

b. $P(2 \mid D) = \dfrac{P(2 \cap D)}{P(D)}$

$\qquad = \dfrac{.39(.02)}{.0418}$

$\qquad = \dfrac{.0078}{.0418}$

$\qquad \approx .1866$

77. a. $P(\text{second class}) = \dfrac{357}{1316} \approx .271$

b. $P(\text{surviving}) = \dfrac{499}{1316} \approx .379$

c. $P(\text{surviving}\mid\text{first class}) = \dfrac{203}{325} \approx .625$

d. $P(\text{surviving}\mid\text{third class child}) = \dfrac{27}{79} \approx .342$

e. $P(\text{female}\mid\text{first class survivor}) = \dfrac{140}{203} \approx .690$

f. $P(\text{third class}\mid\text{male survivor}) = \dfrac{75}{146} \approx .514$

g. Answers vary. No, because third-class men had a slightly lower survival rate than men generally.

78. a.

	Too High	About Right	Too Low	Don't Know	Total
Male	289	192	6	10	497
Female	257	153	3	14	427
Total	546	345	9	24	924

b. 924 were surveyed.

c. 192 men think taxes are about right.

d. 257 women think taxes are about right.

e. 427 women are in the survey

f. 289 of those who think taxes are too high are male.

g. Given that the respondent is male, how many think taxes are too high.

h. P(think taxes are too high | respondent is male)
$$= \frac{289}{497} \approx .581$$

i. P(think taxes are about right | respondent is woman)
$$= \frac{257}{427} \approx .358$$

j. Answers vary.
Possible answer:
There are different conditional probabilities. In (h), the reduced sample space is those who were not satisfied. In (i), it is those who bought used cars.

Chapter 8 Additional Probability Review Exercises

1.

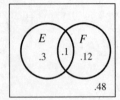

a. $P(E \cup F') = .3 + .1 + .48 = .88$

b. $P(E \cap F') = .30$

c. $P(E' \cup F) = .12 + .48 + .1 = .70$

2. There are a total of 18 $(2 + 3 + 5 + 8)$ marbles in the jar.

a. $P(\text{white}) = \dfrac{2}{18} = \dfrac{1}{9}$

b. $P(\text{orange}) = \dfrac{3}{18} = \dfrac{1}{6}$

c. $P(\text{not black}) = \dfrac{18-8}{18} = \dfrac{10}{18} = \dfrac{5}{9}$

d. $P(\text{orange or yellow}) = \dfrac{3+5}{18} = \dfrac{8}{18} = \dfrac{4}{9}$

3. a. P(Health Care or Energy)
$$= \frac{17.0+9.1}{100} = \frac{26.1}{100} = .261$$

b. P(Financials or Consumer Discretionary)
$$= \frac{19.8+18.1}{100} = \frac{37.9}{100} = .379$$

c. $P(\text{not Energy}) = 1 - P(\text{Energy})$
$$= 1 - \frac{9.1}{100} = \frac{90.9}{100} = .909$$

4. a. P(not in Europe)
$$= 1 - P(\text{Europe})$$
$$= 1 - \frac{36.8}{100} = \frac{63.2}{100} = .632$$

b. P(Europe or North America)
$$= \frac{36.8+46.4}{100} = \frac{83.2}{100} = .832$$

c. P(not Asia nor in Europe)
$$= \frac{46.4+3.8}{100} = \frac{50.2}{100} = .502$$

5. $S = \{1, 2, 3, 4, 5, 6\}$

a. $P(2 \mid \text{odd}) = 0$

b. $P(4 \mid \text{even}) = \dfrac{1}{3}$

c. $P(\text{even} \mid 6) = \dfrac{1}{1} = 1$

6. a. $P(\text{special education}) = \dfrac{6924.6}{24,924.1} \approx .278$

b. $P(\text{Special Education or Vocational and Adult Education})$

$= \dfrac{6924.6 + 1995.0}{24,924.1} = \dfrac{8919.6}{24,924.1} \approx .358$

c. $P(\text{not from Educational Reform})$

$= 1 - P(\text{Educational Reform})$

$= 1 - \dfrac{1792.7}{24,924.1} = \dfrac{23,131.4}{24,924.1} \approx .928$

7. $P(\text{age } 45\text{ - }64 \,|\, 2000) = \dfrac{62,440}{282,125} = .221$

8. $P(\text{age} < 65 \,|\, 2000)$

$= \dfrac{19,218 + 61,331 + 104,075 + 62,440}{282,125}$

$= \dfrac{247,064}{282,125} \approx .876$

9. $P(\text{age} < 5 \,|\, 2020) = \dfrac{22,932}{335,804} \approx .068$

10. $P(\text{age} \geq 65 \,|\, 2020) = \dfrac{47,363 + 7269}{335,804}$

$= \dfrac{54,632}{335,804} \approx .163$

11. No, because being a CPA increases one's likelihood of driving a luxury car.

12.

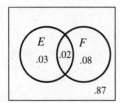

a. $P(E' \cap F) = .08$

b. $P(E' \cup F') = .98$

c. $P(E \cap F') = .03$

Use this diagram for Exercises 13–14.

$P(\text{first box}) = \tfrac{3}{8}$ — $P(\text{orange}) = \tfrac{1}{5}$ — $\tfrac{3}{40}$

$P(\text{red}) = \tfrac{4}{5}$ — $\tfrac{3}{10}$

$P(\text{second box}) = \tfrac{5}{8}$ — $P(\text{orange}) = \tfrac{3}{5}$ — $\tfrac{3}{8}$

$P(\text{red}) = \tfrac{2}{5}$ — $\tfrac{1}{4}$

13. $P(\text{first box} \,|\, \text{orange}) = \dfrac{\frac{3}{40}}{\frac{3}{40} + \frac{3}{8}}$

$= \dfrac{\frac{3}{40}}{\frac{18}{40}}$

$= \dfrac{3}{18}$

$= \dfrac{1}{6}$

14. $P(\text{second box} \,|\, \text{red}) = \dfrac{\frac{1}{4}}{\frac{3}{10} + \frac{1}{4}}$

$= \dfrac{\frac{5}{20}}{\frac{11}{20}}$

$= \dfrac{5}{11}$

15.

$P(\text{good}) = .7$

$P(\text{pass} \,|\, \text{good}) = .8$ — .56

$P(\text{fail} \,|\, \text{good}) = .2$ — .14

$P(\text{poor}) = .3$

$P(\text{pass} \,|\, \text{poor}) = .4$ — .12

$P(\text{fail} \,|\, \text{poor}) = .6$ — .18

$P(\text{good} \,|\, \text{pass}) = \dfrac{.56}{.56 + .12}$

$= \dfrac{.56}{.68}$

$= .824$

$= 82.4\%$

For Exercises 16–19, use the following diagram.

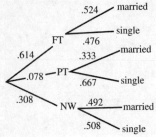

16. $P(\text{married})$

$= .614(.524) + .078(.333) + .308(.492)$

$= .499$

17. $P(\text{full time} \mid \text{married})$

$$\frac{.614(.524)}{.614(.524) + .078(.333) + .308(.492)}$$

$$= \frac{.321736}{.499246} \approx .644$$

18. $P(\text{part-time} \mid \text{not married})$

$$\frac{.078(.667)}{.614(1-.524) + .078(1-.333) + .308(1-.492)}$$

$$= \frac{.052026}{.5008} \approx .104$$

19. $P(\text{not full nor part-time} \mid \text{not married})$

$$= \frac{.308(.508)}{.614(.524) + .078(.333) + .308(.492)}$$

$$= \frac{.156464}{.5008} \approx .312$$

For Exercises 20–21, use the following diagram.

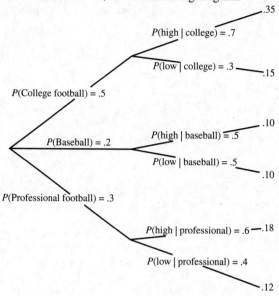

20. $P(\text{college} \mid \text{high rating}) = \dfrac{.35}{.35 + .10 + .18} = \dfrac{.35}{.63} \approx .556$

21. $P(\text{professional} \mid \text{high rating}) = \dfrac{.18}{.35 + .10 + .18} = \dfrac{.18}{.63} \approx .286$

22. $P(\text{individual does not use a seat belt}) = \dfrac{1}{1 + 19}$

$$= \dfrac{1}{20} = .05$$

23. $P(\text{driver used a seatbelt}) = \dfrac{17}{17 + 8} = \dfrac{17}{25} = .68$

24. There is a total of 10 marbles: 2 yellow, 5 red, and 3 blue.

a. $P(\text{marble is red}) = \dfrac{5}{10} = \dfrac{1}{2}$

b. $P(\text{marble is yellow or blue}) = \dfrac{2 + 3}{10}$

$$= \dfrac{5}{10}$$

$$= \dfrac{1}{2}$$

c. $P(\text{marble is yellow or red}) = \dfrac{2 + 5}{10} = \dfrac{7}{10}$

d. Since there are no green marbles,

$P(\text{marble is green}) = \dfrac{0}{10} = 0$

25. Let A be Alam, B be Bartolini, C be Chinn, D be Dickinson, and E be Ellsberg. The sample space is {A and B, A and C, A and D, A and E, B and C, B and D, B and E, C and D, C and E, D and E}. There are 10 equally likely outcomes.

a. $P(\text{Chinn is selected}) = \dfrac{4}{10} = \dfrac{2}{5}$

b. $P(\text{Ellsberg is not selected}) = \dfrac{6}{10} = \dfrac{3}{5}$

c. $P(\text{Alam and Dickinson are selected}) = \dfrac{1}{10}$

d. $P(\text{At least 1 senior partner, Alam or Bartolini, is selected}) = \dfrac{7}{10}$

Case 8 Medical Diagnosis

1.

$$P(H_2 \mid C_1) = \frac{P(C_1 \mid H_2)P(H_2)}{P(C_1 \mid H_1)P(H_1) + P(C_1 \mid H_2)P(H_2) + P(C_1 \mid H_3)P(H_3)}$$

$$= \frac{(.4)(.15)}{(.9)(.8) + (.4)(.15) + (.1)(.05)}$$

$$= \frac{.06}{.72 + .06 + .005}$$

$$\approx .076$$

2.

$$P(H_1 \mid C_2) = \frac{P(C_2 \mid H_1)P(H_1)}{P(C_2 \mid H_1)P(H_1) + P(C_2 \mid H_2)P(H_2) + P(C_2 \mid H_3)P(H_3)}$$

$$= \frac{(.2)(.8)}{(.2)(.8) + (.8)(.15) + (.3)(.05)}$$

$$= \frac{.16}{.16 + .12 + .015}$$

$$\approx .542$$

3.

$$P(H_3 \mid C_2) = \frac{P(C_2 \mid H_3)P(H_3)}{P(C_2 \mid H_1)P(H_1) + P(C_2 \mid H_2)P(H_2) + P(C_2 \mid H_3)P(H_3)}$$

$$= \frac{(.3)(.05)}{(.2)(.8) + (.8)(.15) + (.3)(.05)}$$

$$= \frac{.015}{.16 + .12 + .015}$$

$$\approx .051$$

Chapter 9: Counting, Probability Distributions, and Further Topics in Probability

Section 9.1 Probability Distributions and Expected Value

1. The number of possible samples is 16.

$$P(0) = \frac{{}_4C_0}{16} = \frac{1}{16}$$

$$P(1) = \frac{{}_4C_1}{16} = \frac{1}{4}$$

$$P(2) = \frac{{}_4C_2}{16} = \frac{3}{8}$$

$$P(3) = \frac{{}_4C_3}{16} = \frac{1}{4}$$

$$P(4) = \frac{{}_4C_4}{16} = \frac{1}{16}$$

Number of Boys	0	1	2	3	4
$P(x)$	$\frac{1}{16}$	$\frac{1}{4}$	$\frac{3}{8}$	$\frac{1}{4}$	$\frac{1}{16}$

3. Let x be the number of queens drawn. Then x can take on values 0, 1, 2, or 3. The probabilities follow.

$$P(x = 0) = \frac{{}_4C_0 \cdot {}_{48}C_3}{{}_{52}C_3} = \frac{4324}{5525} \approx .783$$

$$P(x = 1) = \frac{{}_4C_1 \cdot {}_{48}C_2}{{}_{52}C_3} = \frac{1128}{5525} \approx .204$$

$$P(x = 2) = \frac{{}_4C_2 \cdot {}_{48}C_1}{{}_{52}C_3} = \frac{72}{5525} \approx .013$$

$$P(x = 3) = \frac{{}_4C_3 \cdot {}_{48}C_0}{{}_{52}C_3} = \frac{1}{5525} \approx .0002$$

Number of queens	0	1	2	3
$P(x)$	.783	.204	.013	.0002

5.

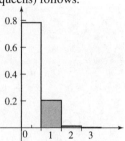

7. The histogram for Exercise 3 with P(at least one queens) follows.

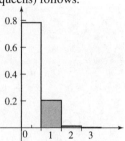

9. expected value $= 1(.2) + 3(.4) + 5(.3) + 7(.1)$

$$= 3.6$$

11. expected value
$$= 0(.14) + 2(.22) + 4(.36) + 8(.18) + 16(.10)$$
$$= 4.92$$

13. $E(x) = 1(.2) + 2(.3) + 3(.1) + 4(.4) = 2.7$

15. $E(x) = 1(.3) + 2(.25) + 3(.2) + 4(.15) + 5(.1)$
$$= 2.5$$

17. $E(x) = 1\left(\dfrac{18}{38}\right) - 1\left(\dfrac{20}{38}\right)$

$$= -\frac{2}{38} = -\frac{1}{19} \approx -0.05$$

19. You have one chance in a thousand of winning $500 on a $1 bet for a net return of $499. In the 999 other outcomes you lose your dollar

$$E(x) = 499\left(\frac{1}{1000}\right) + (-1)\left(\frac{999}{1000}\right)$$

$$= -\frac{500}{1000} = -\$.50 \text{ or } -50¢$$

21. Let x denote the winnings.

$$E(x) = 49,999 \left(\frac{1}{2,000,000} \right) +$$

$$9,999 \left(\frac{2}{2,000,000} \right) -$$

$$1 \left(\frac{1,999,997}{2,000,000} \right) =$$

$$-\$.965 = -96.5¢$$

23. $E(x) = 25(-.87) = -21.75$

25. $E(x) = 0(.15) + 1(.25) + 2(.18)$
$$+3(.15) + 4(.13) + 5(.09) + 6(.03)$$
$$+7(.02) = 2.35$$

27. The distribution is not valid since no probability can be less than 0.

29. The distribution is not valid since no probability can be less than 0.

31. Let x denote the missing probability. Then,
$$.01 + .09 + .25 + .45 + .05 + x = 1.0$$
$$.85 + x = 1.0$$
$$x = 1.0 - .85 = .15$$

33. Let x denote the missing probability. Then
$$.20 + x + .25 + .30 = 1.0$$
$$x + .75 = 1.0$$
$$x = 1.0 - .75 = .25$$

35. Let x and y denote missing probabilities. Then
$$.10 + .10 + .20 + .25 + .05 + x + y = 1.0$$
$$.70 + x + y = 1.0$$
$$x + y = 1.0 - .7 = .3$$
Answers may vary. One possible answer would be .1 and .2.

37. $E(x) = .0007[100(15,000) + 250(10,000) +$
$$500(5,000)] = .0007(6,500,000) = \$4550$$

39. $E(x) - 0(.2707) + 1(.4043) + 2(.2415)$
$$+3(.0721) + 4(.0108) + 5(.0006) = 1.1498$$

41.

Account Number	Expected value	Exist. vol. + exp. value	Class
3	2000	22,000	C
4	1000	51,000	B
5	25,000	30,000	C
6	60,000	60,000	A
7	16,000	46,000	B

43. a. Let x denote the cost of using each antibiotic. For amoxicillin, $E(x) = .75(59.30) + .25(96.15) \approx \$68.51$.
For cefaclor, $E(x) = .90(69.15) + .10(106.00) \approx \$72.84$.

b. Amoxicillin, since the total expected cost is less.

45. a. $E(x) = 630,000(.5) + 315,000(.5)$
$$= 472,500 \text{ pounds}$$

b. $E(x) = 630,000 \left(\frac{1}{9} \right) + 315,000 \left(\frac{8}{9} \right)$
$$= 350,000 \text{ pounds}$$

Section 9.2 The Multiplication Principle, Permutations and Combinations

1. $_4P_2 = \frac{4!}{(4-2)!} = \frac{4!}{2!} = 4 \cdot 3 = 12$

3. $_8C_5 = \frac{8!}{5!(8-5)!} = \frac{8!}{3!5!} = \frac{8 \cdot 7 \cdot 6}{3 \cdot 2 \cdot 1} = 56$

5. $_8P_1 = \frac{8!}{(8-1)!} = \frac{8!}{7!} = 8$

7. $4! = 4 \cdot 3 \cdot 2 \cdot 1 = 24$

9. $_9C_6 = \frac{9!}{6!(9-6)!}$
$$= \frac{9!}{6!3!} = \frac{9 \cdot 8 \cdot 7}{3 \cdot 2 \cdot 1} = 84$$

11. $_{13}P_2 = \dfrac{13!}{(13-2)!} = \dfrac{13!}{11!} = 13 \cdot 12 = 156$

13. $_{25}P_5 = 6,375,600$

15. $_{14}P_5 = 240,240$

17. $_{18}C_5 = \dfrac{18!}{5!13!} = \dfrac{18 \cdot 17 \cdot 16 \cdot 15 \cdot 14}{5 \cdot 4 \cdot 3 \cdot 2 \cdot 1} = 8568$

19. $_{25}C_{16} = 2,042,975$

21. If $0! = 0$, $_4P_4 = \dfrac{4!}{(4-4)!} = \dfrac{24}{0} =$ undefined

23. a. There are two possibilities for each line, and, by the multiplication principle, $2 \cdot 2 \cdot 2 = 8$ trigrams.

b. Since each hexagram is made up of two trigrams, and there are 8 possible trigrams, it follows that there are $8 \cdot 8 = 64$ possible hexagrams.

25. $6 \cdot 8 \cdot 4 \cdot 3 = 576$
There are 576 varieties of autos available.

27. Yes; Since a social security number has 9 digits with no restrictions, there are $10^9 = 1,000,000,000$ (1 billion) different social security numbers. This is enough for every one of the people in the United States to have a social security number.

29. Since a zip code has nine digits with no restrictions, there are 10^9 or $1,000,000,000$ different 9-digit zip codes.

31. a. If the first piece is a traditional piece, the total number of ways they can arrange the program is $6 \cdot 9! = 2,177,280$.

b. If the last piece is an original piece, the total number of ways they can arrange the program is $9! \cdot 4 = 1,451,520$.

33. a. There are 8 possibilities for the first digit, 2 possibilities for the second digit, and 10 possibilities for the last digit. The total number of possible area codes is
$8 \cdot 2 \cdot 10 = 160$
There are 8 possibilities for the first digit and 10 possibilities for each of the next six digits. The total number of phone numbers is
$8 \cdot 10 \cdot 10 \cdot 10 \cdot 10 \cdot 10 \cdot 10$
$= 8 \times 10^6$
$= 8,000,000$

b. Some numbers, like 911, 800, and 900, are reserved for special purposes.

35. In this new plan, there would be 8 possibilities for the first digit, 2 possibilities for the second digit, and 10 possibilities for each of the last 2 digits. The total number of area codes is $8 \cdot 2 \cdot 10 \cdot 10 = 1600$.

37. Answers vary.
Possible answer:
A permutation of a elements ($a \geq 1$) from a set of b elements is any arrangement, without repetition, of the a elements.

39. Since order makes a difference, the number of arrangements is
$_{10}P_9 = \dfrac{10!}{(10-9)!} = \dfrac{10!}{1!} = 3,628,800$.

41. The number of different arrangements of the 4 candidates for the one office is $_4P_4 = 24$. The number of different arrangements for the 3 candidates would be $_3P_3 = 6$. If the 4-candidate office is listed first, there would be $6 \cdot 24 = 144$ different ballots. If the 3-candidate office is listed first, there would be $24 \cdot 6 = 144$ different ballots. The total number of different ballots would be $144 + 144 = 288$.

43. Four people are being selected. Since each receives a different job, order is important. The total number of different officer selections is
$_{32}P_4 = \dfrac{32!}{(32-4)!} = \dfrac{32!}{28!}$
$= 32 \cdot 31 \cdot 30 \cdot 29$
$= 863,040$.

45. $_{17}P_2 = \dfrac{17!}{(17-2)!} = \dfrac{17!}{15!}$

$= 17 \cdot 16$

$= 272$

47. This is a combinations problem.

a. $_{10}C_4 = \dfrac{10!}{4!6!} = 210$

b. $_{10}C_6 = \dfrac{10!}{6!4!} = 210$

49. a. Since there are only four 7's in a deck, it is impossible to get five 7's. The answer is 0.

b. There are 4 each of the 2's, 3's, and 4's, or 12 in all. Thus, there are $_{12}C_5 = 792$ ways to get 2's, 3's or 4's.

c. There are 40 cards that are not 2's, 3's, or 4's in the deck. There are $_{40}C_5 = 658,008$ ways to get 5 cards that are not 2's, 3's nor 4's.

d. To get exactly 2 kings or queens out of the 8 kings and queens in the deck, can be accomplished in $_8C_2 = 28$ ways. Then, there are $_{44}C_3 = 13,244$ ways to get the remaining cards. By the multiplication principle, there are $28 \cdot 13,244 = 370,832$ ways to get exactly 2 kings or queens.

e. There are $_{13}C_2 = 78$ ways to get 2 hearts. There are $_{13}C_3 = 286$ ways to get 3 diamonds. By the multiplication principle, there are $78 \cdot 286 = 22,308$ such hands.

51. Answers vary.

53. a. With repetition permitted, the tree diagram shows 9 different pairs

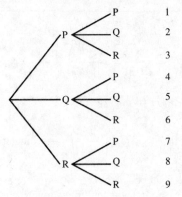

b. If repetition is not permitted, one branch is missing from each of the clusters of second branches, for a total of 6 different pairs.

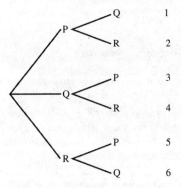

c. Find the number of combinations of 3 elements taken 2 at a time.

$_3C_2 = \dfrac{3!}{2!1!} = 3$

No repetitions are allowed, so the answer cannot equal that for part a. However, since order does not matter, our answer is only half of the answer for part b. For example PQ and QP are distinct in b. but not in c. Thus, the answer differs from both a. and b.

55. This is a combinations problem.

    a. $_{10}C_3 = 120$ delegations

    b. $_6C_3 = 20$ delegations

    c. $_6C_2 \cdot _4C_1 = 15 \cdot 4 = 60$ delegations

    d.   At least one means one or two or three.
$$_6C_2 \cdot _4C_1 + _6C_1 \cdot _4C_2 + _6C_0 \cdot _4C_3$$
$$= 60 + 36 + 4 = 100$$

57. There are no restrictions as to whether the scoops have to be different flavors.

    a. The number of different double-scoops will be $21 \cdot 21 = 441$.

    b. The number of different triple-scoops will be $21 \cdot 21 \cdot 21 = 21^3 = 9261$.

59. a. Since order is not important, this is a combination problem.
$$_{99}C_6 = 1,120,529,256$$

    b. Since order is important, this is a permutations problem.
$$_{99}P_6 = 806,781,064,300$$

61. $3 \cdot 3 \cdot 3 \cdot 3 = 81$

63. It is not possible since $26^3 = 17,576$ (the number of different 3-initial names) and the biologist needs 52,000 names. Actually, 4-initial names would do the biologist's job since $26^4 = 456,976$.

65. This is a combinations problem.

    a. $_5C_3 = 10$

    b. Since you are taking 3 Diet Coke and only 1 Diet Coke exists, this situation is impossible. The answer is 0.

    c. $_3C_3 = 1$

    d. $_5C_2 \cdot _1C_1 = 10 \cdot 1 = 10$

    e. $_5C_2 \cdot _3C_1 = 10 \cdot 3 = 30$

    f. $_3C_2 \cdot _5C_1 = 3 \cdot 5 = 15$

    g. Again, since you are picking 2 Diet Coke and only 1 Diet Coke exists, this situation is impossible. The answer is 0.

67. a. Since order matters, this is a permutation problem. The number of choices for the first bed is 12, for the second there are 11 choices, for the third there are 10 choices, and for the fourth there are 9 choices, so there are $_{12}P_4 = 11,880$ ways the 4 beds can be planted assuming each bed is planted differently.

    b. If order does not matter, this is a combination. The number of ways we can select 4 plant-color combinations from a set of 12 is $_{12}C_4 = 495$.

69. $\left(_{15}P_5\right)^4 \cdot \left(_{15}P_4\right) = 5.524 \times 10^{26}$

71. a. martini
There are 7 total letters, $1m$, $1a$, $1r$, $1t$, $2i$'s, and $1n$.
$$\frac{7!}{1!1!1!1!2!1!} = 2520$$

    b. nunnery
There are 7 letters $3n$'s, $1u$, $1e$, $1r$, and $1y$.
$$\frac{7!}{3!1!1!1!1!} = 840$$

    c. grinding
There are 8 letters $2g$'s, $1r$, $2i$'s, $2n$'s, and $1d$.
$$\frac{8!}{2!1!2!2!1!} = 5040$$

73. a. Total of 12 dinners:
$$_{12}P_{12} = 479,001,600 \text{ ways}$$

b. Since dinners of the same company are considered identical, the problem is to find the number of different arrangements of the three colors.
$$_3P_3 = 6$$

c. $\dfrac{12!}{4!3!5!} = 27,720$

Section 9.3 Applications of Counting

1. A sample containing 4 engines is used to test the shipment.

$P(\text{no defectives}) = \dfrac{_2C_0 \cdot {}_{10}C_4}{_{12}C_4}$

$= \dfrac{1(210)}{495} = .424$

3. There are 8 computers, 5 non-defective, and a sample of size 1 is chosen.

$P(\text{no defectives}) = \dfrac{_5C_1}{_8C_1} = \dfrac{5}{8}$

5. There are 8 computers, 5 non-defective, and a sample of size 3 is chosen.

$P(\text{no defectives}) = \dfrac{_5C_3}{_8C_3} = \dfrac{10}{56} = \dfrac{5}{28}$

7. There are 42 prizes, 10 of which are $100 prizes; 3 are drawn.

$P(\text{all are \$100 prizes}) = \dfrac{_{10}C_3}{_{242}C_3}$

$= \dfrac{120}{2,332,880} \approx .00005$

9. There are 42 prizes, 20 of which are $25 prizes and 22 which are other prizes; 3 are drawn.

$P(\text{two \$25 prizes}) = \dfrac{_{20}C_2 \cdot {}_{22}C_0 \cdot {}_{200}C_1}{_{242}C_3}$

$= \dfrac{190 \cdot 1 \cdot 200}{2,332,880} \approx .0163$

11. There are 200 dummy tickets, and all 3 must come from that group.
$P(\text{no winning ticket})$

$= \dfrac{_{200}C_3}{_{242}C_3} = \dfrac{1,313,400}{2,332,880} \approx .5630$

13. Since order does not make a difference, the number of 2-card hands is $_{52}C_2 = 1326$.

15. There are 48 non-deuces (2's) in a deck of cards.

$P(\text{no deuces}) = \dfrac{_{48}C_2}{_{52}C_2} = \dfrac{1128}{1326} \approx .851$

17. From Exercise 13, we know there are 1326 different 2-card hands. To see how many hands have the same suit, there are $_{13}C_2 = 78$ ways to get a 2-card hand of one particular suit. Because there are 4 different suits, there are $4 \cdot 78 = 312$ hands of the same suit. Therefore, there are $1326 - 312 = 1014$ hands of different suits.

$P(\text{different suits}) = \dfrac{1014}{1326} \approx .765$

19. $P(\text{no more than 1 diamond})$
$= P(\text{no diamonds}) + P(1 \text{ diamond})$
Because there are 39 non-diamonds in a deck,

$P(\text{no more than 1 diamond}) = \dfrac{741}{1326} + \dfrac{39 \cdot 13}{1326}$

$= \dfrac{1248}{1326} \approx .941.$

21. Answer varies
Possible answer:
The advantage of using this rule is that it is many times easier to calculate the probability of the complement of an event than of the given event. For example in dealing a hand of 5 cards, if E is the event to get at least one heart, it is much easier to calculate $P(E)'$, the probability of getting no hearts and subtracting it from 1.

23. In a deck there are 4 ten's, 4 jacks, and 44 other cards.
$P(2 \text{ ten's and 3 jacks})$

$= \dfrac{_4C_2 \cdot {}_4C_3 \cdot {}_{44}C_8}{_{52}C_{13}} \approx .0067$

25. a. We must pick all 6 of our numbers from the 99 total numbers.

$P(\text{all } 6) = \dfrac{_6C_6}{_{99}C_6} \approx 8.9 \times 10^{-10}$

b. $P(\text{all } 6)\ \dfrac{1}{_{99}P_6} \approx 1.2 \times 10^{-12}$

27. The probability of two individuals independently selecting the winning numbers is

$$\left(\frac{1}{120,526,770}\right)^2 \approx 6.9 \times 10^{-17}.$$

29. a. $P(3 \text{ women and 1 man})$

$$= \frac{_3C_3 \cdot {}_{11}C_1}{_{14}C_4}$$

$$= \frac{1 \cdot 11}{1001} \approx .011$$

b. $P(\text{all men}) = \frac{_{11}C_4}{_{14}C_4} = \frac{330}{1001} \approx .330$

c. $P(\text{at least 1 woman})$
$= 1 - P(\text{no women})$
$= 1 - P(\text{all men})$
$= 1 - .330$
$= .670$

31. a. $P(2 \text{ English and 3 Russians})$

$$= \frac{_{10}C_2 \cdot {}_4C_3}{_{25}C_5}$$

$$= \frac{45 \cdot 4}{53,130} \approx .003$$

b. $P(\text{all English speaking})$

$$= \frac{_{10}C_5}{_{25}C_5}$$

$$= \frac{252}{53,130} \approx .005$$

c. There are 15 non-English-speaking children.
$P(\text{no English speaking})$

$$= \frac{_{15}C_5}{_{25}C_5}$$

$$= \frac{3003}{53,130} \approx .057$$

31. Continued

d. $P(\text{at least 2 Vietnamese or Hmong})$

$$= P(2) + P(3) + P(4) + P(5)$$

$$= \frac{_5C_2 \cdot {}_{20}C_3}{_{25}C_5} + \frac{_5C_3 \cdot {}_{20}C_2}{_{25}C_5} + \frac{_5C_4 \cdot {}_{20}C_1}{_{25}C_5}$$

$$+ \frac{_5C_5 \cdot {}_{20}C_0}{_{25}C_5}$$

$$= \frac{10 \cdot 1140}{53,130} + \frac{10 \cdot 190}{53,130} + \frac{5 \cdot 20}{53,130} + \frac{1 \cdot 1}{53,130}$$

$$= \frac{13,401}{53,130} \approx .252$$

This probability can also be found by using the complementary event:
$P(\text{at least 2}) = 1 - [P(0) + P(1)].$

33. The probability that at least 2 of the 100 U.S. Senators have the same birthday is

$$1 - \frac{_{365}P_{100}}{(365)^{100}} \approx 1.$$

35. There are $_{20}C_4 = 4845$ ways to pick the correct 4 numbers out of the 20 the state picks. There are $_{80}C_4 = 1,581,580$ ways to pick 4 numbers out of 80. The probability of winning \$55 is

$$\frac{_{20}C_4}{_{80}C_4} = \frac{4845}{1581580} \approx .003 .$$

37. There are 60 losing numbers from which all 4 must be picked. There are 80 numbers from which 4 are picked. The probability of losing is

$$\frac{_{20}C_0 \cdot {}_{60}C_4}{_{80}C_4} = \frac{1 \cdot 487,635}{1,581,580} \approx .3083$$

39. Answers vary.
Theoretical answers:
This exercise should be solved by computer methods. The solution will vary according to the computer program that is used. The answers are (a) .0399, (b) .5191, (c) .0226.

Section 9.4 Binomial Probability

1. $n = 10, p = .7, x = 6, 1 - p = .3$

$P(\text{exactly } 6) = {}_{10}C_6 (.7)^6 (.3)^4$

$\approx .200$

3. $n = 10, p = .7, x = 0,$

$1 - p = 1 - .7 = .3$

$P(\text{none}) = {}_{10}C_0 (.7)^0 (.3)^{10}$

$\approx .000006$

5. $P(\text{at least } 1) = 1 - P(\text{none})$

From Exercise 3, we have $P(\text{none}) = .000006$, so

$P(\text{at least } 1) = 1 - .000006 = .999994.$

7. We have $n = 10$, $p = .05$, $x = 2$, and

$1 - p = .95$. $P(\text{exactly } 2) =$

${}_{10}C_2 (.05)^2 (.95)^8 \approx .075$.

9. $P(\text{none}) = {}_{10}C_0 (.05)^0 (.95)^{10} \approx .599$

11. $P(\text{at least } 1) = 1 - P(\text{none})$. From Exercise 9, we have $P(\text{none}) = .599$, so $P(\text{at least } 1) = 1 - .599 \approx .401$.

13. $P(\text{all heads}) = {}_5C_5 \left(\dfrac{1}{2}\right)^5 \left(\dfrac{1}{2}\right)^0 = \dfrac{1}{32}$

15. "No more than 3 heads" means 0 heads, 1 head, 2 heads, or 3 heads.

$P(0 \text{ heads}) = {}_5C_0 \left(\dfrac{1}{2}\right)^0 \left(\dfrac{1}{2}\right)^5 = \dfrac{1}{32}$

$P(1 \text{ head}) = {}_5C_1 \left(\dfrac{1}{2}\right)^1 \left(\dfrac{1}{2}\right)^4 = \dfrac{5}{32}$

$P(2 \text{ heads}) = {}_5C_2 \left(\dfrac{1}{2}\right)^2 \left(\dfrac{1}{2}\right)^3 = \dfrac{10}{32}$

$P(3 \text{ heads}) = {}_5C_3 \left(\dfrac{1}{2}\right)^3 \left(\dfrac{1}{2}\right)^2 = \dfrac{10}{32}$

$P(\text{no more than 3 heads})$

$= \dfrac{1}{32} + \dfrac{5}{32} + \dfrac{10}{32} + \dfrac{10}{32} = \dfrac{26}{32} = \dfrac{13}{16}$

17. Answer varies

Possible answer:

A problem involves a binomial experiment if the experiment is repeated several times, there are only two possible outcomes, and the repeated trials are independent.

19. $n = 12$, $p = .03$, $1 - p = .97$

$P(3) = {}_{12}C_3 (.03)^3 (.97)^9 \approx .0045$

21. $P(\text{at most } 2) = P(0) + P(1) + P(2)$

$= {}_{12}C_0 (.03)^0 (.97)^{12} + {}_{12}C_1 (.03)^1 (.97)^{11} +$

${}_{12}C_2 (.03)^2 (.97)^{10}$

$\approx .6938 + .2575 + .0438 = .9952$

23. Since 3% of Americans 65 to 74 have the disease, $6 = .03 \times 200$ would be expected to have the disease in a sample of 200.

25. $P(1) = {}_{15}C_1 (.12)^1 (.88)^{14} \approx .3006$

27. $P(\text{at most } 4) =$

$P(0) + P(1) + P(2) + P(3) + P(4)$

$= {}_{15}C_0 (.12)^0 (.88)^{15} + {}_{15}C_1 (.12)^1 (.88)^{14} +$

${}_{15}C_2 (.12)^2 (.88)^{13} + {}_{15}C_3 (.12)^3 (.88)^{12} +$

${}_{15}C_4 (.12)^4 (.88)^{11}$

$\approx .1470 + .3006 + .2870 + .1696 + .0694$

$= .9736$

$n = 100$, $p = .027$, and $1 - p = .973$

29. $P(\text{exactly 2 sets of twins})$

$= {}_{100}C_2 (.027)^2 (.973)^{98} \approx .247$

31. $n = 9$, $p = .11$, $1 - p = .89$

$P(2) = {}_9C_2 (.11)^2 (.89)^7 \approx .193$

33. $P(\text{none}) = {}_9C_0 (.11)^0 (.89)^9 \approx .350$

35. Since 11% of Americans are left-handed, $3.85 = .11 \times 35$ would be expected to be left-handed from the sample of 35.

37. $P(\text{at most } 3) = P(0) + P(1) + P(2) + P(3)$

$= {}_{16}C_0 (.08)^0 (.92)^{16} + {}_{16}C_1 (.08)^1 (.92)^{15} + {}_{16}C_2 (.08)^2 (.92)^{14} + {}_{16}C_3 (.08)^3 (.92)^{13}$

$\approx .263 + .366 + .239 + .097$

$= .966$

39. Since 66% of those having sexually transmitted disease are younger than 25, $330 = .66 \times 500$ would be expected to be younger than 25.

41. a. $P(\text{fewer than } 15) = {}_{40}C_{14}(.152)^{14}(.848)^{26} + {}_{40}C_{13}(.152)^{13}(.848)^{27} + {}_{40}C_{12}(.152)^{12}(.848)^{28}$

$+ \ {}_{40}C_{11}(.152)^{11}(.848)^{29} + {}_{40}C_{10}(.152)^{10}(.848)^{30} + {}_{40}C_9(.152)^9(.848)^{31}$

$+ \ {}_{40}C_8(.152)^8(.848)^{32} + {}_{40}C_7(.152)^7(.848)^{33} + {}_{40}C_6(.152)^6(.848)^{34}$

$+ \ {}_{40}C_5(.152)^5(.848)^{35} + {}_{40}C_4(.152)^4(.848)^{36} + {}_{40}C_3(.152)^3(.848)^{37}$

$+ \ {}_{40}C_2(.152)^2(.848)^{38} + {}_{40}C_1(.152)^1(.848)^{39} + {}_{40}C_0(.848)^{40}$

$\approx .9995$

b. Answers vary.

43. a. $P(\text{all 5 of the bands match}) = {}_5C_5(.25)^5 = \dfrac{1}{1024}$ or 1 chance in 1024

b. $P(\text{all 20 of the bands match}) = {}_{20}C_{20}(.25)^{20} \approx$ about 1 chance in 1.1×10^{12}

c. $P(\text{16 or more bands match}) = {}_{20}C_{16}(.25)^{16}(.75)^4 + {}_{20}C_{17}(.25)^{17}(.75)^3 + {}_{20}C_{18}(.25)^{18}(.75)^2$

$+ \ {}_{20}C_{19}(.25)^{19}(.75) + {}_{20}C_{20}(.25)^{20}$

$\approx$ about 1 chance in 2.6×10^6

d. Answers vary.

Section 9.5 Markov Chains

1. $\begin{bmatrix} \dfrac{1}{3} & \dfrac{2}{3} \end{bmatrix}$ could be a probability vector because it is a matrix with only one row containing only nonnegative entries whose sum is $\dfrac{1}{3} + \dfrac{2}{3} = 1$.

3. $\begin{bmatrix} 0 & 1 \end{bmatrix}$ could be a probability vector because it has only one row, the entries are nonnegative, and $0 + 1 = 1$.

5. $\begin{bmatrix} .3 & -.1 & .6 \end{bmatrix}$ cannot be a probability vector because it has a negative entry, $-.1$.

7. $\begin{bmatrix} .7 & .1 \\ .5 & .5 \end{bmatrix}$ cannot be a transition matrix because the sum of the entries in the first row is $.7 + .1 = .8 \neq 1$.

9. $\begin{bmatrix} \frac{4}{9} & \frac{1}{3} \\ \frac{1}{5} & \frac{7}{10} \end{bmatrix}$

This could not be a transition matrix because the

sum of the entries in row 1 is $\frac{4}{9} + \frac{1}{3} = \frac{7}{9}$ and the

sum of the entries in row 2 is $\frac{1}{5} + \frac{7}{10} = \frac{9}{10}$.

11. $\begin{bmatrix} \frac{1}{2} & \frac{1}{4} & 1 \\ \frac{2}{3} & 0 & \frac{1}{3} \\ \frac{1}{3} & 1 & 0 \end{bmatrix}$

This could not be a transition matrix because the
sum of the entries in row 1 is

$$\frac{1}{2} + \frac{1}{4} + 1 = \frac{7}{4} \neq 1,$$

and the sum of the entries in row 3 is

$$\frac{1}{3} + 1 = \frac{4}{3} \neq 1.$$

13. This is not a transition diagram because the sum of
the probabilities for changing from state A to
states A, B, and C is

$$\frac{1}{3} + \frac{1}{2} + 1 \neq 1.$$

15. This is a transition diagram. The information
given in this diagram can also be given by the
following matrix.

$$\begin{array}{c} \\ A \\ B \\ C \end{array} \begin{array}{ccc} A & B & C \\ \begin{bmatrix} .6 & .2 & .2 \\ .9 & .02 & .08 \\ .4 & .0 & .6 \end{bmatrix} \end{array}$$

17. Let $A = \begin{bmatrix} .2 & .8 \\ .9 & .1 \end{bmatrix}$

A is a regular transition matrix since $A^2 = A$
contains all positive entries.

19. Let $P = \begin{bmatrix} 0 & 1 & 0 \\ .3 & .3 & .4 \\ 1 & 0 & 0 \end{bmatrix}$.

$$P^2 = \begin{bmatrix} 0 & 1 & 0 \\ .3 & .3 & .4 \\ 1 & 0 & 0 \end{bmatrix} \begin{bmatrix} 0 & 1 & 0 \\ .3 & .3 & .4 \\ 1 & 0 & 0 \end{bmatrix}$$

$$= \begin{bmatrix} .3 & .3 & .4 \\ .49 & .39 & .12 \\ 0 & 1 & 0 \end{bmatrix}$$

$$P^3 = \begin{bmatrix} 0 & 1 & 0 \\ .3 & .3 & .4 \\ 1 & 0 & 0 \end{bmatrix} \begin{bmatrix} .3 & .3 & .4 \\ .49 & .39 & .12 \\ 0 & 1 & 0 \end{bmatrix}$$

$$= \begin{bmatrix} .49 & .39 & .12 \\ .237 & .607 & .156 \\ .3 & .3 & .4 \end{bmatrix}$$

P is a regular transition matrix since P^3 contains
all positive entries.

21. Let $A = \begin{bmatrix} .10 & .70 & 0 & .20 \\ 0 & .36 & .39 & .25 \\ 0 & 0 & 1 & 0 \\ .54 & 0 & .42 & .04 \end{bmatrix}$

$$A^2 = \begin{bmatrix} .1180 & .322 & .357 & .203 \\ .135 & .1296 & .6354 & .10 \\ 0 & 0 & 1 & 0 \\ .0756 & .378 & .4368 & .1096 \end{bmatrix}$$

A is not regular. Any power of A will have zero
entries in the 1st, 2nd, and 4th column of row 3;
thus, it cannot have all positive entries.

23. Let $P = \begin{bmatrix} .66 & .34 \\ .12 & .88 \end{bmatrix}$, and let **V** be the probability

vector $[v_1 \quad v_2]$.

$[v_1 \quad v_2] \begin{bmatrix} .66 & .34 \\ .12 & .88 \end{bmatrix} = [v_1 \quad v_2]$.

$.66v_1 + .12v_2 = v_1$

$.34v_1 + .88v_2 = v_2$

Simplify these equations to get the system

$-.34v_1 + .12v_2 = 0$

$.34v_1 - .12v_2 = 0$.

These equations are dependent. Since **V** is a probability vector,

$v_1 + v_2 = 1$.

Solve the system

$-.34v_1 + .12v_2 = 0$.

$v_1 + \quad v_2 = 1$

by the substitution method.

$-.34(1 - v_2) + .12v_2 = 0$

$-.34 + .34v_2 + .12v_2 = 0$

$.46v_2 = .34$

$v_2 = \dfrac{.34}{.46} = \dfrac{17}{23}$

$v_1 = 1 - \dfrac{17}{23} = \dfrac{6}{23}$

Thus, the equilibrium vector is

$\begin{bmatrix} \dfrac{6}{23}, & \dfrac{17}{23} \end{bmatrix}$.

25. Let $P = \begin{bmatrix} \dfrac{2}{3} & \dfrac{1}{3} \\ \dfrac{1}{8} & \dfrac{7}{8} \end{bmatrix}$, and let **V** be the probability

vector $\begin{bmatrix} v_1 & v_2 \end{bmatrix}$. We want to find **V** such that

VP = V,

or $\begin{bmatrix} v_1 & v_2 \end{bmatrix} \begin{bmatrix} \dfrac{2}{3} & \dfrac{1}{3} \\ \dfrac{1}{8} & \dfrac{7}{8} \end{bmatrix} = \begin{bmatrix} v_1 & v_2 \end{bmatrix}$.

Use matrix multiplication on the left.

$\begin{bmatrix} \dfrac{2}{3}v_1 + \dfrac{1}{8}v_2 & \dfrac{1}{3}v_1 + \dfrac{7}{8}v_2 \end{bmatrix} = \begin{bmatrix} v_1 & v_2 \end{bmatrix}$

Set corresponding entries from the two matrices equal to get

$\dfrac{2}{3}v_1 + \dfrac{1}{8}v_2 = v_1$

$\dfrac{1}{3}v_1 + \dfrac{7}{8}v_2 = v_2$.

Multiply both equations by 24 to eliminate fractions.

$16v_1 + 3v_2 = 24v_1$

$8v_1 + 21v_2 = 24v_2$

Simplify both equations.

$-8v_1 + 3v_2 = 0$

$8v_1 - 3v_2 = 0$

These equations are dependent. To find the values of v_1 and v_2, an additional equation is needed.

Since $\mathbf{V} = \begin{bmatrix} v_1 & v_2 \end{bmatrix}$. is a probability vector

$v_1 + v_2 = 1$.

To find v_1 and v_2, solve the system

$-8v_1 + 3v_2 = 0$ *(1)*

$v_1 + v_2 = 1.$ *(2)*

From equation (2), $v_1 = 1 - v_2$. Substitute $1 - v_2$ for v_1 in equation (1) to get

$-8(1 - v_2) + 3v_2 = 0$

$-8 + 8v_2 + 3v_2 = 0$

$11v_2 = 8$

$v_2 = \dfrac{8}{11}$

and $v_1 = 1 - v_2 = \dfrac{3}{11}$.

The equilibrium vector is

$\begin{bmatrix} \dfrac{3}{11} & \dfrac{8}{11} \end{bmatrix}$.

27. Let $P = \begin{bmatrix} .16 & .28 & .56 \\ .43 & .12 & .45 \\ .86 & .05 & .09 \end{bmatrix}$, and let **V** be the

probability vector $\begin{bmatrix} v_1 & v_2 & v_3 \end{bmatrix}$.

$$\begin{bmatrix} v_1 & v_2 & v_3 \end{bmatrix}\begin{bmatrix} .16 & .28 & .56 \\ .43 & .12 & .45 \\ .86 & .05 & .09 \end{bmatrix} = \begin{bmatrix} v_1 & v_2 & v_3 \end{bmatrix}$$

$.16v_1 + .43v_2 + .86v_3 = v_1$

$.28v_1 + .12v_2 + .05v_3 = v_2$

$.56v_1 + .45v_2 + .09v_3 = v_3$

Simplify these equations by the
Gauss-Jordan method to obtain

$$v_1 = \frac{7783}{16,799}, v_2 = \frac{2828}{16,799}, \text{and } v_3 = \frac{6188}{16,799}$$

The equilibrium vector is

$$\begin{bmatrix} \dfrac{7783}{16,799} & \dfrac{2828}{16,799} & \dfrac{6188}{16,799} \end{bmatrix},$$

or $\begin{bmatrix} .4633 & .1683 & .3684 \end{bmatrix}$ in decimal form.

29. Let $P = \begin{bmatrix} .44 & .31 & .25 \\ .80 & .11 & .09 \\ .26 & .31 & .43 \end{bmatrix}$, and let **V** be the

probability vector $\begin{bmatrix} v_1 & v_2 & v_3 \end{bmatrix}$.

$$\begin{bmatrix} v_1 & v_2 & v_3 \end{bmatrix}\begin{bmatrix} .44 & .31 & .25 \\ .80 & .11 & .09 \\ .26 & .31 & .43 \end{bmatrix} = \begin{bmatrix} v_1 & v_2 & v_3 \end{bmatrix}$$

$.44v_1 + .80v_2 + .26v_3 = v_1$

$.31v_1 + .11v_2 + .31v_3 = v_2$

$.25v_1 + .09v_2 + .43v_3 = v_3$

Simplify these equations to get the system

$-.56v_1 + .80v_2 + .26v_3 = 0$

$.31v_1 - .89v_2 + .31v_3 = 0$

$.25v_1 + .09v_2 - .57v_3 = 0$

Since **V** is the probability vector,

$v_1 + v_2 + v_3 = 1$.

This gives us a system of four equations in three
variables.

$$v_1 + v_2 + v_3 = 1$$

$-.56v_1 + .80v_2 + .26v_3 = 0$

$.31v_1 - .89v_2 + .31v_3 = 0$

$.25v_1 + .09v_2 - .57v_3 = 0$

This system can be solved by the Gauss-Jordan
method. Start with the augmented matrix

$$\begin{bmatrix} 1 & 1 & 1 \\ -.56 & .80 & .26 \\ .31 & -.89 & .31 \\ .25 & .09 & -.57 \end{bmatrix}\begin{bmatrix} 1 \\ 0 \\ 0 \\ 0 \end{bmatrix}.$$

The solution of this system is
$v_1 = .4872, v_2 = .2583,$ and $v_3 = .2545,$ so the

equilibrium vector is $\begin{bmatrix} .4872 & .2583 & .2545 \end{bmatrix}$.

31. The following solution presupposes the use of a TI-82 graphics calculator. similar results can be obtained from other graphics calculators.

Store the given transition matrix as matrix [A].

$$A = \begin{bmatrix} .3 & .2 & .3 & .1 & .1 \\ .4 & .2 & .1 & .2 & .1 \\ .1 & .3 & .2 & .2 & .2 \\ .2 & .1 & .3 & .2 & .2 \\ .1 & .1 & .4 & .2 & .2 \end{bmatrix}$$

$$A^2 = \begin{bmatrix} .23 & .21 & .24 & .17 & .15 \\ .26 & .18 & .26 & .16 & .14 \\ .23 & .18 & .24 & .19 & .16 \\ .19 & .19 & .27 & .18 & .17 \\ .17 & .2 & .26 & .19 & .18 \end{bmatrix}$$

$$A^3 = \begin{bmatrix} .226 & .192 & .249 & .177 & .156 \\ .222 & .196 & .252 & .174 & .156 \\ .219 & .189 & .256 & .177 & .159 \\ .213 & .192 & .252 & .181 & .162 \\ .213 & .189 & .252 & .183 & .163 \end{bmatrix}$$

$$A^4 = \begin{bmatrix} .2205 & .1916 & .2523 & .1774 & .1582 \\ .2206 & .1922 & .2512 & .1778 & .1582 \\ .2182 & .1920 & .2525 & .1781 & .1592 \\ .2183 & .1909 & .2526 & .1787 & .1595 \\ .2176 & .1906 & .2533 & .1787 & .1598 \end{bmatrix}$$

$$A^5 = \begin{bmatrix} .21932 & .19167 & .25227 & .17795 & .15897 \\ .21956 & .19152 & .25226 & .17794 & .15872 \\ .21905 & .19152 & .25227 & .17818 & .15898 \\ .21880 & .19144 & .25251 & .17817 & .15908 \\ .21857 & .19148 & .25253 & .17824 & .15918 \end{bmatrix}$$

The entry in row 2, column 4 of A^5 is .17794, which gives the probability that state 2 changes to state 4 after 5 repetitions.

33. The transition matrix is

$$\begin{bmatrix} \frac{3}{7} & \frac{1}{7} & \frac{2}{7} & \frac{1}{7} \\ \frac{1}{2} & 0 & \frac{1}{4} & \frac{1}{4} \\ \frac{1}{3} & \frac{1}{3} & 0 & \frac{1}{3} \\ \frac{1}{4} & \frac{1}{2} & 0 & \frac{1}{4} \end{bmatrix}$$

Let **V** be the probability vector $\begin{bmatrix} v_1 & v_2 & v_3 \end{bmatrix}$

$$\begin{bmatrix} v_1 & v_2 & v_3 & v_4 \end{bmatrix}\begin{bmatrix} \frac{3}{7} & \frac{1}{7} & \frac{2}{7} & \frac{1}{7} \\ \frac{1}{2} & 0 & \frac{1}{4} & \frac{1}{4} \\ \frac{1}{3} & \frac{1}{3} & 0 & \frac{1}{3} \\ \frac{1}{4} & \frac{1}{2} & 0 & \frac{1}{4} \end{bmatrix}=\begin{bmatrix} v_1 & v_2 & v_3 & v_4 \end{bmatrix}$$

$\frac{3}{7}v_1 + \frac{1}{2}v_2 + \frac{1}{3}v_3 + \frac{1}{4}v_4 = v_1$

$\frac{1}{7}v_1 + 0v_2 + \frac{1}{3}v_3 + \frac{1}{2}v_4 = v_2$

$\frac{2}{7}v_1 + \frac{1}{4}v_2 + 0v_3 + 0v_4 = v_3$

$\frac{1}{7}v_1 + \frac{1}{4}v_2 + \frac{1}{3}v_3 + \frac{1}{4}v_4 = v_4$

Simplify these equations to get the dependent system

$-\frac{4}{7}v_1 + \frac{1}{2}v_2 + \frac{1}{3}v_3 + \frac{1}{4}v_4 = 0$

$\frac{1}{7}v_1 - v_2 + \frac{1}{3}v_3 + \frac{1}{2}v_4 = 0$

$\frac{2}{7}v_1 + \frac{1}{4}v_2 - v_3 + 0v_4 = 0$

$\frac{1}{7}v_1 + \frac{1}{4}v_2 + \frac{1}{3}v_3 - \frac{3}{4}v_4 = 0$

Also, $v_1 + v_2 + v_3 + v_4 = 1$

Solve this system by the Gauss-Jordan method to obtain

$$v_1 = \frac{7}{18}, \; v_2 = \frac{2}{9}, \; v_3 = \frac{1}{6}, \; v_4 = \frac{2}{9}.$$

Therefore, 2.5-magnitude earthquakes will occur 38.9% of the time, 2.6-magnitude 22.2% of the time, 2.7-magnitude 16.7% of the time, and 2.8-magnitude 22.2% of the time.

35. From the transition matrix, we obtain

$.81v_1 + .77v_2 = v_1$ and $.19v_1 + .23v_2 = v_2$.

Collecting like terms, we find that

$-.19v_1 + .77v_2 = 0$ or $v_1 = \frac{77}{19}v_2$.

Substitute this into $v_1 + v_2 = 1$, which yields

$$\frac{96}{19}v_2 = 1.$$

Therefore, $v_2 = \frac{19}{96}$ and $v_1 = \frac{77}{96}$. The line works $\frac{77}{96}$ of the time.

37. Cross pink with color on left.

Resulting Color

Red   Pink   White

$$\begin{array}{c} \text{Red} \\ \text{Pink} \\ \text{White} \end{array}\begin{bmatrix} \frac{1}{2} & \frac{1}{2} & 0 \\ \frac{1}{4} & \frac{1}{2} & \frac{1}{4} \\ 0 & \frac{1}{2} & \frac{1}{2} \end{bmatrix}$$

$\frac{1}{2}v_1 + \frac{1}{4}v_2 = v_1$

$\frac{1}{2}v_1 + \frac{1}{2}v_2 + \frac{1}{2}v_3 = v_2$

$\frac{1}{4}v_2 + \frac{1}{2}v_3 = v_3$

Also, $v_1 + v_2 + v_3 = 1$.

Solving this system, we obtain

$$v_1 = \frac{1}{4}, v_2 = \frac{1}{2}, v_3 = \frac{1}{4}.$$

The equilibrium vector is

$$\begin{bmatrix} \frac{1}{4} & \frac{1}{2} & \frac{1}{4} \end{bmatrix}.$$

The long range prediction is $\frac{1}{4}$ red, $\frac{1}{2}$ pink, and $\frac{1}{4}$ white snapdragons.

39. $\begin{bmatrix} v_1 & v_2 & v_3 \end{bmatrix}\begin{bmatrix} .94 & .06 & 0 \\ .12 & .879 & .001 \\ 0 & .32 & .68 \end{bmatrix}=\begin{bmatrix} v_1 & v_2 & v_3 \end{bmatrix}$

$.94v_1 + .12v_2 + 0v_3 = v_1$

$.06v_1 + .879v_2 + .32v_3 = v_2$

$0v_1 + .001v_2 + .68v_3 = v_3.$

Therefore, we have the system

$v_1 + v_2 + v_3 = 1$

$-.06v_1 + .12v_2 + 0v_3 = 0$

$.06v_1 - .121v_2 + .32v_3 = 0$

$0v_1 + .001v_2 - .32v_3 = 0.$

Solve this system by the Gauss-Jordan method to obtain $v_1 = .666, v_2 = .333, v_3 = .001$. Therefore, 66.6% will be homeowners, 33.3% will be renters, and there will be 0.1% homeless in the long-range.

41. The transition matrix is

$$\begin{bmatrix} .85 & .10 & .05 \\ .15 & .75 & .10 \\ .10 & .30 & .60 \end{bmatrix}.$$

The square of the transition matrix is

$$\begin{bmatrix} .85 & .10 & .05 \\ .15 & .75 & .10 \\ .10 & .30 & .60 \end{bmatrix}\begin{bmatrix} .85 & .10 & .05 \\ .15 & .75 & .10 \\ .10 & .30 & .60 \end{bmatrix}$$

$$= \begin{bmatrix} .7425 & .175 & .0825 \\ .25 & .6075 & .1425 \\ .19 & .415 & .395 \end{bmatrix}.$$

The cube of the transition matrix is

$$\begin{bmatrix} .85 & .10 & .05 \\ .15 & .75 & .10 \\ .10 & .30 & .60 \end{bmatrix}\begin{bmatrix} .7425 & .175 & .0825 \\ .25 & .6075 & .1425 \\ .19 & .415 & .395 \end{bmatrix}$$

$$= \begin{bmatrix} .665625 & .23025 & .104125 \\ .317875 & .523375 & .15875 \\ .26325 & .44875 & .288 \end{bmatrix}.$$

a. $\begin{bmatrix} 50,000 & 0 & 0 \end{bmatrix}\begin{bmatrix} .85 & .10 & .05 \\ .15 & .75 & .10 \\ .10 & .30 & .60 \end{bmatrix}$

$= \begin{bmatrix} 42,500 & 5000 & 2500 \end{bmatrix}$

The numbers in the groups after 1 year are 42,500, 5000, and 2500.

b. $\begin{bmatrix} 50,000 & 0 & 0 \end{bmatrix}\begin{bmatrix} .7425 & .175 & .0825 \\ .25 & .6075 & .1425 \\ .19 & .415 & .395 \end{bmatrix}$

$= \begin{bmatrix} 37,125 & 8750 & 4125 \end{bmatrix}$

The numbers in the groups after 2 years are 37,125, 8750, and 4125.

c. $\begin{bmatrix} 50,000 & 0 & 0 \end{bmatrix}$

$\begin{bmatrix} .665625 & .23025 & .104125 \\ .317875 & .523375 & .15875 \\ .26325 & .44875 & .288 \end{bmatrix}$

$= \begin{bmatrix} 33,281 & 11,513 & 5206 \end{bmatrix}$

The numbers in the groups after 3 years are 33,281, 11,513, and 5206.

41. Continued

d. The system of equations is

$$.85v_1 + .15v_2 + .10v_3 = v_1$$
$$.10v_1 + .75v_2 + .30v_3 = v_2$$
$$.05v_1 + .10v_2 + .60v_3 = v_3$$

After collecting like terms, we have the matrix

$$\begin{bmatrix} -.15 & .15 & .10 & | & 0 \\ .10 & -.25 & .30 & | & 0 \\ .05 & .10 & -.40 & | & 0 \end{bmatrix}$$

Clear the first column.

$$\begin{bmatrix} 1 & -1 & -\dfrac{2}{3} & | & 0 \\ 0 & -\dfrac{3}{20} & \dfrac{11}{30} & | & 0 \\ 0 & \dfrac{3}{20} & -\dfrac{11}{30} & | & 0 \end{bmatrix}$$

Clear the second column.

$$\begin{bmatrix} 1 & 0 & -\dfrac{28}{9} & | & 0 \\ 0 & 1 & -\dfrac{22}{9} & | & 0 \\ 0 & 0 & 0 & | & 0 \end{bmatrix}$$

Hence, $v_1 = \dfrac{28}{9}v_3$ and $v_2 = \dfrac{22}{9}v_3$,

which we substitute into

$$v_1 + v_2 + v_3 = 1.$$

The result is

$$\frac{28}{9}v_3 + \frac{22}{9}v_3 + \frac{9}{9}v_3 = 1$$

$$\frac{59}{9}v_3 = 1$$

$$v_3 = \frac{9}{59}.$$

Then

$$v_2 = \frac{22}{59}, v_1 = \frac{28}{59}, \text{ and }$$

$$\mathbf{V} = \begin{bmatrix} \dfrac{28}{59} & \dfrac{22}{59} & \dfrac{9}{59} \end{bmatrix} \text{ or } \begin{bmatrix} .475 & .373 & .152 \end{bmatrix}.$$

In the long run, the probabilities of no accidents, one accident, and more than one accident are .475, .373, and .152, respectively.

43. a.
$$\begin{bmatrix} .443 & .364 & .193 \\ .277 & .436 & .287 \\ .266 & .304 & .430 \end{bmatrix}\begin{bmatrix} .443 & .364 & .193 \\ .277 & .436 & .287 \\ .266 & .304 & .430 \end{bmatrix}$$

$$= \begin{bmatrix} .348 & .379 & .273 \\ .320 & .378 & .302 \\ .316 & .360 & .323 \end{bmatrix}$$

b. The desired probability is found in row 1, column 1. The probability that if England won the last game, England will win the game after the next one is .348.

c. The desired probability is found in row 2, column 1. The probability that if Australia won the last game, England will win the game after the next one is .320.

45. a. Answers vary.
Possible answer:
If there is no one in line, then after 1 min there will be either 0, 1, or 2 people in line with probabilities
$p_{00} = .4, p_{01} = .3,$ and $p_{02} = .3.$ If there is one person in line, then that person will be served and either 0, 1 or 2 new people will join the line, with probabilities
$p_{10} = .4, p_{11} = .3,$ and $p_{12} = .3.$ If there are two people in line, then one of them will be served and either 1 or 2 new people will join the line, with probabilities $p_{21} = .5$ and
$p_{22} = .5;$ it is impossible for both people in line to be served, so $p_{20} = 0.$ Therefore, the transition matrix is

$$A = \begin{bmatrix} .4 & .3 & .3 \\ .4 & .3 & .3 \\ 0 & .5 & .5 \end{bmatrix}.$$

b. The transition matrix for a two-minute period is

$$A^2 = \begin{bmatrix} .4 & .3 & .3 \\ .4 & .3 & .3 \\ 0 & .5 & .5 \end{bmatrix}\begin{bmatrix} .4 & .3 & .3 \\ .4 & .3 & .3 \\ 0 & .5 & .5 \end{bmatrix}$$

$$= \begin{bmatrix} .28 & .36 & .36 \\ .28 & .36 & .36 \\ .20 & .40 & .40 \end{bmatrix}.$$

c. The probability that a queue with no one in line has two people in line 2 min later is .36 since that is the entry in row 1, column 3 of A^2.

47. The following solution presupposes the use of a TI-82 graphics calculator. Similar results can be obtained from other graphics calculators. Store the given transition matrix as matrix $[A]$ and the probability vector $\begin{bmatrix} .5 & .5 & 0 & 0 \end{bmatrix}$ as matrix $[B]$.
Multiply $[B]$ by successive powers of $[A]$ to obtain $\begin{bmatrix} 0 & 0 & .102273 & .897727 \end{bmatrix}$.
This gives the long-range prediction for the percent of employees in each state for the company training program.

Section 9.6 Decision Making

1. a. An optimist should choose the coast; $150,000 is the largest profit.

b. A pessimist should choose the highway; the worst case of $30,000 is better than −$40,000 if the coast is chosen.

c. If the possibility of heavy opposition is .8, the probability of light opposition is .2. Find his expected profit for each strategy.
Highway:
70,000(.2) + 30,000(.8) = $38,000
Coast:
150,000(.2) + −(40,000)(.8) = −$2000
He should choose the highway for an expected profit of $38,000

d. If the probability of heavy opposition is .4, the probability of light opposition is .6. Find his expected profit for each strategy.
Highway:
70,000(.6) + 30,000(.4) = $54,000
Coast:
150,000(.6) + (−40,000)(.4) = $74,000
He should choose the coast.

3. Note that the costs given in the payoff matrix are given in hundreds of dollars.

a. An optimist should make no upgrade; minimum cost is $2800.

b. A pessimist should make the upgrade; a worst case of $13,000 is better than a possible cost of $45,000 if no upgrade is made.

3. Continued

c. Find the expected cost of each strategy.
Make upgrade:
.7(130) + .2(130) + .1(130) = 130,
or $13,000
Make no upgrade:
.7(28) + .2(180) + .1(450) = 100.6
or $10,600
He should not upgrade. The expected cost to the company if this strategy is chosen is $10,060.

	Fails	Doesn't Fail

5. a.

$$\begin{array}{c} \text{Overhaul} \\ \text{Don't Overhaul} \end{array} \begin{bmatrix} -\$8600 & -\$2600 \\ -\$6000 & \$0 \end{bmatrix}$$

b. Find the expected cost under each strategy.
Overhaul:
.1(−8600) + .9(−2600) = −$3200
Don't Overhaul:
.3(−6000) + .7(0) = −$1800
To minimize his expected costs, the business should not overhaul the machine before shipping.

7. a. An optimistic city council should choose no campaign; the smallest cost that appears in the matrix is $0, which corresponds to not running a campaign.

b. A pessimistic city council should choose the campaign for all; its worst case is the least costly ($800,000 is less than $2,820,000 or $3,100,100).

c. Find the expected cost for each strategy.
Campaign for all:
.8(100,000) + .2(800,000) = $240,000
Campaign for youth:
.8(2,820,000) + .2(20,000) = $2,260,000
No campaign:
.8(3,100,100) + .2(0) = $2,480,080
The city council should choose the campaign for all.

9. Find the expected utility under each strategy.
Jobs:
(.35)(40) + (.65)(−10) = 7.5
Environment:
(.35)(−12) + (.65)(30) = 15.3
She should emphasize the environment. The expected utility of this strategy is 15.3.

Chapter 9 Review Exercises

1.
$$E(x) = 0(.22) + 1(.54) + 2(.16) + 3(.08)$$
$$= 1.1$$

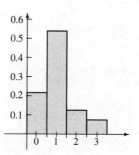

2.
$$E(x) = -3(.15) - 2(.20) - 1(.25) + 0(.18)$$
$$+1(.12) + 2(.06) + 3(.04) = -.74$$

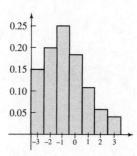

3. $E(x) = -10(.333) + 0(.333) + 10(.333) = 0$

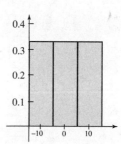

4. $E(x) = 1\left(\dfrac{18}{38}\right) - 1\left(\dfrac{20}{38}\right) = -.052$

x	−1	+1
$P(x)$	.526	.474

5. a. In flipping a coin 4 times, by the multiplication principle, there are $2^4 = 16$ possible outcomes.

$$P(0 \text{ tails}) = \frac{{}_4C_0}{16} = \frac{1}{16}$$

$$P(1 \text{ tail}) = \frac{{}_4C_1}{16} = \frac{4}{16} = \frac{1}{4}$$

$$P(2 \text{ tails}) = \frac{{}_4C_2}{16} = \frac{6}{16} = \frac{3}{8}$$

$$P(3 \text{ tails}) = \frac{{}_4C_3}{16} = \frac{4}{16} = \frac{1}{4}$$

$$P(4 \text{ tails}) = \frac{{}_4C_4}{16} = \frac{1}{16} = \frac{1}{16}$$

Let x be the number of tails.

x	0	1	2	3	4
$P(x)$	$\frac{1}{16}$	$\frac{1}{4}$	$\frac{3}{8}$	$\frac{1}{4}$	$\frac{1}{16}$

b. The expected value is

$$0\left(\frac{1}{16}\right) + 1\left(\frac{1}{4}\right) + 2\left(\frac{3}{8}\right) + 3\left(\frac{1}{4}\right) + 4\left(\frac{1}{16}\right) = 2.$$

6. a. In selecting 2 bouquets from a group of 10, there are ${}_{10}C_2 = 45$ ways.

$$P(0 \text{ roses}) = \frac{{}_3C_0 \cdot {}_7C_2}{{}_{10}C_2} = \frac{21}{45} = \frac{7}{15}$$

$$P(1 \text{ rose}) = \frac{{}_3C_1 \cdot {}_7C_1}{{}_{10}C_2} = \frac{21}{45} = \frac{7}{15}$$

$$P(2 \text{ roses}) = \frac{{}_3C_2 \cdot {}_7C_0}{{}_{10}C_2} = \frac{3}{45} = \frac{1}{15}$$

x	0	1	2
$P(x)$	$\frac{7}{15}$	$\frac{7}{15}$	$\frac{1}{15}$

b. $E(x) = 0\left(\frac{7}{15}\right) + 1\left(\frac{7}{15}\right) + 2\left(\frac{1}{15}\right) = .6$

7. a. Three of the 10 members did not do their homework, and 3 members of the 10 are selected. Three members of the 10 can be selected in ${}_{10}C_3 = 120$ ways. The probability of selecting

$$P(0) = \frac{{}_3C_0 \cdot {}_7C_3}{{}_{10}C_3} = \frac{35}{120} = .292$$

$$P(1) = \frac{{}_3C_1 \cdot {}_7C_2}{{}_{10}C_3} = \frac{63}{120} = .525$$

$$P(2) = \frac{{}_3C_2 \cdot {}_7C_1}{{}_{10}C_3} = \frac{21}{120} = .175$$

$$P(3) = \frac{{}_3C_3 \cdot {}_7C_0}{{}_{10}C_3} = \frac{1}{120} = .008$$

x	0	1	2	3
$P(x)$	.292	.525	.175	.008

b. $E(x) = 0(.292) + 1(.525) + 2(.175) + 3(.008) = .899$

8. Probability of getting 3 hearts

$$= \frac{{}_{13}C_3}{{}_{52}C_3} = \frac{286}{22,100} \approx .0129$$

Probability of not getting 3 hearts
$= 1 - .0129 = .9871$.
Let $x =$ the amount to pay for the game.
If you win, you get $100 - x$.
If you lose, you get $-x$. The expected value must be 0.

$$(100 - x)(.0129) + (-x)(.9871) = 0$$
$$1.29 - .0129x - .9871x = 0$$
$$1.29 = x$$

You must pay $1.29.

9. The probability of winning if you bet "under" is $\frac{15}{36}$ and has a net return of $2. The probability of 7 is $\frac{6}{36}$ and has a net return of $4. Betting "over" has a probability of $\frac{15}{36}$ with a net return of $2.

$$P(\text{sum} < 7) = P(2) + P(3) + P(4) +$$
$$P(5) + P(6)$$
$$= \frac{1}{36} + \frac{2}{36} + \frac{3}{36} + \frac{4}{36} + \frac{5}{36} = \frac{15}{36}$$
$$P(\text{sum} = 7) = \frac{6}{36}$$
$$P(\text{sum} > 7) = P(8) + P(9) + P(10) +$$
$$P(11) + P(12)$$
$$= \frac{5}{36} + \frac{4}{36} + \frac{3}{36} + \frac{2}{36} + \frac{1}{36} = \frac{15}{36}$$

The expected return for each type of bet is as follows:

Under:
$$E(x) = 2\left(\frac{15}{36}\right) - 2\left(\frac{21}{36}\right) = -\$.33$$

Exactly 7:
$$E(x) = 4\left(\frac{6}{36}\right) - 2\left(\frac{30}{36}\right) = -\$1.00$$

Over:
$$E(x) = 2\left(\frac{15}{36}\right) - 2\left(\frac{21}{36}\right) = -\$.33$$

10. $E(x) = 9,998\left|\frac{1}{10,000}\right| + 998\left|\frac{2}{10,000}\right|$
$$+ 98\left(\frac{2}{10,000}\right) - 2\left(\frac{9995}{10,000}\right) = -\$.78$$

11. The probability of having a girl is $\frac{1}{2}$.

The probability of not having a girl is $1 - \frac{1}{2} = \frac{1}{2}$.

Number	Probability
0	$\binom{5}{0}\left(\frac{1}{2}\right)^0\left(\frac{1}{2}\right)^5 = \frac{1}{32}$
1	$\binom{5}{1}\left(\frac{1}{2}\right)^1\left(\frac{1}{2}\right)^4 = \frac{5}{32}$
2	$\binom{5}{2}\left(\frac{1}{2}\right)^2\left(\frac{1}{2}\right)^3 = \frac{10}{32}$
3	$\binom{5}{3}\left(\frac{1}{2}\right)^3\left(\frac{1}{2}\right)^2 = \frac{10}{32}$
4	$\binom{5}{4}\left(\frac{1}{2}\right)^4\left(\frac{1}{2}\right)^1 = \frac{5}{32}$
5	$\binom{5}{5}\left(\frac{1}{2}\right)^5\left(\frac{1}{2}\right)^0 = \frac{1}{32}$

$$E(x) = 0\left|\frac{1}{32}\right| + 1\left|\frac{5}{32}\right| + 2\left|\frac{10}{32}\right| + 3\left|\frac{10}{32}\right|$$
$$+ 4\left(\frac{5}{32}\right) + 5\left(\frac{1}{32}\right) = \frac{80}{32} = 2.5$$

12. a. Three cards can be drawn from a standard deck $_{52}C_3 = 22,100$ ways. If three cards are drawn,

$$P(0 \text{ kings}) = \frac{_4C_0 \cdot {}_{48}C_3}{22,100} = \frac{1 \cdot 17,296}{22,100} \approx .7826$$

$$P(1 \text{ king}) = \frac{_4C_1 \cdot {}_{48}C_2}{22,100} = \frac{4 \cdot 1128}{22,100} \approx .2042$$

$$P(2 \text{ kings}) = \frac{_4C_2 \cdot {}_{48}C_1}{22,100} = \frac{6 \cdot 48}{22,100} = \frac{288}{22,100} \approx .0130$$

$$P(3 \text{ kings}) = \frac{_4C_3 \cdot {}_{48}C_0}{22,100} = \frac{4}{22,100} = .0002$$

$$E(x) = 0(.7826) + 1(.2042) + 2(.0130)$$
$$+ 3(.0002) = .23$$

b.

$$P(0 \text{ diamonds}) = \frac{_{13}C_0 \times {}_{39}C_3}{22,100} = \frac{1 \times 9139}{22,100} \gg .4135$$

$$P(1 \text{ diamonds}) = \frac{_{13}C_1 \times {}_{39}C_2}{22,100} = \frac{13 \times 741}{22,100} \gg .4359$$

$$P(2 \text{ diamonds}) = \frac{_{13}C_2 \times {}_{39}C_1}{22,100} = \frac{78 \times 39}{22,100} = .1376$$

$$P(3 \text{ diamonds}) = \frac{_{13}C_3 \times {}_{39}C_0}{22,100} = \frac{286 \times 1}{22,100} = .0129$$

$$E(3 \text{ diamonds}) = 0(.4135) + (.4359) + 2(.1376)$$
$$+ 3(.0129) = .75$$

13. Since order is important, this is a permutation. Five shuttle vans can line up $_5P_5 = 5! = 120$ ways

14. Since order is important, this is a permutation. There are 7 ways to select the first runner, 6 ways to select the second runner, and 5 ways to select the third runner, or

$$_7P_3 = \frac{7!}{(7-3)!} = \frac{7!}{4!} = 7 \cdot 6 \cdot 5 = 210 \; .$$

15. Since order is not important, this is a combination. Three monitors can be selected from 12 monitors

in $_{12}C_3 = \dfrac{12!}{3!9!} = \dfrac{12 \cdot 11 \cdot 10}{3 \cdot 2 \cdot 1} = 220$ ways.

16. a. One of the 4 broken monitors can be selected in $_4C_1 = 4$ ways. The remaining 2 must come from the 8 nonbroken monitors, and can be selected in $_8C_2 = 28$ ways. By the multiplication principle, the selection can be made in $112 = 4 \times 28$ ways.

b. All 3 monitors must come from the nonbroken group of 8. This can be accomplished in $_8C_3 = 56$ ways.

c. At least one broken monitor can be accomplished by selecting 1, 2, or 3 defective monitors. If 1 monitor is defective, 2 must be nondefective. If 2 are defective, 1 must be nondefective. If 3 are defective, then 0 must be nondefective. The number of ways to select

1 defective: $_4C_1 \cdot _8C_2 = 4 \times 28 = 112$

2 defective: $_4C_2 \cdot _8C_1 = 6 \times 8 = 48$

3 defective: $_4C_3 \cdot _8C_0 = 4 \times 1 = 4$

is then $112 + 48 + 4 = 164$.

17. Since order is important, this is a permutation. There are 30 choices for the first seat, 29 for the second, 28 for the third, 27 for the fourth, 26 for the fifth, and 25 for the sixth, or

$_{30}P_6 = 427,518,000$ ways.

18. The first seat will be occupied by the given student. That leaves 29 choices for the second seat, 28 choices for the third seat, 27 choices for the fourth seat, 26 choices for the fifth seat, and 25 for the sixth seat, or $_{29}P_5 = 14,250,600$ possible ways.

19. a. Since there are 15 students in each major, there are $_{15}P_3 = 2730$ ways to arrange the students within each major. There are also $_2P_2 = 2$ ways to arrange the two different groups. Thus, the total number of arrangements is
$2 \times 2730 \times 2730 = 14,905,800$.

b. Assume the odd seats are occupied by science majors, then because order is important, there are $_{15}P_3 = 2730$ possible arrangements for that group. Then the even seats would be occupied by the business majors with $_{15}P_3 = 2730$ possible arrangements By the multiplication principle, there would be $2730 \times 2730 = 7,452,900$ possible arrangements. However, if the odd seats were occupied by the business majors and the even by the science majors, there would also be $2730 \times 2730 = 7,452,900$ possible arrangements. So, there must be $7,452,900 + 7,452,900 = 14,905,800$ possibilities.

20. Answers vary.

21. Answers vary.

22. There are 26 black cards in the deck.
$$P(\text{both black}) = \frac{_{26}C_2}{_{52}C_2} = \frac{325}{1326} \approx .245$$

23. There are 13 hearts in a deck.
$$P(\text{both hearts}) = \frac{_{13}C_2}{_{52}C_2} = \frac{78}{1326} \approx .059$$

24. To get exactly one face card, you must have one non-face card. There are 12 face cards and 40 non-face cards in a deck.
$P(\text{exactly one face card})$
$$= \frac{_{12}C_1 \cdot _{40}C_1}{_{52}C_2} = \frac{480}{1326} \approx .362$$

25. There are 4 aces in a deck of cards.
$P(\text{at most one ace})$
$$= \frac{_4C_0 \cdot _{48}C_2}{_{52}C_2} + \frac{_4C_1 \cdot _{48}C_1}{_{52}C_2}$$
$$= \frac{1128}{1326} + \frac{192}{1326}$$
$$= \frac{1320}{1326} \approx .995$$

26. There are 12 possible selections, 6 of them being ice cream, and 6 not ice cream.

$$P(3 \text{ ice cream}) = \frac{{}_6C_3 \cdot {}_6C_0}{{}_{12}C_3} = \frac{20 \cdot 1}{220} \approx .091$$

27. There are 12 possible selections, 4 custard and 8 noncustard.

$$P(4 \text{ custard}) = \frac{{}_4C_3 \cdot {}_8C_0}{{}_{12}C_3} = \frac{4 \cdot 1}{220} \approx .018$$

28. There are 12 possible selections, 2 frozen yogurt and 10 nonyogurt.

$$P(\text{at least 1 yogurt}) = 1 - P(\text{no yogurt})$$

$$= 1 - \frac{{}_2C_0 \cdot {}_{10}C_3}{{}_{12}C_3} = 1 = \frac{1 \cdot 120}{220} = \frac{100}{220} \approx .455$$

29. There are 12 possible selections, 4 custard, 6 ice cream, and 2 frozen yogurt.

$$P(1 \text{ custard, 1 ice cream, 1 yogurt})$$

$$= \frac{{}_4C_1 \cdot {}_6C_1 \cdot {}_2C_1}{{}_{12}C_3} = \frac{4 \cdot 6 \cdot 2}{220} = \frac{48}{220} \approx .218$$

30. There are 12 possible selections, 6 ice cream, and 6 not ice cream.

$$P(\text{at most 1 ice cream}) = P(0) + P(1)$$

$$= \frac{{}_6C_0 \cdot {}_6C_3}{{}_{12}C_3} + \frac{{}_6C_1 \cdot {}_6C_2}{{}_{12}C_3}$$

$$= \frac{20}{220} + \frac{90}{220} = \frac{110}{220} = .5$$

31. a. The number of subsets of size 0 is ${}_nC_0$ or 1.

The number of subsets of size 1 is ${}_nC_1$ or n.

The number of subsets of size 2 is ${}_nC_2$ or

$$\frac{n(n-1)}{2}.$$

The number of subsets of size n is ${}_nC_n$ or 1.

b. The total number of subsets of a set with n elements is ${}_nC_0 + {}_nC_1 + {}_nC_2 + \cdots + {}_nC_n$.

c. Answers vary.

31. Continued

d. Let $n = 4$.

$${}_4C_0 + {}_4C_1 + {}_4C_2 + {}_4C_3 + {}_4C_4$$

$$= 1 + 4 + 6 + 4 + 1$$

$$= 16$$

Since $2^4 = 16$, the equation from part c holds.

Let $n = 5$.

$${}_5C_0 + {}_5C_1 + {}_5C_2 + {}_5C_3 + {}_5C_4 + {}_5C_5$$

$$= 1 + 5 + 10 + 10 + 5 + 1$$

$$= 32$$

Since $2^5 = 32$, the equation from part c holds.

32. $n = 20, p = .01, x = 4, 1 - p = .99$

$$P(4) = \binom{20}{4}(.01)^4(.99)^{16} \approx .000041$$

33. $n = 20, p = .01, 1 - p = .99$

$P(\text{no more than 3})$

$= P(0) + P(1) + P(2) + P(3)$

$= {}_{20}C_0(.01)^0(.99)^{20} + {}_{20}C_1(.01)^1(.99)^{19}$

$+ {}_{20}C_2(.01)^2(.99)^{18} + {}_{20}C_3(.01)^3(.99)^{17}$

$\approx .817907 + .165234 + .015856 + .000961$

$= .99996$

34. $n = 6, p = .25, 1 - p = .75$

x	$P(x)$	
0	.1780	$= {}_6C_0(.25)^0(.75)^6$
1	.3560	$= {}_6C_1(.25)^1(.75)^5$
2	.2966	$= {}_6C_2(.25)^2(.75)^4$
3	.1318	$= {}_6C_3(.25)^3(.75)^3$
4	.0330	$= {}_6C_4(.25)^4(.75)^2$
5	.0044	$= {}_6C_5(.25)^5(.75)^1$
6	.0002	$= {}_6C_6(.25)^6(.75)^0$

35. $E(x) = 0(.1780) + 1(.3560) + 2(.2966)$

$+ 3(.1318) + 4(.0330) + 5(.0044) + 6(.0002)$

$= 1.5$

36. a. $n = 4$, $p = .81$, $1 - p = .19$

$$P(4 \text{ invested in U.S.}) = {}_4C_4 (.81)^4 (.19)^0$$
$$= 1 \cdot .4305 \cdot 1 = .4305$$

b. $P(\text{at least 2 invested in U.S.}) =$
$$P(2) + P(3) + P(4) =$$
$${}_4C_2 (.81)^2 (.19)^2 + {}_4C_3 (.81)^3 (.19)^1$$
$$+ {}_4C_4 (.81)^4 (.19)^0$$
$$= .1421 + .4039 + .4305 = .9765$$

c. $P(\text{at most 1 invested in U.S.})$
$$= P(0) + P(1)$$
$$= {}_4C_0 (.81)^0 (.19)^4 + {}_4C_1 (.81)^1 (.19)^3$$
$$= .0013 + .0222 = .0235$$

37. a. If the odds that the fatality occurred in China is 1:4, then the probability that the fatality

occurred in China $= \dfrac{1}{1+4} = \dfrac{1}{5} = .2$.

b. The distribution is as follows:

x	$P(x)$	
0	.1678	$= {}_8C_0 (.2)^0 (.8)^8$
1	.3355	$= {}_8C_1 (.2)^1 (.8)^7$
2	.2936	$= {}_8C_2 (.2)^2 (.8)^6$
3	.1468	$= {}_8C_3 (.2)^3 (.8)^5$
4	.0459	$= {}_8C_4 (.2)^4 (.8)^4$
5	.0092	$= {}_8C_5 (.2)^5 (.8)^3$
6	.0011	$= {}_8C_6 (.2)^6 (.8)^2$
7	.0000	$= {}_8C_7 (.2)^7 (.8)^1$
8	0000	$= {}_8C_8 (.2)^8 (.8)^0$

c. $E(x) = 0(.1678) + 1(.3355)$
$$+ 2(.2936) + 3(.1468) + 4(.0459)$$
$$+ 5(.0092) + 6(.0011) + 7(.0000)$$
$$+ 8(.0000) = 1.599$$

38. $\begin{bmatrix} 0 & 1 \\ .77 & .23 \end{bmatrix}$

This is a regular transition matrix because
$$\begin{bmatrix} 0 & 1 \\ .77 & .23 \end{bmatrix} \begin{bmatrix} 0 & 1 \\ .77 & .23 \end{bmatrix} = \begin{bmatrix} .77 & .23 \\ .1771 & .8229 \end{bmatrix},$$
in which all entries are positive.

39. $\begin{bmatrix} -.2 & .4 \\ .3 & .7 \end{bmatrix}$

This is not a regular transition matrix because there is a negative entry in the first row and first column. In fact, this makes it not even a transition matrix.

40. $\begin{bmatrix} .21 & .15 & .64 \\ .50 & .12 & .38 \\ 1 & 0 & 0 \end{bmatrix}$

This is a regular transition matrix because
$$\begin{bmatrix} .21 & .15 & .64 \\ .50 & .12 & .38 \\ 1 & 0 & 0 \end{bmatrix} \begin{bmatrix} .21 & .15 & .64 \\ .50 & .12 & .38 \\ 1 & 0 & 0 \end{bmatrix}$$
$$= \begin{bmatrix} .7591 & .0495 & .1914 \\ .545 & .0894 & .3656 \\ .21 & .15 & .64 \end{bmatrix},$$
in which all entries are positive.

41. $\begin{bmatrix} .22 & 0 & .78 \\ .40 & .33 & .27 \\ 0 & .61 & .39 \end{bmatrix}$

This is a regular transition matrix because
$$\begin{bmatrix} .22 & 0 & .78 \\ .40 & .33 & .27 \\ 0 & .61 & .39 \end{bmatrix} \begin{bmatrix} .22 & 0 & .78 \\ .40 & .33 & .27 \\ 0 & .61 & .39 \end{bmatrix}$$
$$= \begin{bmatrix} .0484 & .4758 & .4758 \\ .220 & .2736 & .5064 \\ .244 & .4392 & .3168 \end{bmatrix},$$
in which all entries are positive.

42. $P = \begin{bmatrix} .35 & .15 & .50 \\ .30 & .35 & .35 \\ .15 & .30 & .55 \end{bmatrix}$

$I = \begin{bmatrix} .2 & .4 & .4 \end{bmatrix}$

a. The distribution after one month is
$$\begin{bmatrix} .2 & .4 & .4 \end{bmatrix} \begin{bmatrix} .35 & .15 & .50 \\ .30 & .35 & .35 \\ .15 & .30 & .55 \end{bmatrix}$$
$$= \begin{bmatrix} .250 & .290 & .460 \end{bmatrix}.$$

42. Continued

b. $I = \begin{bmatrix} .2 & .4 & .4 \end{bmatrix}$

$$A = \begin{bmatrix} .35 & .15 & .50 \\ .30 & .35 & .35 \\ .15 & .30 & .55 \end{bmatrix}$$

$$A^2 = \begin{bmatrix} .2425 & .2550 & .5025 \\ .2625 & .2725 & .4650 \\ .2250 & .2925 & .4825 \end{bmatrix}$$

The distribution after 2 months is given by
$IA^2 = \begin{bmatrix} .2435 & .2770 & .4795 \end{bmatrix}$.

c. To find the long-range distribution, we use the system

$$v_1 + v_2 + v_3 = 1$$
$$.35v_1 + .3v_2 + .15v_3 = v_1$$
$$.15v_1 + .35v_2 + .3v_3 = v_2$$
$$.5v_1 + .35v_2 + .55v_3 = v_3.$$

Simplify these equations to obtain the system

$$v_1 + v_2 + v_3 = 1$$
$$-.65v_1 + .3v_2 + .15v_3 = 0$$
$$.15v_1 - .65v_2 + .3v_3 = 0$$
$$.5v_1 + .35v_2 - .45v_3 = 0.$$

Solve this system by the Gauss-Jordan method to obtain

$$v_1 = \frac{75}{313}, v_2 = \frac{87}{313}, \text{ and } v_3 \frac{151}{313}.$$

The long-range distribution is
$\begin{bmatrix} .240 & .278 & .482 \end{bmatrix}$.

43. a. $I = \begin{bmatrix} .15 & .60 & .25 \end{bmatrix}$

$$A = \begin{bmatrix} .80 & .14 & .06 \\ .04 & .85 & .11 \\ .03 & .13 & .84 \end{bmatrix}$$

The distribution after one month is

$$\begin{bmatrix} .15 & .60 & .25 \end{bmatrix} \begin{bmatrix} .80 & .14 & .06 \\ .04 & .85 & .11 \\ .03 & .13 & .84 \end{bmatrix}$$

$$= \begin{bmatrix} .1515 & .5635 & .2850 \end{bmatrix}$$

43. Continued

b. $A^2 = \begin{bmatrix} .80 & .14 & .06 \\ .04 & .85 & .11 \\ .03 & .13 & .84 \end{bmatrix} \begin{bmatrix} .80 & .14 & .06 \\ .04 & .85 & .11 \\ .03 & .13 & .84 \end{bmatrix}$

$$= \begin{bmatrix} .6474 & .2388 & .1138 \\ .0693 & .7424 & .1883 \\ .0544 & .2239 & .7217 \end{bmatrix}$$

$$A^3 = \begin{bmatrix} .80 & .14 & .06 \\ .04 & .85 & .11 \\ .03 & .13 & .84 \end{bmatrix}$$

$$\begin{bmatrix} .6474 & .2388 & .1138 \\ .0693 & .7424 & .1883 \\ .0544 & .2239 & .7217 \end{bmatrix}$$

$$= \begin{bmatrix} .530886 & .308410 & .160704 \\ .090785 & .665221 & .243994 \\ .074127 & .291752 & .634121 \end{bmatrix}$$

The distribution after 3 years is given by
$IA^3 = \begin{bmatrix} .1526 & .5183 & .3290 \end{bmatrix}$

c. To find the long range distribution, we use

the system

$$v_1 + v_2 + v_3 = 1$$
$$.80v_1 + .04v_2 + .03v_3 = v_1$$
$$.14v_1 + .85v_2 + .13v_3 = v_3$$
$$.06v_1 + .11v_2 + .84v_3 = v_3$$

This system yields

$$v_1 + v_2 + v_3 = 1$$
$$-.20v_1 + .04v_2 + .03v_3 = 0$$
$$.14v_1 - .15v_2 + .13v_3 = 0$$
$$.06v_1 + .11v_2 - .16v_3 = 0$$

Solving this system yields $v_1 = \frac{97}{643}$,

$v_2 = \frac{302}{643}$ and $v_3 = \frac{244}{643}$. Thus the long

range distribution is $\begin{bmatrix} .1509 & .4697 & .3795 \end{bmatrix}$.

44. a. Since the candidate is an optimist, look for the biggest value in the matrix, which is 5000. Hence, she should oppose it.

b. A pessimistic candidate wants to find the best of the worst things that can happen. If she favors, the worst is –4000. If she waffles the worst is –500. If she opposes, then worst is 0. Since the best of these is 0, she should oppose.

c. Since there is a 40% chance the opponent favors the plant and a 35% that he will waffle, the chance he will oppose is
$1 - .4 - .35 = .25$
Expected gain if she favors:
$(0)(.4) + (-1000)(.35) + (-4000)(.25)$
$= -1350$
Expected gain if she waffles:
$(1000)(.4) + 0(.35) + (-500)(.25) = 275$
Expected gain if she opposes:
$(5000)(.4) + (2000)(.35) + (0)(.25) = 2700$
she should oppose and get 2700 additional votes.

d. Now the opponent has 0 probability of favoring, .7 of waffling and .3 of opposing.
Expected gain if she favors:
$(0)(0) + (.7)(-1000) + (.3)(-4000)$
$= -1900$
Expected gain if she waffles:
$(0)(1000) + (.7)(0) + (.3)(-500)$
$= -150$
Expected gain if she opposes:
$(0)(5000) + (.7)(2000) + (.3)(0)$
$= 1400$
She should oppose and get 1400 additional votes.

45. a. Since the candidate is an optimist, look for the biggest value in the matrix, which is 100. Hence, she should use active learning.

b. A pessimistic wants to find the best of the worst things that can happen. If she lectures, the worst is –80. If she uses active learning, the worst is –30. Since the best of these is –30, she should use active learning.

c. Since there is a 75% chance the class will prefer the lecture format, there is a 25% chance the class will support the active learning format.
Expected gain for lecture format:
$50(.75) - 80(.25) = 17.5$ points.

45. Continued

Expected gain for active learning format:
$-30(.75) + 100(.25) = 2.5$ points. She should use the lecture format where the expected gain is 175.

d. Since there is a 60% chance the class will prefer the active learning format, there is a 40% chance the class will support the lecture format.
Expected gain for lecture format:
$50(.4) - 80(.6) = -28$ points.
Expected gain for active learning format:
$-30(.4) + 100(.6) = 48$ points. She should use the active learning format where the expected gain is 48.

46. $P(\text{product is successful}) = .5$
$P(\text{product is unsuccessful}) = .5$
$P(\text{successful product passing quality control})$
$= .8$
$P(\text{unsuccessful product passing quality control})$
$= .25$
$P(\text{successful product and passes quality control})$
$= (.5)(.8) = .4$
$P(\text{unsuccessful product and passes quality control}) = (.5)(.25) = .125$
$P(\text{passes quality control}) = .4 + .125 = .525$
$P(\text{successful product passes quality control})$
$= \dfrac{.4}{.525} = .7619$
$P(\text{unsuccessful product passes quality control}) = \dfrac{.125}{.525} = .2381$
$E(x) = (40,000,000)(.7619)$
$\qquad + (-15,000,000)(.2381)$
$\qquad \approx 27,000,000$
The expected net profit is (e) $27 million.

47. If a box is good (probability .9) and the merchant samples an excellent piece of fruit from that box (probability .80), then he will accept the box and earn a $200 profit on it. If a box is bad (probability .1) and he samples an excellent piece of fruit from the box (probability .30), then he will accept the box and earn a –$1000 profit on it. If the merchant ever samples a non-excellent piece of fruit, he will not accept the box. In this case he pays nothing and earns nothing, so the profit will be $0.

Let x denote the merchant's earnings.

Note that $.9(.80) = .72$,

$.1(.30) = .03$,

and $1 - (.72 + .03) = .25$.

The probability distribution is as follows.

x	200	−1000	0
$P(x)$	.72	.03	.25

The expected value when the merchant samples the fruit is

$E(x) = 200(.72) + (-1000)(.03)$

$+ 0(.25)$

$= 144 - 30 + 0$

$= 114.$

We must also consider the case in which the merchant does not sample the fruit. Let x again denote the merchant's earnings. The probability distribution is as follows.

x	200	−1000
$P(x)$	.9	.1

The expected value when the merchant does not sample the fruit is

$E(x) = 200(.9) + (-1000)(.1)$

$= 180 - 100$

$= \$80.$

Combining these two results, the expected value of the right to sample is $114 – $80 = $34, which corresponds to choice (c).

48. Let $I(x)$ represent the airline's net income if x people show up.

$I(0) = 0$

$I(1) = 100$

$I(2) = 2 \cdot 100 = 200$

$I(3) = 3 \cdot 100 = 300$

$I(4) = 3 \cdot 100 - 100 = 200$

$I(5) = 3 \cdot 100 - 2 \cdot 100 = 100$

$I(6) = 3 \cdot 100 - 3 \cdot 100 = 0$

Let $P(x)$ represent the probability that x people will show up. Use the binomial probability formula to find the values of $P(x)$.

$$P(0) = \binom{6}{0}(.6)^0(.4)^6 = .004$$

$$P(1) = \binom{6}{1}(.6)^1(.4)^5 = .037$$

$$P(2) = \binom{6}{2}(.6)^2(.4)^4 = .138$$

$$P(3) = \binom{6}{3}(.6)^3(.4)^3 = .276$$

$$P(4) = \binom{6}{4}(.6)^4(.4)^2 = .311$$

$$P(5) = \binom{6}{5}(.6)^5(.4)^1 = .187$$

$$P(6) = \binom{6}{6}(.6)^6(.4)^0 = .047$$

x	0	1	2	3	4	5	6
Income	0	100	200	300	200	100	0
$P(x)$	.004	.037	.138	.276	.311	.187	.047

a. $E(I) = 0(.004) + 100(.037) + 200(.138) + 300(.276) + 200(.311) + 100(.187) + 0(.047)$

$= \$195$

b. $n = 3$

x	0	1	2	3
Income	0	100	200	300
$P(x)$	.064	.288	.432	.216

$E(I) = 0(.064) + 100(.288)$

$+ 200(.432) + 300(.216)$

$= \$180$

48. Continued

On the basis of all calculations, the table given in the exercise is completed as follows.

x	0	1	2	3	4	5	6
Income	0	100	200	300	200	100	0
$P(x)$	.004	.037	.138	.276	.311	.187	.047

$n = 4$

x	0	1	2	3	4
Income	0	100	200	300	200
$P(x)$	.0256	.1536	.3456	.3456	.1296

$$E(I) = 0(.0256) + 100(.1536) + 200(.3456) + 300(.3456) + 200(.1296)$$
$$= \$214.08$$

$n = 5$

x	0	1	2	3	4	5
Income	0	100	200	300	200	100
$P(x)$	.01024	.0768	.2304	.3456	.2592	.07776

$$E(I) = 0(.01024) + 100(.0768) + 200(.2304) + 300(.3456) + 200(.2592) + 100(.07776)$$
$$= \$217.06$$

Since $E(I)$ is greatest when $n = 5$, the airlines should book 5 reservations to maximize revenue.

Case 9 Optimal Inventory for a Service Truck

1. a.
$$C(M_0) = NL\left[1 - (1 - p_1)(1 - p_2)(1 - p_3)\right]$$
$$= 3(54)\left[1 - (1 - .09)(1 - .24)(1 - .17)\right]$$
$$= \$69.01$$

b.
$$C(M_2) = H_2 + NL\left[1 - (1 - p_1)(1 - p_3)\right]$$
$$= 40 + 3(54)\left[1 - (1 - .09)(1 - .17)\right]$$
$$= \$79.64$$

c.
$$C(M_3) = H_3 + NL\left[1 - (1 - p_1)(1 - p_2)\right]$$
$$= 9 + 3(54)\left[1 - (1 - .09)(1 - .24)\right]$$
$$= \$58.96$$

d.
$$C(M_{12}) = H_1 + H_2 + NL\left[1 - (1 - p_3)\right]$$
$$= 15 + 40 + 3(54)\left[1 - (1 - .17)\right]$$
$$= \$82.54$$

e.
$$C(M_{13}) = H_1 + H_3 + NL\left[1 - (1 - p_2)\right]$$
$$= 15 + 9 + 3(54)\left[1 - (1 - .24)\right]$$
$$= \$62.88$$

f.
$$C(M_{123}) = H_1 + H_2 + H_3$$
$$= 15 + 40 + 9$$
$$\$64$$

2. Stock only part 3 on the trunk because $C(M_3)$ has the lowest expected cost.

3. The events of needing parts 1, 2, and 3 are not the only events in the sample space.

4. There would be 2^n different policies to evaluate.

Chapter 10: Introduction to Statistics

Section 10.1 Frequency Distributions and Measures of Central Tendency

1. a–b. Since 15–17 is to be the first interval and there are 3 numbers between 15 and 17 inclusive, we will let all six intervals be of size 3. The other five intervals are 15–17, 18–20, 21–23, 24–26, 27–29, 30–32, 33–35, 36–38, and 39–41. Keeping a tally of how many data values lie in each interval leads to the following frequency distribution.

Interval	Frequency
15–17	4
18–20	7
21–23	8
24–26	10
27–29	7
30–32	7
33–35	3
36–38	3
39–41	1

c. Draw the histogram. It consists of 9 bars of equal width, having heights as determined by the frequency of each interval.

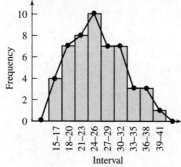

d. To construct the frequency polygon, join consecutive midpoints of the tops of the histogram bars with straight line segments. See the histogram in part (c).

3. a–b. Since 0–9 is to be the first interval, we let all the intervals be of size 10. The largest data value is 71, so the last interval that will be needed is 70–79. The frequency distribution is as follows.

Interval	Frequency
0–9	2
10–19	12
20–29	14
30–39	11
40–49	7
50–59	3
60–69	0
70–79	1

c. Draw the histogram. It consists of 8 bars of equal width and having heights as determined by the frequency of each interval.

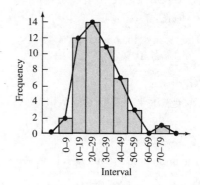

d. Construct the frequency polygon by joining consecutive midpoints of the tops of the histogram bars with straight line segments. See histogram in part (c).

5. The data ranges from 16 to 41.

STEM	LEAVES
1	
1	66678889
2	0001122233344444
2	556667889999
3	001112234
3	5777
4	1

Units: $4|1 = 41$ years

7. Answers vary. Possible answer: A frequency polygon is another form of graph that illustrates a grouped frequency distribution. The polygon is formed by joining consecutive midpoints of the tops of the histogram bars with straight line segments.

9. $\Sigma x = 21,900 + 22,850 + 24,930$
$+ 29,710 + 28,340 + 40,000$
$= 167,730$

The mean of the 6 numbers is
$\bar{x} = \dfrac{167,730}{6} = 27,955$.

11. $\Sigma x = 3.5 + 4.2 + 5.8 + 6.3 + 7.1 +$
$+ 2.8 + 3.7 + 4.2 + 4.2 + 5.7$
$= 47.5$

The mean of the 10 numbers is
$\bar{x} = \dfrac{47.5}{10} = 7.75 \approx 4.8$.

13. $\Sigma x = 9.2 + 10.4 + 13.5 + 8.7 + 9.7 = 51.5$

The mean is $\bar{x} = \dfrac{51.5}{5} = 10.3$.

15.

Value	Frequency	Value × Frequency	
19	3	$19 \cdot 3$	57
20	5	$20 \cdot 5$	100
21	25	$21 \cdot 25$	525
22	8	$22 \cdot 8$	176
23	2	$23 \cdot 2$	46
24	1	$24 \cdot 1$	24
28	1	$28 \cdot 1$	28
	Total: 45	Total:	956

The mean is $\bar{x} = \dfrac{956}{45} \approx 21.2$.

17.

x	f	xf
9	5	45
11	10	110
15	12	180
17	9	153
20	6	120
28	1	28
Totals	43	636

$\bar{x} = \dfrac{636}{43} = 14.8$

19. 28458, 29679, 33679, 38400, 39720
There are 5 numbers here; the median is the middle term, in this case 33,679.

21. First arrange the numbers in numerical order, from smallest to largest.
94.1, 96.8, 97.4, 98.6, 98.4, 98.7, 99.2, 99.9
There are 8 numbers here; the median is the mean of the 2 middle numbers, which is
$\dfrac{98.4 + 98.6}{2} = \dfrac{197}{2} = 98.5$.

23. 1, 2, 2, 1, 2, 2, 1, 1, 2, 2, 3, 4, 2, 3, 4, 2, 3, 2, 3,
The mode is the number that occurs most often. Here, the mode is 2.

25. 62, 65, 71, 74, 71, 76, 71, 63, 59, 65, 65, 64, 72, 71, 77, 63, 65
The mode is the number that occurs most often. Here, there are two modes, 65 and 71, since they both appear four times.

27. 5.7, 5.7, 5.8, 5.6, 5.8, 5.8, 5.8, 5.8, 5.8
The number 5.8 occurs six times; this is more than all others, so the mode is 5.8.

29. Answers vary. Possible answer: The median is the most appropriate measure of central tendency when most of the data are similar, but there are a few extreme values which influence the value of the mean significantly.

31.

Interval	Midpoint, x	Frequency, f	Product, xf
0–9	4.5	2	9.0
10–19	14.5	12	174.0
20–29	24.5	14	343.0
30–39	34.5	11	379.5
40–49	44.5	7	311.5
50–59	54.5	3	163.5
60–69	64.5	0	0
70–79	74.5	1	74.5
		Total: 50	Total: 1455

The mean of this collection of grouped data is

$$\bar{x} = \frac{1455}{50} = 29.1 .$$

The interval 20–29 contains the most data values, 14, so it is the modal class.

33. No. Answers vary.

35. a. Mean $\bar{x}$

$$= 1000 \left(\frac{147,970 + 70,527 + 67,682 + 60,023 + 58,105 + 40,589 + 37,842 + 36,089 + 33,925 + 32,812}{10} \right)$$

$$= \$58,556,400$$

b. The two middle compensations are \$58,105,000 and \$40,589,000. The median compensation is:
$$\frac{58,105,000 + 40,589,000}{2} = \$49,347,000.$$

c. The largest salary of Reuben Mark is more than twice the amount of the next largest salary that affects significantly the value of the mean.

d. Silverman's salary is \$60,023,000.

37. Mean $\bar{x} = \dfrac{16 + 12 + 11 + 9 + 8 + 7 + 7 + 6 + 5 + 5 + 4 + 4 + 2}{13}$

$$\approx 7.38$$

The middle data when the data is in ascending order is 7. So the median number of types is 7.
There are three modes, 7, 5 and 4. Each mode appears twice.

39. The maximum temperatures in ascending order: 39, 39, 40, 44, 47, 50, 51, 60, 69, 70, 78, 79

$$\sum x = 666$$

$$\bar{x} = \frac{\sum x}{n} = \frac{666}{12} = 55.5°F$$

$$\text{Median} = \frac{50 + 51}{12} = 50.5°F$$

41. Arrange the price per bushel data in ascending order: 2.48, 2.62, 2.65, 2.78, 3.38, 3.40, 3.45, 3.56, 4.30, 4.55,

$$\sum x = 33.17$$

$$\bar{x} = \frac{\sum x}{n} = \frac{33.17}{10} \approx \$3.32$$

$$\text{Median} = \frac{3.38 + 3.40}{2} = \$3.39$$

43.

Income Range	Midpoint Salary	Frequency (in thousands)	Product (in thousands)
Under $5,000	$2500	905	2,262,500
$5,000–$9,999	$7500	1451	10,882,500
$10,000–$14,999	$12,500	1158	14,475,000
$15,000–$24,999	$20,000	2197	43,940,000
$25,000–$34,999	$30,000	1904	57,120,000
$35,000–$49,999	$42,500	1984	84,320,000
$50,000–$74,999	$62,500	2051	128,187,500
$75,000–$99,999	$87,500	865	75,687,500
		12,515	416,875,000

$$\bar{x} = \frac{416,875,000}{12,515} \approx \$33,310$$

45. a. According to the graph, about 14% of the population is in the 10–19 age group.

b. About 7% of the population is in the 60–69 age group.

c. The 30–39 and 40–49 age ranges makes up the largest percent of the population at about 15%.

Section 10.2 Measures of Variation

1. Answers vary. Possible answer: The standard deviation of a sample of numbers is the square root of the variance of the sample.

3. 32, 21, 31, 35, 20

   Range $= 35 - 20 = 15$

   $\overline{x} = \dfrac{32 + 21 + 31 + 35 + 20}{5} = \dfrac{139}{5} = 27.8$

Number	Deviation from mean	Square of deviation
32	4.2	17.64
21	−6.8	46.24
31	3.2	10.24
25	7.2	51.84
20	−7.8	60.84
Total		186.80

   $s = \sqrt{\dfrac{186.80}{5-1}} = \sqrt{46.7} \approx 6.83$

5. 40, 25, 25, 36, 35

   Range $= 40 - 25 = 15$

   $\overline{x} = \dfrac{40 + 25 + 25 + 36 + 35}{5} = \dfrac{161}{5} = 32.2$

Number	Deviation from mean	Square of deviation
40	7.8	60.84
25	−7.2	51.84
25	−7.2	51.84
36	3.8	14.44
35	2.8	7.84
Total		186.80

   $s = \sqrt{\dfrac{186.80}{5-1}} = \sqrt{\dfrac{186.80}{4}} = \sqrt{46.7} \approx 6.83$

7. 38, 15, 25, 48, 43, 7, 25, 28

   Range $= 48 - 7 = 41$

   $\overline{x} = \dfrac{38 + 15 + 25 + 48 + 43 + 7 + 25 + 28}{8}$

   $= \dfrac{229}{8} = 28.625$

Number	Deviation from mean	Square of deviation
38	9.375	87.89
15	−13.625	185.64
25	−3.625	13.14
48	19.375	375.39
43	14.375	206.64
7	−21.625	467.64
25	−3.625	13.14
28	−.625	.39
Total		1349.87

   $s = \sqrt{\dfrac{1349.87}{8-1}} = \sqrt{\dfrac{1349.87}{7}} = \sqrt{192.84} \approx 13.89$

9. 3147, 3572, 1559, 3544, 2183, 3119, 3799, 3147, 3232

   Range: $3799 - 1559 = 2240$

   $\overline{x} = (3147 + 3572 + 1559 + 3544 + 2183$
   $+ 3119 + 3799 + 3147 + 3232)/9$

   $= \dfrac{27,302}{9} = 3033.6$

Number	Deviation from mean	Square of deviation
3147	113.4	12,859.6
3572	538.4	289,874.6
1559	−1474.6	2,174,445.2
3544	510.4	260,508.2
2183	−850.6	723,520.4
3119	85.4	7293.2
3799	765.4	585,837.2
3147	113.4	12,859.6
3232	198.4	39,362.6
Total		4,106,560

9. Continued

$$s = \sqrt{\frac{4,106,560}{9-1}} = \sqrt{\frac{4,106,560}{8}}$$

$$= \sqrt{513,320} \approx 716.46$$

11. Expand the table to include columns for the midpoint x of each interval, and for fx, x^2, and fx^2.

Interval	f	x	fx	x^2	fx^2
0–24	4	12	48	144	576
25–49	3	37	111	1369	4107
50–74	6	62	372	3844	23,064
75–99	3	87	261	7569	22,707
100–124	5	112	560	12,544	62,720
125–129	9	137	1233	18,769	168,921
Totals	30		2585		282,095

The mean of the grouped data is
$$\bar{x} = \frac{\Sigma fx}{n} = \frac{2585}{30} \approx 86.2 \,.$$
The standard deviation for the grouped data is

$$s = \sqrt{\frac{\Sigma fx^2 - n\bar{x}^2}{n-1}}$$

$$= \sqrt{\frac{282,095 - 30(86.2)^2}{30-1}}$$

$$\approx \sqrt{2040.7} \approx 45.2 \,.$$

13. This Exercise should be completed using a computer or calculator. The solution may vary according to the computer program or calculator that is used. The answer is given below.
$\bar{x} = 5.0876$; $s = .1087$

15. Use Chebyshev's theorem with $k = 2$.
$$1 - \frac{1}{2^2} = 1 - \frac{1}{4} = \frac{3}{4}$$

So, at least $\frac{3}{4}$ of the numbers lie within 2 standard deviations of the mean.

17. Use Chebyshev's theorem with $k = 1.5$.
$$1 - \frac{1}{(1.5)^2} = 1 - \frac{4}{9} = \frac{5}{9}, \text{ so at least } \frac{5}{9} \text{ of the}$$
numbers lie within 1.5 standard deviations of the mean.

19. Between 26 and 74, $\bar{x} = 50$; so s = 6; we have
$$26 = 50 - 4 \cdot 6 \text{ and}$$
$$74 = 50 + 4 \cdot 6,$$
so $k = 4$.
At least
$$1 - \frac{1}{k^2} = \frac{15}{16} = 93.75\%$$
of the numbers lie between 26 and 74.

21. Less than 32 or more than 68
From Exercise 18, 88.9% of the data lie between 32 and 68. So, no more than
$100\% - 88.9\% = 11.1\%$ of the data are less then 32 or more than 68.

23. a. Male first-year college student heights:
182, 178, 179, 182, 173, 170, 167, 171
$$\bar{x} = \frac{1}{8}(182 + 178 + 179 + 182 + 173$$
$$+ 170 + 167 + 171)$$
$$= \frac{1}{8}(1402) = 175.25$$

The mean height for male first-year college students is 175.25 cm..

x	$x - \bar{x}$	$(x - \bar{x})^2$
182	6.75	45.5625
178	2.75	7.5625
179	3.75	14.0625
182	6.75	45.5625
173	-2.25	5.0625
170	-5.25	27.5625
167	-8.25	68.0625
171	-4.25	18.0625

Total: 231.5

23. Continued

$$s = \sqrt{\frac{231.5}{8-1}} \approx 5.75$$

The standard deviation of the male first-year college students' heights is 5.75 cm.

Female first-year college students
174, 162, 157, 172, 164, 162, 168, 163

$$\bar{x} = \frac{1}{8}(174 + 162 + 157 + 175 + 164$$

$$+ 162 + 168 + 163)$$

$$= \frac{1}{8}(1322) = 165.25$$

The mean life of the female first-year college students is 165.25.

x	$x - \bar{x}$	$(x - \bar{x})^2$
174	8.75	76.5625
162	−3.25	10.5625
157	−8.25	68.0625
172	6.75	45.5625
164	−1.25	1.5625
162	−3.25	10.5625
168	2.75	7.5625
163	−2.25	5.0625
		Total: 225.5

$$s = \sqrt{\frac{225.5}{8-1}} \approx 5.68$$

The standard deviation of the female first-year college students' heights is 5.68 cm.

b. The female first-year college students have a smaller standard deviation, which indicates less variability.

c. The male first-year college students have a higher average height. This is not a surprise because in the general population, males are generally taller.

25.

Year	Number Unemployed	Square of the deviation
1997	6.7	.052258
1998	6.2	.530858
1999	5.9	1.058018
2000	5.7	1.509458
2001	6.8	.016538
2002	8.4	2.165018
2003	8.8	3.502138
	48.5	8.834286

a. $\bar{x} = \dfrac{\sum x}{n} = \dfrac{48.5}{7} = 6.9286$

Unemployment was closest to the mean in 2001.

b. First, the variance:

$$s^2 = \frac{8.834286}{8-1} \approx 1.2620409$$

The standard deviation is:

$$s \approx \sqrt{1.2620409} \approx 1.12$$

c. One standard deviation from the mean consists of the values 5.8086 to 8.0486. Unemployment is within 1 standard deviation of the mean for 4 of these years.

d. Three standard deviations from the mean consists of the values 3.5686 to 10.2886. In all 7 years the unemployment falls within 3 standard deviations of the mean.

27.

Number of blood types	Square of the number
16	256
12	144
11	121
9	81
8	64
7	49
7	49
6	36
5	25
5	25
4	16
4	16
2	4
	886

a. $s^2 = \dfrac{886 - 13(7.38)^2}{13 - 1} \approx 14.8$

$s \approx \sqrt{14.8} \approx 3.8$

b. One standard deviation from the mean consists of the values 3.58 to 11.18. Ten animals have blood types within 1 standard deviation of the mean.

29.

x (days)	x^2
84	7056
91	8281
128	16,384
131	17,161
143	20,449
153	23,409
164	26,896
894	119,636

a. $\bar{x} = \dfrac{894}{7} \approx 127.71$ days

$s = \sqrt{\dfrac{119,636 - 7(\bar{x})^2}{7 - 1}} \approx 30.19$ days

b. Doubling times within 2 standard deviations of the mean range from 67.33 to 188.09 days. All of these cancers have doubling times within 2 standard deviations of the mean.

c. Answers vary. Possible answer: A doubling time of 200 days is more than 2 standard deviations from the mean of these slow growing cancers. The tumor is not growing as expected for these slow-growing cancers.

31.

Sample Number

	1	2	3	4	5	6	7	8	9	10
	2	3	−2	−3	−1	3	0	−1	2	0
	−2	−1	0	1	2	2	1	2	3	0
	1	4	1	2	4	2	2	3	2	2
a. $\bar{x}$	$\dfrac{1}{3}$	2	$-\dfrac{1}{3}$	0	$\dfrac{5}{3}$	$\dfrac{7}{3}$	1	$\dfrac{4}{3}$	$\dfrac{7}{3}$	$\dfrac{2}{3}$
b. s	2.1	2.6	1.5	2.6	2.5	.6	1	2.1	.6	1.2

c. $\bar{X} = \dfrac{\Sigma \bar{x}}{n} \approx \dfrac{11.3}{10} = 1.13$

31. Continued

    d. $\overline{s} = \dfrac{\Sigma s}{n} = \dfrac{16.8}{10} = 1.68$

    e. The upper control limit for the sample means is
$$\overline{X} + 1.954\overline{s} = 1.13 + (1.954)(1.68)$$
$$\approx 4.41$$
The lower control limit for the sample means is
$$\overline{X} - 1.954\overline{s} = 1.13 - (1.954)(1.68)$$
$$\approx -2.15$$
One of the measurements, -3, is outside of these limits, so the process is out of control.

33.

Interval	f	x	xf	x^2	$x^2 f$
50–59	2	54.5	109	2970.25	5940.5
60–69	4	64.5	258	4160.25	16,641
70–79	7	74.5	521.5	5550.25	38,851.75
80–89	9	84.5	760.5	7140.25	64,262.25
90–99	8	94.5	756	8930.25	71,442
Totals:	30		2405		197,137.5

$$\overline{x} = \frac{2405}{30} \approx 80.17$$
$$s = \sqrt{\frac{197,137.5 - 30(80.17)^2}{30 - 1}} \approx 12.2$$

For individualized instruction, the mean is 80.17 and the standard deviation is 12.2.

35. Answers vary. Possible answer: You would expect the mean of the traditional instruction to be smaller, since
13 of the 34 students lie in the first two intervals while only 6 of 30 students lie in the same intervals for individualized instruction. As far as standard deviation is concerned, there is a small difference between the two, 1.1. This gives the impression that the data is pretty much spread out in nearly the same fashion in both types of instruction.

Section 10.3 Normal Distributions

1. The peak in a normal curve occurs directly above the mean.

3. Answers vary. Possible answer:
   If a normal distribution has mean μ and standard deviation σ, then the z-score for the number x is
   $$z = \frac{x - \mu}{\sigma}.$$

5. By looking up 1.75 in Table 2, we get .4599 = 45.99%.

7. −.43 indicates that we are below the mean. Looking up .43 in Table 2, we get .1664 = 16.64%.

9. The entry corresponding to $z = 1.41$ is .4207, and the entry for $z = 2.83$ is .4977, so the area between $z = 1.41$ and $z = 2.83$ is .4977 − .4207 = .0770 or 7.7%.

11. For $z = 2.48$, the entry is .4934, and for $z = .05$, the entry is .0199. The area between $z = -2.48$ and $z = -.05$ is .4934 − .0199 = .4735 or 47.35%.

13. To find the area between $z = -3.05$ and $z = 1.36$, add the area between $z = -3.05$ and $z = 0$ to the area between $z = 1.36$ and $z = 0$. The area is .4989 + .4131 = .9120 = 91.20%.

15. 5% of the total area is to the right of z. The mean divides the area in half or .5. The area from the mean to z standard deviations is .5 − .05 = .45. Using Table 2 backwards, we get the z-score corresponding to the area .45 as 1.64 or 1.65 (approximately).

17. 15% of the total area is to the left of z. The area from the mean to z standard deviations is .5 − .15 = .35. The z-score from the table is 1.04. As the area is to the left of the mean, $z = -1.04$.

19. To find $P(x \le \mu)$ and $P(x \ge \mu)$ consider the normal curve. The curve is symmetric about the vertical line through μ, so $P(x \le \mu)$ represents half the area under the curve. Since the area under the curve is 1, $P(x \le \mu) = .5$. Similarly, $P(x \ge \mu) = .5$.

21. Using Chebyshev's theorem with $k = 3$:
    $$1 - \frac{1}{3^2} = 1 - \frac{1}{9} = \frac{8}{9} \approx .889$$
    Using the normal distribution, the area between $z = -3$ and $z = 3$:
    .4987 + .4987 = .9974
    The probability a number will lie within 3 standard deviations of the mean is greater as indicated by the normal distribution at .9974 than Chebyshev's theorem shows at .889.

23.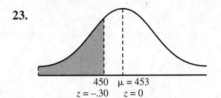

    To find the area to the left of $z = -.30$, find the area between $z = 0$ and $z = -.30$ and subtract that answer from .5. The area is .5 − .1179 = .3821.

25.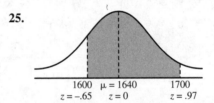

    To find the area between $z = -.65$ and $z = .97$, find the area between $z = 0$ and $z = -.65$. Then find the area between $z = 0$ and $z = .97$. Add these two answers together. The area is .2422 + .3340 = .5762.

27. Let x represent clotting time.
    $\mu = 51.6$, $\sigma = 14.3$

    For $x = 60$, $z = \dfrac{60 - 51.6}{14.3} \approx .59$

    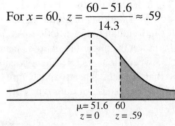

    To find the area to the right of $z = .59$, find the area between $z = .59$ and $z = 0$ and subtract that from 0.5. The area is 0.5 − .2224 = .2776.

29. Let x represent the starting salaries for accounting majors.

$\mu = 41,0000, \ \sigma = 3200$

For $x = 50,000$,

$$z = \frac{50,000 - 41,000}{3200} \approx 2.8125 .$$

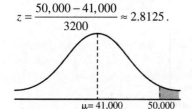

$\mu = 41,000 \quad 50,000$
$z = 0 \qquad z = 2.8125$

To find the area to the right of $z = 2.8125$, find the area between $z = 0$ and $z = 2.8125$ and subtract it from .5. The area is $.5 - .4975 = .0025$. Thus, the probability that an individual will have a starting salary above \$50,000 is .0025.

31. For $x = 500, \ z = \dfrac{500 - 500}{100} = 0.$

The area to the right of $z = 0$ is .5.

$.5 \cdot 10,000 = 5000$ light bulbs are expected to last at least 500 hr.

33. For $x = 650, \ z = \dfrac{650 - 500}{100} = 1.5.$

For $x = 780, \ z = \dfrac{780 - 500}{100} = 2.8.$

The area between $z = 1.5$ and $z = 2.8$ is $.4974 - .4332 = .0642.$

$.0642 \cdot 10,000 = 642$ light bulbs are expected to last between 650 and 780 hr.

35. For $x = 740, \ z = \dfrac{740 - 500}{100} = 2.4.$

The area to the left of $z = 2.4$ is $.4918 + .5 = .9918.$

$.9981 \cdot 10,000 = 9918$ light bulbs are expected to last less than 740 hr.

37. If 80% of the total area straddles the mean, 40% of the area is between z and the mean. Look for a value of z in Table 2 with an area of .4. The closest match is $z = 1.28$.

Let x represent the life of a light bulb.

For the shortest life of the middle 80% of bulbs, $z < 0$.

$$\frac{x - 500}{100} = -1.28$$

$$x - 500 = -128$$

$$x = 372$$

For the longest life of the middle 80% of bulbs, $z > 0$.

$$\frac{x - 500}{100} = 1.28$$

$$x - 500 = 128$$

$$x = 628$$

80% of the light bulbs have lives between 372 and 628 hr.

39. If 85% of the total area is to the left of z, then $z > 0$ and 35% of the area is between z and the mean. Look for a value of z in Table 2 with an area of .35. The closest match is $z = 1.04$. Let x stand for the speed.

$$\frac{x - 40}{5} = 1.04$$

$$x - 40 = 5.2$$

$$x = 45.2$$

The 85th percentile speed is 45.2 mph.

41.

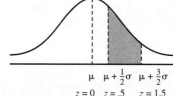

$\mu \quad \mu + \frac{1}{2}\sigma \quad \mu + \frac{3}{2}\sigma$
$z = 0 \quad z = .5 \quad z = 1.5$

To find the area between $z = .5$ and $z = 1.5$, first find the area between $z = 1.5$ and $z = 0$. Then subtract the area between $z = 0$ and $z = .5$. The area is $.4332 - .1915 = .2417 = 24.17\%$, which means 24.17% of the students receive a B.

43. Answers vary. Possible answer: This system would be more fair in a large freshman class in psychology than in a graduate seminar of five students since the large class is more apt to have grades ranging the entire spectrum. The graduate students might have all grades in the 90's which would mean that the graduate student with the lowest score in the 90's would get an F.

45. Mean = 550 units;
Standard deviation = 45 units
The recommended daily allowance is
$\mu + 2.5\sigma = 550 + (2.5)(46)$
$= 665$ units

47. Mean = 155 units; standard
deviation = 14 units
The recommended daily allowance is
$\mu + 2.5\sigma = 155 + (2.5)(14)$
$= 190$ units.

49. To find the area to the right of $z = 1$, find the area
between $z = 0$ and $z = 1$, and subtract that from .5.
The area is $.5 - .3413 = .1587$, which means
15.87% of the students had scores more than 1
standard deviation above the mean.

51. Less than $400
$\mu = 500$, $\sigma = 65$
For $x = 400$, $z = \dfrac{400 - 500}{65} \approx -1.54$
To find the area to the left of $z = -1.54$, subtract
the area between $z = -1.54$ and $z = 0$ from .5.
The area is $.5 - .4382 = .0618$.

53. Between $350 and $600
For $x = 350$,
$$z = \frac{350 - 500}{65}$$
$$= -2.31.$$
For $x = 600$,
$$z = \frac{600 - 500}{65}$$
$$= 1.54.$$
To find the area between $z = -2.31$ and
$z = 1.54$, find the area between $z = 0$ and
$z = 1.54$ and add it to the area between $z = -2.31$
and $z = 0$. The area is $.4382 + .4896 = .9278$.

55. Arrange the 2002 data in ascending order:
2.3, 2.6, 3.8, 3.9, 3.9, 4.2, 5.8, 6.0, 7.0, 8.3
Minimum = 2.3
Maximum = 8.3
Median = $Q_2 = 4.05$
$n = 10$
$.25 \cdot 10 = 25$, which rounds up to 3.
$Q_1 = 3.8$
$.75 \cdot 10 = 7.5$, which rounds up to 8.
$Q_3 = 6.0$

57.

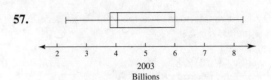

59. Arrange the 1990 data in ascending order:
5.1, 5.1, 5.3, 5.4, 5.9, 6.6, 7.0, 7.1, 7.1, 7.4, 8.0,
8.0, 8.2, 8.4, 9.2
Minimum = 5.1
Maximum = 9.2
Median = $Q_2 = 7.1$
$n = 15$
$.25 \cdot 15 = 3.75$, which rounds up to 4.
$Q_1 = 5.4$
$.75 \cdot 15 = 11.25$, which rounds up to 12.
$Q_3 = 8.0$

61.

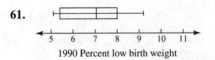

1990 Percent low birth weight

63. Yes. Explanations vary.

Section 10.4 Normal Approximation to the Binomial Distribution

1. Answers vary. Possible answer: To find the mean and standard deviation of a binomial distribution, you must know n, the number of independent repeated trials, and p, the probability of a success in a single trial.

3. $n = 16, p = .5, (1 - p) = .5$

   a. Using the binomial distribution
   $$P(x = 8) = \binom{16}{8}(.5)^8(.5)^8 \approx .1964$$

   b. Since we are using a normal curve approximation.
   $$\mu = np = 16\left(\frac{1}{2}\right) = 8$$
   and
   $$\sigma = \sqrt{np(1-p)}$$
   $$= \sqrt{16\left(\frac{1}{2}\right)\left(1 - \frac{1}{2}\right)} = 2.$$
   The required probability is the area of the region corresponding to the x-values of 7.5 and 8.5.
   $$z = \frac{x - \mu}{\sigma} = \frac{8.5 - 8}{2} = .25$$
   From Table 2, the corresponding area is .0987. Therefore, the total area is $2(.0987) = .1974$.
   This is the probability of getting exactly 8 heads.

5. $n = 16, p = .5, (1 - p) = .5$

   a. Using the binomial distribution,
   $P(x > 13) = P(x = 14) + P(x = 15) + P(x = 16)$
   $$P(x = 14) = \binom{16}{14}(.5)^{14}(.5)^2 \approx .00183$$
   $$P(x = 15) = \binom{16}{15}(.5)^{15}(.5)^1 \approx .00024$$
   $$P(x = 16) = \binom{16}{16}(.5)^{16}(.5)^0 \approx .00002$$
   $$P(x > 13) \approx .00209$$

5. Continued

   b. For more than 13 tails, the desired area is to the right of 13.5.
   $$\mu = np = 16(.5) = 8$$
   $$\sigma = \sqrt{16(.5)(.5)} = 2$$
   For $x = 13.5$, $z = \dfrac{13.5 - 8}{2} = 2.75$.
   $$P(x > 13) = .5 - .4970 = .0030$$

7. Exactly 500 heads
   $$n = 1000, \ p = \frac{1}{2}$$
   $$\mu = np = 1000\left(\frac{1}{2}\right) = 500$$
   $$\sigma = \sqrt{np(1-p)} = \sqrt{1000\left(\frac{1}{2}\right)\left(\frac{1}{2}\right)} = 15.8$$
   We need to find the area of the region corresponding to the x-values of 499.5 and 500.5.
   $$z = \frac{500.5 - 500}{15.8} = .03$$
   The area is .0120. Since the region is symmetric about the mean $\mu = 500$, the probability (area) is $2(.0120) = .0240$.

9. 475 heads or more
   $$p = \frac{1}{2}, \mu = 500, \text{ and } \sigma = 15.8.$$
   The required area is to the right of the x-value 474.5.
   $$z = \frac{474.5 - 500}{15.8} = -1.61$$
   The area is .4463. This area is to the left of the mean. So, the required area (probability) is $.4463 + .5 = .9463$.

11. Exactly 20 fives
    The probability of a five is $p = \dfrac{1}{6}$, and $n = 120$.
    $$\mu = np = 120\left(\frac{1}{6}\right) = 20$$
    $$\sigma = \sqrt{120\left(\frac{1}{6}\right)\left(\frac{5}{6}\right)} = 4.08$$
    $$z = \frac{20.5 - 20}{4.08} = .12$$
    The area is .0478. So, the area between the x-values 19.5 and 20.5 is $2(.0478) = .0956$. This is the required probability.

13. More than 17 threes

$p = \dfrac{1}{6}$, and the required area is to the right of

$x = 17.5$.

$z = \dfrac{17.5 - 20}{4.08} = -.61$

The area is .2291. So, for more than 17 threes, the probability is
.2291 + .5 = .7291.

15. $n = 130$, $p = \dfrac{1}{6}$, $1 - p = \dfrac{5}{6}$

$\mu = np = 130\left(\dfrac{1}{6}\right) \approx 21.6667$

$\sigma = \sqrt{np(1-p)}$

$= \sqrt{130\left(\dfrac{1}{6}\right)\left(\dfrac{5}{6}\right)}$

≈ 4.2492

The required area is to the right of $x = 25.5$.

$z = \dfrac{25.5 - 21.6667}{4.2492} \approx .90$

The area for $z = .90$ is .3159, so the probability is
.5 − .3159 = .1841.

17. $p = .3$ and $n = 26$

$\mu = np = 26(.3) = 7.8$

$\sigma = \sqrt{26(.3)(.7)} = 2.34$

Since at least half of 26 nests escape predation, the area is to the right of $x = 12.5$.

$z = \dfrac{12.5 - 7.8}{2.34} = 2.01$

The area from $x = 12.5$ to the mean is .4778, so the area to the right of $x = 12.5$ is
.5 − .4778 = .0222,
which gives the required probability.

19. $p = .114$, $1 - p = .886$, and $n = 600$

$\mu = np = 600(.114) = 68.4$

$\sigma = \sqrt{np(1-p)}$

$= \sqrt{600(.114)(.886)}$

≈ 7.78

For $x = 80.5$

$z = \dfrac{80.5 - 68.4}{7.78}$

$z \approx 1.56$

The area between $z = 0$ and $z = 1.56$ is .4406, so the probability of $z \geq 1.56$ is .5 − .4406 = .0594.

21. $n = 40$, $p = .245$, $1 - p = .755$

$\mu = np = 40(.245) = 9.8$

$\sigma = \sqrt{np(1-p)} = \sqrt{40(.245)(.755)} \approx 2.7201$

For 15 or fewer, let

$x = 15.5$

$z = \dfrac{15.5 - 9.8}{2.7201} \approx 2.10$

The area is .4821, so the probability is
.5 + .4821 = .9821.

23. $\mu = np = (134)(.20) = 26.8$

$\sigma = \sqrt{np(1-p)} = \sqrt{(134)(.20)(.80)}$

$= 4.63$

a. $P(x = 12) = P(11.5 < x < 12.5)$

If $x = 11.5$,

$z = \dfrac{11.5 - 26.8}{4.63} = -3.305$.

If $x = 12.5$,

$z = \dfrac{12.5 - 26.8}{4.63} = -3.089$.

$P(11.5 < x < 12.5)$

Using a computer with appropriate software, we find that
$= P(-3.305 < z < -3.089) = .0005$

b. $P(\text{no more than } 12) = P(x < 12.5)$

If $x = 12.5$, $z = -3.089$, from (a).

$P(x < 12.5) = P(z < -3.089)$

Using a computer with appropriate software, we find that
$P(z \leq -3.089) = .001$.

c. $P(x = 0) = P(x < .5)$

If $x = .5$,

$z = \dfrac{.5 - 26.8}{4.63} = -5.680$.

$P(x \leq .5) = P(z < -5.680)$

Using a computer with appropriate software, we find that
$P(z \leq -5.680) = .0000$.

25. $\mu = np = (75)(.05) = 3.75$

$\sigma = \sqrt{np(1-p)} = \sqrt{(75)(.05)(.95)}$

$= 1.887$

a. $P(x = 7) = P(6.5 < x < 7.5)$

If $x = 6.5$,

$z = \dfrac{6.5 - 3.75}{1.887} = 1.457$.

If $x = 7.5$,

$z = \dfrac{7.5 - 3.75}{1.887} = 1.987$.

$P(6.5 < x < 7.5) = P(1.46 < z < 1.99)$

Using Appendix B, we then have

$.4767 - .4279 = .0488$.

b. $P(x = 0) = P(x < .5)$

If $x = .5$,

$z = \dfrac{.5 - 3.75}{1.887} = -1.722$.

$P(x < .5) = P(z < -1.722)$

Using a computer with appropriate software, we find that

$P(z < -1.722) = .0425$.

c. $P(\text{at least } 1) = P(x > .5)$

If $x = .5$, $z = -1.722$ from (b).

Using a computer with appropriate software, we see that

$P(z > -1.722) = .9575$.

27. a. $n = 1000, p = .006, 1 - p = .994$

$\mu = np = 6$

$\sigma = \sqrt{np(1-p)}$

$= \sqrt{1000(.006)(.994)}$

≈ 2.4421

The required area is to the right of $x = 9.5$.

$z = \dfrac{9.5 - 6}{2.4421} \approx 1.43$

The area for $z = 1.43$ is .4236, so the probability is $.5 - .4236 = .0764$.

27. Continued

b. $n = 1000, p = .015, 1 - p = .985$

$\mu = np = 15$

$\sigma = \sqrt{np(1-p)}$

$= \sqrt{1000(.015)(.985)}$

≈ 3.8438

The area required is between $x = 19.5$ and $x = 40.5$.

If $x = 19.5$, $z = \dfrac{19.5 - 15}{3.8438} \approx 1.17$.

If $x = 40.5$, $z = \dfrac{40.5 - 15}{3.8438} \approx 6.63$.

The area for $z = 1.17$ is .3790 and for $z = 6.63$ is .5000.

$P(20 \leq x \leq 40) = .5000 - .3790 = .121$

c. $n = 500, p = .015, 1 - p = .985$

$\mu = np = 7.5$

$\sigma = \sqrt{np(1-p)}$

$= \sqrt{500(.015)(.985)}$

≈ 2.7180

For $x = 14.5$, $z = \dfrac{14.5 - 7.5}{2.7180} \approx 2.58$.

The area for $z = 2.58$ is .4951, so the probability is $.5 - .4951 = .0049$.

Yes, this town does appear to have a higher than normal number of B– donors, since the probability of getting these kinds of results from a normal town is only .49%.

29. $n = 1400, p = .55, 1 - p = .45$

$\mu = np = 1400(.55) = 770$

$\sigma = \sqrt{np(1-p)} = \sqrt{1400(.55)(.45)} \approx 18.6145$

The required area is to the right of $x = 749.5$.

$z = \dfrac{749.5 - 770}{18.6145} \approx -1.10$

The area for $z = -1.10$ is .3643, so the probability is $.5 + .3643 = .8643$.

31. a. Using the binomial distribution to find the probability of 4 or more holes in one, $P(x > 3.5)$, is equivalent to finding the probability $1 - P(x < 3.5)$.

$1 - P(x < 3.5) = 1 - (P(x = 0) + P(x = 1) + P(x = 2) + P(x = 3))$

$$P(x = 0) = \binom{156}{0}\left(\frac{1}{3709}\right)^0 \left(\frac{3708}{3709}\right)^{156} \approx .95881$$

$$P(x = 1) = \binom{156}{1}\left(\frac{1}{3709}\right)^1 \left(\frac{3708}{3709}\right)^{155} \approx .04034$$

$$P(x = 2) = \binom{156}{2}\left(\frac{1}{3709}\right)^2 \left(\frac{3708}{3709}\right)^{154} \approx .00084$$

$$P(x = 3) = \binom{156}{3}\left(\frac{1}{3709}\right)^3 \left(\frac{3708}{3709}\right)^{153} \approx .00001$$

$1 - (.95881 + .04034 + .00084 + .00001) \approx 1.2139 \times 10^{-7}$

The probability that 4 or more of 156 golf pros shoot a hole in one is 1.2139×10^{-7}.

b. $n = 156$, $p = \dfrac{1}{3709}$, $1 - p = \dfrac{3708}{3709}$

$$\mu = np = \frac{156}{3709} \approx .04206$$

$$\sigma = \sqrt{np(1-p)} = \sqrt{156\left(\frac{1}{3709}\right)\left(\frac{3708}{3709}\right)} \approx .20506$$

The required area is to the right of $x = 3.5$.

$$z = \frac{3.5 - .04206}{.20506} \approx 16.86$$

The area for $z = 16.86$ is essentially .5000. The probability is then $.5 - .5000 = 0$.
(The probability is actually very small.)
We must be cautious in using this approximation in this application because the data doesn't follow the rule of thumb that $np \geq 5$.

c. $n = 20,000$, $p = \dfrac{1}{3709}$, $1 - p = \dfrac{3708}{3709}$

$$\mu = np = 20,000\left(\frac{1}{3709}\right) \approx 5.3923$$

$$\sigma = \sqrt{np(1-p)} = \sqrt{20,000\left(\frac{1}{3709}\right)\left(\frac{3708}{3709}\right)} \approx 2.3218$$

The required area is to the right of $x = 3.5$.

$$z = \frac{3.5 - 5.3923}{2.3218} = -.82$$

The area for $z = -.82$ is .2939, so the probability is $.5 + .2939 = .7939$.
Discussion answers vary.

Chapter 10 Review Exercises

1. Answers vary. Possible answer: Some reasons for organizing data into a grouped frequency distribution are that
    a. it organizes data so that one can visually understand data;
    b. it allows for graphical representation like histogram;
    c. it provides easier computation of mean and standard deviation when there is a large number of data.

2. Answers vary. Possible answer: In a grouped frequency distribution, there should be from 6 to 15 intervals.

3. a. Since 450–474 is to be the first interval, we will let all the intervals be of size 25. The largest data value is 566, so the last interval that will be needed is 550–574. The frequency distribution is as follows:

Interval	Frequency
450–474	5
475–499	6
500–524	5
525–549	2
550–574	2

    b. Draw the histogram. It consists of 5 bars of equal width and having heights as determined by the frequency of each interval.

    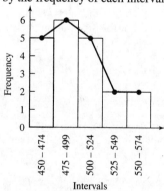

    c. Construct the frequency polygon by joining consecutive midpoints of the tops of the histogram bars with straight line segments. See the histogram in part (b).

4. a.

Interval	Frequency
9–10	3
11–12	6
13–14	6
15–16	7

    b–c.

    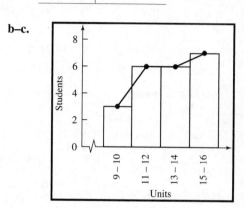

    d.

STEM	LEAVES
0	9
1	00
1	122222
1	333344
1	5555566

5. $\sum x = 480 + 451 + 501 + 478 + 512 + 473 + 509 + 515 + 458 + 566 + 516 + 535 + 492 + 558 + 488 + 547 + 461 + 475 + 492 + 471 = 9978$
    The mean of the 20 numbers is
    $$\bar{x} = \frac{\sum x}{20} = \frac{9978}{20} = 498.9 .$$

6. 10, 9, 16, 12, 13, 15, 13, 16, 15, 11, 13, 12, 12, 15, 12, 14, 10, 12, 14, 15, 15, 13
    $$\bar{x} = \frac{\sum x}{n}$$
    $$= \frac{287}{22}$$
    $$\approx 13.0$$

7.

Interval	Midpoint, x	Frequency, f	Product, xf
60–69	64.5	10	645
70–79	74.5	24	1788
80–89	84.5	6	507
90–99	94.5	3	283.5
100–109	104.5	1	104.5
	Total:	44	3328

The mean of this collection of grouped data is

$$\overline{x} = \frac{3328}{44} = 75.6 \text{ cm} .$$

8.

Interval	Midpoint, x	Freq, f	Product xf
0–999	499.5	1	499.5
1000–1999	1499.5	12	17,994
2000–2999	2499.5	14	34,993
3000–3999	3499.5	11	38,494.5
4000–4999	4499.5	5	22,497.5
5000–5999	5499.5	1	5499.5
	Total:	44	119,978

Use the formula for the mean of a grouped frequency distribution.

$$\overline{x} = \frac{\Sigma xf}{n} = \frac{119,978}{44} \approx 2726.8$$

9. Answers vary. Possible answer: Mean, median, and mode are all types of averages. The mode is the value that occurs the most. The median is the middle value when the data are ranked from highest to lowest. The mean is the sum of the data divided by the number of data items.

10. 65, 68, 71, 72, 72, 73, 73, 73, 78, 80, 84, 89
There are 12 numbers here; the median is the mean of the 2 middle numbers, which is
$\frac{73+73}{2} = 73$. The mode = 73, which occurs 3 times.

11. Arrange the numbers in numerical order, from smallest to largest.
66, 67, 68, 70, 71, 72, 72, 72, 73, 74, 76, 77, 80
The median is the middle number; in this case it is 72.
The mode is the number that occurs most often; in this case it is 72.

12. The modal class is the interval with the greatest frequency. For the distribution of Exercise 7, the modal class is 70–79.

13. The modal class for the distribution of Exercise 8 is the interval 2000–2999, since it contains more data values than any of the other intervals.

14. Answers vary. Possible answer: The range of a distribution is the difference between the largest and smallest data values.

15. Answers vary. Possible answer: The standard deviation is the square root of the variance. The standard deviation measures how spread out the data are from the mean.

16. 14, 17, 18, 19, 30

The range is the difference between the largest and smallest numbers. For this distribution, the range is 30 − 14 = 16.

To find the standard deviation, the first step is to find the mean.

$$\bar{x} = \frac{\Sigma x}{n} = \frac{14+17+18+19+30}{5}$$

$$= \frac{98}{5} = 19.6$$

Now complete the following chart.

x	x^2
14	196
17	289
18	324
19	361
30	900
Total:	2070

$$s = \sqrt{\frac{\Sigma x^2 - n\bar{x}^2}{n-1}} = \sqrt{\frac{2070 - 5(19.6)^2}{4}}$$

$$= \sqrt{37.3} \approx 6.11$$

17. The range is 57 − = 37, the difference of the highest and lowest numbers in the distribution.

The mean is $\bar{x} = \dfrac{\Sigma x}{n} = \dfrac{289}{10} = 28.9$.

Construct a table with the values of x, $x - \bar{x}$, and $(x - \bar{x})^2$.

x	$x - \bar{x}$	$(x - \bar{x})^2$
26	−2.9	8.41
43	14.1	198.81
17	−11.9	141.61
20	−8.9	79.21
25	−3.9	15.21
37	8.1	65.61
54	25.1	630.01
28	−.9	.81
20	−8.9	79.21
19	−9.9	98.01
Totals: 289		1316.90

The standard deviation is

$$s = \sqrt{\frac{1316.90}{10-1}} \approx \sqrt{143.6} \approx 12.10.$$

18. Recall that when working with grouped data, x represents the midpoint of each interval. Complete the following table, which extends the table from Exercise 7.

Interval	f	x	xf	x^2	fx^2
60–69	10	64.5	645	4160.25	41,602.50
70–79	24	74.5	1788	5550.25	133,206.00
80–89	6	84.5	507	7140.25	42,841.50
90–99	3	94.5	283.5	8930.25	26,790.75
100– 09	1	104.5	104.5	10,920.25	10,920.25
Totals:	44		3328		255,361.00

Use the formulas for grouped frequency distributions to find the mean and then the standard deviation. (The mean was also calculated in Exercise 7.)

$$\bar{x} = \frac{\Sigma xf}{n} = \frac{3328}{44} = 75.\overline{63}$$

$$s = \sqrt{\frac{\Sigma fx^2 - n\bar{x}^2}{n-1}}$$

$$= \sqrt{\frac{255,361 - 44(75.63)^2}{43}} \approx 9.20$$

19. Start with the frequency distribution that was the answer to Exercise 8, and expand the table to include columns for the midpoint x of each interval, and for xf, x^2, and f^2 x .

Interval	f	x	xf	x^2	fx^2
0–999	1	499.5	499.50	249,500.25	249,500.25
1000–1999	12	1499.5	17,994.00	2,248,500.25	26,982,003.00
2000–2999	14	2499.5	34,993.00	6,247,500.25	87,465,003.50
3000–3999	11	3499.5	38,494.50	12,246,500.25	134,711,502.75
4000–4999	5	4499.5	22,497.50	20,245,500.25	101,227,501.25
5000–5999	1	5499.5	5,499.50	30,244,500.25	30,244,500.25
Totals:	44		119,978.00		380,880,011.00

The mean of the grouped data is
$$\bar{x} = \frac{\Sigma xf}{n} = \frac{119,978}{44} = 2726.77 .$$
The standard deviation for the grouped data is
$$s = \sqrt{\frac{\Sigma fx^2 - n\bar{x}^2}{n-1}}$$
$$= \sqrt{\frac{380,880,011 - 44(2726.77)^2}{44-1}} \approx \sqrt{1,249,471}$$
$$\approx 1117.80$$

20. Answers vary. Possible answer: A normal distribution is a continuous distribution with the following properties:
    a. The highest frequency is at the mean;
    b. The graph is symmetric about a vertical line through the mean; and
    c. The total area under the curve, above the x-axis, is 1.

21. Answers vary. Possible answer: A distribution in which the peak is not at the center, or mean, is called skewed.

22. Between $z = 0$ and $z = 1.35$
    By Table 2, the area between $z = 0$ and $z = 1.35$ is .4115.

23. To the left of $z = .38$
    The area between $z = 0$ and $z = .38$ is .1480, so the area to the left of $z = .38$ is
    $.5 + .1480 = .6480$.

24. Between $z = -1.88$ and $z = 2.41$
    The area between $z = -1.88$ and $z = 0$ is .4699. The area between $z = 0$ and $z = 2.41$ is .4920. The total area between $z = -1.88$ and $z = 2.41$ is
    $.4699 + .4920 = .9619$.

25. Between $z = 1.53$ and $z = 2.82$
    The area between $z = 2.82$ and $z = 1.53$ is
    $.4976 - .4370 = .0606$.

26. Since 8% of the area is to the right of the z-score, 42% or .4200 is between $z = 0$ and the appropriate z-score. Use Table 2 to find an appropriate z-score whose area is .4200. The closest approximation is $z = 1.41$.

27. Answers vary. Possible answer: The normal distribution is not a good approximation of a binomial distribution that has a value of p close to 0 or 1 because the histogram of such a binomial distribution is skewed and therefore not close to the shape of a normal distribution.

28. a.

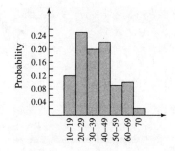

b. No, this is not a normal distribution, it is not mound-shaped, but rather skewed to the left.

29. a. $n = 6, p = .12, 1 - p = .88$

$$P(x = 2) = \binom{6}{2}(.12)^2(.88)^4 \approx .1295$$

b. $\mu = np = 6(.12) = .72$

$$\sigma = \sqrt{np(1-p)} = \sqrt{6(.12)(.88)} \approx .7960$$

On average, one would expect .72 of the 6 ages to be in the interval $10 - 19$.

30. a. For Stock I,

$$\bar{x} = \frac{11 + (-1) + 14}{3} = 8,$$

so the mean (average return) is 8%.

$$s = \sqrt{\frac{\Sigma x^2 - n\bar{x}^2}{n-1}} = \sqrt{\frac{318 - 3(8)^2}{2}}$$

$$= \sqrt{63} \approx 7.9,$$

so the standard deviations is 7.9%.
For Stock II,

$$\bar{x} = \frac{9 + 5 + 10}{3} = 8,$$

so the mean is also 8%.

$$s = \sqrt{\frac{\Sigma x^2 - n\bar{x}^2}{n-1}} = \sqrt{\frac{206 - 3(8)^2}{2}}$$

$$= \sqrt{7} \approx 2.6,$$

so the standard deviation is 2.6%.

b. Both stocks offer an average (mean) return of 8%. The smaller standard deviation for Stock II indicates a more stable return and thus greater security.

31.

Diet	A	B
$\bar{x}$	2.7	1.3
s	2.26	.95

a. Diet A produced the greater mean gain.

b. "Most consistent" means least variable. Diet B has the smaller standard deviation and so produced the most consistent gain.

32. a. Arrange the data for diet A in ascending order: 0, 1, 1, 1, 1, 3, 4, 4, 5, 7

Minimum = 0

Maximum = 7

$$\text{Median} = \frac{1 + 3}{2} = 2$$

$n = 10$

$.25 \cdot 10 = 2.5$ rounds up to 3.

$Q_1 = 1$

$.75 \cdot 10 = 7.5$ rounds up to 8.

$Q_3 = 4$

Diet A

Arrange the data for diet B in ascending order: 0, 0, 1, 1, 1, 1, 2, 2, 2, 3

Minimum = 0

Maximum = 3

Median = 1

$n = 10$

$.25 \cdot 10 = 2.5$ rounds up to 3.

$Q_1 = 1$

$.75 \cdot 10 = 7.5$ rounds up to 8.

$Q_3 = 2$

Diet B

b. Answers vary. Diet A had a greater mean gain but Diet B had more consistent results.

33. a.

x	x^2
67.2	4515.85
69.8	4872.04
71.2	5069.44
74.4	5535.36
74.3	5520.49
77.9	6068.41
79.8	6368.04
514.6	37,949.63

$$\overline{x} = \frac{514.6}{7} = .7351 = \$73.51 \text{ billion}$$

$$s = \sqrt{\frac{37,949.62 - 7(73.51)^2}{6}} \approx 4.46$$

b. Greatest gain: $z = \frac{79.8 - 73.51}{4.54} \approx 1.42$

The greatest gain is 1.42 standard deviations from the mean.

Smallest gain: $z = \frac{67.3 - 73.51}{4.54} \approx -1.37$

The smallest gain is -1.37 standard deviations from the mean.

34. a.

x	x^2
265	70,225
217	47,089
192	36,864
1072	1,149,184
1824	3,326,976
1910	3,648,100
2514	6,320,196
2202	4,848,804
150	22,500
438	191,844
294	86,436
475	225,625
333	110,889
11,886	20,084,732

34. Continued

$$\overline{x} = \frac{11,886}{13} = 914.30$$

$$s = \sqrt{\frac{\Sigma x^2 - n\overline{x}^2}{n-1}} = \sqrt{\frac{20,084,732 - 13(914.30)^2}{13-1}}$$

$$= \sqrt{768,121.14} \approx 876.4$$

x	x^2
285	81,225
315	99,225
559	312,481
1458	2,125,764
4553	20,729,809
2028	4,112,784
3372	11,370,384
2495	6,225,025
469	219,961
850	722,500
337	113,569
539	290,521
401	160,801
17,661	46,564,049

$$\overline{x} = \frac{17,661}{13} = 1358.5$$

$$s = \sqrt{\frac{\Sigma x^2 - n\overline{x}^2}{n-1}} = \sqrt{\frac{46,564,049 - 13(1358.5)^2}{13-1}}$$

$$= \sqrt{1,881,021.6} \approx 1371.5$$

b. The camping industry is closest to the mean in 1990. The camping industry is closest to the mean in 2003.

35. a. Arrange the 1990 data in ascending order.
150, 192, 217, 265, 294, 333, 438, 475,
1072, 1824, 1910, 2202, 2514
Minimum = 150
Maximum = 2514
Median = $Q_2 = 438$
n = 13
$.25 \cdot 13 = 3.25$, which rounds up to 4
$Q_1 = 265$
$.75 \cdot 13 = 9.75$, which rounds up to 10
$Q_3 = 1824$

b. Arrange the 2003 data in ascending order
285, 315, 337, 401, 469, 539, 559, 850,
1458, 2028, 2495, 3372, 4553
Minimum = 285
Maximum = 4553
Median = $Q_2 = 559$
n = 13
$.25 \cdot 13 = 3.25$, which rounds up to 4
$Q_1 = 401$
$.75 \cdot 13 = 9.75$, which rounds up to 10
$Q_3 = 2028$

36. a. 1990 Sales

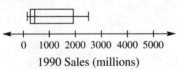

1990 Sales (millions)

2003 Sales

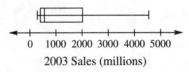

2003 Sales (millions)

b. Sales have increased most dramatically in exercise equipment.

37. a. $\mu = 100$ and $\sigma = 15$
More than 130
We need to find the area to the right of
x = 130.
$$z = \frac{130 - 100}{15} = 2$$
The area is .4772. The area to the right of
x = 130 is .5 − .4772 = .0228.
Therefore, 2.28% of the people score more
than 130.

37. Continued

b. Less than 85
For x = 85,
$$z = \frac{85 - 100}{15} = -1.$$
The area to the left of z = −1 is equal to the
area to the right of z = 1, which is
.5 − .3413 = .1587.
Therefore, 15.87% of the people score less
than 85.

c. Between 85 and 115
For x = 115,
$$z = \frac{115 - 100}{15} = 1.$$
The area is .3413. Since 85 and 115 are
equidistant from the mean, the total required
area is
2(.3413) = .6826.
Thus, 68.26% of the people score between 85
and 115.

38. Let x represent the number of ounces of juice in a
can.
$\mu = 32.1$, $\sigma = .1$
Find the z-score for x = 32.
$$z = \frac{x - \mu}{\sigma} = \frac{32 - 32.1}{.1} = -1$$
As shown in the solution for Exercise 37(b), the
area to the left of z = −1 is .5 − .3413 = .1587.
Therefore, 15.87% of the cartons contain less than
a quart.

39. a. n = 100, p = .98
$\mu = np = 100(.98) = 98$
$\sigma = \sqrt{100(.98)(.02)} = 1.4$
If 95% of the flies are killed, that means 95
flies are killed, which means we must find the
area between x = 94.5 and x = 95.5.
For x = 94.5,
$$z = \frac{94.5 - 98}{1.4} = -2.5.$$
For x = 95.5,
$$z = \frac{95.5 - 98}{1.4} = -1.79.$$
To find the area between z = −2.5 and
z = −1.79, find the area between z = 0 and
z = 1.79 and subtract it from the area between
z = 0 and z = 2.5. The area is
.4938 − .4633 = .0305,
which is the required probability.

39. Continued

   b. Using information from part (a), if at least 95% of the flies are killed, that means 95 or more flies are killed, which means we must find the area to the right of $x = 94.5$, which is to the right of $z = -2.5$. To do this, find the area between $z = 0$ and $z = 2.5$ and add it to .5. The area is $.4938 + .5 = .9938$, which gives the required probability.

   c. If at least 90% of the flies are killed, this means 90 or more flies, which is the area to the right of $x = 89.5$.

$$z = \frac{89.5 - 98}{1.4} = -6.07$$

To find the area to the right of $z = -6.07$, find the area between $z = 0$ and $z = 6.07$ and add it to .5. The area is $.5 + .5 = 1.000$.

   d. If all 100 flies are killed, we need the area between $x = 99.5$ and $x = 100.5$.
For $x = 99.5$,

$$z = \frac{99.5 - 98}{1.4} = 1.07.$$

For $x = 100.5$,

$$z = \frac{100.5 - 98}{1.4} = 1.79.$$

To find the area between $z = 1.07$ and $z = 1.79$, find the area between $z = 0$ and $z = 1.07$, and subtract it from the area between $z = 0$ and $z = 1.79$. The area is $.4633 - .3577 = .1056$.
The binomial distribution gives

$$\binom{100}{100}(.98)^{100}(.02)^0 = .1326.$$

Case 10 Statistics in the Law—The Castaneda Decision

1. $n = 870$, $p = .791$, $1 - p = .209$

$$\sigma = \sqrt{np(1-p)} = \sqrt{870(.791)(.209)} \approx 11.9928$$

$$\mu = np = 870(.791) = 688.17$$

$$z = \frac{339 - 688.17}{11.9928} \approx -29.1$$

2. Answers vary. Possible answer: The courts' figure of 1 in 10^{140} probably comes from the normal approximation to the binomial distribution.

3. **a.** $n = 220$, $p = .791$
     $\mu = np = 220(.791) \approx 174$
     Of 220 jurors from this population, we should expect 174 Mexican-American jurors.

   b. $n = 220$, $p = .791$, $1 - p = .209$
     $\sigma = \sqrt{np(1-p)} = \sqrt{220(.791)(.209)} \approx 6.03$

   c. $z = \dfrac{100 - 174}{6.03} \approx -12.3$

   d. The probability at -12.3 standard deviations from the mean is less than .004, the smallest probability in the table.

4. **a.** $n = 112$, $p = \dfrac{88}{294}$, $(1-p) = \dfrac{206}{294}$

$$\mu = np = 112\left(\frac{88}{294}\right) \approx 33.5238$$

$$\sigma = \sqrt{np(1-p)}$$

$$= \sqrt{112\left(\frac{88}{294}\right)\left(\frac{206}{294}\right)}$$

$$\approx 4.8466$$

$$z = \frac{6 - 33.5238}{4.8466} \approx -5.7$$

The expected number of women in management is about -5.7 standard deviations from the mean.

   b. Answers vary. Possible answer: Yes, it appears to be purposeful discrimination. The women in management do not represent the women in the employee body.

Chapter 11: Differential Calculus

Section 11.1 Limits

1. a. By reading the graph, as x gets closer to 3 from the left or the right, $f(x)$ gets closer to 3, so
$$\lim_{x \to 3} f(x) = 3.$$

b. As x gets closer to -1.5 from the left or right, $f(x)$ gets closer to 0, so
$$\lim_{x \to -1.5} f(x) = 0.$$

3. a. By reading the graph, as x gets closer to -2 from the left $f(x)$ approaches -1. As x gets closer to -2 from the right, $f(x)$ approaches $-\dfrac{1}{2}$. Since these two values of $f(x)$ are not equal, $\lim_{x \to 2} f(x)$ does not exist.

b. By reading the graph, as x gets closer to 1 from the left or right, $f(x)$ gets closer to $-\dfrac{1}{2}$, so
$$\lim_{x \to 1} f(x) = -\frac{1}{2}.$$

5. a. By reading the graph, as x gets closer to 0 from the left or right, $f(x)$ gets closer to 2, so
$$\lim_{x \to 0} f(x) = 2.$$

b. By reading the graph, as x gets closer to -1 from the right, $f(x)$ becomes infinitely large, and from the left, $f(x)$ becomes infinitely small. Since $f(x)$ has no limit as x approaches -1 from both the right and left, $\lim_{x \to -1} f(x)$ does not exist.

7. a. By reading the graph, as x gets closer to 1 from the left or right, $g(x)$ gets closer to 1, so
$$\lim_{x \to 1} g(x) = 1.$$

b. By reading the graph, as x gets closer to -1 from the left or right, $g(x)$ get closer to -1, so
$$\lim_{x \to -1} g(x) = -1.$$

9. $\lim_{x \to 2} F(x)$ in Exercise 2(a) exists because, as x gets closer to 2 from the left or the right, $F(x)$ gets closer to 4, a single number. On the other hand, $\lim_{x \to -2} f(x)$ in Exercise 3(a) does not exist because, as x gets closer to -2 from the left, $f(x)$ gets closer to -1, but, as x gets closer to -2 from the right, $f(x)$ gets closer to another number, $-\dfrac{1}{2}$.

11. $\lim_{x \to 1} \dfrac{\ln x}{x - 1}$

x	$\dfrac{\ln x}{x-1}$
.9	1.0536
.99	1.0050
.999	1.0005
1	
1.001	.9995
1.01	.9950
1.1	.9531

As x gets closer to 1 from the left or right, the value of $\dfrac{\ln x}{x - 1}$ gets closer to 1, so
$$\lim_{x \to 1} \frac{\ln x}{x - 1} = 1.$$

13. $\lim_{x \to 0} \dfrac{e^{3x} - 1}{x}$

x	$\dfrac{e^{3x} - 1}{x}$
$-.1$	2.5918
$-.01$	2.9554
$-.001$	2.9955
0	
.001	3.0045
.01	3.0455
.1	3.4986

As x gets closer to 0 from the left or right, the value of $\dfrac{e^{3x} - 1}{x}$ gets closer to 3, so
$$\lim_{x \to 0} \frac{e^{3x} - 1}{x} = 3.$$

15. $\lim\limits_{x\to 0}(x\cdot\ln|x|)$

x	$x\cdot\ln|x|$
$-.1$	$.2303$
$-.01$	$.0461$
$-.001$	$.0069$
0	
$.001$	$-.0069$
$.01$	$-.0461$
$.1$	$-.2303$

As x gets closer to 0 from the left or right, the value of $x\cdot\ln|x|$ gets closer to 0, so

$\lim\limits_{x\to 0}(x\cdot\ln|x|)=0.$

17. $\lim\limits_{x\to 3}\dfrac{x^3-3x^2-4x+12}{x-3}$

x	$\dfrac{x^3-3x^2-4x+12}{x-3}$
2.9	4.4100
2.99	4.9401
2.999	4.9940
3	
3.001	5.0060
3.01	5.0601
3.1	5.6100

As x gets closer to 3 from the left or right, the value of $\dfrac{x^3-3x^2-4x+12}{x-3}$ gets closer to 5, so

$\lim\limits_{x\to 3}\dfrac{x^3-3x^2-4x+12}{x-3}=5.$

19. $\lim\limits_{x\to -2}\dfrac{x^4+2x^3-x^2+3x+1}{x+2}$

x	$\dfrac{x^4+2x^3-x^2+3x+1}{x+2}$
-2.1	87.839
-2.01	898.8894
-2.001	8998.9889
-2	
-1.999	-9000.989
-1.99	-900.8906
-1.9	-89.959

19. Continued

As x gets closer to -2 from the left, the value of $\dfrac{x^4+2x^3-x^2+3x+1}{x+2}$ becomes infinitely large, and from the right, the value becomes infinitely small, so

$\lim\limits_{x\to -2}\dfrac{x^4+2x^3-x^2+3x+1}{x+2}$ does not exist.

21. $\lim\limits_{x\to 4}[f(x)-g(x)]$
Use property 2.
$\lim\limits_{x\to 4}[f(x)-g(x)]=\lim\limits_{x\to 4}f(x)-\lim\limits_{x\to 4}g(x)$
$=25-10$
$=15$

23. $\lim\limits_{x\to 4}\dfrac{f(x)}{g(x)}$
Use property 4.
$\lim\limits_{x\to 4}\dfrac{f(x)}{g(x)}=\dfrac{\lim\limits_{x\to 4}f(x)}{\lim\limits_{x\to 4}g(x)}$
$=\dfrac{25}{10}$
$=\dfrac{5}{2}$

25. $\lim\limits_{x\to 4}\sqrt{f(x)}=\lim\limits_{x\to 4}[f(x)]^{1/2}$
Use property 5.
$\lim\limits_{x\to 4}\sqrt{f(x)}=\left[\lim\limits_{x\to 4}f(x)\right]^{1/2}$
$=25^{1/2}$
$=5$

27. $\lim\limits_{x\to 4}\dfrac{f(x)+g(x)}{2g(x)}$
$=\dfrac{\lim\limits_{x\to 4}[f(x)+g(x)]}{\lim\limits_{x\to 4}[2g(x)]}$ Property 4
$=\dfrac{\lim\limits_{x\to 4}f(x)+\lim\limits_{x\to 4}g(x)}{2\lim\limits_{x\to 4}g(x)}$ Property 1 and L.C.F.
$=\dfrac{25+10}{2(10)}$
$=\dfrac{35}{20}$
$=\dfrac{7}{4}$

29. a. $f(x) \begin{cases} 3-x & \text{if } x < -2 \\ x+2 & \text{if } -2 \le x < 2 \\ 1 & \text{if } x \ge 2 \end{cases}$

The graph of $f(x) = 3 - x$ if $x < -2$ is the ray through $(-4, 7)$ and $(-2, 5)$, with excluded (open) endpoint $(-2, 5)$. The graph of
$f(x) = x + 2$ if $-2 \le x < 2$ is the segment with endpoint $(-2, 0)$ and open endpoint $(2, 4)$. The graph of $f(x)$ if $x \ge 2$ is the horizontal ray to the right with endpoint $(2, 1)$.

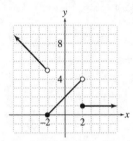

b. $\lim\limits_{x \to -2} f(x)$

As x gets closer to -2 from the left,
$f(x) = 3 - x$ gets closer to $3 - (-2) = 5$
because $f(x) = 3 - x$ if $x < -2$. As x gets
closer to -2 from the right, $f(x + 2)$ gets
closer to 0 because $f(x) = x + 2$ if
$-2 \le x < 2$. $f(x)$ does not get closer to a single
real number. Therefore $\lim\limits_{x \to -2} f(x)$ does not
exist.

c. $\lim\limits_{x \to 1} f(x)$

$f(x)$ is the polynomial function
$f(x) = x + 2$ if $-2 \le x < 2$, so
$\lim\limits_{x \to 1} f(x) = \lim\limits_{x \to 1} f(1)$
$\qquad\qquad = 1 + 2 = 3.$

d. $\lim\limits_{x \to 2} f(x)$

As x gets closer to 2 from the left,
$f(x) = x + 2$ gets closer to $2 + 2 = 4$ because
$f(x) = x + 2$ if $-2 \le x < 2$. As x gets closer to 2
from the right, $f(x) = 1$ gets closer to 1
because $f(x) = 1$ if $x \ge 2$. $f(x)$ does not get
closer to a single real number. Therefore,
$\lim\limits_{x \to 2} f(x)$ does not exist.

31. $\lim\limits_{x \to 2}(2x^3 + 5x^2 + 2x + 2)$

$= (2)(2)^3 + (5)(2)^2 + 2(2) + 2$

Polynomial limits

$= 42.$

33. $\lim\limits_{x \to 3} \dfrac{5x - 5}{2x + 1}$

$= \dfrac{\lim\limits_{x \to 3}(5x - 5)}{\lim\limits_{x \to 3}(2x + 1)}$

$= \dfrac{5(3) - 5}{2(3) + 1}$

$= \dfrac{10}{7}$

35. $\lim\limits_{x \to 3} \dfrac{x^2 - 9}{x - 3}$

$= \lim\limits_{x \to 3} \dfrac{(x - 3)(x + 3)}{x - 3}$

$= \lim\limits_{x \to 3}(x + 3)$ Limit Theorem

$= 3 + 3$

$= 6$

37. $\lim\limits_{x \to -2} \dfrac{x^2 - x - 6}{x + 2}$

$= \lim\limits_{x \to -2} \dfrac{(x - 3)(x + 2)}{x + 2}$

$\lim\limits_{x \to -2}(x - 3)$ Limit Theorem

$= -2 - 3$

$= -5$

39. $\lim\limits_{x \to 2} \dfrac{x^2 - 5x + 6}{x^2 - 6x + 8} = \lim\limits_{x \to 2} \dfrac{(x - 2)(x - 3)}{(x - 2)(x - 4)}$

$= \lim\limits_{x \to 2} \dfrac{x - 3}{x - 4}$

$= \dfrac{\lim\limits_{x \to 2}(x - 3)}{\lim\limits_{x \to 2}(x - 4)}$

$= \dfrac{2 - 3}{2 - 4}$

$= \dfrac{1}{2}$

41. $\lim\limits_{x \to 4} \dfrac{(x + 4)^2(x - 5)}{(x - 4)(x + 4)^2} = \lim\limits_{x \to 4} \dfrac{x - 5}{x - 4}$

$= \dfrac{-1}{0}$ Undefined

The limit does not exist.

43. $\lim\limits_{x \to 3} \sqrt{x^2 - 3} = \sqrt{\lim\limits_{x \to 3}(x^2 - 3)}$

$= \sqrt{3^2 - 3}$

$= \sqrt{6}$

45. $\lim\limits_{x \to 4} \dfrac{-6}{(x-4)^2} = \dfrac{\lim\limits_{x \to 4}(-6)}{\lim\limits_{x \to 4}(x-4)^2}$

$\qquad\qquad = \dfrac{-6}{0}$ Undefined

The limit does not exist.

47. $\lim\limits_{x \to 0} \dfrac{\frac{1}{x+3} - \frac{1}{3}}{x} = \lim\limits_{x \to 0} \left(\dfrac{1}{x+3} - \dfrac{1}{3} \right)\left(\dfrac{1}{x} \right)$

$\qquad\qquad = \lim\limits_{x \to 0} \dfrac{3-x-3}{3(x+3)(x)}$

$\qquad\qquad = \lim\limits_{x \to 0} \dfrac{-x}{3(x+3)x}$

$\qquad\qquad = \lim\limits_{x \to 0} \dfrac{-1}{3(x+3)}$

$\qquad\qquad = -\dfrac{1}{9}$

49. $\lim\limits_{x \to 25} \dfrac{\sqrt{x}-5}{x-25} =$

$\lim\limits_{x \to 25} \dfrac{\sqrt{x}-5}{x-25} \cdot \dfrac{\sqrt{x}+5}{\sqrt{x}+5}$

$= \lim\limits_{x \to 25} \dfrac{x-25}{(x-25)\left(\sqrt{x}+5\right)}$

$= \lim\limits_{x \to 25} \dfrac{1}{\sqrt{x}+5}$

$= \dfrac{1}{\sqrt{25}+5} = \dfrac{1}{5+5} = \dfrac{1}{10}$

51. $\lim\limits_{x \to 5} \dfrac{\sqrt{x}-\sqrt{5}}{x-5} = \lim\limits_{x \to 5} \dfrac{\sqrt{x}-\sqrt{5}}{x-5} \cdot \dfrac{\sqrt{x}+\sqrt{5}}{\sqrt{x}+\sqrt{5}}$

$\qquad\qquad = \lim\limits_{x \to 5} \dfrac{x-5}{(x-5)(\sqrt{x}+\sqrt{5})}$

$\qquad\qquad = \lim\limits_{x \to 5} \dfrac{1}{\sqrt{x}+\sqrt{5}}$

$\qquad\qquad = \dfrac{1}{2\sqrt{5}}$ or $\dfrac{\sqrt{5}}{10}$

53. $P(s) = \dfrac{105s}{s+7}$

a. $P(1) = \dfrac{105(1)}{1+7} = \dfrac{105}{8} = 13.125$

b. $P(13) = \dfrac{105(13)}{13+7} = \dfrac{1365}{20} = 68.25$

c. $\lim\limits_{s \to 13} P(s) = \lim\limits_{s \to 11} \dfrac{105s}{s+7}$

$\qquad\qquad = \dfrac{105(13)}{13+7} = 68.25$

55. $c(x) = 150,000 + 3x$

$\overline{c}(x) = \dfrac{c(x)}{x}$

a. $\overline{c}(1000) = \dfrac{150,000 + 3(1000)}{1000}$

$\qquad\qquad = \dfrac{153,000}{1000}$

$\qquad\qquad = 153$

b. $\overline{c}(10,000) = \dfrac{150,000 + 3(10,000)}{10,000}$

$\qquad\qquad = \dfrac{180,000}{10,000}$

$\qquad\qquad = 18$

c. $\lim\limits_{x \to 100,000} \overline{c}(x) = \lim\limits_{x \to 100,000} \dfrac{150,000 + 3x}{x}$

$\qquad\qquad = \dfrac{150,000 + 3(100,000)}{100,000}$

$\qquad\qquad = \dfrac{450,000}{100,000}$

$\qquad\qquad = 4.50$

57. The curves are continuous.

a. $\lim\limits_{x \to 2030} C(x) = C(2030) = 1.5$

b. $\lim\limits_{x \to 2015} I(x) = I(2015) = 1.2$

c. $\lim\limits_{x \to 2045} C(x) - I(x)$

$= \lim\limits_{x \to 2045} C(x) - \lim\limits_{x \to 2045} I(x)$

$= C(2045) - I(2045)$

$= 1.5 - 1.5$

$= 0$

d. $\lim\limits_{x \to 2045} C(x) + I(x)$

$= \lim\limits_{x \to 2045} C(x) + \lim\limits_{x \to 2045} I(x)$

$= C(2045) + I(2045)$

$= 1.5 + 1.5$

$= 3$

59. a. $P(x) = \begin{cases} .99 & \text{if} \quad 0 \le x \le 20 \\ 1.06 & \text{if} \quad 20 < x \le 21 \\ 1.13 & \text{if} \quad 21 < x \le 22 \\ 1.20 & \text{if} \quad 22 < x \le 23 \end{cases}$

b.

c. The $\lim\limits_{x \to 10} P(x) = .99$ since 10 is between 0 and 20.

d. The $\lim\limits_{x \to 20} P(x)$ does not exist because the limit as x approaches 20 from the left is not equal to the limit as x approaches 20 from the right.

e. The $\lim\limits_{x \to 22.5} P(x) = 1.20$ since 22.5 is between 22 and 23.

Section 11.2 One-Sided Limits and Limits Involving Infinity

1. As x approaches 2 from the right, f(x) approaches 2. Thus, $\lim\limits_{x \to 2^+} f(x) = 2$.

3. As x approaches –2 from the left, f(x) approaches –1. Thus, $\lim\limits_{x \to -2^-} f(x) = -1$.

5. As x approaches 2 from the left, g(x) approaches –1. Thus, $\lim\limits_{x \to 2^-} g(x) = -1$.

7. As x approaches –2 from the right, g(x) approaches 3. Thus, $\lim\limits_{x \to -2^+} g(x) = 3$.

9. a. As x approaches –2 from the left, $f(x)$, approaches 0. Thus $\lim\limits_{x \to -2^-} f(x) = 0$.

b. As x approaches 0 from the right, $f(x)$ approaches 1. Thus, $\lim\limits_{x \to 0^+} f(x) = 1$.

c. As x approaches 3 from the left and right, $f(x)$ approaches –1. Thus, $\lim\limits_{x \to 3} f(x) = -1$.

d. As x approaches 3 from the right, $f(x)$ approaches –1. Thus, $\lim\limits_{x \to 3^+} f(x) = -1$.

11. a. As x approaches –2 from the left, $f(x)$, approaches 2. Thus $\lim\limits_{x \to -2^-} f(x) = 2$.

b. As x approaches 0 from the right, $f(x)$ is undefined. Thus, $\lim\limits_{x \to 0^+} f(x)$ is undefined.

c. As x approaches 3 from the left and right, $f(x)$ approaches 0. Thus, $\lim\limits_{x \to 3} f(x) = 0$.

d. As x approaches 3 from the right, $f(x)$ approaches 2. Thus, $\lim\limits_{x \to 3^+} f(x) = 2$.

13. $\lim\limits_{x \to 2^+} \sqrt{x^2 - 4} = \lim\limits_{x \to 2^+} \left(x^2 - 4\right)^{\frac{1}{2}}$

$= \left[\lim\limits_{x \to 2^+} \left(x^2 - 4\right)\right]^{\frac{1}{2}} = 0^{\frac{1}{2}} = 0$

15. $\lim\limits_{x \to 0^+} \sqrt{x} + x + 1$

$= \lim\limits_{x \to 0^+} \sqrt{x} + \lim\limits_{x \to 0^+} x + \lim\limits_{x \to 0^+} 1$

$= 0 + 0 + 1 = 1$

15. $\lim\limits_{x \to 0^+} \sqrt{x} + x + 1$

$= \lim\limits_{x \to 0^+} \sqrt{x} + \lim\limits_{x \to 0^+} x + \lim\limits_{x \to 0^+} 1$

$= 0 + 0 + 1 = 1$

17. $\lim\limits_{x \to -2^+} \left(x^3 - x^2 - x + 1 \right)$

$= \lim\limits_{x \to -2^+} x^3 - \lim\limits_{x \to -2^+} x^2 - \lim\limits_{x \to -2^+} x + \lim\limits_{x \to -2^+} 1$

$= -8 - 4 + 2 + 1 = -9$

19. $\lim\limits_{x \to -5^+} \dfrac{\sqrt{x+5}+5}{x^2-5} = \lim\limits_{x \to -5^+} \left[\dfrac{\sqrt{x+5}+5}{x^2-5} \cdot \dfrac{\sqrt{x+5}-5}{\sqrt{x+5}-5} \right]$

$= \lim\limits_{x \to -5^+} \left[\dfrac{x-20}{x^2\sqrt{x+5} - 5\sqrt{x+5} - 5x^2 + 25} \right]$

$= \lim\limits_{x \to -5^+} \left[\dfrac{\dfrac{x-20}{x^2}}{\sqrt{x+5} - \dfrac{5\sqrt{x+5}}{x^2} - 5 + \dfrac{25}{x^2}} \right] = \dfrac{-1}{-5+1} = \dfrac{1}{4}$

21. a. As x approaches -2 from the left, $f(x)$ approaches $3 - (-2) = 5$. Thus,

$\lim\limits_{x \to -2^-} f(x) = 5$.

b. As x approaches -2 from the right, $f(x)$ approaches $-2 + 2 = 0$. Thus,

$\lim\limits_{x \to -2^+} f(x) = 0$.

c. Since $f(x)$ does not approach the same value as x approaches -2 from the left and right, $\lim\limits_{x \to -2} f(x)$ doe not exist.

23. a. As x approaches 2 from the left and the right, $f(x)$ approaches positive infinity. Thus,

$\lim\limits_{x \to 2} f(x) = \infty$.

b. As x approaches -2 from the right, $f(x)$ approaches positive infinity. Thus,

$\lim\limits_{x \to -2^-} f(x) = \infty$.

c. As x approaches -2 from the left, $f(x)$ approaches negative infinity. Thus,

$\lim\limits_{x \to -2^-} f(x) = -\infty$.

25. $\lim\limits_{x \to 4} \dfrac{-6}{(x-4)^2} = -\infty$

27. $\lim\limits_{x \to -1^+} \dfrac{2}{1+x} = \infty$

29. $\lim\limits_{x \to \infty} f(x) = \infty$ and $\lim\limits_{x \to -\infty} f(x) = 0$

31. $\lim\limits_{x \to \infty} f(x) = 2$ and $\lim\limits_{x \to -\infty} f(x) = -1$

33. $\lim\limits_{x \to \infty} f(x) = \infty$ and $\lim\limits_{x \to -\infty} f(x) = -\infty$

35. $\lim\limits_{x \to \infty} \left[\sqrt{x^2+1} - (x+1) \right] = -1$

37. $\lim\limits_{x \to -\infty} \dfrac{x^{2/3} - x^{4/3}}{x^3} = 0$

39. $\lim\limits_{x \to -\infty} e^{1/x} = 1$

41. $\lim\limits_{x \to \infty} \dfrac{\ln x}{x} = 0$

43. $\lim\limits_{x \to \infty} \dfrac{3x^2+5}{4x^2-6x+2} = \lim\limits_{x \to \infty} \dfrac{3 + \dfrac{4}{x^2}}{4 - \dfrac{6}{x} + \dfrac{2}{x^2}}$

$= \dfrac{3+0}{4-0+0} = \dfrac{3}{4}$

45. $\lim\limits_{x \to -\infty} \dfrac{2x^2-6x+1}{2+x-x^2} = \lim\limits_{x \to -\infty} \dfrac{2 - \dfrac{6}{x} + \dfrac{1}{x^2}}{\dfrac{2}{x^2} + \dfrac{1}{x} - 1}$

$= \dfrac{2+0+0}{0-0-1} = -2$

47. $\lim\limits_{x \to -\infty} \dfrac{2x^5 - x^3 + 2x - 9}{5 - x^5} = \lim\limits_{x \to -\infty} \dfrac{2 - \dfrac{1}{x^2} + \dfrac{2}{x^4} - \dfrac{9}{x^5}}{\dfrac{5}{x^5} - 1}$

$= \dfrac{2 - 0 + 0 + 0}{-0 - 1} = -2$

49. $\lim\limits_{x\to-\infty} \dfrac{(x-3)(x+2)}{2x^2+x+1} = \lim\limits_{x\to-\infty} \dfrac{x^2-x-6}{2x^2+x+1}$

$= \lim\limits_{x\to-\infty} \dfrac{1-\dfrac{1}{x}-\dfrac{6}{x^2}}{2+\dfrac{1}{x}+\dfrac{1}{x^2}} = \dfrac{1+0-0}{2-0+0} = \dfrac{1}{2}$

51. $\lim\limits_{x\to\infty}\left(3x-\dfrac{1}{x^2}\right) = \lim\limits_{x\to\infty}3x - \lim\limits_{x\to\infty}\dfrac{1}{x^2}$

$= \infty - 0 = \infty$

53. $\lim\limits_{x\to-\infty}\left(\dfrac{3x}{x+2}+\dfrac{2x}{x-1}\right)$

$= \lim\limits_{x\to-\infty}\left(\dfrac{3x^2-3x+2x^2+4x}{x^2+x-2}\right)$

$= \lim\limits_{x\to-\infty}\dfrac{5x^2+x}{x^2+x-2}$

$= \lim\limits_{x\to-\infty}\dfrac{5+\dfrac{1}{x}}{1+\dfrac{1}{x}-\dfrac{2}{x^2}} = \dfrac{5-0}{1-0-0} = 5$

55. $\lim\limits_{x\to\infty}\dfrac{x}{|x|}$

Since x approaches positive infinity, $|x|=x$.

Thus, $\lim\limits_{x\to\infty}\dfrac{x}{|x|} = \lim\limits_{x\to\infty}\dfrac{x}{x} = \lim\limits_{x\to\infty}1 = 1$.

57. $\lim\limits_{x\to-\infty}\dfrac{x}{|x|+1}$

Since x approaches negative infinity, $|x|=-x$.

Thus, $\lim\limits_{x\to-\infty}\dfrac{x}{|x|+1} = \lim\limits_{x\to-\infty}\dfrac{x}{-x+1}$

$= \lim\limits_{x\to\infty}\dfrac{1}{-1+\dfrac{1}{x}} = \dfrac{1}{-1} = -1$

59. a. $N(t) = 71.8e^{-8.96e^{-.0685t}}$

$N(65) = 71.8e^{-8.96e^{-.0685(65)}}$

$= 71.8e^{-8.96e^{-4.4525}}$

$= 71.8e^{-8.96(.0116494)}$

$= 71.8e^{-.1043787}$

$= 71.8(.9008841)$

≈ 64.68

b. $\lim\limits_{t\to\infty}N(t) =$

$\lim\limits_{t\to\infty}71.8e^{-8.96e^{-.0685t}}$

$= \lim\limits_{t\to\infty}71.8e^{-8.96e^{-\infty}}$

$= \lim\limits_{t\to\infty}71.8e^{-8.96(0)} = \lim\limits_{t\to\infty}71.8e^{0}$

$= \lim\limits_{t\to\infty}71.8 \approx 71$ teeth

61. $\lim\limits_{h\to\infty}A(h) = \lim\limits_{h\to\infty}\dfrac{.17h}{h^2+2}$

$= \lim\limits_{h\to\infty}\dfrac{.17}{h+\dfrac{2}{h}} = \dfrac{.17}{\infty+0} = \dfrac{.17}{\infty} = 0$

The concentration of the drug approaches 0 as time increases.

Section 11.3 Rates of Change

1. The average rate of change for $f(x) = x^2+2x$

between $x=0$ and $x=6$ is $\dfrac{f(6)-f(0)}{6-0}$.

$f(6) = 6^2+2(6)$

$= 36+12$

$= 48$

$f(0) = 0^2+2(0)$

$= 0$

The average rate of change is

$\dfrac{48-0}{6-0} = \dfrac{48}{6} = 8$.

3. $f(x) = 2x^3-4x^2+6$

between $x=-1$ and $x=2$.

The average rate of change is

$\dfrac{f(2)-f(-1)}{2-(-1)} = \dfrac{6-(-0)}{3} = \dfrac{6}{3} = 2$.

5. $f(x) = \sqrt{x}$ between $x = 1$ and $x = 9$.

The average rate of change is

$$\frac{f(9) - f(1)}{9 - 1} = \frac{3 - 1}{9 - 1} = \frac{2}{8} = \frac{1}{4}.$$

7. $f(x) = \dfrac{1}{x - 1}$ between $x = -2$ and $x = 0$

The average rate of change is

$$\frac{f(0) - f(-2)}{0 - (-2)} = \frac{-1 - \left(-\frac{1}{3}\right)}{2}$$

$$= \frac{-1 + \frac{1}{3}}{2}$$

$$= \frac{-\frac{2}{3}}{2}$$

$$= -\frac{1}{3}$$

9. Let $S(x)$ be the function representing sales in thousands of dollars of x thousand catalogs.

a. $S(20) = 40$

$S(10) = 30$

$$\text{Average rate of change} = \frac{S(20) - S(10)}{20 - 10}$$

$$= \frac{40 - 30}{10}$$

$$= 1$$

As catalog distribution changes from 10,000 to 20,000, sales will have an average increase of \$1000 for each additional 1000 catalogs distributed.

b. $S(30) = 46$

$S(20) = 40$

$$\text{Average rate of change} = \frac{S(30) - S(20)}{30 - 20}$$

$$= \frac{46 - 40}{10}$$

$$= \frac{6}{10}$$

$$= \frac{3}{5}$$

As catalog distribution changes from 20,000 to 30,000, sales will have an average increase of $\left(\dfrac{3}{5}\right)(1000)$ or \$600 for each additional 1000 catalogs distributed.

9. Continued

c. $S(40) = 50$

$S(30) = 46$

$$\text{Average rate of change} = \frac{S(40) - S(30)}{40 - 30}$$

$$= \frac{50 - 46}{10}$$

$$= \frac{4}{10}$$

$$= \frac{2}{5}$$

As catalog distribution changes from 30,000 to 40,000, sales will have an average increase

of $\left(\dfrac{2}{5}\right)(1000)$ or \$400 for each additional

1000 catalogs distributed.

d. As more catalogs are distributed, sales increase at a smaller and smaller rate.

e. It might be that the market for items in the catalog is becoming saturated.

11. Let $S(t)$ represent the money remaining in the fund after t years.

a. Average rate of change in trust funds is

$$\frac{S(1998) - S(1994)}{1998 - 1994} \approx \frac{124 - 151}{4}$$

$$\approx -\$6.75 \text{ billion}$$

b. Average rate of change in trust funds is

$$\frac{S(2010) - S(1998)}{2010 - 1998} \approx \frac{294 - 124}{12}$$

$$\approx \$14.17 \text{ billion}$$

c. Average rate of change in trust funds is

$$\frac{S(2005) - S(1999)}{2005 - 1999} \approx \frac{250 - 150}{6}$$

$$\approx \$16.67 \text{ billion}$$

13. Let $r(t)$ represent the number of restaurants in year t.

   a. Average rate of change $= \dfrac{r(1975) - r(1965)}{1975 - 1965}$

$$= \dfrac{3352 - 738}{10}$$

$$= 261.4$$

That is an increase of 261.4 restaurants per year.

   b. Average rate of change $= \dfrac{r(1985) - r(1975)}{1985 - 1975}$

$$= \dfrac{6972 - 3352}{10}$$

$$= 362$$

That is an increase of 362 restaurants per year.

   c. Average rate of change $= \dfrac{r(1995) - r(1985)}{1995 - 1985}$

$$= \dfrac{11,368 - 6972}{10}$$

$$= 439.6$$

That is an increase of 439.6 restaurants per year.

   d. Average rate of change $= \dfrac{r(2005) - r(1995)}{2005 - 1995}$

$$= \dfrac{13,609 - 11,368}{10}$$

$$= 224.1$$

That is an increase of 224.1 restaurants per year.

   e. Average rate of change $= \dfrac{r(2005) - r(1965)}{2005 - 1965}$

$$= \dfrac{13,609 - 738}{40}$$

$$\approx 321.8$$

That is an increase of 321.8 restaurants per year.

   f. After increasing for a number of years, the rate of change is now decreasing.

15. Let $R(t)$ represent the annual number of trips after t years.

   a. Average rate of change

$$= \dfrac{R(1995) - R(1992)}{1995 - 1992}$$

$$= \dfrac{7.8 - 8.5}{3}$$

$$\approx -.233$$

About $-.233$ billion per year, which means that during this period, use of mass transportation was decreasing at a rate of about 233,000,000 trips per year.

   b. Average rate of change $= \dfrac{r(2000) - r(1995)}{2000 - 1995}$

$$= \dfrac{9.4 - 7.8}{5}$$

$$= .32$$

During this period, use of mass transportation was increasing at a rate of 320,000,000 trips per year.

   c. Average rate of change

$$= \dfrac{r(1997.75) - r(1992)}{1997.75 - 1992}$$

$$= \dfrac{8.5 - 8.5}{5.75}$$

$$= 0$$

This does not accurately reflect the fact that transportation use decreased significantly and then increased again during this period.

17. The average rate of change of y as x changes from a to b is found by

$$\dfrac{y(b) - y(a)}{b - a}.$$

The instantaneous rate of change of y as $x = a$ is found by

$$\lim_{h \to 0} \dfrac{y(a + h) - y(a)}{h}.$$

19. $s(t) = 2.2t^2$

Time Interval	Average Speed
$t = 5$ to $t = 5.1$	$\frac{s(5.1)-s(5)}{5.1-5} = \frac{57.22-55}{.1} = 22.2$
$t = 5$ to $t = 5.01$	$\frac{s(5.01)-s(5)}{5.01-5} = 22.022$
$t = 5$ to $t = 5.001$	$\frac{s(5.001)-s(5)}{5.001-5} = 22.0022$

This chart suggests that the instantaneous velocity at $t = 5$ is 22 ft/sec.

21. $s(t) = 2.2t^2$

$$\text{Average speed} = \frac{s(30)-s(0)}{30-0}$$
$$= \frac{1980-0}{30}$$
$$= 66 \text{ ft/sec.}$$

23. $s(t) = t^2 + 4t + 3$

For $t = 1$, the instantaneous velocity is

$$\lim_{h \to 0} \frac{s(1+h)-s(1)}{h}.$$

We have

$$s(1+h) = (1+h)^2 + 4(1+h) + 3$$
$$= 1 + 2h + h^2 + 4 + 4h + 3$$
$$= h^2 + 6h + 8,$$

and

$$s(1) = 1^2 + 4(1) + 3 = 8.$$

Thus,

$$s(1+h) - s(1) = (h^2 + 6h + 8) - 8$$
$$= h^2 + 6h,$$

and the instantaneous velocity at $t = 1$ is

$$\lim_{h \to 0} \frac{h^2 + 6h}{h} = \lim_{h \to 0} \frac{h(h+6)}{h}$$
$$= \lim_{h \to 0}(h+6)$$
$$= 6 \text{ ft/sec.}$$

25. a. Average velocity $= \dfrac{s(2)-s(0)}{2-0}$

$$= \frac{10-0}{2}$$
$$= 5 \text{ ft/sec}$$

b. Average velocity $= \dfrac{s(4)-s(2)}{4-2}$

$$= \frac{14-10}{2}$$
$$= 2 \text{ ft/sec}$$

c. Average velocity $= \dfrac{s(6)-s(4)}{6-4}$

$$= \frac{20-14}{2}$$
$$= 3 \text{ ft/sec}$$

d. Average velocity $= \dfrac{s(8)-s(6)}{8-6}$

$$= \frac{30-20}{2}$$
$$= 5 \text{ ft/sec}$$

e. i. Let $t = 4$ and $h = 2$
instantaneous velocity

$$\approx \frac{s(4+2)-s(4)}{2} = \frac{20-14}{2} = 3 \text{ ft/sec}$$

ii. instantaneous velocity $\approx \dfrac{2+3}{2}$

$$= 2.5 \text{ ft/sec}$$

f. i. Let $t = 6$ and $h = 2$
instantaneous velocity

$$\approx \frac{s(6+2)-s(6)}{2} = \frac{30-20}{2} = 5 \text{ ft/sec}$$

ii. instantaneous velocity $\approx \dfrac{5+3}{2}$

$$= 4 \text{ ft/sec}$$

27. $f(x) = x^2 - x - 1$

    a. $f(a+h) = (a+h)^2 - (a+h) - 1$
$$= a^2 + 2ah + h^2 - a - h - 1$$

    b. $\dfrac{f(a+h) - f(a)}{h}$

$$= \frac{a^2 + 2ah + h^2 - a - h - 1 - (a^2 - a - 1)}{h}$$

$$= \frac{2ah + h^2 - h}{h}$$

$$= \frac{h(2a + h - 1)}{h}$$

$$= 2a + h - 1$$

    c. The instantaneous rate of change is

$$\lim_{h \to 0} \frac{f(a+h) - f(a)}{h} = \lim_{h \to 0} 2a + h - 1$$
$$= 2a - 1,$$

so the instantaneous rate of change when $a = 5$ is $2(5) - 1 = 9$.

29. $f(x) = x^3$

    a. $f(a+h) = (a+h)^3$
$$= a^3 + 3a^2h + 3ah^2 + h^3$$

    b.

$$\frac{f(a+h) - f(a)}{h}$$

$$= \frac{a^3 + 3a^2h + 3ah^2 + h^3 - a^3}{h}$$

$$= \frac{h(3a^2 + 3ah + h^2)}{h}$$

$$= 3a^2 + 3ah + h^2$$

    c. The instantaneous rate of change is

$$\lim_{h \to 0} \frac{f(a+h) - f(a)}{h} = \lim_{h \to 0} 3a^2 + 3ah + h^2$$
$$= 3a^2,$$

the instantaneous rate of change when $a = 5$ is $3(5)^2 = 75$.

31. $R = 10x - .002x^2$

$R(1000) = 10(1000) - .002(1000)^2$

$\qquad = 8000$

$R(1001) = 10(1001) - .002(1001)^2$

$\qquad = 8005.998$

a. Average rate of change $= \dfrac{R(1001) - R(1000)}{1001 - 1000}$

$\qquad\qquad = \dfrac{8005.998 - 8000}{1}$

$\qquad\qquad = 5.998$

Since the revenue is given in thousands of dollars, the average rate of change is \$5998 per unit.

b. Marginal revenue $= \lim\limits_{h \to 0} \dfrac{R(1000 + h) - R(1000)}{h}$

$\qquad = \lim\limits_{h \to 0} \dfrac{10(1000 + h) - .002(1000 + h)^2 - 8000}{h}$

$\qquad = \lim\limits_{h \to 0} \dfrac{10,000 + 10h - .002(1,000,000 + 2000h + h^2) - 8000}{h}$

$\qquad = \lim\limits_{h \to 0} \dfrac{6h - .002h^2}{h}$

$\qquad = \lim\limits_{h \to 0} \dfrac{h(6 - .002h)}{h}$

$\qquad = \lim\limits_{h \to 0} 6 - .002h$

$\qquad = 6$

The marginal revenue is \$6000 per unit.

c. Additional revenue $= R(1001) - R(1000)$

$\qquad\qquad = 8005.998 - 8000$

$\qquad\qquad = 5.998$

The additional revenue is \$5998.

d. The answers in (a) and (c) are the same.

33. $p(t) = t^2 + t$ for $0 \le t \le 5$

a. Average rate of change $= \dfrac{p(4) - p(1)}{4 - 1}$

$$= \frac{(4^2 + 4) - (1^2 + 1)}{3}$$

$$= \frac{18}{3} = 6\% \text{ per day}$$

b. Instantaneous rate of change

$$= \lim_{h \to 0} \frac{f(t + h) - f(t)}{h}$$

$$= \lim_{h \to 0} \frac{(t + h)^2 + t + h - t^2 - t}{h}$$

$$= \lim_{h \to 0} \frac{t^2 + 2th + h^2 + t + h - t^2 - t}{h}$$

$$= \lim_{h \to 0} \frac{2th + h^2 + h}{h}$$

$$= \lim_{h \to 0} 2t + h + 1$$

$$= 2t + 1$$

When $t = 3$, the instantaneous rate of change is $2(3) + 1 = 7\%$ per day.

35. a. $\lim_{h \to 0} \dfrac{f(30 + h) - f(30)}{h} = \lim_{h \to 0} \dfrac{\frac{76.7}{1 + 16(.8444^{30+h})} - \frac{76.7}{1 + 16(.8444^{30})}}{h}$

X	Y1
-.001	1.0733
-1E⁻4	1.0732
1E⁻6	1.0732
0	ERROR
1E⁻7	1.0732
1E⁻5	1.0732
1E⁻4	1.0732

X=1E⁻7

In 2000, the number of subscribers was increasing by about 1,073,227 subscribers per year.

b. $\lim_{h \to 0} \dfrac{f(36 + h) - f(36)}{h} = \lim_{h \to 0} \dfrac{\frac{76.7}{1 + 16(.8444^{36+h})} - \frac{76.7}{1 + 16(.8444^{36})}}{h}$

X	Y1
-.001	.43847
-1E⁻4	.43844
1E⁻6	.43843
0	ERROR
1E⁻7	.43838
1E⁻5	.43843
1E⁻4	.43843

X=1E⁻7

In 2006, the number of subscribers was increasing by about 438,432 subscribers per year.

c. Answers vary.

37. Let $L(t)$ represent the crown length in millimeters t weeks after conception.

a. Average rate of growth for $L(t) = -0.01t^2 + .788t - 7.048$ between 22 and 28 weeks is $\dfrac{L(28) - L(22)}{28 - 22}$.

$$L(28) = -.01(28)^2 + .788(28) - 7.048$$
$$= 7.176$$
$$L(22) = -.01(22)^2 + .788(22) - 7.048$$
$$= 5.448$$

The average rate of change is $\dfrac{7.176 - 5.448}{28 - 22} = \dfrac{1.728}{6} = .288$ mm/wk.

b.
$$\lim_{h \to 0} \frac{L(22 + h) - L(22)}{h}$$
$$= \lim_{h \to 0} \frac{-.01(22 + h)^2 + .788(22 + h) - 7.048 - \left(.01(22)^2 + .788(22) - 7.048\right)}{h}$$

X	Y₁	
-.001	.34801	
-1E-4	.348	
1E-6	.348	
0	ERROR	
1E-7	.348	
1E-5	.348	
1E-4	.348	

X=1E-7

At exactly 22 weeks the mesiodistal crown is increasing in length by .348 mm/wk.

c. Answers vary.

$L(t) = -.01t^2 + .788t - 7.048$

39. Let $M(t)$ represent the body mass in kilograms of a yearling bighorn sheep t days since May 25.

a. Average rate of change in body mass between 105 and 115 days past May 25 is $\dfrac{M(115)-M(105)}{115-105}$.

$M(115) = 27.5 + .3(115) - .001(115)^2 = 48.775$

$M(105) = 27.5 + .3(105) - .001(105)^2 = 47.975$

The average rate of change is $\dfrac{48.775-47.975}{115-105} \approx 0.08$ kg/day .

b. $\displaystyle\lim_{h\to 0}\frac{M(105+h)-M(105)}{h} = \lim_{h\to 0}\frac{27.5 + .3(105+h) - .001(105)^2 - \left(27.5 + .3(105) - .001(105)^2\right)}{h}$

X	Y₁
-.001	.09
-1E⁻⁴	.09
1E⁻6	.09
0	ERROR
1E⁻7	.09
1E⁻5	.09
1E⁻4	.09

X=1E⁻7

The instantaneous rate of change for a big horn sheep yearling whose age is 105 days past May 25 is 0.9 kg/day.

c. $M(t) = 27.5 + .3t - .001t^2$

$M(t) = 27.5 + .3t - .001t^2$

d. Answers vary

Section 11.4 Tangent Lines and Derivatives

1. $f'(x) = 3x^2 - 4$

$f'(2) = 3(2)^2 - 4(1)$

$= 12 - 4$

$= 8$

The slope of the tangent line at $(2, f(2)) = (2, 0)$ is 8.

$y - y_1 = m(x - x_1)$

$y - (0) = 8(x - 2)$

$y = 8x - 16$

3. $f'(x) = -\dfrac{1}{x^2}$

$f'(-2) = -\dfrac{1}{(-2)^2} = -\dfrac{1}{4}$

The slope of the tangent line at

$(-2, f(-2)) = \left(-2, -\dfrac{1}{2}\right)$ is $-\dfrac{1}{4}$.

$y - y_1 = m(x - x_1)$

$y + \dfrac{1}{2} = -\dfrac{1}{4}(x + 2)$

$y = -\dfrac{1}{4}x - 1$

5. $f(x) = x^2 - 2$ at $x = 1$.

a. $f(1 + h) = (1 + h)^2 - 2$

$= -1 + 2h + h^2$

$f(1 + h) - f(1) = -1 + 2h + h^2 + 1$

$= 2h + h^2$

$\dfrac{f(1 + h) - f(1)}{h} = 2 + h$

Letting $h \to 0$, slope of tangent $= 2$

b. The slope of the tangent line at $(1, f(1)) = (1, -1)$ is 2.

$y - y_1 = m(x - x_1)$

$y + 1 = 2(x - 1)$

$y = 2x - 3$

7. $f(x) = \dfrac{5}{x}$ at $x = 4$.

a. $f(4 + h) = \dfrac{5}{4 + h}$

$f(4 + h) - f(4) = \dfrac{5}{4 + h} - \dfrac{5}{4}$

$= \dfrac{-5h}{4(4 + h)}$

$\dfrac{f(4 + h) - f(4)}{h} = \dfrac{-5}{4(4 + h)}$

Letting $h \to 0$,

slope of tangent $= \dfrac{-5}{4(4)} = -\dfrac{5}{16}$

b. The slope of the tangent line at

$(4, f(4)) = \left(4, \dfrac{5}{4}\right)$ is $-\dfrac{5}{16}$.

$y - \dfrac{5}{4} = -\dfrac{5}{16}(x - 4)$

$y = -\dfrac{5}{16}x + \dfrac{5}{2}$

9. $f(x) = 4\sqrt{x}$ at $x = 4$.

a. $f(4 + h) = 4\sqrt{4 + h}$

$f(4 + h) - f(4) = 4\sqrt{4 + h} - 4\sqrt{4}$

$= 4\sqrt{4 + h} - 8$

$= 4\left(\sqrt{4 + h} - 2\right)$

$\dfrac{f(4 + h) - f(4)}{h} = 4 \cdot \dfrac{\sqrt{4 + h} - 2}{h}$

$= \dfrac{4\sqrt{4 + h} - 2}{h} \cdot \dfrac{\left(\sqrt{4 + h} + 2\right)}{\left(\sqrt{4 + h} + 2\right)}$

$= \dfrac{4\left[(4 + h) - 4\right]}{h\left(\sqrt{4 + h} + 2\right)} = \dfrac{4h}{h\left(\sqrt{4 + h} + 2\right)}$

$= \dfrac{4}{\sqrt{4 + h} + 2}$

Letting $h \to 0$,

slope of tangent $= \dfrac{4}{\sqrt{4} + 2} = 1$

b. The slope of the tangent line at $(4, f(4)) = (4, 8)$ is 1

$y - 8 = 1(x - 4)$

$y = x + 4$

11. a. From the graph, $f(x)$ is largest at $x = x_5$.

b. $f(x)$ is smallest at $x = x_4$.

c. $f'(x)$ is smallest at $x = x_3$ because x_3 is the only labeled value at which the tangent line has a negative slope.

d. $f'(x)$ is closest to 0 at $x = x_2$ because the tangent line is closer to being horizontal (where the slope is 0) at x_2 than at any other labeled value.

13. If $g'(x) > 0$ for $x < 0$, the slope of the tangent line must be positive, so $g(x)$ must increase for $x < 0$. If $g'(x) < 0$ for $x > 0$, the slope of the tangent line must be negative, so $g(x)$ must decrease for $x > 0$. One of the many possible graphs follows.

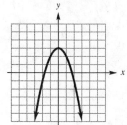

15. a. The derivative is positive because the tangent line is rising from left to right at $x = 100$ and thus has a positive slope.

b. The derivative is negative because the tangent line is falling from left to right at $x = 200$ and thus has a negative slope.

17. Since velocity is the rate of change of distance from a starting point, we must decide which graph represents the function (distance over time) and which represents the derivative of that function (velocity over time).
For $t > 0$, the graph of (a) is positive. If (a) is the derivative, the graph of (b) would have no high or low points when (a) would be 0, which is not the case.
Furthermore, assuming (b) is the derivative of (a), we see that (b) is positive when (a) is increasing ($0 < t < 2$ and $t > 4$) and (b) is negative when (a) is decreasing ($2 < t < 4$).
Therefore, graph (a) is distance and graph (b) is velocity.

19. $f(x) = -4x^2 + 11x$

Step 1
$$f(x+h) = -4(x+h)^2 + 11(x+h)$$
$$= -4(x^2 + 2xh + h^2) + 11x + 11h$$
$$= -4x^2 - 8xh - 4h^2 + 11x + 11h$$

Step 2
$$f(x+h) - f(x)$$
$$= (-4x^2 - 8xh - 4h^2 + 11x + 11h)$$
$$-(-4x^2 + 11x)$$
$$= -8xh - 4h^2 + 11h$$

Step 3
$$\frac{f(x+h) - f(x)}{h}$$
$$= \frac{-8xh - 4h^2 + 11h}{h}$$
$$= \frac{h(-8x - 4h + 11)}{h}$$
$$= -8x - 4h + 11$$

Step 4
$$f'(x) = \lim_{h \to 0} \frac{f(x+h) - f(x)}{h}$$
$$= \lim_{h \to 0} (-8x - 4h + 11)$$
$$= -8x + 11$$
$$f'(2) = -8(2) + 11 = -5$$
$$f'(0) = -8(0) + 11 = 11$$
$$f'(-3) = -8(-3) + 11 = 35$$

21. $f(x) = 8x + 6$
We condense the steps as follows
$$\frac{f(x+h) - f(x)}{h} = \frac{[8(x+h) + 6] - (8x + 6)}{h}$$
$$= \frac{(8x + 8h + 6) - (8x + 6)}{h}$$
$$= \frac{8h}{h}$$
$$= 8$$
$$f'(x) = \lim_{h \to 0} 8 = 8$$
$$f'(2) = 8; f'(0) = 8; f(-3) = 8$$

23. $f(x) = -\dfrac{2}{x}$

$$\frac{f(x+h)-f(x)}{h} = \frac{\frac{-2}{x+h}-\left(\frac{-2}{x}\right)}{h}$$

$$= \frac{\frac{-2x+2(x+h)}{(x+h)x}}{h}$$

$$= \frac{2h}{h(x+h)(x)}$$

$$= \frac{2}{(x+h)(x)}$$

$$f'(x) = \lim_{h\to 0}\frac{2}{(x+h)(x)} = \frac{2}{x^2}$$

$$f'(2) = \frac{2}{2^2} = \frac{1}{2}$$

$$f'(0) = \frac{2}{0^2} \ \text{Undefined}$$

The derivative does not exist at $x = 0$.

$$f'(-3) = \frac{2}{(-3)^2} = \frac{2}{9}$$

25. $\qquad f(x) = \dfrac{4}{x-1}$

$$\frac{f(x+h)-f(x)}{h}$$

$$= \frac{\frac{4}{x+h-1}-\frac{4}{x-1}}{h}$$

$$= \frac{\frac{4(x-1)-4(x+h-1)}{(x+h-1)(x-1)}}{h}$$

$$= \frac{-4h}{h(x-1+h)(x-1)}$$

$$= \frac{-4}{(x-1+h)(x-1)}$$

$$f'(x) = \lim_{h\to 0}\frac{-4}{(x-1+h)(x-1)}$$

$$= \frac{-4}{(x-1)^2}$$

$$f'(2) = \frac{-4}{(2-1)^2} = -4$$

$$f'(0) = \frac{-4}{(0-1)^2} = -4$$

$$f'(-3) = \frac{-4}{(-3-1)^2} = \frac{-4}{16} = -\frac{1}{4}$$

27. For $x = \pm 6$, the graph of $f(x)$ has sharp points. Therefore, there is no derivative for $x = 6$ or $x = -6$.

29. The derivative does not exist at the following x-values.

x	Reason
-5	Function is not defined.
-3	Function has a sharp point.
0	Function is not defined.
2	Function has a sharp point.
4	Function has a vertical tangent

31. $R(x) = 20x - \dfrac{x^2}{500}$

a. The marginal revenue is the rate of change of the revenue, or $R'(x)$, so find $R'(x)$.

$$\frac{R(x+h)-R(x)}{h}$$

$$= \frac{20(x+h)-\frac{(x+h)^2}{500}-20x+\frac{x^2}{500}}{h}$$

$$= \frac{20h-\frac{2xh+h^2}{500}}{h}$$

$$= 20 - \frac{2x+h}{500}$$

$$R'(x) = \lim_{h\to 0}\frac{R(x+h)-R(x)}{h}$$

$$= \lim_{h\to 0}\left(20 - \frac{2x+h}{500}\right)$$

$$= 20 - \frac{2x}{500}$$

At $x = 1000$,

$$R'(x) = 20 - \frac{2(1000)}{500} = 16$$

The marginal revenue is $16/table.

31. Continued

b. The actual revenue from the sale of the 1001st item is

$R(1001) - R(1000)$

$= 20(1001) - \dfrac{1001^2}{500}$

$- \left[20(1000) - \dfrac{1000^2}{500} \right]$

$= 18,015.998 - 18,000$

$= 15.998$ or 16.

The actual revenue is $15.998 or $16.

c. The marginal revenue found in part (a) approximates the actual revenue from the sale of the 1001st item found in part (b).

33. a. Given that the demand for the items is

$D(p) = -2p^2 + 4p + 6,$

where p represents the price, and

$D'(p) = -4p + 4,$

the rate of change of demand with respect to price is $D'(p)$ or $-4p + 4$.

b. When $p = 10$,

$D'(10) = -4(10) + 4 = -36.$

This means that the demand is decreasing at the rate of about 36 items for each increase in price of $1.

35. 1000; the population is increasing at a rate of 1000 shellfish per time unit. 570; the population is increasing more slowly at 570 shellfish per time unit. 250; the population is increasing at a much slower rate of 250 shellfish per time unit.

37. $f(x) = x^2 - 5x + 2$

$f'(x) = 2x - 5$

Numerical derivative:

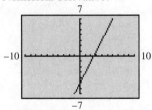

39. $f(x) = \ln x + x$

$f'(x) = \dfrac{1}{x} + 1$

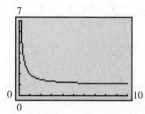

41. $f(t) = 10,000\left(1 - e^{-2t}\right) + 2000$

a. $t = 2$:

$\text{nDeriv}\left(10000(1 - e^{-2t}) + 2000, t, 2\right)$

≈ 1340.640

$t = 10$:

$\text{nDeriv}\left(10000)(1 - e^{-2t}) + 2000, t, 10\right)$

≈ 270.671

$t = 30$:

$\text{nDeriv}\left(10000(1 - e^{-2t}) + 2000, t, 30\right)$

≈ 4.958

$t = 60$:

$\text{nDeriv}\left(10000(1 - e^{-2t}) + 2000, t, 60\right)$

$\approx .012$

b. The rate of increase is getting smaller and smaller as time goes on, which means that the population is almost stable after 60 months.

43. $C(t) = 25e^{-2t} + 45$

a. $t = 0$:

$C'(0) = \text{nDeriv}(C(t), t, 0) \approx -5$

Costs are decreasing by about $5000 per month.

$t = 12$:

$C'(12) = \text{nDeriv}(C(t), t, 12) \approx -.45359$

Costs are decreasing by about $453.59 per month.

$t = 24$:

$C'(24) = \text{nDeriv}(C(t), t, 24) \approx -.04115$

Costs are decreasing by about $41.15 per month.

$t = 36$:

$C'(36) = \text{nDeriv}(C(t), t, 36) \approx -.00373$

Costs are decreasing by about $3.75 per month.

b. Costs are decreasing at a slower and slower rate and are almost constant at the end of three years.

45. a. $f(x) = .5x^5 - 2x^3 + x^2 - 3x + 2;\ -3 \le x \le 3$

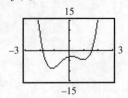

b. $g(x) = 2.5x^4 - 6x^2 + 2x - 3$

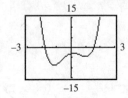

c. The graphs appear identical, which suggests that $f'(x) = g(x)$.

47. $y = \dfrac{4x^2 + x}{x^2 + 1}$

Graph the derivative of y and each possible derivative function on the same screen.

a. $f(x) = \dfrac{2x+1}{2x}$

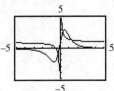

b. $g(x) = \dfrac{x^2 + x}{2x}$

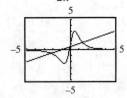

c. $h(x) = \dfrac{2x+1}{x^2+1}$

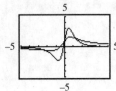

47. Continued

d. $k(x) = \dfrac{-x^2 + 8x + 1}{(x^2 + 1)^2}$

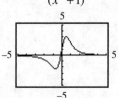

The derivative of y could possibly be

$$k(x) = \dfrac{-x^2 + 8x + 1}{(x^2 + 1)^2}.$$

Section 11.5 Techniques for Finding Derivatives

1. $f(x) = 4x^2 - 3x + 5$

$f'(x) = 4(2x^{2-1}) - 3(1x^{1-1}) + 0$

$\quad = 8x - 3$

3. $y = 2x^3 + 3x^2 - 6x + 2$

$y' = 2(3x^{3-1}) + 3(2x^{2-1}) - 6(1x^{1-1}) + 0$

$\quad = 6x^2 + 6x - 6$

5. $g(x) = x^4 + 3x^3 - 8x - 7$

$g'(x) = 4x^{4-1} + 3(3x^{3-1}) - 8(1x^{1-1}) - 0$

$\quad = 4x^3 + 9x^2 - 8$

7. $f(x) = 6x^{1.5} - 4x^5$

$f'(x) = 6(1.5x^{1.5-1}) - 4(.5x^{-5})$

$\quad = 9x^5 - 2x^{-5}$ or $9x^5 - \dfrac{2}{x^5}$

9. $y = -15x^{3/2} + 2x^{1.9} + x$

$y' = -15\left(\dfrac{3}{2}x^{3/2-1}\right) + 2(1.9x^{1.9-1}) + 1x^{1-1}$

$\quad = -22.5x^{1/2} + 3.8x^9 + 1$

11. $y = 24t^{3/2} + 4t^{1/2}$

$y' = 24\left(\dfrac{3}{2}t^{1/2}\right) + 4\left(\dfrac{1}{2}t^{-1/2}\right)$

$\quad = 36t^{1/2} + 2t^{-1/2}$ or $36t^{1/2} + \dfrac{2}{t^{1/2}}$

13. $y = 8\sqrt{x} + 6x^{3/4}$

$\quad = 8x^{1/2} + 6x^{3/4}$

$\quad y' = 8\left(\dfrac{1}{2}x^{-1/2}\right) + 6\left(\dfrac{3}{4}x^{-1/4}\right)$

$\quad\quad = 4x^{-1/2} + \dfrac{9}{2}x^{-1/4}$ or $\dfrac{4}{x^{1/2}} + \dfrac{9}{2x^{1/4}}$

15. $g(x) = 6x^{-5} - 2x^{-1}$

$\quad g'(x) = 6(-5x^{-6}) - 2(-1)x^{-2}$

$\quad\quad = -30x^{-6} + 2x^{-2}$ or $-\dfrac{30}{x^6} + \dfrac{2}{x^2}$

17. $y = 10x^{-2} + 8x^{-4} - 6x$

$\quad y' = 10(-2x^{-3}) + 8(-4x^{-5}) - 6$

$\quad\quad = -20x^{-3} - 32x^{-5} - 6$

$\quad\quad$ or $-\dfrac{20}{x^3} - \dfrac{32}{x^5} - 6$

19. $f(t) = \dfrac{6}{t} - \dfrac{6}{t^2}$

$\quad\quad = 6t^{-1} - 6t^{-2}$

$\quad f'(t) = 6(-1t^{-2}) - 6(-2t^{-3})$

$\quad\quad = -6t^{-2} + 12t^{-3}$

$\quad\quad$ or $-\dfrac{6}{t^2} + \dfrac{12}{t^3}$

21. $y = \dfrac{9 - 8x + 2x^3}{x^4}$

$\quad y = \dfrac{9}{x^4} - \dfrac{8}{x^3} + \dfrac{2}{x}$

$\quad\quad = 9x^{-4} - 8x^{-3} + 2x^{-1}$

$\quad y' = 9(-4x^{-5}) - 8(-3x^{-4}) + 2(-1x^{-2})$

$\quad\quad = -36x^{-5} + 24x^{-4} - 2x^{-2}$

$\quad\quad$ or $-\dfrac{36}{x^5} + \dfrac{24}{x^4} - \dfrac{2}{x^2}$

23.

$\quad g(x) = 8x^{-1/2} - 5x^{1/2} + x$

$\quad g'(x) = 8\left(-\dfrac{1}{2}x^{-3/2}\right) - 5\left(\dfrac{1}{2}x^{-1/2}\right) + 1$

$\quad\quad = -4x^{-3/2} - \dfrac{5}{2}x^{-1/2} + 1$

$\quad\quad$ or $-\dfrac{4}{x^{3/2}} - \dfrac{5}{2x^{1/2}} + 1$

25. $y = 4x^{-3/2} + 8x^{-1/2} + x^2 - x$

$\quad y' = 4\left(-\dfrac{3}{2}x^{-5/2}\right) + 8\left(-\dfrac{1}{2}x^{-3/2}\right) + 2x - 1$

$\quad\quad = -6x^{-5/2} - 4x^{-3/2} + 2x - 1$

$\quad\quad$ or $-\dfrac{6}{x^{5/2}} - \dfrac{4}{x^{3/2}} + 2x - 1$

27. $y = \dfrac{5}{\sqrt[4]{x}} = 5x^{-1/4}$

$\quad y' = 5\left(-\dfrac{1}{4}x^{-5/4}\right)$

$\quad\quad = -\dfrac{5}{4}x^{-5/4}$ or $-\dfrac{5}{4x^{5/4}}$

29. $y = \dfrac{-6t}{\sqrt[3]{t^2}}$

$\quad\quad = -6t(t^{-2/3})$

$\quad y = -6t^{1/3}$

$\quad y' = -6\left(\dfrac{1}{3}t^{-2/3}\right)$

$\quad\quad = -2t^{-2/3}$ or $-\dfrac{2}{t^{2/3}}$

31. $y = 8x^{-5} - 9x^{-4} + 9x^4$

$\quad \dfrac{dy}{dx} = 8(-5x^{-6}) - 9(-4x^{-5}) + 9\left(4x^{4-1}\right)$

$\quad\quad = -40x^{-6} + 36x^{-5} + 36x^3$

$\quad\quad$ or $-\dfrac{40}{x^6} + \dfrac{36}{x^5} + 36x^3$

33. $D_x\left(9x^{-1/2} + \dfrac{2}{x^{3/2}}\right)$

$\quad = D_x(9x^{-1/2} + 2x^{-3/2})$

$\quad = 9\left(-\dfrac{1}{2}x^{-3/2}\right) + 2\left(-\dfrac{3}{2}x^{-5/2}\right)$

$\quad = -\dfrac{9}{2}x^{-3/2} - 3x^{-5/2}$

$\quad$ or $-\dfrac{9}{2x^{3/2}} - \dfrac{3}{x^{5/2}}$

35. $f(x) = 6x^2 - 2x$

$\quad f'(x) = 12x - 2$

$\quad f'(-2) = 12(-2) - 2$

$\quad\quad = -24 - 2$

$\quad\quad = -26$

37.
$$f(t) = 2\sqrt{t} - \frac{3}{\sqrt{t}}$$
$$= 2t^{1/2} - 3t^{-1/2}$$
$$f'(t) = 2\left(\frac{1}{2}t^{-1/2}\right) - 3\left(-\frac{1}{2}t^{-3/2}\right)$$
$$= t^{-1/2} + \frac{3}{2}t^{-3/2}$$
$$= \frac{1}{t^{1/2}} + \frac{3}{2t^{3/2}}$$
$$f'(4) = \frac{1}{4^{1/2}} + \frac{3}{2(4^{3/2})}$$
$$= \frac{1}{2} + \frac{3}{16} = \frac{11}{16}$$

39.
$$f(x) = -\frac{(3x^2 + x)^2}{7}$$
$$= -\frac{9x^4 + 6x^3 + x^2}{7}$$
$$= -\frac{9}{7}x^4 - \frac{6}{7}x^3 - \frac{1}{7}x^2$$
$$f'(x) = -\frac{9}{7}(4x^3) - \frac{6}{7}(3x^2) - \frac{1}{7}(2x)$$
$$= -\frac{36}{7}x^3 - \frac{18}{7}x^2 - \frac{2}{7}x$$
$$f'(1) = -\frac{36}{7} - \frac{18}{7} - \frac{2}{7}$$
$$= -\frac{56}{7} = -8$$

Of the given choices, –9 is closest to –8, so the answer is (b).

41.
$$f(x) = x^4 - 2x^2 + 1; x = 1$$
$$f'(x) = 4x^3 - 4x$$
The slope of the tangent line at $x = 1$ is
$$f'(1) = 4(1^3) - 4(1) = 4 - 4 = 0.$$
The tangent line at $x = 1$ is the horizontal line through
$$f(1) = 1^4 - 2(1^2) + 1 = 0.$$
The equation of the tangent line is $y = 0$.

43. Let $f(x) = y = 4x^{1/2} + 2x^{3/2} + 1; x = 4$.
$$f'(x) = 4\left(\frac{1}{2}x^{-1/2}\right) + 2\left(\frac{3}{2}x^{1/2}\right)$$
$$= \frac{2}{x^{1/2}} + 3x^{1/2}$$
The slope of the tangent line at $x = 4$ is
$$f'(4) = \frac{2}{4^{1/2}} + 3(4^{1/2})$$
$$= 7.$$
Because
$$f(4) = 4(4^{1/2}) + 2(4^{3/2}) + 1$$
$$= 25,$$
the tangent line a $x = 4$ goes through (4, 25).
Using the point-slope form, the equation of the tangent line at $x = 4$ is
$$y - 25 = 7(x - 4)$$
$$y = 7x - 28 + 25$$
$$y = 7x - 3.$$

45. $P(x) = .03x^2 - 4x - 3x^8 + -5000$
Marginal profit
$$= P'(x) = .03(2x) - 4 + 3(.8x^{-2})$$
$$= .06x - 4 + 2.4x^{-2}$$

a. $P'(100) = .06(100) - 4 + 2.4(100)^{-2}$
$$= \$2.96$$

b. $P'(1000) = .06(1000) - 4 + 2.4(1000)^{-2}$
$$= \$56.60$$

c. $P'(5000) = .06(5000) - 4 + 2.4(5000)^{-2}$
$$= \$296.44$$

d. $P'(10,000) = .06(10,000) - 4 +$
$$2.4(10,000)^{-2} = \$596.38$$

47. $S(t) = 10,000 - 10,000t^{-2} + 100t^1$
$$S'(t) = -10,000(-.2t^{-1.2}) + 100(.1t^{-.9})$$
$$= \frac{2000}{t^{1.2}} + \frac{10}{t^{.9}}$$

a. $S'(1) = \frac{2000}{(1)^{1.2}} + \frac{10}{1^{.9}} = \frac{100}{1} = 2010$

b. $S'(10) = \frac{2000}{(10)^{1.2}} + \frac{10}{10^{.9}} = 127.45$

49. $M'(x) = 3.044(3x^2) - 379.6(2x) + 14,274.5$

$\quad = 9.132x^2 - 759.2x + 14,274.5$

a. For 1920, $x = 20$

$M'(20) = 9.132(20)^2 - 759.2(20) + 14,274.5$

$= 2743$

b. For 1960, $x = 60$

$M'(60) = 9.132(60)^2 - 759.2(60) + 14,274.5$

$= 1598$

c. For 1980, $x = 80$

$M'(80) = 9.132(80)^2 - 759.2(80) + 14,274.5$

$= 11,983$

d. for 2000, $x = 100$

$M'(100) = 9.132(100)^2 - 759.2(100) + 14,274.5$

$= 29,675$

e. For 2002, $x = 102$

$M'(60) = 9.132(102)^2 - 759.2(102) + 14,274.5$

$= 31,845$

f. The amount of money in circulation was increasing at a rate of \$2743 million per year in 1920; it was increasing more slowly in 1960 (\$1598 million per year), and by 2000 it was increasing at a rate of \$29,675 million per year. Finally, in 2002, it was increasing at \$31,845 per year.

51. Demand is $x = 5000 - 100p$

$$p = \frac{5000 - x}{100}$$

$$R(x) = x\left(\frac{5000 - x}{100}\right) = \frac{5000x - x^2}{100}$$

$$R'(x) = \frac{5000 - 2x}{100}$$

a. $R'(1000) = \dfrac{5000 - 2000}{100} = 30$

b. $R'(2500) = \dfrac{5000 - 2(2500)}{100} = 0$

c. $R'(3000) = \dfrac{5000 - 2(3000)}{100} = -10$

53. $V = \pi r^2 h$

Since $h = 80$ mm, $V = \pi r^2(80) = 80\pi r^2$.

a. $\dfrac{dV}{dr} = 80\pi(2r) = 160\pi r$

b. When $r = 4$,

$\dfrac{dV}{dr} = 160\pi(4) = 640\pi$.

The volume of blood is increasing at a rate of 640π mm^3/mm of change in the radius.

c. When $r = 6$,

$\dfrac{dV}{dr} = 160\pi(6) = 960\pi$.

The volume of blood is increasing at a rate of 960π mm^3/mm of change in the radius.

d. When $r = 8$,

$\dfrac{dV}{dr} = 160\pi(8) = 1280\pi$.

The volume of blood is increasing at a rate of 1280π mm^3/mm of change in the radius.

55. $f'(x) = .002(4x^3) - .04(3x^2) + .33(2x) - .475$

$\quad = .008x^3 - .12x^2 + .66x - .475$

a. In 1990, $x = 0$

$f'(0) = .008(0)^3 - .12(0)^2 + .66(0) - .475$

$= -.475$

b. In 1993, $x = 3$

$f'(3) = .008(3)^3 - .12(3)^2 + .66(3) - .475$

$= .641$

c. In 1994, $x = 4$

$f'(4) = .008(4)^3 - .12(4)^2 + .66(4) - .475$

$= .757$

d. In 1997, $x = 7$

$f'(7) = .008(7)^3 - .12(7)^2 + .66(7) - .475$

$= 1.009$

e. In 1999, $x = 9$

$f'(9) = .008(9)^3 - .12(9)^2 + .66(9) - .475$

$= 1.577$

f. Living standards were decreasing in 1990, but then began to increase. The rate of increase steadily rose from 1991–1999, with the rate in 1999 more than double the rate in 1994.

57. a. When $x = 5$

$y_1 = 4.13(5) + 14.63 \approx 35$

$y_2 = -.033(5)^2 + 4.647(5) + 13.347 \approx 36$

b. $\dfrac{dy_1}{dx} = 4.13$; when $x = 5$, $\dfrac{dy_1}{dx} = 4.13$

$\dfrac{dy_2}{dx} = -.033(2x) + 4.647 = -.066x + 4.647$

When $x = 5$, $\dfrac{dy_2}{dx} = -.066(5) + 4.647 \approx 4.317$

c. Use points (15, 76) and (5, 36) to find the slope.

$m = \dfrac{76 - 36}{15 - 5} = \dfrac{40}{10} = 4$

$y - 36 = 4(x - 5)$

$y = 4x + 16$

(Actually, all points except the first lie on this line.)

d. Answers vary.

59. a. $V = C(R_0 - R)R^2 = CR_0 R^2 - CR^3$

$\dfrac{dV}{dR} = 2CR_0 R - 3CR^2$

b. Let $\dfrac{dV}{dR} = 0$.

$2CR_0 R - 3CR^2 = 0$

$CR(2R_0 - 3R) = 0$

$2R_0 - 3R = 0$

$2R_0 = 3R$

$\dfrac{2}{3}R_0 = R$

61. $s(t) = 8t^2 + 3t + 1$

a. $v(t) = 8(2t) + 3 = 16t + 3$

b. $v(0) = 16(0) + 3 = 3$

$v(5) = 16(5) + 3 = 83$

$v(10) = 16(10) + 3 = 163$

63. $s(t) = 2t^3 + 6t^2$

a. $v(t) = 2(3t^2) + 6(2t)$

$= 6t^2 + 12t$

b. $v(0) = 6(0^2) + 12(0) = 0$

$v(5) = 6(5^2) + 12(5) = 210$

$v(10) = 6(10^2) + 12(10) = 720$

65. $s(t) = -16t^2 + 144$

velocity $= s'(t)$

$= -32t$

a. $s'(1) = -32$ ft/sec

$s'(2) = -32 \cdot 2$

$= -64$ ft/sec

b. The rock will hit the ground when $s(t) = 0$.

$-16t^2 + 144 = 0$

$t = \sqrt{\dfrac{144}{16}}$

$= 3$ sec

c. The velocity at impact is the velocity at 3 sec.

$v = -32 \cdot 3$

$= -96$ ft/sec

67. Proof that if $y = x^n, y' = n \cdot x^{n-1}$:

a. $y = (x + h)^n$

By the binomial theorem,

$y = x^n + n \cdot x^{n-1}h$

$+ \dfrac{n(n-1)}{2!}x^{n-2}h^2 + \ldots + h^n$

b. $\dfrac{(x+h)^n - x^n}{h}$

$= n \cdot x^{n-1} + \dfrac{n(n-1)}{2!}x^{n-2}h + \ldots + h^{n-1}$

c. $\displaystyle\lim_{h \to 0} \dfrac{(x+h)^n - x^n}{h} = n \cdot x^{n-1} = y'$

69. $g(x) = 6 - 4x + 3x^2 - x^3$

$g'(x) = -4 + 6x - 3x^2$

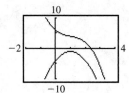

a. $g'(x) > 0$ for no value of x.

b. $g'(x) = 0$ for no value of x.

c. $g'(x) < 0$ on $(-\infty, \infty)$.

d. The derivative is always negative, so the graph of $g(x)$ is always decreasing.

Section 11.6 Derivatives of Products and Quotients

1. $y = (x^2 - 2)(3x + 2)$

$y' = (x^2 - 2)D_x(3x + 2)$

$\quad + (3x + 2)D_x(x^2 - 2)$

$\quad = (x^2 - 2)(3) + (3x + 2)(2x)$

$\quad = 3x^2 - 6 + 6x^2 + 4x$

$\quad = 9x^2 + 4x - 6$

3. $y = (6x^3 + 2)(5x - 3)$

$y' = (6x^3 + 2)D_x(5x - 3)$

$\quad + (5x - 3)D_x(6x^3 + 2)$

$\quad = (6x^3 + 2)(5) + (5x - 3)(18x^2)$

$\quad = 30x^3 + 10 + 90x^3 - 54x^2$

$\quad = 120x^3 - 54x^2 + 10$

5. $y = (x^4 - 2x^3 + 2x)(4x^2 + x - 3)$

$y' = (x^4 - 2x^3 + 2x)[D_x(4x^2 + x - 3)]$

$\quad + (4x^2 + x - 3)[D_x(x^4 - 2x^3 + 2x)]$

$\quad = (x^4 - 2x^3 + 2x)(8x + 1)$

$\quad + (4x^2 + x - 3)(4x^3 - 6x^2 + 2)$

$\quad = 8x^5 - 16x^4 + 16x^2 + x^4 - 2x^3 + 2x$

$\quad + 16x^5 - 20x^4 - 18x^3 + 26x^2 + 2x - 6$

$\quad = 24x^5 - 35x^4 - 20x^3 + 42x^2 + 4x - 6$

7. $y = (6x^2 + 4x)^2$

$\begin{aligned} y' &= (6x^2 + 4x)[D_x(6x^2 + 4x)] \\ &\quad + (6x^2 + 4x)[D_x(6x^2 + 4x)] \\ &= (6x^2 + 4x)(12x + 4) \\ &\quad + (6x^2 + 4x)(12x + 4) \\ &= 72x^3 + 24x^2 + 48x^2 + 16x \\ &\quad + 72x^3 + 24x^2 + 48x^2 + 16x \\ &= 144x^3 + 144x^2 + 32x \end{aligned}$

9. $y = (3x^3 + x^2)^2$

$\begin{aligned} y' &= (3x^3 + x^2)[D_x(3x^3 + x^2)] \\ &\quad + (3x^3 + x^2)[D_x(3x^3 + x^2)] \\ &= (3x^3 + x^2)(9x^2 + 2x) \\ &\quad + (3x^3 + x^2)(9x^2 + 2x) \\ &= 27x^5 + 6x^4 + 9x^4 + 2x^3 \\ &\quad + 27x^5 + 6x^4 + 9x^4 + 2x^3 \\ &= 54x^5 + 30x^4 + 4x^3 \end{aligned}$

11. $y = \dfrac{3x - 5}{x - 3}$

$\begin{aligned} y' &= \frac{(x-3)D_x(3x-5) - (3x-5)D_x(x-3)}{(x-3)^2} \\ &= \frac{(x-3)(3) - (3x-5)(1)}{(x-3)^2} \\ &= \frac{3x - 9 - 3x + 5}{(x-3)^2} \\ &= \frac{-4}{(x-3)^2} \end{aligned}$

13. $f(t) = \dfrac{t^2 - 4t}{t + 3}$

$D_t(t^2 - 4t) = 2t - 4$ and $D_t(t + 3) = 1$

$\begin{aligned} f'(t) &= \frac{(t+3)(2t-4) - (t^2-4t)(1)}{(t+3)^2} \\ &= \frac{2t^2 - 4t + 6t - 12 - t^2 + 4t}{(t+3)^2} \\ &= \frac{t^2 + 6t - 12}{(t+3)^2} \end{aligned}$

15. $g(x) = \dfrac{3x^2 + x}{2x^3 - 2}$

$D_x(3x^2 + x) = 6x + 1$ and $D_x(2x^3 - 2) = 6x^2$

$g'(x) = \dfrac{(2x^3 - 2)(6x + 1)}{(2x^3 - 1)^2} - \dfrac{(3x^2 + x)(6x^2)}{(2x^3 - 2)^2}$

$\quad = \dfrac{12x^4 + 2x^3 - 12x - 2 - 18x^4 + 6x^3}{(2x^3 - 2)^2}$

$\quad = \dfrac{-6x^4 + 8x^3 - 12x - 2}{(2x^3 - 2)^2}$

17. $y = \dfrac{x^2 - 4x + 2}{x + 3}$

$D_x(x^2 - 4x + 2) = 2x - 4$ and $D_x(x + 3) = 1$

$y' = \dfrac{(x + 3)(2x - 4)}{(x + 3)^2} - \dfrac{(x^2 - 4x + 2)(1)}{(x + 3)^2}$

$\quad = \dfrac{2x^2 - 4x + 6x - 12 - x^2 + 4x - 2}{(x + 3)^2}$

$\quad = \dfrac{x^2 + 6x - 14}{(x + 3)^2}$

19. $r(t) = \dfrac{\sqrt{t}}{3t + 4} = \dfrac{t^{1/2}}{3t + 4}$

$D_t(t^{1/2}) = \dfrac{1}{2}t^{-1/2}$ and $D_t(3t + 4) = 3$

$r'(t) = \dfrac{(3t + 4)\left(\frac{1}{2}t^{-1/2}\right) - (t^{1/2})(3)}{(3t + 3)^2}$

$\quad = \dfrac{\frac{3}{2}t^{1/2} + 2t^{-1/2} - 3t^{1/2}}{(3t + 4)^2}$

$\quad = \dfrac{-\frac{3}{2}t^{1/2} + \frac{2}{t^{1/2}}}{(3t + 4)^2}$

$\quad = \dfrac{-\frac{3}{2}\sqrt{t} + \frac{2}{\sqrt{t}}}{(3t + 4)^2}$ or $\dfrac{-3t + 4}{2\sqrt{t}(3t + 4)^2}$

21. $y = \dfrac{9x - 8}{\sqrt{x}} = \dfrac{9x - 8}{x^{1/2}}$

$D_x(9x - 8) = 9$ and $D_x(x^{1/2}) = \dfrac{1}{2}x^{-1/2}$

$y' = \dfrac{x^{1/2}(9) - (9x - 8)\left(\frac{1}{2}x^{-1/2}\right)}{(x^{1/2})^2}$

$\quad = \dfrac{9x^{1/2} - \frac{9}{2}x^{1/2} + 4x^{-1/2}}{x}$

$\quad = \dfrac{\frac{9x^{1/2}}{2} + 4x^{-1/2}}{x}$

$\quad = \dfrac{\frac{9x^{1/2}}{2} + \frac{4}{x^{1/2}}}{x} = \dfrac{\frac{9x + 2}{2x^{1/2}}}{x} = \dfrac{9x + 8}{2x^{3/2}} = \dfrac{9x + 8}{2x\sqrt{x}}$

23. $y = \dfrac{9 - 7x}{1 - x}$

$D_x(9 - 7x) = -7$ and $D_x(1 - x) = -1$

$y' = \dfrac{(1 - x)(-7) - (9 - 7x)(-1)}{(1 - x)^2}$

$\quad = \dfrac{-7 + 7x + 9 - 7x}{(1 - x)^2} = \dfrac{2}{(1 - x)^2}$

25. $f(p) = \dfrac{(2p + 3)(4p - 1)}{3p + 2}$

$f'(p)$

$= \dfrac{(3p + 2)D_p[(2p + 3)(4p - 1)]}{(3p + 2)^2}$

$\quad - \dfrac{(2p + 3)(4p - 1)D_p(3p + 2)}{(3p + 2)^2}$

$= \dfrac{(3p + 2)[(2)(4p - 1) + (2p + 3)(4)]}{(3p + 2)^2}$

$\quad - \dfrac{(2p + 3)(4p - 1)(3)}{(3p + 2)^2}$

$= \dfrac{(3p + 2)(8p - 2 + 8p + 12)}{(3p + 2)^2}$

$\quad - \dfrac{(8p^2 + 10p - 3)(3)}{(3p + 2)^2}$

$= \dfrac{(3p + 2)(16p + 10)}{(3p + 2)^2} - \dfrac{24p^2 + 30p - 9}{(3p + 2)^2}$

$= \dfrac{48p^2 + 62p + 20 - 24p^2 - 30p + 9}{(3p + 2)^2}$

$= \dfrac{24p^2 + 32p + 29}{(3p + 2)^2}$

27. $g(x) = \dfrac{x^3 + 1}{(2x+1)(5x+2)}$

$g'(x) = \dfrac{(2x+1)(5x+2)D_x(x^3+1)}{[(2x+1)(5x+2)]^2} - \dfrac{(x^3+1)D_x[(2x+1)(5x+2)]}{[(2x+1)(5x+2)]^2}$

$\quad = \dfrac{(2x+1)(5x+2)(3x^2)}{(2x+1)^2(5x+2)^2} - \dfrac{(x^3+1)[2(5x+2)+(2x+1)(5)]}{(2x+1)^2(5x+2)^2}$

$\quad = \dfrac{(10x^2+4x+5x+2)(3x^2)}{(2x+1)^2(5x+2)^2} - \dfrac{(x^3+1)(10x+4+10x+5)}{(2x+1)^2(5x+2)^2}$

$\quad = \dfrac{30x^4+27x^3+6x^2-20x^4-9x^3-20x-9}{(2x+1)^2(5x+2)^2}$

$\quad = \dfrac{10x^4+18x^3+6x^2-20x-9}{(2x+1)^2(5x+2)^2}$

29. In the first step, the numerator should be
$(x^2-1)(2) - (2x+5)(2x)$.

31. $f(x) = \dfrac{x}{x-2}$

$f'(x) = \dfrac{(x-2)[D_x(x)]-(x)[D_x(x-2)]}{(x-2)^2} = \dfrac{(x-2)(1)-(x)(1)}{(x-2)^2} = \dfrac{x-2-x}{(x-2)^2} = \dfrac{-2}{(x-2)^2}$

$f'(3) = \dfrac{-2}{(3-2)^2} = -2$

The slope is equal to $f'(3) = -2$.

$(y-3) = -2(x-3)$

$\quad y = -2x+9$

33. $f(x) = \dfrac{(x-1)(2x+3)}{x-5}$

$f(x) = \dfrac{2x^2+x-3}{x-5}$

$f'(x) = \dfrac{(x-5)[D_x(2x^2+x-3)]-(2x^2+x-3)[D_x(x-5)]}{(x-5)^2}$

$f'(x) = \dfrac{(x-5)(4x+1)-(2x^2+x-3)(1)}{(x-5)^2}$

$\quad = \dfrac{4x^2-19x-5-2x^2-x+3}{(x-5)^2}$

$\quad = \dfrac{2x^2-20x-2}{(x-5)^2}$

$f'(9) = \dfrac{2(9)^2-20(9)-2}{(9-5)^2} = -1.25$

The slope is equal to $f'(9) = -1.25$.

$y-42 = -1.25(x-9)$

$\quad y = -1.25x+53.25$

35. $C(x) = \dfrac{3x+2}{x+4}$

a. $C(10) = \dfrac{3(10)+2}{10+4} = \dfrac{32}{14}$

$= 2.286$ hundred dollars

$\dfrac{C(10)}{10} = \dfrac{2.286}{10}$

$= .2286$ hundred dollars per unit

$= \$22.86$ per unit

b. $C(20) = \dfrac{3(20)+2}{20+4} = \dfrac{62}{24}$

$= 2.583$ hundred dollars

$\dfrac{C(20)}{20} = \dfrac{2.583}{20}$

$= .1292$ hundred dollars per unit

$= \$12.92$ per unit

35. Continued

c. $\dfrac{C(x)}{x} = \dfrac{\frac{3x+2}{x+4}}{x} = \dfrac{3x+2}{x(x+4)}$

$= \dfrac{3x+2}{x^2+4x}$ hundred dollars per unit

d. $\bar{C}(x) = \dfrac{3x+2}{x^2+4x}$

Marginal average cost function:

$\bar{C}'(x) = \dfrac{d}{dx}\left(\dfrac{3x+2}{x^2+4x}\right)$

$= \dfrac{(x^2+4x)(3)-(3x+2)(2x+4)}{(x^2+4x)^2}$

$= \dfrac{3x^2+12x-6x^2-16x-8}{(x^2+4x)^2}$

$= \dfrac{-3x^2-4x-8}{(x^2+4x)^2}$

37.a. Let $x = 70$

$C(70) = \dfrac{75,000(70)}{100-70} = \$175,000$

b. $C'(x) = \dfrac{(100-x)[D_x(75,000x)]-(75,000x)[D_x(100-x)]}{(100-x)^2}$

$= \dfrac{(100-x)(75,000)-(75,000x)(-1)}{(100-x)^2}$

$= \dfrac{7,500,000-75,000x+75,000x}{(100-x)^2}$

$C'(x) = \dfrac{7,500,000}{(100-x)^2}$

c. $C'(75) = \dfrac{7,500,000}{(100-75)^2} = \$12,000$ for each 1% removed

d. $C'(90) = \dfrac{7,500,000}{(100-90)^2} = \$75,000$ for each 1% removed

e. $C'(95) = \dfrac{7,500,000}{(100-95)^2} = \$300,000$ for each 1% removed

39. $T(x) = \dfrac{10x}{x^2+5} + 98.6$

a. $\dfrac{dT}{dx}$

$= \dfrac{(x^2+5)[D_x(10x)] - (10x)[D_x(x^2+5)]}{(x^2+5)^2}$

$= \dfrac{(x^2+5)(10) - (10x)(2x)}{(x^2+5)^2}$

$= \dfrac{10x^2 + 50 - 20x^2}{(x^2+5)^2}$

$= \dfrac{-10x^2 + 50}{(x^2+5)^2}$

b. $T'(0) = \dfrac{-10(0)^2 + 50}{((0)^2+5)^2} = 2$

The temperature is increasing at 2 degrees per hour.

c. $T'(1) = \dfrac{-10(1)^2 + 50}{((1)^2+5)^2} \approx 1.1111$

The temperature is increasing at about 1.1111 degrees per hour.

d. $T'(3) = \dfrac{-10(3)^2 + 50}{((3)^2+5)^2} \approx -.2041$

The temperature is decreasing at about .2041 degrees per hour.

e. $T'(9) = \dfrac{-10(9)^2 + 50}{((9)^2+5)^2} \approx -.1028$

The temperature is decreasing at about .1028 degrees per hour.

41. $R(w) = 30\,\dfrac{w-4}{w-1.5}$

a. $R(5) = 30\left(\dfrac{5-4}{5-1.5}\right) \approx 8.57$ minutes

b. $R(7) = 30\left(\dfrac{7-4}{7-1.5}\right) \approx 16.36$ minutes

c. $R'(x) = 30\left(\dfrac{(w-1.5)(1) - (w-4)(1)}{(w-1.5)^2}\right)$

$= 30\left(\dfrac{w-1.5-w+4}{(w-1.5)^2}\right)$

$= \dfrac{75}{(w-1.5)^2}$

$R'(5) = \dfrac{75}{(5-1.5)^2} \approx 6.12$ min/(kcal/min)

$R'(7) = \dfrac{75}{(7-1.5)^2} \approx 2.48$ min/(kcal/min)

43. $f(x) = \dfrac{x^2}{2(1-x)}$

The rate of change of the waiting time is

$f'(x) = \dfrac{2(1-x)(2x) - x^2(-2)}{[2(1-x)]^2}$

$= \dfrac{(2-2x)(2x) + 2x^2}{(2-2x)^2}$

$= \dfrac{4x - 4x^2 + 2x^2}{(2-2x)^2}$

$= \dfrac{-2x^2 + 4x}{(2-2x)^2}$

a. $f'(.1) = \dfrac{-2(.1)^2 + 4(.1)}{[2-2(.1)]^2} = \dfrac{.38}{3.24} = .1173$

The rate of change of the number of vehicles waiting when the traffic intensity is .1 is .1173.

b. $f'(.6) = \dfrac{-2(.6)^2 + 4(.6)}{[2-2(.6)]^2} = \dfrac{1.68}{.64} = 2.625$

The rate of change of the number of vehicles waiting when the traffic intensity is .6 is 2.625.

45. $N(t) = \dfrac{70t^2}{30 + t^2}$

a. $N'(t) = \dfrac{(30 + t^2)(140t) - (70t^2)(2t)}{(30 + t^2)^2}$

$= \dfrac{4200t + 140t^3 - 140t^3}{(30 + t^2)^2}$

$= \dfrac{4200t}{(30 + t^2)^2}$

b. $N'(3) = \dfrac{4200(3)}{(30 + 3^2)^2}$

$= 8.28$ words per min

$N'(5) = \dfrac{4200(5)}{(30 + 5^2)^2}$

$= 6.94$ wpm/hr of instruction

$N'(7) = \dfrac{4200(7)}{(30 + 7^2)^2}$

$= 4.71$ wpm/hr of instruction

$N'(10) = \dfrac{4200(10)}{(30 + 10^2)^2}$

$= 2.49$ wpm/hr of instruction

$N'(15) = \dfrac{4200(15)}{(30 + 15^2)^2}$

$= .97$ wpm/hr of instruction

c. The rate of improvement decreases as the time increases.

Section 11.7 The Chain Rule

In Exercises 1–4, $f(x) = 2x^2 + 3x$ and $g(x) = 4x - 1$.

1. $f[g(5)]$
Find $g(5)$ first.
$g(5) = 4(5) - 1 = 19$
$f[g(5)] = f(19)$
$\qquad = 2(19^2) + 3(19)$
$\qquad = 779$

3. $g[f(5)]$
Find $f(5)$ first.
$f(5) = 2(5^2) + 3(5)$
$\qquad = 65$
$g[f(5)] = g(65)$
$\qquad = 4(65) - 1$
$\qquad = 259$

5. $f(x) = 8x + 12; g(x) = 2x + 3$
$f[g(x)] = 8(2x + 3) + 12$
$\qquad = 16x - 12$
$g[f(x)] = 2(8x + 12) + 3$
$\qquad = 16x + 27$

7. $f(x) = -x^3 + 2; g(x) = 4x + 2$
$f[g(x)] = -(4x + 2)^3 + 2$
$\qquad = -64x^3 + 96x^2 - 48x - 6$
$g[f(x)] = 4(-x^3 + 2) + 2$
$\qquad = -4x^3 + 10$

9. $f(x) = \dfrac{1}{x}; g(x) = x^2$
$f[g(x)] = \dfrac{1}{x^2}$
$g[f(x)] = \left(\dfrac{1}{x}\right)^2$
$\qquad = \dfrac{1}{x^2}$

11. $f(x) = \sqrt{x + 2}; g(x) = 8x^2 - 6x$
$f[g(x)] = \sqrt{8x^2 - 6x + 2}$
$g[f(x)] = 8(\sqrt{x + 2})^2 - 6$
$\qquad = 8x + 16 - 6$
$\qquad = 8x + 10$

13. $y = (4x+3)^5$

If $f(x) = x^5$ and $g(x) = 4x+3$, then

$y = f[g(x)] = (4x+3)^5$.

15. $y = \sqrt{6+3x^2}$

If $f(x) = \sqrt{x}$ and $g(x) = 6+3x^2$, then

$y = f[g(x)] = \sqrt{6+3x^2}$.

17. $y = \dfrac{\sqrt{x}+3}{\sqrt{x}-3}$

If $f(x) = \dfrac{x+3}{x-3}$ and $g(x) = \sqrt{x}$, then

$y = f[g(x)] = \dfrac{\sqrt{x}+3}{\sqrt{x}-3}$.

19. $y = (x^{1/2}-3)^2 + (x^{1/2}-3) + 5$

If $f(x) = x^2 + x + 5$ and

$g(x) = x^{1/2} - 3$, then

$y = f[g(x)]$

$\quad = (x^{1/2}-3)^2 + (x^{1/2}-3) + 5$.

21. $y = (3x+4)^3$

$y' = 3(3x+4)^{3-1}\dfrac{d}{dx}(3x+4)$

$\quad = 3(3x+4)^2(3)$

$\quad = 9(3x+4)^2$

23. $y = 6(3x+2)^5$

$y' = 6[5(3x+2)^4](3)$

$\quad = 90(3x+2)^4$

25. $y = -2(8x^2+6)^4$

$y' = -2[4(8x^2+6)^3](16x)$

$\quad = -128x(8x^2+6)^3$

27. $y = 12(2x+5)^{3/2}$

$y' = 12\left(\dfrac{3}{2}\right)(2x+5)^{1/2}(2)$

$\quad = 36(2x+5)^{1/2}$

29. $y = -7(4x^2+9x)^{3/2}$

$y' = -7\left(\dfrac{3}{2}\right)(4x^2+9x)^{1/2}(8x+9)$

$\quad = -\dfrac{21}{2}(8x+9)(4x^2+9x)^{1/2}$

31. $y = 8\sqrt{4x+7} = 8(4x+7)^{1/2}$

$y' = 8\left(\dfrac{1}{2}\right)(4x+7)^{-1/2}(4)$

$\quad = 16(4x+7)^{-1/2}$

or $\dfrac{16}{\sqrt{4x+7}}$

33. $y = -2\sqrt{x^2+4x} = -2(x^2+4x)^{1/2}$

$y' = -2\left(\dfrac{1}{2}\right)(x^2+4x)^{-1/2}(2x+4)$

$\quad = -(2x+4)(x^2+4x)^{-1/2}$

or $\dfrac{-(2x+4)}{\sqrt{x^2+4x}}$

35. $y = (x+1)(x-3)^2$

$y' = (x+1)D_x(x-3)^2 + (x-3)^2 D_x(x+1)$

$\quad = (x+1)(2)(x-3)(1) + (x-3)^2(1)$

$\quad = 2(x+1)(x-3) + (x-3)^2$

37. $y = 5(x+3)^2(2x-1)^5$

$y' = 5(x+3)^2(5)(2x-1)^4(2) + (5)(2x-1)^5(2)(x+3)$

$\quad = 50(x+3)^2(2x-1)^4 + 10(2x-1)^5(x+3)$

$\quad = 10(x+3)(2x-1)^4 \cdot [5(x+3) + (2x-1)]$

$\quad = 10(x+3)(2x-1)^4(7x+14)$

$\quad = 70(x+3)(2x-1)^4(x+2)$

39. $y = (3x+1)^3\sqrt{x} = (3x+1)^3(x^{1/2})$

$$y' = (3x+1)^3\left(\frac{1}{2}\right)(x^{-1/2})$$
$$+ (x^{1/2})(3)(3x+1)^2(3)$$
$$= \frac{1}{2}(x^{-1/2})(3x+1)^3 + 9(x^{1/2})(3x+1)^2$$
$$= (x^{-1/2})(3x+1)^2\left[\frac{1}{2}(3x+1)+9x\right]$$
$$= (x^{-1/2})(3x+1)^2\left(\frac{3x+1+18x}{2}\right)$$
$$= (x^{-1/2})(3x+1)^2\left(\frac{21x+1}{2}\right)$$
$$= \frac{(3x+1)^2(21x+1)}{2\sqrt{x}}$$

41. $y = \dfrac{1}{(x-4)^2} = (x-4)^{-2}$

$$y' = -2(x-4)^{-3}$$
$$\text{or } \frac{-2}{(x-4)^3}$$

43. $y = \dfrac{(4x+3)^3}{2x-1}$

$$y' = \frac{(2x-1)(3)(4x+3)^2(4)-(4x+3)^3(2)}{(2x-1)^2}$$
$$= \frac{12(2x-1)(4x+3)^2-(2)(4x+3)^3}{(2x-1)^2}$$
$$= \frac{(4x+3)^2[12(2x-1)-2(4x+3)]}{(2x-1)^2}$$
$$= \frac{(4x+3)^2(24x-12-8x-6)}{(2x-1)^2}$$
$$= \frac{(4x+3)^2(16x-18)}{(2x-1)^2}$$
$$= \frac{2(4x+3)^2(8x-9)}{(2x-1)^2}$$

45. $y = \dfrac{x^2+4x}{(5x+2)^2}$

$$y' = \frac{(2x+4)(5x+2)^2}{(5x+2)^4} - \frac{2(5x+2)(5)(x^2+4x)}{(5x+2)^4}$$
$$= \frac{(2x+4)(5x+2)^2}{(5x+2)^4} - \frac{10(5x+2)(x^2+4x)}{(5x+2)^4}$$
$$= \frac{(5x+2)}{(5x+2)^4}\cdot[(2x+4)(5x+2)-10(x^2+4x)]$$
$$= \frac{(10x^2+24x+8-10x^2-40x)}{(5x+2)^3}$$
$$= \frac{-16x+8}{(5x+2)^3}$$

47. $y = (x^{1/2} + 1)(x^{1/2} - 1)^{1/2}$

$y' = (x^{1/2} + 1)\left(\dfrac{1}{2}\right)(x^{1/2} - 1)^{-1/2}\left(\dfrac{1}{2}x^{-1/2}\right) + (x^{1/2} - 1)^{1/2}\left(\dfrac{1}{2}x^{-1/2}\right)$

$\quad = \dfrac{x^{-1/2}(x^{1/2} + 1)(x^{1/2} - 1)^{-1/2}}{4} + \dfrac{x^{-1/2}(x^{1/2} - 1)^{1/2}}{2}$

$\quad = \dfrac{x^{-1/2}(x^{1/2} - 1)^{-1/2}}{2} \cdot \left[\dfrac{x^{1/2} + 1}{2} + (x^{1/2} - 1)\right]$

$\quad = \dfrac{x^{-1/2}(x^{1/2} - 1)^{-1/2}(3x^{1/2} - 1)}{4}$

$\quad = \dfrac{3x^{1/2} - 1}{4x^{1/2}(x^{1/2} - 1)^{1/2}}$

49. Use the table of values in the textbook.

    a. $x = 1$

$\qquad D_x(f[g(x)]) = f'[g(x)] \cdot g'(x)$

$\qquad\qquad = [f'(2)] \cdot \left(\dfrac{2}{7}\right)$

$\qquad\qquad = (-7)\left(\dfrac{2}{7}\right)$

$\qquad\qquad = -2$

    b. $x = 2$

$\qquad D_x(f[g(x)]) = f'[g(x)] \cdot g'(x)$

$\qquad\qquad [f'(3)] \cdot \dfrac{3}{7}$

$\qquad\qquad = (-8)\left(\dfrac{3}{7}\right)$

$\qquad\qquad = -\dfrac{24}{7}$

51. $f(x) = (2x^2 + 3x + 1)^{50}$

$f'(x) = 50(2x^2 + 3x + 1)^{49}(4x + 3)$

$f'(0) = 50(2 \cdot 0^2 + 3 \cdot 0 + 1)^{49}(4 \cdot 0 + 3)$

$\quad = 50(1)(3) = 150$

Since (d) is 150, the correct choice is (d).

53. $C(x) = 500 + \sqrt{100 + 20x^2 - x}$

$\quad = 500 + (100 + 20x^2 - x)^{1/2}$

The marginal cost function is

$C'(x) = \left(\dfrac{1}{2}\right)(100 + 20x^2 - x)^{-1/2}(40x - 1)$

$\quad = \dfrac{40x - 1}{2\sqrt{100 + 20x^2 - x}}.$

55. $R(x) = 10\sqrt{300x - 2x^2} = 10(300x - 2x^2)^{1/2}$

    a.   The marginal revenue function is

$$R'(x) = 10\left(\frac{1}{2}\right)(300x - 2x^2)^{-1/2} \cdot (300 - 4x)$$

$$= \frac{5(300 - 4x)}{\sqrt{300x - 2x^2}}.$$

    b. $R'(30) = \dfrac{5(300 - 4 \cdot 30)}{\sqrt{300 \cdot 30 - 2 \cdot 30^2}}$

$$= \$10.61$$

        $R'(60) = \dfrac{5(300 - 4 \cdot 60)}{\sqrt{300 \cdot 60 - 2 \cdot 60^2}}$

$$= \$2.89$$

        $R'(90) = \dfrac{5(300 - 4 \cdot 90)}{\sqrt{300 \cdot 90 - 2 \cdot 90^2}}$

$$= -\$2.89$$

        $R'(120) = \dfrac{5(300 - 4 \cdot 120)}{\sqrt{300 \cdot 120 - 2 \cdot 120^2}}$

$$= -\$10.61$$

    c.   As the number of items sold increases, the rate of change of revenue per item sold decreases.

57. The marginal revenue function is $R'(x) = (225,000)\left(\dfrac{1}{3}\right)\left(\dfrac{x}{200}\right)^{-2/3}\left(\dfrac{1}{200}\right)$

$$= 375\left(\frac{x}{200}\right)^{-2/3}$$

    a. $R'(600) = 375\left(\dfrac{600}{200}\right)^{-2/3} = \$180.28$

    b. $R'(1000) = 375\left(\dfrac{1000}{200}\right)^{-2/3} = \$128.25$

    c. $R'(1500) = 375\left(\dfrac{1500}{200}\right)^{-2/3} = \$97.87$

    d.   The average revenue from the sale of x sets is $\dfrac{R(x)}{x}$.

$$\overline{R}(x) = \frac{225,000}{x}\sqrt[3]{\frac{x}{200}}$$

57. Continued

e. To find the marginal average revenue, take the derivative of $\bar{R}(x)$.

$$\frac{d\bar{R}(x)}{dx} = \frac{225,000}{x}\left[\left(\frac{1}{3}\right)\left(\frac{x}{200}\right)^{-\frac{2}{3}}\left(\frac{1}{200}\right)\right] + \sqrt[3]{\frac{x}{200}}\left[\frac{-225,000}{x^2}\right]$$

$$= \frac{375}{x}\left(\frac{x}{200}\right)^{-\frac{2}{3}} - \frac{225,000}{x^2}\sqrt[3]{\frac{x}{200}}$$

59. $p = \dfrac{200}{x^{1/2}} = 200x^{-1/2}$

$$\frac{dp}{dx} = 100x^{-3/2} = \frac{-100}{x^{3/2}}$$

$$x = 15n$$

$$\frac{dx}{dn} = 15$$

The marginal revenue product is

$$\frac{dR}{dn} = \left(p + x\frac{dp}{dx}\right)\frac{dx}{dn}$$

$$= \left(\frac{200}{x^{1/2}} + x\left[\frac{-100}{x^{3/2}}\right]\right)(15)$$

$$= \left(\frac{200}{x^{1/2}} - \frac{100}{x^{1/2}}\right)(15)$$

$$= \left(\frac{100}{x^{1/2}}\right)(15)$$

$$= \frac{1500}{x^{1/2}}$$

When $n = 25$,

$$x = 15n = 15(25) = 375.$$

Thus,

$$\frac{dR}{dn} = \frac{1500}{x^{1/2}} = \frac{1500}{(375)^{1/2}}$$

$$= \$77.46 \text{ per additional employee.}$$

61. $r(t) = 2t; A(r) = \pi r^2$

$$A[r(t)] = A(2t) = \pi(2t)^2 = 4\pi t^2$$

$A = 4\pi t^2$ gives the area of the pollution in terms of the time since the pollutants were first emitted.

63. $D(p) = \dfrac{-p^2}{100} + 500; p(c) = 2c - 10$

The demand in terms of the cost is

$$D(c) = D[p(c)]$$

$$= \frac{-(2c-10)^2}{100} + 500$$

$$= \frac{-4(c-5)^2}{100} + 500$$

$$= \frac{-c^2 + 10c - 25}{25} + 500.$$

65. a. $x = \sqrt{15,000 - 1.5p}$ Solve for p.

$$x^2 = 15,000 - 1.5p$$

$$x^2 - 15,000 = -1.5p$$

$$\frac{x^2 - 150,000}{-1.5} = p$$

$$\frac{-2(x^2 - 15,000)}{3} = p$$

$$\frac{30,000 - 2x^2}{3} = p$$

Use the Revenue formula, $R = px$.

$$R = \left(\frac{30,000 - 2x^2}{3}\right)x = \frac{30,000x - 2x^3}{3}$$

b. $P(x) = R(x) - C(x)$

$$= \left(\frac{30,000x - 2x^3}{3}\right) - (2000x + 3500)$$

$$= 10,000x - \frac{2x^3}{3} - 2000x - 3500$$

$$= 8000x - \frac{2x^3}{3} - 3500$$

c. $P'(x) = 8000 - \dfrac{2}{3}(3x^2)$

$$= 8000 - 2x^2$$

65. Continued

   d. Find x by plugging $p = 25$ into the demand equation.

$$x = \sqrt{15,000 - 1.5(25)} \approx 122.3213$$

$$P'(122.3213) \approx 8000 - 2(122.3213)^2$$

$$\approx -\$21,925$$

67. a. To find the life expectancy of a "jawbreaker," set $r(t) = 0$ and solve for t.

$$6 - \frac{3}{17}t = 0$$

$$6 = \frac{3}{17}t$$

$$34 = t$$

The life expectancy is 34 minutes.

   b. Replace r with $6 - \frac{3}{17}t$.

$$V(t) = \frac{4}{3}\pi\left(6 - \frac{3}{17}t\right)^3$$

$$V'(t) = 3\left[\frac{4}{3}\pi\left(6 - \frac{3}{17}t\right)^2\right]\left(-\frac{3}{17}\right)$$

$$= -\frac{12\pi}{17}\left(6 - \frac{3}{17}t\right)^2$$

$$V'(17) = -\frac{12\pi}{17}\left(6 - \frac{3}{17}(17)\right)^2$$

$$= -\frac{12\pi}{17}(3)^2$$

$$= -\frac{108\pi}{17} \text{ mm}^3/\text{min}$$

Replace r with $6 - \frac{3}{17}t$.

$$S(t) = 4\pi\left(6 - \frac{3}{17}t\right)^2$$

$$S'(t) = 2\left[4\pi\left(6 - \frac{3}{17}t\right)\right]\left(-\frac{3}{17}\right)$$

$$= -\frac{24\pi}{17}\left(6 - \frac{3}{17}t\right)$$

$$S'(17) = -\frac{24\pi}{17}\left(6 - \frac{3}{17}(17)\right)$$

$$= -\frac{24\pi}{17}(3)$$

$$= -\frac{72\pi}{17} \text{ mm}^2/\text{min}$$

67. Continued

In a person's mouth, at the 17-minute mark a jawbreaker's volume is decreasing at $\frac{108\pi}{17} \approx 20 \text{ mm}^3/\text{min}$ and its surface area is decreasing at $\frac{72\pi}{17} \approx 13 \text{ mm}^2/\text{min}$.

   c. Answers vary.

69. $F(t) = 60 - \dfrac{150}{\sqrt{8+t^2}}$

$$= 60 - 150(8 + t^2)^{-1/2}$$

   a. $F'(t) = -150\left(-\dfrac{1}{2}\right)(8 + t^2)^{-3/2}(2t)$

$$= 150t(8 + t^2)^{-3/2}$$

$$= \frac{150t}{(8 + t^2)^{3/2}}$$

   b. The rate at which the cashier's speed is increasing after x hr is $F'(x)$.

After 5 hr,

$$F'(5) = \frac{150(5)}{(8 + 5^2)^{3/2}}$$

$$\approx 3.96 \text{ items/min per hr}$$

After 10 hr,

$$F'(10) = \frac{150(10)}{(8 + 10^2)^{3/2}}$$

$$\approx 1.34 \text{ items/min per hr}$$

After 20 hr,

$$F'(20) = \frac{150(20)}{(8 + 20^2)^{3/2}}$$

$$\approx .36 \text{ items/min per hr}$$

After 40 hr,

$$F'(40) = \frac{150(40)}{(8 + 40^2)^{3/2}}$$

$$\approx .09 \text{ items/min per hr}$$

   c. The answers are decreasing with time. This is reasonable because there is an upper limit to how fast a cashier can work, so $F(t)$ has to level off as t increases.

71. The following solution presupposes the use of a TI-83 graphics calculator. Similar results can be obtained from other graphics calculators. Enter

$$K(x) = \sqrt[3]{(2x-1)^2} \quad \text{as } y_1 \text{ and}$$

$$K'(x) = \frac{4}{3\sqrt[3]{2x-1}} \quad \text{as } y_2. \text{ Use } -3 \le x \le 3 \text{ and}$$

$-3 \le y \le 3$. Graph the functions.

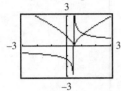

a. From the graph, $K'(x)$ is positive on the interval $(.5, \infty)$.

b. From the graph, $K'(x) = 0$ for no values of x, i.e., the derivative is never zero.

c. From the graph, $K'(x)$ is negative on the interval $(-\infty, .5)$.

d. The derivative does not exist at $x = .5$, which corresponds to a sharp point on the graph of $K(x)$. The derivative is positive when $K(x)$ is increasing and negative when $K(x)$ is decreasing.

Section 11.8 Derivatives of Exponential and Logarithmic Functions

1. $y = e^{5x}$

Let $g(x) = 5x$. Then $y = e^{g(x)}$, so

$$y' = g'(x) \cdot e^{g(x)}.$$

$$g'(x) = 5$$

$$y' = 5e^{5x}$$

3. $f(x) = 5e^{2x}$

$$f'(x) = 5(2e^{2x})$$

$$= 10e^{2x}$$

5. $g(x) = -4e^{-7x}$

$$g'(x) = -4(-7)e^{-7x}$$

$$= 28e^{-7x}$$

7. $y = e^{x^2}$

$$y' = 2xe^{x^2}$$

9. $f(x) = e^{x^3/3}$

$$f'(x) = \frac{1}{3}(3x^2)e^{x^3/3}$$

$$= x^2 e^{x^3/3}$$

11. $y = -3e^{3x^2+5}$

$$y' = (-3)(6x)e^{3x^2+5}$$

$$= -18xe^{3x^2+5}$$

13. $y = \ln(-8x^2 + 6x)$

$$y' = \frac{D_x(-8x^2 + 6x)}{-8x^2 + 6x}$$

$$= \frac{-16x + 6}{-8x^2 + 6x}$$

$$= \frac{-8x + 3}{-4x^2 + 3x}$$

15. $y = \ln\sqrt{3x+2} = \ln(3x+2)^{1/2}$

Use a property of logarithms.

$$y = \frac{1}{2}\ln(3x+2)$$

$$y' = \frac{1}{2} \cdot \frac{D_x(3x+2)}{3x+2}$$

$$= \frac{1}{2} \cdot \frac{3}{3x+2}$$

$$= \frac{3}{2(2x+1)}$$

17. $f(x) = \ln[(2x-3)(x^2+4)]$

$$f'(x) = \frac{(2x-3)(2x) + (x^2+4)(2)}{(2x-3)(x^2+4)}$$

$$= \frac{4x^2 - 6x + 2x^2 + 8}{(2x-3)(x^2+4)}$$

$$= \frac{6x^2 - 6x + 8}{(2x-3)(x^2+4)}$$

19. $y = x^3 e^{-2x}$

$$y' = 3x^2 e^{-2x} + x^3(-2)e^{-2x}$$

$$= 3x^2 e^{-2x} - 2x^3 e^{-2x}$$

$$= (3x^2 - 2x^3)e^{-2x}$$

21. $y = (3x^2 - 4x)e^{-3x}$

$y' = (6x - 4)e^{-3x}$

$\quad + (3x^2 - 4x)(-3)(e^{-3x})$

$= (6x - 4)e^{-3x}$

$\quad + (-9x^2 + 12x)(e^{-3x})$

$= e^{-3x}(6x - 4 - 9x^2 + 12x)$

$= e^{-3x}(18x - 4 - 9x^2)$

$(-9x^2 + 18x - 4)e^{-3x}$

23. $y = \ln\dfrac{6-x}{3x+5}$

Use a property of logarithms.

$y = \ln(6 - x) - \ln(3x + 5)$

$y' = \dfrac{-1}{6-x} - \dfrac{3}{3x+5}$

$= \dfrac{-(3x+5) - 3(6-x)}{(6-x)(3x+5)}$

$= \dfrac{-3x - 5 - 18 + 3x}{(6-x)(3x+5)}$

$= \dfrac{-23}{(6-x)(3x+5)}$

25. $y = \ln[(5x^3 - 2x)^{3/2}]$

$= \dfrac{3}{2}\ln(5x^3 - 2x)$

$= \dfrac{3}{2}\ln[x(5x^2 - 2)]$

$= \dfrac{3}{2}[\ln x + \ln(5x^2 - 2)]$

$y' = \dfrac{3}{2}\left[\dfrac{1}{x} + \dfrac{10x}{5x^2 - 2}\right]$

$= \dfrac{3}{2}\left[\dfrac{5x^2 - 2 + 10x^2}{x(5x^2 - 2)}\right]$

$= \dfrac{3(15x^2 - 2)}{2x(5x^2 - 2)}$

27. $y = x\ln(2 - x^3)$

$y' = x \cdot \dfrac{-3x^2}{2 - x^3} + \ln(2 - x^3) \cdot 1$

$= \dfrac{-3x^3}{2 - x^3} + \ln(2 - x^3)$

29. $y = \dfrac{\ln|x|}{x^4}$

$y' = \dfrac{x^4\left(\frac{1}{x}\right) - (\ln|x|)(4x^3)}{(x^4)^2}$

$= \dfrac{x^3 - 4x^3\ln|x|}{x^8}$

$= \dfrac{x^3(1 - 4\ln|x|)}{x^8}$

$= \dfrac{1 - 4\ln|x|}{x^5}$

31. $y = \dfrac{-5\ln|x|}{5 - 2x}$

$y' = \dfrac{(5 - 2x)\left(\frac{-5}{x}\right) - (-5\ln|x|)(-2)}{(5 - 2x)^2}$

$= \dfrac{\frac{-25}{x} + 10 - 10\ln|x|}{(5 - 2x)^2}$

33. $y = \dfrac{x^3 - 1}{2\ln|x|}$

$y' = \dfrac{2\ln|x|(3x^2) - (x^3 - 1)\left(\frac{2}{x}\right)}{4[\ln|x|]^2}$

$= \dfrac{6x^2\ln|x| - \frac{2(x^3-1)}{x}}{4(\ln|x|)^2}$

$= \dfrac{6x^3\ln|x| - 2(x^3 - 1)}{4x(\ln|x|)^2}$

$= \dfrac{3x^3\ln|x| - (x^3 - 1)}{2x(\ln|x|)^2}$

35. $y = \sqrt{\ln(x-3)} = [\ln(x-3)]^{1/2}$

$y' = \dfrac{1}{2}[\ln(x-3)]^{-1/2} \cdot D_x[\ln(x-3)]$

$= \dfrac{1}{2}[\ln(x-3)]^{-1/2}\dfrac{1}{(x-3)}$

$= \dfrac{1}{2(x-3)\sqrt{\ln(x-3)}}$

37. $y = \dfrac{e^x - 1}{\ln|x|}$

$y' = \dfrac{\ln|x| \cdot e^x - (e^x - 1)\frac{1}{x}}{(\ln|x|)^2}$

$= \dfrac{xe^x\ln|x| - e^x + 1}{x(\ln|x|)^2}$

39. $y = \dfrac{e^x - e^{-x}}{x}$

$y' = \dfrac{x[e^x - (-1)e^{-x}] - (e^x - e^{-x})(1)}{x^2}$

$= \dfrac{xe^x + xe^{-x} - e^x + e^{-x}}{x^2}$

or $\dfrac{e^x(x-1) + e^{-x}(x+1)}{x^2}$

41. $f(x) = e^{3x+2} \ln(4x-5)$

$f'(x) = e^{3x+2}\left(\dfrac{4}{4x-5}\right)$

$\quad + [\ln(4x-5)(e^{3x+2})(3)]$

$= \dfrac{4e^{3x+2}}{4x-5} + 3e^{3x+2} \ln(4x-5)$

43. $y = \dfrac{700}{7 - 10e^{4x}} = 700(7 - 10e^{4x})^{-1}$

$y' = -700(7 - 10e^{4x})^{-2}[-10e^{4x}(.4)]$

$= \dfrac{2800e^{4x}}{(7 - 10^{4x})^2}$

45. $y = \dfrac{500}{12 + 5e^{-.5x}}$

$= 500(12 + 5e^{-.5x})^{-1}$

$y' = -500(12 + 5e^{-.5x})^{-2}[(5)(-.5)e^{-.5x}]$

$= -500(12 + 5e^{-.5x})^{-2}(-2.5e^{-.5x})$

$= 1250e^{-.5x}(12 + 5e^{-.5x})^{-2}$

$= \dfrac{1250e^{-.5x}}{(12 + 5e^{-.5x})^2}$

47. $y = 8^x$

$y = \left(e^{\ln 8}\right)^x = e^{(\ln 8)x}$

$y' = (\ln 8)e^{(\ln 8)x} = (\ln 8)8^x$

49. $y = 15^{2x}$

$y = \left(e^{\ln 15}\right)^{2x} = e^{(\ln 15)(2x)}$

$y' = 2(\ln 15)e^{(\ln 15)(2x)}$

$= 2(\ln 15) \cdot 15^{2x}$

51. $g(x) = \log 6x = \dfrac{\ln 6x}{\ln 10}$

$= \dfrac{1}{\ln 10} \cdot \ln 6x$

$g'(x) = \dfrac{1}{\ln 10} \cdot \dfrac{d}{dx}(\ln 6x)$

$= \dfrac{1}{\ln 10} \cdot \left(\dfrac{1}{6x}\right) \cdot (6) = \dfrac{1}{(\ln 10)x}$

53. $f(x) = \dfrac{e^x}{2x+1}$

$f'(x) = \dfrac{(2x+1)(e^x) - (e^x)(2)}{(2x+1)^2} = \dfrac{e^x(2x-1)}{(2x+1)^2}$

$f'(0) = \dfrac{e^0(2(0) - 1)}{(2(0) + 1)^2} = \dfrac{-1}{1} = -1$

The slope of the tangent line to the graph of f is
$f'(0) = -1$.

$(y - 1) = -1(x - 0)$

$\quad y = -x + 1$

55. $f(x) = \dfrac{x^2}{e^x}$

$f'(x) = \dfrac{(e^x)(2x) - (x^2)(e^x)}{(e^x)^2} = \dfrac{-x^2 + 2x}{e^x}$

$f'(1) = \dfrac{-(1)^2 + 2(1)}{e^{(1)}} = \dfrac{1}{e}$

The slope of the tangent line to the graph of f is

$f'(1) = \dfrac{1}{e}$. When $x = 1$, $f(1) = \dfrac{(1)^2}{e^1} = \dfrac{1}{e}$.

$\left(y - \dfrac{1}{e}\right) = \dfrac{1}{e}(x - 1)$

$\quad y = \dfrac{1}{e}x$

57. $f(x) = e^{2x}$

$f'(x) = 2e^{2x}$

$f'\left[\ln\left(\dfrac{1}{4}\right)\right] = 2e^{2\ln(1/4)}$

$= 2e^{\ln(1/4)^2}$

$= 2\left(\dfrac{1}{4}\right)^2$

$= 2\left(\dfrac{1}{16}\right) = \dfrac{1}{8}$

59. a. Let $x = 2$ (for fall 2001).

$$f(2) = 54e^{.1513(2)} \approx 73$$

The number of Internet users for fall 2001 was about 73 million.

$$f'(x) = .1513(54e^{.1513x})$$

$$f'(2) = .1513(54e^{.1513(2)}) \approx 11.06$$

The rate of increase is approximately 11.06 million people per year.

b. Let $x = 7$ (for fall 2006).

$$f(7) = 54e^{.1513(7)} \approx 155.724$$

The number of Internet users for fall 2006 will be about 155.724 million.

$$f'(x) = .1513(54e^{.1513x})$$

$$f'(7) = .1513(54e^{.1513(7)}) \approx 23.561$$

The rate of increase is approximately 23.56` million people per year.

61. Let $y = .0631e^{.6007x}$

$$y' = (.0631)(.6007)e^{.6007x}$$

$$= .0379e^{.6007x}$$

a. For 2005, $x = 5$

$$y = .0631e^{.6007(5)} \approx 1.27 \text{ million}$$

$$y' = .0379e^{(.6007)(5)} \approx 763,995 \text{ per year}$$

b. For 2007, $x=7$

$$y = .0631e^{.6007(7)} \approx 4.23 \text{ million}$$

$$y' = .0379e^{(.6007)(7)} \approx 2.54 \text{ million per year}$$

c. For 2010, $x = 10$

$$y = .0631e^{.6007(10)} \approx 25.64 \text{ million}$$

$$y' = .0379e^{(.6007)(5)} \approx 15.4 \text{ million per year}$$

63. $H(N) = 1000(1 - e^{-kN})$

If $k = .1$, $H(N) = 1000(1 - e^{-.1N})$.

$$H'(N) = -1000[(-.1)e^{-.1N}]$$

$$= 100e^{-.1N}$$

a. $H'(10) = 100e^{-1} \approx 36.8$

b. $H'(100) = 100e^{-10} \approx .00454$

c. $H'(1000) = 100e^{-100}$

$$\approx 3.72 \times 10^{-42} \approx 0$$

d. $H'(N) = \dfrac{100}{e^{.1N}}$

$e^{.1N}$ is always positive because powers of e are never negative. So, $H'(N)$ is always positive. This means a habit always gets stronger, never weaker, as it is repeated.

65. a. $M(200) = 3102e^{-e^{-.022(200-56)}} \approx 2974.15 \text{ g}$

b. $M'(t) = 3102(-e^{-.022(t-56)})(-.022)\left(e^{-e^{-.022(t-56)}}\right)$

$= 68.244e^{-.022(t-56)}\left(e^{-e^{-.022(t-56)}}\right)$

$M'(200) = 68.244e^{-.022(200-56)}e^{-e^{-.022(200-56)}}$

$\approx 2.75 \text{ g/day}$

c.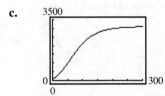

The growth pattern for $M(t)$ is increasing for 0 to 200 days and staying approximately the same for t greater than 200.

d.

Weight	Day	Rate
990.9797	50	24.87793
2121.673	100	17.72981
2733.571	150	7.603823
2974.153	200	2.753855
3058.845	250	.942782
3087.568	300	.316772

67. $f(t) = 4295.5e^{-.2294t}$

$f'(t) = 4295.5\left(e^{-.2294t}\right)(-.2294)$

$= -985.39e^{-.2294t}$

a. $f'(6) = -985.39e^{-.2294(6)}$ In 2006, the average cost is decreasing by \$248.80 per year.

$\approx -\$248.80$

b. $f'(9) = -985.39e^{-.2294(9)}$ In 2009, the average cost is decreasing by \$125.02 per year.

$\approx -\$125.02$

69. a. $V(240) = \dfrac{1100}{(1+1023e^{-.02415(240)})^4}$

$\approx 3.857 \text{ cm}^3$

b. $\dfrac{4}{3}\pi r^3 = 3.857$

$r^3 \approx .92079$

$r \approx .973 \text{ cm}$

c. $\dfrac{1100}{(1+1023e^{-.02415t})^4} = .5$

$\dfrac{(1+1023e^{-.02415t})^4}{1100} = \dfrac{1}{.5}$

$(1+1023e^{-.02415t})^4 = 2200$

$1+1023e^{-.02415t} \approx 6.84866$

$1023e^{-.02415t} \approx 5.84866$

$e^{-.02415t} \approx .00572$

$\ln e^{-.02415t} \approx \ln .00572$

$-.02415t \approx -5.16428$

$t \approx 214$

According to the formula, a tumor of size 0.5 cm^3 has been growing for about 214 months.

d. $V'(t) = \dfrac{(1+1023e^{-.02415t})^4(0) - (1100)(4)(1+1023e^{-.02415t})^3(1023)(-.02415)(e^{-.02415t})}{(1+1023e^{-.02415t})^8}$

$= \dfrac{108,703.98(1+1023e^{-.02415t})^3(e^{-.02415t})}{(1+1023e^{-.02415t})^8}$

$= \dfrac{108,703.98e^{-.02415t}}{(1+1023e^{-.02415t})^5}$

$V'(240) = \dfrac{108,703.98e^{-.02415(240)}}{(1+1023e^{-.02415(240)})^5} \approx .282 \text{ cm}^3 / \text{month}$

When the tumor is 240 months old, it is increasing in volume at the instantaneous rate of $.282 \text{ cm}^3 / \text{month}$.

71. Let $p = 100 + \dfrac{50}{\ln x}$.

   a. The revenue function is given by

$$R(x) = xp = x\left(100 + \frac{50}{\ln x}\right). \text{ The marginal}$$

revenue is

$$R'(x) = x\frac{d}{dx}\left(100 + \frac{50}{\ln x}\right) + \left(100 + \frac{50}{\ln x}\right)\frac{d}{dx}x$$

$$= \left(50(-\ln x)^{-2}\frac{1}{x}\right) + \left(100 + \frac{50}{\ln x}\right)$$

$$= \frac{-50x}{(\ln x)^2} + \frac{50}{\ln x} + 100 = \frac{50(\ln x - 1)}{(\ln x)^2} + 100$$

   b. The revenue from the next thousand items when x = 8 is

$$R'(8) = \frac{50(\ln 8 - 1)}{(\ln 8)^2} + 100 = \$112.48$$

73.

$$f(x) = \frac{67,338 - 12,595\ln x}{x}$$

$$f'(x) = \frac{x(-12,595 \cdot \frac{1}{x}) - (67,338 - 12,595\ln x)}{x^2}$$

$$= \frac{-12,595 - 67,338 + 12,595\ln x}{x^2}$$

$$= \frac{-79,933 + 12,595\ln x}{x^2}$$

   a. $f(30) = \dfrac{67,338 - 12,595\ln(30)}{30} \approx 816.66$

$$f'(30) = \frac{-79,933 + 12,595\ln(30)}{30^2} \approx -41.2$$

When the street is 30 feet wide, the maximum traffic flow is about 817 cars per hour, and the rate of change of the maximum traffic flow is about –41.2 cars per hour per foot.

   b. $f(40) = \dfrac{67,338 - 12,595\ln(40)}{40} \approx 521.91$

$$f'(40) = \frac{-79,933 + 12,595\ln(40)}{40^2} \approx -20.9$$

When the street is 40 feet wide, the maximum traffic flow is about 522 cars per hour, and the rate of change of the maximum traffic flow is about –20.9 cars per hour per foot.

75. $g(x) = -4964.2 + 6284\ln x$

   a. In 2003, x = 2003 – 1990 + 10 = 23, thus
$g(23) = -4964.2 + 6284\ln 23 = 14,739$. In
2006, x = 2006 – 1990 + 10 = 26, thus
$g(26) = -4964.2 + 6284\ln 26 = 15,509$.

   b. $g'(x) = 6284\left(\dfrac{1}{x}\right) = \dfrac{6284}{x}$

$$g'(23) = \frac{6284}{23} \approx 273$$

In 2003, kidney transplants are increasing at a rate of about 273 per year.

$$g'(26) = \frac{6284}{26} \approx 242. \text{ In 2006, kidney}$$

transplants are increasing at a rate of about 242 per year.

77. a. $\dfrac{2\ln E - 2\ln .007}{3\ln 10} = 8.9$

$$2\ln E - 2\ln .007 \approx 61.479$$

$$2\ln E \approx 51.555$$

$$\ln E \approx 25.778$$

$$E \approx 1.567 \times 10^{11} \text{ kWh}$$

   b. 247 3 10,000,000 = 2,470,000,000
10 million households use
2,470,000,000 kWh/month.

$$\frac{1.567 \times 10^{11}}{2,470,000,000} \approx 63.4$$

The energy released by this earthquake would power 10 million households for about 63.4 months.

   c. $M'(E) = \dfrac{(3\ln 10)\left(\frac{2}{E}\right) - (0)(2\ln E - 2\ln .007)}{(3\ln 10)^2}$

$$= \frac{\frac{6\ln 10}{E}}{(3\ln 10)^2}$$

$$= \frac{2}{E(3\ln 10)}$$

$$M'(70,000) = \frac{2}{70,000(3\ln 10)}$$

$$\approx 4.14 \times 10^{-6} \text{ Munit/kWh}$$

   d. $\displaystyle\lim_{E \to \infty} \frac{2}{E(3\ln 10)} = 0$

As E increases, $\dfrac{dM}{dE}$ approaches zero.

Section 11.9 Continuity and Differentiability

1. $\lim\limits_{x \to 0} f(x)$ does not exist and $f(x)$ is not defined at $x = 0$. The open circle at $x = 2$ means $f(2)$ does not exist. Thus, the function is discontinuous at $x = 0$ and $x = 2$.

3. The limit of $h(x)$ as x approaches 1 is -2. The value of $h(x)$ at $x = 1$ is 2 (solid circle). Since the limit of $h(x)$ and the value of $h(x)$ are not equal, $h(x)$ is discontinuous at $x = 1$.

5. $\lim\limits_{x \to -3} y$ does not exist because y approaches 2 from the left and 3 from the right at $x = -3$, so y is not continuous at $x = -3$. Likewise, $\lim\limits_{x \to 0} y$ and $\lim\limits_{x \to 3} y$ do not exist because y approaches different values from the left and right at both $x = 0$ and $x = 3$.
    Therefore, y is discontinuous at $x = -3$, $x = 0$, and $x = 3$.

7. $\lim\limits_{x \to 0} y$ and $\lim\limits_{x \to 6} y$ do not exist because y approaches different values from the left and from the right at both $x = 0$ and $x = 6$. Thus, y is discontinuous at $x = 0$ and $x = 6$.

In Exercises 9–19, the value of the limit of a function is found by the methods established earlier in the chapter. Details are omitted here.

9. $f(x) = \dfrac{4}{x-2}; x = 0, x = 2$

    $f(0) = \dfrac{4}{-2} = -2$

    $f(x)$ is continuous at $x = 0$ because $f(0)$ is defined, $\lim\limits_{x \to 0} f(x)$ exists, and $\lim\limits_{x \to 0} f(x) = f(0)$.

    $f(2) = \dfrac{4}{2-2} = \dfrac{4}{0}$

    so $f(x)$ is not defined at $x = 2$. Hence, $f(x)$ is not continuous at $x = 2$.

11. $h(x) = \dfrac{1}{x(x-3)}; x = 0; x = 3; x = 5$

    $h(0) = \dfrac{1}{0(0-3)} = \dfrac{1}{0}$

    so $h(x)$ is not continuous at $x = 0$.

    $h(3) = \dfrac{1}{3(3-3)} = \dfrac{1}{0}$

    so $h(x)$ is not continuous at $x = 3$.

    $h(5) = \dfrac{1}{5(5-3)} = \dfrac{1}{10}$

    $\lim\limits_{x \to 5} \dfrac{1}{x(x-3)} = \dfrac{1}{10}$, so $h(x)$ is continuous at $x = 5$.

13. $g(x) = \dfrac{x+2}{x^2-x-2} = \dfrac{x+2}{(x-2)(x+1)};$
    $x = 1, x = 2, x = -2$

    $g(1) = \dfrac{1+2}{(1-2)(1+1)} = -\dfrac{3}{2}$

    $\lim\limits_{x \to 1} \dfrac{x+2}{x^2-x-2} = -\dfrac{3}{2}$, so $g(x)$ is continuous at $x = 1$.

    $g(2) = \dfrac{2+2}{(2-2)(2+1)} = \dfrac{4}{0(3)} = \dfrac{4}{0}$

    $g(x)$ is not defined at $x = 2$, so $g(x)$ is not continuous at $x = 2$.

    $g(-2) = \dfrac{-2+2}{(-2-2)(-2+1)} = \dfrac{0}{-4(-1)}$
    $= 0$

    $\lim\limits_{x \to -2} \dfrac{x+2}{x^2-x-2} = 0$, so $g(x)$ is continuous at $x = -2$.

15. $g(x) = \dfrac{x^2-4}{x-2}; x = 0, x = 2, x = -2$

    $g(0) = \dfrac{0-4}{0-2} = 2$

    $\lim\limits_{x \to 0} g(x)$, so $g(x)$ is continuous at $x = 0$.

    $g(2) = \dfrac{4-4}{2-2} = \dfrac{0}{0}$

    so $g(x)$ is not continuous at $x = 2$.

    $g(-2) = \dfrac{4-4}{-4} = 0$

    $\lim\limits_{x \to -2} g(x) = 0$, so $g(x)$ is continuous at $x = -2$.

17. $f(x) = \begin{cases} x-2 & \text{if } x \le 3 \\ 2-x & \text{if } x > 3 \end{cases}$; $x = 2, x = 3$

$p(2) = 2 - 2 = 0$

$\lim\limits_{x \to 2} f(x) = 0$

so $f(x)$ is continuous at $x = 2$.

$f(3) = 3 - 2 = 1$

$\lim\limits_{x \to 3} f(x)$ does not exist

so, $f(x)$ is not continuous at $x = 3$.

19. $f(x) = \dfrac{5+x}{x(x-2)}$; $x = 0, x = 2$ $f(x)$ is not

defined at $x = 0$, so $f(x)$ is not continuous at

$x = 0$. $f(x)$ is not defined at $x = 2$; so $f(x)$ is

not continuous at $x = 2$.

21. $p(x) = x^2 - 4x + 11$

$f(x)$ is continuous everywhere.

23. $k(x) = e^{\sqrt{x-1}}$

Since $\sqrt{x-1}$ is not defined for $x - 1 < 0$, or

$x < 1$, $k(x)$ is not defined for $x < 1$. Thus,

$k(x)$ is not continuous for $x < 1$.

25. a.

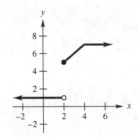

b. Since $\lim\limits_{x \to 2} f(x)$ does not exist, $f(x)$ is

discontinuous at $x = 2$.

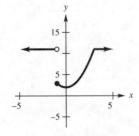

27. $f(x) = \begin{cases} x+k & \text{if } x \le 2 \\ 5-x & \text{if } x > 2 \end{cases}$

We wish to find k, such that

$2 + k = 5 - 2$,

in order for $\lim\limits_{x \to 2} f(x)$ to be the same number from

the left or right.

$2 + k = 5 - 2$

$2 + k = 3$

$k = 1$

29. a. $\lim\limits_{x \to 98} T(x) = 7.25$ cents

b. $\lim\limits_{x \to 02^-} T(x) = 7$ cents

c. $\lim\limits_{x \to 02^+} T(x) = 7.25$ cents

d. $\lim\limits_{x \to 02} T(x)$ does not exists since $T(x)$ does

not have the same value as x approaches from
the left and right.

e. $T(2) = 7.25$ cents

f. The graph of $T(x)$ is discontinuous for years
1935, 1943, 1949, and 1967.

31. The function is discontinuous only at $t = m$, where
$\lim\limits_{t \to m} f(t) = 45\%$ from the left and $\lim\limits_{t \to m} f(t) = 70\%$
from the right.
The function is differentiable everywhere except
at $t = m$ and at the endpoints.

33. a. $\lim\limits_{x \to 3^-} C(x) = \$2.60$

b. $\lim\limits_{x \to 3^+} C(x) = \$3.50$

c. $\lim\limits_{x \to 3} C(x)$ does not exists since $C(x)$ does

not have the same value as x approaches from
the left and right.

d. $C(3) = .80 + 2(.90) = \$2.60$

e.

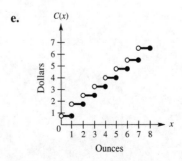

Ounces

35. $C(t)$ represents the total cost.

    a. $C(4) = 4(30) = 120$
        The cost for 4 days is $120.

    b. $C(5) = 5(30) = 150$
        The cost for 5 days is $150.

    c. $C(6) = C(5) = 150$
        The cost for 6 days is $150.

    d. $C(7) = C(5) = 150$
        The cost for 7 days is $150.

    e. Because 6 and 7 are free, 8 days is equivalent
        to 6 days for which a fee is charged.
        $C(8) = 6(30) = 180$
        The cost for 8 days is $180.

    f. $\lim\limits_{t \to 5} C(t) = C(5) = \$150$

    g. $\lim\limits_{t \to 6} C(t) = C(6) = \$150$

37. $[-3, 16]$ since x must be greater than or equal to -3 and less than or equal to 16.

39. $(12, \infty)$ since x must be greater than 12.

41. From the graph in Exercise 6, the function is continuous on $(-6, 0)$, $(4, 8)$. Because the function is not continuous as x approaches 0 from the right, the function is not continuous on $[0, 3]$.

Chapter 11 Review Exercises

1. The graph of $f(x)$ approaches 4 from the left and from the right of $x = -3$. Therefore,
$$\lim\limits_{x \to -3} f(x) = 4.$$

2. $\lim\limits_{x \to -1} g(x)$ does not exist, since the graph shows that $g(x)$ approaches 2 as x approaches -1 from the right but $g(x)$ approaches -2 as x approaches -1 from the left.

3. $\lim\limits_{x \to 1} \dfrac{x^3 - 1.1x^2 - 2x + 2.1}{x - 1}$

x	$\dfrac{x^3 - 1.1x^2 - 2x + 2.1}{x - 1}$
.9	-1.38
.99	-1.2189
.999	-1.2019
1	
1.001	-1.1981
1.01	-1.1809
1.1	-1

As x gets closer to 1 from the left or right, the value of $\dfrac{x^3 - 1.1x^2 - 2x + 2.1}{x - 1}$ gets closer to -1.2, so $\lim\limits_{x \to 1} \dfrac{x^3 - 1.1x^2 - 2x + 2.1}{x - 1} = -1.2$.

4. $\lim\limits_{x \to 2} \dfrac{x^4 + .5x^3 - 4.5x^2 - 2.5x + 3}{x - 2}$

x	$\dfrac{x^4 + .5x^3 - 4.5x^2 - 2.5x + 3}{x - 2}$
1.9	15.334
1.99	17.2758
1.999	17.4775
2	
2.001	17.5225
2.01	17.7259
2.1	19.836

As x gets closer to 2 from the left or right, the value of $\dfrac{x^4 + .5x^3 - 4.5x^2 - 2.5x + 3}{x - 2}$ gets closer to 17.5, so
$$\lim\limits_{x \to 2} \dfrac{x^4 + .5x^3 - 4.5x^2 - 2.5x + 3}{x - 2} = 17.5.$$

5. $\lim\limits_{x \to 0} \dfrac{\sqrt{2-x} - \sqrt{2}}{x}$

x	$\dfrac{\sqrt{2-x} - \sqrt{2}}{x}$
$-.1$	$-.3492$
$-.01$	$-.3531$
$-.001$	$-.3535$
0	
$.001$	$-.3536$
$.01$	$-.3540$
$.1$	$-.3580$

As x gets closer to 0 from the left or right, $\dfrac{\sqrt{2-x} - \sqrt{2}}{x}$ gets closer to $-.35$, so

$$\lim\limits_{x \to 0} \dfrac{\sqrt{2-x} - \sqrt{2}}{x} = -.35 \,.$$

6. $\lim\limits_{x \to -1} \dfrac{10^x - .1}{x+1}$

x	$\dfrac{10^x - .1}{x+1}$
-1.1	$.2057$
-1.01	$.2276$
-1.001	$.2300$
-1	
$-.999$	$.2305$
$-.99$	$.2329$
$-.9$	$.2589$

As x approaches -1 from the left or right, $\dfrac{10^x - .1}{x+1}$ approaches $.23$, so

$$\lim\limits_{x \to -1} \dfrac{10^x - .1}{x+1} = .23 \,.$$

7. $x^2 - 3x + 1$ is a polynomial, so

$$\lim\limits_{x \to 2}(x^2 - 3x + 1) = 2^2 - 3(2) + 1$$
$$= -1.$$

8. $-2x^2 + x - 5$ is a polynomial, so

$$\lim\limits_{x \to -1}(-2x^2 + x - 5) = -2(-1)^2 + (-1) - 5$$
$$= -8.$$

9. $\dfrac{3x+1}{x-2}$ is continuous except at 2, where the expression is not defined.

$$\lim\limits_{x \to 4} \dfrac{3x+1}{x-2} = \dfrac{3(4)+1}{4-2} = \dfrac{13}{2}$$

10. $\dfrac{4x+7}{x-3}$ is continuous except at 3, where the expression is not defined. We must investigate the expression as x approaches 3 from the left and right.

x	$\dfrac{4x+7}{x-3}$
2.9	-186
2.99	-1896
2.999	$-18{,}996$
3	
3.001	$19{,}004$
3.01	1904
3.1	194

Because the expression approaches values that are not the same $(-19{,}000)$ and $(19{,}000)$ as x approaches 3 from the left and from the right,

$$\lim\limits_{x \to 3} \dfrac{4x+7}{x-3} \text{ does not exist.}$$

11. $\lim\limits_{x \to 2} \dfrac{x^2 - 4}{x-2}$

Although $f(x) = \lim\limits_{x \to 2} \dfrac{x^2 - 4}{x-2}$ is discontinuous at $x = 2$, this limit exists. We rewrite the fraction and use the Limit Theorem from Section 11.1.

$$\lim\limits_{x \to 2} \dfrac{x^2 - 4}{x-2} = \lim\limits_{x \to 2} \dfrac{(x+2)(x-2)}{x-2}$$
$$= \lim\limits_{x \to 2}(x+2)$$
$$= 2 + 2 = 4$$

12. $\lim\limits_{x\to-3}\dfrac{x^2+2x-3}{x+3}$

Although the function is discontinuous at
$x=-3$, this limit exists.

$\lim\limits_{x\to-3}\dfrac{x^2+2x-3}{x+3}=\lim\limits_{x\to-3}\dfrac{(x+3)(x-1)}{x+3}$

$\qquad=\lim\limits_{x\to-3}(x-1)$

$\qquad=-3-1=-4$

13. $\lim\limits_{x\to-4}\dfrac{2x^2+3x-20}{x+4}$

$=\lim\limits_{x\to-4}\dfrac{(x+4)(2x-5)}{x+4}$

$=\lim\limits_{x\to-4}(2x-5)$

$=2(-4)-5$

$=-13$

14. $\lim\limits_{x\to3}\dfrac{3x^2-2x-21}{x-3}$

$=\lim\limits_{x\to3}\dfrac{(x-3)(3x+7)}{x-3}$

$=\lim\limits_{x\to3}(3x+7)$

$=3(3)+7$

$=16$

15. $\lim\limits_{x\to9}\dfrac{\sqrt{x}-3}{x-9}$

$=\lim\limits_{x\to9}\dfrac{(\sqrt{x}-3)(\sqrt{x}+3)}{(x-9)(\sqrt{x}+3)}$

$=\lim\limits_{x\to9}\dfrac{x-9}{(x-9)(\sqrt{x}+3)}$

$=\lim\limits_{x\to9}\dfrac{1}{\sqrt{x}+3}$

$=\dfrac{1}{\sqrt{9}+3}=\dfrac{1}{6}$

16. $\lim\limits_{x\to16}\dfrac{\sqrt{x}-4}{x-16}$

$=\lim\limits_{x\to16}\dfrac{(\sqrt{x}-4)(\sqrt{x}+4)}{(x-16)(\sqrt{x}+4)}$

$=\lim\limits_{x\to16}\dfrac{x-16}{(x-16)(\sqrt{x}+4)}$

$=\lim\limits_{x\to16}\dfrac{1}{\sqrt{x}+4}$

$=\dfrac{1}{\sqrt{16}+4}=\dfrac{1}{8}$

17. $\lim\limits_{x\to-1^+}\sqrt{9+8x-x^2}=\lim\limits_{x\to-1^+}\sqrt{(9-x)(1+x)}$

$=\left(\lim\limits_{x\to-1^+}\sqrt{9-x}\right)\left(\lim\limits_{x\to-1^+}\sqrt{1+x}\right)$

$=10\cdot0=0$

18. $\lim\limits_{x\to8^+}\dfrac{x^2-64}{x-8}=\lim\limits_{x\to8^+}\dfrac{(x-8)(x+8)}{x-8}$

$=\lim\limits_{x\to8^+}(x+8)=\lim\limits_{x\to8^+}x+\lim\limits_{x\to8^+}8$

$=8+8=16$

19. $\lim\limits_{x\to-5^+}\dfrac{|x+5|}{x+5}$

As x approaches -5 from the right,
$|x+5|=(x+5)$, so

$\lim\limits_{x\to-5^+}\dfrac{|x+5|}{x+5}=\lim\limits_{x\to-5^+}\dfrac{(x+5)}{x+5}$

$=\lim\limits_{x\to-5^+}1=1$

20. $\lim\limits_{x\to7^-}\left(\sqrt{7-x^2+6x}+2\right)$

$=\lim\limits_{x\to7^-}\sqrt{7-x^2+6x}$

$=\lim\limits_{x\to7^-}2=0+2=2$

21. $\lim\limits_{x\to\infty}\dfrac{2x^3-3x^2+5x-1}{4x^3+2x^2-x+10}$

$=\lim\limits_{x\to\infty}\dfrac{2-\dfrac{3}{x}+\dfrac{5}{x^2}-\dfrac{1}{x^3}}{4+\dfrac{2}{x}-\dfrac{1}{x^2}+\dfrac{10}{x^3}}=\dfrac{\lim\limits_{x\to\infty}2-\dfrac{3}{x}+\dfrac{5}{x^2}-\dfrac{1}{x^3}}{\lim\limits_{x\to\infty}4+\dfrac{2}{x}-\dfrac{1}{x^2}+\dfrac{10}{x^3}}$

$=\dfrac{2}{4}=\dfrac{1}{2}$

22. $\displaystyle\lim_{x\to-\infty}\frac{4-3x-2x^2}{x^3+2x+5}=\lim_{x\to-\infty}\frac{\dfrac{4}{x^3}-\dfrac{3}{x^2}-\dfrac{2}{x}}{1+\dfrac{2}{x^2}+\dfrac{5}{x^3}}$

$\displaystyle=\frac{\lim\limits_{x\to-\infty}\dfrac{4}{x^3}-\dfrac{3}{x^2}-\dfrac{2}{x}}{\lim\limits_{x\to-\infty}1+\dfrac{2}{x^2}+\dfrac{5}{x^3}}=\frac{0}{1}=0$

23. $\displaystyle\lim_{x\to-\infty}\left(\frac{2x+1}{x-3}+\frac{4x-1}{3x}\right)$

$\displaystyle=\lim_{x\to-\infty}\left(\frac{6x^2+2x+4x^2-13x+3}{3x^2-9x}\right)$

$\displaystyle=\lim_{x\to-\infty}\frac{10x^2-10x+2}{3x^2-9x}=\lim_{x\to-\infty}\frac{10-\dfrac{10}{x}+\dfrac{3}{x^2}}{3-\dfrac{9}{x}}$

$\displaystyle=\frac{\lim\limits_{x\to-\infty}10-\dfrac{10}{x}+\dfrac{3}{x^2}}{\lim\limits_{x\to-\infty}3-\dfrac{9}{x}}=\frac{10}{3}$

24. $\displaystyle\lim_{x\to\infty}\frac{x}{\left(\ln x\right)^2}$

By constructing a table using a calculator,

X	Y1
5	1.9303
50	3.2671
500	12.946
5000	68.925
50000	427.1
500000	2903.7
5E6	21015

X=5000000

it is obvious that as x approaches infinity,

$\dfrac{x}{\left(\ln x\right)^2}$ gets larger and larger. Thus,

$\displaystyle\lim_{x\to\infty}\frac{x}{\left(\ln x\right)^2}=\infty$.

25. $x=0$ to $x=4$

From the graph, $f(0)=0$ and $f(4)=1$.

Average rate of change

$=\dfrac{f(4)-f(0)}{4-0}=\dfrac{1-0}{4-0}=\dfrac{1}{4}$

26. $x=2$ to $x=8$

From the graph, $f(8)=4$ and $f(2)=4$.

Average rate of change

$=\dfrac{f(8)-f(2)}{8-2}=\dfrac{4-4}{8-2}=0$

27. $f(x)=3x^2-5$, from $x=1$ to $x=5$

$f(1)=3(1^2)-5=-2$

$f(5)=3(5^2)-5=70$

Average rate of change

$=\dfrac{f(6)-f(1)}{5-1}=\dfrac{70-(-2)}{4}=18$

28. $g(x)=-x^3+2x^2+1$,

from $x=-2$ to $x=3$

$g(-2)=-(-2)^3+2(-2)^2+1=17$

$g(3)=-3^3+2(3^2)+1=-8$

Average rate of change

$=\dfrac{g(3)-g(-2)}{3-(-2)}=\dfrac{-8-17}{5}=\dfrac{-25}{5}=-5$

29. $h(x)=\dfrac{6-x}{2x+3}$, from $x=0$ to $x=6$

$h(0)=\dfrac{6-0}{2(0)+3}=\dfrac{6}{3}=2$

$h(6)=\dfrac{6-6}{2(6)+3}=\dfrac{0}{15}=0$

Average rate of change

$=\dfrac{h(6)-h(0)}{6-0}=\dfrac{0-2}{6}=-\dfrac{1}{3}$

30. $f(x)=e^{2x}+5\ln x$

from $x=1$ to $x=7$

$f(1)=e^2+5\ln 1$

≈ 7.389056099

$f(7)=e^{14}+5\ln 7$

$\approx 1,206,214$

Average rate of change

$=\dfrac{f(7)-f(1)}{7-1}=\dfrac{1,206,206.6}{6}$

$\approx 200,434.44$

31. $y=2x+3=f(x)$

$f(x+h)=2(x+h)+3=2x+2h+3$

$\qquad f(x+h)-f(x)$

$\qquad=(2x+2h+3)-(2x+3)$

$\qquad=2h$

$\dfrac{f(x+h)-f(x)}{h}=\dfrac{2h}{h}=2$

$f'(x)=y'=\lim\limits_{h\to 0}\dfrac{f(x+h)-f(x)}{h}$

$\qquad=\lim\limits_{h\to 0}2=2$

32. $y = x^2 + 2x$

$y' = \lim_{h \to 0} \dfrac{[(x+h)^2 + 2(x+h)] - (x^2 + 2x)}{h}$

$= \lim_{h \to 0} \dfrac{x^2 + 2xh + h^2 + 2x + 2h - x^2 - 2x}{h}$

$= \lim_{h \to 0} \dfrac{2xh + h^2 + 2h}{h}$

$= \lim_{h \to 0} (2x + h + 2)$

$= 2x + 2$

33. $y = 2x^2 - x - 1$

$y' = \lim_{h \to 0} \dfrac{[2(x+h)^2 - (x+h) - 1] - (2x^2 - x - 1)}{h}$

$= \lim_{h \to 0} \dfrac{2x^2 + 4xh + 2h^2 - x - h - 1 - 2x^2 + x + 1}{h}$

$= \lim_{h \to 0} \dfrac{4xh + 2h^2 - h}{h}$

$= \lim_{h \to 0} (4x + 2h - 1)$

$= 4x - 1$

34. $y = x^3 + 5$

$y' = \lim_{h \to 0} \dfrac{[(x+h)^3 + 5] - (x^3 + 5)}{h}$

$= \lim_{h \to 0} \dfrac{x^3 + 3x^2h + 3xh^2 + h^3 + 5 - x^3 - 5}{h}$

$= \lim_{h \to 0} \dfrac{3x^2h + 3xh^2 + h^3}{h}$

$= \lim_{h \to 0} (3x^2 + 3xh + h^2) = 3x^2$

35. $y = x^2 - 6x$; at $x = 2$

Let $f(x) = y$.

$f(2) = 2^2 - 6(2) = -8$

$f'(x) = 2x - 6$

The slope of the tangent line at $(2, -8)$ is

$f'(2) = 2(2) - 6 = -2$.

Use point-slope form with $x_1 = 2$, $y_1 = -8$ and

$m = -2$ to find equation of tangent line.

$y - y_1 = m(x - x_1)$

$y - (-8) = -2(x - 2)$

$y + 8 = -2x + 4$

$y = -2x - 4$ or

$y + 2x = -4$

36. $y = 8 - x^2$; at $x = 1$

Let $f(x) = y$.

$f(1) = 8 - 1^2 = 7$

$f'(x) = -2x$

The slope of the tangent line at $(1, 7)$ is

$f'(1) = -2(1) = -2$

The equation is

$y - y_1 = m(x - x_1)$

$y - 7 = -2(x - 1)$

$y - 7 = -2x + 2$

$2x + y = 9$.

37. $y = \dfrac{-2}{x + 5}$; at $x = -2$

Let $f(x) = y$.

$f(-2) = \dfrac{-2}{-2 + 5} = -\dfrac{2}{3}$

$f(x) = -2(x + 5)^{-1}$

$f'(x) = -2[-1(x + 5)^{-2}(1)]$

$\qquad = \dfrac{2}{(x + 5)^2}$

slope $= f'(-2) = \dfrac{2}{(-2 + 5)^2}$

$\qquad = \dfrac{2}{(3)^2}$

$\qquad = \dfrac{2}{9}$

The equation is

$y - y_1 = m(x - x_1)$

$y + \dfrac{2}{3} = \dfrac{2}{9}(x + 2)$

$9y + 6 = 2(x + 2)$

$9y + 6 = 2x + 4$

$2x - 9y = 2$.

38. $y = \sqrt{6x-2}$; at $x = 3$

Let $f(x) = y$.

$$f(3) = \sqrt{6(3)-2} = 4$$

$$f(x) = (6x-2)^{1/2}$$

$$f'(x) = \frac{1}{2}(6x-2)^{-1/2}(6)$$

$$= 3(6x-2)^{-1/2}$$

slope $= y'(3)$

$$= 3(6 \cdot 3 - 2)^{-1/2}$$

$$= 3(16)^{-1/2}$$

$$= \frac{3}{16^{1/2}} = \frac{3}{4}$$

The equation is

$$y - 4 = \frac{3}{4}(x-3)$$

$$4y - 16 = 3x - 9$$

$$4y = 3x + 7, \text{ or}$$

$$3x - 4y = -7.$$

39. $T(p) = .06p^4 - 1.25p^3 + 6.5p^2 - 18p + 200$

$(0 < p \leq 11)$

$$T(8) = .06(8)^4 - 1.25(8)^3 + 6.5(8)^2 - 18(8) + 200$$

$$= 77.76$$

$$T(5) = .06(5)^4 - 1.25(5)^3 + 6.5(5)^2 - 18(5) + 200$$

$$= 153.75$$

a. Average rate of change in demand for price change from \$5 to \$8 is

$$\frac{T(8)-T(5)}{8-5} = \frac{77.76-153.75}{3}$$

$$= -25.33 \text{ per dollar.}$$

b. $T'(p) = .24p^3 - 3.75p^2 + 13p - 18$

Instantaneous rate of change in demand when price is \$5 is

$$T'(5) = .24(5)^3 - 3.75(5)^2 + 13(5) - 18$$

$$= -16.75 \text{ per dollar}$$

c. Instantaneous rate of change in demand when price is \$8 is

$$T'(8) = .24(8)^3 - 3.75(8)^2 + 13(8) - 18$$

$$= -31.12 \text{ per dollar}$$

40. The average rate of change of a function $f(x)$ from $x = 0$ to $x = 4$ is

$$\frac{f(4)-f(0)}{4-0}.$$

If this quotient equals 0, it means that

$$f(4) - f(0) = 0,$$

$$\text{or } f(4) = f(0).$$

It is not necessary for f to be constant between $x = 0$ and $x = 4$, although that could be the case. For example, suppose we have the quadratic function

$$f(x) = (x-2)^2.$$

Then $f(0) = 4$ and $f(4) = 4$, and

$$\frac{f(4)-f(0)}{4-0} = \frac{4-4}{4} = 0,$$

although $f(x)$ is not constant from $x = 0$ to $x = 4$.

41. $y = 5x^2 - 8x - 9$

$$y' = 5(2x) - 8$$

$$= 10x - 8$$

42. $y = x^3 - 3x^2$

$$y' = 3x^2 - 3(2x)$$

$$= 3x^2 - 6x$$

43. $y = 6x^{7/3}$

$$y' = 6\left(\frac{7}{3}\right)(x^{4/3})$$

$$= \frac{42}{3}x^{4/3}$$

$$= 14x^{4/3}$$

44. $y = -3x^{-2}$

$$y' = (-3)(-2)x^{-3}$$

$$= 6x^{-3} \text{ or } \frac{6}{x^3}$$

45.

$$f(x) = x^{-5} + \sqrt{x}$$

$$= x^{-5} + x^{1/2}$$

$$f'(x) = -5x^{-6} + \left(\frac{1}{2}\right)x^{-1/2}$$

$$\text{or } -\frac{5}{x^6} + \frac{1}{2x^{1/2}}$$

46. $f(x) = 6x^{-2} - 2\sqrt{x}$

$\quad\quad = 6x^{-2} - 2(x)^{1/2}$

$\quad f'(x) = 6(-2)x^{-3} - 2\left(\dfrac{1}{2}\right)x^{-1/2}$

$\quad\quad\quad = -12x^{-3} - x^{-1/2}$

$\quad\quad\quad$ or $-\dfrac{12}{x^3} - \dfrac{1}{x^{1/2}}$

47. $y = (3t^2 + 7)(t^3 - t)$

Use the product rule.

$\quad y' = (3t^2 + 7)(3t^2 - 1) + (t^3 - t)(6t)$

$\quad\quad = 9t^4 + 18t^2 - 7 + 6t^4 - 6t^2$

$\quad\quad = 15t^4 + 12t^2 - 7$

48. $y = (-5t + 4)(t^3 - 2t^2)$

Use the product rule.

$\quad y' = (-5t + 4)(3t^2 - 4t) + (t^3 - 2t^2)(-5)$

$\quad\quad = -15t^3 + 20t^2 + 12t^2 - 16t - 5t^3 + 10t^2$

$\quad\quad = -20t^3 + 42t^2 - 16t$

49. $y = 8x^{3/4}(3x + 2)$

Use the product rule.

$\quad y' = (8x^{3/4})(3) + 8(3x + 2)\left(\dfrac{3}{4}\right)(x^{-1/4})$

$\quad\quad = 24x^{3/4} + 18x^{3/4} + 12x^{-1/4}$

$\quad\quad = 42x^{3/4} + 12x^{-1/4}$

50. $y = 25x^{-3/5}(x^2 + 5)$

Use the product rule.

$\quad y' = 25x^{-3/5}(2x) + 25(x^2 + 5)\left(-\dfrac{3}{5}\right)x^{-8/5}$

$\quad\quad = 50x^{2/5} - 15x^{-8/5}(x^2 + 5)$

$\quad\quad = 50x^{2/5} - 15x^{2/5} - 75x^{-8/5}$

$\quad\quad = 35x^{2/5} - 75x^{-8/5}$

51. $f(x) = \dfrac{2x}{x^2 + 2}$

Use the quotient rule.

$\quad f'(x) = \dfrac{(x^2 + 2)(2) - 2x(2x)}{(x^2 + 2)^2}$

$\quad\quad\quad = \dfrac{2x^2 + 4 - 4x^2}{(x^2 + 2)^2}$

$\quad\quad\quad = \dfrac{-2x^2 + 4}{(x^2 + 2)^2}$

52. $g(x) = \dfrac{-3x^2}{3x + 4}$

Use the quotient rule.

$\quad g'(x) = \dfrac{(3x + 4)(-3)(2x) - (-3x^2)(3)}{(3x + 4)^2}$

$\quad\quad\quad = \dfrac{-18x^2 - 24x + 9x^2}{(3x + 4)^2}$

$\quad\quad\quad = \dfrac{-9x^2 - 24x}{(3x + 4)^2}$

53. $y = \dfrac{\sqrt{x} - 1}{2x + 2} = \dfrac{x^{1/2} - 1}{2x + 2}$

Use the quotient rule.

$\quad y' = \dfrac{(2x + 2)\left(\dfrac{1}{2}x^{-1/2}\right) - (x^{1/2} - 1)2}{(x + 2)^2}$

$\quad\quad = \dfrac{x^{1/2} + x^{-1/2} - 2x^{1/2} + 2}{(x + 2)^2}$

$\quad\quad = \dfrac{-x^{1/2} + x^{-1/2} + 2}{2(x + 2)^2} \cdot \dfrac{x^{1/2}}{x^{1/2}}$

$\quad\quad = \dfrac{-x + 1 + 2x^{1/2}}{x^{1/2}(x + 2)^2}$

$\quad\quad = \dfrac{-x + 1 + 2\sqrt{x}}{\sqrt{x}(x + 2)^2}$

54. $y = \dfrac{\sqrt{x} + 6}{x - 3} = \dfrac{x^{1/2} + 6}{x - 3}$

Use the quotient rule.

$\quad y' = \dfrac{(x - 3)\left(\dfrac{1}{2}x^{-1/2}\right) - (x^{1/2} + 6)(1)}{(x - 3)^2}$

$\quad\quad = \dfrac{\frac{1}{2}x^{1/2} - \frac{3}{2}x^{-1/2} - x^{1/2} - 6}{(x - 3)^2}$

$\quad\quad = \dfrac{-\frac{1}{2}x^{1/2} - \frac{3}{2}x^{-1/2} - 6}{(x - 3)^2}$

$\quad\quad = \dfrac{\frac{1}{2}x^{-1/2}(-x - 3 - 12x^{1/2})}{(x - 3)^2}$

$\quad\quad = \dfrac{-x - 3 - 12x^{1/2}}{2x^{1/2}(x - 3)^2}$

$\quad\quad = \dfrac{-x - 3 - 12\sqrt{x}}{2\sqrt{x}(x - 3)^2}$

55. $y = \dfrac{x^2 - x + 1}{x - 1}$

Use the quotient rule.

$$y' = \frac{(x-1)(2x-1) - (x^2 - x + 1)(1)}{(x-1)^2}$$

$$= \frac{2x^2 - 3x + 1 - x^2 + x - 1}{(x-1)^2}$$

$$= \frac{x^2 - 2x}{(x-1)^2}$$

56. $y = \dfrac{2x^3 - 5x^2}{x + 2}$

Use the quotient rule.

$$y' = \frac{(x+2)(6x^2 - 10x) - (2x^3 - 5x^2)(1)}{(x+2)^2}$$

$$= \frac{6x^3 - 10x^2 + 12x^2 - 20x - 2x^3 + 5x^2}{(x+2)^2}$$

$$= \frac{4x^3 + 7x^2 - 20x}{(x+2)^2}$$

57. $f(x) = (4x - 2)^4$

Use the chain rule.

$$f'(x) = 4(4x-2)^3(4)$$

$$= 16(4x-2)^3$$

58. $k(x) = (5x - 1)^6$

Use the chain rule.

$$k'(x) = 6(5x-1)^5(5)$$

$$= 30(5x-1)^5$$

59. $y = \sqrt{2t - 5}$

$$= (2t - 5)^{1/2}$$

Use the chain rule.

$$y' = \frac{1}{2}(2t-5)^{-1/2}(2)$$

$$= (2t-5)^{-1/2} \text{ or } \frac{1}{(2t-5)^{1/2}}$$

$$= \frac{1}{\sqrt{2t-5}}$$

60. $y = -3\sqrt{8t - 1} = -3(8t - 1)^{1/2}$

$$y' = -3\left[\frac{1}{2}(8t-1)^{-1/2}\right](8)$$

$$= -12(8t-1)^{-1/2}$$

$$= \frac{-12}{(8t-1)^{1/2}} = \frac{-12}{\sqrt{8t-1}}$$

61. $y = 2x(3x - 4)^3$

Use the product rule and the chain rule.

$$y' = 2x(3)(3x-4)^2(3) + (3x-4)^3(2)$$

$$= 18x(3x-4)^2 + 2(3x-4)^3$$

62. $y = 5x^2(2x + 3)^5$

$$y' = 5x^2(5)(2x+3)^4(2) + (2x+3)^5(10x)$$

$$= 50x^2(2x+3)^4 + 10x(2x+3)^5$$

63. $f(u) = \dfrac{3u^2 - 4u}{(2u + 3)^3}$

Use the quotient rule and the chain rule.

$$f'(u) = \frac{(2u+3)^3(6u-4) - (3u^2-4u)(3)(2u+3)^2(2)}{(2u+3)^6}$$

$$= \frac{(2u+3)^3(6u-4) - 6(3u^2-4u)(2u+3)^2}{(2u+3)^6}$$

$$= \frac{(2u+3)(6u-4) - 6(3u^2-4u)}{(2u+3)^4}$$

64. $g(t) = \dfrac{t^3 + t - 2}{(2t - 1)^5}$

$g'(t)$

$$= \frac{(2t-1)^5(3t^2+1) - (t^3+t-2)(5)(2t-1)^4(2)}{(2t-1)^{10}}$$

$$= \frac{(2t-1)(3t^2+1) - 10(t^3+t-2)}{(2t-1)^6}$$

65. $y = e^{-2x^3}$

$$y' = -6x^2 e^{-2x^3}$$

66. $y = -5e^{x^2}$

$$y' = -5(2xe^{x^2}) = -10xe^{x^2}$$

67. $y = 5x \cdot e^{2x}$

$$y' = 5x(2)e^{2x} + e^{2x}(5) \quad \text{Product Rule}$$

$$= 10xe^{2x} + 5e^{2x}$$

$$\text{or } 5e^{2x}(2x+1)$$

68. $y = -7x^2 \cdot e^{-3x}$

$$y' = -7x^2(-3e^{-3x}) + e^{-3x}(-14x)$$

$$= 21x^2 e^{-3x} - 14xe^{-3x}$$

$$\text{or } 7xe^{-3x}(3x-2)$$

69. $y = \ln(x^2 + 4x - 1)$

$$y' = \frac{D_x(x^2 + 4x + 1)}{x^2 + 4x - 1} = \frac{2x + 4}{x^2 + 4x - 1}$$

70. $y = \ln(4x^3 + 2x)$

$$y' = \frac{12x^2 + 2}{4x^3 + 2x}$$

$$= \frac{2(6x^2 + 1)}{2(2x^3 + x)}$$

$$= \frac{6x^2 + 1}{2x^3 + x}$$

71. $y = \frac{\ln 6x}{x^2 - 1}$

$$y' = \frac{(x^2 - 1)\left(\frac{6}{6x}\right) - (\ln 6x)(2x)}{(x^2 - 1)^2}$$

$$= \frac{x - \frac{1}{x} - 2x(\ln 6x)}{(x^2 - 1)^2}$$

72. $y = \frac{\ln(3x + 5)}{x^2 + 5x}$

$$y' = \frac{(x^2 + 5x)\left(\frac{3}{3x+5}\right) - [\ln(3x+5)](2x+5)}{(x^2 + 5x)^2}$$

$$= \frac{3}{(x^2 + 5x)(3x + 5)} - \frac{(2x + 5)\ln(3x + 5)}{(x^2 + 5x)^2}$$

73. $y = \frac{x^2 + 3x - 10}{x - 3}$

$$y' = \frac{(x - 3)(2x + 3) - (x^2 + 3x - 10)(1)}{(x - 3)^2}$$

$$= \frac{2x^2 - 3x - 9 - x^2 - 3x + 10}{(x - 3)^2}$$

$$= \frac{x^2 - 6x + 1}{(x - 3)^2}$$

74. $y = \frac{x^2 - x - 6}{x - 2}$

$$y' = \frac{(x - 2)(2x - 1) - (x^2 - x - 6)(1)}{(x - 2)^2}$$

$$= \frac{2x^2 - 5x + 2 - x^2 + x + 6}{(x - 2)^2}$$

$$= \frac{x^2 - 4x + 8}{(x - 2)^2}.$$

75. $y = -6e^{2x}$

$$y' = -6(2)e^{2x} = -12e^{2x}$$

76. $y = 8e^{.5x}$

$$y' = 8(.5e^{.5x}) = 4e^{.5x}$$

77. $D_x\left(\frac{\sqrt{x} + 1}{\sqrt{x} - 1}\right)$

$$= D_x\left(\frac{x^{1/2} + 1}{x^{1/2} - 1}\right)$$

$$= \frac{(x^{1/2} - 1)\left(\frac{1}{2}x^{-1/2}\right) - (x^{1/2} + 1)\left(\frac{1}{2}x^{-1/2}\right)}{(x^{1/2} - 1)^2}$$

$$= \frac{(x^{1/2} - 1 - x^{1/2} - 1)\left(\frac{1}{2}x^{-1/2}\right)}{(x^{1/2} - 1)^2}$$

$$= \frac{-2\left(\frac{1}{2}\right)x^{-1/2}}{(x^{1/2} - 1)^2}$$

$$= \frac{-1}{x^{1/2}(x^{1/2} - 1)^2}$$

78. $D_x\left(\frac{2x + \sqrt{x}}{1 - x}\right) = D_x\left(\frac{2x + x^{1/2}}{1 - x}\right)$

$$= \frac{(1 - x)\left(2 + \frac{1}{2}x^{-1/2}\right) - (2x + x^{1/2})(-1)}{(1 - x)^2}$$

$$= \frac{(1 - x)\left(2 + \frac{1}{2x^{1/2}}\right) + (2x + x^{1/2})}{(1 - x)^2}$$

$$= \frac{(1 - x)\left(\frac{4x^{1/2} + 1}{2x^{1/2}}\right) + (2x + x^{1/2})}{(1 - x)^2}$$

$$= \frac{(1 - x)(4x^{1/2} + 1) + 2x^{1/2}(2x + x^{1/2})}{2x^{1/2}(1 - x)^2}$$

$$= \frac{4x^{1/2} + 1 - 4x^{3/2} - x + 4x^{3/2} + 2x}{2x^{1/2}(1 - x)^2}$$

$$= \frac{4x^{1/2} + x + 1}{2x^{1/2}(1 - x)^2}$$

79. $y = \sqrt{t^{1/2} + t} = (t^{1/2} + t)^{1/2}$

$$\frac{dy}{dt} = \frac{1}{2}(t^{1/2} + t)^{-1/2}\left(\frac{1}{2}t^{-1/2} + 1\right)$$

$$= \frac{\left(\frac{1}{2}t^{-1/2} + 1\right)(2t^{1/2})}{2(t^{1/2} + t)^{1/2}(2t^{1/2})}$$

$$= \frac{1 + 2t^{1/2}}{4t^{1/2}(t^{1/2} + t)^{1/2}}$$

80. $y = \dfrac{\sqrt{x-1}}{x} = \dfrac{(x-1)^{1/2}}{x}$

$\dfrac{dy}{dx} = \dfrac{x\left[\frac{1}{2}(x-1)^{-1/2}(1)\right] - (x-1)^{1/2}(1)}{x^2}$

$= \dfrac{\frac{x}{2(x-1)^{1/2}} - (x-1)^{1/2}}{x^2}$

$= \dfrac{x - 2(x-1)}{2x^2(x-1)^{1/2}}$

$= \dfrac{2-x}{2x^2(x-1)^{1/2}}$

81. $f(x) = \dfrac{\sqrt{8+x}}{x+1} = \dfrac{(8+x)^{1/2}}{x+1}$

$f'(x) = \dfrac{(x+1)\left(\dfrac{1}{2}\right)(8+x)^{-1/2}(1)}{(x+1)^2}$

$\qquad - \dfrac{(8+x)^{1/2}(1)}{(x+1)^2}$

$= \dfrac{\frac{1}{2}(x+1)(8+x)^{-1/2} - (8+x)^{1/2}}{(x+1)^2}$

$f'(1) = \dfrac{\frac{1}{2}(1+1)(8+1)^{-1/2} - (8+1)^{1/2}}{(1+1)^2}$

$= \dfrac{\frac{1}{2}(2)(9)^{-1/2} - (9)^{1/2}}{2^2}$

$= \dfrac{\frac{1}{3} - 3}{4} = \dfrac{-\frac{8}{3}}{4} = -\dfrac{2}{3}$

82. $f(t) = \dfrac{2-3t}{\sqrt{2+t}}$

$f'(-2)$ does not exist since $f(-2)$ is undefined.

83. The graph is continuous throughout, because the function is defined for every x and the limit of the function is equal to the value of the function for every x. Thus, there is no point of discontinuity for this function.

84. The graph is discontinuous at $x = -4$ and $x = 2$, since the limit of the function fails to exist at each point.

85. $f(x) = \dfrac{2x-3}{2x+3}; x = -\dfrac{3}{2}, x = 0, x = \dfrac{3}{2}$

$f\left(-\dfrac{3}{2}\right) = \dfrac{2\left(-\frac{3}{2}\right)-3}{2\left(-\frac{3}{2}\right)+3} = \dfrac{-6}{0}$,

so $f(x)$ is not defined at $x = -\dfrac{3}{2}$. Thus, $f(x)$ is not

continuous at $x = -\dfrac{3}{2}$.

$f(0) = \dfrac{2(0)-3}{2(0)+3} = -1$

$\lim\limits_{x\to 0} f(x) = -1$, so $f(x)$ is continuous at $x = 0$.

$f\left(\dfrac{3}{2}\right) = \dfrac{2\left(\frac{3}{2}\right)-3}{2\left(\frac{3}{2}\right)+3} = \dfrac{0}{6} = 0$

$\lim\limits_{x\to 3/2} f(x) = 0$, so $f(x)$ is continuous at $x = \dfrac{3}{2}$.

86. $g(x) = \dfrac{2x-1}{x^3+x^2}; x = -1, x = 0, x = \dfrac{1}{2}$

$g(-1) = \dfrac{2(-1)-1}{(-1)^3+(-1)^2} = \dfrac{-3}{0}$,

so $g(x)$ is not continuous at $x = -1$.

$g(0) = \dfrac{2(0)-1}{0^3+0^2} = \dfrac{-1}{0}$,

so $g(x)$ is not continuous at $x = 0$.

$g\left(\dfrac{1}{2}\right) = \dfrac{2\left(\frac{1}{2}\right)-1}{\left(\frac{1}{2}\right)^3+\left(\frac{1}{2}\right)^2} = \dfrac{0}{\frac{3}{8}} = 0$

$\lim\limits_{x\to 1/2} g(x) = 0$, so $g(x)$ is continuous at

$x = \dfrac{1}{2}$.

87. $h(x) = \dfrac{2-3x}{2-x-x^2}$

$; x = -2, x = \dfrac{2}{3}, x = 1$

$h(-2) = \dfrac{2-3(-2)}{2-(-2)-(-2)^2} = \dfrac{8}{0}$,

so $h(x)$ is not continuous at $x = -2$.

$h\left(\dfrac{2}{3}\right) = \dfrac{2-3\left(\frac{2}{3}\right)}{2-\frac{2}{3}-\left(\frac{2}{3}\right)^2} = \dfrac{0}{\frac{8}{9}} = 0$

$\lim\limits_{x\to 2/3} h(x) = 0$, so $h(x)$ is continuous at $x = \dfrac{2}{3}$.

$h(1) = \dfrac{2-3\cdot 1}{2-1-1^2} = \dfrac{-1}{0}$

so $h(x)$ is not continuous at $x = 1$.

88. $f(x) = \dfrac{x^2-4}{x^2-x-6}; x=2, x=3, x=4$

$f(2) = \dfrac{2^2-4}{2^2-2-6} = \dfrac{0}{-4} = 0,$

$\lim\limits_{x\to 2} f(x) = 0,$ so $f(x)$ is continuous at $x=2$.

$f(3) = \dfrac{3^2-4}{3^2-3-6} = \dfrac{5}{0},$

so $f(x)$ is not continuous at $x=3$.

$f(4) = \dfrac{4^2-4}{4^2-4-6} = \dfrac{12}{6} = 2$

$\lim\limits_{x\to 4} f(x) = 2,$ so $f(x)$ is continuous at $x=4$.

89. $f(x) = \dfrac{x-6}{x+5}; x=6, x=-5, x=0$

Since $f(x)$ is a rational function, it will not be continuous at any x for which $x+5=0$. Then, if $x=-5$, $f(x)$ will not be continuous, but it is continuous at all other points, including $x=6$ and $x=0$.

90. $f(x) = \dfrac{x^2-9}{x+3}; x=3, x=-3, x=0$

$f(-3)$ is not defined, but $f(x)$ will be continuous for all other values of x.

Thus, $f(x)$ is continuous at $x=3$ and $x=0$ and discontinuous at $x=-3$.

91. a. In 2010, $x=10$

$g(10) = -.00096(10)^3 - .1(10)^2 + 11.3(10) + 1274$

$\approx 1,376$ million

In 2020, $x=20,$ $g(20) = -.00096(20)^2 - .1(20)^2 + 11.3(20) + 1274$

$\approx 1,452.32$ million

b. The average rate of change from 2000 to 2006 is

$g(6) = -.00096(6)^3 - .1(6)^2 + 11.3(6) + 1274$

≈ 1337.99264

$g(0) = -.00096(0)^3 - .1(0)^2 + 11.3(0) + 1274$

≈ 1274

$\dfrac{g(6)-g(0)}{6-0} = \dfrac{1337.99264-1274}{6-0}$

$= 10,665,440$ people per year

The average rate of change from 1010 to 2020 is

$\dfrac{g(20)-g(10)}{20-10} = \dfrac{1452.32-1376.04}{10}$

$= 7,628,000$ people per year

c. $g'(x) = -.00096(3)x^2 - .1(2x) + 11.3$

$= -.00288x^2 - .2x + 11.3$

In 2015, the population was increasing at

$g'(15) = -.00288(15)^2 - .2(15) + 11.3$

$= 7.652$ or $7,652,000$ people per year

In 2025, the population was increasing at

$g'(25) = -.00288(25)^2 - .2(25) + 11.3$

$= 4.5$ or $4,500,000$ people per year

92. Let $g(x) = .010068x^4 - .2736x^3 - 3.6598x^2 + 138.2894x + 12.8694$

    a. For 1983, $x = 3$

        $g(3) = 388.23$ or \$388,230,000

        For 1989, $x = 9$

        $g(9) = 827.63$ or \$827,630,000

        For 2004, $x = 24$

        $g(24) = 781.84$ or \$781,840,000

    b. The average rate of lottery sales from 1983 to 1989 is

        $\dfrac{g(9) - g(3)}{9 - 3} = 73.234040$ or \$73,234,040 per year .

        The average rate of lottery sales from 1989 to 2004 is

        $\dfrac{g(24) - g(9)}{24 - 9} = -3.052492$ or -\$3,052,492 per year

    c. $g'(x) = .010068(4)x^3 - .2736(3)x^2 - 3.6598(2)x + 138.2894$

              $= 0.040272x^3 - 0.8208x^2 - 7.3196x + 138.2894$

        Lottery sales in 1988 were changing at a rate of

        $g'(8) = 47.821$ or \$47,821,000 per year

        Lottery sales in 200 were changing at a rate of

        $g'(20) = -14.250$ or $-\$14,250,000$ per year

93. Let $f(x) = 23.46e^{.088x}$

    a. For 2001, $x = 1$

        $f(1) = 23.46e^{.088(1)} = 25.618$ or \$25,618,000,000

        For 2004, $x = 4$

        $f(4) = 23.46e^{.088(4)} = 33.358$ or \$33,358,000,000

    b. Marginal revenue is $f'(x) = 23.46(.088)e^{.088x} = 2.06448e^{.088x}$

        In 2001, marginal revenue is $f'(1) = 2.2544$ or \$2,254,400,000 per year

        In 2004, marginal revenue is $f'(4) = 2.9355$ or \$2,935,500,000 per year

94. Let $g(x) = 2.267x^4 - 20.035x^3 + 66.5x^2 - 91.02x + 21.02$

    a. For 2001, $x = 1$ $g(1) = -21.268$ or $-\$21,268,000,000$

        For 2004, $x = 4$ $g(4) = 19.052$ or \$19,052,000,000

    b. Marginal revenue is $g'(x) = 2.267(4)x^3 - 20.035(3)x^2 - 91.02$

              $= 9.068x^3 - 60.105x^2 + 133x - 91.02$

        For 2001, marginal revenue is $g'(1) = -9.057$ or $-\$9,057,000,000$ per year .

        For 2004, marginal revenue is $g'(4) = 59.652$ or \$59,652,000,000 per year .

95. Let $f(x) = 2.9 + 2.63 \ln x$

   a. For 2006, $x = 6$ $f(6) = 2.9 + 2.63 \ln 6 = 7.6123$ or 7,612,300 passengers

   For 2010, $x = 10$ $f(10) = 2.9 + 2.63 \ln 10 = 8.9558$ or 8,955,800 passengers

   b. The rate at which passengers are boarding is $f'(x)\dfrac{2.63}{x}$

   In 2006, $f'(6) = \dfrac{2.63}{6} = .438333$ or 438,333 passengers per year

   In 1010, $f'(x) = \dfrac{2.63}{10} = .263$ or 263,000 passengers per year

96. Let $f(x) = 7311e^{-.00823x}$

   a. In 1990, $x = 15$ and $f(15) = 7311e^{-.00823(15)} = 6462$.

   In 2000, $x = 25$ and $f(25) = 7311e^{-.00823(25)} = 5951$

   In 2006, $x = 31$ and $f(31) = 7311e^{-.00823(31)} = 5665$

   b. The rate at which hospital numbers are changing is
   $f'(x) = 7311(-.00823)e^{-.00823x} = -60.16953e^{-.00823x}$

   In 2000, $f'(25) \approx -49$ or decreasing by about 49 hospitals per year.

   In 2006, $f'(31) \approx -47$ or decreasing by about 47 hospitals per year.

97. Let $g(x) = \dfrac{20.3}{1 + 2.95e^{-.129x}}$

   a. In 2002, $x = 2$ and $g(2) = \dfrac{20.3}{1 + 2.95e^{-.129(2)}} = 6.191$ or \$6191   In 2005, $x = 5$ and

   $g(5) = \dfrac{20.3}{1 + 2.95e^{-.129(5)}} = 7.968$ or \$7968

   b. Costs are changing at a rate of $g'(x) = \dfrac{(-20.3)(2.95e^{-.129x})(-.129)}{(1 + 2.95e^{-.129x})^2} = \dfrac{7.725165e^{-.129x}}{(1 + 2.95e^{-.129x})^2}$

   In 2002, costs were changing at a rate of $g'(2) = .555$ or increasing at \$555 per year.

   In 2005, costs were changing at a rate of $g'(5) = .624$ or increasing at \$624 per year.

98. Let $f(x) = 15.76e^{.18x}$

   a. In 2000, $x = 10$ and $f(10) = 15.76e^{.18(10)} = 95.342$ or 95,342,000.

   b. Cellular phone numbers are changing at a rate of $f'(x) = 15.76(e^{.18x})(.18) = 2.8368e^{.18x}$.

   In 2000, numbers were changing at a rate of $f'(10) = 17.162$ or 17,162,000 cellular phones per year.

99. From the graph, the slope of the tangent line to the Hands curve where it crosses the Bat curve is 0, and the slope of the tangent line to the Bat curve where it crosses the Hands curve is approximately 650. So when the Hands and Bat functions are equal, the derivative of the Hands function is 0 mph per sec, and the derivative of the Bat function is approximately 640 mph per sec. This represents the acceleration of the hands and the bat at the moment when their velocities are equal.

100. $L = 71.5(1 - e^{-.1t})$

$W = .01289 \cdot L^{2.9}$

a.   The approximate length of a 5-year-old monkeyface is
$L(5) = 71.5(1 - e^{-.1(5)}) = 28.1$ cm.

b.   The rate at which the length of a 5-year-old is increasing is $L'(5)$.

$L'(t) = 71.5(.1e^{-.1t}) = 7.15e^{-.1t}$.

$L'(5) = 7.15e^{-.1(5)} = 4.34$ cm/year.

c.   The approximate weight of a 5-year-old monkeyface is
$W(L(5)) = W(28.1) = .01289 \cdot 28.1^{2.9}$
$= 205$ g.

d.   The rate of change of the weight with respect to length for a 5-year-old monkeyface is $W'(L(5))$.

$W' = .01289(2.9)L^{1.9}$

$= .037381 \cdot L^{1.9}$

$W'(L(5)) = W'(28.1) = .037381(28.1^{1.9})$

$= 21.19$ g/cm.

e.   The rate at which the weight of a 5-year-old monkeyface is increasing is $W'(5)$.

$W'(t) = W'(t) \cdot L'(t)$.

So $W'(5) = W'(L(5)) \cdot L'(5)$

$= 21.1$ g/cm $\cdot$ 4.34 cm/year

$= 91.6$ g/year.

Case 11 Price Elasticity of Demand

1. $q = -3.003p + 675.23$

   $$\frac{dq}{dp} = -3.003$$

   $$E = -\frac{p}{q} \cdot \frac{dq}{dp} = \frac{3.003p}{-3.003p + 675.23}$$

   For $p = 70$,

   $$E = \frac{3.003(70)}{-3.003(70) + 675.23}$$

   $$\approx .452$$

2. $q = -1000p + 70{,}000$

   $$\frac{dq}{dp} = -1000$$

   $$E = -\frac{p}{q} \cdot \frac{dq}{dp} = \frac{1000p}{-1000p + 70{,}000}$$

   a. For $p = 30$,

   $$E = \frac{1000(30)}{-1000(30) + 70{,}000}$$

   $$= .75$$

   b. For $p = 40$,

   $$E = \frac{1000(40)}{-1000(40) + 70{,}000}$$

   $$\approx 1.33$$

   Since the demand is elastic at $p = \$40$, a smaller price increase would be better.

3. $q = -2481.52p + 472{,}191.2$

   $$\frac{dq}{dp} = -2481.52$$

   $$E = -\frac{p}{q} \cdot \frac{dq}{dp} = \frac{2481.52p}{-2481.52p + 472{,}191.2}$$

   a. For $p = 100$,

   $$E = \frac{2481.52(100)}{-2481.52(100) + 472{,}191.2}$$

   $$\approx 1.1$$

   For $p = 75$,

   $$E = \frac{2481.52(75)}{-2481.52(75) + 472{,}191.2}$$

   $$\approx .65$$

   b. Let $1 = \dfrac{2481.52p}{-2481.52p + 472{,}191.2}$

   $$-2481.52p + 472{,}191.2 = 2481.52p$$

   $$472{,}191.2 = 4963.04p$$

   $$p \approx 95.14$$

   The demand has unit elasticity at $p \approx \$95.14$.

4. $q = -2.35p + 28.26$

   $$\frac{dq}{dp} = -2.35$$

   $$E = -\frac{p}{q} \cdot \frac{dq}{dp} = \frac{2.35p}{-2.35p + 28.26}$$

   For $p = 3.00$,

   $$E = \frac{2.35(3)}{-2.35(3) + 28.26}$$

   $$\approx .33$$

   Demand is inelastic at a price of $3.00. Even though demand drops as price rises, at this price level a small price increase leads to an overall revenue increase.

5. $E = 0$ means that

   $$-\frac{p}{q} \cdot \frac{dq}{dp} = 0.$$

   Because the price $p \neq 0$, $\dfrac{dq}{dp} = 0$. Changes in price produce no change at all in demand, i.e., the demand is constant.

Chapter 12: Applications of the Derivative

Section 12.1 Derivatives and Graphs

1. The function is increasing on $(1, \infty)$, and decreasing on $(-\infty, 1)$. The lowest point on the graph has coordinates $(1, -4)$, so the local minimum of -4 occurs at $x = 1$.

3. The function is increasing on $(-\infty, -2)$ and decreasing on $(-2, \infty)$. The highest point on the graph has coordinates $(-2, 3)$, so a local maximum of 3 occurs at $x = -2$.

5. The function is increasing on $(-\infty, -4)$ and $(-2, \infty)$, and decreasing on $(-4, -2)$. A relatively high point occurs at $(-4, 3)$ and a relatively low point occurs at $(-2, 1)$, so a local maximum of 3 occurs at $x = -4$ and a local minimum of 1 occurs at $x = -2$.

7. The function is increasing on $(-7, -4)$ and $(-2, \infty)$ and decreasing on $(-\infty, -7)$ and $(-4, -2)$. Relatively low points occur at $(-7, -2)$ and $(-2, -2)$, and relatively high point occurs at $(-4, 3)$. Thus a local minimum of -2 occurs at $x = -7$ and $x = -2$, and a local maximum of 3 occurs at $x = -4$.

9. $f(x) = 2x^3 - 5x^2 - 4x + 2$

$f'(x) = 6x^2 - 10x - 4$

$6x^2 - 10x - 4 = 0$

$2(3x^2 - 5x - 2) = 0$

$2(3x + 1)(x - 2) = 0$

$x = -\dfrac{1}{3}$ or $x = 2$

Test $f'(x)$ at $x = -2, x = 0, x = 3$.

$f'(-2) = 6(-2)^2 - 10(-2) - 4 = 40 > 0$

$f'(0) = 6(0)^2 - 10(0) - 4 = -4 < 0$

$f'(3) = 6(3)^2 - 10(3) - 4 = 20 > 0$

$f'(x)$ is positive on $\left(-\infty, -\dfrac{1}{3}\right)$, so $f(x)$ is increasing.

$f'(x)$ is negative on $\left(-\dfrac{1}{3}, 2\right)$, so $f(x)$ is decreasing.

$f'(x)$ is positive on $(2, \infty)$, so $f(x)$ is increasing.

11. $f(x) = \dfrac{x+1}{x+3}$

$f'(x) = \dfrac{(x+3)(1) - (x+1)(1)}{(x+3)^2} = \dfrac{3}{(x+3)^2}$

so $f'(x) = \dfrac{3}{(x+3)^2} = 0$ has no solution.

$f'(x)$ does not exist if

$(x+3)^2 = 0$

$x = -3$.

Test $f'(x)$ at $x = -4$ and $x = 0$.

$f'(-4) = \dfrac{3}{(-4+3)^2} = 3$

$f'(0) = \dfrac{3}{(0+3)^2} = \dfrac{3}{9} = \dfrac{1}{3}$

$f'(x)$ is positive on $(-\infty, -3)$ and $(-3, \infty)$, so $f(x)$ is increasing on those intervals.

13. $f(x) = \sqrt{6-x} = (6-x)^{\frac{1}{2}}$, so $f(x)$ is defined on $x \leq 6$.

$f'(x) = \dfrac{1}{2}(6-x)^{-\frac{1}{2}}(-1) = \dfrac{-1}{2\sqrt{6-x}}$

so $f'(x) = \dfrac{-1}{2\sqrt{6-x}} = 0$ has no solution.

$f'(x)$ does not exist if $2\sqrt{6-x} = 0$

$6 - x = 0$

$x = 6$.

Test $f'(x)$ at $x = 0$,

$f'(0) = \dfrac{-1}{2\sqrt{6-0}} = \dfrac{-1}{2\sqrt{6}} < 0$

$f'(x)$ is negative on $(-\infty, 6)$, so $f(x)$ is decreasing.

15. $f(x) = 2x^3 + 3x^2 - 12x + 5$

$f'(x) = 6x^2 + 6x - 12$

$6x^2 + 6x - 12 = 0$

$6(x^2 + x - 2) = 0$

$6(x + 2)(x - 1) = 0$

$x = -2$ or $x = 1$

Test $f'(x)$ at $x = -3$, $x = 0$, $x = 3$

$f'(-3) = 6(-3)^2 - 6(-3) - 12 = 60 > 0$

$f'(0) = 6(0)^2 - 6(0) - 12 = -12 < 0$

$f'(3) = 6(3)^2 - 6(3) - 12 = 24 > 0$

$f'(x)$ is positive on $(-\infty, -2)$ and $(1, \infty)$, so $f(x)$ is increasing.

$f'(x)$ is negative on $(-2, 1)$, so $f(x)$ is decreasing.

17.   The graph of f' is 0 when $x = -2$, $x = -1$, $x = 2$, so these are the critical numbers of f.

19. $f(x) = x^3 + 3x^2 - 3$

$f'(x) = 3x^2 + 6x$

$\quad\quad\; = 3x(x + 2)$

Find the critical numbers by solving the equation $f'(x) = 0$

$3x(x + 2) = 0$

$3x = 0$ or $x + 2 = 0$

$x = 0$ or $\quad\; x = -2$

Use the first derivative test by testing $f'(x)$ at $x = -3$, $x = 1$, and $x = 3$.

$f'(-3) = 3(-3)^2 + 6(-3) = 9 > 0$

$f'(-1) = 3(-1)^2 + 6(-1) = -3 < 0$

$f'(3) = 3(3)^2 + 6(3) = 45 > 0$

$f'(x)$ is positive on $(-\infty, -2)$ and negative on $(-2, 0)$, so a local maximum occurs at $x = -2$.

$f'(x)$ is negative on $(-2, 0)$ and positive on $(0, \infty)$, so a local minimum occurs at $x = 0$.

Values of extrema:

$f(0) = 0^3 + 3(0)^2 - 3 = -3$

$f(-2) = (-2)^3 + 3(-2)^2 - 3 = 1$

A local maximum of 1 occurs at $x = -2$ and a local minimum of -3 occurs at $x = 0$.

21. $f(x) = x^3 + 6x^2 + 9x + 2$

$f'(x) = 3x^2 + 12x + 9$

$\quad\quad\; = 3(x^2 + 4x + 3)$

$\quad\quad\; = 3(x + 3)(x + 1) = 0$

$x + 3 = 0$ or $x + 1 = 0$

$x = -3$ or $\quad\; x = -1$

Test $f'(x)$ at $x = -4$, $x = -2$, and $x = 0$.

$f'(-4) = 3(-4)^2 + 12(-4) + 9 = 9 > 0$

$f'(-2) = 3(-2)^2 + 12(-2) + 9 = -3 < 0$

$f'(0) = 3(0) + 12(0) + 9 = 9 > 0$

$f'(x)$ is positive on $(-\infty, -3)$ and negative on $(-3, -1)$, so a local maximum occurs at $x = -3$.

$f'(x)$ is negative on $(-3, -1)$ and positive on $(-1, \infty)$, so a local minimum occurs at $x = -1$.

$f(-3) = (-3)^3 + 6(-3)^2 + 9(-3) + 2 = 2$

$f(-1) = (-1)^3 + 6(-1)^2 + 9(-1) + 2 = -2$

Thus, f has a local maximum of 2 at $x = -3$ and a local minimum of -2 at $x = -1$.

23. $f(x) = \dfrac{4}{3}x^3 - \dfrac{21}{2}x^2 + 5x + 3$

$f'(x) = 4x^2 - 21x + 5$

$4x^2 - 21x + 5 = 0$

$(4x - 1)(x - 5) = 0$

$x = 5$ or $x = \dfrac{1}{4}$

Test $f'(x)$ at 0, 1, and 6.

$f'(0) = 5 > 0$

$f'(1) = -12 < 0$

$f'(6) = 23 > 0$

$f'(x)$ is positive on $(-\infty, \dfrac{1}{4})$ and positive on

$(5, \infty)$. $f'(x)$ is negative on $\left(\dfrac{1}{4}, 5\right)$.

By the first derivative test, $f(x)$ has a local

maximum at $x = \dfrac{1}{4}$, and a local minimum at

$x = 5$.

$f(-5) = -\dfrac{377}{6}$

$f\left(-\dfrac{1}{4}\right) = \dfrac{827}{96}$

Thus, f has a local maximum of $\dfrac{539}{96}$ at $x = \dfrac{1}{4}$

and a local minimum of $-\dfrac{395}{6}$ at $x = 5$.

25.

$$f(x) = \frac{2}{3}x^3 - x^2 - 12x + 2$$

$$f'(x) = 2x^2 - 2x - 12$$

$$2(x^2 - x - 6) = 0$$

$$2(x + 2)(x - 3) = 0$$

$$x + 2 = 0 \quad \text{or} \quad x - 3 = 0$$

$$x = -2 \quad \text{or} \quad x = 3$$

Test $f'(x)$ at $x = -3$, $x = 0$, and $x = 4$.

$$f'(-3) = 2(-3)^2 - 2(-3) - 12 = 12 > 0$$

$$f'(0) = 2(0)^2 - 2(0) - 12 = -12 < 0$$

$$f'(4) = 2(4)^2 - 2(4) - 12 = 12 > 0$$

$f'(x)$ is positive on $(-\infty, -2)$ and negative on $(-2, 3)$, so $f(x)$ has a local maximum at $x = -2$.

$f'(x)$ is negative on $(-2, 3)$ and positive on $(3, \infty)$, so $f(x)$ has a local minimum at $x = 3$.

$$f(-2) = \frac{2}{3}(-2)^3 - (-2)^2 - 12(-2) + 2$$

$$= \frac{50}{3}$$

$$f(3) = \frac{2}{3}(3)^3 - 3^2 - 12(3) + 2$$

$$= -25$$

Thus, f has a local maximum of $\frac{50}{3}$ at $x = -2$ and a local minimum of -25 at $x = 3$.

27.

$$f(x) = x^5 + 20x^2 + 8$$

$$f'(x) = 5x^4 + 40x$$

$$5x^4 + 40x = 0$$

$$5x(x^3 + 8) = 0$$

$$5x(x + 2)(x^2 - 2x + 4) = 0$$

$$5x = 0 \quad \text{or} \quad x + 2 = 0 \quad \text{or} \quad x^2 + 2x + 4 = 0$$

$$x = 0 \quad \text{or} \quad x = -2 \qquad \text{No real solution}$$

Test $f'(x)$ at $x = -3$, $x = -1$, and $x = 3$.

$$f'(-3) = 5(-3)^4 + 40(-3) = 285 > 0$$

$$f'(-1) = 5(-1)^4 + 40(-1) = -35 < 0$$

$$f'(3) = 5(3)^4 + 40(3) = 285 > 0$$

$f'(x)$ is positive on $(-\infty, -2)$ and negative on $(-2, 0)$, so $f(x)$ has a local maximum at $x = -2$.

$f'(x)$ is negative on $(-2, 0)$ and positive on $(0, \infty)$, so $f(x)$ has a local minimum at $x = 0$.

$$f(-2) = (-2)^5 + 20(-2)^2 + 8 = 56$$

$$f(0) = 0^5 + 20(0)^2 + 8 = 8$$

Thus, f has a local maximum of 56 at $x = -2$ and a local minimum of 8 at $x = 0$.

29.

$$f(x) = x^{\frac{11}{5}} - x^{\frac{6}{5}} + 1$$

$$f'(x) = \frac{11}{5}x^{\frac{6}{5}} - \frac{6}{5}x^{\frac{1}{5}}$$

$$= \frac{1}{5}x^{\frac{1}{5}}(11x - 6)$$

Critical numbers:

$$\frac{1}{5}x^{\frac{1}{5}}(11x - 6) = 0$$

$$\frac{1}{5}x^{\frac{1}{5}} = 0 \quad \text{or} \quad 11x - 6 = 0$$

$$x = 0 \quad \text{or} \qquad x = \frac{6}{11}$$

Test $f'(x)$ at $x = -1$, $x = \frac{1}{2}$, and $x = 1$.

$$f'(-1) = \frac{11}{5}(-1)^{\frac{6}{5}} - \frac{6}{5}(-1)^{\frac{1}{5}}$$

$$= \frac{17}{5} > 0$$

$$f'\left(\frac{1}{2}\right) = \frac{11}{5}\left(\frac{1}{2}\right)^{\frac{6}{5}} - \frac{6}{5}\left(\frac{1}{2}\right)^{\frac{1}{5}}$$

$$\approx -.087 < 0$$

$$f'(1) = \frac{11}{5}(1)^{\frac{6}{5}} - \frac{6}{5}(1)^{\frac{1}{5}} = 1 > 0$$

$f'(x)$ is positive on $(-\infty, 0)$ and negative on $\left(0, \frac{6}{11}\right)$, so a local maximum occurs at $x = 0$.

$f'(x)$ is negative on $\left(0, \frac{6}{11}\right)$ and positive on $\left(\frac{6}{11}, \infty\right)$, so a local minimum occurs at $x = \frac{6}{11}$.

$$f(0) = 0^{\frac{11}{5}} - 0^{\frac{6}{5}} + 1 = 1$$

$$f\left(\frac{6}{11}\right) = \left(\frac{6}{11}\right)^{\frac{11}{5}} - \left(\frac{6}{11}\right)^{\frac{6}{5}} + 1$$

$$\approx .7804$$

A local maximum of 1 occurs at $x = 0$, and a local minimum of approximately .7804 occurs at $x = \frac{6}{11}$.

31. $f(x) = -(3-4x)^{\frac{2}{5}} + 4$

$$f'(x) = -\frac{2}{5}(3-4x)^{-\frac{3}{5}}(-4)$$

$$= \frac{8}{5}(3-4x)^{-\frac{3}{5}}$$

$$= \frac{8}{5(3-4x)^{\frac{3}{5}}}$$

Critical numbers:

$f'(x)$ is never zero, but $f'(x)$ does not exist if

$3 - 4x = 0$, or $x = \frac{3}{4}$. Test $f'(x)$ at $x = 0$ and $x = 1$.

$$f'(0) = \frac{8}{5(3-4\cdot 0)^{\frac{3}{5}}} \approx .828 > 0$$

$$f'(1) = \frac{8}{5(3-4\cdot 1)^{\frac{3}{5}}} \approx -\frac{8}{5} < 0$$

$f'(x)$ is positive on $\left(-\infty, \frac{3}{4}\right)$ and negative on

$\left(\frac{3}{4}, \infty\right)$, so a local maximum occurs at $x = \frac{3}{4}$.

$$f\left(\frac{3}{4}\right) = \left[3 - 4\left(\frac{3}{4}\right)\right]^{\frac{2}{5}} + 4 = 4$$

A local maximum of 4 occurs at $x = \frac{3}{4}$. There is

no local minimum.

33. $f(x) = \dfrac{x^3}{x^3 + 1}$

$$f'(x) = \frac{\left(x^3 + 1\right)3x^2 - x^3\left(3x^2\right)}{\left(x^3 + 1\right)^2}$$

$$= \frac{3x^2}{\left(x^3 + 1\right)^2}$$

$f'(x) = 0$ when $x = 0$.

Note that both $f(x)$ and $f'(x)$ do not exist at

$x = -1$, so -1 is not a critical number.

Test $f'(x)$ at $x = -2$ and $x = 1$.

$$f'(-2) = \frac{12}{49} > 0$$

$$f'(1) = \frac{3}{4} > 0$$

$f'(x)$ is positive on $(-\infty, 0)$ and $(0, \infty)$.

f has no local extrema.

35. $f(x) = -xe^x$

$$f'(x) = (-1)\left(e^x\right) + (-x)\left(e^x\right)$$

$$= -e^x - xe^x = -e^x(1 + x)$$

$f'(x) = 0$ when

$-e^x = 0$ or $1 + x = 0$

No solution $x = -1$

The only critical number is -1. Test $f'(x)$ at $x = -2$ and $x = 0$.

$$f'(-2) = -e^{-2}(1 - 2) = -e^{-2}(-1)$$

$$= e^{-2} = .135 > 0$$

$$f'(0) = -e^0(1) = -1 < 0$$

$f'(x)$ is positive on $(-\infty, -1)$ and negative on $(-1, \infty)$.

$$f(-1) = -(-1)e^{-1} = \frac{1}{e}$$

f has a local maximum of $\dfrac{1}{e}$ at $x = -1$.

37. $f(x) = x \cdot \ln|x|$

$$f'(x) = (1)\ln|x| + x\left(\frac{1}{x}\right) = \ln|x| + 1$$

$\ln|x| + 1 = 0$

$\ln|x| = -1$

If $x > 0$,

$$x = e^{-1} = \frac{1}{e} \approx .3678$$

Test $f'(x)$ at $x = .2$ and $x = 1$.

$$f'(.2) = \ln.2 + 1 \approx -.6094 < 0$$

$$f'(1) = \ln 1 + 1 = 1 > 0$$

$f'(x)$ is negative on $\left(-\infty, \frac{1}{e}\right)$ and positive on

$\left(\frac{1}{e}, \infty\right)$.

$$f\left(\frac{1}{e}\right) = \frac{1}{e}\ln\left(\frac{1}{e}\right) = \frac{1}{e}(-1) = -\frac{1}{e}$$

f has a local minimum of $-\dfrac{1}{e}$ at $x = \dfrac{1}{e}$. By

symmetry and absolute value, there is a local

maximum of $\dfrac{1}{e}$ at $x = -\dfrac{1}{e}$.

39. $f(x) = xe^{3x} - 2$

$f'(x) = (1)e^{3x} + (3)xe^{3x}$

$\qquad = e^{3x}(1 + 3x)$

Critical numbers:

$e^{3x}(1 + 3x) = 0$

$e^{3x} = 0 \quad$ or $\quad 1 + 3x = 0$

No solution $\qquad x = -\dfrac{1}{3}$

The critical number is $x = -\dfrac{1}{3}$.

$f'(-1) = e^{-3}(1 - 3) = -2e^{-3} < 0$

$f'(0) = e^{3(0)}(1 + 3(0)) = 1 > 0$

Values of extremum:

$f\left(-\dfrac{1}{3}\right) = \dfrac{-1}{3}e^{3\left(-\frac{1}{3}\right)} - 2$

$\qquad = \dfrac{-1}{3e} - 2 \approx -2.1226$

A local minimum of -2.1226 occurs at $x = -\dfrac{1}{3}$.

41. $f(x) = e^x + e^{-x}$

$f'(x) = e^x - e^{-x}$

Find the critical numbers.

$e^x - e^{-x} = 0$

$e^x\left(1 - e^{-2x}\right) = 0$

$e^x = 0 \quad$ or $\quad 1 - e^{-2x} = 0$

No solution $\qquad e^{-2x} = 1$

$\qquad\qquad\qquad -2x = 0$

$\qquad\qquad\qquad x = 0$

Test $f'(x)$ at $x = -1$ and $x = 1$.

$f'(-1) = e^{-1} - e^1 = \dfrac{1}{e} - e \approx -2.35 < 0$

$f'(1) = e^1 - e^{-1} = e - \dfrac{1}{e} \approx 2.35 > 0$

$f'(x)$ is negative on $(-\infty, 0)$ and positive on $(0, \infty)$.

$f(0) = e^0 + e^0 = 2$

f has a local minimum of 2 at $x = 0$.

43. Graph the function

$h(x) = .2x^4 - x^3 - 12x^2 + 99x - 5$

by entering it as y_1.

Use $-10 \le x \le 15$ and $-800 \le y \le 400$.
Under the CALC menu use, "minimum" and
"maximum" to find a local minimum of -563.42
at $x = -5.5861$.

45. Graph the function

$f(x) = .01x^5 + .2x^4 - x^3 - 6x^2 + 5x + 40$

by entering it as y_1. Use $-20 \le x \le 10$ and
$-15 \le y \le 7000$.

Under the CALC menu, use "minimum" and
"maximum" to find local maxima of 5982.75 at x
$= -18.5239$ and 40.9831 at $x = .3837$, and local
minima of 11.4750 at $x = -2.8304$, and -53.7683
$x = 4.9706$.

47. a. The derivative is undefined at any point
where the graph shows a "peak", that is, a
sharp change from increasing to decreasing.

b. The graph of the function for particulates is
increasing from April to July, decreasing
from July to November, and constant from
January to April and from November to
December.

c. All four lower graphs indicate the
corresponding functions are constant from
January to April and November to
December. Air pollution is greatly reduced
when the temperature is low, as is the case
during these months.

49. $P(x) = -x^3 + 3x^2 + 72x$

$P'(x) = -3x^2 + 6x + 72$

a. Solve $P'(x) = 0$ to find the critical numbers.

$-3x^2 + 6x + 72 = 0$

$-3\left(x^2 - 2x - 24\right) = 0$

$-3(x - 6)(x + 4) = 0$

$x = 6 \quad$ or $\quad x = -4$

The number of units sold must be positive,
so determine whether P is a maximum for
$x = 6$.

$P'(1) = 75 > 0$

$P'(10) = -168 < 0$

Thus, 6 units must be sold to maximize
profit.

b. $P(6) = -6^3 + 3(6)^2 + 72(6)$

$\qquad = 324$

Thus, \$324 is the maximum profit.

51. $f(x) = \dfrac{1}{\sqrt{2\pi}}e^{-\frac{x^2}{2}}$

$f'(x) = \dfrac{-x}{\sqrt{2\pi}}e^{-\frac{x^2}{2}}$

$\dfrac{-x}{\sqrt{2\pi}}e^{-\frac{x^2}{2}} = 0$ when $x = 0$.

Test $f'(x)$ at $x = -1, x = 1$.

$f'(-1) = \dfrac{-(-1)}{\sqrt{2\pi}}e^{-\frac{(-1)^2}{2}} = \dfrac{1}{\sqrt{2\pi}}e^{-\frac{1}{2}} > 0$

$f'(1) = \dfrac{-1}{\sqrt{2\pi}}e^{-\frac{(1)^2}{2}} = \dfrac{-1}{\sqrt{2\pi}}e^{-\frac{1}{2}} < 0$

$f'(x)$ is positive on $(-\infty, 0)$, so $f(x)$ is increasing.

$f'(x)$ is negative on $(0, \infty)$, so $f(x)$ is decreasing.

53. $C(x) = 4000 - 4x \quad 0 \le x \le 17{,}000$

$R(x) = 20x - \dfrac{x^2}{1000} \quad 0 \le x \le 17{,}000$

Profit function,
$P(x) = R(x) - C(x)$

$= 20x - \dfrac{x^2}{1000} - (4000 - 4x)$

$= -\dfrac{x^2}{1000} + 24x - 4000$

$P'(x) = -\dfrac{x}{500} + 24 = 0$

$x = 12{,}000$

Test $P'(x)$ at $x = 0, x = 17{,}000$

$P'(0) = -\dfrac{0}{500} + 24 = 24 > 0$

$P'(17{,}000) = -\dfrac{17{,}000}{500} + 24 = -10 < 0$

$P'(x)$ is positive on $(0, 12{,}000)$ so $P(x)$ is increasing.

55. $A(x) = .004x^3 - .05x^2 + .16x + .05$,
$0 \le x \le 6$
$A'(x) = .012x^2 - .1x + .16$

Set the derivative equal to zero and solve for x.
$.012x^2 - .1x + .16 = 0$
$x \approx 2.16 \quad$ or $\quad x \approx 6.17$
since $0 \le x \le 6$, $x \approx 2.16$ is the only critical number.
Test $A'(x)$ for $x = 0$, and $x = 3$.
$A'(0) = .012(0)^2 - .1(0) + .16 = .16 > 0$
$A'(3) = .012(3)^2 - .1(3) + .16 = -.032 < 0$

55. Continued

a. $A'(x)$ is positive on $(0, 2.16)$, so $A(x)$, the alcohol concentration, is increasing.

b. $A'(x)$ is negative on $(2.16, 6)$, so $A(x)$ is decreasing.

57. a. The number of accidents is increasing on $(0, 1)$, $(2, 5)$, $(10, 11)$, $(12, 14)$, $(18, 19)$ and $(20, 23)$.

b. The number accidents is decreasing on $(1, 2)$, $(5, 10)$, $(11, 12)$, $(14, 18)$, and $(19, 20)$.

c. Local maxima occur at $t = 1, 5, 11, 14$, and 19; local minima occur at $t = 2, 10, 12, 18$, and 20.

59. a. The minimum wage was about \$8.00 in 1969, \$6.00 in 1982, \$4.65 in 1989, \$5.44 in 1998 and \$4.90 in 2003.

b. Between 1965 and 2003 there are 6 local maxima; they correspond approximately to the years 1969, 1975, 1977, 1979, 1992 and 1998.

c. Between 1980 and 2003 local minima occur around the years 1989 and 1995.

d. Between 1980 and 2000 the minimum wage decreases in the intervals (1980, 1989), (1992, 1995), (1997, 2003).

61. $f(x) = .0361x^3 - .853x^2 + 5.117x + .45$

$f'(x) = .1083x^2 - 1.706x + 5.117$

Set $f'(x) = 0$ and solve for x

$0 = .1083x^2 - 1.706x + 5.117$

$x = \dfrac{-(-1.706) \pm \sqrt{(-1.706)^2 - 4(.1083)(5.117)}}{2(.1083)}$

$x \approx 4.031, 11.722$

$f(1) = 4.750$

$f(4.031) = 9.581$

$f(11.722) = 1.370$

$f(14) = 3.958$

Hence a local maximum occurred in 1944, and a local minimum occurred in 2002.

63. $g(x) = .0058x^4 - .2314x^3 + 3.2917x^2$
$- 19.9399x + 49.0718$

Using a graphing calculator with $5 \leq x \leq 15$, the function has a local maximum when $x = 8.98$, and local minima when $x = 6.74$

and $x = 14.21$. This means that interest rates peaked in late 1998, and reached a low point in late 1996 and early 2004

65. $g(x) = -.00096x^3 - .1x^2 + 11.3x + 1274$

$g'(x) = -.00288x^2 - .2x + 11.3$. $g'(x) = 0$ when

$x = 36.8965$. $g'(30) = 2.708 > 0$,

$g'(40) = -1.308 < 0$. Hence, $x = 36.8965$ is where a local maximum occurs. This corresponds to late 2036, and the population will be

$g(36.8965) \approx 1506.5753$. The population will be about 1,506,575,300.

67. $D(x) = -x^4 + 8x^3 + 80x^2$, $0 \leq x \leq 13$

$D'(x) = -4x^3 + 24x^2 + 160x$

$-4x^3 + 24x^2 + 160x = 0$

$-4x(x^2 - 6x - 40) = 0$

$-4x(x+4)(x-10) = 0$

$x = 0$ or $x = -4$ or $x = 10$

Reject $x = -4$, which is not the interval.

$D'(-1) < 0$, $D'(9) > 0$, $D'(11) < 0$

$D(0)$ is a minimum and $D(10)$ is a maximum. A speaker should aim for a discrepancy of 10 to maximize the attitude change.

69. $R(t) = \dfrac{20t}{t^2 + 100}$, $R'(t) = \dfrac{(t^2 + 100)(20) - 20t(2t)}{(t^2 + 100)^2}$

$= \dfrac{20t^2 - 2000 - 40t^2}{(t^2 + 100)^2}$

$= \dfrac{-20(t^2 - 100)}{(t^2 + 100)^2}$

$R'(t) = 0$ when $t = 10$. This is the only critical number. $R'(9) \approx .0116 > 0$ and

$R'(11) \approx -.0086 < 0$. Hence a local maximum occurs at $t = 10$. The film length that received the highest rating is 10 minutes.

Section 12.2 The Second Derivative

1. $f(x) = x^3 - 6x^2 + 1$

$f'(x) = 3x^2 - 12x$

$f''(x)$ is the derivative of $f'(x)$.

$f''(x) = 6x - 12$

$f''(0) = 6(0) - 12 = -12$

$f''(2) = 6(2) - 12 = 0$

$f''(-3) = 6(-3) - 12 = -30$

3. $f(x) = (x+3)^4$

$f'(x) = 4(x+3)^3(1) = 4(x+3)^3$

$f''(x) = 12(x+3)^2$

$f''(0) = 12(0+3)^2 = 108$

$f''(2) = 12(2+3)^2 = 300$

$f''(-3) = 12(-3+3)^2 = 0$

5. $f(x) = \dfrac{x^2}{1+x}$

$f'(x) = \dfrac{(1+x)(2x) - x^2(1)}{(1+x)^2}$

$= \dfrac{2x + x^2}{(1+x)^2}$

$f''(x) = \dfrac{(1+x)^2(2+2x) - (2x+x^2)(2)(1+x)}{(1+x)^4}$

$= \dfrac{(1+x)(2+2x) - (2x+x^2)(2)}{(1+x)^3}$

$= \dfrac{2}{(1+x)^3}$

$f''(0) = 2$

$f''(2) = \dfrac{2}{27}$

$f''(-3) = -\dfrac{1}{4}$

7. $f(x) = \sqrt{x+4} = (x+4)^{\frac{1}{2}}$

$f'(x) = \frac{1}{2}(x+4)^{-\frac{1}{2}}$

$f''(x) = \left(-\frac{1}{2}\right)\frac{1}{2}(x+4)^{-\frac{3}{2}}$

$\quad = -\frac{(x+4)^{-\frac{3}{2}}}{4}$

or $-\frac{1}{4(x+4)^{\frac{3}{2}}}$

$f''(0) = -\frac{1}{4(0+4)^{\frac{3}{2}}} = -\frac{1}{4(4)^{\frac{3}{2}}}$

$\quad = -\frac{1}{4(8)}$

$\quad = -\frac{1}{32}$

$f''(2) = -\frac{1}{4(2+4)^{\frac{3}{2}}}$

$\quad = -\frac{1}{4(6)^{\frac{3}{2}}}$

$f''(-3) = -\frac{1}{4(-3+4)^{\frac{3}{2}}} = -\frac{1}{4(1)^{\frac{3}{2}}}$

$\quad = -\frac{1}{4}$

9. $f(x) = 5x^{\frac{4}{5}}$

$f'(x) = 4x^{-\frac{1}{5}}$

$f''(x) = -\frac{4}{5}x^{-\frac{6}{5}}$ or $-\frac{4}{5x^{\frac{6}{5}}}$

$f''(0)$ does not exist.

$f''(2) = -\frac{4}{5\left(2^{\frac{6}{5}}\right)} = -\frac{4}{5(2)\left(2^{\frac{1}{5}}\right)} = -\frac{2^{\frac{4}{5}}}{5}$

$f''(-3) = -\frac{4}{5}(-3)^{-\frac{6}{5}} = -\frac{4}{5(-3)(-3)^{\frac{1}{5}}} = \frac{4}{15(-3)^{\frac{1}{5}}}$

11. $f(x) = 2e^x$

$f'(x) = 2e^x$

$f''(x) = 2e^x$

$f''(0) = 2e^0 = 2(1) = 2$

$f''(2) = 2e^2$

$f''(-3) = 2e^{-3}$ or $\frac{2}{e^3}$

13. $f(x) = 6e^{2x}$

$f'(x) = 6e^{2x}(2) = 12e^{2x}$

$f''(x) = 12e^{2x}(2) = 24e^{2x}$

$f''(0) = 24e^0 = 24$

$f''(2) = 24e^4$

$f''(-3) = 24e^{-6} = \frac{24}{e^6}$

15. $f(x) = \ln|x|$

$f'(x) = \frac{1}{x} = x^{-1}$

$f''(x) = -x^{-2} = -\frac{1}{x^2}$

$f''(0) = -\frac{1}{0^2}$ does not exist

$f''(2) = -\frac{1}{2^2} = -\frac{1}{4}$

$f''(-3) = -\frac{1}{(-3)^2} = -\frac{1}{9}$

17. $f(x) = x\ln|x|$

$f'(x) = (1)\ln|x| + x\left(\frac{1}{x}\right)$

$\quad = \ln|x| + 1$

$f''(x) = \frac{1}{x}$

$f''(0) = \frac{1}{0}$ does not exist

$f''(2) = \frac{1}{2}$

$f''(-3) = -\frac{1}{3}$

19. As the price $P(t)$ decreases, the slope of the graph is negative, so $P'(t)$ is negative. Since the price is decreasing faster and faster, the graph resembles the right-hand side of the graph in Figure 12.23(a). Here the slope is decreasing, so the second derivative $P''(t)$ is negative.

21. $s(t) = 6t^2 + 5t$

$v(t) = s'(t) = 12t + 5$

$a(t) = v'(t) = s''(t) = 12$

$v(0) = 12(0) + 5 = 5$ cm/sec

$v(4) = 12(4) + 5 = 53$ cm/sec

$a(0) = 12$ cm/sec^2

$a(4) = 12$ cm/sec^2

23. $s(t) = 3t^3 - 4t^2 + 8t - 9$

$v(t) = s'(t) = 9t^2 - 8t + 8$

$a(t) = v'(t) = s''(t) = 18t - 8$

$v(0) = 9(0)^2 - 8(0) + 8 = 8$ cm/sec

$v(4) = 9(4)^2 - 8(4) + 8 = 120$ cm/sec

$a(0) = 18(0) - 8 = -8$ cm/sec^2

$a(4) = 18(4) - 8 = 64$ cm/sec^2

25. $f(x) = x^3 + 3x - 5$

$f'(x) = 3x^2 + 3$

$f''(x) = 6x$

$f''(x) > 0$ if $x > 0$

$f''(x) < 0$ if $x < 0$

so $f''(x) > 0$ on $(0, \infty)$ and $f(x)$ is concave upward.

$f''(x) < 0$ on $(-\infty, 0)$ and $f(x)$ is concave downward. $f''(x) = 0$ when $x = 0$ and so there is a point of inflection at $(0, -5)$.

27. $f(x) = x^3 + 4x^2 - 6x + 3$

$f'(x) = 3x^2 + 8x - 6$

$f''(x) = 6x + 8$

f is concave upward when

$f''(x) = 6x + 8 > 0$

$x > -\dfrac{4}{3}$

f is concave downward when

$f''(x) = 6x + 8 < 0$

$x < -\dfrac{4}{3}$

point of inflection at $x = -\dfrac{4}{3}$

$f\left(-\dfrac{4}{3}\right) = \left(-\dfrac{4}{3}\right)^3 + 4\left(-\dfrac{4}{3}\right)^2 - 6\left(-\dfrac{4}{3}\right) + 3$

$= \dfrac{425}{27}$

so f is concave upward on $\left(-\dfrac{4}{3}, \infty\right)$, concave

downward on $\left(-\infty, -\dfrac{4}{3}\right)$, and has a point of

inflection at $\left(-\dfrac{4}{3}, \dfrac{425}{27}\right)$.

29. $f(x) = \dfrac{2}{x - 4}$

$f'(x) = \dfrac{-2}{(x - 4)^2}$

$f''(x) = \dfrac{4}{(x - 4)^3}$

$\dfrac{4}{(x - 4)^3} > 0$ when $(x - 4)^3 > 0$

$x > 4$

$\dfrac{4}{(x - 4)^3} < 0$ when $(x - 4)^3 < 0$

$x < 4$

$f(4) = \dfrac{2}{4 - 4} = \dfrac{2}{0}$ which does not exist.

So f is concave upward on $(4, \infty)$, concave downward on $(-\infty, 4)$, and has no points of inflection.

31. $f(x) = x^4 + 8x^3 - 30x^2 + 24x - 3$

$f'(x) = 4x^3 + 24x^2 - 60x + 24$

$f''(x) = 12x^2 + 48x - 60$

$f''(x) > 0$ when $12x^2 + 48x - 60 > 0$

$12\left(x^2 + 4 - 5\right) > 0$

$12(x + 5)(x - 1) > 0$

$x + 5 > 0$ and $x - 1 > 0$, or $x + 5 < 0$ and $x - 1 < 0$

$x > -5$ and $x > 1$, or $x < -5$ and $x < 1$

$x > 1$ or $x < -5$

$f''(x) < 0$ when $12x^2 + 48x - 60 < 0$

$12\left(x^2 + 4 - 5\right) < 0$

$12(x + 5)(x - 1) < 0$

$x + 5 > 0$ and $x - 1 < 0$, or $x + 5 < 0$ and $x - 1 > 0$

$x > -5$ and $x < 1$, or $x < -5$ and $x > 1$

$-5 < x < 1$

$f(-5) = (-5)^4 + 8(-5)^3 - 30(-5)^2 + 24(-5) - 3$

$= -1248$

$f(1) = (1)^4 + 8(1)^3 - 30(1)^2 + 24(1) - 3 = 0$

So f is concave upward on $(-\infty, -5)$ and $(1, \infty)$, concave downward on $(-5, 1)$, and has points of inflection at $(-5, -1248)$ and $(1, 0)$.

33. $R(x) = 10,000 - x^3 + 42x^2 + 800x, \ 0 \le x \le 20$

$R'(x) = -3x^2 + 84x + 800$

$R''(x) = -6x + 84$

$R''(x) = 0$ when $-6x + 84 = 0$

$\qquad\qquad x = 14$

$R(14) = 10,000 - (14)^3 + 42(14)^2 + 800(14)$

$\qquad = 26,688$

So the point of diminishing returns is (14, 26,688).

35. $f(x) = -2x^3 - 3x^2 - 72x + 1$

$f'(x) = -6x^2 - 6x - 72$

$f''(x) = -12x - 6$

$f'(x) = 0$ means

$-6x^2 - 6x - 72 = 0$

$-6(x^2 + x + 12) = 0$

but the solutions to $x^2 + x + 12 = 0$ are not real numbers. Because there are no critical numbers, there are no local maxima or minima.

37. $f(x) = x^3 + \dfrac{3}{2}x^2 - 60x + 100$

$f'(x) = 3x^2 + 3x - 60$

Solve $f'(x) = 0$.

$3x^2 + 3x - 60 = 0$

$3(x^2 + x - 20) = 0$

$3(x + 5)(x - 4) = 0$

$x + 5 = 0 \quad$ or $\quad x - 4 = 0$

$\quad x = -5 \quad$ or $\qquad x = 4$

Use the second derivative test.

$f''(x) = 6x + 3$

$f''(-5) = 6(-5) + 3 = -27 < 0$

$f''(4) = 6(4) + 3 = 27 > 0$

Thus, $x = -5$ gives a local maximum and $x = 4$ gives a local minimum.

39. $f(x) = x^4 - 8x^2$

$f'(x) = 4x^3 - 16x$

Solve for $f'(x) = 0$.

$4x^3 - 16x = 0$

$4x(x^2 - 4) = 0$

$4x = 0 \quad$ or $\quad x^2 - 4 = 0$

$\qquad\qquad\qquad x^2 = 4$

$x = 0 \quad$ or $\quad x = 2 \quad$ or $\quad x = -2$

Use the second derivative test.

$f''(x) = 12x^2 - 16$

$f''(0) = -16 < 0$

$f''(2) = 48 - 16 = 32 > 0$

$f''(-2) = 48 - 16 = 32 > 0$

Thus, $x = 0$ gives a local maximum, and $x = 2$ and $x = -2$ give local minimum.

41. $f(x) = x + \dfrac{4}{x}$

$\qquad = x + 4x^{-1}$

$f'(x) = 1 - 4x^{-2}$

Solve $f'(x) = 0$.

$1 - 4x^{-2} = 0$

$1 - \dfrac{4}{x^2} = 0$

$1 = \dfrac{4}{x^2}$

$x^2 = 4$

$x = \pm 2$

Use the second derivative test.

$f''(x) = 8x^{-3} = \dfrac{8}{x^3}$

$f''(2) = \dfrac{8}{(2)^3} = \dfrac{8}{8} = 1 > 0$

$f''(-2) = \dfrac{8}{(-2)^3} = \dfrac{8}{-8} = -1 < 0$

Thus, $x = 2$ gives a local minimum and $x = -2$ gives a local maximum.

43. $f(x) = \dfrac{x^2 + 9}{2x}$

$f'(x) = \dfrac{(2x)(2x) - (2)\left(x^2 + 9\right)}{(2x)^2}$

$= \dfrac{4x^2 - 2x^2 - 18}{4x^2}$

$= \dfrac{2x^2 - 18}{4x^2}$

$= \dfrac{x^2 - 9}{2x^2}$

Solve $f'(x) = 0$.

$\dfrac{x^2 - 9}{2x^2} = 0$

$\dfrac{(x+3)(x-3)}{2x^2} = 0$

$(x+3)(x-3) = 0$

$x = 3 \quad \text{or} \quad x = -3$

Use the second derivative test.

$f''(x) = \dfrac{(4x)\left(4x^2\right) - (8x)\left(2x^2 - 18\right)}{\left(4x^2\right)^2}$

$= \dfrac{144x}{16x^4} = \dfrac{9}{x^3}$

$f''(3) = \dfrac{9}{(3)^3} = \dfrac{9}{27} = \dfrac{1}{3} > 0$

$f''(-3) = \dfrac{9}{(-3)^3} = -\dfrac{9}{27} = -\dfrac{1}{3} < 0$

Thus $x = 3$ gives a local minimum and $x = -3$ gives a local maximum.

45. $f(x) = \dfrac{2 - x}{2 + x}$

$f'(x) = \dfrac{(-1)(2 + x) - (1)(2 - x)}{(2 + x)^2}$

$= \dfrac{-2 - x - 2 + x}{(2 + x)^2}$

$= \dfrac{-4}{(2 + x)^2}$

$f'(x)$ does not equal 0 for any value. $f'(-2)$ does not exist, but -2 is not a critical number because $f(-2)$ is undefined. Thus, there are no critical numbers and, consequently, there can be no local maxima or minima.

47. $f'(x) = (x - 1)(x - 2)(x - 4)$

$= x^3 - 7x^2 + 14x - 8$

$f'(x) = 0$ when $x = 1$ or $x = 2$ or $x = 4$

$f''(x) = 3x^2 - 14x + 14$

$f''(1) = 3(1)^2 - 14(1) + 14 = 3 > 0$

$f''(2) = 3(2)^2 - 14(2) + 14 = -2 < 0$

$f''(4) = 3(4)^2 - 14(4) + 14 = 6 > 0$

$f''(x) = 0$ when $3x^2 - 14x + 14 = 0$

$x = \dfrac{-(-14) \pm \sqrt{\left(-14^2\right) - 4(3)(14)}}{2(3)}$

$= \dfrac{14 \pm \sqrt{28}}{6} = \dfrac{7 \pm \sqrt{7}}{3}$

So f has a local maximum at $x = 2$, local minimum at $x = 1$ and $x = 4$, and points of inflection at $x = \dfrac{7 + \sqrt{7}}{3}$ and $x = \dfrac{7 - \sqrt{7}}{3}$.

49. $f'(x) = (x - 2)^2(x - 1) = x^3 - 5x^2 + 8x - 4$

$f'(x) = 0$ when $x = 2$ or $x = 1$

$f''(x) = 3x^2 - 10x + 8$

$f''(2) = 3(2)^2 - 10(2) + 8 = 0$

$f''(1) = 3(1)^2 - 10(1) + 8 = 1 > 0$

$f''(x) = 0$ when

$3x^2 - 10x + 8 = 0$

$(3x - 4)(x - 2) = 0$

$x = \dfrac{4}{3} \quad \text{or} \quad x = 2$

So f has a local minimum at $x = 1$, and points of inflection at $x = \dfrac{4}{3}$ and $x = 2$.

51. a. $f'(x) > 0$ and $f''(x) > 0$

These conditions indicate a point where the function is increasing and the graph is concave up. Point E satisfies these conditions.

b. $f'(x) < 0$ and $f''(x) > 0$

These conditions indicate a point where the function is decreasing and the graph concave up. Point A satisfies these conditions.

c. $f'(x) = 0$ and $f''(x) < 0$

These conditions indicate a critical point of the function where the graph is concave down. Point C satisfies these conditions.

51. Continued

d. $f'(x) = 0$ and $f''(x) > 0$

These conditions indicate a critical point of the function where the graph is concave up. Point B satisfies these conditions.

e. $f'(x) < 0$ and $f''(x) = 0$

These conditions indicate a point where the function is decreasing and the same point is an inflection point. Point D satisfies these conditions.

52. a. Since the second derivative repeatedly changes sign, the curve is concave up and concave down throughout, which means that the first derivative is alternately increasing and decreasing. However, "decline" means that the first derivative is generally negative.

b. Answers vary.
Possible answer: As age increases, the decrease in ability is not a smooth one.

53. $s(t) = -16t^2$
$v(t) = s'(t) = -32t$

a. $s'(3) = -32(3) = -96$ ft/sec

b. $s'(5) = -32(5) = -160$ ft/sec

c. $s'(8) = -32(8) = -256$ ft/sec

d. $a(t) = s''(t) = -32$ ft/sec^2

55. $N(x) = -3x^3 + 135x^2 + 3600x + 12,000$
$N'(x) = -9x^2 + 270x + 3600$

$N''(x) = -18x + 270$
Solve $N''(x) = 0$ for x.
$-18x + 270 = 0$
$270 = 18x$
$15 = x$
Now $N(15) = 86,250$
Since $N''(1) = 252 > 0$ and $N''(20) = -90 < 0$, the graph changes from concave upward to concave downward at $x = 15$. Therefore there is a point of diminishing returns at the inflection point $(15, 86,250)$. This indicates 86,250 cameras are sold when $15,000 is spent on advertising.

57. $f(x) = -.0013x^3 + .0324x^2 + .198x + 5.8$
$(1 \le x \le 15)$

$f'(x) = -.0039x^2 + .0648x + .198$
$f''(x) = -.0078x + .0648$
$0 = -.0078x + .0648$
$-.0648 = -.0078x$
$\dfrac{-.0648}{-.0078} = x$
$x \approx 8.3$
$f'(8.3) \approx .47 > 0$
$f''(8.0) \approx .0024 > 0$
$f''(8.5) \approx -.0015 < 0$
The point of inflection occurs when $x = 8.3$. In late 1998 the rate of spending begins to slow down (although it is still increasing).

59. $f(x) = .124x^3 - 4.587x^2 + 44.146x + 259.9$ $(0 \le x \le 25)$

a. $f'(x) = .372x^2 - 9.174x + 44.146$
$0 = .372x^2 - 9.174x + 44.146$
$x = \dfrac{-(-9.174) \pm \sqrt{(-9.174)^2 - 4(.372)(44.146)}}{2(.372)}$
$= \dfrac{9.174 \pm \sqrt{18.473}}{.744}$
$x \approx 6.5537$ or $x \approx 18.1076$ are the critical numbers.

b. Test $f'(x)$ at $x = 0$, $x = 10$, and $x = 20$.
$f'(0) = .372(0)^2 - 9.174(0) + 44.146$
$= 44.146 > 0$
$f'(10) = .372(10)^2 - 9.174(10) + 44.146$
$= -10.394 < 0$
$f'(20) = .372(20)^2 - 9.174(20) + 44.146$
$= 9.466 > 0$
Defense spending was at a local maximum in mid-1986 ($x = 6.5537$) and at a local minimum in mid-1998 ($x = 18.1076$).

c. $f''(x) = .744x - 9.174$
$0 = .744x - 9.174$
$9.174 = .744x$
$12.331 = x$
$f(x)$ has an inflection point at $x = 12.331$. This indicates that the rate of change in spending began to increase in early 1992.

61. $f(x) = .315x^3 - 13.78x^2 + 143.2x + 887.2$
$(4 \leq x \leq 24)$
$f'(x) = .945x^2 - 27.56x + 143.2$
Solve $f'(x) = 0$.
$.945x^2 - 27.56x + 143.2 = 0$

$x = \dfrac{-(-27.56) \pm \sqrt{(27.56)^2 - 4(.945)(143.2)}}{2(.945)}$

$= \dfrac{27.56 \pm \sqrt{218.258}}{1.89}$

$x \approx 6.77$ or $x \approx 22.40$ are the critical numbers.
Test $f'(x)$ at $x = 0$,
$x = 10$, and $x = 23$
$f'(0) = .945(0)^2 - 27.56(0) + 143.2$
$= 143.2 > 0$
$f'(10) = .945(10)^2 - 27.56(10) + 143.2$
$= -37.9 < 0$
$f'(23) = .945(23)^2 - 27.56(23) + 143.2$
$= 9.2 > 0$
local maximum occurs at
$x = 6.77$. $f(6.77) \approx 1322.8$ and hence
burglaries peaked in late 1986.

63. $C(x) = 250x + \dfrac{16,000}{x} + 1000$

$(1 \leq x \leq 30)$

$C'(x) = 250 - \dfrac{16,000}{x^2}$

Solve $C'(x) = 0$.

$250 - \dfrac{16,000}{x^2} = 0$

$250x^2 = 16,000$

$x^2 = 64$

$x = \pm 8$

8 is the only critical number in the interval (1, 30).

$C''(x) = 2\left(\dfrac{16,000}{x^3}\right) = \dfrac{32,000}{x^3}$

$C''(8) = \dfrac{32,000}{8^3} = 62.5 > 0$

$x = 8$ gives a local minimum. Check the endpoints of the interval.

$C(1) = 250 + 16,000 + 1000 = 17,250$

$C(8) = 250(8) + \dfrac{16,000}{8} + 1000 = 5000$

$C(30) = 250(30) + \dfrac{16,000}{30} + 1000$

≈ 9033

The company should employ 8 employees full time.

65. $P(x) = \ln\left(-x^3 + 3x^2 + 72x + .0001\right)$

$P'(x) = \dfrac{-3x^2 + 6x + 72}{-x^3 + 3x^2 + 72x - .0001}$

$= \dfrac{3\left(x^2 - 2x - 24\right)}{x^3 - 3x^2 - 72x - .0001}$

Find the critical numbers. Solve $P'(x) = 0$.

$\dfrac{3\left(x^2 - 2x - 24\right)}{x^3 - 3x^2 - 72x - .0001} = 0$

$\dfrac{3(x - 6)(x + 4)}{x^3 - 3x^2 - 72x - .0001} = 0$ from which we the

only critical number $x = 6$. Test $P'(x)$ at $x = 5$ and

$P'(5) = \dfrac{3\left(5^2 - 2 \cdot 5 - 24\right)}{5^3 - 3 \cdot 5^2 - 72 \cdot 5 - .0001}$

$= .0871 > 0$

$x = 7$.

$P'(7) = \dfrac{3\left(7^2 - 2 \cdot 7 - 24\right)}{7^3 - 3 \cdot 7^2 - 72 \cdot 7 - .0001}$

$= -.1071 < 0$

A local maximum occurs at $x = 6$ and the number of units that must be sold are 60,000. The maximum

profit is $P(6) = \ln\left(-6^3 + 3 \cdot 6^2 + 72 \cdot 6 + .0001\right)$

$= 5.780744$

The maximum profit is \$5,780,744.

67. $p(t) = \dfrac{20t^3 - t^4}{1000}$ $(0 \leq t \leq 20)$

$p'(t) = \dfrac{60t^2 - 4t^3}{1000}$

a. $p'(t) = 0$ when
$60t^2 - 4t^3 = 0$
$4t^2(15 - t) = 0$
$t = 0$ or $t = 15$

$p''(t) = \dfrac{120t - 12t^2}{1000}$

$p''(0) = \dfrac{120(0) - 12(0)^2}{1000} = 0$

$p''(15) = \dfrac{120(15) - 12(15)^2}{1000} = -.9 < 0$

So p has a local maximum at 15, thus the percent of the population infected is a maximum at 15 days.

67. **Continued**

b. $p(15) = \dfrac{20(15)^3 - (15)^4}{1000}$

$= \dfrac{16,875}{1000} = 16.875$

so the maximum percent of the population infected is 16.875%.
So the maximum percent of the population infected is about 17.

Section 12.3 Optimization Applications

1. On $[0, 4]$ f has an absolute maximum at $x = 4$ and an absolute minimum at $x = 1$.

3. On $[-4, 2]$ f has an absolute maximum at $x = 2$ and an absolute minimum at $x = -2$.

5. On $[-8, 0]$ f has an absolute maximum at $x = -4$, and an absolute minimum at $x = -7$ and $x = -2$.

7. $f(x) = x^4 - 32x^2 - 7$ on $[-5, 6]$

$f'(x) = 4x^3 - 64x = 0$

$4x(x^2 - 16) = 0$

$4x(x - 4)(x + 4) = 0$

$x = 0, 4,$ or -4

x	$f(x)$
-5	-182
-4	-263
0	-7
4	-263
6	137

So on $[-5, 6]$ f has an absolute maximum at $x = 6$, and an absolute minimum at $x = -4$ and $x = 4$.

9. $f(x) = \dfrac{8 + x}{8 - x}$ on $[4, 6]$

$f'(x) = \dfrac{(8 - x)(1) - (8 + x)(-1)}{(8 - x)^2}$

$= \dfrac{16}{(8 - x)^2}$

$f'(x)$ is never 0, but is undefined when $x = 8$. Since 8 is not $[4, 6]$, check only 4 and 6.

x	$f(x)$
4	3
6	7

So on $[4, 6]$, f has an absolute maximum at $x = 6$ and an absolute minimum at $x = 4$.

11. $f(x) = \dfrac{x}{x^2 + 2}$ on $[-1, 4]$

$f'(x) = \dfrac{(x^2 + 2)(1) - x(2x)}{(x^2 + 2)^2}$

$= \dfrac{-x^2 + 2}{(x^2 + 2)^2} = 0$

when $x = \sqrt{2}$ or $-\sqrt{2}$.
$-\sqrt{2}$ is not in $[0, 4]$, so check $x = -1, \sqrt{2}, 4$.

x	$f(x)$
-1	$-\dfrac{1}{3}$
$\sqrt{2}$	$\dfrac{\sqrt{2}}{4}$
4	$\dfrac{2}{9}$

So on $[-1, 4]$ f has an absolute maximum at $x = \sqrt{2}$ and an absolute minimum at $x = -1$.

13. $f(x) = (x^2 + 18)^{\frac{2}{3}}$ on [−3, 2]

$$f'(x) = \frac{2}{3}(x^2 + 18)^{-\frac{1}{3}}(2x) = \frac{4x}{3(x^2 + 18)^{\frac{1}{3}}}$$

$f'(x) = 0$ when $x = 0$.

x	$f(x)$
−3	9
0	$3\sqrt[3]{12} \approx 6.87$
2	$22^{\frac{2}{3}} \approx 7.85$

So on [−3, 2] f has an absolute maximum at $x = -3$ and an absolute minimum at $x = 0$.

15. $f(x) = \dfrac{1}{\sqrt{x^2 + 1}}$ on [−1, 1]

$$f(x) = (x^2 + 1)^{-\frac{1}{2}}$$

$$f'(x) = -\frac{1}{2}(x^2 + 1)^{-\frac{3}{2}}(2x)$$

$$= \frac{-x}{(x^2 + 1)^{\frac{3}{2}}} = 0 \qquad \text{when } x = 0$$

x	$f(x)$
−1	$\dfrac{1}{\sqrt{2}}$
0	1
1	$\dfrac{1}{\sqrt{2}}$

So on [−1, 1] f has an absolute maximum at $x = 0$, and an absolute minimum at $x = -1$ and $x = 1$.

17. $f(x) = 2x^3 - 3x^2 - 12x + 1$ on (0, 4)

$f'(x) = 6x^2 - 6x - 12 = 0$. Solve $f'(x) = 0$.
$6(x - 2)(x + 1) = 0$
$x = 2$, or −1. Only $x = 2$ is the only critical number that lies in the given interval.

$f''(x) = 12x - 6$
$f''(2) = 12 > 0$

Hence, f has a local minimum at $x = 2$. By the Critical Point Theorem, the absolute minimum of f on the interval (0,4) occurs at $x = 2$.

19. $g(x) = \dfrac{1}{x} = x^{-1}$ on $(0, \infty)$

$$f'(x) = -x^{-2} = -\frac{1}{x^2}$$

Solve $f'(x) = 0$. $f'(x) = \dfrac{1}{x^2} = 0$ Since $f'(x)$ is never equal to 0 and $f'(x)$ is defined on $(0, \infty)$ there are no absolute extrema.

21. $g(x) = 6x^{\frac{2}{3}} - 4x$ on $(0, \infty)$

$$f'(x) = \frac{4}{x^{\frac{1}{3}}} - 4$$

Solve $f'(x) = 0$. $f'(x) = \dfrac{4}{x^{\frac{1}{3}}} - 4 = 0$.

$\dfrac{4 - 4x^{\frac{1}{3}}}{x^{\frac{1}{3}}} = 0$, $4 = 4x^{\frac{1}{3}}$ and $x = 1$ is the only critical number that lies in the given interval.

$$f''(x) = \frac{-4}{3x^{\frac{4}{3}}}$$

$$f''(1) = \frac{-4}{3} < 0$$

Hence, f has a local maximum at $x = 1$. By the Critical Point Theorem, the absolute maximum of f on the interval $(0, \infty)$ occurs at $x = 1$.

23. The average cost per monitor is $\dfrac{C(x)}{x}$.

$$\frac{C(x)}{x} = \frac{.13x^3 - 70x^2 + 10,000x}{x}$$

$$= .13x^2 - 70x + 10,000$$

$$\left(\frac{C(x)}{x}\right)' = .26x - 70$$

$$0 = .26x - 70$$

$$70 = .26x$$

$$x \approx 269.23, \text{ round down to } 269.$$

$\dfrac{C(x)}{x}$ has a minimum at $x = 269$ monitors.

$$\frac{C(269)}{269} \approx 577$$

Producing 269 monitors will give the lowest average cost per monitor of $577.

25. a. Let $R(x)$ be the revenue generated from selling x radiators.
$$R(x) = 350x$$

b. If $P(x)$ is the profit from x radiators then
$$P(x) = R(x) - C(x)$$
$$= 350x - (600,000 - 25x + .01x^2)$$
$$P(x) = -0.1x^2 + 375x - 600,000$$

c. $P'(x) = 375 - .02x$
$$0 = 375 - .02x$$
$$.02x = 375$$
$$x = 18,750$$

$P(x)$ has one critical value, $x = 18,750$.
$P''(x) = -.02 < 0$
By the Second Derivative Test, $P(x)$ has a maximum at $x = 18,750$.
$P(18,750) = 2,915,625$
The manufacturer should make 18,750 radiators for a maximum profit of $2,915,625.

27. $p(t) = 10te^{-t/8}$ $(0 \le t \le 40)$

a. $p'(t) = 10t\left(-\dfrac{1}{8}\right)e^{-t/8} + 10e^{-t/8}$

$$= e^{-t/8}\left(-\frac{5}{4}t + 10\right)$$

$$0 = e^{-t/8}\left(-\frac{5}{4}t + 10\right)$$

$$0 = -\frac{5}{4}t + 10$$

$$-10 = -\frac{5}{4}t$$

$$8 = t$$

Test $p(t)$ at $x = 0$, $x = 8$, and $x = 40$.
$p(0) = 0$
$p(8) \approx 29.43$
$p(40) \approx 2.70$
The maximum number of people are infected after 8 days.

b. $p(8) = 10(8)e^{-8/8} = 29.43$
The maximum percent of people infected is 29.43%.

29. $d(x) = -.0000038x^3 - .0008x^2$
$$+ .069x + 1.55 \quad (0 \le x \le 50)$$
$$d'(x) = -.0000114x^2 - .00174x + .069$$

Set $d'(x) = 0$ and use the quadratic formula to solve for the critical numbers.

$$x = \frac{-(-.00174) \pm \sqrt{(-.00174)^2 - 4(-.0000114)(.069)}}{2(-.0000114)}$$

$$\approx -185.30 \text{ or } 32.66$$

$x = 32.6646$ is the only critical number that lies in the given interval.

$$d''(x) = -.0000228x - .0017$$
$$d''(32.6646) = -.00248475 < 0$$

Hence, f has a local maximum at $x = 32.6646$. The number of doctors was highest in mid-2002.

31. $g(x) = -.0069x^3 + .617x^2 - 12.6x + 188.4$

$5 \le x \le 34$

$g'(x) = -.0207x^2 + 1.234x - 12.6$

Find the critical numbers.

Set $g'(x) = 0$.

$-.0207x^2 + 1.234x - 12.6 = 0$

$$x = \frac{-(1.234) \pm \sqrt{(1.234)^2 - 4(-.0207)(-12.6)}}{2(-.0207)}$$

$x = 13.08$ or $x = 46.52$

$f(5) = 139.96$

$f(13.08) = 113.71$

Notice that 46.52 is not in the domain. Since $x = 5$ corresponds to the year 1975, $x = 13.03$ corresponds to the year 1983. Hence, the horsepower was lowest in 1983.

33. Let x = the length of a side of the square that is cut from each corner, in feet. Then the width of the box is $3 - 2x$, the length is $8 - 2x$, and the height is x. The height must not be more than $\dfrac{3}{2}$ ft because the width of the cardboard is 3 ft.

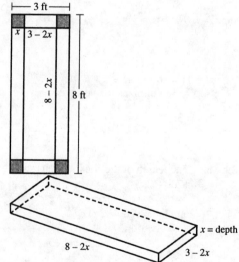

$V(x) = (8 - 2x)(3 - 2x)x$

$\quad = 24x - 22x^2 + 4x^3$

The domain is $\left(0, \dfrac{3}{2}\right)$.

$V'(x) = 24 - 44x + 12x^2$

Solve $V'(x) = 0$.

$24 - 44x + 12x^2 = 0$

$4\left(6 - 11x + 3x^2\right) = 0$

$(3 - x)(2 - 3x) = 0$

33. Continued

$3 - x = 0$ or $2 - 3x = 0$

$x = 3$ or $\qquad x = \dfrac{2}{3}$

Because 3 is not in the domain, the only critical number of interest is $x = \dfrac{2}{3}$. Evaluate $V(x)$ at $x = \dfrac{2}{3}$ and at the endpoints of the domain.

x	$V(x)$	
0	0	
$\dfrac{2}{3}$	$\dfrac{200}{27} = 7\dfrac{11}{27}$	← Maximum
$\dfrac{3}{2}$	0	

The square should be $\dfrac{2}{3}$ ft by $\dfrac{2}{3}$ ft or 8 inches by 8 inches.

35. Let x = a side of the square base. The volume is $16,000$ cm^3 and $V = LWH$, or $H = \dfrac{V}{LW}$, so the height of the box is $\dfrac{16,000}{x^2}$.

The area of each base (top or bottom) is x^2. The area of each of the four sides is

$$x\left(\frac{16,000}{x^2}\right).$$

The cost of both the top and bottom is $3 \cdot 2x^2$. The cost of the other four sides is

$$1.50 \cdot 4x\left(\frac{16,000}{x^2}\right).$$

The total cost is

$$C(x) = 3 \cdot 2x^2 + 1.50 \cdot 4x\left(\frac{16,000}{x^2}\right)$$

$$= 6x^2 + \frac{96,000}{x}$$

$$C'(x) = 12x - \frac{96,000}{x^2} = \frac{12x^3 - 96,000}{x^2}$$

When $C'(x) = 0$,

$$12x^3 - 96,000 = 0$$

$$x = 20.$$

$$C''(x) = 12 + \frac{192,000}{x^3} > 0 \text{ if } x > 0.$$

So $x = 20$ leads to a minimum cost. The dimensions of the box of minimum total cost are 20 cm by 20 cm by

$$\frac{16,000}{20^2} = 40 \text{ cm.}$$

$$C(20) = 6 \cdot 20^2 + \frac{96,000}{20} = 7200$$

The minimum total cost is 7200 cents or $72.

37. Let x = width.
Then
$2x$ = length, and
h = height.
An equation for the volume is

$$36 = (2x)(x)h$$

$$36 = 2x^2 h$$

$$\frac{18}{x^2} = h.$$

The surface area is

$$S(x) = 2x(x) + 2xh + 2(2x)h$$

$$= 2x^2 + 6xh$$

$$= 2x^2 + 6x\left(\frac{18}{x^2}\right)$$

$$= 2x^2 + \frac{108}{x}.$$

$$S'(x) = 4x - \frac{108}{x^2}$$

When $S'(x) = 0$,

$$\frac{4x^3 - 108}{x^2} = 0$$

$$4\left(x^3 - 27\right) = 0$$

$$x = 3.$$

$$S''(x) = 4 + \frac{108(2)}{x^3}$$

$$= 4 + \frac{216}{x^3} > 0$$

since $x > 0$.
So $x = 3$ minimizes the volume.

If $x = 3$, $h = \dfrac{18}{x^2} = \dfrac{18}{9} = 2$.

The dimensions are 3 ft by 6 ft by 2 ft.

39. a. The length of the field in meters is $1200 - 2x$.

b. $A(x) = x(1200 - 2x)$
$\qquad = 1200x - 2x^2$

c. $A'(x) = 1200 - 4x$
$\quad$ When $\quad 1200 - 4x = 0$
$$4(300 - x) = 0$$
$$x = 300.$$
$A''(x) = -4 < 0$ for all x.
The maximum area occurs at $x = 300$ m.

d. $A(300) = 1200(300) - 2(300)^2$
$\qquad = 360,000 - 180,000$
$\qquad = 180,000$

The maximum area is 180,000 sq m.

41. Let x = length of sides that cost $6 per foot.
y = length of sides that cost $3 per foot.

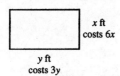

x ft
costs 6x

y ft
costs 3y

An equation for the cost of the fencing is
$2(3y) + 2(6x) = 2400$
$$6y = 2400 - 12x$$
$$y = 400 - 2x.$$

$A = xy$
$A(x) = x(400 - 2x)$
$$= 400x - 2x^2$$
$A'(x) = 400 - 4x$
When $400 - 4x = 0$,
$$x = 100.$$
$A''(x) = -4 < 0$ for all x.
$A(x)$ is maximum when $x = 100$.
If $x = 100$, $y = 400 - 2(100) = 200$.
$A(100) = 100[400 - 2(100)]$
$$= 100(200) = 20,000$$
The maximum area is 20,000 sq ft with 200 ft on the $3 sides and 100 ft on the $6 sides.

43. Let x = the length of the side opposite the existing fence.

The area is 15,625 m^2, so $\dfrac{15,625}{x}$ = the length of each end. The cost of each end is $2\left(\dfrac{15,625}{x}\right)$, and the cost of the side opposite the existing fence is $4x$. The total cost of the new fence for three sides of the rectangular area is

$C(x) = 4x + 2(2)\left(\dfrac{15,625}{x}\right)$

$\qquad = 4x + \dfrac{62,500}{x}$

$C'(x) = 4 - \dfrac{62,500}{x^2}$

When $4 - \dfrac{62,500}{x^2} = 0$,

$4x^2 = 62,500$

$x^2 = 15,625$

$x = \pm 125.$

Only $x = 125$ is in the domain.

$C''(x) = \dfrac{125,000}{x^3} > 0$,

so $x = 125$ gives a minimum cost.

$C(125) = 4(125) + \dfrac{62,500}{125} = 1000$

The least expensive fence costs $1000.

45. a. Price in cents for x thousand bars is

$P(x) = 100 - \dfrac{x}{10}$.

Revenue (in cents) is

$1000x \cdot p(x) = 1000x\left(100 - \dfrac{x}{10}\right)$.

$R(x) = 100,000x - 100x^2$

b. $R'(x) = 100,000 - 200x$
When $100,000 - 200x = 0$,
$$x = 500.$$
$R''(x) = -200 < 0$ for all x.
The maximum revenue is attained when $x = 500$.

c. $R(500) = 100,000(500) - 100(500)^2$
$$= 25,000,000$$

The maximum revenue is 25,000,000 cents, or $250,000.

47. Let x = speed in miles per hour and G = gallons burned per mile.

$$G(x) = \frac{1}{32}\left(\frac{64}{x} + \frac{x}{50}\right)$$

The cost of fuel on the 400-mi trip is $2.88/gal. The total cost in dollars is

$$C(x) = 2.88\left(\frac{1}{32}\right)\left(\frac{64}{x} + \frac{x}{50}\right)(400)$$

$$= 36\left(\frac{64}{x} + \frac{x}{50}\right)$$

$$= \frac{2304}{x} + \frac{18}{25}x$$

a. $C'(x) = \frac{-2304}{x^2} + \frac{18}{25}$

Solve $C'(x) = 0$.

$$\frac{-2304}{x^2} + \frac{18}{25} = 0$$

$$\frac{2304}{x^2} = \frac{18}{25}$$

$$x^2 = 3200$$

$$x = \pm\sqrt{3200}$$

$$x = \pm 40\sqrt{2}$$

$$x \approx 56.6 \text{ or } -56.6$$

Discard the negative solution since speed must be nonnegative.

$$C''(x) = \frac{-2304(-2)}{x^3} > 0 \text{ for all } x.$$

(Recall x must be positive.)

$C(x)$ is minimum at $x = 40\sqrt{2} \approx 56.6$ mph.

b. $C\left(40\sqrt{2}\right) = \frac{2304}{40\sqrt{2}} + \frac{18}{25}\left(40\sqrt{2}\right)$

$$= \frac{288\sqrt{2}}{5}$$

$$\approx 81.46$$

The minimum total cost is $81.46.

49. Distance on shore: $9 - x$ miles
Cost on shore: $400 per mile

Distance underwater: $\sqrt{x^2 + 36}$
Cost underwater: $500 per mile
Find the distance from A, that is, $(9 - x)$, to minimize cost, $C(x)$.

$$C(x) = (9 - x)400 + \left(\sqrt{x^2 + 36}\right)500$$

$$= 3600 - 400x + 500\left(x^2 + 36\right)^{\frac{1}{2}}$$

$$C'(x) = -400 + 500\left(\frac{1}{2}\right)\left(x^2 + 36\right)^{-\frac{1}{2}}(2x)$$

$$= -400 + \frac{500x}{\sqrt{x^2 + 36}}$$

If $C'(x) = 0$,

$$\frac{500x}{\sqrt{x^2 + 36}} = 400$$

$$\frac{5x}{4} = \sqrt{x^2 + 36}$$

$$\frac{25}{16}x^2 = x^2 + 36$$

$$\left(\frac{25}{16} - 1\right)x^2 = 36$$

$$\frac{9}{16}x^2 = 36$$

$$x^2 = \frac{36 \cdot 16}{9}$$

$$x = 64$$

$$x = \pm 8.$$

A distance cannot be negative, so we discard -8. Use the second derivative test to be determined whether the critical number 8 yields a maximum or minimum.

$$C''(x) =$$

$$\frac{\left(x^2 + 36\right)^{\frac{1}{2}}(500) - 500x\left(\frac{1}{2}\right)\left(x^2 + 36\right)^{-\frac{1}{2}}(2x)}{x^2 + 36}$$

$$= \frac{500\left(x^2 + 36\right)^{\frac{1}{2}} - 500x^2\left(x^2 + 36\right)^{-\frac{1}{2}}}{x^2 + 36}$$

$C''(8) > 0$, so $x = 8$ produces the minimum total cost. To yield the minimum total cost, the distance is $9 - x = 9 - 8 = 1$ mile from point A.

51. x = number of batches per year

$M = 100,000$ units produced annually

$k = \$1$, cost to store one unit for one year

$a = \$500$, cost to set up, or fixed cost

$$x = \sqrt{\frac{kM}{2a}} = \sqrt{\frac{(1)(100,000)}{2(500)}}$$

$$= \sqrt{100} = 10$$

Each year, 10 batches should be produced.

53. $k = \$3$, cost to store for one year

$a = \$7$, cost to set up, or fixed cost

$M = 16,800$ units produced annually

$$x = \sqrt{\frac{kM}{2a}} = \sqrt{\frac{(3)(16,800)}{(2)(7)}}$$

$$= \sqrt{3600} = 60$$

Each year, 60 orders should be placed.

55. x = number of orders per year

$k = \$.50$, cost to store for one year

$M = 100,000$, annual demand

$a = \$60$, cost to place an order (fixed cost)

$$x = \sqrt{\frac{kM}{2a}}$$

$$= \sqrt{\frac{.50(100,000)}{2(60)}} \approx 20 \text{ orders}$$

Since the annual demand is 100,000, and the number of orders is 20, the optimum number of copies per order is $\frac{100,000}{20} = 5000$ copies per order.

57. The formula for the economic order quantity (which is another name for the economic lot size) finds the number of batches to be ordered depending on the cost of storing one unit per year, a fixed cost of placing an order, and a known number of units demanded. The best answer is (c). Periodic demand for the goods is known.

Section 12.4 Curve Sketching

1. $f(x) = -x^2 - 10x - 25$

The y-intercept is $f(0) = -25$, but $(0, -25)$ is not a convenient point to plot. To find any x-intercepts, let $y = 0$.

$$0 = -x^2 - 10x - 25$$

$$0 = -1\left(x^2 + 10x + 25\right)$$

$$0 = -1(x + 5)^2$$

$$0 = (x + 5)^2$$

$$0 = x + 5$$

$$x = -5$$

The only x-intercept is -5. This is a polynomial function, so there are no asymptotes.

Find $f'(x)$ and set it equal to zero to find critical numbers.

$$f'(x) = -2x - 10$$

$$= -2(x + 5) = 0$$

Critical number: $x = -5$

$f''(x) = -2 < 0$ for all x

$f(-5) = 0$

The graph is concave downward on $(-\infty, \infty)$. Hence there is a local maximum of 0 at $x = -5$.

There are no points of inflection. The graph is a parabola opening downward. The local maximum (which is also the absolute maximum) occurs at the vertex $(-5, 0)$. Plot some additional points and sketch the graph.

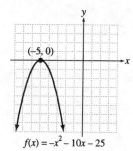

$f(x) = -x^2 - 10x - 25$

3. $f(x) = 3x^3 - 3x^2 + 1$

The y-intercept is $f(0) = 1$. It is not convenient to solve $f(x) = 0$ to find any x-intercepts.

$$f'(x) = 9x^2 - 6x$$
$$= 3x(3x - 2) = 0$$

Critical numbers: $x = 0$ or $x = \dfrac{2}{3}$

$$f''(x) = 18x - 6$$
$$f''(0) = -6 < 0$$
$$f''\left(\frac{2}{3}\right) = 6 > 0$$

There is a local maximum of 1 at 0 and a local minimum of $\dfrac{5}{9}$ at $\dfrac{2}{3}$.

$$f''(x) = 18x - 6 = 0$$
$$x = \frac{1}{3}$$
$$f''(0) = -6 < 0$$
$$f''(1) = 12 > 0$$

There is a point of inflection at $\left(\dfrac{1}{3}, \dfrac{7}{9}\right)$.

Plot the critical points, point of inflection, and a few additional points to sketch the graph.

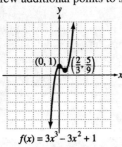

$f(x) = 3x^3 - 3x^2 + 1$

5. $f(x) = -2x^3 - 9x^2 + 108x - 10$

The y-intercept is $f(0) = -10$. It is not convenient to find the x-intercepts.

$$f'(x) = -6x^2 - 18x + 108$$
$$= -6\left(x^2 + 3x - 18\right) = 0$$
$$(x + 6)(x - 3) = 0$$

Critical numbers: $x = -6$ or $x = 3$

$$f''(x) = -12x - 18$$
$$f''(-6) = 54 > 0$$
$$f''(3) = -54 < 0$$

There is a local maximum of 179 at 3 and a local minimum of -550 at -6.

$$f''(x) = -12x - 18 = 0$$
$$x = -\frac{3}{2}$$
$$f''(-2) = 6 > 0$$
$$f''(-1) = -6 < 0$$

There is a point of inflection at $\left(-\dfrac{3}{2}, -\dfrac{371}{2}\right)$.

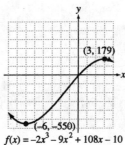

$f(x) = -2x^3 - 9x^2 + 108x - 10$

7. $f(x) = 2x^3 + \dfrac{7}{2}x^2 - 5x + 3$

The y-intercept is $f(0) = 3$. It is not convenient to find any possible x-intercepts.

$f'(x) = 6x^2 + 7x - 5 = 0$

$(2x - 1)(3x + 5) = 0$

Critical numbers: $x = \dfrac{1}{2}$ or $x = -\dfrac{5}{3}$

$f''(x) = 12x + 7$

$f''\left(\dfrac{1}{2}\right) = 13 > 0$

$f''\left(-\dfrac{5}{3}\right) = -13 < 0$

There is a local maximum of $\dfrac{637}{54}$ at $-\dfrac{5}{3}$ and a local minimum of $\dfrac{13}{8}$ and $\dfrac{1}{2}$.

$f''(x) = 12x + 7 = 0$

$x = -\dfrac{7}{12}$

$f''(-1) = -5 < 0$

$f''(0) = 7 > 0$

There is a point of inflection at $\left(-\dfrac{7}{12}, \dfrac{2899}{432}\right)$.

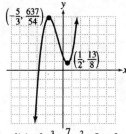

$f(x) = 2x^3 + \dfrac{7}{2}x^2 - 5x + 3$

9. $f(x) = (x + 3)^4$

The y-intercept is $f(0) = 3^4 = 81$, but $(0, 81)$ is not a convenient point to plot. Because the right-hand side is in factored form, it is easy to find any x-intercepts.

$0 = (x + 3)^4$

$0 = x + 3$

$x = -3$

The only x-intercept is -3.

$f'(x) = 4(x + 3)^3 (1)$

$\qquad = 4(x + 3)^3 = 0$

Critical number: $x = -3$

$f''(x) = 12(x + 3)^2$

$f''(-3) = 12(-3 + 3)^2 = 0$

The second derivative test fails, so use the first derivative test.

$f'(-2) = 4(-2 + 3)^3 = 4(1)^3 = 4 > 0$

$f'(-4) = 4(-4 + 3)^3 = 4(-1)^3 = -4 < 0$

$f(x)$ is increasing on $(-3, \infty)$ and decreasing on $(-\infty, -3)$. Thus, there is a local minimum at $x = -3$, which is also an absolute minimum.

$f''(x) = 12(x + 3)^2 > 0$ for all x, so the graph is always concave upward and there is no point of inflection.

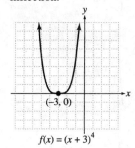

$f(x) = (x + 3)^4$

11. $f(x) = x^4 - 18x^2 + 5$

The y-intercept is $f(0) = 5$. It is not convenient to find the x-intercepts.

$f'(x) = 4x^3 - 36x$

$\qquad = 4x(x^2 - 9) = 0$

The critical numbers are 0, 3, and –3.

$f''(x) = 12x^2 - 36$

$f''(0) = -36 < 0$

$f''(3) = 72 > 0$

$f''(-3) = 72 > 0$

There is a local maximum of 5 at $x = 0$ and a local minimum of –76 at $x = 3$ and $x = -3$.

$f''(x) = 12x^2 - 36 = 0$

$\qquad x^2 = 3$

$\qquad x = \pm\sqrt{3}$

$f''(-2) = 12 > 0$

$f''(-1) = -24 < 0$

There is a point of inflection at $\left(-\sqrt{3}, -40\right)$.

$f''(1) = -24$

$f''(2) = 12$

There is another point of inflection at $\left(\sqrt{3}, -40\right)$.

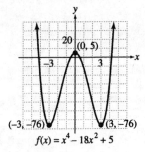

$f(x) = x^4 - 18x^2 + 5$

13. $f(x) = x - \dfrac{1}{x}$

Step 1 There is no y-intercept since $f(0)$ is undefined. Find any x-intercepts by solving the equation $f(x) = 0$.

$f(x) = x - \dfrac{1}{x} = 0$

$x\left(x - \dfrac{1}{x}\right) = x(0)$

$\qquad x^2 - 1 = 0$

$(x+1)(x-1) = 0$

$x = -1 \quad \text{or} \quad x = 1$

The x-intercepts are –1 and 1.

Step 2 f is a rational function. To find any vertical asymptotes, rewrite the function rule so that we have a single fraction.

$f(x) = x - \dfrac{1}{x} = \dfrac{x^2 - 1}{x}$

When $x = 0$, the denominator is 0 but the numerator is not, so the line $x = 0$ (the y-axis) is a vertical asymptote.

Step 3 Find the first derivative and use it to look for critical numbers.

$f(x) = x - \dfrac{1}{x} = x - x^{-1}$

$f'(x) = 1 - (-1)x^{-2}$

$\qquad = 1 + x^{-2}$

$\qquad = 1 + \dfrac{1}{x^2}$

The equation

$f'(x) = 1 + \dfrac{1}{x^2} = 0$

has no real solution.

Although $f'(0)$ does not exist, 0 is not a critical number because $f(0)$ also does not exist. Thus, there are no critical numbers and consequently no local extrema.

The vertical asymptote at $x = 0$ (the y-axis) separates the graph into two branches. Note that

$f'(x) = 1 + \dfrac{1}{x^2}$

is always positive, so the graph is increasing on both $(-\infty, 0)$ and $(0, \infty)$.

13. Continued

Step 4 We now consider the second derivative.

$$f''(x) = -2x^{-3}$$

$$= -\frac{2}{x^3}$$

The equation

$$f''(x) = -\frac{2}{x^3} = 0$$

has no solution. Although $f''(x)$ does not exist at $x = 0$, there cannot be a point of inflection because $f(0)$ does not exist. (Recall that $x = 0$ is a vertical asymptote.) Thus, the graph has no point of inflection. Although there is no point of inflection, concavity may also change at a vertical asymptote, so we examine the second derivative on the two open intervals determined by the asymptote. We will test $x = -1$ in $(-\infty, 0)$ and $x = 1$ in $(0, \infty)$.

$$f''(-1) = -\frac{2}{(-1)^3} = 2 > 0$$

$$f''(1) = -\frac{2}{1^3} = -2 < 0$$

Thus, the graph is concave upward on $(-\infty, 0)$ and concave downward on $(0, \infty)$.

Step 5 To find any horizontal asymptote, consider the following limit.

$$\lim_{x \to \infty} x - \frac{1}{x} = \lim_{x \to \infty} \frac{x^2 - 1}{x}$$

$$= \lim_{x \to \infty} \frac{\frac{x^2}{x^2} - \frac{1}{x^2}}{\frac{x}{x^2}}$$

$$= \lim_{x \to \infty} \frac{1 - \frac{1}{x^2}}{\frac{1}{x}}$$

$$= \frac{1 - 0}{0} = \frac{1}{0}$$

Because division by 0 is undefined, this limit does not exist. Similarly, $\lim_{x \to \infty} f(x)$ does not exist.

Therefore, the graph has no horizontal asymptote.

Notice that as $x \to \infty$, $\frac{1}{x} \to 0$, so that

$$f(x) = x - \frac{1}{x} \approx x.$$

Likewise, as $x \to -\infty$,

$$f(x) = x - \frac{1}{x} \approx x.$$

This means that as x gets larger and larger in absolute value, the graph gets closer and closer to the line $y = x$, which we call an oblique or slant asymptote.

Step 6 All of the information we have gathered about this function will guide us in sketching the graph. Because there are no local extrema or points of inflection and also no y-intercept, the only points we have found are the x-intercepts -1 and 1. A few additional points should be plotted on each branch of the curve. The asymptotes are very helpful because they serve as guidelines for the graph.

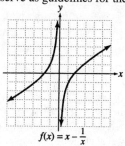

$$f(x) = x - \frac{1}{x}$$

15. $f(x) = \dfrac{x^2 + 25}{x}$

$f(0)$ is undefined, so there is no y-intercept.

$f(x) = 0$ has no real solution because $x^2 + 25$ is always positive. Thus, there are no x-intercepts. When $x = 0$, the denominator is 0 but the numerator is not, so the line $x = 0$ (the y-axis) is a vertical asymptote. We now consider the first derivative and look for critical numbers.

$$f'(x) = \frac{x(2x) - (x^2 + 25)}{x^2}$$

$$= \frac{x^2 - 25}{x^2}$$

$$f'(x) = \frac{(x + 5)(x - 5)}{x^2} = 0$$

$$(x + 5)(x - 5) = 0$$

$$x = -5 \quad \text{or} \quad x = 5$$

Although $f'(0)$ does not exist, 0 is not a critical number because $f(0)$ also does not exist.

Use the second derivative test to determine if there are local maxima or minima at the critical numbers -5 and 5.

$$f''(x) = \frac{x^2(2x) - 2x(x^2 - 25)}{x^4}$$

$$= \frac{50}{x^3}$$

$$f''(-5) = -\frac{2}{5} < 0$$

$$f''(5) = \frac{2}{5} > 0$$

15. Continued

Thus, there is a local maximum of –10 at $x = -5$ and a local minimum of 10 at $x = 5$. f is increasing on $(-\infty, -5)$ and $(5, \infty)$ and decreasing on $(-5, 0)$ and $(0, 5)$.

$$f''(x) = \frac{50}{x^3} \neq 0 \text{ for any } x,$$

so there are no points of inflection. This graph changes concavity at the vertical asymptote $x = 0$. Earlier we found that $f''(-5) < 0$ and $f''(5) > 0$. This tells us that the graph is concave downward on $(-\infty, 0)$ and concave upward on $(0, \infty)$.

Following the method shown in Example 1 and the solution for Exercise 13, we determine that there is no horizontal asymptote. However, if we rewrite the function rule as

$$f(x) = x + \frac{25}{x},$$

we see that as x gets very large in absolute value, $\frac{25}{x} \to 0$, and the graph approaches the line $y = x$, which is an oblique or slant asymptote.

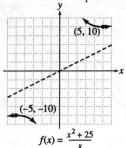

$$f(x) = \frac{x^2 + 25}{x}$$

17. $f(x) = \dfrac{x-1}{x+1}$

$f(0) = -1$, so the y-intercept is -1. $f(x) = 0$ when $x - 1 = 0$ or $x = 1$, so there is one x-intercept, 1. When $x = -1$, the denominator is 0 but the numerator is not, so there is one vertical asymptote, the line $x = -1$.

17. Continued

$$f'(x) = \frac{(x+1)-(x-1)}{(x+1)^2}$$

$$= \frac{2}{(x+1)^2}$$

$f'(x)$ is never zero.

$f'(x)$ fails to exist for $x = -1$, but -1 is not a critical number because $f(-1)$ does not exist. Thus, there are no critical points and there can be no local extrema.

Because $f'(x)$ is always positive where it exists, the function is increasing on $(-\infty, -1)$ and $(-1, \infty)$.

$$f''(x) = \frac{(x+1)^2(0) - 2(2)(x+1)^3}{(x+1)^4}$$

$$= \frac{-4(x+1)^3}{(x+1)^4} = \frac{-4}{x+1}$$

$f''(x)$ is never equal to 0, so there are no points of inflection. The concavity of the graph may change at the vertical asymptote, so we test numbers in the intervals $(-\infty, -1)$ and $(-1, \infty)$.

$$f''(-2) = \frac{-4}{-2+1} = 4 > 0$$

$$f''(0) = \frac{-4}{0+1} = -4 < 0$$

Thus, f is concave upward on $(-\infty, -1)$ and concave downward on $(-1, \infty)$. Because this is a rational function, we now look for a horizontal asymptote.

$$\lim_{x\to\infty} \frac{x-1}{x+1} = \lim_{x\to\infty} \frac{\frac{x}{x} - \frac{1}{x}}{\frac{x}{x} + \frac{1}{x}}$$

$$= \lim_{x\to\infty} \frac{1 - \frac{1}{x}}{1 + \frac{1}{x}}$$

$$= \frac{1-0}{1+0} = 1$$

Likewise, $\lim_{x\to-\infty} f(x) = 1$. Thus, there is a horizontal asymptote at $x = 1$. The vertical and horizontal asymptotes act as guidelines for the graph.

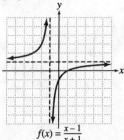

$$f(x) = \frac{x-1}{x+1}$$

19. $y = x - \ln|x|$

$$y' = 1 - \frac{1}{x} = \frac{x-1}{x}$$

$$y'' = \frac{1}{x^2}$$

$y' = 0$ when $x = 1$. There is a vertical asymptote at $x = 0$. The critical numbers are 0 and 1. Test y' at

$$x = -1, \frac{1}{2}, \text{ and } \frac{3}{2}.$$

$y'(-1) = 2$, $y'\left(\frac{1}{2}\right) = -1$, $y'\left(\frac{3}{2}\right) = \frac{1}{3}$. So y is

increasing on $(-\infty, 0)$ and $(1, \infty)$, and decreasing on $(0, 1)$. Since $y'' \neq 0$, there are no points of inflection. Also, y'' is always positive. Hence, the graph is always concave upward. Here is the graph.

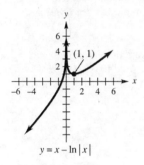

$y = x - \ln|x|$

21. $y = xe^{-x}$

$$y' = (1-x)e^{-x}$$
$$y'' = (x-2)e^{-x}$$

Set $y' = 0$ and solve for x to get the critical numbers. Since $e^{-x} \neq 0$, the only critical number is $x = 1$. Test y' at $x = 0$ and $x = 2$.

$$y'(0) = 1$$
$$y'(2) \approx -e^{-2}$$

y is increasing on $(-\infty, 1)$ and decreasing on $(1, \infty)$.

Set $y'' = 0$ to find points of inflection.

$y'' = 0$ implies $(x-2)e^{-x} = 0$. An inflection point occurs at $x = 2$. Here is the graph.

21. Continued

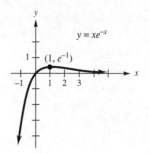

$y = xe^{-x}$

23.

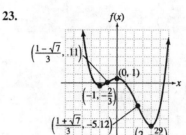

25.

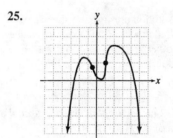

27.

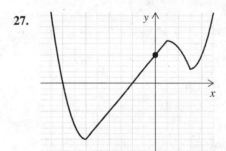

29. a. The first derivative is always positive because the function is increasing.

b. The second derivative is positive because the graph is concave upward. This means that the risk is increasing at a faster and faster rate.

31. The following solution presupposes the use of a TI-83 graphics calculator. Similar results can be obtained from other graphics calculators.

Enter the function $f(x) = .1x^3 - .1x^2 - .005x + 1$ as y_1. Use $-.3 \le x \le 1$ and $.95 \le y \le 1.05$. Under the CALC menu, use "minimum" and "maximum" to find a local maximum of 1.0001 at $x = -.0241$ and a local minimum of .9818 at $x = .6908$. For this function,

$$f'(x) = .3x^2 - .2x - .005$$
$$f''(x) = .6x - .2.$$

If $.6x - .2 = 0$, $x = \dfrac{1}{3}$ and we see an inflection point occurs when $x = \dfrac{1}{3}$.

$$f\left(\frac{1}{3}\right) = \frac{1}{270} - \frac{1}{90} - \frac{5}{3000} + 1 = \frac{5351}{5400}.$$

There is an inflection point at $\left(\dfrac{1}{3}, \dfrac{5351}{5400}\right)$.

33. The following solution presupposes the use of TI-83 graphics calculator. Similar results can be obtained from other graphics calculators.

Enter the function

$f(x) = .01x^5 + x^4 - x^3 - 6x^2 + 5x + 4$ as y_1. First use $-100 \le x \le 100$ and $-20 \le y \le 10,000,000$. Under the CALC menu, use "maximum" to find a local maximum of 8,671,701.6 at $x = -80.7064$. Then change the window to $-3 \le x \le 3$ and $-10 \le y \le 10$. Under the CALC menu, use "minimum" and "maximum" to find another local maximum of 5.0017 at $x = .3982$ and local minima of -8.8191 at $x = -1.6166$ and -1.7461 at $x = 1.9249$.

For this function,

$f'(x) = .05x^4 + 4x^3 - 3x^2 - 12x + 5$ and

$f''(x) = .2x^3 + 12x^2 - 6x - 12$. Enter f'' as y_1. First use $-80 \le x \le 10$ and $-10,000 \le y \le 10,000$. Under the CALC menu use "root" to find that f'' changes sign at $x = -60.4796$. Then change the window to $-2 \le x \le 2$ and $-50 \le y \le 50$. Again, under the CALC menu use "root" to find where f'' changes sign. This occurs at $x = -.7847$ and at $x = 1.2643$. There are points of inflection at these three x-values. Since

$f(-60.4796) = 5,486,563.4,$

$f(-.7847) = -2.7584,$

and $f(1.2643) = 1.2972,$

there are inflection points at $(-60.4796, 5,486,563.4)$, $(-.7847, -2.7584)$, and $(1.2643, 1.2972)$.

Chapter 12 Review Exercises

1. If $f'(x) > 0$ for each x in an interval, then f is increasing on the interval; if $f'(x) < 0$ for each x in an interval, then f is decreasing on the interval.

2. If either $f'(c) = 0$ or $f'(c)$ does not exist, there may be a local extremum at c.
The first derivative test says that $f(c)$ is a local maximum if $f'(a) > 0$ and $f'(b) < 0$. $f(c)$ is a local minimum of $f'(a) < 0$ and $f'(b) > 0$. If $f''(c)$ exists, the second derivative test says that if $f''(c) < 0$, then $f(c)$ is a local maximum. If $f''(c) > 0$, then $f(c)$ is a local minimum.

3. A local extremum is a maximum or minimum value of the function on an open interval. An absolute extremum is the largest or smallest possible value of the function. A local extremum can be an absolute extremum, but not necessarily.

4. The first derivative can be used to determine where a graph is increasing or decreasing. The second derivative can be used to determine where a graph is concave upward or concave downward.

5. $f(x) = x^2 + 9x - 9$
$f'(x) = 2x + 9$

When $f'(x) = 0$,

$$2x + 9 = 0$$
$$x = -\frac{9}{2}.$$

$f''(x) = 2 > 0$ for all x. A local minimum occurs at $x = -\dfrac{9}{2}$. f is increasing on $\left(-\dfrac{9}{2}, \infty\right)$ and decreasing on $\left(-\infty, -\dfrac{9}{2}\right)$.

6. $f(x) = -3x^2 - 3x + 11$
$f'(x) = -6x - 3$

When $f'(x) = 0$,

$$-6x - 3 = 0$$
$$x = -\frac{1}{2}.$$

$f''(x) = -6 < 0$ for x. A local maximum occurs at $x = -\dfrac{1}{2}$. f is increasing on $\left(-\infty, -\dfrac{1}{2}\right)$ and decreasing on $\left(-\dfrac{1}{2}, \infty\right)$.

7. $g(x) = 2x^3 - x^2 - 4x + 7$

$g'(x) = 6x^2 - 2x - 4$

When $g'(x) = 0$

$6x^2 - 2x - 4 = 0$

$2(3x^2 - x - 2) = 0$

$2(3x + 2)(x - 1) = 0$

$3x + 2 = 0$ or $x - 1 = 0$

$x = -\dfrac{2}{3}$ or $x = 1$

Use the second derivative test.

$g''(x) = 12x - 2$

$g''\left(-\dfrac{2}{3}\right) = -10 < 0$

$g''(1) = 10 > 0$

A local maximum occurs at $x = -\dfrac{2}{3}$. A local

minimum occurs at $x = 1$. f is increasing

on $\left(-\infty, -\dfrac{2}{3}\right)$ and $(1, \infty)$, and decreasing on

$\left(-\dfrac{2}{3}, 1\right)$.

8. $g(x) = -4x^3 - 5x^2 + 8x + 1$

$g'(x) = -12x^2 - 10x + 8$

When $g'(x) = 0$,

$-12x^2 - 10x + 8 = 0$

$-2(6x^2 + 5x - 4) = 0$

$-2(3x + 4)(2x - 1) = 0$

$x = -\dfrac{4}{3}$ or $x = \dfrac{1}{2}$

Use the second derivative test.

$g''(x) = -24x - 10$

$g''\left(-\dfrac{4}{3}\right) = 22 < 0$

$g''\left(\dfrac{1}{2}\right) = -22 > 0$

A local maximum occurs at $x = \dfrac{1}{2}$.

A local minimum occurs at $x = -\dfrac{4}{3}$. f is

increasing on $\left(-\dfrac{4}{3}, \dfrac{1}{2}\right)$, and decreasing on both

$\left(-\infty, -\dfrac{4}{3}\right)$ and $\left(\dfrac{1}{2}, \infty\right)$.

9. $f(x) = \dfrac{4}{x - 4} = 4(x - 4)^{-1}$

$f'(x) = -4(x - 4)^{-2} = \dfrac{-4}{(x - 4)^2}$

$f'(x) = 0$ has no solution.

$f'(4)$ does not exist. Although 4 is not a critical number because $f(4)$ is undefined, this number determines the open intervals $(-\infty, 4)$ and $(4, \infty)$ we need to look at. (The graph will have a vertical asymptote at $x = 4$.)

$f'(2) = -1 < 0$

$f'(5) = -4 < 0$

f is never increasing. It is decreasing on $(-\infty, 4)$ and $(4, \infty)$.

10. $f(x) = \dfrac{-6}{3x - 5} = -6(3x - 5)^{-1}$

$f'(x) = -6(-1)(3x - 5)^{-2}(3)$

$= 18(3x - 5)^{-2} = \dfrac{18}{(3x - 5)^2}$

$f'(x) = 0$ has no solution.

$f'(x)$ does not exist if $x = \dfrac{5}{3}$, but this is not a

critical number because $f\left(\dfrac{5}{3}\right)$ is undefined.

There are no critical numbers and there is no local extrema.

Test -1 in $\left(-\infty, \dfrac{5}{3}\right)$ and 2 in $\left(\dfrac{5}{3}, \infty\right)$.

$f'(-1) = \dfrac{9}{2} > 0$

$f'(2) = 18 > 0$

f is never decreasing. It is increasing on $\left(-\infty, \dfrac{5}{3}\right)$

and $\left(\dfrac{5}{3}, \infty\right)$.

11. $f(x) = 2x^3 - 3x^2 - 36x + 10$

$f'(x) = 6x^2 - 6x - 36 = 0$

$6(x^2 - x - 6) = 0$

$(x-3)(x+2) = 0$

$x = 3$ or $x = -2$

Critical numbers: 3, –2

$f''(x) = 12x - 6$

$f''(3) = 30 > 0$, so there is a local minimum at $x = 3$.

$f''(-2) = -30 < 0$, so there is a local maximum at $x = -2$.

$f(-2) = 54$

$f(3) = -71$

f has a local maximum of 54 at $x = -2$ and a local minimum of –71 at $x = 3$.

12. $f(x) = 2x^3 - 3x^2 - 12x + 2$

$f'(x) = 6x^2 - 6x - 12 = 0$

$x^2 - x - 2 = 0$

$(x-2)(x+1) = 0$

–1 and 2 are the critical numbers.

$f''(x) = 12x - 6$

$f''(-1) = -18 < 0$

$f''(2) = 18 > 0$

$f(-1) = 9$

$f(2) = -18$

f has a local maximum of 9 at $x = -1$ and a local minimum of –18 at $x = 2$.

13. $f(x) = x^4 - \dfrac{8}{3}x^3 - 6x^2 + 2$

$f'(x) = 4x^3 - 8x^2 - 12x$

$\qquad = 4x(x^2 - 2x - 3)$

$\qquad = 4x(x-3)(x+1)$

$f'(x) = 0$ when $x = -1, 0,$ or 3, so these are the critical numbers.

$f(-1) = -\dfrac{1}{3}; f(0) = 2; f(3) = -43$

$f''(x) = 12x^2 - 16x - 12$

$f''(3) = 48$

$f''(0) = -12$

$f''(-1) = -16$

A local minimum of $-\dfrac{1}{3}$ occurs at $x = -1$.

A local maximum of 2 occurs at $x = 0$.

A local minimum of –43 occurs at $x = 3$.

14. $f(x) = x \cdot e^x$

Find the derivative and set it equal to 0.

$f'(x) = xe^x + e^x = 0$

$e^x(x+1) = 0$

$x = -1$

The only critical number is –1.

$f''(x) = xe^x + e^x + e^x$

$\qquad = xe^x + 2e^x$

$f''(-1) = -1e^{-1} + 2e^{-1}$

$\qquad = -\dfrac{1}{e} + \dfrac{2}{e}$

$\qquad = \dfrac{1}{e} > 0$

$f(-1) = -1e^{-1}$

$\qquad = -\dfrac{1}{e} \approx -.368$

f has a local minimum of $-\dfrac{1}{e}$ at $x = -1$.

15. $f(x) = 3x \cdot e^{-x}$

$f'(x) = 3x(-e^{-x}) + 3e^{-x}$

$\qquad = (3 - 3x)e^{-x}$

$f''(x) = (3 - 3x)(-e^{-x}) + (-3)e^{-x}$

$\qquad = (3x - 6)e^{-x}$

$f'(x) = 0$ means $(3 - 3x)e^{-x} = 0$ or $3 - 3x = 0$. (e^{-x} is never 0.) $x = 1$ is the critical number.

$f(1) = 3e^{-1} = \dfrac{3}{e}$

$f''(1) = -3e^{-1} < 0$, so $x = 1$ is the location of a local maximum. A local maximum of $\dfrac{3}{e}$ occurs at $x = 1$.

16. $f(x) = \dfrac{e^x}{x-1}$

Find the first derivative and set it equal to zero.

$f'(x) = \dfrac{(x-1)e^x - e^x}{(x-1)^2} = 0$

$\dfrac{e^x(x-1-1)}{(x-1)^2} = 0$

$\dfrac{e^x(x-2)}{(x-1)^2} = 0$

$x - 2 = 0$

$x = 2$

$f''(x) = \dfrac{(x-1)^2\left[e^x + (x-2)e^x\right]}{(x-1)^4}$

$\qquad - \dfrac{e^x(x-2)(2)(x-1)}{(x-1)^4}$

$f''(2) = \dfrac{1^2\left(e^2\right) - 0}{1^4} = e^2 > 0$

$f(2) = \dfrac{e^2}{1} = e^2$

A local minimum of e^2 occurs at $x = 2$.

17. $f(x) = 2x^5 - 5x^3 + 3x - 1$

$f'(x) = 10x^4 - 15x^2 + 3$

$f''(x) = 40x^3 - 30x$

$f''(1) = 40(1)^3 - 30(1) = 10$

$f''(-2) = 40(-2)^3 - 30(-2) = -260$

18. $f(x) = \dfrac{3-2x}{x+3}$

$f'(x) = \dfrac{(x+3)(-2) - (3-2x)(1)}{(x+3)^2}$

$\qquad = \dfrac{-9}{(x+3)^2} = -9(x+3)^{-2}$

$f''(x) = -9(-2)(x+3)^{-3}$

$\qquad = \dfrac{18}{(x+3)^3}$

$f''(1) = \dfrac{18}{(1+3)^3} = \dfrac{18}{64} = \dfrac{9}{32}$

$f''(-2) = \dfrac{18}{(-2+3)^3} = \dfrac{18}{1} = 18$

19. $f(x) = -5e^{2x}$

$f'(x) = -5e^{2x}(2) = -10e^{2x}$

$f''(x) = -10e^{2x}(2) = -20e^{2x}$

$f''(1) = -20e^{2(1)} = -20e^2$

$f''(-2) = -20e^{2(-2)} = -20e^{-4}$

20. $f(x) = \ln|5x+2|$

$f'(x) = \dfrac{1}{5x+2}(5) = 5(5x+2)^{-1}$

$f''(x) = 5(-1)(5x+2)^{-2}(5)$

$\qquad = -\dfrac{25}{(5x+2)^2}$

$f''(1) = -\dfrac{25}{[5(1)+2]^2} = -\dfrac{25}{49}$

$f''(-2) = -\dfrac{25}{[5(-2)+2]^2} = -\dfrac{25}{64}$

21. $f(x) = -2x^3 - \dfrac{1}{2}x^2 - x - 3$

$f'(x) = -6x^2 - x - 1$

Solve $-6x^2 - x - 1 = 0$ for x:

$x = \dfrac{-b \pm \sqrt{b^2 - 4ac}}{2a}$

$\quad = \dfrac{-(-1) \pm \sqrt{(-1)^2 - 4(-6)(-1)}}{2(-6)}$

$\quad = \dfrac{1 \pm \sqrt{1-24}}{-12}$

$\quad = \dfrac{1 \pm \sqrt{-23}}{-12},$

which has no real-number solution. Therefore there are no local extrema.

$f''(x) = -12x - 1$

When $f''(x) = 0$,

$-12x - 1 = 0$

$x = -\dfrac{1}{12}.$

$f(x)$ is concave upward on $\left(-\infty, -\dfrac{1}{12}\right)$ and

concave downward on $\left(-\dfrac{1}{12}, \infty\right)$.

$f\left(-\dfrac{1}{12}\right) = -2\left(-\dfrac{1}{12}\right)^3 - \dfrac{1}{2}\left(-\dfrac{1}{12}\right)^2 - \left(-\dfrac{1}{12}\right) - 3$

$\qquad \approx -2.92$

21. Continued

$\left(-\dfrac{1}{12}, -2.92\right)$ is an inflection point.

When $x = 0$,

$$f(x) = -2(0)^3 - \frac{1}{2}(0)^2 - 0 - 3 = -3 .$$

$f(0) = -3$, so the y-intercept is -3. It is not convenient to find the x-intercept. The function is a polynomial, so there are no asymptotes. Use all of the above information and plot some additional points to sketch the graph.

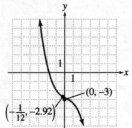

22. $f(x) = -\dfrac{4}{3}x^3 + x^2 + 30x - 7$

$f'(x) = -4x^2 + 2x + 30$

When $f'(x) = 0$,

$-2\left(2x^2 - x - 15\right) = 0$

$(2x + 5)(x - 3) = 0.$

$x = -\dfrac{5}{2}$ and $x = 3$ are critical numbers.

$f\left(-\dfrac{5}{2}\right) = -\dfrac{4}{3}\left(-\dfrac{5}{2}\right)^3 + \left(-\dfrac{5}{2}\right)^2 + 30\left(-\dfrac{5}{2}\right) - 7$

≈ -54.9

$f(3) = -\dfrac{4}{3}(3)^3 + 3^2 + 30(3) - 7$

$= 56$

$\left(-\dfrac{5}{2}, -54.9\right)$ and $(3, 56)$ are critical points.

$f''(x) = -8x + 2$

$f''\left(-\dfrac{5}{2}\right) = 22 > 0$

22. Continued

There is a local minimum at $x = -\dfrac{5}{2}$.

$f''(3) = -22 < 0$

There is a local maximum at $x = 3$. $f(x)$ is increasing on $\left(-\dfrac{5}{2}, 3\right)$ and decreasing on $\left(-\infty, -\dfrac{5}{2}\right)$ and $(3, \infty)$.

When $f''(x) = 0$,

$-8x + 2 = 0$

$x = \dfrac{1}{4} .$

$f(x)$ is concave upward on $\left(-\infty, \dfrac{1}{4}\right)$ and concave downward on $\left(-\dfrac{1}{4}, \infty\right)$.

$f\left(\dfrac{1}{4}\right) = \dfrac{4}{3}\left(\dfrac{1}{4}\right)^3 + \left(\dfrac{1}{4}\right)^2 + 30\left(\dfrac{1}{4}\right) - 7$

$\approx .54$

$\left(\dfrac{1}{4}, .54\right)$ is an inflection point. $f(0) = -7$, so the y-intercept is 0. It is not convenient to find the x-intercept.

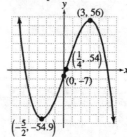

23.

$$f(x) = x^4 - \frac{4}{3}x^3 - 4x^2 + 1$$

$$f'(x) = 4x^3 - 4x^2 - 8x$$

When $f'(x) = 0$,

$$4x(x^2 - x - 2) = 0$$

$$x(x+1)(x-2) = 0$$

-1, 0, and 2 are critical numbers.

$$f(-1) = (-1)^4 - \frac{4}{3}(-1)^3 - 4(-1)^2 + 1$$

$$= -\frac{2}{3}$$

$$f(0) = 1$$

$$f(2) = 2^4 - \frac{4}{3}(2)^3 - 4(2)^2 + 1$$

$$= -\frac{29}{3}$$

$\left(-1, -\frac{2}{3}\right)$, $(0, 1)$, and $\left(2, -\frac{29}{3}\right)$ are critical points.

$$f''(x) = 12x^2 - 8x - 8$$

$$f''(-1) = 12 > 0$$

There is a local minimum at $x = -1$.

$$f''(0) = -8 < 0$$

There is a local maximum at $x = 0$.

$$f''(2) = 24 > 0$$

There is a local minimum at $x = 2$. $f(x)$ is increasing on $(-1, 0)$ and $(2, \infty)$ and decreasing on $(-\infty, -1)$ and $(0, 2)$.

When $f''(x) = 0$,

$$4(3x^2 - 2x - 2) = 0.$$

Solve this equation by the quadratic formula.

$$x = \frac{2 \pm \sqrt{2^2 + 24}}{6} = \frac{2 \pm 2\sqrt{7}}{6} = \frac{1 \pm \sqrt{7}}{3}$$

$f(x)$ is concave upward on

$\left(-\infty, \frac{1-\sqrt{7}}{3}\right)$ and $\left(\frac{1+\sqrt{7}}{3}, \infty\right)$ and concave

downward on $\left(\frac{1-\sqrt{7}}{3}, \frac{1+\sqrt{7}}{3}\right)$.

23. Continued

$$f\left(\frac{1-\sqrt{7}}{3}\right) \approx .11$$

$$f\left(\frac{1+\sqrt{7}}{3}\right) \approx -5.12$$

$\left(\frac{1-\sqrt{7}}{3}, .11\right)$ and $\left(\frac{1+\sqrt{7}}{3}, -5.12\right)$ are

inflection points.

$f(0) = 1$, so the y-intercept is 1. The x-intercepts are not convenient to find but can be estimated from the graph. This is a polynomial function, so there are no asymptotes.

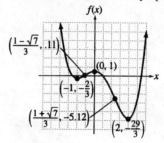

24.

$$f(x) = -\frac{2}{3}x^3 + \frac{9}{2}x^2 + 5x + 1$$

$$f'(x) = -2x^2 + 9x + 5$$

When $f'(x) = 0$,

$$2x^2 - 9x - 5 = 0$$

$$(2x+1)(x-5) = 0.$$

$-\frac{1}{2}$ and 5 are critical numbers.

$$f\left(-\frac{1}{2}\right) \approx -.29$$

$$f(5) \approx 55.17$$

$\left(-\frac{1}{2}, -.29\right)$ and $(5, 55.17)$ are critical points.

$$f''(x) = -4x + 9$$

$$f''\left(-\frac{1}{2}\right) = 11 > 0$$

There is a local minimum at $x = -\frac{1}{2}$.

$$f''(5) = -11 < 0$$

24. Continued

There is a local maximum at $x = 5$.

$f(x)$ is increasing on $\left(-\dfrac{1}{2}, 5\right)$ and decreasing on

$\left(-\infty, -\dfrac{1}{2}\right)$ and $(5, \infty)$.

When $f''(x) = 0$,

$-4x + 9 = 0$

$x = \dfrac{9}{4}.$

$f(x)$ is concave upward on $\left(-\infty, \dfrac{9}{4}\right)$ and concave

downward on $\left(\dfrac{9}{4}, \infty\right)$.

$f\left(\dfrac{9}{4}\right) \approx 27.44$

$\left(\dfrac{9}{4}, 27.44\right)$ is an inflection point. $f(0) = 1$, so the

y-intercept is 1. It is not convenient to find the
x-intercepts.

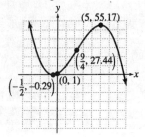

25.

$f(x) = \dfrac{x-1}{2x+1}$

$f'(x) = \dfrac{(2x+1) - (x-1)(2)}{(2x+1)^2}$

$= \dfrac{2}{(2x+1)^2} = 3(2x+1)^{-2}$

$f'(x)$ is never zero. $f'\left(-\dfrac{1}{2}\right)$ does not exist, but

since $f\left(-\dfrac{1}{2}\right)$ is undefined, $-\dfrac{1}{2}$ is not a critical

number. (There is a vertical asymptote at

$x = -\dfrac{1}{2}$.) This function has no critical numbers,

so there can be no local extrema. To determine
where the function is increasing and decreasing,

test points in the intervals $\left(-\infty, -\dfrac{1}{2}\right)$ and

$\left(-\dfrac{1}{2}, \infty\right)$.

$f'(-1) = \dfrac{2}{[2(-1)+1]^2} = 2 > 0$

$f'(0) = \dfrac{2}{[2(0)+1]^2} = 2 > 0$

$f(x)$ is increasing on $\left(-\infty, -\dfrac{1}{2}\right)$ and on

$\left(-\dfrac{1}{2}, \infty\right)$.

$f''(x) = 3(-2)(2x+1)^{-3}(2)$

$= -\dfrac{12}{(2x+1)^3}$

Test $x = -1$ and $x = 0$ in $f''(x)$.

$f''(-1) = -\dfrac{12}{[2(-1)+1]^3} = 12 > 0$

$f(x)$ is concave upward on $\left(-\infty, -\dfrac{1}{2}\right)$.

$f''(0) = -\dfrac{12}{[2(0)+1]^3} = -12 < 0$

25. Continued

$f(x)$ is concave downward on $\left(-\dfrac{1}{2}, \infty\right)$.

$f''(x)$ is never zero, so there are no inflection points.
When $f(x) = 0$,

$$\frac{x-1}{2x+1} = 0$$

$$x - 1 = 0$$

$$x = 1.$$

The only x-intercept is 1.

When $x = 0$,

$$f(x) = \frac{0-1}{2(0)+1} = -1.$$

The y-intercept is -1.
This rational function has a vertical asymptote

where the denominator is 0, that is, at $x = -\dfrac{1}{2}$.

To find the horizontal asymptote, find the limit at infinity.

$$\lim_{x \to \infty} f(x) = \lim_{x \to \infty} \frac{x-1}{2x+1}$$

$$= \lim_{x \to \infty} \frac{\frac{x}{x} - \frac{1}{x}}{\frac{2x}{x} + \frac{1}{x}}$$

$$= \lim_{x \to \infty} \frac{1 - \frac{1}{x}}{2 + \frac{1}{x}}$$

$$= \frac{1-0}{2+0} = \frac{1}{2}$$

$y = \dfrac{1}{2}$ is a horizontal asymptote.

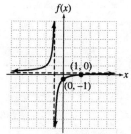

26.

$$f(x) = \frac{2x-5}{x+3}$$

$$f'(x) = \frac{(x+3)(2) - (2x-5)}{(x+3)^2}$$

$$= \frac{11}{(x+3)^2} = 11(x+3)^{-2}$$

$f'(x)$ is never zero. -3 is not a critical number because $f(x)$ does not exist at $x = -3$.
There are no local extrema. Test $x = -4$ and $x = -2$ in $f'(x)$.

$$f'(-4) = \frac{11}{(-4+3)^2} = 11 > 0$$

$$f(-2) = \frac{11}{(-2+3)^2} = 11 > 0$$

$f(x)$ is increasing on $(-\infty, -3)$ and on $(-3, \infty)$.

$$f''(x) = 11(-2)(x+3)^{-3}$$

$$= -22(x+3)^{-3}$$

Test $x = -4$ and $x = -2$ in $f''(x)$.

$$f''(-4) = -22(-4+3)^{-3} = 22 > 0$$

$f(x)$ is concave upward on $(-\infty, -3)$.

$$f''(-2) = -22(-2+3)^{-3} = -22 < 0$$

$f(x)$ is concave downward on $(-3, \infty)$.

$f''(x)$ is never zero, so there are no inflection points.
If $f(x) = 0$,

$$\frac{2x-5}{x+3} = 0$$

$$2x - 5 = 0$$

$$x = \frac{5}{2}.$$

The x-intercept is $\dfrac{5}{2}$.

If $x = 0$,

$$f(x) = \frac{2 \cdot 0 - 5}{0+3}$$

$$= -\frac{5}{3}.$$

The y-intercept is $-\dfrac{5}{3}$.

26. Continued

This rational function has a vertical asymptote at $x = -3$.

$$\lim_{x\to\infty} f(x) = \lim_{x\to\infty}\frac{2x-5}{x+3}$$

$$= \lim_{x\to\infty}\frac{\frac{2x}{x}-\frac{5}{x}}{\frac{x}{x}+\frac{3}{x}}$$

$$= \lim_{x\to\infty}\frac{2-\frac{5}{x}}{1+\frac{3}{x}}$$

$$= \frac{2-0}{1+0} = 2$$

$y = 2$ is a horizontal asymptote.

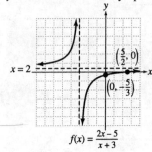

$$f(x) = \frac{2x-5}{x+3}$$

27. $f(x) = -4x^3 - x^2 + 4x + 5$

$f'(x) = -12x^2 - 2x + 4$

When $f'(x) = 0$,

$-2(6x^2 + x - 2) = 0$

$(3x+2)(2x-1) = 0.$

$-\dfrac{2}{3}$ and $\dfrac{1}{2}$ are critical numbers.

$f\left(-\dfrac{2}{3}\right) \approx 3.07$

$f\left(\dfrac{1}{2}\right) = 6.25$

$\left(-\dfrac{2}{3}, 3.07\right)$ and $\left(\dfrac{1}{2}, 6.25\right)$ are critical points.

$f''(x) = -24x - 2$

$f''\left(-\dfrac{2}{3}\right) = 16 > 0$

There is a local minimum at $x = -\dfrac{2}{3}$.

$f''\left(\dfrac{1}{2}\right) = -14 < 0$

27. Continued

There is a local maximum at $x = \dfrac{1}{2}$.

f is increasing on $\left(-\dfrac{2}{3}, \dfrac{1}{2}\right)$ and decreasing on

$\left(-\infty, -\dfrac{2}{3}\right)$ and $\left(\dfrac{1}{2}, \infty\right)$.

When $f''(x) = 0$,

$-24x - 2 = 0$

$x = -\dfrac{1}{12}.$

f is concave upward on $\left(-\infty, -\dfrac{1}{12}\right)$ and concave

downward on $\left(-\dfrac{1}{12}, \infty\right)$.

$f\left(-\dfrac{1}{12}\right) \approx 4.66$

$\left(-\dfrac{1}{12}, 4.66\right)$ is an inflection point. $f(0) = 5$, so

the y-intercept is 5. It is not convenient to find the x-intercept.

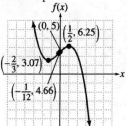

28. $f(x) = x^3 + \dfrac{5}{2}x^2 - 2x - 3$

$f'(x) = 3x^2 + 5x - 2$

$\qquad = (3x-1)(x+2) = 0$

$x = \dfrac{1}{3}$ or $x = -2$

$\dfrac{1}{3}$ and -2 are critical numbers.

$f(-2) = 3$

$f\left(\dfrac{1}{3}\right) \approx -3.35$

$(-2, 3)$ and $\left(\dfrac{1}{3}, -3.35\right)$ are critical points.

28. Continued

$$f''(x) = 6x + 5$$

If $f''(x) = 6x + 5 = 0$

$$x = -\frac{5}{6}$$

$f''\left(\frac{1}{3}\right) = 7 > 0$, so $f(x)$ has a local minimum at

$x = \frac{1}{3}$.

$f''(-2) = -7 < 0$, so $f(x)$ has a local maximum at $x = -2$.

f is increasing on $(-\infty, -2)$ and $\left(\frac{1}{3}, \infty\right)$ and

decreasing on $\left(-2, \frac{1}{3}\right)$. Thus, f is concave

downward on $\left(-\infty, -\frac{5}{6}\right)$ and concave upward on

$\left(-\frac{5}{6}, \infty\right)$.

$f\left(-\frac{5}{6}\right) \approx -.18$

There is a point of inflection at $\left(-\frac{5}{6}, -.18\right)$. The

y-intercept is -3. It is not convenient to find the x-intercepts.

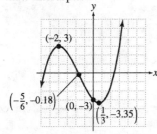

29. $f(x) = x^4 + 2x^2$

$f'(x) = 4x^3 + 4x$

If $f'(x) = 0$,

$4x(x^2 + 1) = 0$.

$x^2 + 1 = 0$ has no real solution, so $x = 0$ is the only critical number. $f(0) = 0$, so $(0, 0)$ is a critical number.

$f''(x) = 12x^2 + 4$

$f''(0) = 4 > 0$

There is a local minimum at $x = 0$. $f(x)$ is increasing on $(0, \infty)$ and decreasing on $(-\infty, 0)$.

$f''(x) = 12x^2 + 4 > 0$ for all x, so $f(x)$ is concave upward on $(-\infty, \infty)$.

$f''(x)$ is never zero, so there are no inflection points.

$f(0) = 0$, so the y-intercept is 0. To find any x-intercepts, let $f(x) = 0$.

$f(x) = x^4 + 2x^2 = 0$

$x^2(x^2 + 2) = 0$

The only real solution is $x = 0$, so the only x-intercept is 0.

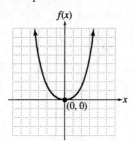

30. $f(x) = 6x^3 - x^4$
$f'(x) = 18x^2 - 4x^3$

When $f'(x) = 0$,
$2x^2(9 - 2x) = 0$.

0 and $\dfrac{9}{2}$ are critical numbers.

$f(0) = 0$

$f\left(\dfrac{9}{2}\right) \approx 136.7$

$(0, 0)$ and $\left(\dfrac{9}{2}, 136.7\right)$ are critical points.

$f''(x) = 36x - 12x^2$
$f''(0) = 0$
There is neither a local minimum nor a local maximum at $x = 0$.

$f''\left(\dfrac{9}{2}\right) = -81 < 0$

There is a local maximum at $x = \dfrac{9}{2}$.

f is increasing on $\left(-\infty, \dfrac{9}{2}\right)$ and decreasing on

$\left(\dfrac{9}{2}, \infty\right)$.

When $f''(x) = 0$,
$36x - 12x^2 = 0$
$12x(3 - x) = 0$
$x = 0$ or $x = 3$
Test $f''(x)$ at $x = -1$, $x = 1$, and $x = 4$.
$f''(-1) = -48 < 0$
f is concave downward on $(-\infty, 0)$.
$f''(1) = 24 > 0$
f is concave upward on $(0, 3)$.
$f''(4) = -48 < 0$
f is concave downward on $(3, \infty)$.
$f(3) = 81$
There are inflection points at $(0, 0)$ and $(3, 81)$.
$f(0) = 0$, so the y-intercept is 0.

To find any x-intercepts, let $f(x) = 0$.
$f(x) = 6x^3 - x^4 = 0$
$x^3(6 - x) = 0$
$x = 0$ or $x = 6$

30. Continued

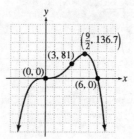

31. $f(x) = \dfrac{x^2 + 4}{x} = x + \dfrac{4}{x}$

$f'(x) = 1 - \dfrac{4}{x^2}$

When $f'(x) = 0$,

$1 - \dfrac{4}{x^2} = 0$

$\dfrac{4}{x^2} = 1$

$4 = x^2$.

-2 and 2 are critical numbers. Notice that $f(x)$ does not exist at $x = 0$. The line $x = 0$ is a vertical asymptote.
$f(-2) = -4$
$f(2) = 4$
$(-2, -4)$ and $(2, 4)$ are critical points.

$f''(x) = \dfrac{8}{x^3}$

$f''(-2) = -1 < 0$
There is a local maximum at $x = -2$.
$f''(2) = 1 > 0$
There is a local minimum at $x = 2$. $f(x)$ is increasing on $(-\infty, -2)$ and $(2, \infty)$ and decreasing on $(-2, 0)$ and $(0, 2)$.
$f''(x)$ is never zero, so there are no points of inflection. $f(x)$ is concave upward on $(0, \infty)$ and concave downward on $(-\infty, 0)$. Because $f(0)$ does not exist, there is no y-intercept. To find any x-intercepts, let
$f(x) = 0$.

$f(x) = \dfrac{x^2 + 4}{x} = 0$

$x^2 + 4 = 0$
This equation has no real solution, so there are no x-intercepts. When $x = 0$, the denominator of

$\dfrac{x^2 + 4}{x}$ is equal to 0 but the numerator is not, so

$x = 0$ (the y-axis) is a vertical asymptote.

31. Continued

We now determine whether this rational function has a horizontal asymptote.

$$\lim_{x \to \infty} \frac{x^2+4}{x} = \lim_{x \to \infty} \frac{\frac{x^2}{x^2}+\frac{4}{x^2}}{\frac{x}{x^2}}$$

$$= \lim_{x \to \infty} \frac{1+\frac{4}{x^2}}{\frac{1}{x}}$$

$$= \frac{1+0}{0} = \frac{1}{0}$$

This expression is undefined, which indicates that there is no horizontal asymptote. Rewrite the function rule as

$$f(x) = \frac{x^2+4}{x} = x + \frac{4}{x}.$$

As x gets larger and larger in absolute value, $\frac{4}{x} \to 0$, and the graph approaches the line $y = x$, which is an oblique or slant asymptote.

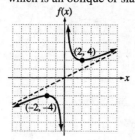

32.
$$f(x) = x + \frac{8}{x} = x + 8x^{-1}$$

$$f'(x) = 1 - 8x^{-2} = 1 - \frac{8}{x^2} = 0$$

$$x^2 = 8$$

$$x = \pm 2\sqrt{2} \quad \text{Critical numbers}$$

Notice that $f(x)$ does not exist at $x = 0$. The line $x = 0$ is a vertical asymptote.

$$f\left(2\sqrt{2}\right) = 2\sqrt{2} + \frac{8}{2\sqrt{2}}$$

$$= 2\sqrt{2} + 2\sqrt{2} = 4\sqrt{2}$$

$$f\left(-2\sqrt{2}\right) = -4\sqrt{2}$$

32. Continued

Thus, $\left(2\sqrt{2}, 4\sqrt{2}\right)$ and $\left(-2\sqrt{2}, -4\sqrt{2}\right)$ are critical points.

$$f''(x) = \frac{16}{x^3}$$

$f''\left(2\sqrt{2}\right) > 0$, so $f(x)$ has a local minimum at $x = 2\sqrt{2}$.

$f''\left(-2\sqrt{2}\right) < 0$, so $f(x)$ has a local maximum at $x = -2\sqrt{2}$.

$f(x)$ is increasing on $\left(-\infty, -2\sqrt{2}\right)$ and $\left(2\sqrt{2}, \infty\right)$ and decreasing on $\left(-2\sqrt{2}, 0\right)$ and $\left(0, 2\sqrt{2}\right)$.

$f''(x) > 0$ when $x > 0$.

$f''(x) < 0$ when $x < 0$.

Thus, $f(x)$ is concave upward on $(0, \infty)$ and concave downward on $(-\infty, 0)$.

There is no point of inflection at $x = 0$, even though $f(x)$ changes concavity because $f(0)$ is undefined. $f(0)$ is undefined, so there is no y-intercept.

To any x-intercepts, let $f(x) = 0$.

$$f(x) = x + \frac{8}{x} = 0$$

$$x\left(x + \frac{8}{x}\right) = x(0)$$

$$x^2 + 8 = 0$$

This equation has no real solution, so there are no x-intercepts. To look for vertical and horizontal asymptotes, we rewrite the function rule so that the right-hand side is a single fraction.

$$f(x) = x + \frac{8}{x} = \frac{x^2+8}{x}$$

When $x = 0$, the denominator is 0 but the numerator is not, so $x = 0$ (the y-axis) is a vertical asymptote. Using the method shown in the solution for Exercise 31, we find that

$$\lim_{x \to \infty} \frac{x^2+8}{x} = \frac{1}{0},$$

which is undefined. This indicates that there is no horizontal asymptote.

32. Continued

To find a possible slant asymptote, use the function rule in its original form. As $x \cdot \infty$ or $x \cdot -\infty$, $\dfrac{8}{x} \to 0$, so the graph will get closer and closer to the line $y = x$, which is a slant asymptote.

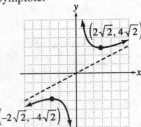

33. $f(x) = -x^2 + 6x + 1; \ [2, 4]$

$f'(x) = -2x + 6 = 0$ when $x = 3$

The absolute extrema must occur at the critical number 3 or at the endpoints.

$f(2) = 9$ Absolute minimum

$f(3) = 10$ Absolute maximum

$f(4) = 9$ Absolute minimum

The absolute maximum of 10 occurs at $x = 3$. The absolute minimum of 9 occurs at $x = 2$ and $x = 4$.

34. $f(x) = 4x^2 - 4x - 7; \ [-1, 3]$

$f'(x) = 8x - 4 = 0$

$x = \dfrac{1}{2}$

x	-1	$\dfrac{1}{2}$	3
$f(x)$	1	-8	17

The absolute maximum of 17 occurs at $x = 3$ and the absolute minimum of -8 occurs at $x = \dfrac{1}{2}$.

35. $f(x) = x^3 + 2x^2 - 15x + 3; \ [-.5, 3.3]$

$f'(x) = 3x^2 + 4x - 15 = 0$ when

$(3x - 5)(x + 3) = 0$

$x = \dfrac{5}{3}$ or $x = -3$.

Find the values of the function at the critical numbers and at the endpoints.

$f(-.5) = 10.875$

$f(3.3) = 11.217$ Absolute maximum

$f\left(\dfrac{5}{3}\right) = -\dfrac{319}{27}$ Absolute minimum

$f(2) = -11$

The absolute maximum of 11.217 occurs at $x = 3.3$.

The absolute minimum of $-\dfrac{319}{27}$ occurs at $x = \dfrac{5}{3}$.

36. $f(x) = -2x^3 - x^2 + 4x - 1; \ [-3, 1]$

$f'(x) = -6x^2 - 2x + 4$

$\quad\quad = -2\left(3x^2 + x - 2\right)$

$\quad\quad = -2(3x - 2)(x + 1)$

$f'(x) = 0$ when $x = -1$ or $x = \dfrac{2}{3}$,

so the critical numbers are -1 and $\dfrac{2}{3}$.

Find the values of the function at the critical numbers and at the endpoint.

x	-3	-1	$\dfrac{2}{3}$	1
$f(x)$	32	-4	$\dfrac{17}{27}$	0

The absolute maximum of 32 occurs at $x = -3$ and the absolute minimum of -4 occurs at $x = -1$.

37. $S(x) = -x^3 + 3x^2 + 360x + 5000, \ 6 \le x \le 20$

$$S'(x) = -3x^2 + 6x + 360$$
$$0 = -3x^2 + 6x + 360$$
$$x = \frac{-6 \pm \sqrt{6^2 - 4(-3)(360)}}{2(-3)}$$
$$= \frac{-6 \pm \sqrt{4356}}{-6}$$
$$x = -10 \ \text{or} \ x = 12$$

The point $x = -10$ is out of range. Test $S(x)$ at $x = 6$, $x = 12$, and $x = 20$.
$S(6) = 7052$
$S(12) = 8024$
$S(20) = 5400$
A temperature of 12°C produces the maximum number of bacteria.

38. $f(x) = .0205x^3 - .619x^2 - 5.44x - 6.04$
$(9 \le x \le 15)$

$$f'(x) = .0614x^2 - 1.238x + 5.44$$
$$0 = .0614x^2 - 1.238x + 5.44$$
$$x = \frac{-(-1.238) \pm \sqrt{(-1.238)^2 - 4(.0614)(5.44)}}{2(.0614)}$$
$$= \frac{1.238 \pm \sqrt{.1966}}{.1228}$$
$$x \approx 6.4707 \ \text{or} \ x \approx 13.6921$$

The point $x = 6.4707$ is out of range.
Test $f(x)$ at $x = 9$, $x = 13.6921$, and $x = 15$.
$f(9) = 7.7255$
$f(13.6921) \approx 5.0205$
$f(15) = 5.4725$

Interest rates were highest in 1999, and lowest in 2003.

39. $g(x) = -.025x^3 + .24x^2 - .32x + 3.9$
$(0 \le x \le 4)$

$$g'(x) = -.075x^2 + .48x - .32$$
$$0 = -.075x^2 + .48x - .32$$
$$x = \frac{-.48 \pm \sqrt{.48^2 - 4(-.075)(-.32)}}{2(-.075)}$$
$$= \frac{-.48 \pm \sqrt{.1344}}{-.150}$$
$$x \approx 5.6440 \ \text{or} \ x \approx .7560$$

The point $x = 5.6440$ is out of range.

Test $f(x)$ at $x = 0$, $x = .7560$, and $x = 4$.
$f(0) = 3.9$
$f(.7560) \approx 3.7844$
$f(4) = 4.86$
The least amount was spent on research and development in 2001 and the most was spent in 2004.

40. $f(x) = \dfrac{1041.2}{1 + 4.05e^{-.6x}}$
$(0 \le x \le 13)$

$$f'(x) = \frac{\left(1 + 4.05e^{-.6x}\right)(0) - 1041.2\left((4.05)(-.6)e^{-.6x}\right)}{\left(1 + 4.05e^{-.6x}\right)^2}$$
$$= \frac{2530.116\left(e^{-.6x}\right)}{\left(1 + 4.05e^{-.6x}\right)^2}$$

Notice that $f'(x) \ne 0$, and so we test $f(x)$ at $x = 0$, and $x = 13$.
$f(0) = 206.1782$
$f(13) = 1039.4751$
Most lung transplants were performed in 2003 $(x = 13)$ and the least transplants occurred in 1990 $(x = 0)$.

41. $f(x) = -.0083x^3 - .15x^2 + 1.18x + 3.92$
$(0 \le x \le 5)$

$$f'(x) = -.0249x^2 - .3x + 1.18$$
$$0 = -.0249x^2 - .3x + 1.18$$
$$x = \frac{-(-.3) \pm \sqrt{(-.3)^2 - 4(-.0249)(1.18)}}{2(-.0249)}$$
$$x \approx 3.1235 \ \text{or} \ x \approx -15.1717$$

Test $f'(x)$ at $x = 0$, $x = 3.1235$, and $x = 5$.
$f(0) = 3.92$
$f(3.1235) \approx 5.8894$
$f(5) = 5.0325$

Unemployment was lowest in 2000, and highest in 2003.

42. Let x = length and width of a side of base;
h = height.
Volume = 27 cubic meters with a square base and top. Find height, length, and width for minimum surface area.

Volume $= x^2 h$

$$x^2 h = 27$$

$$h = \frac{27}{x^2}$$

Surface area $= 2x^2 + 4xh$

$$A(x) = 2x^2 + 4x\left(\frac{27}{x^2}\right)$$

$$= 2x^2 + 108x^{-1}$$

$$A'(x) = 4x - 108x^{-2}$$

If $A' = 0$,

$$4x - \frac{108}{x^2} = 0$$

$$\frac{4x^3 - 108}{x^2} = 0$$

$$4x^3 = 108$$

$$x^3 = 27$$

$$x = 3.$$

$$A''(x) = 4 + 2(108)x^{-3}$$

$$= 4 + \frac{216}{x^3}$$

$$A''(3) = 12 > 0$$

The minimum occurs at $x = 3$ where

$$h = \frac{27}{3^2} = 3.$$

The dimensions are 3 m by 3 m by 3 m.

43. Let x = width of play area:
y = length of play area.

Building

An equation describing the amount of fencing is

$$900 = 2x + y$$

$$y = 900 - 2x.$$

$$A = xy$$

$$A(x) = x(900 - 2x)$$

$$= 900x - 2x^2.$$

If $A'(x) = 900 - 4x = 0$

$$x = 225.$$

Then $y = 900 - 2(225) = 450$.
$A''(x) = -4 < 0$, so $A(225)$ is a maximum.
Dimensions for maximum area are 225 m by 450 m.

44. Volume of open cylinder = 27π cubic inches. Find radius of bottom to minimize cost of material.

Volume of cylinder $= \pi r^2 h$
Surface area of cylinder open at one end $= 2\pi rh + \pi r^2$,

$$V = \pi r^2 h = 27\pi$$

$$h = \frac{27\pi}{\pi r^2}$$

$$A = 2\pi r\left(\frac{27\pi}{\pi r^2}\right) + \pi r^2$$

$$= 54\pi r^{-1} + \pi r^2$$

$$A' = -54\pi r^{-2} + 2\pi r$$

If $A' = 0$,

$$2\pi r = \frac{54\pi}{r^2}$$

$$r^3 = 27$$

$$r = 3.$$

$A'' = 108\pi r^{-3} + 2\pi > 0$, so for minimum cost, the radius should be 3 inches.

45. x = number of batches per year
M = 240,000 units produced annually
k = \$2, cost to store one unit for one year
a = \$15 cost to set up, or fixed cost

$$x = \sqrt{\frac{kM}{2a}} = \sqrt{\frac{2(240,000)}{2(15)}}$$

$$\approx 126$$

Since the number of batches is a whole number, the answer is 126 batches.

46. M = 128,000 cases sold per year
k = 1, cost to store 1 case for 1 year
a = 10, fixed cost for order
x = number of orders per year

$$x = \sqrt{\frac{kM}{2a}}$$

$$= \sqrt{\frac{(1)(128,000)}{2(10)}}$$

$$= \sqrt{6400} = 80$$

The company should produce 80 lots annually.

47. Let x = width of play area;
y = length of play area.
An equation describing the amount of fencing uses the formula for the perimeter of a rectangle.

$$900 = 2x + 2y$$

$$900 - 2x = 2y$$

$$450 - x = y$$

Substitute this expression for y in the formula for area of a rectangle.

$$A = xy$$

$$A = x(450 - x)$$

$$= 450x - x^2$$

If $A'(x) = 450 - 2x = 0$,

$$x = 225.$$

Then $y = 450 - 225 = 225$.

$A''(x) = -2 < 0$, which confirms that the maximum occurs at $x = 225$.
The dimensions for maximum area are 225 m by 225 m.

Case 12   A Total Cost Model for a Training Program

1. $$Z(m) = \frac{C_1}{m} + DtC_2 + DC_3\left(\frac{m-1}{2}\right)$$

$$= C_1 m^{-1} + DtC_2 + \frac{DC_3 m}{2} - \frac{DC_3}{2}$$

$$Z'(m) = -C_1 m^{-2} + \frac{DC_3}{2}$$

$$= -\frac{C_1}{m^2} + \frac{DC_3}{2}$$

2. $$Z'(m) = 0$$

$$-\frac{C_1}{m^2} + \frac{DC_3}{2} = 0$$

$$-\frac{C_1}{m^2} = -\frac{DC_3}{2}$$

$$-\frac{C_1}{m^2} = -\frac{DC_3}{2}$$

$$2C_1 = DC_3 m^2$$

$$\sqrt{\frac{2C_1}{DC_3}} = m \qquad (m \text{ is positive})$$

3. Let $D = 3$, $t = 12$, $C_1 = 15,000$, $C_3 = 900$.

$$m = \sqrt{\frac{2(15,000)}{(3)(900)}} = \sqrt{\frac{100}{9}} = \frac{10}{3}$$

4. $m^+ = 4$ Whole number larger than $\dfrac{10}{3}$

$m^- = 3$ Whole number smaller than $\dfrac{10}{3}$

5. Let $C_1 = 15,000$, $m^+ = 4$, $m^- = 3$, $D = 3$,
$C_2 = 100$, $C_3 = 900$, $t = 12$.

$$Z(m) = \frac{C_1}{m} + DtC_2 + DC_3\left(\frac{m-1}{2}\right)$$

$$Z(m^+) = Z(4) = \frac{15,000}{4} + (3)(12)(100)$$

$$+ (3)(900)\left(\frac{4-1}{2}\right)$$

$$= 3750 + 3600 + 4050$$

$$= \$11,400$$

$$Z(m^-) = Z(3) = \frac{15,000}{3} + (3)(12)(100)$$

$$+ (3)(900)\left(\frac{3-1}{2}\right)$$

$$= 5000 + 3600 + 2700$$

$$= \$11,300$$

6. Since $Z(m^-)$ is smaller than $Z(m^+)$, the optimum value of Z is \$11,300. This occurs when $t = 3$ months. Now $N = mD = (3)(3) = 9$ so the number of trainees in a batch is 9.

Chapter 13: Integral Calculus

Section 13.1 Antiderivatives

1. If $F(x)$ and $G(x)$ are both antiderivatives of $f(x)$, then they differ only by a constant.

3. Answers vary.

5. $\int 12x \, dx = 12 \int x \, dx$

$$= 12 \left(\frac{x^2}{2} \right) + C$$

$$= 6x^2 + C$$

7. $\int 8p^2 \, dp = 8 \int p^2 \, dp$

$$= 8 \left(\frac{p^3}{3} \right) + C$$

$$= \frac{8p^3}{3} + C$$

9. $\int 105 \, dx = 105 \int 1 \, dx$

$$= 105 \int x^0 \, dx$$

$$= 105 \left(\frac{x^1}{1} \right) + C$$

$$= 105x + C$$

11. $\int (5z - 1) \, dz = 5 \int z \, dz - \int 1 \, dz$

$$= 5 \left(\frac{z^2}{2} \right) - \int z^0 \, dz$$

$$= \frac{5z^2}{2} - \frac{z^1}{1} + C$$

$$= \frac{5z^2}{2} - z + C$$

13. $\int (z^2 - 4z + 2) \, dz$

$$= \int z^2 \, dz - 4 \int z \, dz + 2 \int 1 \, dz$$

$$= \frac{z^3}{3} - 4 \left(\frac{z^2}{2} \right) + 2 \int z^0 \, dz$$

$$= \frac{z^3}{3} - 2z^2 + 2 \left(\frac{z^1}{1} \right) + C$$

$$= \frac{z^3}{3} - 2z^2 + 2z + C$$

15. $\int (x^3 - 14x^2 + 22x + 3) \, dx$

$$= \int x^3 \, dx - 14 \int x^2 \, dx + 22 \int x \, dx + 3 \int 1 \, dx$$

$$= \frac{x^4}{4} - 14 \left(\frac{x^3}{3} \right) + 22 \left(\frac{x^2}{2} \right) + 3 \int x^0 \, dx$$

$$= \frac{x^4}{4} - \frac{14x^3}{3} + 11x^2 + 3 \left(\frac{x^1}{1} \right) + C$$

$$= \frac{x^4}{4} - \frac{14x^3}{3} + 11x^2 + 3x + C$$

17. $\int 6\sqrt{y} \, dy = 6 \int y^{\frac{1}{2}} \, dy$

$$= 6 \left(\frac{y^{\frac{3}{2}}}{\frac{3}{2}} \right) + C$$

$$= 6 \left(\frac{2}{3} y^{\frac{3}{2}} \right) + C$$

$$= 4y^{\frac{3}{2}} + C$$

19. $\int \left(6t\sqrt{t} + 3\sqrt[4]{t} \right) dt$

$$= 6 \int t \cdot t^{\frac{1}{2}} \, dt + 3 \int t^{\frac{1}{4}} \, dt$$

$$= 6 \int t^{\frac{3}{2}} \, dt + 3 \int t^{\frac{1}{4}} \, dt$$

$$= 6 \left(\frac{t^{\frac{5}{2}}}{\frac{5}{2}} \right) + 3 \left(\frac{t^{\frac{5}{4}}}{\frac{5}{4}} \right) + C$$

$$= 6 \left(\frac{2}{5} t^{\frac{5}{2}} \right) + 3 \left(\frac{4}{5} t^{\frac{5}{4}} \right) + C$$

$$= \frac{12t^{\frac{5}{2}}}{5} + \frac{12t^{\frac{5}{4}}}{5} + C$$

21. $\int \left(56t^{\frac{1}{2}} + 18t^{\frac{7}{2}} \right) dt$

$$= 56 \int t^{\frac{1}{2}} \, dt + 18 \int t^{\frac{7}{2}} \, dt$$

$$= 56 \left(\frac{t^{\frac{3}{2}}}{\frac{3}{2}} \right) + 18 \left(\frac{t^{\frac{9}{2}}}{\frac{9}{2}} \right) + C$$

$$= 56 \left(\frac{2}{3} t^{\frac{3}{2}} \right) + 18 \left(\frac{2}{9} t^{\frac{9}{2}} \right) + C$$

$$= \frac{112t^{\frac{3}{2}}}{3} + 4t^{\frac{9}{2}} + C$$

23. $\int \dfrac{24}{x^3} dx = 24 \int \dfrac{1}{x^3} dx$

$= 24 \int x^{-3} dx$

$= 24 \left(\dfrac{x^{-2}}{-2} \right) + C$

$= -12 x^{-2} + C$

$= -12 \left(\dfrac{1}{x^2} \right) + C$

$= -\dfrac{12}{x^2} + C$

25. $\int \left(\dfrac{1}{y^3} - \dfrac{2}{\sqrt{y}} \right) dy = \int \left(y^{-3} - 2y^{-\frac{1}{2}} \right) dy$

$= \int y^{-3} dy - 2 \int y^{-\frac{1}{2}} dy$

$= \dfrac{y^{-2}}{-2} - 2 \left(\dfrac{y^{\frac{1}{2}}}{\frac{1}{2}} \right) + C$

$= \dfrac{1}{-2y^2} - 2 \left(2y^{\frac{1}{2}} \right) + C$

$= -\dfrac{1}{2y^2} - 4y^{\frac{1}{2}} + C$

$= -\dfrac{1}{2y^2} - 4\sqrt{y} + C$

27. $\int \left(6x^{-3} + 2x^{-1} \right) dx$

$= 6 \int x^{-3} dx + 2 \int x^{-1} dx$

$= 6 \left(\dfrac{x^{-2}}{-2} \right) + 2 \ln|x| + C$

$= -3x^{-2} + 2 \ln|x| + C$

29. $\int 4e^{3u} du = 4 \int e^{3u} du$

$= 4 \left(\dfrac{1}{3} e^{3u} \right) + C$

$= \dfrac{4e^{3u}}{3} + C$

31. $\int 3e^{-.8x} dx = \dfrac{3\left(e^{-.8x}\right)}{-.8} + C$

$= \dfrac{-15e^{-.8x}}{4} + C$

33. $\int \left(\dfrac{6}{x} + 4e^{-5x} \right) dx$

$= 6 \ln|x| + \dfrac{4e^{-5x}}{-.5} + C$

$= 6 \ln|x| - 8e^{-5x} + C$

35. $\int \dfrac{1+2t^4}{t} dt = \int \left(\dfrac{1}{t} + \dfrac{2t^4}{t} \right) dt$

$= \int \left(\dfrac{1}{t} + 2t^3 \right) dt$

$= \ln|t| + \dfrac{2t^4}{4} + C$

$= \ln|t| + \dfrac{t^4}{2} + C$

37. $\int \left(e^{2u} + \dfrac{4}{u} \right) du = \int e^{2u} du + \int \dfrac{4}{u} du$

$= \dfrac{1}{2} e^{2u} + 4 \ln|u| + C$

39. $\int (5x+1)^2 dx = \int \left(25x^2 + 10x + 1 \right) dx$

$= 25 \int x^2 dx + 10 \int x dx + \int 1 dx$

$= 25 \left(\dfrac{x^3}{3} \right) + 10 \left(\dfrac{x^2}{2} \right) + x + C$

$= 25 \left(\dfrac{x^3}{3} \right) + 5x^2 + x + C$

41. $\int \dfrac{\sqrt{x}+1}{\sqrt[3]{x}} dx = \int \left(\dfrac{\sqrt{x}}{\sqrt[3]{x}} + \dfrac{1}{\sqrt[3]{x}} \right) dx$

$= \int \left(x^{\left(\frac{1}{2} - \frac{1}{3}\right)} + x^{-\frac{1}{3}} \right) dx$

$= \int \left(x^{\frac{1}{6}} + x^{-\frac{1}{3}} \right) dx$

$= \dfrac{x^{\frac{7}{6}}}{\frac{7}{6}} + \dfrac{x^{\frac{2}{3}}}{\frac{2}{3}} + C$

$= \dfrac{6x^{\frac{7}{6}}}{7} + \dfrac{3x^{\frac{2}{3}}}{2} + C$

43. $f'(x) = 6x^2 - 4x + 3$

$f(x) = \int (6x^2 - 4x + 3) dx$

$\qquad = 6\left(\dfrac{x^3}{3}\right) - 4\left(\dfrac{x^2}{2}\right) + 3x + C$

$\qquad = 2x^3 - 2x^2 + 3x + C$

Since $(0, 1)$ is on the curve, $f(0) = 1$. Thus,
$2(0)^3 - 2(0)^2 + 3(0) + C = 1$ and $C = 1$. Therefore,
the equation of the curve is
$f(x) = 2x^3 - 2x^2 + 3x + 1$.

45. a. $G'(x) = g(x)$ and $G(0) = 144$

$G(x) = \int (4.8x^2 - 16.4x + 20.6) dx$

$\qquad = 4.8 \int x^2 dx - 16.4 \int x\, dx + 20.6 \int 1\, dx$

$\qquad = 4.8\left(\dfrac{x^3}{3}\right) - 16.4\left(\dfrac{x^2}{2}\right) + 20.6x + C$

$G(x) = 1.6x^3 - 8.2x^2 + 20.6x + C$

$144 = 1.6(0)^3 - 8.2(0)^2 + 20.6(0) + C$

$144 = C$

$G(x) = 1.6x^3 - 8.2x^2 + 20.6x + 144$

b. $x = 2005 - 1995 = 10$

$G(4) = 1.6(10)^3 - 8.2(10)^2 + 20.6(10) + 144$

$\qquad = 1130$

There were \$1130 billion in imports from Canada in 2005.

47. a. $R'(x) = r(x)$ and $R(0) = 1952$

$R(x) = \int (-1.72x^3 + 42x^2 - 262x + 917) dx$

$\qquad = -1.72 \int x^3 dx + 42 \int x^2 dx - 262 \int x\, dx + 971 \int 1\, dx$

$\qquad = -1.72\left(\dfrac{x^4}{4}\right) + 42\left(\dfrac{x^3}{3}\right) - 262\left(\dfrac{x^2}{2}\right) + 917x + C$

$R(x) = -.43x^4 + 14x^3 - 131x^2 + 917x + C$

$R(0) = -.43(0)^4 + 14(0)^3 - 131(0)^2 + 917(0) + C$

$1952 = C$

$R(x) = -.43x^4 + 14x^3 - 131x^2 + 917x + 1952$

b. $x = 2000 - 1980 = 20$

$R(20) = -.43(20)^4 + 14(20)^3 - 131(20)^2 + 917(20) + 1952$

$\qquad \approx 11,092$

The revenue in 2000 was about \$11,092 million.

49. $P'(x) = 4 - 6x + 3x^2$ and $P(0) = -40$

$P(x) = \int \left(4 - 6x + 3x^2 \right) dx$

$\quad = 4 \int 1\, dx - 6 \int x\, dx + 3 \int x^2 dx$

$\quad = 4x - 6\left(\dfrac{x^2}{2} \right) + 3\left(\dfrac{x^3}{3} \right) + C$

$P(x) = 4x - 3x^2 + x^3 + C$

$-40 = 4(0) - 3(0)^2 + (0)^3 + C$

$-40 = C$

Therefore,

$\quad P(x) = x^3 - 3x^2 + 4x - 40$.

51. $C'(x) = x^{\frac{2}{3}} + 5$, $C(8) = 58$

$C(x) = \int \left(x^{\frac{2}{3}} + 5 \right) dx$

$\quad = \dfrac{3x^{\frac{5}{3}}}{5} + 5x + C$

$58 = \dfrac{3(8)^{\frac{5}{3}}}{5} + 5(8) + C$

$58 = \dfrac{3(32)}{5} + 40 + C$

$58 = \dfrac{96}{5} + 40 + C$

$\dfrac{-6}{5} = -1.2 = C$

So,

$C(x) = \dfrac{3x^{\frac{5}{3}}}{5} + 5x - 1.2$.

53.

$C'(x) = .2x^2 + .4x + .8$, $C(6) = 32.50$

$C(x) = \int (.2x^2 + .4x + .8) dx$

$\quad = \dfrac{x^3}{15} + \dfrac{x^2}{5} + .8x + k$

$C(6) = \dfrac{6^3}{15} + \dfrac{6^2}{5} + .8(6) + k$

Since $C(6) = 32.50$

$\quad 32.50 = \dfrac{216}{15} + \dfrac{36}{5} + 4.8 + k$

$\quad\quad k = 6.1$

$C(x) = \dfrac{x^3}{15} + \dfrac{x^2}{5} + .8x + 6.1$

55. $C'(x) = -\dfrac{40}{e^{.05x}} + 100$, $C(5) = 1400$

$C(x) = \int \left(-\dfrac{40}{e^{.05x}} + 100 \right) dx$

$\quad = \dfrac{800}{e^{.05x}} + 100x + k$

$C(5) = \dfrac{800}{e^{.05(5)}} + 100(5) + k$

Since $C(10) = 25$

$25 = \dfrac{15}{4} + 11 + 2.4 + 2.5 + k$

$k = 5.35$

$C(x) = \dfrac{15x^4}{40,000} + \dfrac{11x^3}{1000} + \dfrac{6x^2}{250} + .25x + 5.35$

Since $C(5) = 1400$

$1400 = \dfrac{800}{e^{.25}} + 500 + k$

$k \approx 276.96$

$C(x) \approx \dfrac{800}{e^{.05x}} + 100x + 276.96$

57. a. $F'(t) = f(t)$, $F(0) = 4.52$

$F(t) = \int .064184 e^{.0142t}\, dt$

$\quad = .064184 \int e^{.0142t}\, dt$

$\quad = .064184 \cdot \dfrac{1}{.0142} e^{.0142t} + C$

$F(t) = 4.52 e^{.0142t} + C$

$F(0) = 4.52 e^{.0142(0)} + C = 4.52 + C$

$4.52 = 4.52 + C$

$\quad 0 = C$

$F(t) = 4.52 e^{.0142t}$

b. $t = 2005 - 1980 = 25$

$F(25) = 4.52 e^{.0142(25)} \approx 6.45$

The function estimates the world population in 2005 to be 6.45 billion.

59. a. $G'(x) = g(x)$, $G(1) = 2.04$

$G(x) = \int -\dfrac{.28}{x}\, dx = -.28 \int \dfrac{1}{x}\, dx$

$G(x) = -.28 \ln x + C$

$G(1) = -.28 \ln 1 + C$

$2.04 = C$

$G(x) = -.28 \ln x + 2.04$

59. Continued

    b. $x = 2004 - 1990 + 1 = 15$
          $G(15) = -.28\ln 15 + 2.04 \approx 1.28$
          There were about 1.28 million military
          personnel on active duty in 2004.

Section 13.2 Integration by Substitution

1.   Integration by substitution is related to the chain rule for derivatives, but in reverse.
Difficult integrals in which u replaces an expression in the integrand and the derivative of u also appears to suggest using integration by substitution.

3. $\int 3(12x - 1)^2 \, dx$

Let $u = 12x - 1$. Then $du = 12 \, dx$.

$\int 3(12x - 1)^2 \, dx$

$= 3\int (12x - 1)^2 \, dx$

$= 3 \cdot \dfrac{1}{12} \cdot 12 \int (12x - 1)^2 \, dx$

$= \dfrac{3}{12} \int (12x - 1)^2 \, 12 \, dx$

$= \dfrac{1}{4} \int u^2 \, du$

$= \dfrac{1}{4}\left(\dfrac{u^3}{3}\right) + C$

$= \dfrac{u^3}{12} + C$

$= \dfrac{(12x - 1)^3}{12} + C$

5. $\int \dfrac{4}{(3t + 6)^2} \, dt$

Let $u = 3t + 6$. Then $du = 3 \, dt$.

$\int \dfrac{4}{(3t + 6)^2} \, dt$

$= 4\int (3t + 6)^{-2} \, dt$

$= 4 \cdot \dfrac{1}{3} \int (3t + 6)^{-2} (3) \, dt$

$= \dfrac{4}{3} \int (3t + 6)^{-2} (3) \, dt$

$= \dfrac{4}{3} \int u^{-2} \, du$

$= \dfrac{4}{3}\left(\dfrac{u^{-1}}{-1}\right) + C$

$= -\dfrac{4}{3u} + C = -\dfrac{4}{3(3t + 6)} + C$

7. $\int \dfrac{x + 1}{\left(x^2 + 2x - 4\right)^{\frac{3}{2}}} \, dx$

Let $u = x^2 + 2x - 4$. Then $du = (2x + 2) \, dx$.

$\int \dfrac{(x + 1)}{\left(x^2 + 2x - 4\right)^{\frac{3}{2}}} \, dx$

$= \int \left(x^2 + 2x - 4\right)^{-\frac{3}{2}} (x + 1) \, dx$

$= \dfrac{1}{2} \cdot 2 \int \left(x^2 + 2x - 4\right)^{-\frac{3}{2}} (x + 1) \, dx$

$= \dfrac{1}{2} \int \left(x^2 + 2x - 4\right)^{-\frac{3}{2}} (2x + 2) \, dx$

$= \dfrac{1}{2} \int u^{-\frac{3}{2}} \, du$

$= \dfrac{1}{2}\left(\dfrac{u^{-\frac{1}{2}}}{-\frac{1}{2}}\right) + C$

$= -\dfrac{1}{u^{\frac{1}{2}}} + C$

$= -\dfrac{1}{\left(x^2 + 2x - 4\right)^{\frac{1}{2}}} + C$

$= -\dfrac{1}{\sqrt{x^2 + 2x - 4}} + C$

9. $\int r^2 \sqrt{r^3 + 3} \, dr$

Let $u = r^3 + 3$. Then $du = 3r^2 dr$.

$\int r^2 \sqrt{r^3 + 3} \, dr$

$= \int \left(r^3 + 3\right)^{\frac{1}{2}} r^2 dr$

$= \frac{1}{3} \cdot 3 \int \left(r^3 + 3\right)^{\frac{1}{2}} r^2 dr$

$= \frac{1}{3} \int \left(r^3 + 3\right)^{\frac{1}{2}} 3r^2 dr$

$= \frac{1}{3} \int u^{\frac{1}{2}} du$

$= \frac{1}{3} \left(\frac{u^{\frac{3}{2}}}{\frac{3}{2}}\right) + C$

$= \frac{1}{3} \left(\frac{2}{3} u^{\frac{3}{2}}\right) + C$

$= \frac{2}{9} \left(r^3 + 3\right)^{\frac{3}{2}} + C = \frac{2\left(r^3 + 3\right)^{\frac{3}{2}}}{9} + C$

11. $\int \left(-3e^{7k}\right) dk$

Let $u = 7k$. Then $du = 7 \, dk$.

$\int \left(-3e^{7k}\right) dk = -3 \int e^{7k} dk$

$= -3 \cdot \frac{1}{7} \cdot 7 \int e^{7k} dk$

$= -\frac{3}{7} \int e^{7k} 7 \, dk$

$= -\frac{3}{7} \int e^u du$

$= -\frac{3}{7} e^u + C$

$= -\frac{3}{7} e^{7k} + C$

$= -\frac{3e^{7k}}{7} + C$

13. $\int 4w^3 e^{2w^4} dw$

Let $u = 2w^4$. Then $du = 8w^3 dw$.

$\int 4w^3 e^{2w^4} dw$

$= 4 \int e^{2w^4} \cdot w^3 dw$

$= 4 \cdot \frac{1}{8} \cdot 8 \int e^{2w^4} \cdot w^3 dw$

$= \frac{4}{8} \int e^{2w^4} \cdot 8w^3 dw$

$= \frac{1}{2} \int e^u du$

$= \frac{1}{2} e^u + C$

$= \frac{e^{2w^4}}{2} + C$

15. $\int (2 - t) e^{4t - t^2} dt$

Let $u = 4t - t^2$. Then $du = (4 - 2t) \, dt$.

$\int (2 - t) e^{4t - t^2} dt$

$= \frac{1}{2} \cdot 2 \int e^{4t - t^2} \cdot (2 - t) dt$

$= \frac{1}{2} \int e^{4t - t^2} (4 - 2t) dt$

$= \frac{1}{2} \int e^u du$

$= \frac{1}{2} e^u + C$

$= \frac{e^{4t - t^2}}{2} + C$

17. $\int \frac{e^{\sqrt{y}}}{\sqrt{y}} dy$

$\int \frac{e^{\sqrt{y}}}{\sqrt{y}} dy = \int e^{y^{\frac{1}{2}}} \cdot y^{-\frac{1}{2}} dy$

Let $u = y^{\frac{1}{2}}$. Then $du = \frac{1}{2} y^{-\frac{1}{2}} dy$.

$\int \frac{e^{\sqrt{y}}}{\sqrt{y}} dy = 2 \cdot \frac{1}{2} \int e^{y^{\frac{1}{2}}} \cdot y^{-\frac{1}{2}} dy$

$= 2 \int e^{y^{\frac{1}{2}}} \cdot \frac{1}{2} y^{-\frac{1}{2}} dy$

$= 2 \int e^u du$

$= 2 e^u + C$

$= 2 e^{\sqrt{y}} + C$

19. $\int \dfrac{-4}{12+5x}\,dx$

Let $u = 12 + 5x$. Then $du = 5\,dx$.

$$\int \frac{-4}{12+5x}\,dx = -4\int \frac{1}{12+5x}\,dx$$

$$= -4\cdot\frac{1}{5}\cdot 5\int \frac{1}{12+5x}\,dx$$

$$= -\frac{4}{5}\int \frac{1}{12+5x}\,5\,dx$$

$$= -\frac{4}{5}\int \frac{1}{u}\,du$$

$$= -\frac{4}{5}\ln|u| + C$$

$$= -\frac{4\ln|12+5x|}{5} + C$$

21. $\int \dfrac{e^{2t}}{e^{2t}+1}\,dt$

Let $u = e^{2t} + 1$. Then $du = e^{2t}(2)\,dt$.

$$\int \frac{e^{2t}}{2^{2t}+1}\,dt = \frac{1}{2}\cdot 2\int \frac{1}{e^{2t}+1}\cdot e^{2t}\,dt$$

$$= \frac{1}{2}\int \frac{1}{e^{2t}+1}\cdot 2e^{2t}\,dt$$

$$= \frac{1}{2}\int \frac{1}{u}\,du$$

$$= \frac{1}{2}\ln|u| + C$$

$$= \frac{1}{2}\ln|e^{2t}+1| + C$$

23. $\int \dfrac{x+2}{\left(2x^2+8x\right)^3}\,dx$

$$= \int \left(2x^2+8x\right)^{-3}(x+2)\,dx$$

Let $u = 2x^2 + 8x$. Then $du = (4x + 8)\,dx$.

$$\int \frac{x+2}{\left(2x^2+8x\right)^3}\,dx$$

$$= \frac{1}{4}\cdot 4\int \left(2x^2+8x\right)^{-3}(x+2)\,dx$$

$$= \frac{1}{4}\int \left(2x^2+8x\right)^{-3}(4x+8)\,dx$$

$$= \frac{1}{4}\int u^{-3}\,du$$

$$= \frac{1}{4}\cdot \frac{u^{-2}}{-2} + C$$

$$= -\frac{1}{8u^2} + C$$

$$= -\frac{1}{8\left(2x^2+8x\right)^2} + C$$

25. $\int 5\left(\dfrac{1}{r}+r\right)\left(1-\dfrac{1}{r^2}\right)dr$

Let $u = \dfrac{1}{r} + r$. Then $du = \left(-\dfrac{1}{r^2}+1\right)dr$.

$$\int 5\left(\frac{1}{r}+r\right)\left(1-\frac{1}{r^2}\right)dr = 5\int u\,du$$

$$= \frac{5u^2}{2} + C$$

$$= \frac{5\left(\frac{1}{r}+r\right)^2}{2} + C$$

27. $\int \dfrac{x^2+1}{\left(x^3+3x\right)^{\frac{2}{3}}}dx$

$= \dfrac{1}{3}\int \dfrac{3\left(x^2+1\right)dx}{\left(x^3+3x\right)^{\frac{2}{3}}}$.

Let $u=x^3+3x$. Then

$du=\left(3x^2+3\right)dx$

$\quad = 3\left(x^2+1\right)dx$

$\int \dfrac{x^2+1}{\left(x^3+3x\right)^{\frac{2}{3}}}dx = \dfrac{1}{3}\int \dfrac{du}{u^{\frac{2}{3}}}$

$\quad\quad = \dfrac{1}{3}\int u^{-\frac{2}{3}}du$

$\quad\quad = \dfrac{1}{3}\left(\dfrac{u^{\frac{1}{3}}}{\frac{1}{3}}\right)+C$

$\quad\quad = u^{\frac{1}{3}}+C$

$\quad\quad = \left(x^3+3x\right)^{\frac{1}{3}}+C$

29. $\int \dfrac{6x+7}{3x^2+7x+8}dx$

$= \int \dfrac{1}{3x^2+7x+8}(6x+7)dx$

Let $u=3x^2+7x+8$. Then $du=(6x+7)dx$.

$\int \dfrac{6x+7}{3x^2+7x+8}dx$

$= \int \dfrac{1}{u}du = \ln|u|+C$

$= \ln\left|3x^3+7x+8\right|+C$

31. $\int 2x\left(x^2+5\right)^3dx$

Let $u=x^2+5$. Then $du=2x\,dx$.

$\int 2x\left(x^2+5\right)^3dx = \int u^3du$

$\quad\quad = \dfrac{u^4}{4}+C$

$\quad\quad = \dfrac{\left(x^2+5\right)^4}{4}+C$

33. $\int \left(\sqrt{x^2+12x}\right)(x+6)dx$

$= \int \left(x^2+12x\right)^{\frac{1}{2}}(x+6)dx$

Let $u=x^2+12x$. Then $du=2x+12dx$ or

$(x+6)dx=\dfrac{du}{2}$.

$\int \left(\sqrt{x^2+12x}\right)(x+6)dx$

$= \dfrac{1}{2}\int u^{\frac{1}{2}}du$

$= \dfrac{1}{2}\left(\dfrac{2}{3}\right)u^{\frac{3}{2}}+C$

$= \dfrac{\left(x^2+12x\right)^{\frac{3}{2}}}{3}+C$

35. $\int \dfrac{(10+\ln x)^2}{x}dx$

$\int \dfrac{(10+\ln x)^2}{x}dx = \int (10+\ln x)^2 \cdot \dfrac{1}{x}dx$

Let $u=10+\ln x$. Then $du=\dfrac{1}{x}dx$.

$\int \dfrac{(10+\ln x)^2}{x}dx = \int u^2du$

$\quad\quad = \dfrac{u^3}{3}+C$

$\quad\quad = \dfrac{(10+\ln x)^3}{3}+C$

37. $\int \dfrac{5u}{\sqrt{u-1}}du$

$= \int 5u(u-1)^{-\frac{1}{2}}du$

Let $w=u-1$. Then $dw=du$ and $u=w+1$.

$= 5\int (w+1)w^{-\frac{1}{2}}dw$

$= 5\int w^{\frac{1}{2}}+w^{-\frac{1}{2}}dw$

$= \dfrac{5w^{\frac{3}{2}}}{\frac{3}{2}}+\dfrac{5w^{\frac{1}{2}}}{\frac{1}{2}}+C$

$= \dfrac{10}{3}w^{\frac{3}{2}}+10w^{\frac{1}{2}}+C$

$= \dfrac{10}{3}(u-1)^{\frac{3}{2}}+10(u-1)^{\frac{1}{2}}+C$

39. $\int t\sqrt{5t-1}\,dt$

$= \frac{1}{5}\int 5t(5t-1)^{\frac{1}{2}}\,dt$

Let $u = 5t - 1$. Then $du = 5\,dt$ and $t = \frac{u+1}{5}$.

$\int t\sqrt{5t-1}\,dt$

$= \frac{1}{5}\int\left(\frac{u+1}{5}\right)u^{\frac{1}{2}}\,du$

$= \frac{1}{25}\int\left(u^{\frac{3}{2}}+u^{\frac{1}{2}}\right)du$

$= \frac{1}{25}\left[\frac{u^{\frac{5}{2}}}{\frac{5}{2}}+\frac{u^{\frac{3}{2}}}{\frac{3}{2}}\right]+C$

$= \frac{1}{25}\left[\frac{2}{5}(5t-1)^{\frac{5}{2}}+\frac{2}{3}(5t-1)^{\frac{3}{2}}\right]+C$

$= \frac{2(5t-1)^{\frac{5}{2}}}{125}+\frac{2(5t-1)^{\frac{3}{2}}}{75}+C$

41. a. $C(x) = \int C'(x)\,dx = \int \frac{60x}{5x^2+1}\,dx$

Let $u = 5x^2 + 1$. Then $du = 10x\,dx$.

$C(x) = \int (5x^2+1)^{-1}6\cdot 10x\,dx$

$= 6\int u^{-1}\,du$

$= 6\ln|u| + k$

Notice that for any x, $u \geq 0$.

$C(x) = 6\ln(5x^2+1) + k$

$C(0) = 6\ln(5(0)^2+1) + k$

$10 = k$

$C(x) = 6\ln(5x^2+1) + 10$

b. $C(5) = 6\ln(5(5)^2+1) + 10 \approx 39.02$

Since this function is always increasing, only the cost at the 5th year is about \$39.02 thousand, so yes, they should add the new line.

43. a. $G'(x) = g(x),\ G(0) = 22.5$

$G(x) = \int -\frac{2.41}{x+1}\,dx = -2.41\int\frac{1}{x+1}\,dx$

$G(x) = -2.41\ln|x+1| + C$

$G(0) = -2.41\ln|0+1| + C$

$22.5 = C$

(Since only years after 1980 will be considered, we can do away with the absolute value signs.)

$G(x) = 22.5 - 2.41\ln(x+1)$

b. $x = 2005 - 1980 = 25$

$G(25) = 22.5 - 2.41\ln(25+1) \approx 14.6$

This function estimates 14.6 deaths per 100,000 population.

45. a. $G'(x) = g(x),\ G(0) = 18.6$

$G(x) = \int 1.11e^{.06x}\,dx = 1.11\frac{1}{.06}e^{.06x} + C$

$G(x) = 18.5e^{.06x} + C$

$G(0) = 18.5e^{.06(0)} + C$

$18.6 = 18.5 + C$

$.1 = C$

$G(x) = 18.5e^{.06x} + .1$

b. $x = 2006 - 1990 = 16$

$G(16) = 18.5e^{.06(16)} + .1 \approx 48.4$

This function estimates tourism revenues in Spain in 2006 at about \$48.4 billion.

Section 13.3 Area and the Definite Integral

1. The total usage of electricity is the total area between the graph of the rate function and the x-axis from $x = 0$ (midnight) to $x = 24$ (the next midnight). Approximate this area by using 12 rectangles each with base of length 2 and heights determined by the graph. Estimate the area to be the sum

$2\cdot 3 + 2\cdot 3 + 2\cdot 3.5 + 2\cdot 4 + 2\cdot 5 + 2\cdot 6 + 2\cdot 8$
$+ 2\cdot 11 + 2\cdot 11.5 + 2\cdot 10 + 2\cdot 6 + 2\cdot 4.5 = 151$

The total usage of electricity is about 151 kWh. (Your answer may vary depending on how you interpreted the height of each rectangle from the graph.)

3. Approximate the alcohol in the bloodstream by finding the sum of the areas of 8 rectangles, each with base of length 1 and heights determined by the graph.

$1\cdot 0 + 1\cdot 1 + 1\cdot 2 + 1\cdot 3 + 1\cdot 3.5 + 1\cdot 3.75$
$+ 1\cdot 3.5 + 1\cdot 2 = 19.25$

The total amount of alcohol in the bloodstream is about 20 units.

5. The indefinite integral $\int f(x)dx$ denotes a set of functions, whereas the definite integral represents a number.

7. **a.** $f(x) = 3x + 8; [0, 4]$
two rectangles:
$$\Delta x = \frac{4-0}{2} = 2$$

i	x_i	$f(x_i)$
1	0	8
2	2	14

$$\sum_{i=1}^{2} f(x_i)\Delta x = 8(2) + 14(2)$$
$$= 16 + 28$$
$$= 44$$

b. four rectangles:
$$\Delta x = \frac{4-0}{4} = 1$$

i	x_i	$f(x_i)$
1	0	8
2	1	11
3	2	14
4	3	17

$$\sum_{i=1}^{4} f(x_i)\Delta x$$
$$= 8(1) + 11(1) + 14(1) + 17(1)$$
$$= 8 + 11 + 14 + 17$$
$$= 50$$

c. 40 rectangles:
$$\Delta x = \frac{4-0}{40} = .1$$
Using a graphing calculator
$$\sum_{i=1}^{40} f(x_i)\Delta x = 55.4$$

9. **a.** $f(x) = 4 - x^2; [-2, 2]$
two rectangles:
$$\Delta x = \frac{2-(-2)}{2} = 2$$

i	x_i	$f(x_i)$
1	-2	0
2	0	4

$$\sum_{x=1}^{2} f(x_i)\Delta x = 0(2) + 4(2)$$
$$= 8$$

b. four rectangles:
$$\Delta x = \frac{2-(-2)}{4} = 1$$

i	x_i	$f(x_i)$
1	-2	0
2	-1	3
3	0	4
4	1	3

$$\sum_{i=1}^{4} f(x_i)\Delta x$$
$$= 0(1) + 3(1) + 4(1) + 3(1)$$
$$= 0 + 3 + 4 + 3$$
$$= 10$$

c. 40 rectangles:
$$\Delta x = \frac{2-(-2)}{40} = .1$$
Using a graphing calculator
$$\sum_{i=1}^{40} f(x_i)\Delta x = 10.66$$

11. a. $f(x) = e^{2x} - .5$; $[0, 2]$

two rectangles:

$$\Delta x = \frac{2-0}{2} = 1$$

i	x_i	$f(x_i)$
1	0	.5
2	1	6.89

$$\sum_{i=1}^{2} f(x_i)\Delta x = .5(1) + 6.89(1)$$

$$= 7.39$$

b. four rectangles:

$$\Delta x = \frac{2-0}{4} = 0.5$$

i	x_i	$f(x_i)$
1	0	.5
2	.5	2.22
3	1	6.89
4	1.5	19.59

$$\sum_{i=1}^{4} f(x_i)\Delta x = .25 + 1.110 + 3.445 + 9.795$$

$$= 14.6$$

c. 40 rectangles:

$$\Delta x = \frac{2-0}{40} = .05$$

Using a graphing calculator

$$\sum_{i=1}^{40} f(x_i)\Delta x = 24.48$$

13. $f(x) = \frac{x}{2}$ between $x = 0$ and $x = 4$

four rectangles:

a. $\Delta x = \frac{4-0}{4} = 1$

i	x_i	$f(x_i)$
1	0	0
2	1	.5
3	2	1
4	3	1.5

$$\sum_{i=1}^{4} f(x_i)\Delta x = 0 + .5 + 1 + 1.5$$

$$= 3$$

b. eight rectangles:

$$\Delta x = \frac{4-0}{8} = .5$$

i	x_i	$f(x_i)$
1	0	0
2	.5	.25
3	1.0	.50
4	1.5	.75
5	2.0	1.00
6	2.5	1.25
7	3.0	1.50
8	3.5	1.75

$$\sum_{i=1}^{8} f(x_i)\Delta x$$

$$= 0(.5) + .25(.5) + .5(.5) + .75(.5)$$

$$+ 1(.5) + 1.25(.5)$$

$$+ 1.5(.5) + 1.75(.5)$$

$$= 3.5$$

c. $\int_{0}^{4} f(x)dx$

$$= \int_{0}^{4} \frac{x}{2}dx$$

$$= \frac{1}{2} \text{ (base)(height)}$$

$$= \frac{1}{2}(4)(2) = 4$$

15. $\int_{-5}^{0}\left(x^3+6x^2-10x+2\right)dx$

Under the MATH menu, use fnInt.
Key in the following:

$\text{fnInt}\left(x^3+6x^2-10x+2,\ x,\ -5,\ 0\right)$

The value returned is 228.75. Therefore, the area is approximately 229.

17. $\int_{2}^{7}5\ln\left(2x^2+1\right)dx$

Under the MATH menu, use fnInt.
Key in the following:

$\text{fnInt}\left(5\ln\left(2x^2+1\right),\ x,\ 2,\ 7\right).$

The value returned is 90.55. Therefore the area is approximately 91.

19. $\int_{0}^{3}4x^2e^{-3x}\,dx$

Under the MATH menu use fnInt and key in the following:

$\text{fnInt}\left(4x^2e(-3x),\ x,\ 0,\ 3\right).$

The value returned is .294.
Therefore, the area is approximately .3.

21. $MR(x)=.04x^3-.5x^2+2x,\ [0,\ 12]$

Total Revenue $=\int_{0}^{12}(.04x^3-.5x^2+2x)dx$

Under the MATH menu use fnInt and key in

$\text{fnInt}(.04x^3-.5x^2+2x,\ x,\ 0,\ 12).$

The value returned is 63.36.
Therefore, the total revenue over the period is $6636.

23. $MR(x)=.26x^4-6.25x^3+30.2x^2+87.5x,\ [0,\ 5]$

Total Revenue

$=\int_{0}^{5}(.26x^4-6.25x^3+30.2x^2+87.5x)dx$

Under the MATH menu use fnInt and key in

$\text{fnInt}(.26x^4-6.25x^3+30.2x^2+87.5x,\ x,\ 0,\ 5).$

The value returned 1538.020833.
Therefore the total revenue over the period is about $153,802.

25. a. 2000 hours at an hourly wage of $g(x)$ gives a yearly wage of

$f(x)=2000(13.42x+.356x+.0063\,x^2)$

$f(x)=26,840+712x+12.6x^2$

b.

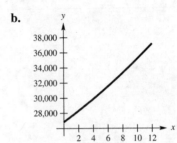

c. Estimate the total wages from January 1, 2000 through December 31, 2002 by summing the area of 3 rectangles, each with a base of length 1 and heights determined by the function f.

$1(35,220)+1(36,196.6)+1(37,198.4)$

$=108,615$

Using rectangles, the total wages over this time period are about $108,615. Because there is area under the graph of the function and not accounted for in the area of the rectangles, this estimate is less than the actual total.

d. Under the MATH menu select fnInt and key in $\text{fnInt}(26,840+712x+12.6x^2,\ x,\ 10,\ 13).$
The value returned is 110,111.
The actual total wages are $110,111.

27. Read values of the function on the graph for every 5 sec from $t = 0$ to $t = 25$. These are speeds in miles per hour, so multiply these values by $\dfrac{5280}{3600}$ to get the speeds in feet per second. Then to estimate the distance traveled, find the sum of areas of rectangles with heights the sides of the rectangles and with widths of 5 sec. The last rectangle has a width of 3 sec.

The total distance traveled is approximately

$$\dfrac{5280}{3600}[0(5) + 40(5) + 64(5) + 77(5) + 88(5) + 95(3)]$$

$$= \dfrac{5280}{3600}(1630)$$

$$\approx 2391.$$

The total distance traveled is approximately 2400 ft.

29. a. At 2 seconds, the area under the curve for car A is $\dfrac{1}{2}(1)(6) + 1(6) = 9$, so car A has traveled 9 ft.

b. Car A is furthest ahead of car B when the graphs intersect, at 2 sec.

c. At 2 seconds, car B has traveled approximately
.5(.3 + 1 + 2.9 + 5) = 4.6
So car A is $9 - 4.6 \approx 4$ feet ahead of car B.

d. Car B catches up with car A when the areas under the curves are equal, which is roughly somewhere between 3 and 3.5 seconds.

Section 13.4 The Fundamental Theorem of Calculus

1. $\displaystyle\int_{-1}^{3}\left(6x^2 - 7x + 3\right)dx$

$$= \left(2x^3 - \dfrac{7x^2}{2} + 3x\right)\Bigg|_{-1}^{3}$$

$$= \left[2(3)^3 - \dfrac{7(3)^2}{2} + 3(3)\right]$$

$$\quad - \left[2(-1)^3 - \dfrac{7(-1)^2}{2} + 3(-1)\right]$$

$$= (54 - \dfrac{63}{2} + 9) - (-2 - \dfrac{7}{2} - 3)$$

$$= \dfrac{63}{2} - \left(\dfrac{17}{2}\right)$$

$$= \dfrac{80}{2} = 40$$

3. $\displaystyle\int_{0}^{2} 3\sqrt{4u + 1}\,du$

Let $w = 4u + 1$. Then $dw = 4\,du$. If $u = 2$, $w = 9$. If $u = 0$, $w = 1$.

$$\int_{0}^{2} 3\sqrt{4u + 1}\,du$$

$$= \dfrac{3}{4}\int_{0}^{2}(4u + 1)^{\frac{1}{2}}\,4\,du$$

$$= \dfrac{3}{4}\int_{1}^{9} w^{\frac{1}{2}}\,dw$$

$$= \dfrac{3}{4}\left(\dfrac{w^{\frac{3}{2}}}{\frac{3}{2}}\right)\Bigg|_{1}^{9}$$

$$= \dfrac{3}{4}\cdot\dfrac{2}{3}w^{\frac{3}{2}}\Bigg|_{1}^{9}$$

$$= \dfrac{1}{2}(9)^{\frac{3}{2}} - \dfrac{1}{2}(1)^{\frac{3}{2}}$$

$$= \dfrac{27}{2} - \dfrac{1}{2}$$

$$= 13$$

5. $\int_0^1 2\left(t^{\frac{1}{2}}-9t\right)dt$

$= 2\left(\dfrac{t^{\frac{3}{2}}}{\frac{3}{2}}+\dfrac{9t^2}{2}\right)\Bigg|_0^1$

$= 2\left(\dfrac{2}{3}t^{\frac{3}{2}}+\dfrac{9}{2}t^2\right)\Bigg|_0^1$

$= 2\left[\dfrac{2}{3}(1)^{\frac{3}{2}}+\dfrac{9}{2}(1)^2\right]-2\left[\dfrac{2}{3}(0)^{\frac{3}{2}}+\dfrac{9}{2}(0)^2\right]$

$= 2\left(\dfrac{2}{3}+\dfrac{9}{2}\right)-2(0-0)$

$= 2\left(\dfrac{31}{6}\right)=10\dfrac{1}{3}$

7. $\int_1^4 \left(5y\sqrt{y}+3\sqrt{y}\right)dy$

$= \int_1^4 \left(5y^{\frac{3}{2}}+3y^{\frac{1}{2}}\right)dy$

$= \left(5\cdot\dfrac{y^{\frac{5}{2}}}{\frac{5}{2}}+3\cdot\dfrac{y^{\frac{3}{2}}}{\frac{3}{2}}\right)\Bigg|_1^4$

$= \left(5\cdot\dfrac{2}{5}y^{\frac{5}{2}}+3\cdot\dfrac{2}{3}y^{\frac{3}{2}}\right)\Bigg|_1^4$

$= \left(2y^{\frac{5}{2}}+2y^{\frac{3}{2}}\right)\Bigg|_1^4$

$= \left[2(4)^{\frac{5}{2}}+2(4)^{\frac{3}{2}}\right]-\left[2(1)^{\frac{5}{2}}+2(1)^{\frac{3}{2}}\right]$

$= (64+16)-(2+2)$

$= 80-4=76$

9. $\int_4^7 \dfrac{11}{(x-3)^2}\,dx = \int_4^7 11(x-3)^{-2}\,dx$

Let $u = x-3$. Then $du = dx$. If $x=7$, $u=4$.
If $x=4$, $u=1$.

$\int_4^7 \dfrac{11}{(x-3)^2}\,dx = \int_1^4 11u^{-2}\,du$

$= 11\cdot\dfrac{u^{-1}}{-1}\Bigg|_1^4$

$= -\dfrac{11}{u}\Bigg|_1^4$

$= -\dfrac{11}{4}-\left(-\dfrac{11}{1}\right)$

$= -\dfrac{11}{4}+\dfrac{44}{4}=\dfrac{33}{4}=8\dfrac{1}{4}$

11. $\int_1^5 \left(5n^{-1}+n^{-3}\right)dn$

$= \int_1^5 5n^{-1}\,dn+\int_1^5 n^{-3}\,dn$

$= 5\int_1^5 \dfrac{1}{n}\,dn+\int_1^5 n^{-3}\,dn$

$= 5\ln n\Big|_1^5+\dfrac{n^{-2}}{-2}\Bigg|_1^5$

$= 5(\ln 5-\ln 1)-\dfrac{1}{2}\left(\dfrac{1}{n^2}\right)\Bigg|_1^5$

$= 5\ln 5-\dfrac{1}{2}\left(\dfrac{1}{25}-1\right)$

$= 5\ln 5-\dfrac{1}{2}\left(-\dfrac{24}{25}\right)$

$= 5\ln 5+\dfrac{12}{25}$

≈ 8.527

13. $\int_2^3 \left(.1e^{-.1A}+\dfrac{3}{A}\right)dA$

$= 0.1\int_2^3 e^{-.1A}\,dA+3\int_2^3 \dfrac{1}{A}\,dA$

$= 0.1\cdot\dfrac{e^{-.1A}}{-.1}\Bigg|_2^3+3\ln|A|\Big|_2^3$

$= -e^{-.1A}\Big|_2^3+3\ln|A|\Big|_2^3$

$= -e^{-.3}-\left(-e^{-.2}\right)+3\ln 3-3\ln 2$

$= e^{-.2}-e^{-.3}+3\ln 3-3\ln 2$

≈ 1.294

15. $\int_1^2 \left(e^{6u}-\dfrac{1}{u^2}\right)du = \int_1^2 e^{6u}\,du-\int_1^2 \dfrac{1}{u^2}\,du$

$= \dfrac{e^{6u}}{6}\Bigg|_1^2+\dfrac{1}{u}\Bigg|_1^2$

$= \dfrac{e^{12}}{6}-\dfrac{e^6}{6}+\dfrac{1}{2}-1$

$= \dfrac{e^{12}}{6}-\dfrac{e^6}{6}-\dfrac{1}{2}$

$\approx 27,058.06$

17. $\int_{-1}^{0} y\left(2y^2 - 3\right)^5 dy$

Let $u = 2y^2 - 3$. Then $du = 4y\,dy$. If $y = 0$, then
$u = -3$. If $y = -1$, then $u = -1$.

$\int_{-1}^{0} y\left(2y^2 - 3\right)^5 dy$

$= \frac{1}{4}\int_{-1}^{0}\left(2y^2 - 3\right)^5 4y\,dy$

$= \frac{1}{4}\int_{-1}^{-3} u^5\,du$

$= \frac{1}{4}\cdot\frac{u^6}{6}\Big|_{-1}^{-3}$

$= \frac{1}{24}(-3)^6 - \frac{1}{24}(-1)^6$

$= \frac{729}{24} - \frac{1}{24} = \frac{728}{24}$

$= \frac{91}{3} = 30.\overline{3}$

19. $\int_{1}^{64}\frac{\sqrt{z} - 2}{\sqrt[3]{z}}dz$

$= \int_{1}^{64}\left(\frac{z^{\frac{1}{2}}}{z^{\frac{1}{3}}} - 2z^{-\frac{1}{3}}\right)dz$

$= \int_{1}^{64} z^{\frac{1}{6}}dz - 2\int_{1}^{64} z^{-\frac{1}{3}}dz$

$= \frac{z^{\frac{7}{6}}}{\frac{7}{6}}\Big|_{1}^{64} - 2\cdot\frac{z^{\frac{2}{3}}}{\frac{2}{3}}\Big|_{1}^{64}$

$= \frac{6z^{\frac{7}{6}}}{7}\Big|_{1}^{64} - 3z^{\frac{2}{3}}\Big|_{1}^{64}$

$= \frac{6(64)^{\frac{7}{6}}}{7} - \frac{6(1)^{\frac{7}{6}}}{7} - 3\left(64^{\frac{2}{3}} - 1^{\frac{2}{3}}\right)$

$= \frac{6(128)}{7} - \frac{6}{7} - 3(16 - 1)$

$= \frac{768 - 6 - 315}{7}$

$= \frac{447}{7} \approx 63.857$

21. $\int_{1}^{2}\frac{\ln x}{3x}dx = \frac{1}{3}\int_{1}^{2}\ln x \cdot \frac{1}{x}dx$

Let $u = \ln x$. Then $du = \frac{1}{x}dx$. If $x = 2$, then
$u = \ln 2$. If $x = 1$, $u = \ln 1 = 0$.

$\int_{1}^{2}\frac{\ln x}{3x}dx = \frac{1}{3}\int_{0}^{\ln 2} u\,du$

$= \left(\frac{1}{3}\right)\cdot\frac{u^2}{2}\Big|_{0}^{\ln 2}$

$= \frac{(\ln 2)^2}{6} - \frac{0^2}{6}$

$= \frac{(\ln 2)^2}{6} \approx .0801$

23. $\int_{0}^{8} x^{\frac{1}{3}}\sqrt{x^{\frac{4}{3}} + 9}\,dx$

Let $u = x^{\frac{4}{3}} + 9$. Then $du = \frac{4}{3}x^{\frac{1}{3}}dx$. If $x = 8$, then
$u = 25$. If $x = 0$, then $u = 9$.

$\int_{0}^{8} x^{\frac{1}{3}}\sqrt{x^{\frac{4}{3}} + 9}\,dx$

$= \frac{3}{4}\int_{0}^{8}\left(x^{\frac{4}{3}} + 9\right)^{\frac{1}{2}}\cdot\frac{4}{3}x^{\frac{1}{3}}dx$

$= \frac{3}{4}\int_{9}^{25} u^{\frac{1}{2}}du$

$= \frac{3}{4}\cdot\frac{u^{\frac{3}{2}}}{\frac{3}{2}}\Big|_{9}^{25}$

$= \frac{3}{4}\cdot\frac{2}{3}u^{\frac{3}{2}}\Big|_{9}^{25}$

$= \frac{1}{2}(25)^{\frac{3}{2}} - \frac{1}{2}(9)^{\frac{3}{2}}$

$= \frac{125}{2} - \frac{27}{2} = \frac{98}{2}$

$= 49$

25. $\int_0^1 \dfrac{4e^t}{\left(3+e^t\right)^2} \, dt$

Let $u = 3 + e^t$. Then $du = e^t dt$. If $t = 1$, then
$u = 3 + e$. If $t = 0$, $u = 4$.

$$\int_0^1 \dfrac{4e^t}{\left(3+e^t\right)^2} \, dt = 4\int_0^1 \left(3+e^t\right)^{-2} e^t dt$$

$$= 4\int_4^{3+e} u^{-2} \, du$$

$$= \dfrac{4u^{-1}}{-1}\Bigg|_4^{3+e}$$

$$= -\dfrac{4}{u}\Bigg|_4^{3+e}$$

$$= -\dfrac{4}{3+e} - \left(-\dfrac{4}{4}\right)$$

$$= 1 - \dfrac{4}{3+e}$$

$$\approx .3005$$

27. $\int_1^{49} \dfrac{\left(1+\sqrt{x}\right)^{\frac{4}{3}}}{\sqrt{x}} \, dx$

$$= \int_1^{49} \left(1+x^{\frac{1}{2}}\right)^{\frac{4}{3}} x^{-\frac{1}{2}} \, dx$$

Let $u = 1 + x^{\frac{1}{2}}$.

Then $du = \dfrac{1}{2} x^{-\frac{1}{2}} dx$. If $x = 49$, $u = 8$. If

$x = 1$, $u = 2$.

$$\int_1^{49} \dfrac{\left(1+\sqrt{x}\right)^{\frac{4}{3}}}{\sqrt{x}} \, dx$$

$$= 2\int_1^{49} \left(1+x^{\frac{1}{2}}\right)^{\frac{4}{3}} \dfrac{1}{2} x^{-\frac{1}{2}} \, dx$$

$$= 2\int_2^8 u^{\frac{4}{3}} \, du$$

$$= 2 \cdot \dfrac{u^{\frac{7}{3}}}{\frac{7}{3}}\Bigg|_2^8$$

$$= 2 \cdot \dfrac{3}{7} u^{\frac{7}{3}}\Bigg|_2^8$$

$$= \dfrac{6}{7}(8)^{\frac{7}{3}} - \dfrac{6}{7}(2)^{\frac{7}{3}}$$

$$= \dfrac{6}{7}\left(128 - 2^{\frac{7}{3}}\right)$$

$$\approx 105.3946$$

29. $\int_2^3 \dfrac{8x^3+5}{2x^4+5x+9} \, dx$

Let $u = 2x^4 + 5x + 9$. Then $du = \left(8x^3+5\right)dx$.

If $x = 3$, then $u = 2(3)^4 + 5(3) + 9 = 186$.

If $x = 2$, then $u = 2(2)^4 + 5(2) + 9 = 51$.

$$\int_2^3 \dfrac{8x^3+5}{2x^4+5x+9} \, dx$$

$$= \int_{51}^{186} \dfrac{1}{u} \, du$$

$$= \ln|u|\Big\|_{51}^{186} = \ln 186 - \ln 51$$

$$= 1.2939$$

31. A negative definite integral for the first year
and a half would indicate a loss for that period.
The overall profit for the two-year period is
represented by $\int_0^2 \left(6x^2 - 7x - 3\right) dx$.

33. $f(x) = 9 - x^2$; $[0, 4]$
The graph of f is above the x-axis on $[0, 3)$ and
below the x-axis on $(3, 4]$.

$$A = \int_0^3 \left(9 - x^2\right) dx + \left|\int_3^4 \left(9 - x^2\right) dx\right|$$

$$= \left(9x - \dfrac{x^3}{3}\right)\Bigg|_0^3 + \left|\left(9x - \dfrac{x^3}{3}\right)\Bigg|_3^4\right|$$

$$= (27 - 9) - 0 + \left|\left(36 - \dfrac{64}{3}\right) - (27 - 9)\right|$$

$$= 18 + \left|\dfrac{44}{3} - 18\right|$$

$$= 18 + \left|-\dfrac{10}{3}\right|$$

$$= 18 + \dfrac{10}{3}$$

$$= \dfrac{64}{3} = 21.3\overline{3}$$

35. $f(x) = x^3 - 1$; $[-1, 2]$

The graph of f is below the x-axis on $[-1, 1)$ and above the x-axis on $(1, 2]$.

$$A = \left| \int_{-1}^{1} \left(x^3 - 1 \right) dx \right| + \int_{1}^{2} \left(x^3 - 1 \right) dx$$

$$= \left| \left(\frac{x^4}{4} - x \right) \Big|_{-1}^{1} \right| + \left(\frac{x^4}{4} - x \right) \Big|_{1}^{2}$$

$$= \left| \left(\frac{1}{4} - 1 \right) - \left(\frac{1}{4} + 1 \right) \right| + (4 - 2) - \left(\frac{1}{4} - 1 \right)$$

$$= \left| -\frac{3}{4} - \frac{5}{4} \right| + 2 - \left(-\frac{3}{4} \right)$$

$$= \left| -\frac{8}{4} \right| + 2 + \frac{3}{4}$$

$$= 2 + 2 + \frac{3}{4}$$

$$= 4.75$$

37. $f(x) = e^{2x} - 1$; $[-2, 1]$

Solve $f(x) = 0$ to determine where the graph crosses the x-axis.

$$e^{2x} - 1 = 0$$

$$e^{2x} = 1$$

$$2x \ln e = \ln 1$$

$$2x = 0$$

$$x = 0$$

The graph crosses the x-axis at 0 in the given interval $[-2, 1]$. The total area is

$$\left| \int_{-2}^{0} \left(e^{2x} - 1 \right) dx \right| + \int_{0}^{1} \left(e^{2x} - 1 \right) dx$$

$$= \left| \left(\frac{e^{2x}}{2} - x \right) \Big|_{-2}^{0} \right| + \left(\frac{e^{2x}}{2} - x \right) \Big|_{0}^{1}$$

$$= \left| \left(\frac{1}{2} - 0 \right) - \left(\frac{e^{-4}}{2} + 2 \right) \right| + \left(\frac{e^2}{2} - 1 \right) - \left(\frac{1}{2} - 0 \right)$$

$$= \left| -\frac{3}{2} - \frac{e^{-4}}{2} \right| + \frac{e^2}{2} - 1 - \frac{1}{2}$$

$$= \left| -\frac{3}{2} - \frac{e^{-4}}{2} \right| + \frac{e^2}{2} - \frac{3}{2}$$

$$\approx 3.7037$$

39. $f(x) \geq 0$ if $x \in [0, 3]$.

$$\int_{0}^{3} x^2 e^{-x^3/2} \, dx$$

Let $u = -\frac{1}{2} x^3$. Then $du = -\frac{3}{2} x^2 dx$. If $x = 3$,

then $u = -\frac{27}{2}$. If $x = 0$, then $u = 0$.

$$\int_{0}^{3} x^2 e^{-x^3/2} \, dx = -\frac{2}{3} \int_{0}^{-27/2} e^u \, du$$

$$= -\frac{2}{3} e^u \Big|_{0}^{-27/2} = -\frac{2}{3} \left(e^{-27/2} - 1 \right)$$

$$\approx .6667$$

41. $f(x) = \frac{1}{x}$; $[1, e]$

$\frac{1}{x} = 0$ has no solution, so the graph does not cross the x-axis in the given interval $[1, e]$.

$$\int_{1}^{e} \frac{1}{x} \, dx = \ln x \Big|_{1}^{e}$$

$$= \ln e - \ln 1$$

$$= 1$$

43. $\int_{1}^{4} \frac{12 \left(\ln x \right)^3}{x} \, dx$

Let $u = \ln x$. Then $du = \frac{1}{x} dx$. If $x = 4$, then $u = \ln 4$. If $x = 1$, then $u = 0$.

$$\int_{1}^{4} \frac{12 \left(\ln x \right)^3}{x} \, dx = 12 \int_{0}^{\ln 4} u^3 \, du$$

$$= 12 \frac{u^4}{4} \Big|_{0}^{\ln 4} = 3 \left(\left(\ln 4 \right)^4 - 0 \right)$$

$$\approx 11.08$$

45.
$$A = \int_0^2 \left(2 - .5x^2\right) dx$$
$$+ \left| \int_2^3 \left(2 - .5x^2\right) dx \right|$$

$$= \left(2x - \frac{.5x^3}{3}\right)\Big|_0^2 + \left|\left(2x - \frac{.5x^3}{3}\right)\Big|_2^3\right|$$

$$= \left(4 - \frac{4}{3}\right) - 0 + \left|\left(6 - \frac{9}{2}\right) - \left(4 - \frac{4}{3}\right)\right|$$

$$= \frac{8}{3} + \left|\frac{3}{2} - \frac{8}{3}\right|$$

$$= \frac{8}{3} + \left|-\frac{7}{6}\right|$$

$$= \frac{16}{6} + \frac{7}{6}$$

$$= \frac{23}{6} = 3.8\overline{3}$$

47.
$$A = \int_{-1}^0 \left(x^2 - 2x\right) dx + \left| \int_0^2 \left(x^2 - 2x\right) dx \right|$$

$$= \left(\frac{x^3}{3} - x^2\right)\Big|_{-1}^0 + \left|\left(\frac{x^3}{3} - x^2\right)\Big|_0^2\right|$$

$$= 0 - \left(-\frac{1}{3} - 1\right) + \left|\left(\frac{8}{3} - 4\right) - 0\right|$$

$$= \frac{4}{3} + \left|-\frac{4}{3}\right|$$

$$= \frac{8}{3}$$

$$= 2.\overline{6}$$

49. $f(x) = \begin{cases} 2x+3 & \text{if } x \le 2 \\ -.5x+8 & \text{if } x > 2 \end{cases}$

$$\int_1^4 f(x)\,dx$$

$$= \int_1^2 f(x)\,dx + \int_2^4 f(x)\,dx$$

$$= \int_1^2 (2x+3)\,dx + \int_2^4 (-.5x+8)\,dx$$

$$= \left(x^2 + 3x\right)\Big|_1^2 + \left(-.5\frac{x^2}{2} + 8x\right)\Big|_2^4$$

$$= (4+6) - (1+3) + (-4+32) - (-1+16)$$

$$= 10 - 4 + 28 - 15$$

$$= 19$$

51. $E'(x) = 4x + 2$ is the rate of expenditure per day.

a. The total expenditure is hundreds of dollars in 10 days is

$$\int_0^{10} (4x+2)\,dx$$

$$= \left(\frac{4x^2}{2} + 2x\right)\Big|_0^{10}$$

$$= 2(100) + 20 - 0$$

$$= 220.$$

Therefore, since 220(100) = 22,000, the total expenditure is $22,000.

b. From the tenth to the twentyfifth day:

$$\int_{10}^{25} (4x+2)\,dx$$

$$= \left(\frac{4x^2}{2} + 2x\right)\Big|_{10}^{25}$$

$$= [2(625) + 50] - [2(100) + 20]$$

$$= 1080$$

That is, $108,000 is spent.

c. If $76,000, or 760(100), is spent,

$$\int_0^a (4x+2)\,dx = 760 .$$

$$\int_0^a (4x+2)\,dx = \left(2x^2 + 2x\right)\Big|_0^a$$

$$= 2a^2 + 2a$$

Solve $760 = 2a^2 + 2a$ by the quadratic formula.

$$2a^2 + 2a - 760 = 0$$

$$a^2 + a - 380 = 0$$

$$a = \frac{-1 \pm \sqrt{1 - 4(-380)}}{2}$$

Since the number of days must be positive,

$$a = \frac{-1 \pm \sqrt{1521}}{2}$$

$$= 19 \text{ days.}$$

53. $\int_1^{11} \left(584.7 + 70.3x - 1.338x^2 - .3556x^3\right) dx$

$= \left(584.7x + 70.3\frac{x^2}{2} - 1.338\frac{x^3}{3} - .3556\frac{x^4}{4}\right)\Big|_1^{11}$

$= 584.7(11) + 70.3\frac{(11)^2}{2} - 1.338\frac{(11)^3}{3}$

$\quad - .3556\frac{(11)^4}{4}$

$\quad - \left(584.7(1) + 70.3\frac{1^2}{2} - 1.338\frac{1^3}{3} - .3556\frac{1^4}{4}\right)$

$= 8789.6391 - 619.3151$

$= 8170.324 \approx 8170$

CD sales from 1994 to 2003 are about $8,170,000,000.

55. $\int_4^{20} (-.0028x^3 + .168x^2 + 7.8x + 404)\,dx$

$= \left(-.0028\frac{x^4}{4} + .168\frac{x^3}{3} + 7.8\frac{x^2}{2} + 404x\right)\Big|_4^{20}$

$= -.0028\frac{20^4}{4} + .168\frac{20^3}{3} + 7.8\frac{20^2}{2} + 404(20)$

$\quad - \left(-.0028\frac{4^4}{4} + .168\frac{4^3}{3} + 7.8\frac{4^2}{2} + 404(4)\right)$

$= 9976 - 1681.8048$

$= 8294.1952$

World energy consumption from 2004 to 2020 is projected to be about 8294.1952 quadrillion BTUs.

57. $L'(t) = \dfrac{70\ln(t+1)}{t+1}$

Consider $\int \dfrac{70\ln(t+1)}{t+1}\,dt$.

Let $u = \ln(t+1)$.

Then $du = \dfrac{1}{t+1}\,dt$.

$\int \dfrac{70\ln(t+1)}{(t+1)} = 70\int \ln(t+1)\dfrac{1}{t+1}\,dt$

$\qquad = 70\int u\,du$

$\qquad = 70\left(\dfrac{u^2}{2}\right) + C$

$\qquad = 35u^2 + C$

$\qquad = 35[\ln(t+1)]^2 + C$

a. The total number of barrels leaked on the first day is

$\int_0^{24} \dfrac{70\ln(t+1)}{t+1} = 35[\ln(t+1)]^2\Big|_0^{24}$

$\qquad = 35(\ln 25)^2 - 40(\ln 1)^2$

$\qquad = 35(\ln 25)^2$

$\qquad \approx 363$ barrels.

b. The total number of barrels leaked on the second day is

$\int_{24}^{48} \dfrac{70\ln(t+1)}{t+1}\,dt = 35[\ln(t+1)]^2\Big|_{24}^{48}$

$\qquad = 35(\ln 49)^2 - 35(\ln 25)^2$

$\qquad \approx 167$ barrels.

57. Continued

c. The amount of oil leaked from one day to the next is given by

$$\int_a^{a+24} \frac{70\ln(t+1)}{t+1}\,dt = [35\ln(t+1)]^2\Big|_a^{a+24}$$

$$= 35[\ln(a+25)]^2 - 35[\ln(a+1)]^2$$

$$= 35[\ln(a+25) + \ln(a+1)]$$

$$\quad\cdot[\ln(a+25) - \ln(a+1)]$$

$$= 35[\ln(a+25) + \ln(a+1)]\cdot\ln\left(\frac{a+25}{a+1}\right).$$

As a gets larger,

$$\ln\left(\frac{a+25}{a+1}\right) \approx \ln 1 = 0.$$

In the long run, the amount of oil leaked per day is decreasing to 0.

59. a $\displaystyle\int_0^9 f(x)\,dx = \int_0^9 (-.77x^2 + 2.5x + 40)\,dx$

$$= \left(-.77\frac{x^3}{3} + 2.5\frac{x^2}{2} + 40x\right)\Big|_0^9$$

$$= -.77\frac{9^3}{3} + 2.5\frac{9^2}{2} + 40(9) - 0$$

$$= 274.14$$

This integral represents the total number of people from age 0 to 90. There are about 274.14 million people aged 0 to 90 according to the 2000 census.

b. $\displaystyle\int_{3.5}^{5.5} f(x)\,dx$

$$= \left(-.77\frac{x^3}{3} + 2.5\frac{x^2}{2} + 40x\right)\Big|_{3.5}^{5.5}$$

$$= -.77\frac{5.5^3}{3} + 2.5\frac{5.5^2}{2} + 40(5.5)$$

$$\quad - \left(-.77\frac{3.5^3}{3} + 2.5\frac{3.5^2}{2} + 40(3.5)\right)$$

$$\approx 70.8$$

There are about 70.8 million baby boomers.

61. $\displaystyle\int_6^{13} \frac{5298}{62 - .533}\,dx = \frac{5298}{-.533}\ln(62 - .533x)\Big|_6^{13}$

$$= \frac{5298}{-.533}\left(\begin{array}{c}\ln(62 - .533(13)) \\ -\ln(62 - .533(6))\end{array}\right)$$

$$= 651.5904450$$

The integral represents the number of millions of short tons of new paper and paperboard supplies sold from January 1, 1995 to the end of 2001. There were 651.6 million short tons sold.

63. $\displaystyle\int_0^5 \frac{1}{3}e^{-x/3}\,dx =$

$$= -e^{-x/3}\Big|_0^5 = -e^{-5/3} - (-1)$$

$$= .811$$

65. $\displaystyle\int_2^5 \frac{1}{8}e^{-x/8}\,dx =$

$$= -e^{-x/8}\Big|_2^5 = -e^{-5/8} - \left(-e^{-2/8}\right)$$

$$= .244$$

Section 13.5 Applications of Integrals

1. $M(x) = 60\left(1 + x^2\right)$

The total maintenance charge on a two-year lease is given by

$$\int_0^2 60\left(1 + x^2\right)dx$$

$$= 60\int_0^2 \left(1 + x^2\right)dx$$

$$= 60\left(x + \frac{x^3}{3}\right)\Big|_0^2 = 60\left(2 + \frac{2^3}{3} - 0\right)$$

$$= 60\left(2 + \frac{8}{3}\right) = 120 + 160$$

$$= \$280.$$

Monthly addition for maintenance:

$$\frac{280}{24} = \$11.67$$

3. $\int_{15}^{18} (3.95 + .03x + .016x^2)dx$

$$= \left(3.95x + \frac{.03x^2}{2} + .\frac{.016x^3}{3}\right)\Big|_{15}^{18}$$

$$= 3.95(18) + .03\frac{(18)^2}{2} + .016\frac{(18)^3}{3}$$

$$- \left(3.95(15) + .03\frac{(15)^2}{2} + .016\frac{(15)^3}{3}\right)$$

$$= 26.44$$

Total revenue from 2000 through the end of 2002 is $26.44 billion.

5. Rate of savings:

$S(t) = 1000(t + 2)$

During the first year

Total savings:

$$\int_0^1 1000(t+2)dt = 1000\int_0^1 (t+2)dt$$

$$= 1000\left(\frac{t^2}{2} + 2t\right)\Big|_0^1$$

$$= 1000\left(\frac{1}{2} + 2 - 0\right)$$

$$= 1000\left(\frac{5}{2}\right)$$

$$= \$2500$$

5. Continued

During the first 6 years

Total savings:

$$\int_0^6 1000(t+2)dt = 100\int_0^6 (t+2)dt$$

$$= 1000\left(\frac{t^2}{2} + 2t\right)\Big|_0^6$$

$$= 1000\left(\frac{36}{2} + 12 - 0\right)$$

$$= 1000(30)$$

$$= \$30,000$$

7. The rate of production is given by the function $P(x) = 1000e^{.2x}$.

In the first 4 years, the total production will be

$$\int_0^4 1000e^{.2x}dx = 1000\int_0^4 e^{.2x}dx$$

$$= 1000\left(\frac{e^{.2x}}{.2}\right)\Big|_0^4$$

$$= \frac{1000}{.2}\left(e^{.8} - e^0\right)$$

$$\approx 6127.7$$

so a production of 20,000 units in the first 4 years will not be met.

9. $y = 3x$ and $y = x^2 - 4$ from $x = -1$ to $x = 4$

On the interval, $3x \geq x^2 - 4$

$$\int_{-1}^4 (3x - (x^2 - 4))dx$$

$$= \int_{-1}^4 (-x^2 + 3x + 4)dx$$

$$= -\frac{x^3}{3} + \frac{3x^2}{2} + 4x\Big|_{-1}^4$$

$$= -\frac{(4)^3}{3} + \frac{3(4)^2}{2} + 4(4)$$

$$- \left(-\frac{(-1)^3}{3} + \frac{3(-1)^2}{2} + 4(-1)\right)$$

$$= \frac{125}{6} = 20.83$$

11. $y = x^2$ and $y = x^3$ from $x = 0$ to $x = 1$

On the interval, $x^2 \geq x^3$

$$\int_0^1 (x^2 - x^3)\,dx = \left(\frac{x^3}{3} - \frac{x^4}{4}\right)\Big|_0^1$$

$$= \frac{1}{3} - \frac{1}{4} - 0$$

$$= \frac{1}{12}$$

13. The following solution presupposes the use of TI-83 graphics calculator. Similar results can be obtained from other graphics calculators.

$$A = \int_1^4 (2xe^x - \ln x)\,dx$$

Under the MATH menu use fnInt key in the following:

fnInt$(2xe^x - \ln x, x, 1, 4)$.

The value returned is 325.0437. Therefore, the area is approximately 325.04.

15. $A = \int_{-1}^2 \left(\sqrt{9 - x^2} - \sqrt{x+1}\right)dx$

Under the MATH menu we use fnInt key in the following:

fnInt$\left(\sqrt{9 - x^2} - \sqrt{x+1}, x, -1, 2\right)$

The value returned is 4.999220084. The area is approximately 4.999.

17. $S'(x) = 150 - x^2$

$$C'(x) = x^2 + \frac{11}{4}x$$

a. Let a be the last year it will be profitable to use this new machine. Then,

$$\int_a^{a+1} (150 - x^2)\,dx$$

$$-\int_a^{a+1}\left(x^2 + \frac{11}{4}x\right)dx < 0$$

$$\left(150x - \frac{x^3}{3}\right)\Big|_a^{a+1}$$

$$-\left(\frac{x^3}{3} + \frac{11}{4}\cdot\frac{x^2}{2}\right)\Big|_a^{a+1} < 0$$

17. Continued

Multiply by 24 to clear fractions.

$$\left(3600x - 8x^3\right)\Big|_a^{a+1} - \left(8x^3 + 33x^2\right)\Big|_a^{a+1} < 0$$

$$\left[3600(a+1) - 8(a+1)^3\right]$$

$$-\left(3600a - 8a^3\right)$$

$$-\left[8(a+1)^3 + 33(a+1)^2\right]$$

$$+\left(8a^3 + 33a^2\right) < 0$$

$$3600a + 3600 - 8\left(a^3 + 3a^2 + 3a + 1\right)$$

$$-3600a + 8a^3$$

$$-\left[8\left(a^3 + 3a^2 + 3a + 1\right)\right.$$

$$\left. +33\left(a^2 + 2a + 1\right)\right] + 8a^3 + 33a^2 < 0$$

$$3600a + 3600 - 8a^3 - 24a^2 - 24a - 8$$

$$-3600a + 8a^3 - 8a^3 - 24a^2 - 24a - 8$$

$$-33a^2 - 66a - 33 + 8a^3 + 33a^2 < 0$$

$$-48a^2 - 114a + 3351 < 0$$

Multiply by -1.

$$48a^2 + 114a - 3351 > 0$$

Solving the equation

$$48a^2 + 114a - 3351 = 0,$$

we find

$a \approx -9.63$ or $z \approx 7.25$.

The solution of $48a^2 + 114a - 3351 > 0$ is $(-\infty, -9.63)$ or $(7.25, \infty)$.

Thus, the last year it will be profitable is during the 8th yr.

(7.25 yr occurs during year 8.)

b. The total net savings during the first year is given by

$$\int_0^1 (150 - x^2)\,dx - \int_0^1\left(x^2 + \frac{11}{4}x\right)dx$$

$$= \left(150x - \frac{x^3}{3}\right)\Big|_0^1 - \left(\frac{x^3}{3} + \frac{11}{4}\cdot\frac{x^2}{2}\right)\Big|_0^1$$

$$= \left(150 - \frac{1}{3}\right) - 0 - \left[\left(\frac{1}{3} + \frac{11}{8}\right) - 0\right]$$

$$\approx 147.96$$

The total is about $148.

17. Continued

c. The total net savings over the entire period of use is given by

$$\int_0^8 \left(150 - x^2\right) dx - \int_0^8 \left(x^2 + \frac{11}{4}x\right) dx$$

$$= \left(150x - \frac{x^3}{3}\right)\Big|_0^8 - \left(\frac{x^3}{3} + \frac{11}{4} \cdot \frac{x^2}{2}\right)\Big|_0^8$$

$$= \left(1200 - \frac{512}{3}\right) - 0 - \left[\left(\frac{512}{3} + \frac{11}{4} \cdot \frac{64}{2}\right) - 0\right]$$

$$= 770.\overline{6}$$

The total is about $771.

19. $E(x) = e^{.15x}$, $I(x) = 120.3 - e^{.15x}$

a. To find the optimum number of days, solve the equation $E(x) = I(x)$. The solution will give the value of x where the two curves meet.

$$e^{.15x} = 120.3 - e^{.15x}$$

$$2e^{.15x} = 120.3$$

$$e^{.15x} = 60.15$$

$$.15x = \ln 60.15$$

$$x = \frac{\ln 60.15}{.15}$$

$$x \approx 27$$

The optimum number of days is 27.

b. The total income for the optimum number of days is given by

$$\int_0^{27} \left(120.3 - e^{.15x}\right) dx$$

$$= 120.3x - \frac{1}{.15}e^{.15x}\Big|_0^{27}$$

$$= (3248.1 - 382.65) - (0 - 6.67)$$

$$= 2872.12$$

The total is $2872.12.

19. Continued

c. The total expenditure for the optimum number of days is given by

$$\int_0^{27} e^{.15x} dx = \frac{1}{.15}e^{.15x}\Big|_0^{27}$$

$$= 382.65 - 6.67$$

$$= 375.98.$$

The total is $375.98.

d. The maximum profit is
$2872.12 - $375.98 = $2496.14.

21. a. The rate of consumption will equal the rate of production when

$$\frac{20}{1.2t + 1.6} = t + .8$$

$$20 = (1.2t + 1.6)(t + .8)$$

$$20 = 1.2t^2 + 2.56t + 1.28$$

$$0 = 1.2t^2 + 2.56t - 18.72$$

Multiply by 100.

$$120t^2 + 256t - 1872 = 0$$

Applying the quadratic formula, $t = 3.02$ or $t = -5.16$. Reject $t = -5.16$. In 3 yr the rates will be equal.

21. Continued

b. The total excess production before consumption and production are equal is given by

$$\int_0^{3.02} \frac{20}{1.2t+1.6} dt - \int_0^{3.02} (t+.8)dt$$

$$= \frac{20}{1.2} \ln |1.2t+1.6| \Big|_0^{3.02} - \left(\frac{t^2}{2}+.8t\right)\Big|_0^{3.02}$$

$$= \frac{20}{1.2} \ln 5.224 - \frac{20}{1.2} \ln 1.6 - \left[\frac{(3.02)^2}{2}+.8(3.02)-0\right]$$

$$\approx 12.74$$

The total is about 12.74 trillion gallons.

23. If a is the number of years before the net effect is zero, then

$$\int_0^a 140t^{.4} dt = \int_0^a 1.6t^{2.5} dt$$

$$140 \cdot \frac{t^{1.4}}{1.4}\Big|_0^a = 1.6 \cdot \frac{t^{3.5}}{3.5}\Big|_0^a$$

$$100a^{1.4} = \frac{16}{35} a^{3.5}$$

$$3500a^{1.4} = 16a^{3.5}$$

$$\frac{3500}{16} = a^{2.1}$$

$$a = \left(\frac{3500}{16}\right)^{\frac{1}{2.1}}$$

$$a \approx 13.0095.$$

The net effect is zero in about 13 yr.

25. a. We graph the linear functions

$$S(q) = \frac{7}{5}q$$

and

$$D(q) = -\frac{3}{5}q + 10$$

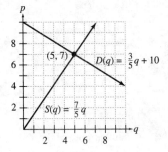

b. $S(q) = D(q)$

$$\frac{7}{5}q = -\frac{3}{5}q + 10$$

$$7q = -3q + 50$$

$$10q = 50$$

$$q = 5$$

$$S(5) = 7$$

The equilibrium point is (5, 7).

25. Continued

c. The consumers' surplus is given by

$$\int_0^5 \left[\left(-\frac{3}{5}q + 10 \right) - 7 \right] dq$$

$$= \int_0^5 \left(-\frac{3}{5}q + 3 \right) dq$$

$$= \left(-\frac{3}{5} \cdot \frac{q^2}{2} + 3q \right) \Big|_0^5$$

$$= \left(-\frac{15}{2} + 15 \right) - 0$$

$$= 7.5.$$

The consumers' surplus is $7.50.

d. The producers' surplus is given by

$$\int_0^5 \left(7 - \frac{7}{5}q \right) dq = \left(7q - \frac{7}{5} \cdot \frac{q^2}{2} \right) \Big|_0^5$$

$$= \left(35 - \frac{35}{2} \right) - 0$$

$$= 17.5.$$

The producers' surplus is $17.50.

27. If $S(q) = 100 + 3q^{\frac{3}{2}} + q^{\frac{5}{2}}$,

$S(9) = 100 + 81 + 243 = 424$.

Therefore, the producers' surplus is given by

$$\int_0^9 \left[424 - \left(100 + 3q^{\frac{3}{2}} + q^{\frac{5}{2}} \right) \right] dq$$

$$= \int_0^9 \left(324 - 3q^{\frac{3}{2}} - q^{\frac{5}{2}} \right) dq$$

$$= \left(324q - 3 \cdot \frac{q^{\frac{5}{2}}}{\frac{5}{2}} - \frac{q^{\frac{7}{2}}}{\frac{7}{2}} \right) \Big|_0^9$$

$$= \left(324q - \frac{6}{5}q^{\frac{5}{2}} - \frac{2}{7}q^{\frac{7}{2}} \right) \Big|_0^9$$

$$\approx (2916 - 291.6 - 624.86) - 0$$

$$= 1999.54.$$

The producers' surplus is $1999.54.

29. If $D(q) = \dfrac{15,500}{(3.2q + 7)^3}$, then

$$D(5) = \frac{15,500}{(16 + 7)^3} = 1.2739 .$$

Therefore, consumers' surplus is given by

$$\int_0^5 \left[\frac{15,500}{(3.2q + 7)^3} - 1.2739 \right] dq .$$

Let $u = 3.2q + 7$. Then $du = 3.2 \, dq$. If $q = 5$, $u = 23$. If $q = 0$, $u = 7$.

$$\int_0^5 \left[\frac{15,500}{(3.2q + 7)^3} - 1.2739 \right] dq$$

$$= \frac{1}{3.2} \int_0^6 \left[15,500(3.2q + 7)^{-3} - 1.2739 \right] 3.2 \, dq$$

$$= \frac{1}{3.2} \int_7^{23} \left(15,500u^{-3} - 1.2739 \right) du$$

$$= \frac{1}{3.2} \left(15,500 \cdot \frac{u^{-2}}{-2} - 1.2739u \right) \Big|_7^{23}$$

$$= \frac{1}{3.2} \left(-\frac{7750}{u^2} - 1.2739u \right) \Big|_7^{23}$$

$$= \frac{1}{3.2} \left(-\frac{7750}{529} - 29.2997 \right) - \frac{1}{3.2} \left(-\frac{7750}{49} - 8.9173 \right)$$

$$= 38.50$$

The consumers' surplus is $30.50.

31. a. We graph the quadratic functions

$$S(q) = q^2 + \frac{11}{4}q \text{ and } D(q) = 150 - q^2 .$$

See the graph in the answer section of the textbook.

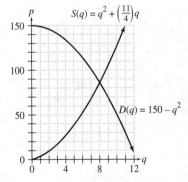

31. Continued

b. $S(q) = D(q)$

$$q^2 + \frac{11}{4}q = 150 - q^2$$

$$2q^2 + \frac{11}{4}q - 150 = 0$$

$$8q^2 + 11q - 600 = 0$$

$$(8q + 75)(q - 8) = 0$$

$$q = -\frac{75}{8} \text{ or } q = 8$$

We discard the negative value.

$$S(8) = 8^2 + \frac{11}{4} \cdot 8 = 86$$

The equilibrium point is (8, 86).

c. The consumers' surplus is given by

$$\int_0^8 \left[(150 - q^2) - 86 \right] dq$$

$$= \int_0^8 (64 - q^2) dq$$

$$= \left(64q - \frac{q^3}{3} \right) \Big|_0^8$$

$$= \left(512 - \frac{512}{3} \right) - 0$$

$$\approx 341.33.$$

The consumers' surplus is $341.33.

d. The producers' surplus is given by

$$\int_0^8 \left[86 - \left(q^2 + \frac{11}{4}q \right) \right] dq$$

$$= \left(86q - \frac{q^3}{3} - \frac{11}{4} \cdot \frac{q^2}{2} \right) \Big|_0^8$$

$$\approx (688 - 170.67 - 88) - 0$$

$$= 429.33.$$

The producers' surplus is $429.33.

33. $y = 3\sqrt{x}$ and $y = 2x$

Determine the points of intersection

$$3\sqrt{x} = 2x$$

$$9x = 4x^2$$

$$4x^2 - 9x = 0$$

$$x(4x - 9) = 0$$

$$x = 0 \text{ and } x = \frac{9}{4}$$

On the interval $x = 0$ and $x = \frac{9}{4}$, $3\sqrt{x} \geq 2x$.

$$\int_0^{9/4} \left(3\sqrt{x} - 2x \right) dx = \left(\frac{3x^{3/2}}{\frac{3}{2}} - x^2 \right) \Big|_0^{9/4}$$

$$= \left(2x^{3/2} - x^2 \right) \Big|_0^{9/4}$$

$$= 2 \left(\frac{9}{4} \right)^{3/2} - \left(\frac{9}{4} \right)^2 - 0$$

$$= 2 \left(\frac{27}{8} \right) - \left(\frac{81}{16} \right)$$

$$= \frac{27}{16}$$

The area of the poster board is $\frac{27}{16}$ square units.

35. $f(x) = 70.71e^{.43x}$

The total increase is given by

$$\int_0^4 70.71e^{.43x} dx = 70.71 \cdot \frac{e^{.43x}}{.43} \Big|_0^4$$

$$= 70.71 \frac{e^{.43(4)}}{.43} - \frac{70.71}{.43}$$

$$= 753.89.$$

The total increase in costs is $753.89.

37. $I(x) = .9x^2 + .1x$

a. $I(.1) = .9(.1)^2 + .1(.1) = .019$
The lower 10% of income producers earn 1.9% of the total income of the population.

b. $I(.5) = .9(.5)^2 + .1(.5) = .275$
The lower 50% of income producers earn 27.5% of the total income of the population.

c. $I(.9) = .9(.9)^2 + .1(.9) = .819$
The lower 90% of income producers earn 81.9% of the total income of the population.

37. Continued

d. Graph $I(x) = x$ and $I(x) = .9x^2 + .1x$ for $0 \le x \le 1$ on the same set of axes.

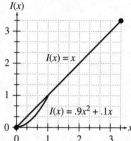

e. The area between the curves is given by

$$\int_0^1 \left[x - \left(.9x^2 + .1x \right) \right] dx$$

$$= \int_0^1 \left(.9x - .9x^2 \right) dx$$

$$= \left(.9 \cdot \frac{x^2}{2} - .9 \cdot \frac{x^3}{3} \right) \bigg|_0^1$$

$$= \left(.45x^2 - .3x^3 \right) \bigg|_0^1$$

$$= (.45 - .3) - 0 = .15.$$

This represents the amount of inequality of income distribution.

39. a. $\int_0^T H(x)dx$ will be a maximum when its derivative, $H(x)$, is equal to 0. This occurs when

$$H(x) = 20 - 2x = 0$$

$$x = 10.$$

Thus, to achieve the maximum production rate, 10 items must be made.

b. $\int_0^T H(x)dx = \int_0^T (20 - 2x)dx$

$$= \left(20x - x^2 \right) \bigg|_0^T$$

$$= 20T - T^2$$

This will be a maximum when its derivative, $H(T)$, is equal to 0.

$$H(T) = 20 - 2T = 0$$

$$T = 10$$

Thus, the maximum production rate per item is 10 hours per item.

Section 13.6 Tables of Integrals (Optional)

1. $\int \dfrac{-7}{\sqrt{x^2 + 36}} dx = -7 \int \dfrac{dx}{\sqrt{x^2 + 36}}$

Use entry 5 from the table with $a = 6$.

$$= -7 \ln \left| x + \sqrt{x^2 + 36} \right| + C$$

3. $\int \dfrac{6}{x^2 - 9} dx = 6 \int \dfrac{1}{x^2 - 9} dx$

Use entry 8 from the table with $a = 3$. Note that $x^2 > 3^2$.

$$= 6 \left(\frac{1}{2(3)} \ln \left| \frac{x-3}{x+3} \right| \right) + C$$

$$= \ln \left| \frac{x-3}{x+3} \right| + C,$$

$$\left(x^2 > 9 \right)$$

5. $\int \dfrac{-4}{x\sqrt{9-x^2}} dx = -4 \int \dfrac{dx}{x\sqrt{9-x^2}}$

Use entry 9 from the table with $a = 3$. Note that $0 < x < 3$.

$$= \frac{4}{3} \ln \left| \frac{3 + \sqrt{9-x^2}}{x} \right| + C,$$

$$(0 < x < 3)$$

7. $\int \dfrac{-5x}{3x+1} dx = -5 \int \dfrac{x}{3x+1} dx$

Use entry 11 from table with $a = 3$, $b = 1$.

$$= -5 \left(\frac{x}{3} - \frac{1}{9} \ln \left| 3x + 1 \right| \right) + C$$

$$= -\frac{5x}{3} + \frac{5}{9} \ln \left| 3x + 1 \right| + C$$

9. $\int \dfrac{13}{3x(3x-5)} dx = \dfrac{13}{3} \int \dfrac{1}{x(3x-5)} dx$

Use entry 13 from the table with $a = 3$, $b = -5$.

$$= \frac{13}{3} \left(-\frac{1}{5} \ln \left| \frac{x}{3x-5} \right| \right) + C$$

$$= -\frac{13}{15} \ln \left| \frac{x}{3x-5} \right| + C$$

11. $\int \frac{4}{4x^2-1}dx$

$= 4\int \frac{dx}{4x^2-1}$

$= 2\int \frac{2\,dx}{4x^2-1}$

Let $u = 2x$. Then $du = 2\,dx$.

$\int \frac{4}{4x^2-1}dx = 2\int \frac{du}{u^2-1}$

Use entry 8 from the table with $a = 1$, $u^2 > 1$.

$= 2 \cdot \frac{1}{2}\ln\left|\frac{u-1}{u+1}\right| + C,\ u^2 > 1,$

Substitute $2x$ for u. Since

$u^2 > 1,\ 4x^2 > 1,\ \text{or}\ x^2 > \frac{1}{4}.$

$\int \frac{4}{4x^2-1}dx = \ln\left|\frac{2x-1}{2x+1}\right| + C,\ \left(x^2 > \frac{1}{4}\right)$

13. $\int \frac{3}{x\sqrt{1-9x^2}}dx = 3\int \frac{3\,dx}{3x\sqrt{1-9x^2}}$

Let $u = 3x$. Then $du = 3\,dx$.

$\int \frac{3}{x\sqrt{1-9x^2}}dx = 3\int \frac{du}{u\sqrt{1-u^2}}$

Use entry 9 from the table with $a = 1$.
Note $0 < u < 1$.

$= -3\ln\left|\frac{1+\sqrt{1-u^2}}{u}\right| + C$

$= -3\ln\left|\frac{1+\sqrt{1-9x^2}}{3x}\right| + C,$

$\left(0 < 3x < 1,\ \text{or}\ 0 < x < \frac{1}{3}\right)$

15. $\int \frac{15x}{2x+3}dx = 15\int \frac{x}{2x+3}dx$

Use entry 11 from the table with $a = 2$,
$b = 3$.

$= 15\left(\frac{x}{2} - \frac{3}{4}\ln|2x+3|\right) + C$

$= \frac{15x}{2} - \frac{45}{4}\ln|2x+3| + C$

17. $\int \frac{-x}{(5x-1)^2}dx = -\int \frac{x\,dx}{(5x-1)^2}$

Use entry 12 form the table with $a = 5$,
$b = -1$.

$= -\left(\frac{-1}{25(5x-1)} + \frac{1}{25}\ln|5x-1|\right) + C$

$= \frac{1}{25(5x-1)} - \frac{\ln|5x-1|}{25} + C$

19. $\int \frac{3x^4 \ln|x|}{4}dx$

Use entry 16, with $n = 4$.

$= \frac{3}{4}x^5\left(\frac{\ln|x|}{5} - \frac{1}{25}\right) + C$

21. $\int \frac{7\ln|x|}{x^2}dx = 7\int x^{-2}\ln|x|\,dx$

Use entry 16 from the table with $n = -2$.

$= 7x^{-1}\left(\frac{\ln|x|}{-1} - \frac{1}{1}\right) + C$

$= \frac{7}{x}\left(-\ln|x| - 1\right) + C$

23. $\int xe^{-2x}dx$

Use entry 17 from the table with
$n = 1$ and $a = -2$.

$= \frac{xe^{-2x}}{-2} - \frac{1}{-2}\int x^0 e^{-2x}dx$

$= -\frac{1}{2}xe^{-2x} + \frac{1}{2}\int e^{-2x}dx$

$= -\frac{1}{2}xe^{-2x} + \frac{1}{2}\left(\frac{e^{-2x}}{-2}\right) + C$

$= -\frac{1}{2}xe^{-2x} - \frac{1}{4}e^{-2x} + C$

$= -\frac{xe^{-2x}}{2} - \frac{e^{-2x}}{4} + C$

25. $R'(x) = \dfrac{1000}{\sqrt{x^2 + 25}}$

The total revenue is given by

$$\int_0^{20} \frac{1000}{\sqrt{x^2 + 25}}\,dx = 1000\int_0^{20} \frac{1}{\sqrt{x^2 + 5^2}}\,dx\ .$$

We use entry 5 in the table with $a = 5$.

$$\int_0^{20} \frac{1000}{\sqrt{x^2 + 25}}\,dx$$

$$= 1000 \ln\left|\frac{x + \sqrt{x^2 + 25}}{5}\right|\Bigg\|_0^{20}$$

$$= 1000 \ln\left(\frac{20 + \sqrt{425}}{5}\right) - 1000 \ln 1$$

$$\approx 2094.71$$

The total revenue is approximately \$2094.71.

27. Rate of growth: $m'(x) = 25xe^{2x}$

The total growth is given by

$$\int_0^3 25xe^{2x}\,dx\ .$$

Use entry 17 with $n = 1$ and $a = 2$.

$$= 25\left(\frac{xe^{2x}}{2}\bigg|_0^3 - \frac{1}{2}\int_0^3 e^{2x}\,dx\right)$$

$$= 25\left(\frac{3e^6}{2} - \frac{e^{2x}}{4}\bigg|_0^3\right)$$

$$= 25\left[\frac{3e^6}{2} - \frac{1}{4}\left(e^6 - e^0\right)\right]$$

$$= \frac{75}{2}e^6 - \frac{25}{4}\left(e^6 - 1\right)$$

$$= 15{,}128.58 - 2515.18 \approx 12{,}613$$

The total accumulated growth after 3 days is about 12,613 microbes.

Section 13.7 Differential Equations

1. $\dfrac{dy}{dx} = -2x + 5x^2$

$$dy = \left(-2x + 5x^2\right)dx$$

$$\int dy = \int\left(-2x + 5x^2\right)dx$$

$$y = \frac{-2x^2}{2} + \frac{5x^3}{3} + C$$

$$y = -x^2 + \frac{5x^3}{3} + C$$

3. $3x^3 - 2\dfrac{dy}{dx} = 0$

$$-2\frac{dy}{dx} = -3x^3$$

$$\frac{dy}{dx} = \frac{3}{2}x^3$$

$$y = \int \frac{3}{2}x^3\,dx$$

$$y = \frac{3}{2}\int x^3\,dx$$

$$y = \frac{3}{2}\cdot\frac{x^4}{4} + C$$

$$y = \frac{3x^4}{8} + C$$

5. $y\dfrac{dy}{dx} = x$

$$y\,dy = x\,dx$$

$$\frac{y^2}{2} = \frac{x^2}{2} + C_1$$

$$y^2 = x^2 + C$$

7. $\dfrac{dy}{dx} = 4xy$

$\dfrac{dy}{y} = 4x\,dx$

$\displaystyle\int \dfrac{dy}{y} = 4\int x\,dx$

$\ln|y| = 4\dfrac{x^2}{2} + C$

$\ln|y| = 2x^2 + C$

$|y| = e^{2x^2 + C} = e^{2x^2} e^c$

$y = Me^{2x^2}$

9. $\dfrac{dy}{dx} = 3x^2 y - 2xy$

$\dfrac{dy}{dx} = y\left(3x^2 - 2x\right)$

$\dfrac{dy}{y} = \left(3x^2 - 2x\right)dx$

$\displaystyle\int \dfrac{dy}{y} = \int \left(3x^2 - 2x\right)dx$

$\ln|y| = \dfrac{3x^3}{3} - \dfrac{2x^2}{2} + C$

$\ln|y| = x^3 - x^2 + C$

$|y| = e^{x^3 - x^2 + C}$

$|y| = e^{x^3 - x^2} e^C$

$y = Me^{x^3 - x^2}$

11. $\dfrac{dy}{dx} = \dfrac{y}{x}, \; x > 0$

$\dfrac{dy}{y} = \dfrac{dx}{x}$

$\displaystyle\int \dfrac{dy}{y} = \int \dfrac{dx}{x}$

$\ln|y| = \ln|x| + C$

$|y| = e^{\ln|x| + C}$

$y = e^{\ln|x|} e^C$

$y = Mx$

13. $\dfrac{dy}{dx} = y - 7$

$\dfrac{dy}{y - 7} = dx$

$\ln|y - 7| = x + C$

$|y - 7| = e^{x + C} = e^x e^C$

$y - 7 = Me^x$

$y = 7 + Me^x$

15. $\dfrac{dy}{dx} = y^2 e^x$

$\dfrac{dy}{y^2} = e^x\,dx$

$\displaystyle\int y^{-2}\,dy = \int e^x\,dx$

$-y^{-1} = e^x + C$

$-\dfrac{1}{y} = e^x + C$

$y = -\dfrac{1}{e^x + C}$

17. $\dfrac{dy}{dx} + 2x = 3x^2 \, ; y = 2$ when $x = 0$.

$\dfrac{dy}{dx} = 3x^2 - 2x$

$y = \displaystyle\int \left(3x^2 - 2x\right)dx$

The general solution is

$y = x^3 - x^2 + C$.

When $x = 0$, $y = 2$.

$2 = 0^3 - 0^2 + C$

$C = 2$

Thus, the particular solution is

$y = x^3 - x^2 + 2$.

19. $\dfrac{dy}{dx}\left(x^3 + 28\right) = \dfrac{6x^2}{y} \, ; y^2 = 6$ when $x = -3$

$y\,dy = \dfrac{6x^2}{x^3 + 28}\,dx$

$\displaystyle\int y\,dy = \int \dfrac{6x^2}{x^3 + 28}\,dx$

$\dfrac{y^2}{2} = \dfrac{6\ln\left|x^3 + 28\right|}{3} + C$

$y^2 = 4\ln\left|x^3 + 28\right| + C$

19. Continued

$y^2 = 6$ when $x = -3$, so

$6 = 4\ln\left|(-3)^3 + 28\right| + C$

$C = 6$

so $y^2 = 4\ln\left|x^3 + 28\right| + 6$

21. $\dfrac{dy}{dx} = \dfrac{x^2}{y}$; $y = 3$ when $x = 0$.

$y\,dy = x^2\,dx$

$\int y\,dy = \int x^2\,dx$

$\dfrac{y^2}{2} = \dfrac{x^3}{3} + C$

Let $x = 0$ and $y = 3$.

$\dfrac{9}{2} = C$

$\dfrac{y^2}{2} = \dfrac{x^3}{3} + \dfrac{9}{2}$

The particular solution is

$y^2 = \dfrac{2}{3}x^3 + 9$.

23. $(5x + 3)y = \dfrac{dy}{dx}$; $y = 1$ when $x = 0$.

$(5x + 3)dx = \dfrac{dy}{y}$

$\int (5x + 3)dx = \int \dfrac{dy}{y}$

$\dfrac{5x^2}{2} + 3x = \ln|y| + C$

If $x = 0$ and $y = 1$,

$0 = \ln|1| + C$

$0 = C$.

We now have

$\dfrac{5x^2}{2} + 3x = \ln|y|$.

Rewrite this equation in exponential form. The particular solution is

$y = e^{2.5x^2 + 3x}$.

25. $\dfrac{dy}{dx} = \dfrac{7x + 1}{y - 3}$; $y = 4$ when $x = 0$.

$(y - 3)dy = (7x + 1)dx$

$\int (y - 3)dy = \int (7x + 1)dx$

$\dfrac{y^2}{2} - 3y = \dfrac{7x^2}{2} + x + C$

If $x = 0$ and $y = 4$,

$\dfrac{16}{2} - 12 = C$

$-4 = C$.

The particular solution is

$\dfrac{y^2}{2} - 3y = \dfrac{7x^2}{2} + x - 4$.

27. Answers vary.

29. $\dfrac{dy}{dx} = -\dfrac{40}{32 - 4x}$

$\dfrac{dy}{dx} = -\dfrac{4(10)}{4(8 - x)}$

$\dfrac{dy}{dx} = -\dfrac{10}{8 - x} = \dfrac{10}{x - 8}$

$y = \int \dfrac{10}{x - 8}\,dx$

$y = 10\int \dfrac{dx}{x - 8}$

$y = 10\ln|x - 8| + C$

Because x represents the investment in thousands of dollars, an investment of \$1000 corresponds to $x = 1$.

When $x = 1$, $y = 100$.

$100 = 10\ln|1 - 8| + C$

$100 = 10\ln 7 + C$

$C = 100 - 10\ln 7$

$y = 10\ln|x - 8| + 100 - 10\ln 7$

$y = 10\left(\ln|x - 8| - \ln 7\right) + 100$

$y = 10\ln\left|\dfrac{x - 8}{7}\right| + 100$

a. \$3000 corresponds to $x = 3$. If $x = 3$,

$y = 10\ln\left|\dfrac{-5}{7}\right| + 100$

≈ 96.64

The productivity is approximately 96.64.

29. Continued

b. $5000 corresponds to $x = 5$. If $x = 5$,

$$y = 10 \ln \left| \frac{-3}{7} \right| + 100$$

$$y \approx 91.53$$

The productivity is approximately 91.53.

c. Expenditures cannot reach $8000 since the fraction would be zero and 0 is not in the domain of the ln function.

31. $\dfrac{dy}{dt} = kt$; $k = 8$ and $y = 50$ when $t = 0$..

a. Thus, $\dfrac{dy}{dt} = 8t$

$$dy = 8t \, dt$$

$$y = \int 8t \, dt$$

$$y = 4t^2 + C.$$

When $t = 0$, $y = 50$, so

$$50 = 4(0)^2 + C$$

$$C = 50,$$

and

$$y = 4t^2 + 50 .$$

b. If $y = 550$, then

$$550 = 4t^2 + 50$$

$$4t^2 = 500$$

$$t^2 = 125$$

$$t = \pm 11.$$

The product should be dated 11 days from time $t = 0$.

33. $y = Me^{kt}$

1995 corresponds to $t = 0$, when the population is 31.4 million, so

$$y = 31.4e^{kt}$$

when $t = 55$, in 2050,

$y = 53.6$, so

$$53.6 = 31.4e^{k(55)}$$

$$k = .009723$$

so $y = 31.4e^{.009723t}$

35. $\dfrac{dy}{dx} = 30 - y$; $y = 1000$ when $x = 0$

$$\frac{dy}{30 - y} = dx$$

$$\int \frac{dy}{30 - y} = \int dx$$

$$-\ln|30 - y| = x + C$$

$$\ln|30 - y| = -x + C$$

$$|30 - y| = e^{-x + C}$$

$$30 - y = Me^{-x}$$

$$y = 30 - Me^{-x}$$

$$1000 = 30 - M$$

$$M = -970$$

$$y = 30 + 970e^{-x}$$

a. If $x = 2$, $y = 30 + 970e^{-2} \approx 161.28$.
About 161.28 thousand bacteria are present.

b. If $x = 7$, $y = 30 + 970e^{-7} \approx 30.88$.
About 30.88 thousand bacteria are present.

c. If $x = 10$, $y = 30 + 970e^{-10} \approx 30.44$.
About 30.44 thousand bacteria are present.

37. $E = 2$

$$-\frac{p}{q} \cdot \frac{dq}{dp} = 2$$

$$\frac{1}{q} dq = -\frac{2}{p} dp$$

$$\int \frac{1}{q} dq = -2 \int \frac{1}{p} dp$$

$$\ln q = -2 \ln p + \ln C$$

$$\ln q = \ln p^{-2} + \ln C$$

$$e^{\ln q} = e^{\ln p^{-2} + \ln C} = e^{\ln p^{-2}} e^{\ln C}$$

$$q = Cp^{-2}$$

$$q = \frac{C}{p^2}$$

39. a. $\dfrac{dy}{dt} = ky$

$\displaystyle\int \dfrac{dy}{y} = \int k\,dt$

$\ln|y| = kt + C$

$e^{\ln|y|} = e^{kt+C}$

$y = \pm\left(e^{kt}\right)\left(e^{C}\right)$

$y = Me^{kt}$

If $y = 1$ when $t = 0$ and $y = 5$ when $t = 2$, we have the following system of equations.

$1 = Me^{k(0)}$ (1)

$5 = Me^{2k}$ (2)

From equation (1),

$1 = M(1)$

$M = 1$.

Substitute 1 for M in equation (2) to find k.

$5 = (1)e^{2k}$

$e^{2k} = 5$

$2k \ln e = \ln 5$

$k = \dfrac{\ln 5}{2}$

$k \approx .8$

b. If $k = .8$ and $M = 1$, then

$y = e^{.8t}$.

When $t = 3$,

$y = e^{.8(3)}$

$y = e^{2.4}$

$y \approx 11$.

In 3 days, 11 people heard the rumor.

c. When $t = 5$,

$y = e^{.8(5)}$

$y = e^{4}$

$y \approx 55$.

In 5 days, 55 people heard the rumor.

d. When $t = 10$,

$y = e^{.8(10)}$

$y = e^{8}$

$y \approx 2981$.

In 10 days, 2981 people heard the rumor.

41. $\dfrac{dT}{dt} = -k(T - C)$

$= -k(T - 68)$

$\dfrac{1}{T-68}\,dT = -k\,dt$

$\displaystyle\int \dfrac{1}{T-68}\,dT = \int -k\,dt$

$\ln|T - 68| = -kt$

$T - 68 = Me^{-kt}$

$T = Me^{-kt} + 68$

$M = 98.6 - 68 = 30.6$, so we have

$T = 30.6e^{-kt} + 68$

when $t = 1$, $T = 90$, so

$90 = 30.6e^{-.33t} + 68$, $k = .33$

a. $T = 68 + 30.6e^{-.33t}$

b. After two hours,

$T = 68 + 30.6e^{-.33(2)} = 83.82$

so the temperature of the body is about $83.8°$.

c. When $T = 75$, we have

$75 = 68 + 30.6e^{-.33t}$

so $t = 4.46$

so the body will be $75°$ in about 4.5 hours.

d. The body will be within $.01°$ of the surrounding air when

$68.01 = 68 + 30.6e^{-.33t}$

$t = 24.32$

so the body will be within $.01°$ in about 24.3 hours.

Chapter 13 Review Exercises

1. $\displaystyle\int \left(x^2 - 3x - 5\right)dx$

$= \dfrac{x^3}{3} - 3\cdot\dfrac{x^2}{2} - 5x + C$

$= \dfrac{x^3}{3} - \dfrac{3x^2}{2} - 5x + C$

2. $\displaystyle\int \left(6 - x^2\right)dx = 6x - \dfrac{x^3}{3} + C$

3. $\int 7\sqrt{x}\,dx = 7\int x^{\frac{1}{2}}\,dx$

$$= 7 \cdot \frac{x^{\frac{3}{2}}}{\frac{3}{2}} + C$$

$$= 7 \cdot \frac{2}{3} x^{\frac{3}{2}} + C$$

$$= \frac{14x^{\frac{3}{2}}}{3} + C$$

4. $\int \frac{\sqrt{x}}{6}\,dx = \frac{1}{6}\int x^{\frac{1}{2}} + C$

$$= \frac{1}{6} \cdot \frac{x^{\frac{3}{2}}}{\frac{3}{2}} + C$$

$$= \frac{1}{6} \cdot \frac{2}{3} x^{\frac{3}{2}} + C$$

$$= \frac{x^{\frac{3}{2}}}{9} + C$$

5. $\int \left(x^{\frac{1}{2}} + 3x^{-\frac{2}{3}} \right) dx$

$$= \frac{x^{\frac{3}{2}}}{\frac{3}{2}} + 3 \cdot \frac{x^{\frac{1}{3}}}{\frac{1}{3}} + C$$

$$= \frac{2}{3} x^{\frac{3}{2}} + 3 \cdot \frac{3}{1} x^{\frac{1}{3}} + C$$

$$= \frac{2x^{\frac{3}{2}}}{3} + 9x^{\frac{1}{3}} + C$$

6. $\int \left(8x^{\frac{4}{3}} + x^{-\frac{1}{2}} \right) dx$

$$= 8 \cdot \frac{x^{\frac{7}{3}}}{\frac{7}{3}} + \frac{x^{\frac{1}{2}}}{\frac{1}{2}} + C$$

$$= 8 \cdot \frac{3}{7} x^{\frac{7}{3}} + 2x^{\frac{1}{2}} + C$$

$$= \frac{24x^{\frac{7}{3}}}{7} + 2x^{\frac{1}{2}} + C$$

7. $\int \frac{-4}{x^3}\,dx = -4\int x^{-3}\,dx$

$$= -4 \cdot \frac{x^{-2}}{-2} + C$$

$$= 2x^{-2} + C$$

8. $\int \frac{9}{x^4}\,dx = 9\int x^{-4}\,dx$

$$= 9 \cdot \frac{x^{-3}}{-3} + C$$

$$= -\frac{3}{x^3} + C$$

9. $\int -3e^{2x}\,dx = -3\int e^{2x}\,dx$

$$= -3 \cdot \frac{1}{2} e^{2x} + C$$

$$= -\frac{3e^{2x}}{2} + C$$

10. $\int 5e^{-x}\,dx = 5\int e^{-x}\,dx$

$$= 5 \cdot \frac{1}{-1} e^{-x} + C$$

$$= -5e^{-x} + C$$

11. $\int \frac{12}{x-1}\,dx = 12\int \frac{1}{x-1}\,dx$

$$= 12\ln|x-1| + C$$

12. $\int \frac{-14}{x+2}\,dx = -14\int \frac{1}{x+2}\,dx$

$$= -14\ln|x+2| + C$$

13. $\int xe^{3x^2}\,dx$

Let $u = 3x^2$. Then $du = 6x\,dx$.

$$\int xe^{3x^2}\,dx = \frac{1}{6}\int e^{3x^2} \cdot 6x\,dx$$

$$= \frac{1}{6}\int e^u\,du$$

$$= \frac{1}{6}e^u + C$$

$$= \frac{1}{6}e^{3x^2} + C$$

$$= \frac{e^{3x^2}}{6} + C$$

14. $\int 4xe^{x^2}\,dx$

Let $u = x^2$. Then $du = 2x\,dx$.

$$\int 4xe^{x^2}\,dx = 2\int e^{x^2} \cdot 2x\,dx$$

$$= 2\int e^u\,du$$

$$= 2e^u + C$$

$$= 2e^{x^2} + C$$

15. $\int \dfrac{3x}{x^2-1}dx$

Let $u = x^2 - 1$. Then $du = 2x\,dx$.

$$\int \dfrac{3x}{x^2-1}dx = 3 \cdot \dfrac{1}{2}\int \dfrac{1}{x^2-1}2x\,dx$$

$$= \dfrac{3}{2}\int \dfrac{1}{u}du$$

$$= \dfrac{3}{2}\ln|u| + C$$

$$= \dfrac{3\ln|x^2-1|}{2} + C$$

16. $\int \dfrac{-6x}{2-x^2}dx$

Let $u = 2 - x^2$. Then $du = -2x\,dx$.

$$\int \dfrac{-6x}{2-x^2}dx = 6 \cdot \dfrac{1}{2}\int \dfrac{1}{2-x^2}(-2x)dx$$

$$= 3\int \dfrac{1}{u}du$$

$$= 3\ln|u| + C$$

$$= 3\ln|2-x^2| + C$$

17. $\int \dfrac{12x^2\,dx}{\left(x^3+5\right)^4} = 12 \cdot \dfrac{1}{3}\int \dfrac{3x^2\,dx}{\left(x^3+5\right)^4}$

Let $u = x^3 + 5$. Then $du = 3x^2\,dx$.

$$\int \dfrac{12x^2\,dx}{\left(x^3+5\right)^4} = 12 \cdot \dfrac{1}{3}\int \dfrac{du}{u^4}$$

$$= 4\int u^{-4}\,du$$

$$= 4\left(\dfrac{u^{-3}}{-3}\right) + C$$

$$= \dfrac{4\left(x^3+5\right)^{-3}}{-3} + C$$

$$\text{or } \dfrac{4}{-3\left(x^3+5\right)^3} + C$$

18. $\int \left(x^2-5x\right)^4(2x-5)dx$

Let $u = x^2 - 5x$. Then $du = (2x-5)dx$.

$$\int \left(x^2-5x\right)^4(2x-5)dx = \int u^4\,du$$

$$= \dfrac{u^5}{5} + C$$

$$= \dfrac{\left(x^2-5x\right)^5}{5} + C$$

19. $\int \dfrac{4x-5}{2x^2-5x}dx$

Let $u = 2x^2 - 5x$. Then $du = (4x-5)dx$.

$$\int \dfrac{4x-5}{2x^2-5x}dx = \int \dfrac{du}{u}$$

$$= \ln|u| + C$$

$$= \ln|2x^2-5x| + C$$

20. $\int \dfrac{8(2x+9)}{x^2+9x+1}dx$

Let $u = x^2 + 9x + 1$. Then $du = (2x+9)\,dx$.

$$\int \dfrac{8(2x+9)}{x^2+9x+1}dx$$

$$= 8\int \dfrac{du}{u}$$

$$= 8\ln|u| + C$$

$$= 8\ln|x^2+9x+1| + C$$

21. $\int \dfrac{x^3}{e^{3x^4}}dx = \int x^3 e^{-3x^4}$

$$= -\dfrac{1}{12}\int -12x^3 e^{-3x^4}\,dx$$

Let $u = -3x^4$. Then $du = -12x^3\,dx$.

$$\int \dfrac{x^3}{e^{3x^4}}dx = -\dfrac{1}{12}\int e^u\,du$$

$$= -\dfrac{1}{12}e^u + C$$

$$= \dfrac{-e^{-3x^4}}{12} + C$$

22. $\int 2e^{3x^2+4} x \, dx$

Let $u = 3x^2 + 4$. Then $du = 6x \, dx$.

$2\int e^{3x^2+4} x \, dx = 2 \cdot \frac{1}{6}\int e^{3x^2+4} \cdot 6x \, dx$

$= \frac{1}{3}\int e^u \, du$

$= \frac{1}{3}e^u + C$

$= \frac{1}{3}e^{3x^2+4} + C$

$= \frac{e^{3x^2+4}}{3} + C$

23. $\int -2e^{-5x} \, dx = \frac{-2}{-5}\int -5e^{-5x} \, dx$

Let $u = -5x$. Then $du = -5 \, dx$.

$\int -2e^{-5x} \, dx = \frac{2}{5}\int e^u \, du$

$= \frac{2}{5}e^u + C$

$= \frac{2e^{-5x}}{5} + C$

24. $\int 11e^{-4x} \, dx = \frac{11e^{-4x}}{-4} + C$

$= -\frac{11e^{-4x}}{4} + C$

25. $\int \frac{3(\ln x)^5}{x} \, dx = 3\int \frac{(\ln x)^5}{x} \, dx$

Let $u = \ln x$. Then $du = \frac{1}{x} \, dx$.

$\int \frac{3(\ln x)^5}{x} \, dx = 3\int u^5 \, du$

$3\left(\frac{u^6}{6}\right) + C = \frac{u^6}{2} + C$

$= \frac{(\ln x)^6}{2} + C$

26. $\int \frac{10(\ln(5x+3))^2}{5x+3} \, dx = 10\int \frac{(\ln(5x+3))^2}{5x+2} \, dx$

Let $u = \ln(5x+3)$. Then $du = \left(\frac{1}{5x+3} \cdot 5\right) dx$.

$\int \frac{10(\ln(5x+3))^2}{5x+3} \, dx = 10 \cdot \frac{1}{5}\int u^2 \, du$

$= 2\frac{u^3}{3} + C = \frac{2(\ln(5x+3))^3}{3} + C$

27. $\int 25e^{-50x} \, dx = 25\int e^{-50x} \, dx$

Let $u = -50x$. Then $du = -50 \, dx$.

$\int 25e^{-50x} \, dx = 25\left(-\frac{1}{50}\right)\int e^u \, du$

$= -\frac{1}{2}e^u + C = -\frac{1}{2}e^{-50x} + C$

28. $\int 4xe^{-3x^2+7} \, dx = 4\int xe^{-3x^2+7} \, dx$

Let $u = -3x^2 + 7$. Then $du = -6x \, dx$.

$\int 4xe^{-3x^2+7} \, dx = 4\left(-\frac{1}{6}\right)\int e^u \, du$

$= -\frac{2}{3}e^u + C = -\frac{2}{3}e^{-3x^2+7} + C$

29. Answers vary. Possible answer: By dividing the area under the curve into rectangles and then finding the sum of the areas of the rectangles.

30. a. $f(x) = 16x^2 - x^4 + 2$

from $x = -2$ to $x = 3$

Area $= f(-2) + f(-1) + f(0) + f(1) + f(2)$

$= 50 + 17 + 2 + 17 + 50 = 136$

b. Using a graphing calculator, under the MATH menu, use fnInt. Key in

fnInt$(16x^2 - x^4 + 2, \, x, \, -2, \, 3)$.

The value returned is 141.667, so the area is approximately 142.

31. $g(x) = -x^4 + 12x^2 + x + 5$

from $x = -3$ to $x = 3$

a. Area $= f(-3) + f(-2) + f(-1)$

$+ f(0) + f(1) + f(2)$

$= 29 + 35 + 15 + 5 + 17 + 39 = 140$

31. Continued

  b. Under the MATH menu, use fnInt.
  Key in
  $\text{fnInt}\left(-x^4 + 12x^2 + x + 5, \, x, \, -3, \, 3\right).$
  The value returned is 148.8, so the area is
  approximately 148.8.

32. $f(x) = 2x + 3$ from $x = 0$ to $x = 4$

$$\Delta x = \frac{4-0}{4} = 1$$

i	x_i	$f\left(x_i\right)$
1	0	3
2	1	5
3	2	7
4	3	9

$$\sum_{i=1}^{4} f\left(x_i\right)\Delta x$$
$$= 3(1) + 5(1) + 7(1) + 9(1)$$
$$= 24$$

33. $\displaystyle\int_0^4 (2x+3)\,dx$

$$A = \frac{1}{2}(B+b)h$$
$$h = 4 - 0 = 4$$
$$B = f(4) = 2(4) + 3 = 11$$
$$b = f(0) = 2(0) + 3 = 3$$
$$\int_0^4 (2x+3) = \frac{1}{2}(11+3)(4)$$
$$= 28$$

The area formula gives the exact answer, 28. The
answer obtained in Exercise 28, which is 24, is an
approximation of the area. Because it uses left
endpoints and the line $y = 2x + 3$ has a positive
slope, the area obtained using the four rectangles
will be less than the exact area.

34. Answers vary. Possible answer: Substitution is
useful in integration with complicated functions
when some expression in the integral can be
replaced by u along with the derivative of u, du.

35. $\displaystyle\int_0^1 (x^3 - x^2)\,dx = \left(\frac{x^4}{4} - \frac{x^3}{3}\right)\Bigg|_0^1$

$$= \frac{1}{4} - \frac{1}{3} - 0$$

$$= -\frac{1}{12}$$

36. $\displaystyle\int_0^1 e^{3t}\,dt = \frac{1}{3}e^{3t}\Bigg|_0^1 = \frac{1}{3}e^{3(1)} - \frac{1}{3}e^{3(0)} = \frac{1}{3}\left(e^3 - 1\right)$

37. $\displaystyle\int_1^5 \left(6x^{-2} + x^{-3}\right)dx$

$$= \left(\frac{6x^{-1}}{-1} + \frac{x^{-2}}{-2}\right)\Bigg|_1^5$$

$$= \left(-\frac{6}{5} - \frac{1}{50}\right) - \left(-6 - \frac{1}{2}\right)$$

$$= \frac{-61}{50} + \frac{13}{2}$$

$$= \frac{-61 + 325}{50}$$

$$= \frac{264}{50}$$

$$= \frac{132}{25} \approx 5.28$$

38. $\displaystyle\int_2^3 \left(5x^{-2} + 7x^{-4}\right)dx$

$$= \left(\frac{5x^{-1}}{-1} + \frac{7x^{-3}}{-3}\right)\Bigg|_2^3$$

$$= \left(\frac{-5}{x} - \frac{7}{3x^3}\right)\Bigg|_2^3$$

$$= \left(\frac{-5}{3} - \frac{7}{81}\right) - \left(\frac{-5}{2} - \frac{7}{24}\right)$$

$$= \frac{-5}{3} - \frac{7}{81} + \frac{5}{2} + \frac{7}{24}$$

$$= \frac{-1080 - 56 + 1620 + 189}{648}$$

$$= \frac{673}{648} \approx 1.039$$

39. $\int_1^3 15x^{-1}dx = 15\int_1^3 \frac{dx}{x}$

$= 15\ln|x|\Big\|_1^3$

$= 15\ln 3 - 15\ln 1$

$= 15\ln 3$

$= 16.479$

40. $\int_1^6 \frac{8x^{-1}}{3}dx = \frac{1}{3}\int_1^6 \frac{8}{x}dx$

$= \frac{8}{3}(\ln x)\Big|_1^6$

$= \frac{8}{3}(\ln 6 - \ln 1)$

$= \frac{8}{3}\ln 6$

≈ 4.778

41. $\int_0^4 2e^{-5x}dx = \frac{2e^{-5x}}{-5}\Big|_0^4$

$= -\frac{2}{5}e^{-20} - \frac{2}{5}e^0$

$= -\frac{2}{5}e^{-20} - \frac{2}{5}$

$\approx .40$

42. $\int_1^2 \frac{5}{2}e^{4x}dx = \frac{5}{2}\cdot\frac{e^{4x}}{4}\Big|_1^2$

$= \frac{5e^{4x}}{8}\Big|_1^2$

$= \frac{5(e^8 - e^4)}{8}$

≈ 1828.97

43. $\int_{\sqrt{5}}^5 2x\sqrt{x^2 - 3}\,dx$

Let $u = x^2 - 3$. Then $du = 2x\,dx$.

If $x = 5$, $u = 22$. If $x = \sqrt{5}$, $u = 2$.

$\int_{\sqrt{5}}^5 2x\sqrt{x^2 - 3}\,dx = \int_{\sqrt{5}}^5 \sqrt{x^2 - 3}\cdot 2x\,dx$

$= \int_2^{22} u^{\frac{1}{2}}\,du$

$= \frac{u^{\frac{3}{2}}}{\frac{3}{2}}\Big|_2^{22}$

$= \frac{2}{3}\left(22^{\frac{3}{2}} - 2^{\frac{3}{2}}\right)$

≈ 66.907

44. $\int_0^1 x\sqrt{5x^2 + 4}\,dx$

Let $u = 5x^2 + 4$. Then $du = 10x\,dx$.

If $x = 1$, $u = 9$. If $x = 0$, $u = 4$.

$\int_0^1 x\sqrt{5x^2 + 4}\,dx$

$= \frac{1}{10}\int_0^1 \left(5x^2 + 4\right)^{\frac{1}{2}}\cdot 10x\,dx$

$= \frac{1}{10}\int_4^9 u^{\frac{1}{2}}\,du$

$= \frac{1}{10}\cdot\frac{u^{\frac{3}{2}}}{\frac{3}{2}}\Big|_4^9$

$= \frac{2}{30}\left(9^{\frac{3}{2}} - 4^{\frac{3}{2}}\right)$

$= \frac{38}{30} = \frac{19}{15} = 1.2\overline{6}$

45. $\int_1^6 \frac{8x + 12}{x^2 + 3x + 9}dx = 4\int_1^6 \frac{2x + 3}{x^2 + 3x + 9}dx$

Let $u = x^2 + 3x + 9$. Then $du = (2x + 3)dx$. If

$x = 6$, then $u = 63$. If $x = 1$, then $u = 13$.

$4\int_1^6 \frac{2x + 3}{x^2 + 3x + 9}dx = 4\int_{13}^{63}\frac{1}{u}du$

$= 4\ln|u|\Big\|_{13}^{63} = 4(\ln 63 - \ln 13)$

$= 4\ln\frac{63}{13} \approx 6.313$

46. $\int_1^4 \frac{3(\ln x)^5}{x}\,dx = 3\int_1^4 \frac{(\ln x)^5}{x}\,dx$

Let $u = \ln x$. Then $du = \frac{1}{x}\,dx$. If $x = 4$, the

$u = \ln 4$. If $x = 1$, then $u = 0$.

$\int_1^4 \frac{3(\ln x)^5}{x}\,dx = 3\int_0^{\ln 4} u^5\,du$

$= \frac{3u^6}{6}\Big|_0^{\ln 4} = \frac{1}{2}\Big[(\ln 4)^6 - 0\Big]$

$= \frac{(\ln 4)^6}{2} \approx 3.549$

47. $f(x) = e^x$; $[0, 2]$

$A = \int_0^2 e^x\,dx$

$= e^x\Big|_0^2$

$= e^2 - e^0$

$= e^2 - 1 \approx 6.3891$

48. $f(x) = 1 + e^{-x}$; $[0, 4]$

$A = \int_0^4 \left(1 + e^{-x}\right)dx$

$= \left(x + \frac{1}{-1}e^{-x}\right)\Big|_0^4$

$= \left(4 - e^{-4}\right) - (0 - 1)$

$= 5 - e^{-4} \approx 4.982$

49. $C'(x) = 10 - 5x$; fixed cost is \$4. The fixed cost tells us that $C(0) = 4$.

$C(x) = \int (10 - 5x)dx$

$C(x) = 10x - \frac{5x^2}{2} + k$

$4 = 10(0) - \frac{5}{2}(0)^2 + k$

$k = 4$

Thus, the cost function is

$C(x) = 10x - \frac{5x^2}{2} + 4$.

50. $C'(x) = 2x + 3x^2$; 2 units cost \$12. Since, 2 units cost \$12, we have $C(2) = 12$.

$C(x) = \int \left(2x + 3x^2\right)dx$

$C(x) = x^2 + x^3 + k$

$12 = 2^2 + 2^3 + k$

$12 = 4 + 8 + k$

$k = 0$

Thus, the cost function is

$C(x) = x^2 + x^3$.

51. $C'(x) = 3\sqrt{2x - 1}$; 13 units cost \$270. Since 13 units cost \$270, we have $C(13) = 270$.

$C(x) = \int 3\sqrt{2x - 1}\,dx$

Let $u = 2x - 1$. Then $du = 2\,dx$.

$C(x) = 3 \cdot \frac{1}{2}\int \sqrt{2x - 1} \cdot 2\,dx$

$C(x) = \frac{3}{2}\int u^{\frac{1}{2}}\,du$

$C(x) = \frac{3}{2} \cdot \frac{u^{\frac{3}{2}}}{\frac{3}{2}} + k$

$C(x) = u^{\frac{3}{2}} + k$

$C(x) = (2x - 1)^{\frac{3}{2}} + k$

$270 = 25^{\frac{3}{2}} + k$

$270 = 125 + k$

$k = 145$

Thus, the cost function is

$C(x) = (2x - 1)^{\frac{3}{2}} + 145$.

52. $C'(x) = \frac{6}{x + 1}$; fixed cost is \$18. Since the fixed cost is \$18, $C(0) = 18$.

$C(x) = \int \frac{6}{x + 1}\,dx$

$C(x) = 6\ln|x + 1| + k$

$18 = 6\ln 1 + k$

$k = 18$

Thus, the cost function is

$C(x) = 6\ln|x + 1| + 18$.

53. $S' = \sqrt{x} + 3$

$$S(x) = \int_0^9 \left(x^{\frac{1}{2}} + 3 \right) dx$$

$$= \left(\frac{x^{\frac{3}{2}}}{\frac{3}{2}} + 3x \right) \Bigg|_0^9$$

$$= \frac{2}{3}(9)^{\frac{3}{2}} + 27$$

$$= 45$$

Total sales are 45(1000), or 45,000 units.

54.

Midpoint	Height of rectangle (function values)
1	11,000
3	9500
5	12,000
7	10,000
9	6500

The total area is approximately
11,000(2) + 9500(2) + 12,000(2)
+ 10,000(2) + 6500(2)
= 22,000 + 19,000 + 24,000 + 20,000 + 13,000
= 98,000.
The total income is approximately $98,000.

55. $\int_0^T 100,000 e^{.03t} dt = 4,000,000$

$$100,000 \cdot \frac{1}{.03} e^{.03t} \Bigg|_0^T = 4,000,000$$

$$\frac{1}{.03} \left(e^{.03T} - 1 \right) = 40$$

$$e^{.03T} - 1 = 1.2$$

$$e^{.03T} = 2.2$$

$$.03T = \ln 2.2$$

$$T = \frac{\ln 2.2}{.03}$$

$$T \approx 26.28$$

The supply will be used up in about 26.3 yr.

56. $\int_0^{10} \left(150 - \sqrt{3.2t + 4} \right) dt$

$$= \int_0^{10} 150 \, dt - \int_0^{10} \sqrt{3.2t + 4} \, dt$$

Let $u = 3.2t + 4$ and $du = 3.2 \, dt$ or $\frac{1}{3.2} du = dt$.

$$= 150t \Big|_0^{10} - \int_0^{10} \sqrt{u} \frac{1}{3.2} du$$

$$= 1500 - \frac{1}{3.2} \frac{u^{1.5}}{1.5}$$

$$= 1500 - .2083(3.2t + 4)^{1.5} \Big|_0^{10}$$

$$= 1500$$
$$- \left(.2083(3.2 \cdot 10 + 4)^{1.5} - .2083(3.2 \cdot 0 + 4)^{1.5} \right)$$

$$\approx 1456.66$$

In the first 10 months there were about
1457 spiders.

57. $\int_{10}^{22} \left(556 - 20.23t + 2.21t^2 - 0.07t^3 \right) dt$

$$= 556t - 20.23 \frac{t^2}{2} + 2.21 \frac{t^3}{3} - 0.07 \frac{t^4}{4} \Bigg|_{10}^{22}$$

$$= 556(22) - \frac{20.23}{2}(22)^2 + \frac{2.21}{3}(22)^3$$

$$- \frac{.07}{4}(22)^4$$

$$- 556(10) - \frac{20.23}{2}(10)^2 + \frac{2.21}{3}(10)^3$$

$$- \frac{.07}{4}(10)^4$$

$$= 11,080.8867 - 5110.1667 = 5970.72$$

The total number of births from 1990 to the end
of 2001 is 5,970,720.

58. $\int_0^{15} 24.5 e^{.063t} dt = 24.5 \frac{1}{.063} e^{.063t} \Bigg|_0^{15}$

$$= 388.89 e^{.063(15)} - 388.89 e^{.063(0)}$$

$$\approx 611.65$$

From 1990 to 2005 about $611.65 billion was
paid out in life insurance proceeds.

59. $\int_0^{35} 550(x+1)^{1/7} dx$

$= 550\int_0^{35} (x+1)^{1/7} dx$

Let $u = x + 1$ and $du = dx$.

$= 550\int u^{1/7} du$

$= 550\dfrac{u^{8/7}}{\frac{8}{7}}$

$= 481.25 u^{8/7}$

$= 481.25(x+1)^{8/7}\Big|_0^{35}$

$= 481.25(35+1)^{8/7} - 481.25(1)^{8/7}$

$\approx 28,906.943 - 481.25$

$\approx 28,425.693$

No, total circulation from 1970 to 2005 reached only 28.4 billion papers.

60. $S'(x) = 225 - x^2$

$C'(x) = x^2 + 25x + 150$

$S'(x) = C'(x)$

$225 - x^2 = x^2 + 25x + 150$

$2x^2 + 25x - 75 = 0$

$(2x - 5)(x + 15) = 0$

$x = \dfrac{5}{2}$ or $x = -15$

Discard the negative solution. The company should use the machinery for 2.5 yr.

$\int_0^{2.5}\Big[(225 - x^2) - (x^2 + 25x + 150)\Big] dx$

$= \int_0^{2.5}(-2x^2 - 25x + 75) dx$

$= \left(\dfrac{-2x^3}{3} - \dfrac{25x^2}{2} + 75x\right)\Big|_0^{2.5}$

$= \dfrac{-2(2.5)^3}{3} - \dfrac{25(2.5)^2}{2} + 75(2.5)$

≈ 98.95833

The net savings are 98,958.33, or about \$99,000.

61. Answers vary.

Possible answer: Consumers' surplus is the total of all differences between the equilibrium price on an item and the higher prices individuals will be willing to pay and is thought of as savings realized by those individuals.

Producers' surplus is the total of all differences between the equilibrium price and the lower prices at which the manufacturers would sell the product and is considered added income for the manufacturers.

62. $S(q) = q^2 + 5q + 100$

$D(q) = 350 - q^2$

$S(q) = D(q)$

$q^2 + 5q + 100 = 350 - q^2$

$2q^2 + 5q - 250 = 0$

$(2q + 25)(q - 10) = 0$

$q = \dfrac{-25}{2}$ or $q = 10$

Discard the negative solution.

$S(10) = 10^2 + 5(10) + 100 = 250$

The equilibrium point is (10, 250).

a. The producers' surplus is given by

$\int_0^{10}\Big[250 - (q^2 + 5q + 100)\Big] dq$

$= \int_0^{10}(150 - q^2 - 5q) dq$

$= \left(150q - \dfrac{q^3}{3} - \dfrac{5q^2}{2}\right)\Big|_0^{10}$

$= 916.\overline{6}$

The producers' surplus is \$916.67.

b. The consumers' surplus is given by

$\int_0^{10}\Big[(350 - q^2) - 250\Big] dq$

$= \int_0^{10}(100 - q^2) dq$

$= \left(100q - \dfrac{q^3}{3}\right)\Big|_0^{10}$

$= \left[100(10) - \dfrac{10^3}{3}\right] - 0$

$= 666.\overline{6}$

The consumers' surplus is \$666.67.

63. $I'(t) = \dfrac{100t}{t^2 + 2}$

The total number of infected people is given by

$\displaystyle\int_0^4 \dfrac{100t}{t^2 + 2}\, dt$.

Let $u = t^2 + 2$. Then $du = 2t\, dt$.
If $t = 4$, $u = 18$. If $t = 0$, $u = 2$.

$\displaystyle\int_0^4 \dfrac{100t}{t^2 + 2}\, dt = 50 \int_0^4 \dfrac{2t\, dt}{t^2 + 2}$

$\qquad = 50 \int_2^{18} \dfrac{du}{u}$

$\qquad = 50 \ln |u| \Big\|_2^{18}$

$\qquad = 50 \ln 18 - 50 \ln 2$

$\qquad = 50 \ln \left(\dfrac{18}{2}\right)$

$\qquad = 50 \ln 9$

$\qquad \approx 109.86$

Approximately 110 people will be infected.

64. a. $\displaystyle\int_0^2 \left(\dfrac{1}{2} e^{-\frac{x}{2}}\right) dx = -e^{-\frac{x}{2}} \Big|_0^2$

$\qquad = -e^{-1} - \left(-e^0\right)$

$\qquad = -.3679 + 1$

$\qquad = .6321$

b. $1 - \displaystyle\int_0^3 \left(\dfrac{1}{2} e^{-\frac{x}{2}}\right) dx = 1 - \left[-e^{-\frac{x}{2}} \Big|_0^3\right]$

$\qquad = 1 - \left[-e^{-\frac{3}{2}} - \left(-e^0\right)\right]$

$\qquad = 1 - \left[-e^{-1.5} + 1\right]$

$\qquad = .2231$

65. a. $\displaystyle\int_1^2 .44 e^{-.44t}\, dt = -e^{-.44t} \Big|_1^2$

$\qquad = -e^{-.88} - \left(-e^{-.44}\right)$

$\qquad = .2293$

b. $1 - \displaystyle\int_0^4 .44 e^{-.44t}\, dt = 1 - \left[-e^{-.44t} \Big|_0^4\right]$

$\qquad = 1 - \left[-e^{-1.76} - \left(e^0\right)\right]$

$\qquad = 1 - \left[-e^{-1.76} - 1\right]$

$\qquad = .1720$

66. $\displaystyle\int \dfrac{1}{\sqrt{x^2 - 64}}\, dx$

Use entry 6 from the table of integrals with $a = 8$.

$= \ln \left| x + \sqrt{x^2 - 64} \right| + C$

67. $\displaystyle\int \dfrac{10}{x\sqrt{25 + x^2}}\, dx = 10 \int \dfrac{1}{x\sqrt{25 + x^2}}\, dx$

Use entry 10 in the table with $a = 5$.

$= -10 \left[\dfrac{1}{5} \ln \left| \dfrac{5 + \sqrt{5^2 + x^2}}{x} \right| \right] + C$

$= -2 \ln \left| \dfrac{5 + \sqrt{25 + x^2}}{x} \right| + C$

68. $\displaystyle\int \dfrac{18}{x^2 - 9}\, dx = 18 \int \dfrac{dx}{x^2 - 9}$

Use entry 8 from the table with $a = 3$.

$= 18 \left(\dfrac{1}{6} \ln \left| \dfrac{x - 3}{x + 3} \right| \right) + C$

$= 3 \ln \left| \dfrac{x - 3}{x + 3} \right| + C \;\; \left(x^2 > 9\right)$

69. $\displaystyle\int \dfrac{15x}{2x - 5}\, dx = 15 \int \dfrac{x}{2x - 5}\, dx$

This matches entry 11 in the table with $a = 2$ and $b = -5$.

$= 15 \left[\dfrac{x}{2} - \dfrac{-5}{2^2} \ln |2x - 5| \right] + C$

$= 15 \left[\dfrac{x}{2} + \dfrac{5 \ln |2x - 5|}{4} \right] + C$

or $\dfrac{15x}{2} + \dfrac{75}{4} \ln |2x - 5| + C$

70. Answers vary.

71. $\dfrac{dy}{dx} = 2x^3 + 6x + 5$

$dy = \left(2x^3 + 6x + 5\right) dx$

$\displaystyle\int dy = \int \left(2x^3 + 6x + 5\right) dx$

$y = \dfrac{2x^4}{4} + \dfrac{6x^2}{2} + 5x + C$

$y = \dfrac{x^4}{2} + 3x^2 + 5x + C$

72. $\dfrac{dy}{dx} = x^2 + \dfrac{5x^4}{8}$

$\quad y = \displaystyle\int\left(x^2 + \dfrac{5x^4}{8}\right)dx$

$\quad y = \dfrac{x^3}{3} + \dfrac{x^5}{8} + C$

73. $\dfrac{dy}{dx} = \dfrac{3x+1}{y}$

$\quad y\,dy = (3x+1)dx$

$\quad \displaystyle\int y\,dy = \int(3x+1)dx$

$\quad \dfrac{y^2}{2} = \dfrac{3x^2}{2} + x + C_1$

$\quad y^2 = 3x^2 + 2x + C$

74. $\dfrac{dy}{dx} = \dfrac{e^x + x}{y-1}$

$\quad (y-1)dy = \left(e^x + x\right)dx$

$\quad \dfrac{y^2}{2} - y = e^x + \dfrac{x^2}{2} + C$

75. $\dfrac{dy}{dx} = 5\left(e^{-x} - 1\right)$; $y = 17$ when $x = 0$

$\quad dy = 5\left(e^{-x} - 1\right)dx$

$\quad \displaystyle\int dy = \int 5\left(e^{-x} - 1\right)dx$

$\quad y = 5\left[\dfrac{e^{-x}}{-1} - x\right] + C$

The general solution is

$y = -5e^{-x} - 5x + C$.

Let $x = 0$ and $y = 17$.

$17 = -5e^0 - 5(0) + C$

$17 = -5 + C$

$22 = C$

The particular solution is

$y = -5e^{-x} - 5x + 22$.

76. $\dfrac{dy}{dx} = \dfrac{x}{x^2-3} + 7$; $y = 52$ when $x = 2$

$\quad y = \displaystyle\int\left(\dfrac{x}{x^2-3} + 7\right)dx$

$\quad y = \dfrac{1}{2}\displaystyle\int\left(\dfrac{2x}{x^2-3} + 7\right)dx$

The general solution is

$y = \dfrac{1}{2}\ln\left|x^2-3\right| + 7x + C$.

Let $x = 2$ and $y = 52$.

$52 = \dfrac{1}{2}\ln\left|4-3\right| + 14 + C$

$38 = \dfrac{1}{2}\ln 1 + C$

$38 = 0 + C$

$38 = C$

The particular solution is

$y = \dfrac{1}{2}\ln\left|x^2-3\right| + 7x + 38$.

77. $(5-2x)y = \dfrac{dy}{dx}$; $y = 2$ when $x = 0$

$\quad (5-2x)dx = \dfrac{dy}{y}$

$\quad \displaystyle\int(5-2x)dx = \int\dfrac{dy}{y}$

The general solution is

$5x - x^2 = \ln|y| + C$.

Let $x = 0$ and $y = 2$.

$\quad 0 = \ln 2 + C$

$\quad C = -\ln 2$

$5x - x^2 = \ln|y| - \ln 2 = \ln\left|\dfrac{y}{2}\right|$

$e^{5x-x^2} = \left|\dfrac{y}{2}\right|$

$2e^{5x-x^2} = |y|$

The particular solution is

$y = 2e^{5x-x^2}$.

78. $\sqrt{x}\dfrac{dy}{dx} = xy$; $y = 4$ when $x = 1$

Write $\sqrt{x}$ as $x^{\frac{1}{2}}$ and separate the variables.

$$y = \frac{x^{\frac{1}{2}}}{x}\frac{dy}{dx}$$

$$\frac{1}{y} = x^{\frac{1}{2}}\frac{dx}{dy}$$

Integrate.

$$\frac{1}{y}dy = x^{\frac{1}{2}}dx$$

$$\ln|y| = \frac{2}{3}x^{\frac{3}{2}} + C$$

Let $x = 1$ and $y = 4$.

$$\ln 4 = \frac{2}{3} + C$$

$$C = \ln 4 - \frac{2}{3} \approx .7196$$

$$\ln|y| \approx \frac{2}{3}x^{\frac{3}{2}} + .7196$$

$$|y| \approx e^{.7196}e^{\frac{2x^{3/2}}{3}}$$

$$y \approx \pm 2.054e^{\frac{2x^{3/2}}{3}}$$

Since $y = 4$ when $x = 1$, we must use the positive solution.

$$y = 2.054e^{\frac{2x^{3/2}}{3}}$$

79. $A = 10{,}000$ when $t = 0$, $r = .05$, $D = -1000$

a. $\dfrac{dA}{dt} = .05A - 1000$

b.
$$\frac{1}{.05A - 1000}dA = dt$$

$$\frac{1}{.05}\ln|.05A - 1000| = t + k$$

$$\ln|.05A - 1000| = .05t + k$$

$$\ln|.05(10{,}000) - 1000| = k$$

$$k = \ln|-500| = \ln 500$$

$$\ln|.05A - 1000| = .05t + \ln 500$$

$$|.05A - 1000| = 500e^{.05t}$$

79. Continued

Since $.05A < 1000$,

$$|.05A - 1000| = 1000 - .05A$$

$$1000 - .05A = 500e^{.05t}$$

$$A = \frac{1}{.05}\left(1000 - 500e^{.05t}\right)$$

$$= 10{,}000\left(2 - e^{.05t}\right)$$

When $t = 1$,

$$A = 10{,}000\left(2 - e^{.05(1)}\right)$$

$$\approx 9487.29.$$

The amount left in the account after 1 yr will be \$9487.29.

80. a. $\dfrac{dy}{dx} = 4.2e^{.3x}$

$$dy = 4.2e^{.3x}dx$$

$$y = \frac{4.2}{.3}e^{.3x} + C$$

$$y = 14e^{.3x} + C$$

When $x = 0$, $y = 0$.

$$0 = 14e^0 + C$$

$$C = -14 \qquad e^0 = 1$$

$$y = 14e^{.3x} - 14$$

When $x = 8$,

$$y = 14e^{2.4} - 14$$

$$\approx 140.3245$$

Sales after 8 mo are \$14,032.45.

b. When $x = 12$,

$$y = 14e^{3.6} - 14$$

$$\approx 498.3753.$$

Sales after 12 mo are \$49,837.53.

81.
$$\frac{dy}{dx} = .2(125 - y)$$

$$\frac{dy}{125 - y} = .2\,dx$$

$$\int \frac{dy}{125 - y} = \int .2\,dx$$

$$-\ln|125 - y| = .2x + C_1$$

$$\ln|125 - y| = -.2x + C_2$$

$$|125 - y| = e^{-.2x + C_2}$$

$$|125 - y| = e^{-.2x} \cdot e^{C_2}$$

$$|125 - y| = C_3 e^{-.2x}$$

$$125 - y = \pm C_3 e^{-.2x}$$

$$125 \pm C_3 e^{-.2x} = y$$

$$y = 125 + M e^{-.2x}$$

When $x = 0$, $y = 20$.

$$20 = 125 + M e^0$$

$$M = -105$$

$$y = 125 - 105 e^{-.2x}$$

a. Find the value of y when $x = 10$.

$$y = 125 - 105 e^{-2}$$

$$y \approx 110.79$$

In 10 days the worker will produce about 111 items.

b. Theoretically, $125 - 105 e^{-.2x}$ will never equal 125, but, for practical purposes, for large values of x it will approximately equal 125, so the worker can produce approximately 125 items a day.

82.
$$\frac{dT}{dt} = k\left(T - T_F\right)$$

$$\frac{dT}{T - T_F} = k\,dt$$

$$\int \frac{dT}{T - T_F} = \int k\,dt$$

$$\ln|T - T_F| = kt + C_1$$

$$|T - T_F| = e^{kt + C_1}$$

$$|T - T_F| = e^{kt} \cdot e^{C_1}$$

$$T - T_F = \pm e^{C_1} e^{kt}$$

$$T = T_F + M e^{kt}$$

82. Continued

$T_F = 300°$, $T(0) = 40°$, and $T(1) = 150°$

$$T = 300 + M e^{kt}$$

$$40 = 300 + M \cdot e^0$$

$$M = -260$$

$$T = 300 - 260 e^{kt}$$

$$150 = 300 - 260 e^{k \cdot 1}$$

$$-150 = -260 e^{k}$$

$$e^k = \frac{150}{260}$$

$$k = \ln\left(\frac{150}{260}\right) \approx -.550$$

$$T = 300 - 260 e^{-.550t}$$

a. After 2 hr, the temperature is

$$T = 300 - 260 e^{-.550(2)}$$

$$T \approx 213.45$$

The temperature is approximately 213°.

b. If $T = 250$,

$$250 = 300 - 260 e^{-.550t}$$

$$-50 = -260 e^{-.550t}$$

$$e^{-.550t} = \frac{50}{260}$$

$$-.550t = \ln\left(\frac{5}{26}\right)$$

$$t = \frac{\ln\left(\frac{5}{26}\right)}{-.550}$$

$$t \approx 2.998.$$

The temperature will be 250° in approximately 3 hr.

83. First note that $\dfrac{dy}{dx}$ is the ratio of $\dfrac{dy}{dt}$ to $\dfrac{dx}{dt}$. Then the variables may be separated and an integration may be performed.

$$\frac{dy}{dx} = \frac{\frac{dy}{dt}}{\frac{dx}{dt}}$$

$$\frac{dy}{dx} = \frac{-.3y + .4xy}{.2x - .5xy}$$

$$\frac{dy}{dx} = \frac{(-.3 + .4x)y}{(.2 - .5y)x}$$

$$(.2 - .5y)\frac{dy}{y} = (-.3 + .4x)\frac{dx}{x}$$

$$\int \frac{.2 - .5y}{y}\,dy = \int \frac{-.3 + .4x}{x}\,dx$$

$$\int \left(\frac{.2}{y} - .5\right)dy = \int \left(-\frac{.3}{x} + .4\right)dx$$

$$.2\int \frac{dy}{y} - .5\int dy = -.3\int \frac{dx}{x} + .4\int dx$$

$.2\ln|y| - .5y = -.3\ln|x| + .4x + C$ This is an equation that relates x to y (we can assume $x > 0$ and $y > 0$). Both growth rates being zero means that

$$\frac{dx}{dt} = 0 \text{ and } \frac{dy}{dt} = 0.$$

The equation would then become

$0 = .2x - .5xy$

$0 = -.3y + .4xy$

Note that $x = 0$ and $y = 0$ is a solution to this system of equations, but this solution would mean that both species were absent. Suppose neither is 0 and see if there are other solutions.
The first equation can be written as

$.5xy = .2x$

$$y = \frac{.2x}{.5x} = \frac{2}{5} \ (x \neq 0)$$

The second equation can be written as

$.3y = .4xy$.

$$x = \frac{.3y}{.4y} = \frac{3}{4} \ (y \neq 0)$$

So both growth rates are 0 when $x = \dfrac{3}{4}$ unit and

$y = \dfrac{2}{5}$ unit.

Case 13 Bounded Population Growth

1. Verify $\dfrac{1}{P(C-P)} = \dfrac{\frac{1}{C}}{P} + \dfrac{\frac{1}{C}}{C-P}$

$$\frac{\frac{1}{C}}{P} + \frac{\frac{1}{C}}{C-P} = \frac{\left(\frac{1}{C}\right)(C-P)}{P(C-P)} + \frac{\left(\frac{1}{C}\right)P}{P(C-P)}$$

$$= \frac{\left(\frac{1}{C}\right)(C-P) + \left(\frac{1}{C}\right)P}{P(C-P)}$$

$$= \frac{\left(\frac{1}{C}\right)C - \left(\frac{1}{C}\right)P + \left(\frac{1}{C}\right)P}{P(C-P)}$$

$$= \frac{1-0}{P(C-P)}$$

$$= \frac{1}{P(C-P)}$$

2. $P(0) = 10$

$$\frac{1}{200}\ln(P) - \frac{1}{200}\ln(200 - P) = .005t + K$$

$\text{fi } \ln(P) - \ln(200 - P) = t + 200K$

$\Rightarrow e^{\ln(P) - \ln(200-P)} = e^{t + 200K}$

$\Rightarrow \dfrac{P}{200 - P} = e^{200K}e^{t}$

$\Rightarrow P = 200e^{200K}e^{t} - Pe^{200K}e^{t}$

$\Rightarrow P + P(e^{200K}e^{t}) = 200e^{200K}e^{t}$

$\Rightarrow P = \dfrac{200e^{200K}e^{t}}{1 + e^{200K}e^{t}}$

$\Rightarrow P = \dfrac{200e^{t}}{e^{-200K} + e^{t}}$

Let $A = e^{-200K}$

$$P(t) = \frac{200e^{t}}{A + e^{t}}$$

Since $P(0) = \dfrac{200}{A+1}$ must equal 10, it follows that

$A = 19$ and $P(t) = \dfrac{200e^{t}}{e^{t} + 19}$.

3. $P(0) = 6$
$C = 20$
$k = .0011$

$$\frac{dP}{dt} = .0011P(20 - P)$$

$$\int \frac{1}{P(20 - P)}\,dP = \int .0011\,dt$$

$$\frac{1}{20}\ln(P) - \frac{1}{20}\ln(20 - P) = .0011t + K$$

3. Continued

fi $\ln(P) - \ln(20 - P) = .022t + 20K$
Performing similar calculations as in Exercise 2,

$$P = \frac{20e^{.022t}}{e^{-20K} + e^{.022t}}$$

$$P(0) = \frac{20}{e^{-20K} + 1}$$

$$6 = \frac{20}{e^{-20K} + 1}$$

$$e^{-20K} + 1 = \frac{10}{3}$$

$$e^{-20K} = \frac{7}{3}$$

$$-20K = \ln\frac{7}{3}$$

$$K = \frac{\ln\frac{7}{3}}{-20}$$

Rewrite P with the value of K.

$$P = \frac{20e^{.022t}}{\frac{7}{3} + e^{.022t}}$$

$$P(t) = \frac{60e^{.022t}}{7 + 3e^{.022t}}$$

Now find t when $P(t) = 18$.

$$18 = \frac{60e^{.022t}}{7 + 3e^{.022t}}$$

$$18(7 + 3e^{.022t}) = 60e^{.022t}$$

$$126 + 54e^{.022t} = 60e^{.022t}$$

$$126 = 6e^{.022t}$$

$$21 = e^{.022t}$$

$$\ln 21 = .022t$$

$$\frac{\ln 21}{.022} = t$$

$$138.387 \approx t$$

The population will reach 18 billion after about 138 years.

4.
No, $P(t)$ will approach but never reach 20 billion. This model was chosen because it agreed with the expected behavior of a bounded population. The population was bounded by assigning the carrying capacity as 20 billion.

Chapter 14: Multivariate Calculus

Section 14.1 Functions of Several Variables

1. $f(x, y) = 5x + 2y - 4$

$f(2, -1) = 5(2) + 2(-1) - 4$

$\qquad = 10 - 2 - 4 = 4$

$f(-4, 1) = 5(-4) + 2(1) - 4$

$\qquad = -20 + 2 - 4 = -22$

$f(-2, -3) = 5(-2) + 2(-3) - 4$

$\qquad = -10 - 6 - 4 = -20$

$f(0, 8) = 5(0) + 2(8) - 4$

$\qquad = 0 + 16 - 4 = 12$

3. $f(x, y) = \sqrt{y^2 + 2x^2}$

$f(2, -1) = \sqrt{(-1)^2 + 2(2)^2} = \sqrt{1 + 8}$

$\qquad = \sqrt{9} = 3$

$f(-4, 1) = \sqrt{(1)^2 + 2(-4)^2} = \sqrt{1 + 32}$

$\qquad = \sqrt{33}$

$f(-2, -3) = \sqrt{(-3)^2 + 2(-2)^2} = \sqrt{9 + 8}$

$\qquad = \sqrt{17}$

$f(0, 8) = \sqrt{(8)^2 + 2(0)^2} = \sqrt{64} = 8$

5. a. $g(-2, 4) = -(-2)^2 - 4(-2)(4) + (4)^3$

$\qquad = -4 + 32 + 64$

$\qquad = 92$

b. $g(-1, -2) = -(-1)^2 - 4(-1)(-2) + (-2)^3$

$\qquad = -1 - 8 - 8$

$\qquad = -17$

c. $g(-2, 3) = -(-2)^2 - 4(-2)(3) + (3)^3$

$\qquad = -4 + 24 + 27$

$\qquad = 47$

d. $g(5, 1) = -(5)^2 - 4(5)(1) + (1)^3$

$\qquad = -25 - 20 + 1$

$\qquad = -44$

7. The xy-, xz- and yz- traces of a graph are the curves that result when a surface that is the graph of a function of two variables, $z = f(x, y)$, is cut by the xy-, xz- and yz- planes, respectively.

9. $3x + 2y + z = 12$

Let $x = 0$ and $y = 0$. Then $z = 12$. The point $(0, 0, 12)$ is on the graph. Let $x = 0$ and $z = 0$. Then $y = 6$. The point $(0, 6, 0)$ is on the graph. Let $y = 0$ and $z = 0$. Then $x = 4$. The point $(4, 0, 0)$ is on the graph. The graph is a plane. We sketch the portion in the first octant through these three points.

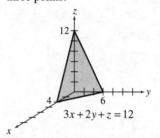

11. $x + y = 4$

We let $x = 0$ and find $y = 4$. The point $(0, 4, 0)$ is on the graph. We let $y = 0$ and find $x = 4$. The point $(4, 0, 0)$ is on the graph. Since there is no z-intercept, the plane is parallel to the z-axis. We sketch the portion in the first octant through these two points.

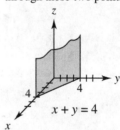

13. $z = 4$

The z-intercept is $(0, 0, 4)$. Since there is no x-intercept or y-intercept, the plane is parallel to the xy-plane. We sketch the portion in the first octant through the point $(0, 0, 4)$.

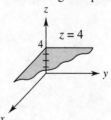

15. $3x + 2y + z = 18$

Value of z	Equation of level curve
$z = 0$	$3x + 2y = 18$
$z = 2$	$3x + 2y = 16$
$z = 4$	$3x + 2y = 14$

We graph the line $3x + 2y = 18$ in the plane $z = 0$. We graph the line $3x + 2y = 16$ in the plane $z = 2$. We graph the line $3x + 2y = 14$ in the plane $z = 4$.

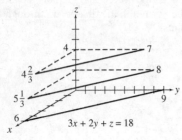

17. $y^2 - x = -z$

Value of z	Equation of level curve
$z = 0$	$x = y^2$
$z = 2$	$x = y^2 + 2$
$z = 4$	$x = y^2 + 4$

We graph the parabola $x = y^2$ (for $y > 0$) in the plane $z = 0$. We graph the parabola $x = y^2 + 2$ (for $y \geq 0$) in the plane $z = 2$. We graph the parabola $x = y^2 + 4$ (for $y \geq 0$) in the plane $z = 4$.

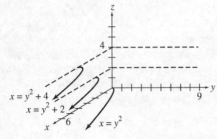

19. $z = x^7 y^3$ and $z = 500 = x^7 y^3 = 500$

$$y^3 = \frac{500}{x^7} \text{ or } y^{\frac{3}{10}} = \frac{500}{x^{\frac{7}{10}}}$$

$$y = \left(\frac{500}{x^{\frac{7}{10}}}\right)^{\frac{10}{3}} = \frac{500^{\frac{10}{3}}}{x^{\frac{7}{3}}} \approx \frac{9.9 \times 10^8}{x^{\frac{7}{3}}}$$

$$\approx \frac{10^9}{x^{\frac{7}{3}}}$$

$y = 500^{10/3} x^{-7/3}$

21. $z = x^7 y^3$ If x is doubled

$z = (2x)^7 y^3 = (2^7) x^7 y^3 \approx 1.6 x^7 y^3$.

The effect of doubling x is to multiply z by $2^7 \approx 1.6$.

If y is doubled,

$z = x^7 (2y)^3 = (2)^3 x^7 y^3 \approx 1.2 x^7 y^3$.

The effect of doubling y is to multiply z by $2^3 \approx 1.2$.

If both x and y are doubled,

$z = (2x)^7 (2y)^3 = (2)^7 x^7 (2)^3 y^3$

$= 2^{7+3} x^7 y^3 = 2 x^7 y^3$.

The effect of doubling both x and y is to double z.

23. a. If $x = 32$ and $y = 1$, then

$$P(32,1) = 100\left(\frac{3}{5}(32)^{-\frac{2}{5}} + \frac{2}{5}(1)^{-\frac{2}{5}}\right)^{-5}$$

$$= 100\left(\frac{3}{5}\left(\frac{1}{4}\right) + \frac{2}{5}(1)\right)^{-5}$$

$$= 100\left(\frac{11}{20}\right)^{-5} = 1986.9482$$

When 32 units of labor and 1 unit of capital are provided, production is about 1987 precision cameras.

23. Continued

b. If $x = 1$ and $y = 32$, then

$$P(1, 32) = 100\left(\frac{3}{5}(1)^{-\frac{2}{5}} + \frac{2}{5}(32)^{-\frac{2}{5}}\right)^{-5}$$

$$= 100\left(\frac{3}{5}(1) + \frac{2}{5}\left(\frac{1}{4}\right)\right)^{-5}$$

$$= 100\left(\frac{3}{5} + \frac{1}{10}\right)^{-5} = 100\left(\frac{7}{10}\right)^{-5}$$

$$= 594.9902$$

When 1 unit of labor and 32 units of capital are provided, production is about 594 precision cameras.

c. If $x = 32$ and $y = 243$, then

$$P(32, 243) = 100\left(\frac{3}{5}(32)^{-\frac{2}{5}} + \frac{2}{5}(243)^{-\frac{2}{5}}\right)^{-5}$$

$$= 100(.1944)^{-5}$$

$$\approx 359,768 \text{ precision cameras.}$$

25. $f(d, w) = 15 + .4(d + w)$

a. If $d = 85$ and $w = 60$,

$$f(85, 60) = 15 + .4(85 + 60)$$

$$= 15 + .4(145)$$

$$= 15 + 58$$

$$= 73$$

A temperature humidity index of 73 would feel comfortable.

b. If $d = 85$ and $w = 70$,

$$f(85, 70) = 15 + .4(85 + 70)$$

$$= 15 + .4(155)$$

$$= 15 + 62$$

$$= 77$$

A temperature humidity index of 77 would feel uncomfortable.

27. $A = .202W^{.425}H^{.725}$

a. If $W = 72$ and $H = 1.78$,
$$A = .2.2(72)^{.425}(1.78)^{.725} = 1.8892 \text{ m}^2$$

b. If $W = 65$ and $H = 1.40$,
$$A = .2.2(65)^{.425}(1.40)^{.725} = 1.5198 \text{ m}^2$$

c. If $W = 70$ and $H = 1.60$,
$$A = .2.2(70)^{.425}(1.60)^{.725} = 1.7278 \text{ m}^2$$

d. Answers vary

29. a. Solve for v. For a ferret,

$$2.56 = \frac{v^2}{(9.81)(.09)}$$

$$2.56 = \frac{v^2}{.8829}$$

$$2.260224 = v^2$$

$$1.5 \approx v$$

The velocity at which this change occurs for a ferret is about 1.5 m/sec.
For a rhinoceros,

$$2.56 = \frac{v^2}{(9.81)(1.2)}$$

$$2.56 = \frac{v^2}{11.772}$$

$$30.13632 = v^2$$

$$5.5 \approx v$$

The velocity at which this change occurs for a rhinoceros is about 5.5 m/sec.

b. Solve for v.

$$.025 = \frac{v^2}{(9.81)(4)}$$

$$.025 = \frac{v^2}{39.24}$$

$$.981 = v^2$$

$$1 \approx v$$

The velocity at which the sauropods were traveling was about 1 m/sec.

31. The distance around the parcel is $2H + 2W$. The length is L. Therefore, if the sum of length and girth is given by f,
$$f(L, W, H) = L + 2H + 2W.$$

33. $z = x^2 + y^2$
Find the traces.
xy-trace: $0 = x^2 + y^2$
xz-trace: $z = x^2$
yz-trace: $z = y^2$
These traces match with graph (c), the correct graph.

35. $x^2 - y^2 = z$
xy-trace: $x^2 - y^2 = 0$
yz-trace: $y^2 = -z$
xz-trace: $x^2 = z$
These traces match with graph (e), the correct graph.

37. $\dfrac{x^2}{16} + \dfrac{y^2}{25} + \dfrac{z^2}{4} = 1$

xy-trace: $\dfrac{x^2}{16} + \dfrac{y^2}{25} = 1$

yz-trace: $\dfrac{y^2}{25} + \dfrac{z^2}{4} = 1$

xz-trace: $\dfrac{x^2}{16} + \dfrac{z^2}{4} = 1$

These traces match with graph (b), the correct graph.

Section 14.2 Partial Derivatives

1. $z = f(x, y) = 8x^3 - 4x^2 y + 9y^2$

a. $\dfrac{\partial z}{\partial x} = 24x^2 - 8xy + 0 = 24x^2 - 8xy$

b. $\dfrac{\partial z}{\partial y} = 0 - 4x^2 + 18y = -4x^2 + 18y$

c. $f_x(2, 3) = 24(2)^2 - 8(2)(3) = 48$

d. $f_y(1, -2) = -4(1)^2 + 18(-2) = -40$

3. $f(x, y) = -x^2 y + 3x^4 - 8$

$f_x = -2xy + 12x^3 - 0$

$\quad = -2xy + 12x^3$

$f_y = -x^2 + 0 - 0 = -x^2$

$f_x(2, -1) = -2(2)(-1) + 12(2)^3 = 100$

$f_y(-4, 3) = -(-4)^2 = -16$

5. $f(x, y) = e^{2x+y}$

$f_x = e^{2x+y}(2) = 2e^{2x+y}$

$f_y = e^{2x+y}(1) = e^{2x+y}$

$f_x(2, -1) = 2e^{2(2)+(-1)} = 2e^3$

$f_y(-4, 3) = e^{2(-4)+3} = e^{-5}$

7. $f(x, y) = \dfrac{-2}{e^{x+2y}}$

$f(x, y) = -2e^{-x-2y}$

$f_x = -2(-1)e^{-x-2y} = 2e^{-x-2y} = \dfrac{2}{e^{x+2y}}$

$f_y = -2(-2)e^{-x-2y} = 4e^{-x-2y} = \dfrac{4}{e^{x+2y}}$

$f_x(2, -1) = \dfrac{2}{e^{2+2(-1)}} = 2$

$f_y(-4, 3) = \dfrac{4}{e^{-4+2(3)}} = \dfrac{4}{e^2}$

9. $f(x, y) = \dfrac{x + 3y^2}{x^2 + y^3}$

Use the quotient rule to find $f_x(x, y)$ and $f_y(x, y)$.

$f_x = \dfrac{\left(x^2 + y^3\right)(1) - \left(x + 3y^2\right)(2x)}{\left(x^2 + y^3\right)^2}$

$\quad = \dfrac{x^2 + y^3 - 2x^2 - 6xy^2}{\left(x^2 + y^3\right)^2}$

$\quad = \dfrac{y^3 - x^2 - 6xy^2}{\left(x^2 + y^3\right)^2}$

$f_y = \dfrac{\left(x^2 + y^3\right)(6y) - \left(x + 3y^2\right)\left(3y^2\right)}{\left(x^2 + y^3\right)^2}$

$\quad = \dfrac{6x^2 y + 6y^4 - 3xy^2 - 9y^4}{\left(x^2 + y^3\right)^2}$

$\quad = \dfrac{6x^2 y - 3xy^2 - 3y^4}{\left(x^2 + y^3\right)^2}$

$f_x(2, -1) = \dfrac{(-1)^3 - 2^2 - 6(2)(-1)^2}{\left[(2)^2 + (-1)^3\right]^2}$

$\quad = \dfrac{-1 - 4 - 12}{3^2}$

$\quad = -\dfrac{17}{9}$

$f_y(-4, 3) = \dfrac{6(-4)^2(3) - 3(-4)(3)^2 - 3(3)^4}{\left[(-4)^2 + (3)^3\right]^2}$

$\quad = \dfrac{288 + 108 - 243}{43^2}$

$\quad = \dfrac{153}{1849} \approx .083$

11. $f(x, y) = \ln\left|2x - x^2 y\right|$

$$f_x = \frac{2 - 2xy}{2x - x^2 y}$$

$$f_y = \frac{0 - x^2}{2x - x^2 y} = \frac{x(-x)}{x(2x - xy)}$$

$$= -\frac{x}{2 - xy}$$

$$f_x(2, -1) = \frac{2 - 2(2)(-1)}{2(2) - (2)^2(-1)}$$

$$= \frac{2 + 4}{4 + 4}$$

$$= \frac{6}{8} = \frac{3}{4}$$

$$f_y(-4, 3) = \frac{-(-4)}{2 - (-4)(3)} = \frac{4}{14} = \frac{2}{7}$$

13. $f(x, y) = x^2 e^{2xy}$

$$f_x = x^2 \cdot 2ye^{2xy} + 2xe^{2xy}$$

$$= 2x^2 ye^{2xy} + 2xe^{2xy}$$

$$f_y = x^2 \cdot 2xe^{2xy} = 2x^3 e^{2xy}$$

$$f_x(2, -1) = 2(2)^2(-1)e^{2(2)(-1)} + 2(2)e^{2(2)(-1)}$$

$$= -8e^{-4} + 4e^{-4}$$

$$= -4e^{-4}$$

$$f_y(-4, 3) = 2(-4)^3 e^{2(-4)(3)}$$

$$= -128e^{-24}$$

15. $f(x, y) = 10x^2 y^3 - 5x^3 - 3y$

$$f_x = 20xy^3 - 15x^2 + 0$$

$$= 20xy^3 - 15x^2$$

$$f_y = 30x^2 y^2 - 0 - 3 = 30x^2 y^2 - 3$$

$$f_{xx} = 20y^3 - 30x$$

$$f_{xy} = 60xy^2 - 0 = 60xy^2$$

$$f_{yy} = 60x^2 y + 0 = 60x^2 y$$

$$f_{yx} = 60xy^2 + 0 = 60xy^2$$

17. $h(x, y) = -3y^2 - 4x^2 y^2 + 7xy^2$

$$h_x = 0 - 8xy^2 + 7y^2 = -8xy^2 + 7y^2$$

$$h_y = -6y - 8x^2 y + 14xy$$

$$h_{xx} = -8y^2 + 0 = -8y^2$$

$$h_{xy} = -16xy + 14y$$

$$h_{yy} = -6 - 8x^2 + 14x$$

$$h_{yx} = 0 - 16xy + 14y$$

$$= -16xy + 14y$$

19. $R(x, y) = \dfrac{3y}{2x + y}$

$$R(x, y) = 3y(2x + y)^{-1}$$

$$R_x(x, y) = -3y(2x + y)^{-2}(2)$$

$$= -\frac{6y}{(2x + y)^2} = -6y(2x + y)^{-2}$$

$$R_y(x, y) = \frac{(2x + y) \cdot 3 - 3y(1)}{(2x + y)^2}$$

$$= \frac{6x}{(2x + y)^2} = 6x(2x + y)^{-2}$$

$$R_{xx}(x, y) = (-6y)(-2)(2x + y)^{-3}(2)$$

$$= \frac{24y}{(2x + y)^3}$$

$$R_{xy}(x, y) = -\frac{(2x + y)^2(6) - 6y(2)(2x + y)^1(1)}{(2x + y)^4}$$

$$= -\frac{(2x + y)(6) - 12y}{(2x + y)^3}$$

$$= -\frac{12x - 6y}{(2x + y)^3}$$

$$= \frac{-12x + 6y}{(2x + y)^3}$$

$$R_{yy}(x, y) = 6x(-2)(2x + y)^{-3}(1)$$

$$= -\frac{12x}{(2x + y)^3}$$

$$R_{yx}(x, y) = \frac{(2x + y)^2(6) - 6x(2)(2x + y)(2)}{(2x + y)^4}$$

$$= \frac{(2x + y)(6) - 24x}{(2x + y)^3}$$

$$= \frac{-12x + 6y}{(2x + y)^3}$$

21. $z = 4xe^y$

$$z_x = 4e^y$$

$$z_y = 4xe^y$$

$$z_{xx} = 0$$

$$z_{yy} = 4xe^y$$

$$z_{xy} = 4e^y = z_{yx}$$

23. $r = \ln(x + y)$

$$r_x = \frac{1}{x + y}$$

$$r_y = \frac{1}{x + y}$$

$$r_{xx} = -\frac{1}{(x + y)^2}$$

$$r_{yy} = -\frac{1}{(x + y)^2}$$

$$r_{xy} = -\frac{1}{(x + y)^2} = r_{yx}$$

25. $z = x \ln(xy)$

$$z_x = \ln xy + 1$$

$$z_y = \frac{x}{y}$$

$$z_{xx} = \frac{1}{x}$$

$$z_{yy} = -xy^{-2} = -\frac{x}{y^2}$$

$$z_{xy} = \frac{1}{y} = z_{yx}$$

27. $f(x, y) = x \ln(xy)$

$$f(x, y) = x(\ln x + \ln y)$$

$$f(x, y) = x \ln x + x \ln y$$

$$f_x(x, y) = x\left(\frac{1}{x}\right) + 1(\ln x) + \ln y$$

$$= 1 + \ln x + \ln y$$

$$f_y(x, y) = 0 + x\left(\frac{1}{y}\right) = \frac{x}{y}$$

$$f_{xy}(x, y) = 0 + 0 + \frac{1}{y} = \frac{1}{y}$$

$$f_{yy}(x, y) = -\frac{x}{y^2}$$

$$f_{xy}(2, 1) = \frac{1}{1} = 1$$

$$f_{yy}(1, 2) = -\frac{1}{2^2} = -\frac{1}{4}$$

29. $f(x, y) = 6x^2 + 6y^2 + 6xy + 36x - 5$

$$f_x(x, y) = 12x + 0 + 6y + 36$$

$$= 12x + 6y + 36$$

$$f_y(x, y) = 0 + 12y + 6x + 0 - 0$$

$$= 12y + 6x$$

Let $f_x(x, y) = 0$ and $f_y(x, y) = 0$. Solve the resulting system of equations.

$$12x + 6y + 36 = 0 \quad (1)$$

$$6x + 12y = 0 \quad (2)$$

To solve the system by the elimination method, multiply equation (2) by -2 and add the result to the first equation.

$$12x + 6y = -36$$

$$\underline{-12x - 24y = 0}$$

$$-18y = -36$$

$$y = 2$$

Substitute 2 for y in equation (2).

$$6x + 12(2) = 0$$

$$6x = -24$$

$$x = -4$$

Thus, $x = -4$ and $y = 2$.

31. $f(x, y) = 9xy - x^3 - y^3 - 6$

$$f_x(x, y) = 9y - 3x^2 - 0 - 0 = 9y - 3x^2$$

$$f_y(x, y) = 9x - 0 - 3y^2 - 0 = 9x - 3y^2$$

If $f_x(x, y) = 0$ and $f_y(x, y) = 0$, we have the nonlinear system of equations

$$9y - 3x^2 = 0$$

$$9x - 3y^2 = 0.$$

To simplify this system divide each equation by 3.

$$3y - x^2 = 0 \quad (1)$$

$$3x - y^2 = 0 \quad (2)$$

We will solve this system by the substitution method, which may be used to solve nonlinear as well as linear systems. First, solve equation (1) for y.

$$3y = x^2$$

$$y = \frac{x^2}{3}$$

Now substitute $\dfrac{x^2}{3}$ for y in equation (2) and solve the resulting equation for x.

31. Continued

$$3x - \left(\frac{x^2}{3}\right)^2 = 0$$

$$3x - \frac{x^4}{9} = 0$$

$$27x - x^4 = 0$$

$$x\left(27 - x^3\right) = 0$$

$$x = 0 \quad \text{or} \quad x = 3$$

If $x = 0$, $y = \dfrac{0^2}{3} = 0$.

If $x = 3$, $y = \dfrac{3^3}{3} = 3$.

Thus, there are two solutions: $x = 0$, $y = 0$ and $x = 3$, $y = 3$.

33. $f(x, y, z) = x^2 + yz + z^4$

$$f_x = 2x$$
$$f_y = z$$
$$f_z = y + 4z^3$$
$$f_{yz} = 1$$
$$f_y(2, -1, 3) = 3$$
$$f_{yz}(-1, 1, 0) = 1$$

35. $f(x, y, z) = \dfrac{6x - 5y}{4z + 5}$

$$f_x = \frac{6}{4z + 5}$$

$$f_y = -\frac{5}{4z + 5}$$

$$f_z = -\frac{4(6x - 5y)}{(4z + 5)^2}$$

$$f_{yz} = \frac{20}{(4z + 5)^2}$$

37. $f(x, y, z) = \ln\left(x^2 - 5xz^2 + y^4\right)$

$$f_x = \frac{2x - 5z^2}{x^2 - 5xz^2 + y^4}$$

$$f_y = \frac{4y^3}{x^2 - 5xz^2 + y^4}$$

$$f_z = -\frac{10xz}{x^2 - 5xz^2 + y^4}$$

$$f_{yz} = \frac{4y^3(10zx)}{\left(x^2 - 5xz^2 + y^4\right)^2}$$

$$= \frac{40xy^3z}{\left(x^2 - 5xz^2 + y^4\right)^2}$$

39. A function with three independent variables has three partial derivatives. For a function $f(x, y, z)$, the 3 partial derivatives are f_x, f_y and f_z. In the second-order partial derivatives, each of the two subscripts may be x, y, or z, so there will be $3 \cdot 3 = 9$ possibilities. For a function $f(x, y, z)$, the 9 second-order partial derivatives are f_{xx}, f_{xy}, f_{xz}, f_{yx}, f_{yy}, f_{yz}, f_{zx}, f_{zy} and f_{zz} .

41. $M(x, y) = 40x^2 + 30y^2 - 10xy + 30$ where x is the cost of electronic chips and y is the cost of labor.

a. $M_y(x, y) = 60y - 10x$

$$M_y(4, 2) = 60(2) - 10(4)$$
$$= 120 - 40$$
$$= 80$$

The manufacturing costs are $80.

b. $M_x(x, y) = 80x - 10y$

$$M_x(3, 6) = 80(3) - 10(6)$$
$$= 240 - 60$$
$$= 180$$

The manufacturing costs are $180.

c. $(\partial M / \partial x) = 80x - 10y$

$$(\partial M / \partial x)(2, 5) = 80(2) - 10(5)$$
$$= 160 - 50$$
$$= 110$$

The manufacturing costs are $110.

41. Continued

   d. $(\partial M / \partial y) = 60y - 10x$ The

   $(\partial M / \partial y)(6,7) = 60(7) - 10(6)$

   $\qquad = 420 - 60$

   $\qquad = 360$

   manufacturing costs are $360.

43. $f(p,i) = 99p - .5pi - .0025p^2$

   a. If $p = 19,400$ and $i = 8$, then

   $f(19400, 8)$

   $= 99(19,400) - .5(19,400)(8) - .0025(19,400)^2$

   $= 1,920,600 - 77,600 - 940,900$

   $= 902,100$

   The weekly sales are $902,100.

   b. $f(p,i) = 99p - .5pi - .0025p^2$

   $f_p(p,i) = 99 - .5i - .005p$

   $f_i(p,i) = -.5p$

   $f_p(p,i)$ is the rate at which weekly sales are changing per unit change in price while the interest rate remains constant.

   $f_i(p,i)$ is the rate at which weekly sales are changing per unit change in interest rate while price remains constant.

45. $f(m,v) = 25.92m^{.68} + \dfrac{3.62m^{.75}}{v}$

   a. $f(300,10) = 25.92(300)^{.68} + \dfrac{3.62(300)^{.75}}{10}$

   $= 1253.36 + 26.09 = 1279.45 \approx 1279$

   The energy expended is 1279 kcal per hour.

   b. $f_m(m,v)$

   $= 25.92(.68)m^{-.32} + \dfrac{3.62}{v}(.75m^{-.25})$

   $= 17.6256m^{-.32} + \dfrac{2.715m^{-.25}}{v}$

   $f_m(300,10)$

   $= 17.625(300)^{-.32} + \dfrac{2.715(300)^{-.25}}{10}$

   $= 2.84 + .07 = 2.91$

   The instantaneous rate of change of energy for a 300-kg animal traveling at 10 km per hour is about 2.91 kcal per hour per g.

47. $A(M,H) = .202M^{.425}H^{.725}$

   a. $A_M(M,H) = .202(.425)M^{-.575}H^{.725}$

   $= .08585M^{-.575}H^{.725}$

   $A_M(72,1.8) = .08585(72)^{-.575}(1.8)^{.725}$

   $= .0112$

   b. $A_H(M,H) = .202M^{.425}(.725)H^{-.275}$

   $A_H(70,1.6) = .14645M^{.425}H^{-.275}$

   $= .14645(70)^{.425}(1.6)^{-.275}$

   $= .783$

49. $f(x,y) = \left[\dfrac{1}{3}x^{-\frac{1}{3}} + \dfrac{2}{3}y^{-\frac{1}{3}}\right]^{-3}$

   a. $f(27, 64)$

   $= \left[\dfrac{1}{3}(27)^{-\frac{1}{3}} + \dfrac{2}{3}(64)^{-\frac{1}{3}}\right]^{-3}$

   $= \left[\dfrac{1}{3}\left(\dfrac{1}{3}\right) + \dfrac{2}{3}\left(\dfrac{1}{4}\right)\right]^{-3}$

   $= \left(\dfrac{1}{9} + \dfrac{1}{6}\right)^{-3}$

   $= \left(\dfrac{5}{18}\right)^{-3}$

   $= \left(\dfrac{18}{5}\right)^{3}$

   $= 46.656$

   b. $f_x = -3\left[\dfrac{1}{3}x^{-\frac{1}{3}} + \dfrac{2}{3}y^{-\frac{1}{3}}\right]^{-4}\left(-\dfrac{1}{9}x^{-\frac{4}{3}}\right)$

   $f_x(27, 64)$

   $= -3\left[\dfrac{1}{3}(27)^{-\frac{1}{3}} + \dfrac{2}{3}(64)^{-\frac{1}{3}}\right]^{-4}\cdot\left(-\dfrac{1}{9}\right)(27)^{-\frac{4}{3}}$

   $= -3\left(\dfrac{5}{18}\right)^{-4}\left(-\dfrac{1}{9}\right)\left(\dfrac{1}{81}\right)$

   $= \dfrac{432}{625}$

   $f_x(27, 64) = .6912$, which represents the rate at which production is changing when labor changes by 1 unit from 27 to 28 and capital remains constant.

49. Continued

$$f_y = -3\left[\frac{1}{3}x^{-\frac{1}{3}} + \frac{2}{3}y^{-\frac{1}{3}}\right]^{-4}\left(-\frac{2}{9}y^{-\frac{4}{3}}\right)$$

$f_y(27, 64)$

$$= -3\left[\frac{1}{3}(27)^{-\frac{1}{3}} + \frac{2}{3}(64)^{-\frac{1}{3}}\right]^{-4} \cdot \left(-\frac{2}{9}\right)(64)^{-\frac{4}{3}}$$

$$= -3\left(\frac{5}{18}\right)^{-4}\left(-\frac{2}{9}\right)\left(\frac{1}{256}\right)$$

$$= \frac{2187}{5000}$$

$$= .4374$$

which represents the rate at which production is changing when capital changes by 1 unit from 64 to 65 and labor remains constant.

c. If labor increases by 1 unit then production would increase by .6912(100) or about 69 units. (See part (b) of this solution.)

51. $z = x^7 y^3$, where x is labor, y is capital.

Marginal productivity of labor is

$$\frac{\partial z}{\partial x} = .7x^{-3}y^3 = \frac{.7y^3}{x^3}.$$

Marginal productivity of capital is

$$\frac{\partial z}{\partial y} = .3x^7 y^{-7} = \frac{.3x^7}{y^7}.$$

53. $C(a, b, v) = \dfrac{b}{a-v} = b(a-v)^{-1}$

a. $C(160, 200, 125) = \dfrac{200}{160-125}$

$$= \frac{200}{35}$$

$$\approx 5.71$$

b. $C(180, 260, 142) = \dfrac{260}{180-142}$

$$= \frac{260}{38}$$

$$\approx 6.84$$

53. Continued

c. $\dfrac{\partial C}{\partial b} = \dfrac{1}{a-v}$

d. $\dfrac{\partial C}{\partial v} = -b(a-v)^{-2}(-1)$

$$= \frac{b}{(a-v)^2}$$

55. $p = f(s, n, a) = .078a + 4(sn)^{\frac{1}{2}}$

a. If $s = 8$, $n = 6$, and $a = 450$, then

$$p = f(8, 6, 450) = .078(450) + 4(8 \cdot 6)^{\frac{1}{2}}$$

$$= 35.1 + 4(48)^{\frac{1}{2}}$$

$$= 35.1 + 27.7128$$

$$= 62.8$$

The probability of passing the course is 62.8%.

b. If $s = 3$, $n = 3$, and $a = 320$, then

$$p = f(3, 3, 320) = .078(320) + 4(3 \cdot 3)^{\frac{1}{2}}$$

$$= 24.96 + 4(9)^{\frac{1}{2}}$$

$$= 24.96 + 12$$

$$= 36.96$$

c. $f_n(s, n, a) = 4\left(s^{\frac{1}{2}}\right)\left(\dfrac{1}{2}n^{-\frac{1}{2}}\right) = 2s^{\frac{1}{2}}n^{-\frac{1}{2}}$

$$f_n(3, 3, 320) = 2\left(3^{\frac{1}{2}}\right)\left(3^{-\frac{1}{2}}\right) = 2\left(3^0\right) = 2$$

The rate of change of probability per additional semester of high school where placement score and SAT score remain constant is 2%.

$$f_a(s, n, a) = 0.078$$

$$f_a(3, 3, 320) = 0.078$$

The rate of change of probability per unit change of SAT score where placement score and number of semesters remain constant is 0.078%.

57. $F = \dfrac{mgR^2}{r^2}$

a. $F_m = \dfrac{gR^2}{r^2}$

This represents the rate of change in force per unit change in mass.

$F_r = \dfrac{-2mgR^2}{r^3}$

This represents the rate of change in force per unit change in distance.

b. Since all variables represent nonnegative

values, $F_m = \dfrac{gR^2}{r^2} > 0$ and

$F_r = \dfrac{-2mgR^2}{r^3} < 0.$

59. a. Let $w = 220$ and $h = 74$.

$B = \dfrac{703(220)}{(74)^2} \approx 28.24$

b. $\dfrac{\partial B}{\partial w} = \dfrac{703}{h^2}$

$\dfrac{\partial B}{\partial h} = -2(703w)h^{-3} = \dfrac{-1406w}{h^3}$

Section 14.3 Extrema of Functions of Several Variables

1. Answers vary.

3. $f(x, y) = 2x^2 + 4xy + 6y^2 - 8x - 10$

$\quad f_x = 4x + 4y + 0 - 8 - 0$

$\quad\quad = 4x + 4y - 8$

$\quad f_y = 0 + 4x + 12y - 0 - 0$

$\quad\quad = 4x + 12y$

Set each of these two partial derivatives equal to zero, forming a system of linear equations.

$4x + 4y - 8 = 0 \quad (1)$

$4x + 12y \quad = 0 \quad (2)$

To solve this system by the elimination method, rewrite equation (1) in the form of $ax + by = c$ and multiply equation (2) by -1; then add the results.

3. Continued

$\begin{array}{r} 4x + 4y = 8 \quad (1) \\ -4x - 12y = 0 \quad (2) \\ \hline -8y = 8 \end{array}$

$y = -1$

Substitute -1 for y in equation (2).

$4x + 12(-1) = 0$

$\quad 4x = 12$

$\quad\quad x = 3$

The critical point is $(3, -1)$.

$f_{xx} = 4,\ f_{yy} = 12,\ \text{and}\ f_{xy} = 4$

$M = f_{xx}(3, -1) \cdot f_{yy}(3, -1) - \left[f_{xy}(3, -1) \right]^2$

$\quad = 4 \cdot 12 - 4^2$

$\quad = 32$

Since $M > 0$ and $f_{xx}(3, -1) > 0$, we have a local minimum at $(3, -1)$.

$f(3, -1) = 2(3)^2 + 4(3)(-1) + 6(-1)^2$

$\quad\quad - 8(3) - 10$

$\quad\quad = 18 - 12 + 6 - 24 - 10$

$\quad\quad = -22$

There is a local minimum of -22 at $(3, -1)$.

5. $f(x, y) = x^2 - xy + 2y^2 + 2x + 6y + 8$

$\quad f_x = 2x - y + 0 + 2 + 0 + 0$

$\quad\quad = 2x - y + 2$

$\quad f_y = 0 - x + 4y + 0 + 6 + 0$

$\quad\quad = -x + 4y + 6$

Set $f_x = 0$ and $f_y = 0$. Solve the resulting system of linear equations by the elimination method.

$2x - y + 2 = 0$

$-x + 4y + 6 = 0$

$\begin{array}{r} 2x - y = -2 \quad (1) \\ -2x + 8y = -12 \quad (2) \\ \hline 7y = -14 \end{array}$

$y = -2$

$2x - (-2) = -2 \quad (1)$

$\quad x = -2$

5. Continued

The critical point is $(-2, -2)$.
$f_{xx} = 2$, $f_{yy} = 4$, and $f_{xy} = -1$
$M = f_{xx}(-2, -2) \cdot f_{yy}(-2, -2)$
$\quad - \left[f_{xy}(-2, -2) \right]^2$
$\quad = 2 \cdot 4 - (-1)^2 = 7$
Since $M > 0$ and $f_{xx}(-2, -2) > 0$, we have a local minimum at $(-2, -2)$.
$f(-2, -2) = (-2)^2 - (-2)(-2)$
$\qquad\qquad + 2(-2)^2 + 2(-2) + 6(-2) + 8$
$\qquad = 4 - 4 + 8 - 4 - 12 + 8$
$\qquad = 0$
There is a local minimum of 0 at $(-2, -2)$.

7. $f(x, y) = 2x^2 + 4xy + y^2 - 4x + 4y$
$\quad f_x = 4x + 4y + 0 - 4 + 0$
$\qquad = 4x + 4y - 4$
$\quad f_y = 0 + 4x + 2y - 0 + 4$
$\qquad = 4x + 2y + 4$

Solve the system of equations.
$\quad 4x + 4y - 4 = 0$
$\quad \underline{4x + 2y + 4 = 0}$
$\quad 4x + 4y = 4 \quad (1)$
$\quad \underline{-4x - 2y = 4 \quad (2)}$
$\qquad 2y = 8$
$\qquad y = 4$
If $y = 4$, then
$\quad 4x + 4y = 4 \qquad (1)$
$\quad 4x + 4(4) = 4$
$\qquad 4x = -12$
$\qquad x = -3$
The critical point is $(-3, 4)$.
$f_{xx} = 4$, $f_{yy} = 2$, and $f_{xy} = 4$
$M = f_{xx}(-3, 4) \cdot f_{yy}(-3, 4) - \left[f_{xy}(-3, 4) \right]^2$
$\quad = 4 \cdot 2 - (4)^2 = -8$
Since $M < 0$, we have a saddle point at $(-3, 4)$.

9. $f(x, y) = 4xy - 10x^2 - 4y^2 + 8x + 8y + 9$
$\quad f_x = 4y - 20x + 8$
$\quad f_y = 4x - 8y + 8$
Solve the system of equations.
$\quad 4y - 20x + 8 = 0$
$\quad \underline{4x - 8y + 8 = 0}$
$\quad 4y - 20x + 8 = 0 \quad (1)$
$\quad \underline{-4y + 2x + 4 = 0 \quad (2)}$
$\qquad -18x + 12 = 0$
$\qquad x = \dfrac{2}{3}$
$\quad 4y - 20\left(\dfrac{2}{3}\right) + 8 = 0 \quad (1)$
$\qquad y = \dfrac{4}{3}$
The critical point is $\left(\dfrac{2}{3}, \dfrac{4}{3}\right)$.
$\quad f_{xx} = -20$
$\quad f_{yy} = -8$
$\quad f_{xy} = 4$
For $\left(\dfrac{2}{3}, \dfrac{4}{3}\right)$,
$M = (-20)(-8) - 16 = 144 > 0$.
Since $f_{xx} < 0$, a local maximum occurs at $\left(\dfrac{2}{3}, \dfrac{4}{3}\right)$.
$f\left(\dfrac{2}{3}, \dfrac{4}{3}\right) = 4\left(\dfrac{2}{3}\right)\left(\dfrac{4}{3}\right) - 10\left(\dfrac{2}{3}\right)^2 - 4\left(\dfrac{4}{3}\right)^2$
$\qquad\qquad + 8\left(\dfrac{2}{3}\right) + 8\left(\dfrac{4}{3}\right) + 9$
$\qquad = 17$
There is a local maximum of 17 at $\left(\dfrac{2}{3}, \dfrac{4}{3}\right)$.

11. $f(x, y) = x^2 + xy - 2x - 2y + 2$

$f_x = 2x + y - 2$

$f_y = x - 2$

Solve the system of equations.

$2x + y - 2 = 0$

$\underline{x - 2 = 0}$

$ x = 2$

$2(2) + y - 2 = 0$

$ y = -2$

This critical point is $(2, -2)$.

$f_{xx} = 2$

$f_{yy} = 0$

$f_{xy} = 1$

For $(2, -2)$,

$M = 2(0) - 1^2 = -1 < 0$

There is a saddle point at $(2, -2)$.

13. $f(x, y) = x^2 - y^2 - 2x + 4y - 7$

$f_x = 2x - 2$

$f_y = -2y + 4$

If $2x - 2 = 0$, $x = 1$.

If $-2y + 4 = 0$, $y = 2$.

The critical point is $(1, 2)$.

$f_{xx} = 2$

$f_{yy} = -2$

$f_{xy} = 0$

For $(1, 2)$,

$M = -4 - 0 < 0$.

There is a saddle point at $(1, 2)$.

15. $f(x, y) = 2x^3 + 2y^2 - 12xy + 15$

$ f_x = 6x^2 + 0 - 12y + 0$

$ = 6x^2 - 12y$

$ f_y = 0 + 4y - 12x + 0$

$ = 4y - 12x$

Setting $f_x = 0$ and $f_y = 0$, we obtain the following nonlinear system.

$6x^2 - 12y = 0$ (1)

$4y - 12x = 0$ (2)

We will solve the system by the substitution method. Simplify equation (1) by dividing both sides by 6. Solve equation (2) for y.

$x^2 - 2y = 0$ (3)

$ y = 3x$ (4)

Now substitute $3x$ for y in equation (3) and solve the resulting equation for x.

15.   Continued

$x^2 - 2(3x) = 0$

$x(x - 6) = 0$

$x = 0$ or $x = 6$

If $x = 0$, $y = 0$.

If $x = 6$, $y = 18$.

There are two critical points:

$(0, 0)$ and $(6, 18)$.

$f_{xx} = 12x$, $f_{yy} = 4$, and $f_{xy} = -12$

At $(0, 0)$,

$M = f_{xx}(0, 0) \cdot f_{yy}(0, 0) - \left[f_{xy}(0, 0) \right]^2$

$ = 0(4) - (-12)^2$

$ = -144$

Since $M < 0$, there is a saddle point at $(0, 0)$. At $(6, 18)$,

$M = f_{xx}(6, 18) \cdot f_{yy}(6, 18) - \left[f_{xy}(6, 18) \right]^2$

$ = 72(4) - (-12)^2$

$ = 144$

Since $M > 0$ and $f_{xx}(6, 18) > 0$, there is a local minimum at $(6, 18)$.

$f(6, 18) = 2(6)^3 + 2(18)^2 - 12(6)(18) + 15 = -201$

There is a local minimum of -201 at $(6, 18)$.

17. $f(x, y) = 3x^2 + 6y^3 - 36xy + 27$

$f_x = 6x + 0 - 36y = 6x - 36y$

$f_y = 0 + 18y^2 - 36x = 18y^2 - 36x$

As in Exercise 15, setting $f_x = 0$ and $f_y = 0$

yields a nonlinear system of equations which we can solve by the substitution method.

$6x - 36y = 0$

$\underline{18y^2 - 36x = 0}$

$ x = 6y$ (1)

$ \underline{y^2 - 2x = 0}$ (2)

$y^2 - 2(6y) = 0$

$y^2 - 12y = 0$

$y(y - 12) = 0$

$y = 0$ or $y = 12$

If $y = 0$, $x = 0$.

If $y = 12$, $x = 72$.

There are two critical points, $(0, 0)$ and $(72, 12)$.

17. Continued

$f_{xx} = 6$, $f_{yy} = 36y$, and $f_{xy} = -36$

At $(0, 0)$,

$$M = f_{xx}(0, 0) \cdot f_{yy}(0, 0) - \left[f_{xy}(0, 0) \right]^2$$
$$= 6(0) - (-36)^2$$
$$= -1296$$

Since $M < 0$, there is a saddle point at $(0, 0)$.

At $(72, 12)$,

$$M = f_{xx}(72,12) \cdot f_{yy}(72, 12) - \left[f_{xy}(72, 12) \right]^2$$
$$= 6(432) - (-36)^2$$
$$= 1296$$

Since $M > 0$ and $f_{xx}(72, 12) > 0$, there is a local minimum at $(72, 12)$.

$$f(72,12) = 3(72)^2 + 6(12)^3 - 36(72)(12) + 27$$
$$= -5157$$

There is a local minimum of -5157 at $(72, 12)$.

19. $f(x, y) = e^{xy}$

$$f_x = ye^{xy}$$
$$f_y = xe^{xy}$$
$$ye^{xy} = 0$$
$$xe^{xy} = 0$$
$$x = y = 0$$

Solve the system of equations.
The critical point is $(0, 0)$.

$$f_{xx} = y^2 e^{xy}$$
$$f_{yy} = x^2 e^{xy}$$
$$f_{xy} = e^{xy} + xye^{xy}$$

For $(0, 0)$,

$$M = 0 \cdot 0 - \left(e^0 \right)^2 = -1 < 0.$$

There is a saddle point at $(0, 0)$.

21. $f(x, y) = -3xy + x^3 - y^3 + \dfrac{1}{8}$

$$f_x = -3y + 3x^2$$
$$f_y = -3x - 3y^2$$

Solve the system $f_x = 0$, $f_y = 0$.

$$-3y + 3x^2 = 0$$
$$-3x - 3y^2 = 0$$

Solve the first equation for y.

$$3y = 3x^2$$
$$y = x^2$$

21. Continued

Substitute into the second equation and solve for x.

$$-3x - 3\left(x^2\right)^2 = 0$$
$$-3x - 3x^4 = 0$$
$$-3x\left(1 + x^3\right) = 0$$
$$x = 0 \quad \text{or} \quad x = -1$$

Then, $y = 0$ or $y = 1$.
The critical points are $(0, 0)$ and $(-1, 1)$.

$$f_{xx} = 6x$$
$$f_{yy} = -6y$$
$$f_{xy} = -3$$

For $(0, 0)$,

$$M = f_{xx}(0, 0) \cdot f_{yy}(0, 0) - \left[f_{xy}(0, 0) \right]^2$$
$$= 0(0) - (-3)^2$$
$$= -9 < 0.$$

So $(0, 0)$ is a saddle point.

For $(-1, 1)$,

$$M = f_{xx}(-1, 1) \cdot f_{yy}(-1, 1) - \left[f_{xy}(-1, 1) \right]^2$$
$$= (-6)(-6) - [-3]^2$$
$$= 36 - 9$$
$$= 27 > 0.$$

So $(-1, 1)$ is a local maximum.

$$f(-1, 1) = -3(-1)(1) + (-1)^3 - (1)^3 + \frac{1}{8} = \frac{9}{8}$$

There is a relative maximum of $\dfrac{9}{8}$ or $1\dfrac{1}{8}$

at $(-1, 1)$. Graph (a).

23. $f(x, y) = y^4 - 2y^2 + x^2 - \dfrac{17}{16}$

$$f_x = 2x$$
$$f_y = 4y^3 - 4y$$

Solve the system $f_x = 0$, $f_y = 0$.

$$2x = 0 \quad (1)$$
$$4y^3 - 4y = 0 \quad (2)$$

Equation (1) gives
$$2x = 0$$
$$x = 0$$

Equation (2) gives
$$4y(y^2 - 1) = 0$$
$$y = 0 \text{ or } y = 1 \text{ or } y = -1$$

23. Continued

The critical points are $(0, 0)$, $(0, 1)$, and $(0, -1)$.
$f_{xx} = 2$
$f_{yy} = 12y^2 - 4$
$f_{xy} = 0$
For $(0, 0)$,
$M = (2)(-4) - 0^2 = -8 < 0$.
There is a saddle point at $(0, 0)$.
For $(0, 1)$,
$M = (2)(8) - 0^2 = 16 > 0$
$f_{xx} = 2 > 0$
So $(0, 1)$ is a local minimum.
For $(0, -1)$,
$M = (2)(8) - 0^2 = 16 > 0$
$f_{xx} = 2 > 0$
So $(0, -1)$ is a local minimum.

$$f(0, 1) = (1)^4 - 2(1)^2 + (0)^2 - \frac{17}{16}$$
$$= -\frac{33}{16}$$
$$f(0, -1) = (-1)^4 - 2(-1)^2 + (0)^2 - \frac{17}{16}$$
$$= -\frac{33}{16}$$

A local minimum of $-\frac{33}{16}$ or $-2\frac{1}{16}$ occurs at $(0, 1)$ and $(0, -1)$.
Graph (b).

25. $f(x, y) = -x^4 + y^4 + 2x^2 - 2y^2 + \frac{1}{16}$

$f_x = -4x^3 + 4x$
$f_y = 4y^3 - 4y$
Solve $f_x = 0$, $f_y = 0$.
$$-4x^3 + 4x = 0 \quad (1)$$
$$+4y^3 - 4y = 0 \quad (2)$$
$$-4x(x^2 - 1) = 0 \quad (1)$$
$$-4x(x+1)(x-1) = 0$$
$x = 0$ or $x = -1$ or $x = 1$
$$4y^3 - 4y = 0 \quad (2)$$
$$4y(y^2 - 1) = 0$$
$$4y(y+1)(y-1) = 0$$
$4y = 0$ or $y + 1 = 0$ or $y - 1 = 0$
$y = 0$ or $y = -1$ or $y = 1$

25. Continued

Because any of the 3 x-values can be paired with any of the 3 y-values, there are 9 critical points.
The critical points are $(0, 0)$, $(0, -1)$, $(0, 1)$, $(-1, 0)$, $(-1, -1)$, $(-1, 1)$, $(1, 0)$, $(1, -1)$, and $(1, 1)$.
$f_{xx} = -12x^2 + 4$
$f_{yy} = 12y^2 - 4$
$f_{xy} = 0$
For $(0, 0)$,
$M = (4)(-4) - 0 = -16 < 0$.
Saddle point
For $(0, -1)$,
$M = (4)(8) - 0 = 32 > 0$
and $f_{xx} = 4$.   Local minimum
For $(0, 1)$,
$M = (4)(8) - 0 = 32 > 0$
and $f_{xx} = 4$.   Local minimum

For $(-1, 0)$,
$M = (-8)(-4) - 0 = 32 > 0$
and $f_{xx} = -8$.   Local maximum
For $(-1, -1)$,
$M = (-8)(8) - 0 = -64 < 0$.
Saddle point
For $(-1, 1)$,
$M = (-8)(8) - 0 = -64 < 0$.
Saddle point
For $(1, 0)$,
$M = (-8)(-4) - 0 = 32 > 0$
and $f_{xx} = -8$ Local maximum
For $(1, -1)$,
$M = (-8)(8) - 0 = -64 < 0$.
Saddle point
For $(1, 1)$,
$M = (-8)(8) - 0 = -64 < 0$.
Saddle point
The saddle points are at $(0, 0)$, $(-1, -1)$, $(-1, 1)$, $(1, -1)$, and $(1, 1)$.
The local maxima are at $(1, 0)$ and $(-1, 0)$.
The local minima are at $(0, 1)$ and $(0, -1)$.
$$f(1, 0) = -(1)^4 - (0)^4 + 2(1)^2 - 2(0)^2 + \frac{1}{16}$$
$$= \frac{17}{16}$$
$$f(-1, 0) = -(-1)^4 - (0)^4 - 2(-1)^2 + 2(0)^2 + \frac{1}{16}$$
$$= \frac{17}{16}$$

A local maximum of $\frac{17}{16}$ or $1\frac{1}{16}$ occurs at $(1, 0)$ and $(-1, 0)$.

25. Continued

$$f(0, 1) = -(0)^4 + (1)^4 + 2(0)^2 - 2(1)^2 + \frac{1}{16}$$

$$= -\frac{15}{16}$$

$$f(0, -1) = -(0)^4 + (-1)^4 + 2(0)^2 - 2(-1)^2 + \frac{1}{16}$$

$$= -\frac{15}{16}$$

A local minimum of $-\dfrac{15}{16}$ occurs at $(0, 1)$ and $(0, -1)$.

Graph (e).

27. $P(x, y) = 1000 + 24x - x^2 + 80y - y^2$

$\qquad P_x = 24 - 2x$

$\qquad P_y = 80 - 2y$

$\qquad 24 - 2x = 0$

$\qquad \underline{80 - 2y = 0}$

$\qquad 12 = x$

$\qquad 40 = y$

Critical point $(12, 40)$

$\qquad P_{xx} = -2$

$\qquad P_{yy} = -2$

$\qquad P_{xy} = 0$

For $(12, 40)$,

$M = -2(-2) - 0^2 = 4 > 0$.

Since $P_{xx} < 0$, $x = 12$ and $y = 40$ maximize the profit.

Therefore,

$$P(12, 40) = 1000 + 24(12) - (12)^2$$

$$+ 80(40) - (40)^2$$

$$= 2744$$

is the maximum profit.

29. $P(x, y) = 800 - 2x^3 + 12xy - y^2$

$\qquad P_x = 0 - 6x^2 + 12y - 0 = -6x^2 + 12y$

$\qquad P_y = 0 - 0 + 12x - 2y = 12x - 2y$

Set $P_x = 0$ and $P_y = 0$. Solve the resulting nonlinear system by the substitution method.

29. Continued

$\qquad -6x^2 + 12y = 0$

$\qquad \underline{12x - 2y = 0}$

$\qquad x^2 - 2y = 0$

$\qquad \underline{\qquad y = 6x}$

$\qquad x^2 - 2(6x) = 0$

$\qquad x(x - 12) = 0$

$\qquad x = 0 \quad$ or $\quad x = 12$

If $x = 0$, $y = 0$.

If $x = 12$, $y = 72$.

The critical points are $(0, 0)$ and $(12, 72)$. We ignore $(0, 0)$ since it makes no sense in the problem.

$P_{xx} = -12x$, $P_{yy} = -2$, and $P_{xy} = 12$

At $(12, 72)$,

$$M = P_{xx}(12, 72) \cdot P_{yy}(12, 72) - \left[P_{xy}(12, 72) \right]^2$$

$$= (-144)(-2) - (12)^2$$

$$= 144$$

Since $M > 0$ and $P_{xx}(12, 72) < 0$, there is a local maximum at $(12, 72)$.

$$P(12, 72) = 800 - 2(12)^3 + 12(12)(72) - (72)^2$$

$$P(12, 72) = 2528$$

The profit function gives the profit in thousands of dollars. The maximum profit of $2,528,000 occurs when the cost of a unit of chips is $12 and the cost of a unit of labor is $72.

31. $C(x, y) = 2x^2 + 3y^2 - 2xy + 2x - 126y + 3800$

$\qquad C_x = 4x - 2y + 2; \ C_y = 6y - 2x - 126$

$\qquad 4x - 2y + 2 = 0 \quad (1)$

$\qquad \underline{6y - 2x - 126 = 0 \quad (2)}$

$\qquad 2x - y + 1 = 0$

$\qquad \underline{-2x + 6y - 126 = 0}$

$\qquad 5y - 125 = 0$

$\qquad \qquad y = 25$

Substitute 25 for y in equation (1).

$\qquad 4x - 2(25) + 2 = 0$

$\qquad \qquad x = 12$

Critical point $(12, 25)$.

$\qquad C_{xx} = 4$, $C_{yy} = 6$, $C_{xy} = -2$

31. Continued

For (12, 25),
$M = 4(6) - 4 = 20 > 0$.
Since $C_{xx} > 0$, 12 units of electrical tape and 25 units of packing tape should be produced to yield a minimum cost.

$$C(12, 25) = 2(12)^2 + 3(25)^2 - 2(12 \cdot 25)$$
$$+ 2(12) - 126(25) + 3800$$
$$= 2237$$

The minimum cost is 2237.

33. The volume is

$$V = xyz = 27, \text{ so } z = \frac{27}{xy}.$$

The surface area is

$$S = 2xy + 2yz + 2xz$$

$$= 2xy + 2y\left(\frac{27}{xy}\right) + 2x\left(\frac{27}{xy}\right)$$

$$= 2xy + \frac{54}{x} + \frac{54}{y}$$

$$S_x = 2y - \frac{54}{x^2}, \quad S_y = 2x - \frac{54}{y^2}$$

Let $S_x = 0$ and $S_y = 0$ to obtain the following system.

$$2y - \frac{54}{x^2} = 0 \quad (1)$$

$$2x - \frac{54}{y^2} = 0 \quad (2)$$

Solve equation (1) for y.

$$2y = \frac{54}{x^2}$$

$$y = \frac{27}{x^2}$$

Substitute this expression for y into equation (2).

$$2x - \frac{54}{\left(\frac{27}{x^2}\right)^2} = 0$$

$$x = \frac{27}{\frac{(27)^2}{x^4}}$$

$$27x = x^4$$

$$0 = x^4 - 27x$$

$$x = x\left(x^3 - 27\right)$$

33. Continued

$$x = 0 \quad \text{or} \quad x^3 - 27 = 0$$
$$x^3 = 27$$
$$x = 3$$

If $x = 0$, $y = \frac{27}{0}$, which is undefined.

(If $x = 0$, there would be no box, so this is not relevant.)

If $x = 3$, $y = \frac{27}{x^2} = \frac{27}{3^2} = \frac{27}{9} = 3$.

Thus, the critical point is (3, 3).

$$S_{xx} = \frac{108}{x^3}, \quad S_{yy} = \frac{108}{y^3}, \quad S_{xy} = 2$$

$$M = (4)(4) - 2^2 = 12 > 0$$

$$S_{xx} = \frac{108}{27} = 4 > 0$$

A minimum surface area will occur when

$x = 3$, $y = 3$, and $z = \frac{27}{3}(3) = 3$.

The dimensions of the box will be
3 m by 3 m by 3 m.

35. $V = LWH$ and $2W + 2H + L = 108$
$$L = 108 - 2W - 2H$$
$$V(H, W) = (108 - 2W - 2H)WH$$
$$V(H, W) = 108WH - 2W^2H - 2WH^2$$
$$V_H = 108W - 2W^2 - 4HW$$
$$V_w = 108H - 4WH - 2H^2$$
$$108W - 2W^2 - 4HW = 0$$
$$\underline{108H - 4WH - 2H^2 = 0}$$
$$2W(54 - W - 2H) = 0$$
$$\underline{2H(54 - 2W - H) = 0}$$
$$2W = 0 \quad \text{or} \quad 54 - W - 2H = 0$$
$$\text{and}$$
$$2H = 0 \quad \text{or} \quad 54 - 2W - H = 0$$

Clearly $W = 0$ or $H = 0$ makes no sense in the problem, so we have the following system.
$$W + 2H = 54$$
$$2W + H = 54$$
Solve this system by the elimination method.
$$-2W - 4H = -108$$
$$\underline{2W + H = 54}$$
$$-3H = -54$$
$$H = 18$$

35. Continued

If $H = 18$, $2W + 18 = 54$ and $W = 18$. The only critical point we need consider is $(18, 18)$.

$V_{HH} = -4W$, $V_{WW} = -4H$

$V_{HW} = 108 - 4W - 4H$

$M = V_{HH}(18, 18) \cdot V_{WW}(18, 18) - \left[V_{HW}(18, 18)\right]^2$

$= (-72)(-72) - [108 - 4(18) - 4(18)]^2$

$= 3888$

Since $M > 0$ and $V_{HH}(18, 18) < 0$, we have a local maximum at $(18, 18)$.

$L = 108 - 2(18) - 2(18) = 36$

The dimensions of the box with maximum volume are 18 inches by 18 inches by 36 inches.

37. $P(x, y) = R(x, y) - C(x, y)$

$= (2xy + 2y + 12) - \left(2x^2 + y^2\right)$

$= 2xy + 2y + 12 - 2x^2 - y^2$

$P_x(x, y) = 2y + 0 + 0 - 4x - 0$

$= 2y - 4x$

$P_y(x, y) = 2x + 2 + 0 - 0 - 2y$

$= 2x + 2 - 2y$

$2y - 4x = 0$ (1)

$2x + 2 - 2y = 0$ (2)

$y = 2x$ (3)

$x - y = -1$ (4)

Substitute $2x$ for y in equation (4).

$x - 2x = -1$

$x = 1$

If $x = 1$, $y = 2$.

The critical point is $(1, 2)$.

$P_{xx} = -4$, $P_{yy} = -2$, and $P_{xy} = 2$

$M = P_{xx}(1, 2) \cdot P_{yy}(1, 2) - \left[P_{xy}(1, 2)\right]^2$

$= (-4)(-2) - (2)^2$

$= 4$

Since $M > 0$ and $P_{xx}(1, 2) < 0$, we have a local maximum at $(1, 2)$.

$P(1, 2) = 2(1)(2) + 2(2) + 12 - 2(1)^2 - (2)^2 = 14$ A maximum profit of \$1400 occurs when 1000 tons of grade A and 2000 tons of grade B ore are used.

39. $P(x, y) = 36xy - x^3 - 8y^3$

$P_x(x, y) = 36y - 3x^2$

$P_y(x, y) = 36x - 24y^2$

$36y - 3x^2 = 0$

$36x - 24y^2 = 0$

$y = \dfrac{x^2}{12}$ (1)

$3x - 2y^2 = 0$ (2)

Substitute $\dfrac{x^2}{12}$ for y in equation (2).

$3x - 2\left(\dfrac{x^2}{12}\right)^2 = 0$

$3x - 2\left(\dfrac{x^4}{144}\right) = 0$

$3x - \dfrac{x^4}{72} = 0$

$x\left(3 - \dfrac{x^3}{72}\right) = 0$

$x = 0$ or $3 - \dfrac{x^3}{72} = 0$

$216 - x^3 = 0$

$x^3 = 216$

$x = 6$

Disregard $x = 0$ since it makes no sense in the problem. If $x = 6$, then $y = \dfrac{6^2}{12} = \dfrac{36}{12} = 3$.

Consider the critical point $(6, 3)$.

$P_{xx}(x, y) = -6x \Rightarrow P_{xx}(6, 3) = -6(6) = -36$

$P_{yy}(x, y) = -48y \Rightarrow P_{yy}(6, 3) = -48(3) = -144$

$P_{xy}(x, y) = 36 \Rightarrow P_{xy}(6, 3) = 36$

$M = (-36) \cdot (-144) - [36]^2 = 3888$

Since $M > 0$ and $P_{xx} < 0$, there is a local maximum at $(6, 3)$.

$P(6, 3) = 36(6)(3) - (6)^3 - 8(3)^3 = 216$.

Profit is a maximum of \$216,000 when 6 tons of steel and 3 tons of aluminum are used.

Section 14.4 Lagrange Multipliers

1. Maximize $f(x,y) = 2xy$ subject to $x + y = 12$.

The Lagrange function is:
$$F(x,y,\lambda) = f(x,y) - \lambda \cdot g(x,y)$$
$$= 2xy - \lambda x - \lambda y + 12\lambda$$

Compute the partial derivatives:
$$F_x(x,y,\lambda) = 2y - \lambda$$
$$F_y(x,y,\lambda) = 2x - \lambda$$
$$F_\lambda(x,y,\lambda) = -x - x + 12$$

Set each partial derivative to zero to obtain the system:
$$2y - \lambda = 0$$
$$2x - \lambda = 0$$
$$-x - y + 12 = 0$$

Then solve the system:
$$\left.\begin{array}{c} 2y - \lambda = 0 \\ 2x - \lambda = 0 \end{array}\right\} \Rightarrow \left.\begin{array}{c} \lambda = 2y \\ \lambda = 2x \end{array}\right\} \Rightarrow \begin{array}{c} 2x = 2y \\ x = y \end{array}$$
$$-x - y + 12 = 0$$
$$-x - x + 12 = 0$$
$$-2x = -12$$
$$x = 6$$

Since $y = x$, $y = 6$ and $\lambda = 2x = 2 \cdot 6 = 12$. So, the maximum value of $f(x,y) = 2xy$ subject to $x + y = 12$ occurs when $x = 6$ and when $y = 6$. The maximum value is $f(6,6) = 2 \cdot 6 \cdot 6 = 72$.

3. Maximize $f(x,y) = x^2 y$ subject to $2x + y = 4$.

The Lagrange function is:
$$F(x,y,\lambda) = f(x,y) - \lambda \cdot g(x,y)$$
$$= x^2 y - 2\lambda x - \lambda y + 4\lambda$$

Compute the partial derivatives:
$$F_x(x,y,\lambda) = 2xy - 2\lambda$$
$$F_y(x,y,\lambda) = x^2 - \lambda$$
$$F_\lambda(x,y,\lambda) = -2x - y + 4$$

Set each partial derivative to zero to obtain the system:
$$2xy - 2\lambda = 0$$
$$x^2 - \lambda = 0$$
$$-2x - y + 4 = 0$$

Then solve the system:

3. **Continued**

$$\left.\begin{array}{c} 2xy - 2\lambda = 0 \\ x^2 - \lambda = 0 \end{array}\right\} \Rightarrow \left.\begin{array}{c} 2\lambda = 2xy \\ \lambda = x^2 \end{array}\right\} \Rightarrow \begin{array}{c} xy = x^2 \\ x = y \end{array}$$
$$-2x - x + 4 = 0$$
$$-3x = -4$$
$$x = \frac{4}{3}$$

Since $y = x$, $y = \frac{4}{3}$ and $\lambda = x^2 = \left(\frac{4}{3}\right)^2 = \frac{16}{9}$.

So, the maximum value of $f(x,y) = x^2 y$ subject to $2x + y = 4$ occurs when $x = \frac{4}{3}$ and when $y = \frac{4}{3}$. The maximum value is

$$f\left(\frac{4}{3}, \frac{4}{3}\right) = \left(\frac{4}{3}\right)^2 \left(\frac{4}{3}\right) = \frac{64}{27}.$$

5. Minimize $f(x,y) = x^2 + 2y^2 - xy$ subject to $x + y = 8$. The Lagrange function is:
$$F(x,y,\lambda) = f(x,y) - \lambda \cdot g(x,y)$$
$$= x^2 + 2y^2 - xy - x\lambda - y\lambda + 8\lambda$$

Compute the partial derivatives:
$$F_x(x,y,\lambda) = 2x - y - \lambda$$
$$F_y(x,y,\lambda) = 4y - x - \lambda$$
$$F_\lambda(x,y,\lambda) = -x - y + 8$$

Set each partial derivative to zero to obtain the system:
$$2x - y - \lambda = 0$$
$$4y - x - \lambda = 0$$
$$-x - y + 8 = 0$$

Then solve the system:
$$2x - y = 4y - x$$
$$\left.\begin{array}{c} \lambda = 2x - y \\ \lambda = 4y - x \end{array}\right\} \Rightarrow \begin{array}{c} 3x = 5y \\ y = \frac{3}{5}x \end{array}$$
$$-x - \frac{3}{5}x + 8 = 0$$
$$-\frac{8}{5}x + 8 = 0$$
$$x = 5$$

Since $y = \frac{3}{5}x$, $y = \frac{3}{5}(5) = 3$. So, the minimum value of $f(x,y) = x^2 + 2y^2 - xy$ subject to $x + y = 8$ occurs when $x = 5$ and when $y = 3$. The maximum value is

$$f(5,3) = 5^2 + 2(3)^2 - (5)(3) = 28.$$

7. Maximize $f(x,y) = x^2 - 10y^2$ subject to $x - y = 18$. The Lagrange function is:

$$F(x,y,\lambda) = f(x,y) - \lambda \cdot g(x,y)$$
$$= x^2 - 10y^2 - x\lambda + y\lambda + 18\lambda$$

Compute the partial derivatives:

$F_x(x,y,\lambda) = 2x - \lambda$

$F_y(x,y,\lambda) = -20y + \lambda$

$F_\lambda(x,y,\lambda) = -x + y + 18$

Set each partial derivative to zero to obtain the system:

$2x - \lambda = 0$

$-20y + \lambda = 0$

$-x + y + 18 = 0$

Then solve the system:

$\left. \begin{array}{l} \lambda = 2x \\ \lambda = 20y \end{array} \right\} \Rightarrow \begin{array}{l} 2x = 20y \\ x = 10y \end{array}$

$-(10y) + y + 18 = 0$

$-9y + 18 = 0$

$y = 2$

Since $x = 10y$, $x = 10(2) = 20$. So, the maximum value of $f(x,y) = x^2 - 10y^2$ subject to $x - y = 18$ occurs when $x = 20$ and when $y = 2$. The maximum value is

$f(20,2) = 20^2 - 10(2)^2 = 360$.

9. Maximize $f(x,y,z) = xyz^2$ subject to $x + y + z = 6$. The Lagrange function is:

$$F(x,y,z,\lambda) = f(x,y,z) - \lambda \cdot g(x,y,z)$$
$$= xyz^2 - x\lambda - y\lambda - z\lambda + 6\lambda$$

Compute the partial derivatives:

$F_x(x,y,z,\lambda) = yz^2 - \lambda$

$F_y(x,y,z,\lambda) = xz^2 - \lambda$

$F_z(x,y,z,\lambda) = 2xyz - \lambda$

$F_\lambda(x,y,z,\lambda) = -x - y - z + 6$

Set each partial derivative to zero to obtain the system:

$yz^2 - \lambda = 0$

$xz^2 - \lambda = 0$

$2xyz - \lambda = 0$

$-x - y - z + 6 = 0$

Then solve the system:

9. Continued

$\left. \begin{array}{l} \lambda = yz^2 \\ \lambda = xz^2 \\ \lambda = 2xyz \end{array} \right\} \Rightarrow \begin{array}{l} yz^2 = xz^2 \Rightarrow y = x \\ xz^2 = 2xyz \Rightarrow z = 2y \end{array}$

$-y - y - 2y + 6 = 0$

$-4y = -6$

$y = \dfrac{3}{2}$

Since $x = y$, $x = \dfrac{3}{2}$ and since $z = 2y$,

$z = 2\left(\dfrac{3}{2}\right) = 3$. So, the maximum value of

$f(x,y,z) = xyz^2$ subject to $x + y + z = 6$ occurs

when $x = \dfrac{3}{2}$, $y = \dfrac{3}{2}$, and when $z = 3$. The

maximum value is

$f\left(\dfrac{3}{2}, \dfrac{3}{2}, 3\right) = \left(\dfrac{3}{2}\right)\left(\dfrac{3}{2}\right)(3)^2 = \dfrac{81}{4}$.

11. Maximize $f(x,y) = xy^2$ subject to $x + y = 18$. The Lagrange function is:

$$F(x,y,\lambda) = f(x,y) - \lambda \cdot g(x,y)$$
$$= xy^2 - \lambda x - \lambda y + 18\lambda$$

Compute the partial derivatives:

$F_x(x,y,\lambda) = y^2 - \lambda$

$F_y(x,y,\lambda) = 2xy - \lambda$

$F_\lambda(x,y,\lambda) = -x - y + 18$

Set each partial derivative to zero to obtain the system:

$y^2 - \lambda = 0$

$2xy - \lambda = 0$

$-x - y + 18 = 0$

Then solve the system:

$\left. \begin{array}{l} \lambda = y^2 \\ \lambda = 2xy \end{array} \right\} \Rightarrow \begin{array}{l} y^2 = 2xy \\ y = 2x \end{array}$

$-x - (2x) + 18 = 0$

$-3x = -18$

$x = 6$

Since $y = 2x$, $y = 2(6) = 12$. So, the maximum value of $f(x,y) = xy^2$ subject to $x + y = 18$ occurs when $x = 6$ and when $y = 12$. The maximum value is $f(6,12) = (6)(12)^2 = 864$.

13. Maximize $f(x,y,z) = xyz$ subject to $x+y+z = 90$. The Lagrange function is:

$$F(x,y,z,\lambda) = f(x,y,z) - \lambda \cdot g(x,y,z)$$
$$= xyz - \lambda x - \lambda y - \lambda z + 90\lambda$$

Compute the partial derivatives:

$F_x(x,y,z,\lambda) = yz - \lambda$

$F_y(x,y,z,\lambda) = xz - \lambda$

$F_z(x,y,z,\lambda) = xy - \lambda$

$F_\lambda(x,y,z,\lambda) = -x - y - z + 90$

Set each partial derivative to zero to obtain the system:

$yz - \lambda = 0$

$xz - \lambda = 0$

$xy - \lambda = 0$

$-x - y - z + 90 = 0$

Then solve the system:

$$\left. \begin{array}{l} \lambda = yz \\ \lambda = xz \\ \lambda = xy \end{array} \right\} \Rightarrow \begin{array}{l} x = y \\ y = z \end{array}$$

$x = y = z$

$-x - x - x + 90 = 0$

$x = 30$

Since $x = y = z$, $y = 30$ and $z = 30$. So, the maximum value of $f(x,y) = xyz$ subject to $x+y+z = 90$ occurs when $x = 30$, $y = 30$, and when $z = 30$. The maximum value is

$f(30,30,30) = (30)(30)(30) = 27,000$.

15. Answers vary.

17. Answers vary.

$F(x,y,\lambda) = f(x,y) - \lambda \cdot g(x,y)$

$F_x(x,y,\lambda) = f_x(x,y) - \lambda \cdot g_x(x,y)$

$F_y(x,y,\lambda) = f_y(x,y) - \lambda \cdot g_y(x,y)$

$0 = f_x(x,y) - \lambda \cdot g_x(x,y)$

$f_x(x,y) = \lambda g_x(x,y)$

$0 = f_y(x,y) - \lambda \cdot g_y(x,y)$

$f_y(x,y) = \lambda g_y(x,y)$

$F_\lambda(x,y,\lambda) = f_\lambda(x,y) - \left[\lambda \cdot g_\lambda(x,y) + g(x,y)\right]$

$0 = 0 - \left[\lambda \cdot 0 + g(x,y)\right]$

$g(x,y) = 0$

19. Let x = length of the ends and y = length of the side opposite. Maximize $A(x,y) = xy$ subject to $2(8x) + 6y = 1200$. The Lagrange function is:

$$F(x,y,\lambda) = f(x,y) - \lambda \cdot g(x,y)$$
$$= xy - 16\lambda x - 6\lambda y + 1200\lambda$$

Compute the partial derivatives:

$F_x(x,y,\lambda) = y - 16\lambda$

$F_y(x,y,\lambda) = x - 6\lambda$

$F_\lambda(x,y,\lambda) = -16x - 6y + 1200$

Set each partial derivative to zero to obtain the system:

$y - 16\lambda = 0$

$x - 6\lambda = 0$

$-16x - 6y + 1200 = 0$

Then solve the system:

$$\left. \begin{array}{l} 16\lambda = y \\ 6\lambda = x \end{array} \right\} \Rightarrow \left. \begin{array}{l} \lambda = \dfrac{y}{16} \\ \lambda = \dfrac{x}{6} \end{array} \right\} \Rightarrow \begin{array}{l} \dfrac{y}{16} = \dfrac{x}{6} \\ x = \dfrac{6y}{16} = \dfrac{3}{8}y \end{array}$$

$-16\left(\dfrac{3y}{8}\right) - 6y + 1200 = 0$

$-6y - 6y + 1200 = 0$

$-12y = -1200$

$y = 100$

Since $x = \dfrac{3}{8}y$, $x = \dfrac{300}{8} = 37.5$. So, the maximum area is produced when $x = 37.5$ feet and $y = 100$ feet. The area is

$A(37.5,100) = 37.5 \cdot 60 = 3750$ ft^2.

21. Maximize $P(x,y) = -x^2 - y^2 + 4x + 8y$ subject to $x+y = 6$. The Lagrange function is:

$$F(x,y,\lambda) = f(x,y) - \lambda \cdot g(x,y)$$
$$= -x^2 - y^2 + 4x + 8y - \lambda x - \lambda y + 6\lambda$$

Compute the partial derivatives:

$F_x(x,y,\lambda) = -2x + 4 - \lambda$

$F_y(x,y,\lambda) = -2y + 8 - \lambda$

$F_\lambda(x,y,\lambda) = -x - y + 6$

Set each partial derivative to zero to obtain the system:

$-2x + 4 - \lambda = 0$

$-2y + 8 - \lambda = 0$

$-x - y + 6 = 0$

Then solve the system:

21. Continued

$$\left.\begin{array}{l}\lambda = -2x+4 \\ \lambda = -2y+8\end{array}\right\} \Rightarrow \begin{array}{l}-2x+4=-2y+8 \\ -2x=-2y+4 \\ x=y-2\end{array}$$

$$-(y-2)-y+6=0$$

$$-y+2-y+6=0$$

$$-2y=-8$$

$$y=4$$

Since $x=y-2$, $x=4-2=2$. Profit is maximized when 2 automobile radiators and 4 generator radiators are sold. Maximum profit is $P(2,4)=-(2)^2-(4)^2+4(2)+8(4)=\$20$.

23. Maximize $f(x,y)=12x^{3/4}y^{1/4}$ subject to $100x+180y=25,200$. The Lagrange function is:

$$F(x,y,\lambda)=f(x,y)-\lambda\cdot g(x,y)$$
$$=12x^{3/4}y^{1/4}-100\lambda x-180\lambda y+25,200\lambda$$

Compute the partial derivatives:

$$F_x(x,y,\lambda)=9x^{-1/4}y^{1/4}-100\lambda$$

$$F_y(x,y,\lambda)=3x^{3/4}y^{-3/4}-180\lambda$$

$$F_\lambda(x,y,\lambda)=-100x-180y+25,200$$

Set each partial derivative to zero to obtain the system:

$$9x^{-1/4}y^{1/4}-100\lambda=0$$

$$3x^{3/4}y^{-3/4}-180\lambda=0$$

$$-100x-180y+25,200=0$$

Then solve the system:

$$\left.\begin{array}{l}\lambda=\dfrac{9x^{-1/4}y^{1/4}}{100} \\[2mm] \lambda=\dfrac{x^{3/4}y^{-3/4}}{60}\end{array}\right\} \Rightarrow \dfrac{9x^{-1/4}y^{1/4}}{100}=\dfrac{x^{3/4}y^{-3/4}}{60} \quad \dfrac{9y^{1/4}}{100x^{1/4}}=\dfrac{x^{3/4}}{60y^{3/4}}$$

$$540y=100x \qquad x=5.4y$$

$$-100(5.4y)-180y+25,200=0$$

$$-540y-180y+25,200=0$$

$$-720y=-25,200$$

$$y=35$$

Since $x=5.4y$, $x=5.4(35)=189$. Maximum production occurs when 189 units of labor and when 35 units of capital are expended.

25. Let x = length of the ends and y = length of the side opposite. Maximize $A(x,y)=xy$ subject to $2x+y=600$. The Lagrange function is:

$$F(x,y,\lambda)=f(x,y)-\lambda\cdot g(x,y)$$
$$=xy-2\lambda x-\lambda y+600\lambda$$

Compute the partial derivatives:

$$F_x(x,y,\lambda)=y-2\lambda$$

$$F_y(x,y,\lambda)=x-\lambda$$

$$F_\lambda(x,y,\lambda)=-2x-y+600$$

Set each partial derivative to zero to obtain the system:

$$y-2\lambda=0$$

$$x-\lambda=0$$

$$-2x-y+600=0$$

Then solve the system:

$$\left.\begin{array}{l}\lambda=\dfrac{y}{2} \\[2mm] \lambda=x \\[2mm] -2x-y+600=0\end{array}\right\} \Rightarrow x=\dfrac{y}{2}$$

$$-2\left(\dfrac{1}{2}y\right)-y+600=0$$

$$-2y=-600$$

$$y=300$$

Since $x=\dfrac{1}{2}y$, $x=\dfrac{1}{2}(300)=150$. The maximum area is enclosed by the fence occurs when the two ends are 150 meters each and the single opposite side is 300 meters, resulting in an area of $A(150,300)=45,000 \text{ m}^2$.

27. Let h = height of cylinder and r = radius. Minimize the surface area, $S(h,r)=2\pi r^2+2hr\pi$ subject to $V(h,r)=h\pi r^2=25$.

The Lagrange function is:

$$F(h,r,\lambda)=f(h,r)-\lambda\cdot g(h,r)$$
$$=2\pi r^2+2\pi hr-\pi\lambda hr^2+25\lambda$$

Compute the partial derivatives:

$$F_h(h,r,\lambda)=2\pi r-\pi r^2\lambda$$

$$F_r(h,r,\lambda)=4\pi r+2\pi h-2\pi hr\lambda$$

$$F_\lambda(h,r,\lambda)=-\pi hr^2+25$$

Set each partial derivative to zero to obtain the system:

$$2\pi r-\pi r^2\lambda=0$$

$$4\pi r+2\pi h-2\pi hr\lambda=0$$

$$-\pi hr^2+25=0$$

27. Continued

Then solve the system:

$$\left. \begin{array}{l} \pi r^2 \lambda = 2\pi r \\ \lambda = \dfrac{2}{r} \\ 2\pi hr\lambda = 4\pi r + 2\pi h \\ hr\lambda = 2r + h \\ \lambda = \dfrac{2r+h}{hr} \end{array} \right\} \Rightarrow \left. \begin{array}{l} \dfrac{2}{r} = \dfrac{2r+h}{hr} \\ \\ 2h = 2r + h \\ h = 2r \end{array} \right.$$

$$-\pi(2r)r^2 + 25 = 0$$

$$-2\pi r^3 = -25$$

$$r^3 = \frac{25}{2\pi}$$

$$r = \sqrt[3]{\frac{25}{2\pi}} \approx 1.58$$

Since $h = 2r$, $h = 2(1.58) = 3.17$. The

cylindrical cal holding $25\ \text{in}^3$ has a minimum surface area if the radius is 1.58 inches and the height is 3.17 inches.

29. Let l = length, w = width, and h = height of the box. Minimize the surface area,

$S(l,w,h) = 2lh + 2wh + 2lw$, subject to the

constraint, $V(l,w,h) = lwh = 185$.

The Lagrange function is:

$$\begin{aligned} F(l,w,h,\lambda) &= f(l,w,h) - \lambda \cdot g(l,w,h) \\ &= 2lh + 2wh + 2lw - \lambda lwh + 185\lambda \end{aligned}$$

Compute the partial derivatives:

$$F_l(l,w,h,\lambda) = 2h + 2w - \lambda wh$$

$$F_w(l,w,h,\lambda) = 2h + 2l - \lambda lh$$

$$F_h(l,w,h,\lambda) = 2l + 2w - \lambda lw$$

$$F_\lambda(l,w,h,\lambda) = -lwh + 185$$

Set each partial derivative to zero to obtain the system:

$$2h + 2w - \lambda wh = 0$$

$$2h + 2l - \lambda lh = 0$$

$$2l + 2w - \lambda lw = 0$$

$$-lwh + 185 = 0$$

29. Continued

Then solve the system:

$$\left. \begin{array}{l} \lambda = \dfrac{2h+2w}{wh} \\ \lambda = \dfrac{2h+2l}{lh} \\ \lambda = \dfrac{2l+2w}{lw} \end{array} \right\} \Rightarrow \left. \begin{array}{l} \dfrac{2h+2w}{wh} = \dfrac{2h+2l}{lh} \\ \dfrac{2h+2l}{lh} = \dfrac{2l+2w}{lw} \end{array} \right\}$$

$$2hl + 2wl = 2hw + 2lw$$

$$\Rightarrow w = l$$

$$2hw + 2lw = 2lh + 2wh$$

$$w = h$$

$$\Rightarrow w = l = h$$

$$-www = -185$$

$$w^3 = 185$$

$$w = \sqrt[3]{185} \approx 5.70$$

Since $w = l = 2h$, $w = l \approx 2(6.45) \approx 12.91$.

A box having a volume of $185\ \text{in}^3$ has minimum surface area when it has dimensions 5.70 in x 5.70 in x 5.70 in.

31. Let l = length, w = width, and h = height of the acquarium. Minimize the surface area,

$S(l,w,h) = 2lh + 2wh + lw$, subject to the

constraint, $V(l,w,h) = lwh = 32$.

The Lagrange function is:

$$\begin{aligned} F(l,w,h,\lambda) &= f(l,w,h) - \lambda \cdot g(l,w,h) \\ &= 2lh + 2wh + lw - \lambda lwh + 32\lambda \end{aligned}$$

Compute the partial derivatives:

$$F_l(l,w,h,\lambda) = 2h + 2w - \lambda wh$$

$$F_w(l,w,h,\lambda) = 2h + l - \lambda lh$$

$$F_h(l,w,h,\lambda) = 2l + 2w - \lambda lw$$

$$F_\lambda(l,w,h,\lambda) = -lwh + 32$$

Set each partial derivative to zero to obtain the system:

$$2h + 2w - \lambda wh = 0$$

$$2h + l - \lambda lh = 0$$

$$2l + 2w - \lambda lw = 0$$

$$-lwh + 32 = 0$$

31. Continued

Then solve the system:

$$\left.\begin{array}{l} \lambda = \dfrac{2h+2w}{wh} \\[2mm] \lambda = \dfrac{2h+l}{lh} \\[2mm] \lambda = \dfrac{2l+2w}{lw} \end{array}\right\} \Rightarrow \left.\begin{array}{l} \dfrac{2h+2w}{wh} = \dfrac{2h+l}{lh} \\[2mm] \dfrac{2h+l}{lh} = \dfrac{2l+2w}{lw} \end{array}\right\}$$

$$2hl + wl = 2hw + lw$$
$$\Rightarrow \quad w = l$$
$$2hw + lw = 2lh + 2wh$$
$$w = 2h$$

$$-(2h)(2h)h = -32$$
$$4h^3 = 32$$
$$h^3 = 8$$
$$h = 2$$

Since $w = 2h$, $w = 2(2) = 4$, and $l = 4$. The minimum surface area of the aquarium occurs when the dimensions are 4 ft x 4 ft x 2 ft.

33. a. $F(r,s,t,\lambda) = P(r,s,t) - \lambda \cdot (r+s+t-\alpha)$
$$= rs(1-t) + (1-r)st + r(1-s)t + rst$$
$$\quad -\lambda \cdot (r+s+t-\alpha)$$
$$= rs - 2rst + st + rt - \lambda r - \lambda s - \lambda t + \lambda \alpha$$

b. Assume $\alpha = .75$

$$F(r,s,t,\lambda)$$
$$= rs - 2rst + st + rt - \lambda r - \lambda s - \lambda t + .75\lambda$$
Partial derivatives of F:
$$F_r(r,s,t,\lambda) = s - 2st + t - \lambda$$
$$F_s(r,s,t,\lambda) = r - 2rt + t - \lambda$$
$$F_t(r,s,t,\lambda) = -2rs + s + r - \lambda$$
$$F_\lambda(r,s,t,\lambda) = -r - s - t + .75$$
Resulting system:
$$s - 2st + t - \lambda = 0$$
$$r - 2rt + t - \lambda = 0$$
$$-2rs + s + r - \lambda = 0$$
$$-r - s - t + .75 = 0$$
Solving the system:

33. Continued

$$\left.\begin{array}{l} \lambda = s - 2st + t \\ \lambda = r - 2rt + t \\ \lambda = -2rs + s + r \end{array}\right\}$$

$$s - 2st + t = r - 2rt + t$$
$$s(1-2t) = r(1-2t) \qquad \Rightarrow r = s$$
$$\Rightarrow$$
$$r - 2rt + t = -2rs + s + r$$
$$t(1-2r) = s(1-2r) \qquad \Rightarrow s = t$$

$$-r - r - r = -.75$$
$$-3r = -.75$$
$$r = .25$$

When $\alpha = .75$, the probability of convicting is minimized when $r = s = t = .25$.

c. If $\alpha = 3$, then $-3r = -3$; so, $r = s = t = 1.0$.

Chapter 14 Review Exercises

1. $f(x, y) = 6y^2 - 5xy + 2x$
$$f(-1, 2) = 6(2)^2 - 5(-1)(2) + 2(-1)$$
$$= 32$$
$$f(6, -3) = 6(-3)^2 - 5(6)(-3) + 2(6)$$
$$= 156$$

2. $f(x, y) = -3x + 2x^2y^2 + 5y$
$$f(-1, 2) = -3(-1) + 2(-1)^2(2)^2 + 5(2)$$
$$= 21$$
$$f(6, -3) = -3(6) + 2(6)^2(-3)^2 + 5(-3)$$
$$= 615$$

3. $f(x, y) = \dfrac{2x-4}{x+3y}$
$$f(-1, 2) = \dfrac{2(-1)-4}{(-1)+3(2)} = \dfrac{-6}{5} = -\dfrac{6}{5}$$
$$f(6, -3) = \dfrac{2(6)-4}{6+3(-3)} = \dfrac{8}{-3} = -\dfrac{8}{3}$$

4. $f(x, y) = x\sqrt{x^2 + y^2}$
$$f(-1, 2) = (-1)\sqrt{(-1)^2 + (2)^2} = -\sqrt{5}$$
$$f(6, -3) = 6\sqrt{(6)^2 + (-3)^2} = 6\sqrt{45}$$
$$= 6(3\sqrt{5}) = 18\sqrt{5}$$

5. Answers vary.

6. Answers vary.

7. $x + 2y + 4z = 4$

Let $x = 0$ and $y = 0$. Then $z = 1$. The point $(0, 0, 1)$ is on the graph. Let $x = 0$ and $z = 0$. Then $y = 2$. The point $(0, 2, 0)$ is on the graph. Let $y = 0$ and $z = 0$. Then $x = 4$. The point $(4, 0, 0)$ is on the graph. The graph is a plane. We sketch the portion in the first octant through these three points.

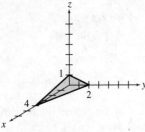

8. $3x + 2y = 6$

We let $x = 0$ and find $y = 3$. The point $(0, 3, 0)$ is on the graph. We let $y = 0$ and find $x = 2$. The point $(2, 0, 0)$ is on the graph. Since there is no z-intercept, the plane is parallel to the z-axis. We sketch the portion in the first octant through these two points.

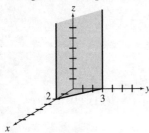

9. $4x + 5y = 20$

We let $x = 0$ and find $y = 4$. The point $(0, 4, 0)$ is on the graph. We let $y = 0$ and find $x = 5$. The point $(5, 0, 0)$ is on the graph. Since there is no z-intercept, the plane is parallel to the z-axis. We sketch the portion in the first octant through these two points.

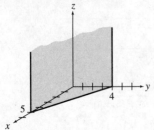

10. $x = 6$

The x-intercept is $(6, 0, 0)$. Since there is no y-intercept or z-intercept, the plane is parallel to the yz-plane. We sketch the portion in the first octant through the point $(6, 0, 0)$.

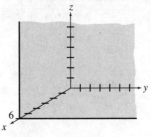

11. $z = f(x, y) = -2x^2 + 5xy + y^2$

a. $\dfrac{\partial z}{\partial x} = -4x + 5y$

b. $\dfrac{\partial z}{\partial y} = 5x + 2y$

$\dfrac{\partial z}{\partial y}(-1, 4) = 5(-1) + 2(4) = 3$

c. $f_{xy} = 5$

$f_{xy}(2, -1) = 5$

12. $z = f(x, y) = \dfrac{2y + x^2}{3y - x}$

Use the quotient rule to find the partial derivatives.

a. $\dfrac{\partial z}{\partial y} = \dfrac{(3y - x)(2) - (2y + x^2)(3)}{(3y - x)^2}$

$= \dfrac{-2x - 3x^2}{(3y - x)^2}$

b. $\dfrac{\partial z}{\partial x} = \dfrac{(3y - x)(2x) - (2y + x^2)(-1)}{(3y - x)^2}$

$= \dfrac{6xy - x^2 + 2y}{(3y - x)^2}$

$\dfrac{\partial z}{\partial x}(0, 2) = \dfrac{6(0)(2) - (0)^2 + 2(2)}{[3(2) - 0]^2}$

$= \dfrac{4}{36} = \dfrac{1}{9}$

12. Continued

c. $f_y = \left(-2x - 3x^2\right)(3y - x)^{-2}$

$f_{yy} = -2\left(-2x - 3x^2\right)(3y - x)^{-3}(3)$

$= \dfrac{6\left(2x + 3x^2\right)}{(3y - x)^3}$

$f_{yy}(-1, 0) = \dfrac{6\left[2(-1) + 3(-1)^2\right]}{[3(0) - (-1)]^3} = \dfrac{6(1)}{1}$

$= 6$

13. Answers vary.

14. $f(x, y) = 3y - 7x^2 y^3$

$f_x = 0 - 14xy^3 = -14xy^3$

$f_y = 3 - 7 \cdot 3x^2 y^2 = 3 - 21x^2 y^2$

15. $f(x, y) = 4x^3 y + 10xy^4$

$f_x = 4 \cdot 3x^2 y + 10 \cdot 1 \cdot y^4$

$= 12x^2 y + 10y^4$

$f_y = 4x^3 \cdot 1 + 10 \cdot 4xy^3$

$= 4x^3 + 40xy^3$

16. $f(x, y) = \sqrt{3x^2 + 2y^2}$

$f(x, y) = \left(3x^2 + 2y^2\right)^{\frac{1}{2}}$

$f_x = \dfrac{1}{2}\left(3x^2 + 2y^2\right)^{-\frac{1}{2}} \cdot 6x$

$= \dfrac{3x}{\sqrt{3x^2 + 2y^2}}$

$f_y = \dfrac{1}{2}\left(3x^2 + 2y^2\right)^{-\frac{1}{2}} \cdot 4y$

$= \dfrac{2y}{\sqrt{3x^2 + 2y^2}}$

17. $f(x, y) = \dfrac{3x - 2y^2}{x^2 + 4y}$

$f_x = \dfrac{\left(x^2 + 4y\right)(3) - \left(3x - 2y^2\right)(2x)}{\left(x^2 + 4y\right)^2}$

$= \dfrac{3x^2 + 12y - 6x^2 + 4xy^2}{\left(x^2 + 4y\right)^2}$

$= \dfrac{12y - 3x^2 + 4xy^2}{\left(x^2 + 4y\right)^2}$

$f_y = \dfrac{\left(x^2 + 4y\right)(-4y) - \left(3x - 2y^2\right)(4)}{\left(x^2 + 4y\right)^2}$

$= \dfrac{-4x^2 y - 16y^2 - 12x + 8y^2}{\left(x^2 + 4y\right)^2}$

$= \dfrac{-4x^2 y - 8y^2 - 12x}{\left(x^2 + 4y\right)^2}$

18. $f(x, y) = x^3 e^{3y}$

$f_x = 3x^2 e^{3y}$

$f_y = x^3 \cdot 3e^{3y} = 3x^3 e^{3y}$

19. $f(x, y) = (y + 1)^2 e^{2x + y}$

$f_x = (y + 1)^2 \cdot 2e^{2x + y}$

$= 2(y + 1)^2 e^{2x + y}$

$f_y = (y + 1)^2 \cdot 1 \cdot e^{2x + y} + 2(y + 1)^1 \cdot 1 \cdot e^{2x + y}$

$= (y + 1)^2 e^{2x + y} + 2(y + 1)e^{2x + y}$

$= (y + 1)e^{2x + y}[(y + 1) + 2]$

$= (y + 1)(y + 3)e^{2x + y}$

20. $f(x, y) = \ln\left|x^2 - 4y^3\right|$

$f_x = \dfrac{2x}{x^2 - 4y^3}$

$f_y = \dfrac{-12y^2}{x^2 - 4y^3}$

21. $f(x, y) = \ln\left|1 + x^3 y^2\right|$

$f_x = \dfrac{0 + 3x^2 y^2}{1 + x^3 y^2} = \dfrac{3x^2 y^2}{1 + x^3 y^2}$

$f_y = \dfrac{0 + 2x^3 y}{1 + x^3 y^2} = \dfrac{2x^3 y}{1 + x^3 y^2}$

22. Answers vary.

23. $f(x, y) = 4x^3y^2 - 8xy$

$f_x = 12x^2y^2 - 8y$

$f_{xx} = 24xy^2$

$f_{xy} = 24x^2y - 8$

24. $f(x, y) = -6xy^4 + x^2y$

$f_x = -6y^4 + 2xy$

$f_{xx} = 2y$

$f_{xy} = -24y^3 + 2x$

25. $f(x, y) = \dfrac{2x}{x - 2y}$

First, use the quotient rule to find f_x.

$f_x = \dfrac{(x - 2y) \cdot 2 - (2x) \cdot 1}{(x - 2y)^2}$

$= \dfrac{-4y}{(x - 2y)^2}$

$= -4y(x - 2y)^{-2}$

$f_{xx} = -4y\left[-2(x - 2y)^{-3}\right]$

$= 8y(x - 2y)^{-3}$

$= \dfrac{8y}{(x - 2y)^3}$

Use the quotient rule on

$f_x = \dfrac{-4y}{(x - 2y)^2}$

to get

$f_{xy} = \dfrac{(x - 2y)^2 \cdot (-4) + 4y[2(x - 2y)(-2)]}{(x - 2y)^4}$

$= \dfrac{-4(x - 2y)[(x - 2y) + 4y)]}{(x - 2y)^4}$

$= \dfrac{-4(x + 2y)}{(x - 2y)^3}$

$= \dfrac{-4x - 8y}{(x - 2y)^3}.$

26. $f(x, y) = \dfrac{3x + y}{x - 1}$

$f_x = \dfrac{(x - 1) \cdot 3 - (3x + y) \cdot 1}{(x - 1)^2}$

$= \dfrac{-3 - y}{(x - 1)^2} = (-3 - y)(x - 1)^{-2}$

$f_{xx} = -2(-3 - y)(x - 1)^{-3}$

$= \dfrac{2(3 + y)}{(x - 1)^3}$

$f_{xy} = \dfrac{-1}{(x - 1)^2}$

27. $f(x, y) = x^2e^y$

$f_x = 2xe^y$

$f_{xx} = 2e^y$

$f_{xy} = 2xe^y$

28. $f(x, y) = ye^{x^2}$

$f_x = 2xye^{x^2}$

$f_{xx} = 2xy \cdot 2xe^{x^2} + e^{x^2} \cdot 2y$

$= 2ye^{x^2}\left(2x^2 + 1\right)$

$f_{xy} = 2xe^{x^2}$

29. $f(x, y) = \ln\left(2 - x^2y\right)$

$f_x = \dfrac{1}{2 - x^2y} \cdot (-2xy)$

$= \dfrac{2xy}{x^2y - 2}$

$f_{xx} = \dfrac{\left(x^2y - 2\right)2y - 2xy(2xy)}{\left(x^2y - 2\right)^2}$

$= \dfrac{2y\left[\left(x^2y - 2\right) - 2x^2y\right]}{\left(x^2y - 2\right)^2}$

$= \dfrac{2y\left(-x^2y - 2\right)}{\left(x^2y - 2\right)^2}$

$= \dfrac{-2x^2y^2 - 4y}{\left(x^2y - 2\right)^2}$

29. Continued

$$f_{xy} = \frac{\left(x^2y - 2\right)2x - (2xy)x^2}{\left(x^2y - 2\right)^2}$$

$$= \frac{2x\left[\left(x^2y - 2\right) - x^2y\right]}{\left(x^2y - 2\right)^2}$$

$$= \frac{2x(-2)}{\left(x^2y - 2\right)^2}$$

$$= -\frac{4x}{\left(x^2y - 2\right)^2}$$

30. $f(x, y) = \ln\left(1 + 3xy^2\right)$

$$f_x = \frac{1}{1 + 3xy^2} \cdot 3y^2$$

$$= \frac{3y^2}{1 + 3xy^2}$$

$$= 3y^2\left(1 + 3xy^2\right)^{-1}$$

$$f_{xx} = 3y^2 \cdot \left(-3y^2\right)\left(1 + 3xy^2\right)^{-2}$$

$$= \frac{-9y^4}{\left(1 + 3xy^2\right)^2}$$

$$f_{xy} = \frac{\left(1 + 3xy^2\right) \cdot 6y - 3y^2(6xy)}{\left(1 + 3xy^2\right)^2}$$

$$= \frac{6y}{\left(1 + 3xy^2\right)^2}$$

31. Total area = area bottom and top + area of ends + area of sides, so:

$F(L, W, H) = 2LW + 2WH + 2LH$.

32. $c(x, y) = 2x + y^2 + 4xy + 25$

a. $\dfrac{\partial c}{\partial x} = 2 + 4y$

$$\frac{\partial c}{\partial x}(64, 6) = 2 + 4(6)$$

$$= 2 + 24 = \$26$$

b. $\dfrac{\partial c}{\partial y} = 2y + 4x$

$$\frac{\partial c}{\partial y}(128, 12) = 2(12) + 4(128)$$

$$= 24 + 512 = \$536$$

33. $z = f(x, y) = x^2 + 2y^2 - 4y$

$f_x = 2x$

$f_y = 4y - 4$

Set each of these two partial derivatives equal to zero forming a system of linear equations.

$2x = 0$

$4y - 4 = 0$

Solving the system, we obtain the critical point $(0, 1)$.

$f_{xx} = 2$, $f_{yy} = 4$, $f_{xy} = 0$

$$M = f_{xx}(0, 1) \cdot f_{yy}(0, 1) - \left[f_{xy}(0, 1)\right]^2$$

$$= 2 \cdot 4 - 0 = 8$$

Since $M > 0$ and $f_{xx}(0, 1) > 0$, we have a local minimum at $(0, 1)$.

$$z = f(0, 1) = 0^2 + 2(1)^2 - 4(1)$$

$$= 0 + 2 - 4 = -2$$

There is a local minimum of -2 at $(0, 1)$.

34. $z = f(x, y) = x^2 + y^2 + 9x - 8y + 1$

$f_x = 2x + 9$

$f_y = 2y - 8$

Set each of these two partial derivatives equal to zero forming a system of linear equations.

$2x + 9 = 0$

$2y - 8 = 0$

Solving the system, we obtain the critical point $\left(-\dfrac{9}{2}, 4\right)$.

$f_{xx} = 2$, $f_{yy} = 2$, $f_{xy} = 0$

$$M = f_{xx}\left(-\frac{9}{2}, 4\right) \cdot f_{yy}\left(-\frac{9}{2}, 4\right) - \left[f_{xy}\left(-\frac{9}{2}, 4\right)\right]^2$$

$$= 2 \cdot 2 - 0 = 4$$

Since $M > 0$ and $f_{xx}\left(-\dfrac{9}{2}, 4\right) > 0$, we have a

local minimum at $\left(-\dfrac{9}{2}, 4\right)$.

$$z = f\left(-\frac{9}{2}, 4\right) = \left(-\frac{9}{2}\right)^2 + (4)^2 + 9\left(-\frac{9}{2}\right) - 8(4) + 1$$

$$= -\frac{141}{4}$$

There is a local minimum of $-\dfrac{141}{4}$ at $\left(-\dfrac{9}{2}, 4\right)$.

35. $f(x,y) = x^2 + 5xy - 10x + 3y^2 - 12y$

$f_x = 2x + 5y - 10$

$f_y = 5x + 6y - 12$

Set each of these two partial derivatives equal to zero forming a system of linear equations.

$2x + 5y - 10 = 0$

$5x + 6y - 12 = 0$

Solving the system using eliminatin, we obtain the critical point $(0,2)$.

$f_{xx} = 2, \ f_{yy} = 6, \ f_{xy} = 5$

$M = f_{xx}(0,2) \cdot f_{yy}(0,2) - \left[f_{xy}(0,2) \right]^2$

$= 2 \cdot 6 - (5)^2 = 12 - 25 = 13$

Since $M < 0$, we have a saddle point at $(0,2)$.

36. $z = f(x,y) = x^3 - 8y^2 + 6xy + 4$

$f_x = 3x^2 + 6y$

$f_y = -16y + 6x$

Set each of these two partial derivatives equal to zero forming a system of linear equations.

$3x^2 + 6y = 0$

$-16y + 6x = 0$

Solving the system

$6x = 16y$

$3x = 8y$

$x = \dfrac{8}{3} y$

$3 \left(\dfrac{8}{3} y \right)^2 + 6y = 0$

$\dfrac{64}{3} y^2 + 6y = 0$

$y \left(\dfrac{64}{3} y + 6 \right) = 0 \Rightarrow y = 0$ or $y = -\dfrac{9}{32}$,

If $y = 0$, then $x = 0$.

If $y = -\dfrac{9}{32}$, then $x = -\dfrac{3}{4}$.

we obtain two critical points: $(0,0)$ and

$\left(-\dfrac{3}{4}, -\dfrac{9}{32} \right)$.

$f_{xx} = 6x, \ f_{yy} = -16, \ f_{xy} = 6$

At $(0,0)$

$M = f_{xx}(0,0) \cdot f_{yy}(0,0) - \left[f_{xy}(0,0) \right]^2$

$= (6)(0) \cdot (-16) - (6)^2 = -36$

Since $M < 0$ there is a saddle point at $(0,0)$.

36. Continued

At $\left(-\dfrac{3}{4}, -\dfrac{9}{32} \right)$

$M = f_{xx} \left(-\dfrac{3}{4}, -\dfrac{9}{32} \right) \cdot f_{yy} \left(-\dfrac{3}{4}, -\dfrac{9}{32} \right)$

$- \left[f_{xy} \left(-\dfrac{3}{4}, -\dfrac{9}{32} \right) \right]^2$

$= (6) \left(-\dfrac{3}{4} \right) \cdot (-16) - (6)^2 = 72 - 36 = 36$

Since $M > 0$ and $f_{xx} \left(-\dfrac{3}{4}, -\dfrac{9}{32} \right) < 0$, there is a

local maximum at $\left(-\dfrac{3}{4}, -\dfrac{9}{32} \right)$.

$f \left(-\dfrac{3}{4}, -\dfrac{9}{32} \right) = \left(-\dfrac{3}{4} \right)^3 - 8 \left(-\dfrac{9}{32} \right)^2$

$+ 6 \left(-\dfrac{3}{4} \right) \left(-\dfrac{9}{32} \right) + 4$

$= \dfrac{539}{128} \approx 4.21$

There is a local maximum of 4.21 at

$\left(-\dfrac{3}{4}, -\dfrac{9}{32} \right)$.

37. $z = f(x,y) = x^3 + y^2 + 2xy - 4x - 3y - 2$

$f_x = 3x^2 + 2y - 4$

$f_y = 2y + 2x - 3$

Set each of these two partial derivatives equal to zero forming a system of nonlinear equations.

$3x^2 + 2y - 4 = 0$

$2y + 2x - 3 = 0$

Solving the system

$2y = 3 - 2x$

$3x^2 + 3 - 2x - 4 = 0$

$3x^2 - 2x - 1 = 0$

$(3x + 1)(x - 1) = 0$

$x = -\dfrac{1}{3}$ or $x = 1$

If $x = -\dfrac{1}{3}$

$2y = 3 - 2 \left(-\dfrac{1}{3} \right) = \dfrac{11}{3} \Rightarrow y = \dfrac{11}{6}$.

If $x = 1$,

$2y = 3 - 2(1) = 1 \Rightarrow y = \dfrac{1}{2}$.

37. Continued

we obtain two critical points: $\left(-\dfrac{1}{3}, \dfrac{11}{6}\right)$ and

$\left(1, \dfrac{1}{2}\right)$.

$f_{xx} = 6x$, $f_{yy} = 2$, $f_{xy} = 2$

At $\left(-\dfrac{1}{3}, \dfrac{11}{6}\right)$

$M = f_{xx}\left(-\dfrac{1}{3}, \dfrac{11}{6}\right) \cdot f_{yy}\left(-\dfrac{1}{3}, \dfrac{11}{6}\right)$

$-\left[f_{xy}\left(-\dfrac{1}{3}, \dfrac{11}{6}\right)\right]^2$

$= (-2)(2) - (2)^2 = -4 - 4 = -8$

Since $M < 0$ there is a saddle point at $\left(-\dfrac{1}{3}, \dfrac{11}{6}\right)$.

At $\left(1, \dfrac{1}{2}\right)$

$M = f_{xx}\left(1, \dfrac{1}{2}\right) \cdot f_{yy}\left(1, \dfrac{1}{2}\right) - \left[f_{xy}\left(1, \dfrac{1}{2}\right)\right]^2$

$= 6(1) \cdot 2 - (2)^2 = 12 - 4 = 8$

$z = f\left(1, \dfrac{1}{2}\right) = (1)^2 + \left(\dfrac{1}{2}\right)^3 + 2(1)\left(\dfrac{1}{2}\right)$

$-4 - 3\left(\dfrac{1}{2}\right) - 2$

$= -\dfrac{21}{4} \approx -5.25$

Since $M > 0$ and $f_{xx}\left(1, \dfrac{1}{2}\right) = 6 > 0$, there is a

local minimum of -5.25 at $\left(1, \dfrac{1}{2}\right)$.

38. $z = f(x, y) = 7x^2 + y^2 - 3x + 6y - 5xy$

$f_x = 14x - 3 - 5y$

$f_y = 2y + 6 - 5x$

Set each of these two partial derivatives equal to zero forming a system of linear equations.

$14x - 3 - 5y = 0$

$2y + 6 - 5x = 0$

Solving the system using the elimination method, we obtain the critical point $(-8, -23)$.

$f_{xx} = 14$, $f_{yy} = 2$, $f_{xy} = -5$

$M = f_{xx}(-8, -23) \cdot f_{yy}(-8, -23) - \left[f_{xy}(-8, -23)\right]^2$

$= (14) \cdot (2) - (-5)^2 = 3$

Since $M > 0$ and $f_{xx}(0,1) > 0$, we have a local

minimum at $(-8, -23)$.

$z = f(-8, -23) = 7(-8)^2 + (-23)^2 - 3(-8)$

$+ 6(-23) - 5(-8)(-23) = -57$

There is a local minimum of -57 at $(-8, -23)$.

39. a. Minimize $c(x, y)$

$= x^2 + 5y^2 + 4xy - 70x - 164y + 1800$

$c_x = 2x + 4y - 70$

$c_y = 10y + 4x - 164$

$2x + 4y - 70 = 0$

$\underline{4x + 10y - 164 = 0}$

$-4x - 8y + 140 = 0$

$\underline{4x + 10y - 164 = 0}$

$2y - 24 = 0$

$y = 12$

$4x + 10(12) - 164 = 0$

$4x = 44$

$x = 11$

$c_{xx} = 2$, $c_{yy} = 10$, $c_{xy} = 4$

At $(11, 12)$,

$M = 2 \cdot 10 - 4^2$

$= 4 > 0$ and $c_{xx} > 0$

A local minimum occurs at $(11, 12)$.

b. The minimum cost is $c(11, 12)$

$= (11)^2 + 5(12)^2 + 4(11)(12)$

$- 70(11) - 164(12) + 1800$

$= 121 + 720 + 528 - 770 - 1968 + 1800$

$= \$431.$

40. $P(x, y)$

$= .01\left(-x^2 + 3xy + 160x - 5y^2 + 200y + 2600\right)$

a. $x + y = 280$

$y = 280 - x$

$P(x, y) = .01[-x^2 + 3x(280 - x) + 160x$

$- 5(280 - x)^2 + 200(280 - x)$

$+ 2600]$

$= .01[-x^2 + 840x - 3x^2 + 160x$

$- 5\left(78,400 - 560x + x^2\right)$

$+ 56,000 - 200x + 2600]$

$= .01\left(-x^2 + 840x - 3x^2 + 160x\right.$

$- 392,000 + 2800x - 5x^2$

$\left. + 56,000 - 200x + 2600\right)$

$P(x, y) = .01\left(-9x^2 + 3600x - 333,400\right)$

$P'(x, y) = .01(-18x + 3600)$

$.01(-18x + 3600) = 0$

$x = 200$

If $x < 200$, $P' > 0$.

If $x > 200$, $P' < 0$.

There is a maximum when $x = 200$.

If $x = 200$, $y = 280 - 200 = 80$.

$P(200, 80) = .01\left[-(200)^2 + 3(200)(80)\right.$

$+ 160(200) - 5(80)^2$

$\left. + 200(80) + 2600\right]$

$= 266$

$200 spent on fertilizer and $80 spent on seed will produce a maximum profit of $266 per acre.

40. Continued

b. $P_x = .01(-2x + 3y + 160)$

$P_y = .01(3x - 10y + 200)$

$-2x + 3y + 160 = 0$

$\underline{3x - 10y + 200 = 0}$

$-2x + 3y = -160$

$\underline{3x - 10y = -200}$

$-6x + 9y = -480$

$\underline{6x - 20y = -400}$

$-11y = -880$

$y = 80$

$-2x + 3(80) = -160$

$-2x = -400$

$x = 200$

The critical point is (200, 80).

$P_{xx} = .01(-2) = -.02$

$P_{yy} = .01(-10) = -.1$

$P_{xy} = .01(3) = .03$

At (200, 80),

$M = P_{xx}(200, 80) \cdot P_{yy}(200, 80)$

$- \left[P_{xy}(200, 80)\right]^2$

$= (-.02)(-.1) - (.03)^2$

$= .0011.$

Since $M > 0$ and $P_{xx}(200, 80) < 0$, there is a local maximum at (200, 80).

c. In (a), we saw that $P(200, 80) = 266$. $200 spent on fertilizer and $80 spent on seed will produce a maximum profit of $266 per acre.

41. Minimize $f(x,y) = x^2 + y^2$ subject to $x = y + 2$.
The Lagrange function is:
$$F(x,y,\lambda) = f(x,y) - \lambda \cdot g(x,y)$$
$$= x^2 + y^2 - \lambda x + \lambda y + 2\lambda$$
Compute the partial derivatives:
$$F_x(x,y,\lambda) = 2x - \lambda$$
$$F_y(x,y,\lambda) = 2y + \lambda$$
$$F_\lambda(x,y,\lambda) = -x + y + 2$$
Set each partial derivative to zero to obtain the system:
$$2x - \lambda = 0$$
$$2y + \lambda = 0$$
$$-x + y + 2 = 0$$

Then solve the system:
$$\left.\begin{array}{l}\lambda = 2x \\ \lambda = -2y\end{array}\right\} \Rightarrow \begin{array}{l} 2x = -2y \\ x = -y \end{array}$$
$$-(-y) + y + 2 = 0$$
$$2y = -2$$
$$y = -1$$
Since $x = -y$, $x = -(-1) = 1$. So, the minimum value of $f(x,y) = x^2 + y^2$ subject to $x = y + 2$ occurs when $x = 1$ and when $y = -1$. The minimum value is $f(1,-1) = (1)^2 + (-1)^2 = 2$.

42. Minimize and maximize $f(x,y) = x^2 y$ subject to $x + y = 4$. The Lagrange function is:
$$F(x,y,\lambda) = f(x,y) - \lambda \cdot g(x,y)$$
$$= x^2 y - \lambda x - \lambda y + 4\lambda$$
Compute the partial derivatives:
$$F_x(x,y,\lambda) = 2xy - \lambda$$
$$F_y(x,y,\lambda) = x^2 - \lambda$$
$$F_\lambda(x,y,\lambda) = -x - y + 4$$
Set each partial derivative to zero to obtain the system:
$$2xy - \lambda = 0$$
$$x^2 - \lambda = 0$$
$$-x - y + 4 = 0$$

42. Conitnued

Then solve the system:
$$\left.\begin{array}{l}\lambda = 2xy \\ \lambda = x^2\end{array}\right\} \Rightarrow \begin{array}{l} 2xy = x^2 \Rightarrow x = 0 \Rightarrow y = 0 \\ 2y = x \end{array}$$
$$-(2y) - y + 4 = 0$$
$$-3x = -4$$
$$y = \frac{4}{3}$$

Since $x = 2y$, $x = 2\left(\frac{4}{3}\right) = \frac{8}{3}$.

So, the maximum value of $f(x,y) = x^2 y$ subject to $x + y = 4$ occurs when $x = \frac{8}{3}$ and when

$y = \frac{4}{3}$. The maximum value is

$f\left(\frac{8}{3}, \frac{4}{3}\right) = \left(\frac{8}{3}\right)^2 \left(\frac{8}{3}\right) = \frac{256}{27}$. The minimum

value occurs when $x = 0$ and $y = 0$.

43. Maximize $f(x,y) = x^2 y$ subject to $x + y = 80$.
The Lagrange function is:
$$F(x,y,\lambda) = f(x,y) - \lambda \cdot g(x,y)$$
$$= x^2 y - \lambda x - \lambda y + 80\lambda$$
Compute the partial derivatives:
$$F_x(x,y,\lambda) = 2xy - \lambda$$
$$F_y(x,y,\lambda) = x^2 - \lambda$$
$$F_\lambda(x,y,\lambda) = -x - y + 80$$
Set each partial derivative to zero to obtain the system:
$$2xy - \lambda = 0$$
$$x^2 - \lambda = 0$$
$$-x - y + 80 = 0$$

Then solve the system:
$$\left.\begin{array}{l}\lambda = 2xy \\ \lambda = x^2\end{array}\right\} \Rightarrow \begin{array}{l} 2xy = x^2 \\ 2y = x \end{array}$$
$$-2y - y + 80 = 0$$
$$-3y = -80$$
$$y = \frac{80}{3}$$

43. Continued

Since $x = 2y$, $x = 2\left(\dfrac{80}{3}\right) = \dfrac{160}{3}$. So, the

maximum value of $f(x, y) = x^2 y$ subject to

$x + y = 80$ occurs when $x = \dfrac{160}{3}$ and when

$y = \dfrac{80}{3}$. The maximum value is

$f\left(\dfrac{160}{3}, \dfrac{80}{3}\right) = \dfrac{2,048,000}{27} \approx 75,852$.

44. Maximize $f(x, y) = xy^2$ subject to $x + y = 50$.

The Lagrange function is:

$F(x, y, \lambda) = f(x, y) - \lambda \cdot g(x, y)$

$\qquad = xy^2 - \lambda x - \lambda y + 50\lambda$

Compute the partial derivatives:

$F_x(x, y, \lambda) = y^2 - \lambda$

$F_y(x, y, \lambda) = 2xy - \lambda$

$F_\lambda(x, y, \lambda) = -x - y + 50$

Set each partial derivative to zero to obtain the system:

$y^2 - \lambda = 0$

$2xy - \lambda = 0$

$-x - y + 50 = 0$

Then solve the system:

$\left. \begin{matrix} \lambda = 2xy \\ \lambda = y^2 \end{matrix} \right\} \Rightarrow \begin{matrix} 2xy = y^2 \\ y = 2x \end{matrix}$

$-x - (2x) + 50 = 0$

$-3x + 50 = 0$

$x = \dfrac{50}{3}$

Since $y = 2x$, $y = 2\left(\dfrac{50}{3}\right) = \dfrac{100}{3}$. So, the

maximum value of $f(x, y) = xy^2$ subject to

$x + y = 50$ occurs when $y = \dfrac{100}{3}$ and when

$x = \dfrac{50}{3}$. The maximum value is

$f\left(\dfrac{50}{3}, \dfrac{100}{3}\right) = \left(\dfrac{50}{3}\right)\left(\dfrac{100}{3}\right)^2$

$= \dfrac{500,000}{27} \approx 18.519$

45. Answers vary

46. Maximize

$P(x, y)$

$= .01\left(-x^2 + 3xy + 160x - 5y^2 + 200y + 2600\right)$

subject to $x + y = 280$.

The Lagrange function is:

$F(x, y, \lambda) = f(x, y) - \lambda \cdot g(x, y)$

$= .01\left(-x^2 + 3xy + 160x - 5y^2 + 200y + 2600\right)$

$-\lambda x - \lambda y + 280\lambda$

$= -.01x^2 + .03xy + 1.6x - .05y^2 + 2y + 26$

$-\lambda x - \lambda y + 280\lambda$

Compute the partial derivatives:

$F_x(x, y, \lambda) = -.02x + .03y + 1.6 - \lambda$

$F_y(x, y, \lambda) = .03x - .1y + 2 - \lambda$

$F_\lambda(x, y, \lambda) = -x - y + 280$

Set each partial derivative to zero to obtain the system:

$-.02x + .03y + 1.6 - \lambda = 0$

$.03x - .1y + 2 - \lambda = 0$

$-x - y + 280 = 0$

Then solve the system using the elimination method (using technology), we obtain $x = 200$ and $y = 80$.

$P(200, 80) =$

$.01(-(200)^2 + 3(200)(80) + 160(200)$

$-5(80)^2 + 200(80) + 2600) = 266$

$P(x, y)$ has a maximum of \$266 when \$200 is spent on fertilizer and when \$80 is spent on hybrid seed.

Case 14 Global Warming and the Method of Least Squares

1.

$f_m(m, b) = -101,033 + 22,825,000m + 11,700b$

$f_b(m, b) = -51.78 + 11,700m + 6b$

$f_{mm}(m, b) = 22,825,000$

$f_{bb}(m, b) = 6$

$f_{mb}(m, b) = 11,700$

$f_{mm}f_{bb} - [f_{mb}]^2 = 60,000$

At $m = .0062, b = -3.46$, we have

$f_m = f_b = 0, f_{mm} > 0$, and $f_{mm}f_{bb} - [f_{mb}]^2 > 0$, so f has a minimum at $(.0062, -3.46)$.

2. $m = \dfrac{3(1900 \cdot 8.47 + 1950 \cdot 8.33 + 2000 \cdot 9.09) - (1900 + 1950 + 2000)(8.47 + 8.33 \cdot 9.09)}{3(1900^2 + 1950^2 + 2000^2) - (1900 + 1950 + 2000)^2}$

$= .0062$

$b = \dfrac{8.47 + 8.33 + 9.09}{3} - .0062 \left(\dfrac{1900 + 1950 + 2000}{3} \right)$

$= -3.46$

3. Using summation notation,

$m = \dfrac{5 \sum xy - \sum x \sum y}{5 \sum x^2 - \left(\sum x \right)^2} = .00532$

$b = \dfrac{\sum y}{5} - m \dfrac{\sum x}{5} = -1.794$

Answers vary. The slope .00532 of the least-squares line still suggests a gradual rise in the neighborhood of .005 – .006 degrees per year, but the pattern of data points suggests that the global land temperature fell during the first half of the century, and then rose during the second half at something like .015°C/year.

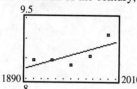